IMPORTANT:

HERE IS YOUR REGISTRATION CODE TO ACCESS
YOUR PREMIUM McGRAW-HILL ONLINE RESOURCES.

For key premium online resources you need THIS CODE to gain access. Once the code is entered, you will be able to use the Web resources for the length of your course.

If your course is using **WebCT** or **Blackboard**, you'll be able to use this code to access the McGraw-Hill content within your instructor's online course.

Access is provided if you have purchased a new book. If the registration code is missing from this book, the registration screen on our Website, and within your WebCT or Blackboard course, will tell you how to obtain your new code.

Registering for McGraw-Hill Online Resources

REGISTRATION CODE

W6PW-LCZM-YDD1-I1ZX-51BM

TO gain access to your McGraw-Hill web resources simply follow the steps below:

(1) USE YOUR WEB BROWSER TO GO TO: **www.mhhe.com/tillery**

(2) CLICK ON **FIRST TIME USER**.

(3) ENTER THE REGISTRATION CODE* PRINTED ON THE TEAR-OFF BOOKMARK ON THE RIGHT.

(4) AFTER YOU HAVE ENTERED YOUR REGISTRATION CODE, CLICK **REGISTER**.

(5) FOLLOW THE INSTRUCTIONS TO SET-UP YOUR PERSONAL UserID AND PASSWORD.

(6) WRITE YOUR UserID AND PASSWORD DOWN FOR FUTURE REFERENCE.
KEEP IT IN A SAFE PLACE.

TO GAIN ACCESS to the McGraw-Hill content in your instructor's **WebCT** or **Blackboard** course simply log in to the course with the UserID and Password provided by your instructor. Enter the registration code exactly as it appears in the box to the right when prompted by the system. You will only need to use the code the first time you click on McGraw-Hill content.

Thank you, and welcome to your McGraw-Hill online Resources!

0-07-291904-3 T/A TILLERY/ENGER/ROSS: INTEGRATED SCIENCE, 2/E

Integrated SCIENCE

Integrated SCIENCE

SECOND EDITION

Bill W. Tillery
ARIZONA STATE UNIVERSITY

Eldon D. Enger
DELTA COLLEGE

Frederick C. Ross
DELTA COLLEGE

Boston   Burr Ridge, IL   Dubuque, IA   Madison, WI   New York   San Francisco   St. Louis
Bangkok   Bogotá   Caracas   Kuala Lumpur   Lisbon   London   Madrid   Mexico City
Milan   Montreal   New Delhi   Santiago   Seoul   Singapore   Sydney   Taipei   Toronto

The McGraw-Hill Companies

Mc Graw Hill Higher Education

INTEGRATED SCIENCE, SECOND EDITION

Published by McGraw-Hill, a business unit of The McGraw-Hill Companies, Inc., 1221 Avenue of the Americas, New York, NY 10020. Copyright © 2004, 2001 by The McGraw-Hill Companies, Inc. All rights reserved. No part of this publication may be reproduced or distributed in any form or by any means, or stored in a database or retrieval system, without the prior written consent of The McGraw-Hill Companies, Inc., including, but not limited to, in any network or other electronic storage or transmission, or broadcast for distance learning.

Some ancillaries, including electronic and print components, may not be available to customers outside the United States.

This book is printed on acid-free paper.

2 3 4 5 6 7 8 9 0 VNH/VNH 0 9 8 7 6 5 4 3

ISBN 0-07-246700-2

Publisher: *Kent A. Peterson*
Sponsoring editor: *Daryl Bruflodt*
Developmental editor: *Mary E. Hurley*
Marketing manager: *Debra B. Hash*
Senior project manager: *Gloria G. Schiesl*
Production supervisor: *Sherry L. Kane*
Lead media project manager: *Judi David*
Senior media technology producer: *Jeffry Schmitt*
Designer: *K. Wayne Harms*
Cover/interior designer: *Christopher Reese*
Cover image: *Michael Javorka/Getty Images*
Senior photo research coordinator: *John C. Leland*
Photo research: *Mary Reeg*
Supplement producer: *Brenda A. Ernzen*
Compositor: *Carlisle Communications, Ltd.*
Typeface: *10/12 Minion*
Printer: *Von Hoffmann Corporation*

The credits section for this book begins on page 637 and is considered an extension of the copyright page.

Library of Congress Cataloging-in-Publication Data

Tillery, Bill W.
    Integrated science / Bill W. Tillery, Eldon D. Enger, Frederick C. Ross.—2nd ed.
        p.   cm.
    Includes index.
    ISBN 0-07-246700-2 (hard copy : alk. paper)
    1. Science.   I. Enger, Eldon D.   II. Ross, Frederick C.   III. Title.

Q161.2.T54   2004
500—dc21

                                                    2003006578
                                                    CIP

www.mhhe.com

CONTENTS

The Earth   275

The Earth's Surface   295

Earth's Weather   319

Earth's Waters   353

Organic and Biochemistry   373

The Nature of Living Things   401

Introduction

Integrated Science is a straightforward, easy-to-read, but substantial introduction to the fundamental behavior of matter and energy in living and nonliving systems. It is intended to serve the needs of nonscience majors who are required to complete one or more science courses as part of a general or basic studies requirement. It introduces basic concepts and key ideas while providing opportunities for students to learn reasoning skills and a new way of thinking about their environment. No prior work in science is assumed. The language, as well as the mathematics, is as simple as can be practical for a college-level science course.

The *Integrated Science* sequence of chapters is flexible, and the instructor can determine topic sequence and depth of coverage as needed. The materials are also designed to support a conceptual approach or a combined conceptual and problem-solving approach. The *Integrated Science* Online Learning Center offers suggestions for integrating the text's topics around theme options. With laboratory studies, the text contains enough material for the instructor to select a sequence for a one- or two-semester course.

> " I would generally describe the writing style of the book as chatty, informal and friendly, and the material flows well. I liked the writing style and felt there was a very low 'fog index'."
> —Stephen C. Brown, University of Texas at San Antonio

> ". . . the topics are segregated enough that the instructors can choose to cover them in any sequence."
> —Tasneem F. Khaleel, Montana State University

> "The use of formulae is useful. If one chooses to use them, they are included, while one can always choose to ignore them if it is not part of the course."
> —David J. DiMattio, St. Bonaventure University

The Goals of Integrated Science

1. **Create an introductory science course aimed at the nonscience major.** The origin of this book is rooted in our concern for the education of introductory-level students in the field of science. Historically, nonscience majors had to enroll in courses intended for science or science-related majors such as premed, architecture, or engineering. Such courses are important for these majors, but are mostly inappropriate for introductory-level nonscience students who are simply not interested in science, and perhaps apprehensive about taking

a science course. To put a nonscience student into such a course is a mistake. Few students will have the time or background to move through the facts, equations, and specialized language to gain any significant insights into the logic or fundamental understandings; instead, they will leave the course with a distaste for science. Today, society has a great need for a few technically trained people, but a much larger need for individuals who understand science.

2. **Introduce a course that presents a coherent and clear picture of all science disciplines—an interdisciplinary approach—to help students confirm and calibrate the big picture with the real world.** Recent studies and position papers have called for an interdisciplinary approach to teaching science to nonmajors. For example, the need is discussed in *Science for All Americans—Project 2061* (American Association for the Advancement of Science), *National Science Education Standards* (National Research Council, 1994), and *Science in the National Interest* (The White House, 1994). Interdisciplinary science is an attempt to broaden and humanize science education by reducing and breaking down the barriers that enclose traditional science disciplines as distinct subjects.

3. **Help instructors build their own mix of descriptive and analytical aspects of science, arousing student interest and feelings as they help students reach the educational goals of their particular course.** The spirit of interdisciplinary science is sometimes found in courses called "General Science," "Combined Science," or "Integrated Science." These courses draw concepts from a wide range of the traditional fields of science, but are not concentrated around certain problems or questions. For example, rather than just dealing with the physics of energy,

an interdisciplinary approach might consider broad aspects of energy—dealing with potential problems of an energy crisis—including social and ethical issues. A number of approaches can be used in interdisciplinary science, including the teaching of science in a *social, historical, philosophical,* or *problem-solving* context, but there is no single best approach. One of the characteristics of interdisciplinary science is that it is not constrained by the necessity of teaching certain facts or by traditions. It likewise cannot be imposed as a formal discipline, with certain facts to be learned. It is justified by its success in attracting and holding the attention and interest of students, making them a little wiser as they make their way toward various careers and callings.

4. **Humanize science for nonscience majors.** Each chapter presents historical background where appropriate, uses every day examples in developing concepts, and follows a logical flow of presentation. The historical chronology, which is of special interest to the humanistically inclined nonscience major, serves to humanize the science being presented. The use of everyday examples appeals to the nonscience major, who is typically accustomed to reading narration, not scientific technical writing, and also tends to bring relevancy to the material being presented. The logical flow of presentation is helpful to students not accustomed to thinking about relationships between what is being read and previous knowledge learned, a useful skill in understanding the sciences.

" . . . *many of our students generally do not select physics or chemistry as their physical science course and graduate without even a basic understanding of these sciences. . . . It is here that the Integrated approach is especially important because physics and chemistry can be taught at a basic level and students can learn to appreciate the central role that they play in everyday life. The authors' approach to teaching science is one that is needed in today's colleges.*"
—Jay R. Yett, Orange Coast College

"*I am very impressed with* Integrated Science's *objectives and philosophy. . . . If more textbooks and teachers took this approach, I think there would be fewer students who are afraid of science. This book, overall, does an excellent job in achieving its stated goals and that is why we use it. What I particularly like is how it 'humanizes' science.*"
—Stephen C. Brown, University of Texas at San Antonio

"*The approach, in terms of goals, as outlined in the preface is right on the money. I share these goals and applaud the attempt, which seems somewhat unique amongst textbooks I have seen. . . . We recognize the need for a more interdisciplinary approach that will help science 'come alive' for our students, particularly in regard to science as a human endeavor with close ties to philosophy.*"
—Joe H. Mehaffey, Francis Marion University

New to This Edition

In general, there's been a concerted effort to make the text even more conceptually oriented and relevant for students:

- A new Concepts Applied feature was added throughout. This new feature merges the old Activities and Q & A features, while adding more applications of relevance for students.
- The more difficult problem exercises and questions were removed, reducing the overall number, and making the end-of-chapter problems and questions more conceptually oriented and student friendly.
- The text material was made more conceptually oriented and student friendly throughout.
- The overall size of the text was reduced by reorganizing and cutting materials, including many Connections and Closer Look features that reviewers indicated were not closely related to the topics presented.
- The number of chapters was reduced to 25.

Several organizational changes were made to make the text more conceptually oriented and to better balance the coverage of physics, chemistry, earth science, astronomy, and biology:

- Chapter 2, "Motion," and Chapter 3, "Patterns of Motion," were merged into one chapter, "Motion," with much of the historical material removed. The chapter is now more relevant to students.
- Old chapter 7 (new chapter 6), "Electricity," was completely rewritten to be more relevant and conceptually attainable with a student friendly approach.
- Old chapter 9 "Atomic Structure" (new chapter 8) was substantially rewritten with a more student friendly approach to atomic theory and was renamed "Atoms and Periodic Properties."
- Old chapters 9, 10, and 11 were merged and condensed into one chapter, "Chemical Reactions" (new chapter 9), which is more conceptually friendly and relevant to students.
- Old chapter 13 (new chapter 10), "Water and Solutions," was substantially rewritten to be more conceptually friendly and relevant to students.
- Old chapter 14, "Organic Chemistry," was integrated with Biochemistry in a new chapter 18, "Organic and Biochemistry," and this material was substantially rewritten.
- Old astronomy chapters 16, 17, and 18 were merged into two chapters (12 and 13) "The Universe" and "The Solar System," with substantially updated content and less historical materials.
- The geologic time section was moved from old earth science chapter 20, "The Earth's Surface," and integrated with materials on human evolution and the classification of organisms into a new chapter 21, "The History of Life on Earth." This material was substantially rewritten to be more conceptually friendly.

- Old chapter 24 (new chapter 20), "The Origin and Evolution of Life" now includes sections on evidence of evolution and misconceptions about evolution. The material on taxonomy and human evolution was moved to the new chapter 21, "The History of Life on Earth."
- Old chapter 25 (new chapter 22), "Ecology and Environment," was changed substantially with new illustrations and the addition of material on aquatic ecosystems. Other organizational changes were made to condense the material.
- The original chapter 26, "Anatomy and Physiology," was divided into two chapters: "Human Biology: Sex and Nutrition" (new chapter 23) and, "Human Biology: Material Exchange and Control Mechanisms" (new chapter 24) to help expand and balance the biology covered in the text.

The Learning System

To achieve the goals stated, this text includes a variety of features that should make your study of *Integrated Science* more effective and enjoyable. These aids are included to help you clearly understand the concepts and principles that serve as the foundation of the integrated sciences.

Overview to Integrated Science

Chapter 1, "What Is Science?" provides an overview or orientation to integrated science in general and to this text in particular. It also describes the fundamental methods and techniques used by scientists to study and understand the world around us.

Chapter Opening Tools

Chapter Outlines

The chapter outline includes all the major topic headings and subheadings within the body of the chapter. It gives you a quick glimpse of the chapter's content and helps you locate sections dealing with particular topics.

Chapter Overviews

Each chapter begins with an introductory overview. The overview previews the content of the chapter and highlights what you can expect to learn from reading it. The overview adds to the general outline of the chapter by introducing you to the concepts that are covered, facilitating in the integration of topics and helping you to stay focused and organized while reading the chapter for the first time. After reading the introduction, browse through the chapter, paying particular attention to the topic headings and illustrations so that you get a feel for the kinds of ideas included within the chapter.

CHAPTER TWELVE

The Universe

This is a planetary nebula in the constellation Aquarius. Planetary nebulae are clouds of ionized gases with no relationship to any planet. They were named long ago, when they appeared similar to the planets Neptune and Uranus when viewed through early telescopes.

OUTLINE

The Night Sky
Origin of Stars
Brightness of Stars
Star Temperature
Star Types
The Life of a Star

Galaxies
The Milky Way Galaxy
Other Galaxies
A Closer Look: Extraterrestrials?
The Life of a Galaxy
A Closer Look: Dark Matter

In chapter 1, you learned some "tools and rules" and some techniques for finding order in your surroundings. You can see motion in natural events such as clouds moving, rain and snow falling, and streams of water, all moving in a never-ending cycle. Order is often found in the form of patterns, or relationships between quantities that are expressed as equations. Equations can be used to (1) describe properties, (2) define concepts, and (3) describe how quantities change relative to one another. In all three uses, patterns are quantified, conceptualized, and used to gain a general understanding about what is happening in nature.

In the study of science, certain parts of nature are often considered and studied together for convenience. One of the more obvious groupings involves *movement*. Most objects around you appear to spend a great deal of time sitting quietly without motion.

Buildings, rocks, utility poles, and trees rarely, if ever, move from one place to another. Even things that do move from time to time sit still for a great deal of time. This includes you, automobiles, and bicycles (figure 2.1). On the other hand, the sun, the moon, and starry heavens always seem to move, never standing still. Why do things stand still? Why do things move?

Questions about motion have captured the attention of people for thousands of years. But the ancient people answered questions about motion with stories of mysticism and spirits that lived in objects. It was during the classic Greek culture, between 600 B.C. and 300 B.C., that people began to look beyond magic and spirits. One particular Greek philosopher, Aristotle, wrote a theory about the universe that offered not only explanations about things such as motion but also offered a sense of beauty, order, and perfection. The theory seemed to fit with other ideas that people had and was held to be correct for nearly two thousand years after it was written. It was not until the work of Galileo and Newton during the 1600s that a new, correct understanding about motion was developed. The development of ideas about motion is an amazing and absorbing story. You will learn in this chapter how to describe and use some properties of motion. This will provide some basic understandings about motion, and will be very helpful in understanding some important aspects of astronomy and the earth sciences, as well as the movement of living things.

DESCRIBING MOTION

Motion is one of the more common events in your surroundings. You can see motion in natural events such as clouds moving, rain and snow falling, and streams of water, all moving in a never-ending cycle. Motion can also be seen in the activities of people who walk, jog, or drive various machines from place to place. Motion is so common that you would think everyone would intuitively understand the concepts of motion, but history indicates that it was only during the past three hundred years or so that people began to understand motion correctly. Perhaps the correct concepts are subtle and contrary to common sense, requiring a search for simple, clear concepts in an otherwise complex situation. The process of finding such order in a multitude of sensory impressions by taking measurable data, and then inventing a concept to describe what is happening, is the activity called *science*. We will now apply this process to motion.

What is motion? Consider a ball that you notice one morning in the middle of a lawn. Later in the afternoon, you notice that the ball is at the edge of the lawn, against a fence, and you wonder if the wind or some person moved the ball. You do not know if the wind blew it at a steady rate, if many gusts of wind moved it, or even if some children kicked it all over the yard. All you know for sure is that the ball has been moved because it is in a different position after some time passed. These are the two important aspects of motion: (1) a change of position and (2) the passage of time.

If you did happen to see the ball rolling across the lawn in the wind, you would see more than the ball at just two locations. You would see the ball moving continuously. You could consider, however, the ball in continuous motion to be a series of individual locations with very small time intervals. Moving involves a change of position during some time period. Motion is the act or process of something changing position.

The motion of an object is usually described with respect to something else that is considered to be not moving. (Such a sta-

tionary object is said to be "at rest.") Imagine that you are traveling in an automobile with another person. You know that you are moving across the land outside the car since your location on the highway changes from one moment to another. Observing your fellow passenger, however, reveals no change of position. You are in motion relative to the highway outside the car. You are not in motion relative to your fellow passenger. Your motion, and the motion of any other object or body, is the process of a change in position *relative* to some reference object or location. Thus *motion* can be defined as the act or process of changing position relative to some reference during a period of time.

MEASURING MOTION

You have learned that objects can be described by measuring certain fundamental properties such as mass and length. Since motion involves (1) a change of position and (2) the passage of *time*, the motion of objects can be described by using combinations of the fundamental properties of length and time. Combinations of these measurements describe three properties of motion: *speed*, *velocity*, and *acceleration*.

Speed

Suppose you are in a car that is moving over a straight road. How could you describe your motion? You need at least two measurements, (1) the distance you have traveled, and (2) the time that has elapsed while you covered this distance. Such a distance and time can be expressed as a ratio that describes your motion. This ratio is a property of motion called **speed**, which is a measure of how fast you are moving. Speed is defined as distance per unit of time, or

$$speed = \frac{distance}{time}$$

The units used to describe speed are usually miles/hour (mi/h), kilometers/hour (km/h), or meters/second (m/s).

22 **Chapter Two** Motion

Applying Science to the Real World

Concepts Applied

As you browse through each chapter, you will find one or more Concepts Applied activities. These are simple exercises that you can perform at home or in the classroom to demonstrate important concepts and reinforce your understanding of them. This feature also describes the application of those concepts to your everyday life.

> "The Tillery book is a comprehensive integrated science book. Ideas on home experiments and applications of knowledge to familiar case scenarios are strengths. The thought provoking questions at the end of the chapter are useful and promote critical thinking."
>
> —Tasneem F. Khaleel, Montana State University

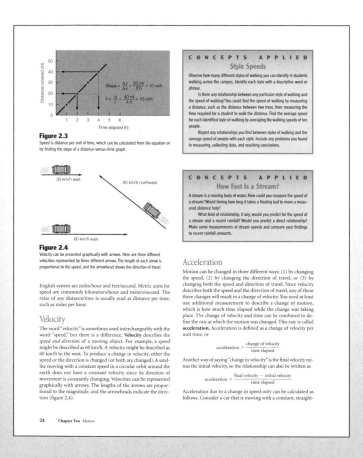

Figure 2.3
Speed is distance per unit of time, which can be calculated from the equation or by finding the slope of a distance-versus-time graph.

Figure 2.4
Velocity can be presented graphically with arrows. Here are three different velocities represented by three different arrows. The length of each arrow is proportional to the speed, and the arrowhead shows the direction of travel.

English system are miles/hour and feet/second. Metric units for speed are commonly kilometers/hour and meters/second. The ratio of any distance/time is usually read as distance per time, such as miles per hour.

Velocity

The word "velocity" is sometimes used interchangeably with the word "speed," but there is a difference. **Velocity** describes the *speed and direction* of a moving object. For example, a speed might be described as 60 km/h. A velocity might be described as 60 km/h to the west. To produce a change in velocity, either the speed or the direction is changed (or both are changed). A satellite moving with a constant speed in a circular orbit around the earth does not have a constant velocity since its direction of movement is constantly changing. Velocities can be represented graphically with arrows. The lengths of the arrows are proportional to the magnitude, and the arrowheads indicate the direction (figure 2.4).

CONCEPTS APPLIED
Style Speeds

Observe how many different styles of walking you can identify in students walking across the campus. Identify each style with a descriptive word or phrase.

Is there any relationship between any particular style of walking and the speed of walking? You could find the speed of walking by measuring a distance, such as the distance between two trees, then measuring the time required for a student to walk the distance. Find the average speed for each identified style of walking by averaging the walking speeds of ten people.

Report any relationships you find between styles of walking and the average speed of people with each style. Include any problems you found in measuring, collecting data, and reaching conclusions.

CONCEPTS APPLIED
How Fast Is a Stream?

A stream is a moving body of water. How could you measure the speed of a stream? Would timing how long it takes a floating leaf to move a measured distance help?

What kind of relationship, if any, would you predict for the speed of a stream and a recent rainfall? Would you predict a direct relationship? Make some measurements of stream speeds and compare your findings to recent rainfall amounts.

Acceleration

Motion can be changed in three different ways: (1) by changing the speed, (2) by changing the direction of travel, or (3) by changing both the speed and the direction of travel. Since velocity describes both the speed and the direction of travel, any of these three changes will result in a change of velocity. You need at least one additional measurement to describe a change of motion, which is how much time elapsed while the change was taking place. The change of velocity and time can be combined to define the *rate* at which the motion was changed. This rate is called **acceleration**. Acceleration is defined as a change of velocity per unit time, or

$$\text{acceleration} = \frac{\text{change of velocity}}{\text{time elapsed}}$$

Another way of saying "change in velocity" is the final velocity minus the initial velocity, so the relationship can also be written as

$$\text{acceleration} = \frac{\text{final velocity} - \text{initial velocity}}{\text{time elapsed}}$$

Acceleration due to a change in speed only can be calculated as follows. Consider a car that is moving with a constant, straight-

Closer Look and Connections

Each chapter of *Integrated Science* also includes one or more **Closer Look** readings that discuss topics of special human or environmental concern, topics about interesting technological applications, or topics on the cutting edge of scientific research. These readings enhance the learning experience by taking a more detailed look at related topics and adding concrete examples to help you better appreciate the real-world applications of science.

All boxed features include informative materials that are supplementary in nature. In addition to the **Closer Look** readings, each chapter contains concrete interdisciplinary **Connec-**

tions that are set aside and highlighted. **Connections** will help you better appreciate the interdisciplinary nature of the sciences. The **Closer Look** and **Connections** readings highlight valuable information beyond the scope of the text and relate intrinsic concepts in the text to real world issues, underscoring the relevance of integrated science in confronting the many issues we face in our daily lives. They are identified with the following icons:

General: This icon identifies interdisciplinary topics that cross over several categories; for example, life sciences and technology.

Life: This icon identifies interdisciplinary life science topics, meaning connections concerning all living organisms collectively: plant life, animal life, marine life, and any other classification of life.

Technology: This icon identifies interdisciplinary technology topics, that is, connections concerned with the application of science for the comfort and well-being of people, especially through industrial and commercial means.

Measurement, Thinking, Scientific Methods: This icon identifies interdisciplinary concepts and understandings concerned with people trying to make sense out of their surroundings by making observations, measuring, thinking, developing explanations for what is observed, and experimenting to test those explanations.

Environmental Science: This icon identifies interdisciplinary concepts and understandings about the problems caused by human use of the natural world and remedies for those problems.

> "I really like the activities and Connections and Closer Look boxes. This really breaks the textbook up into 'easily consumable' chunks."
>
> —Stephen C. Brown, University of Texas at San Antonio

A Closer Look Abnormal Cell Division: Cancer

When cells lose their ability to control mitotic divisions, a group of cells forms a *tumor*. A **tumor** is a mass of undifferentiated cells not normally found in a certain portion of the body. A *benign tumor* is a cell mass that does not heal or spread beyond its original area of growth. A benign tumor can become harmful by growing large enough to interfere with normal body functions. Some tumors are malignant. *Malignant tumors* are nonencapsulated growths of tumor *cells* that are harmful; they may spread or invade other parts of the body. Cells of these tumors move from the original site (**metastasize**) and establish new colonies in other regions of the body. Cells break off from the original tumor and enter the bloodstream. When they get stuck to the inside of a capillary, these cells move through the wall of the blood vessel and invade the tissue, where they begin to reproduce by mitosis. This tumor causes new blood vessels to grow into this new site, which will carry nutrients to this growing mass. These vessels can also bring even more spreading cells to the new tumor site. **Cancer** is the term used to refer to any abnormal growth of cells that has a malignant potential. Agents responsible for causing cancer are called **carcinogens**.

CONCEPTS APPLIED
What Is Cancer?

Cancer is a group of many diseases characterized by the uncontrolled growth and spread of abnormal cells. If the spread is not controlled or checked, it results in death. However, many cancers can be cured if detected and treated promptly.

Cancer's Seven Warning Signals
1. A persistent change in bowel or bladder habits
2. A sore that does not heal
3. Unusual bleeding or discharge
4. A lump or thickening in the breast or elsewhere
5. Persistent indigestion or difficulty in swallowing
6. An obvious change in a wart or mole
7. A nagging cough or hoarseness

What percent of Americans will eventually have cancer? Cancer will strike three out of four families. If you notice one of the preceding symptoms, see your doctor.

To become better informed about cancer, contact your local Cancer Society office or write the American Cancer Society, Inc., 90 Park Ave., New York, NY 10016. Request a copy of the latest publication, *Cancer Facts and Figures*. Go to the Internet and search some of the boldface terms used in this chapter to learn more about this topic. Be certain that the sources you access are "good science."

Connections...
Keep Your Lawn Crabgrass Free with a Mitosis Inhibitor

The crabgrass control agent dithiopyr appears to bind to a cellular protein of the spindle. It is thought that dithiopyr acts on the chemical Microtubule Associated Protein (MAP), altering spindle stability. When this occurs, spindle formation does not occur and the cell dies in metaphase.

SUMMARY

Living things show the characteristics of (1) *metabolic processes*, (2) *generative processes*, (3) *responsive processes*, (4) *control processes*, and (5) a *unique structural organization*. The concept of the *cell* has developed over a number of years. Initially, only two regions, the *cytoplasm* and the *nucleus*, could be identified. At present, numerous *organelles* are recognized as essential components of both *prokaryotic* and *eukaryotic* cell types.

In the process of respiration, organisms convert foods into energy (ATP) and waste materials (carbon dioxide and water). Organisms that have oxygen (O_2) available can employ the Krebs cycle and electron-transport system (ETS). Plants use the waste materials of respiration. Therefore, there is a constant cycling of materials between plants and animals. Sunlight supplies the essential initial energy for making the large organic molecules necessary to maintain the forms of life we know.

In the light-capturing reaction stage of photosynthesis, plants use chlorophyll to trap the energy of sunlight. During the light-dependent stage they manufacture a source of chemical energy, ATP, and a source of hydrogen, NADPH. Atmospheric oxygen is released in this stage. In the light-independent reaction stage of photosynthesis, the ATP energy is used in a series of reactions to join the hydrogen from the NADPH to a molecule of carbon dioxide and form a simple carbohydrate. In subsequent reactions, plants use this molecule as a source of energy and raw materials to make complex carbohydrates, fats, and other organic molecules.

The *cell* is the common unit of life. We study individual cells and their structures to understand how they function as *individual living organisms* and as parts of *many-celled beings*. Knowing how prokaryotic and eukaryotic cell types resemble or differ from each other helps physicians control some organisms dangerous to humans.

Cell division is necessary for growth, repair, and reproduction. Cells go through a *cell cycle* that includes *cell division* (*mitosis* and *cytokinesis*) and *interphase*. *Interphase* is the period of growth and preparation for division. Mitosis is divided into four stages: *prophase, metaphase, anaphase,* and *telophase*. During mitosis, *two daughter nuclei* are formed from *one parent nucleus*. These nuclei have *identical sets of chromosomes and genes that are exact copies* of those of the parent. Although the process of *mitosis* has been presented as a series of phases, you should realize that it is a *continuous, flowing process* from prophase through telophase. Following mitosis, *cytokinesis* divides the *cytoplasm*, and the cell returns to *interphase*.

---

name of a scientist and can be expressed verbally, with an equation, or with a graph.

A *model* is used to help understand something that cannot be observed directly, explaining the unknown in terms of things already understood. Physical models, mental models, and equations are all examples of models that explain how nature behaves. A *theory* is a broad, detailed explanation that guides development and interpretations of experiments in a field of study.

Science and *nonscience* can be distinguished by the kinds of laws and rules that are constructed to unify the body of knowledge. Science involves the continuous *testing* of rules and principles by the collection of new facts. If the rules are not testable, or if no rules are used, it is not science. *Pseudoscience* uses scientific appearances to mislead.

Summary of Equations

1.1
$$\text{mass density} = \frac{\text{mass}}{\text{volume}}$$
$$\rho = \frac{m}{V}$$

KEY TERMS

controlled experiment (p. 13)   quantities (p. 10)
data (p. 8)   referent (p. 3)
density (p. 9)   scientific law (p. 14)
equation (p. 11)   standard unit (p. 4)
hypothesis (p. 13)   theory (p. 16)
measurement (p. 4)   unit (p. 4)
model (p. 14)   variables (p. 11)
properties (p. 3)

APPLYING THE CONCEPTS

1. The process of comparing a property of an object to a well-defined and agreed-upon referent is called the process of
   (a) generalizing.
   (b) measurement.
   (c) graphing.
   (d) scientific investigation.
2. The height of an average person is closest to
   (a) 1.0 m.
   (b) 1.5 m.
   (c) 2.5 m.
   (d) 3.5 m.
3. Which of the following standard units is defined in terms of an object as opposed to an event?
   (a) kilogram
   (b) meter
   (c) second
   (d) none of the above
4. One-half liter of water has a mass of
   (a) 0.5 g.
   (b) 5 g.
   (c) 50 g.
   (d) 500 g.
5. A cubic centimeter (cm^3) of water has a mass of about 1
   (a) mL.
   (b) kg.
   (c) g.
   (d) dm.
6. Measurement information that is used to describe something is called
   (a) referents.
   (b) properties.
   (c) data.
   (d) a scientific investigation.
7. The property of volume is a measure of
   (a) how much matter an object contains.
   (b) how much space an object occupies.
   (c) the compactness of matter in a certain size.
   (d) the area on the outside surface.
8. As the volume of a cube becomes larger and larger, the surface-area-to-volume ratio
   (a) increases.
   (b) decreases.
   (c) remains the same.
   (d) sometimes increases and sometimes decreases.
9. If you consider a very small portion of a material that is the same throughout, the density of the small sample will be
   (a) much less.
   (b) slightly less.
   (c) the same.
   (d) greater.
10. A scientific law can be expressed as
   (a) a written concept.
   (b) an equation.
   (c) a graph.
   (d) any of the above.

Answers
1. b 2. b 3. a 4. d 5. c 6. c 7. b 8. b 9. c 10. d

QUESTIONS FOR THOUGHT

1. What is a concept?
2. What two things does a measurement statement always contain? What do the two things tell you?
3. Other than familiarity, what are the advantages of the English system of measurement?
4. Describe the metric standard units for length, mass, and time.
5. Does the density of a liquid change with the shape of a container? Explain.
6. Does a flattened pancake of clay have the same density as the same clay rolled into a ball? Explain.
7. Compare and contrast a scientific hypothesis and a scientific theory.
8. What is a model? How are models used?
9. Are all theories always completely accepted or completely rejected? Explain.
10. What is pseudoscience and how can you always recognize it?

Connections...
Kidney Machines in Action

When the kidneys do not function properly, the toxic waste urea builds up in the blood because the ailing kidneys cannot filter it out. This eventually leads to a condition known as *uremia*, "urine in the blood." Unless the urea is removed from the body, the concentration of urea increases and will eventually lead to death. Kidney or dialysis machines use tubular cellulose membranes to separate the urea from the blood.

Cellulose is a polymer of glucose molecules and contains openings through which urea molecules pass. In such a machine, cellulose membrane tubes are immersed in a large volume of "cleansing" or dialysis solution. The blood is pumped from the vein of a person with uremia through this tubing, and then returned to the patient. While in the machine, the membrane allows most of the urea in the blood to pass through the tubing walls but not the larger blood proteins and blood cells. Therefore, the patient does not loose vital proteins or blood cells. This system works because the dialysis solution in which the tubing is immersed has a lower amount of urea than does the blood. Because of this concentration gradient, urea in the blood passes from the high concentration in the blood through the cellulose membrane tubing pores into the surrounding dialysis solution. This results in a decrease in urea concentration in the blood. As the concentration of the urea increases in the dialysis solution, the solution is flushed out. This enables the dialysis solution to maintain a constant, low concentration of urea and maintain the movement of urea molecules out of the person's body. To maintain the blood's concentration of other vital ions, the solution is made with the same concentration of those ions as found in the blood. This means that these other ions are in dynamic equilibrium, and so their concentration does not change during kidney dialysis (box figure 19.3).

Blood inlet from patient

Dialysis solution outlet

Tubular cellulose membranes

Dialysis solution inlet

Blood outlet return to patient

Box Figure 19.3

is, chemical bond energy does not have to be expended. For example, animals are constantly using oxygen in various chemical reactions. Consequently, the oxygen concentration in cells always remains low. The cells, then, contain a lower concentration of oxygen than the oxygen level outside of the cells. This creates a diffusion gradient, and the oxygen molecules diffuse from the outside of the cell to the inside of the cell.

the passage of ions. If a particular portion of the membrane has a large number of positive ions on its surface, positively charged ions in the environment will be repelled and prevented from crossing the membrane.

Water molecules easily diffuse through cell membranes. The net movement (diffusion) of water molecules through a selectively permeable membrane is known as **osmosis**. In any

---

End-of-Chapter Features

At the end of each chapter you will find the following materials:

- *Summary:* highlights the key elements of the chapter.
- *Summary of Equations* (chapters 1–8, 11): reinforces your retention of the equations presented.
- *Key Terms:* page references where you will find the terms defined in the chapter.

- *Applying the Concepts:* tests your comprehension of the material covered with a multiple-choice quiz.
- *Questions for Thought:* challenges you to demonstrate your understanding of the topics.
- *Parallel Exercises* (chapters 1–9, 11): reinforces your problem-solving skills. There are two groups of parallel exercises, Group A and Group B. The Group A parallel exercises have complete solutions worked out, along with useful comments in appendix D. The Group B parallel exercises are similar to those in Group A but do not contain answers in the text. By working through the Group A parallel exercises and checking the solutions in appendix D, you will gain confidence in tackling the parallel exercises in Group B.

"The end of chapter material is one of the most useful aspects of this book. I feel that the Applying the Concepts and Questions for Thought are very helpful in preparing students for quizzes and tests."
—Stephen C. Brown, University of Texas at San Antonio

"The chapter review and discussion questions are very useful in our course. We try to let the student see that science is a learning process."
—David J. DiMattio, St. Bonaventure University

End-of-Text Material

At the back of the text you will find appendixes that provide additional background details, charts, and answers to chapter exercises. There is also a glossary of all key terms and an index organized alphabetically by subject matter. Special tables are printed on the inside covers for reference use.

"... many books addressing similar disciplines have a tendency to talk over a student's head, making a student frustrated further in a class they do not want to be attending.... Personally I would admit that Integrated Science has a slight edge. The glossary seems up to date and centers in on words many nonscience majors may not understand."

—David J. DiMattio, St. Bonaventure University

APPENDIX A

Mathematical Review

WORKING WITH EQUATIONS

Many of the problems of science involve an equation, a shorthand way of describing patterns and relationships that are observed in nature. Equations are also used to identify properties and to define certain concepts. But all uses have well-established meanings, symbols that are used by convention, and allowed mathematical operations. This appendix will assist you in better understanding equations and the reasoning that goes with the manipulation of equations in problem-solving activities.

Background

In addition to a knowledge of rules for carrying out mathematical operations, an understanding of certain quantitative ideas and concepts can be very helpful when working with equations. Among these helpful concepts are (1) the meaning of inverse and reciprocal, (2) the concept of a ratio, and (3) fractions.

585

GLOSSARY

A

abiotic factors nonliving parts of an organism's environment

absolute humidity a measure of the actual amount of water vapor in the air at a given time—for example, in grams per cubic meter

absolute magnitude a classification scheme to compensate for the distance differences to stars...

615

Multimedia Supplements

Digital Content Manager

The **Digital Content Manager** contains jpeg files of the four-color illustrations from the text as well as a selection of animations. These digital assets are grouped by chapter on a cross-platform CD-ROM with a user-friendly interface. With the help of these valuable resources, instructors can create customized classroom presentations, visually based tests and quizzes, dynamic course website content, and attractive printed support materials.

Supplements

Integrated Science is accompanied by a variety of supplementary materials. The Online Learning Center is a fully interactive website, including additional resources for the instructor and study and self-testing materials for the student. The instructor's manual for the text can also be found on the Online Learning Center. Also available are an Instructor's Testing and Resource CD-ROM and a Digital Content Manager CD-ROM for the instructor, along with more traditional supplements such as a laboratory manual and transparencies.

Online Learning Center (OLC)

The **Online Learning Center** is an online repository of teaching and learning aids for the *Integrated Science* text. It houses downloadable and printable versions of traditional ancillaries, plus a wealth of online content in both an Instructor's Edition and a Student Edition.

Instructor's Edition of the Online Learning Center The text-specific Online Learning Center includes the fully downloadable instructor's manual. The instructor's manual, written by the text authors, provides chapter outlines, an introduction and summary for each chapter, suggestions for discussion and demonstrations, and multiple-choice questions (with answers) that can be used as resources for cooperative learning. It also includes answers and solutions to all end-of-chapter questions and exercises not provided in the text. Additionally, this instructor's edition of the OLC features integration theme ideas and syllabi contributed by instructors who are presently teaching the Integrated Science course across the country.

The Online Learning Center also contains the figures from the text in jpeg format, over 200 animations, a feedback page, and many other features.

Student Edition of the Online Learning Center Students can use the Online Learning Center to study in a variety of ways that include: scorable practice quizzes, additional self-assessment quizzes, animations, crossword puzzles, and flashcards that use key terms and definitions from the text, an online glossary, a career center, and Web links. Students can also read up-to-the-minute news feeds on integrated science topics.

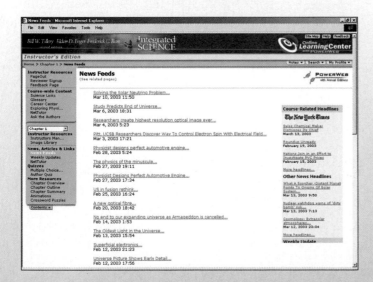

Instructor's Testing and Resource CD-ROM

The **Instructor's Testing and Resource CD-ROM** contains the *Integrated Science* test bank (over 700 test questions in a combination of true/false and multiple-choice formats) within the Brownstone DIPLOMA© test generator. The Brownstone software includes a test generator, an online testing program, Internet testing, and a grade management system. This user-friendly software's testing capability is consistently ranked number one in evaluations over other products. Also located on the Instructor's Testing and Resources CD-ROM are Word and PDF files of the test bank, the instructor's manual, and the quizzes from the Online Learning Center. The Word files for the test bank, instructor's manual, and Online Learning Center quizzes can be used in combination with the Brownstone software or independently.

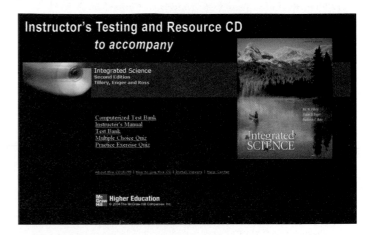

Printed Supplementary Materials

Laboratory Manual

The laboratory manual, written and classroom-tested by the authors, presents a selection of laboratory exercises specifically written for the interest and abilities of nonscience majors. There are laboratory exercises that require measurement, data analysis, and thinking in a more structured learning environment. Alternative open-ended exercises, "Invitations to Inquiry," are provided for instructors who would like a less-structured approach. When the laboratory manual is used with *Integrated Science,* students will have an opportunity to master basic scientific principles and concepts, learn new problem-solving and thinking skills, and understand the nature of scientific inquiry from the perspective of hands-on experiences. There is also an **instructor's edition lab manual** available for professors upon request.

Transparencies

A set of 100 full-color transparencies features images from the text. The images have been modified to ensure maximum readability in both small and large classroom settings.

Acknowledgments

This revision of *Integrated Science* has been made possible by the many users and reviewers of its first edition. The authors are indebted to the first edition reviewers for their critical reviews, comments, and suggestions. The first edition reviewers were:

Stephen C. Brown *University of Texas–San Antonio*
Linda Chamberlin *Valdosta State University*
Kailash Chandra *Savannah State University*
David J. DiMattio *St. Bonaventure University*

Gian S. Ghuman *Savannah State University*
Michael Giazzoni *Point Park College*
Tasneem F. Khaleel *Montana State University–Billings*
Estela Llinás *University of Pittsburgh–Greensburg*
Joe Mehaffey *Francis Marion University*
Kenneth S. Mendelson *Marquette University*
Lee Netherton *John Brown University*
G. A. Nixon *Texas A&M University–Commerce*
Mary Ellen Teasdale *Texas A&M University–Commerce*
Laura Thurlow *Jackson Community College*
Jay R. Yett *Orange Coast College*
Thad Zaleskiewicz *University of Pittsburgh–Greensburg*

Many constructive suggestions, new ideas, and invaluable advice were provided by the initial reviewers through several stages of manuscript development, which resulted in the first edition. Special thanks and appreciation to those who reviewed all or parts of the manuscript:

Douglas Allchin *University of Texas–El Paso*
B. J. Bateman *Troy State University*
William Blaker *Furman University*
Lauretta Buschar *Beaver College*
Tracey Cascadden *University of New Mexico*
Tim Champion *J. C. Smith University*
James Courtright *Marquette University*
David Fawcett *Columbia State Community College*
David Grainger *Colorado State University*
James J. Grant *St. Peter's College*
William Keller *St. Petersburg Junior College*
Mary Hurn Korte *Concordia University*
Theo Koupelis *University of Wisconsin–Marathon*
Bryan Long *Columbia State Community College*
Bruce MacLaren *Eastern Kentucky University*
Donald Miller *University of Michigan–Dearborn*
Scott Mohr *Boston University*
Joaquin Ruiz *University of Arizona*
John Oakes *Marian College*
C. Dianne Phillips *University of Arkansas*
Richard Sapanaro *Contra Costa College*
Herbert Stewart *Florida Atlantic University*
Deborah Tull *University of New Mexico*
Andrew Wallace *Angelo State University*
Jim Westgate *Lamar University*
Robert Wingfield, Jr. *Fisk University*
Yen-Yuan James Wu *Texas A & M University–Commerce*

The authors would also like to thank the following contributors for their theme integration ideas, which appear on the *Integrated Science* Online Learning Center:

Mary Brown *Lansing Community College*
David J. DiMattio *St. Bonaventure University*
Tasneem F. Khaleel *Montana State University–Billings*
G. A. Nixon and Mary Ellen Teasdale *Texas A&M University–Commerce*
Thad Zaleskiewicz and Jennifer Siegert *University of Pittsburgh–Greensburg*

Meet the Authors

Bill W. Tillery

Bill W. Tillery is a professor in the Department of Physics and Astronomy at Arizona State University, where he has been a member of the faculty since 1973. He earned a Bachelors degree at Northeastern State University (1960), and Masters and Doctorate degrees from the University of Northern Colorado (1967). Before moving to Arizona State University, he served as Director of the Science and Mathematics Teaching Center at the University of Wyoming (1969–1973) and as an assistant professor at Florida State University (1967–1969). Bill has served on numerous councils, boards, and committees and was honored as the "Outstanding University Educator" at the University of Wyoming in 1972. He was elected the "Outstanding Teacher" in the Department of Physics and Astronomy at Arizona State University in 1995.

During his time at Arizona State, Bill has taught a variety of courses, including general education courses in science and society, physical science, and introduction to physics. He has received more than 40 grants from the National Science Foundation, the U.S. Office of Education, from private industry (Arizona Public Service), and private foundations (The Flinn Foundation) for science curriculum development and science teacher inservice training. In addition to teaching and grant work, Bill has authored or coauthored more than 60 textbooks, many monographs, and served as editor of three separate newsletters and journals between 1977 and 1996.

Bill also maintains a website dedicated to providing resources for science teachers. This site is named The Science Education Resource Page (SERP) and is funded by a grant from the Flinn Foundation. The URL is http://serp.la.asu.edu.

Eldon D. Enger

Eldon D. Enger is professor emeritus of biology at Delta College, a community college near Saginaw, Michigan. He received his B.A. and M.S. degrees from the University of Michigan. Eldon has over 30 years of teaching experience, during which he has taught biology, zoology, environmental science, and several other courses. He has been very active in curriculum and course development.

Eldon is an advocate for variety in teaching methodology. He feels that if students are provided with varied experiences, they are more likely to learn. In addition to the standard textbook assignments, lectures, and laboratory activities, his classes have included writing assignments, student presentation of lecture material, debates by students on controversial issues, field experiences, individual student projects, and discussions of local examples and relevant current events. He feels that textbooks are very valuable for presenting content, especially if they contain accurate, informative drawings and visual examples. Lectures are best used to help students see themes and make connections, and laboratory activities provide important hands-on activities.

He has been a Fulbright Exchange Teacher to Australia and Scotland; received the Bergstein Award for Teaching Excellence and the Scholarly Achievement Award from Delta College; and participated as a volunteer in Earthwatch Research Programs in Costa Rica, the Virgin Islands, and Australia. During 2001 he was a member of a People to People delegation to South Africa.

Eldon is married and has two adult sons. He enjoys a variety of outdoor pursuits such as cross-country skiing, hiking, hunting, kayaking, fishing, camping, and gardening. Other interests include reading a wide variety of periodicals, beekeeping, singing in a church choir, and preserving garden produce.

Frederick C. Ross

Fred Ross is professor emeritus of biology at Delta College, a community college near Saginaw, Michigan. He received his B.S. and M.S. from Wayne State University, Detroit, Michigan, and has attended several other universities and institutions. Fred has over 30 years of teaching experience, including junior and senior high school, during which he has taught biology, cell biology, biological chemistry, microbiology, environmental science, and zoology. He has been very active in curriculum and course development. These activities included the development of courses in infection control and microbiology, AIDS and infectious diseases, and a PBS ScienceLine course for elementary and secondary education majors in cooperation with Central Michigan University. In addition, he was involved in the development of the wastewater microbiology technician curriculum offered by Delta College.

He was also actively involved in the National Task Force of Two-Year College Biologists (American Institute of Biological Sciences); N.S.F. College Science Improvement Program (COSIP); Evaluator for Science and Engineering Fairs, Michigan Community College Biologists (MCCB); Judge for The Michigan Science Olympiad, Science Bowl; a member of a committee to develop and update Blood-Borne Pathogen Standards protocol; American Society for Microbiology as a member of Topic Outlines in Introductory Microbiology Study Group; and Member Sigma Xi, honor society for scientists and engineers.

Fred involved his students in a variety of learning techniques and has been a prime advocate of writing-to-learn. Besides writing, his students typically engaged in active learning techniques, including use of inquiry-based learning, the Internet, e-mail communications, field experiences, classroom presentation, as well as lab work. The goal of his classroom presentations and teaching was to actively engage the minds of his students in understanding the material, not just memorization of scientific facts.

Fred is married and recently a grandfather. He enjoys sailing, horseback riding, and cross-country skiing.

CHAPTER ONE

What Is Science?

Science is concerned with your surroundings and your concepts and understanding of these surroundings.

OUTLINE

Have you ever thought about your thinking and what you know? On a very simplified level, you could say that everything you know came to you through your senses. You see, hear, and touch things of your choosing and you can also smell and taste things in your surroundings. Information is gathered and sent to your brain by your sense organs. Somehow, your brain processes all this information in an attempt to find order and make sense of it all. Finding order helps you understand the world and what may be happening at a particular place and time. Finding order also helps you predict what may happen next, which can be very important in a lot of situations.

This is a book on thinking about and understanding your surroundings. These surroundings range from the obvious, such as the landscape and the day-to-day weather, to the not so obvious, such as how atoms are put together. Your surroundings include natural things as well as things that people have made and used (figure 1.1). You will learn how to think about your surroundings, whatever your previous experience with thought-demanding situations. This first chapter is about "tools and rules" that you will use in the thinking process.

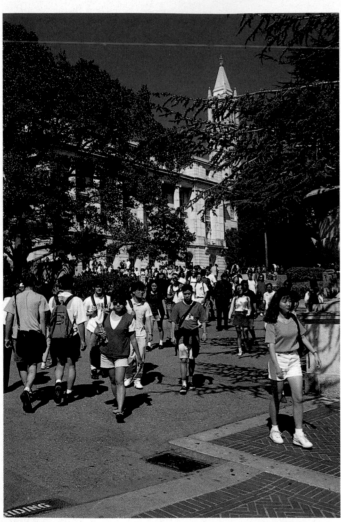

Figure 1.1

Your surroundings include naturally occurring objects and manufactured objects such as sidewalks and buildings.

OBJECTS AND PROPERTIES

Science is concerned with making sense out of the environment. The early stages of this "search for sense" usually involve objects in the environment, things that can be seen or touched. These could be *objects* you see every day, such as a glass of water, a moving automobile, or a running dog. They could be quite large, such as the sun, the moon, or even the solar system, or invisible to the unaided human eye. Objects can be any size, but people are usually concerned with objects that are larger than a pinhead and smaller than a house. Outside these limits, the actual size of an object is difficult for most people to comprehend.

As you were growing up, you learned to form a generalized mental image of objects called a *concept*. Your concept of an object is an idea of what it is, in general, or what it should be according to your idea (figure 1.2). You usually have a word stored away in your mind that represents a concept. The word "chair," for example, probably evokes an idea of "something to sit on." Your generalized mental image for the concept that goes with the word "chair" probably includes a four-legged object with a backrest. Upon close inspection, most of your (and everyone else's) concepts are found to be somewhat vague. For example, if the word "chair" brings forth a mental image of something with four legs and a backrest (the concept), what is the difference between a "high chair" and a "bar stool"? When is a chair a chair and not a stool? These kinds of questions can be troublesome for many people.

Not all of your concepts are about material objects. You also have concepts about intangibles such as time, motion, and relationships between events. As was the case with concepts of material objects, words represent the existence of intangible concepts. For example, the words "second," "hour," "day," and "month" represent concepts of time. A concept of the pushes and pulls that come with changes of motion during an airplane flight might be represented with such words as "accelerate" and "falling." Intangible concepts might seem to be more abstract since they do not represent material objects.

By the time you reach adulthood you have literally thousands of words to represent thousands of concepts. But most, you would find on inspection, are somewhat ambiguous and not at all clear-cut. That is why you find it necessary to talk about certain concepts for a minute or two to see if the other person has the same "concept" for words as you do. That is why when one person says, "Boy, was it hot!" the other person may respond, "How hot was it?" The meaning of "hot" can be quite different for two people, especially if one is from Arizona and the other from Alaska!

The problem with words, concepts, and mental images can be illustrated by imagining a situation involving you and another person. Suppose that you have found a rock that you believe would make a great bookend. Suppose further that you are talk-

Figure 1.2

What is your concept of a chair? Are all of these pieces of furniture chairs? Most people have concepts, or ideas of what things in general should be, that are loosely defined. The concept of a chair is one example of a loosely defined concept.

Figure 1.3

Could you describe this rock to another person over the telephone so that the other person would know *exactly* what you see? This is not likely with everyday language, which is full of implied comparisons, assumptions, and inaccurate descriptions.

ing to the other person on the telephone, and you want to discuss the suitability of the rock as a bookend, but you do not know the name of the rock. If you knew the name, you would simply state that you found a "_____." Then you would probably discuss the rock for a minute or so to see if the other person really understood what you were talking about. But not knowing the name of the rock, and wanting to communicate about the suitability of the object as a bookend, what would you do? You would probably describe the characteristics, or **properties,** of the rock. Properties are the qualities or attributes that, taken together, are usually peculiar to an object. Since you commonly determine properties with your senses (smell, sight, hearing, touch, and taste), you could say that the properties of an object are the effect the object has on your senses. For example, you might say that the rock is a "big, yellow, smooth rock with shiny gold cubes on one side." But consider the mental image that the other person on the telephone forms when you describe these properties. It is entirely possible that the other person is thinking of something very different from what you are describing (figure 1.3)!

As you can see, the example of describing a proposed bookend by listing its properties in everyday language leaves much to be desired. The description does not really help the other person form an accurate mental image of the rock. One problem with the attempted communication is that the description of any property implies some kind of *referent*. The word **referent** means that you *refer to*, or think of, a given property in terms of another, more familiar object. Colors, for example, are sometimes stated with a referent. Examples are "sky blue," "grass green," or "lemon yellow." The referents for the colors blue, green, and yellow are, respectively, the sky, living grass, and a ripe lemon.

Referents for properties are not always as explicit as they are with colors, but a comparison is always implied. Since the comparison is implied, it often goes unspoken and leads to assump-

tions in communications. For example, when you stated that the rock was "big," you assumed that the other person knew that you did not mean as big as a house or even as big as a bicycle. You assumed that the other person knew that you meant that the rock was about as large as a book, perhaps a bit larger.

Another problem with the listed properties of the rock is the use of the word "smooth." The other person would not know if you meant that the rock *looked* smooth or *felt* smooth. After all, some objects can look smooth and feel rough. Other objects can look rough and feel smooth. Thus, here is another assumption, and probably all of the properties lead to implied comparisons, assumptions, and a not very accurate communication. This is the nature of your everyday language and the nature of most attempts at communication.

QUANTIFYING PROPERTIES

Typical day-to-day communications are often vague and leave much to be assumed. A communication between two people, for example, could involve one person describing some person, object, or event to a second person. The description is made by using referents and comparisons that the second person may or may not have in mind. Thus, such attributes as "long" fingernails or "short" hair may have entirely different meanings to different people involved in a conversation. Assumptions and vagueness can be avoided by using **measurement** in a description. Measurement is a process of comparing a property to a well-defined and agreed-upon referent. The well-defined and agreed-upon referent is used as a standard called a **unit**. The measurement process involves three steps: (1) *comparing* the referent unit to the property being described, (2) following a *procedure,* or operation, which specifies how the comparison is made, and (3) *counting* how many standard units describe the property being considered.

As an example of how the measurement process works, consider the property of *length.* Most people are familiar with the concept of the length of something (long or short), the use of length to describe distances (close or far), and the use of length to describe heights (tall or short). The referent units used for measuring length are the familiar inch, foot, and mile from the English system and the centimeter, meter, and kilometer of the metric system. These systems and specific units will be discussed later. For now, imagine that these units do not exist but that you need to measure the length and width of this book. This imaginary exercise will illustrate how the measurement process eliminates vagueness and assumption in communication.

The first requirement in the measurement process is to choose some referent unit of length. You could arbitrarily choose something that is handy, such as the length of a standard paper clip, and you could call this length a "clip." Now you must decide on a procedure to specify how you will use the clip unit. You could define some specific procedures. For example:

1. Place a clip parallel to and on the long edge, or length, of the book so the end of the referent clip is lined up with the bottom edge of the book. Make a small pencil mark at the other end of the clip, as shown in figure 1.4.
2. Move the outside end of the clip to the mark and make a second mark at the other end. Continue doing this until you reach the top edge of the book.
3. Compare how many clip replications are in the book length by counting.
4. Record the length measurements by writing (a) how many clip replications were made and (b) the name of the clip length.

If the book length did not measure to a whole number of clips, you might need to divide the clip length into smaller subunits to be more precise. You could develop a *scale* of the basic clip unit and subunits. In fact, you could use multiples of the basic clip unit for an extended scale, using the scale for measurement rather than moving an individual clip unit. You could call the scale a "clipstick" (as in yardstick or meterstick).

Figure 1.4

As an example of the measurement process, a standard paper-clip length is selected as a referent unit. The unit is compared to the property that is being described. In this example, the property of book length is measured by counting clip lengths.

The measurement process thus uses a defined referent unit, which is compared to a property being measured. The *value* of the property is determined by counting the number of referent units. The name of the unit implies the procedure that results in the number. A measurement statement always contains a *number* and *name* for the referent unit. The number answers the question "How much?" and the name answers the question "Of what?" Thus a measurement always tells you "how much of what." You will find that using measurements will sharpen your communications. You will also find that using measurements is one of the first steps in understanding your physical environment.

MEASUREMENT SYSTEMS

Measurement is a process that brings precision to a description by specifying the "how much" and "of what" of a property in a particular situation. A number expresses the value of the property, and the name of a unit tells you what the referent is as well as implying the procedure for obtaining the number. Referent units must be defined and established, however, if others are to understand and reproduce a measurement. It would be meaningless, for example, for you to talk about a length in "clips" if other people did not know what you meant by a "clip" unit. When standards are established the referent unit is called a **standard unit** (figure 1.5). The use of standard units makes it possible to communicate and duplicate measurements. Standard units are usually defined and established by governments and their agencies that are created for that purpose. In the United States, the agency concerned with measurement standards is named the National Institute of Standards and Technology. In Canada, the Standards Council of Canada oversees the National Standard System.

There are two major *systems* of standard units in use today, the English system and the metric system. The metric system is used throughout the world except in the United States, where both systems are in use. The continued use of the English system

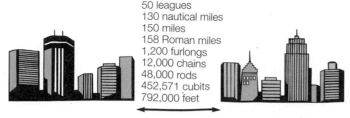

50 leagues
130 nautical miles
150 miles
158 Roman miles
1,200 furlongs
12,000 chains
48,000 rods
452,571 cubits
792,000 feet

Figure 1.5

Any of these units and values could have been used at some time or another to describe the same distance between these hypothetical towns. Any unit could be used for this purpose, but when one particular unit is officially adopted, it becomes known as the *standard unit.*

in the United States presents problems in international trade, so there is pressure for a complete conversion to the metric system. More and more metric units are being used in everyday measurements, but a complete conversion will involve an enormous cost. Appendix A contains a method for converting from one system to the other easily. Consult this section if you need to convert from one metric unit to another metric unit or to convert from English to metric units or vice versa. Conversion factors are listed inside the front cover.

People have used referents to communicate about properties of things throughout human history. The ancient Greek civilization, for example, used units of *stadia* to communicate about distances and elevations. The "stadium" was a unit of length of the racetrack at the local stadium ("stadia" is the plural of stadium), based on a length of 125 paces. Later civilizations, such as the ancient Romans, adopted the stadia and other referent units from the ancient Greeks. Some of these same referent units were later adopted by the early English civilization, which eventually led to the *English system* of measurement. Some adopted units of the English system were originally based on parts of the human body, presumably because you always had these referents with you (figure 1.6). The inch, for example, used the end joint of the thumb for a referent. A foot, naturally, was the length of a foot, and a yard was the distance from the tip of the nose to the end of the fingers on an arm held straight out. A cubit was the distance from the end of an elbow to the fingertip, and a fathom was the distance between the fingertips of two arms held straight out. As you can imagine, there were problems with these early units because everyone was not the same size. Beginning in the 1300s, the sizes of the units were gradually standardized by various English kings (table 1.1). In 1879, the United States, along with sixteen other countries, signed the *Treaty of the Meter,* defining the English units in terms of the metric system. The United States thus became officially metric, but not entirely metric in everyday practice.

The *metric system* was established by the French Academy of Sciences in 1791. The academy created a measurement system that was based on invariable referents in nature, not human body parts. These referents have been redefined over time to make the standard units more reproducible. In 1960, six standard metric units were established by international agreement.

1,000 double paces = 1 mile

Figure 1.6

Many early units for measurement were originally based on the human body. Some of the units were later standardized by governments to become the basis of the English system of measurement.

Table 1.1	Early Conversion Table for English Units of Volume
Two Quantities	**Equivalent Quantity**
2 mouthfuls	= 1 jigger
2 jiggers	= 1 jack
2 jacks	= 1 jill
2 jills	= 1 cup
2 cups	= 1 pint
2 pints	= 1 quart
2 quarts	= 1 pottle
2 pottles	= 1 gallon
2 gallons	= 1 pail
2 pails	= 1 peck
2 pecks	= 1 bushel

The *International System of Units,* abbreviated *SI,* is a modernized version of the metric system. Today, the SI system has seven units that define standards for the properties of length, mass, time, electric current, temperature, amount of substance, and

Table 1.2 The SI Standard Units

Property	Unit	Symbol
Length	meter	m
Mass	kilogram	kg
Time	second	s
Electric current	ampere	A
Temperature	kelvin	K
Amount of substance	mole	mol
Luminous intensity	candela	cd

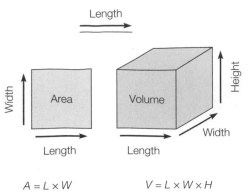

$$A = L \times W \qquad V = L \times W \times H$$

Figure 1.7

Area, or the extent of a surface, can be described by two length measurements. Volume, or the space that an object occupies, can be described by three length measurements. Length, however, can be described only in terms of how it is measured, so it is called a *fundamental property*.

light intensity (table 1.2). The standard units for the properties of length, mass, and time are introduced in this chapter. The remaining units will be introduced in later chapters as the properties they measure are discussed.

STANDARD UNITS FOR THE METRIC SYSTEM

If you consider all the properties of all the objects and events in your surroundings, the number seems overwhelming. Yet, close inspection of how properties are measured reveals that some properties are combinations of other properties (figure 1.7). Volume, for example, is described by the three length measurements of length, width, and height. Area, on the other hand, is described by just the two length measurements of length and width. Length, however, cannot be defined in simpler terms of any other property. There are four properties that cannot be described in simpler terms, and all other properties are combinations of these four. For this reason they are called the *fundamental properties*. A fundamental property cannot be defined in simpler terms other than to describe how it is measured. These four fundamental properties are (1) *length,* (2) *mass,* (3) *time,* and (4) *charge.* Used individually or in combinations, these four properties will describe or measure what you observe in nature. Metric units for measuring the fundamental properties of length, mass, and time will be described next. The fourth fundamental property, charge, is associated with electricity, and a unit for this property will be discussed in Chapter 6.

Length

The standard unit for length in the metric system is the *meter* (the symbol or abbreviation is m) A meter is defined in terms of the distance that light travels in a vacuum during a certain time period, 1/299,792,458 second. The important thing to remember, however, is that the meter is the metric *standard unit* for length. A meter is slightly longer than a yard, 39.3 inches. It is approximately the distance from your left shoulder to the tip of your right hand when your arm is held straight out. Many doorknobs are about one meter above the floor. Think about these distances when you are trying to visualize a meter length.

Mass

The standard unit for mass in the metric system is the *kilogram* (kg). The kilogram is defined as the mass of a certain metal cylinder kept by the International Bureau of Weights and Measures in France. This is the only standard unit that is still defined in terms of an object. The property of mass is sometimes confused with the property of weight since they are directly proportional to each other at a given location on the surface of the earth. They are, however, two completely different properties and are measured with different units. All objects tend to maintain their state of rest or straight-line motion, and this property is called "inertia." The *mass* of an object is a measure of the inertia of an object. The *weight* of the object is a measure of the force of gravity on it. This distinction between weight and mass will be discussed in detail in chapter 2. For now, remember that weight and mass are not the same property.

Time

The standard unit for time is the *second* (s). The second was originally defined as 1/86,400 of a solar day (1/60 × 1/60 × 1/24). The earth's spin was found not to be as constant as thought, so the second was redefined to be the duration required for a certain number of vibrations of a certain cesium atom. A special spectrometer called an "atomic clock" measures these vibrations and keeps time with an accuracy of several millionths of a second per year.

METRIC PREFIXES

The metric system uses prefixes to represent larger or smaller amounts by factors of 10. Some of the more commonly used prefixes, their abbreviations, and their meanings are listed in table 1.3. Suppose you wish to measure something smaller than the standard unit of length, the meter. The meter is subdivided into ten equal-sized subunits called *decimeters*. The prefix *deci-* has a meaning of "one-tenth of," and it takes 10 decimeters to equal the

Most people have heard of a leap year, but not a leap second. A *leap year* is needed because the earth does not complete an exact number of turns on its axis while completing one trip around the sun. Our calendar system was designed to stay in step with the seasons with 365-day years and a 366-day year (leap year) every fourth year.

Likewise, our clocks are occasionally adjusted by a one-second increment known as a *leap second*. The leap second is needed because the earth does not have a constant spin. Coordinated Universal Time is the worldwide scientific standard of timekeeping. It is based upon the earth's rotation and is kept accurate to within microseconds with carefully maintained atomic clocks. A leap second is a second added to Coordinated Universal Time to make it agree with astronomical time to within 0.9 second.

In 1955 astronomers at the U.S. Naval Observatory and the National Physical Laboratory in England measured the relationship between the frequency of the cesium atom (the standard of time) and the rotation of the earth at a particular period of time. The standard atomic clock second was defined to be equivalent to the fraction 1/31,556,925.9747 of the year 1900—or, an average second for that year. This turned out to be the time required for 9,192,631,770 vibrations of the cesium 133 atom. The second was defined in 1967 in terms of the length of time required for 9,192,631,770 vibrations of the cesium 133 atom. So, the atomic second was set equal to an average second of the earth rotation time near the turn of the twentieth century, but defined in terms of the frequency of a cesium atom.

The earth is constantly slowing from the frictional effects of the tides. Evidence of this slowing can be found in records of ancient observations of eclipses. From these records it is possible to determine the slowing of the earth, and one day used to be only 18 hours long 900 million years ago.

It has been over a century since the referent year used for the definition of a second, and the difference is now roughly 2 milliseconds per day. Other factors also affect the earth's spin, such as the wind from hurricanes, so that it is necessary to monitor the earth's rotation continuously and add or subtract leap seconds when needed.

Table 1.3 Some Metric Prefixes

Prefix	Symbol	Meaning
tera-	T	1,000,000,000,000 times the unit
giga-	G	1,000,000,000 times the unit
mega-	M	1,000,000 times the unit
kilo-	k	1,000 times the unit
hecto-	h	100 times the unit
deka-	da	10 times the unit
Unit		
deci-	d	0.1 of the unit
centi-	c	0.01 of the unit
milli-	m	0.001 of the unit
micro-	μ	0.000001 of the unit
nano-	n	0.000000001 of the unit
pico-	p	0.000000000001 of the unit

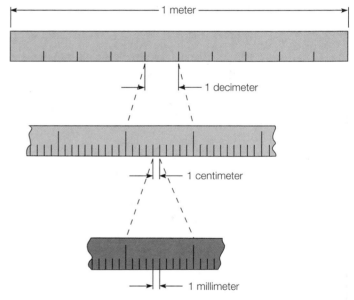

Figure 1.8

Compare the units shown above. How many millimeters fit into the space occupied by one centimeter? How many millimeters fit into the space of one decimeter? Can you express this as multiples of ten?

length of 1 meter. For even smaller measurements, each decimeter is divided into ten equal-sized subunits called *centimeters*. It takes 10 centimeters to equal 1 decimeter and 100 to equal 1 meter. In a similar fashion, each prefix up or down the metric ladder represents a simple increase or decrease by a factor of 10 (figure 1.8).

When the metric system was established in 1791, the standard unit of mass was defined in terms of the mass of a certain volume of water. A cubic decimeter (dm^3) of pure water at 4°C was *defined* to have a mass of 1 kilogram (kg). This definition was convenient because it created a relationship between length, mass, and volume. As illustrated in figure 1.9, a cubic decimeter is 10 cm on each side. The volume of this cube is therefore 10 cm × 10 cm × 10 cm, or 1,000 cubic centimeters (abbreviated as cc or cm^3). Thus, a volume of 1,000 cm^3 of water has a mass of 1 kg. Since 1 kg is 1,000 g, 1 cm^3 of water has a mass of 1 g.

The volume of 1,000 cm^3 also defines a metric unit that is commonly used to measure liquid volume, the *liter* (L). For smaller amounts of liquid volume, the milliliter (mL) is used. The relationship between liquid volume, volume, and mass of water is therefore

$$1.0 \text{ L} \rightarrow 1.0 \text{ dm}^3 \text{ and has a mass of } 1.0 \text{ kg}$$

or, for smaller amounts,

$$1.0 \text{ mL} \rightarrow 1.0 \text{ cm}^3 \text{ and has a mass of } 1.0 \text{ g}$$

Figure 1.9
A cubic decimeter of water (1,000 cm³) has a liquid volume of 1 L (1,000 mL) and a mass of 1 kg (1,000 g). Therefore, 1 cm³ of water has a liquid volume of 1 mL and a mass of 1 g.

UNDERSTANDINGS FROM MEASUREMENTS

One of the more basic uses of measurement is to *describe* something in an exact way that everyone can understand. For example, if a friend in another city tells you that the weather has been "warm," you might not understand what temperature is being described. A statement that the air temperature is 70°F carries more exact information than a statement about "warm weather." The statement that the air temperature is 70°F contains two important concepts: (1) the numerical value of 70 and (2) the referent unit of degrees Fahrenheit. Note that both a numerical value and a unit are necessary to communicate a measurement correctly. Thus, weather reports describe weather conditions with numerically specified units; for example, 70° Fahrenheit for air temperature, 5 miles per hour for wind speed, and 0.5 inch for rainfall (figure 1.10). When such numerically specified units are used in a description, or a weather report, everyone understands *exactly* the condition being described.

Data

Measurement information used to describe something is called **data.** Data can be used to describe objects, conditions, events, or changes that might be occurring. You really do not know if the weather is changing much from year to year until you compare the yearly weather data. The data will tell you, for example, if the weather is becoming hotter or dryer or is staying about the same from year to year.

Let's see how data can be used to describe something and how the data can be analyzed for further understanding. The cubes illustrated in figure 1.11 will serve as an example. Each cube can be described by measuring the properties of size and surface area.

First, consider the size of each cube. Size can be described by *volume,* which means *how much space something occupies.*

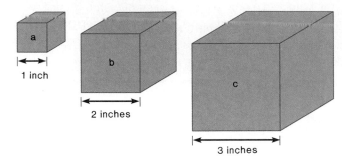

Figure 1.10
A weather report gives exact information, data that describe the weather by reporting numerically specified units for each condition.

Figure 1.11
Cube *a* is 1 inch on each side, cube *b* is 2 inches on each side, and cube *c* is 3 inches on each side. These three cubes can be described and compared with data, or measurement information, but some form of analysis is needed to find patterns or meaning in the data.

The volume of a cube can be obtained by measuring and multiplying the length, width, and height. The data is

volume of cube *a*	1 in³
volume of cube *b*	8 in³
volume of cube *c*	27 in³

Now consider the surface area of each cube. *Area* means *the extent of a surface,* and each cube has six surfaces, or faces (top, bottom, and four sides). The area of any face can be obtained by measuring and multiplying length and width. The data for the three cubes thus describes them as follows:

	Volume	Surface Area
cube *a*	1 in³	6 in²
cube *b*	8 in³	24 in²
cube *c*	27 in³	54 in²

Ratios and Generalizations

Data on the volume and surface area of the three cubes in figure 1.11 describes the cubes, but whether it says anything about a relationship between the volume and surface area of a cube is difficult to tell. Nature seems to have a tendency to camouflage relationships, making it difficult to extract meaning from raw data. Seeing through the camouflage requires the use of mathematical techniques to expose patterns. Let's see how such techniques can be applied to the data on the three cubes and what the pattern means.

One mathematical technique for reducing data to a more manageable form is to expose patterns through a *ratio*. A ratio is a relationship between two numbers obtained when one number is divided by another number. Suppose, for example, that an instructor has 50 sheets of graph paper for a laboratory group of 25 students. The relationship, or ratio, between the number of sheets and the number of students is 50 papers to 25 students, and this can be written as 50 papers/25 students. This ratio is *simplified* by dividing 25 into 50, and the ratio becomes 2 papers/1 student. The 1 is usually understood (not stated), and the ratio is written as simply 2 papers/student. It is read as 2 papers "for each" student, or 2 papers "per" student. The concept of simplifying with a ratio is an important one, and you will see it time and time again throughout science. It is important that you understand the meaning of "per" and "for each" when used with numbers and units.

Applying the ratio concept to the three cubes in figure 1.11, the ratio of surface area to volume for the smallest cube, cube *a*, is 6 in² to 1 in³, or

$$\frac{6 \text{ in}^2}{1 \text{ in}^3} = 6 \frac{\text{in}^2}{\text{in}^3}$$

meaning there are 6 square inches of area *for each* cubic inch of volume.

The middle-sized cube, cube *b*, had a surface area of 24 in² and a volume of 8 in³. The ratio of surface area to volume for this cube is therefore

$$\frac{24 \text{ in}^2}{8 \text{ in}^3} = 3 \frac{\text{in}^2}{\text{in}^3}$$

meaning there are 3 square inches of area *for each* cubic inch of volume.

The largest cube, cube *c*, had a surface area of 54 in² and a volume of 27 in³. The ratio is

$$\frac{54 \text{ in}^2}{27 \text{ in}^3} = 2 \frac{\text{in}^2}{\text{in}^3}$$

or 2 square inches of area *for each* cubic inch of volume. Summarizing the ratio of surface area to volume for all three cubes, you have

small cube	*a*–6:1
middle cube	*b*–3:1
large cube	*c*–2:1

Now that you have simplified the data through ratios, you are ready to generalize about what the information means. You can generalize that the surface-area-to-volume ratio of a cube *decreases* as the volume of a cube becomes larger. Reasoning from this generalization will provide an explanation for a number of related observations. For example, why does crushed ice melt faster than a single large block of ice with the same volume? The explanation is that the crushed ice has a larger surface-area-to-volume ratio than the large block, so more surface is exposed to warm air. If the generalization is found to be true for shapes other than cubes, you could explain why a log chopped into

small chunks burns faster than the whole log. Further generalizing might enable you to predict if 10 lb of large potatoes would require more or less peeling than 10 lb of small potatoes. When generalized explanations result in predictions that can be verified by experience, you gain confidence in the explanation. Finding patterns of relationships is a satisfying intellectual adventure that leads to understanding and generalizations that are frequently practical.

The Density Ratio

The power of using a ratio to simplify things, making explanations more accessible, is evident when you compare the simplified ratio 6 to 3 to 2 with the hodgepodge of numbers that you would have to consider without using ratios. The power of using the ratio technique is also evident when considering other properties of matter. Volume is a property that is sometimes confused with mass. Larger objects do not necessarily contain more matter than smaller objects. A large balloon, for example, is much larger than this book, but the book is much more massive than the balloon. The simplified way of comparing the mass of a particular volume is to find the ratio of mass to volume. This ratio is called mass **density**, which is defined as *mass per unit volume*. The "per" means "for each" as previously discussed, and "unit" means one, or each. Thus "mass per unit volume" literally means the "mass of one volume" (figure 1.12). The relationship can be written as

$$\text{mass density} = \frac{\text{mass}}{\text{volume}}$$

or

$$\rho = \frac{m}{V}$$

equation 1.1

(ρ **is the symbol for the Greek letter** *rho*.)

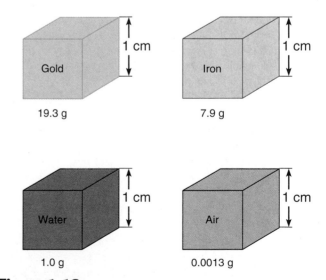

Figure 1.12

Equal volumes of different substances do not have the same mass, as these cube units show. Calculate the densities in g/m³. Do equal volumes of different substances have the same density? Explain.

Connections . . .

Sharks and rays are marine animals that have an internal skeleton made entirely of cartilage. These animals have no swim bladder to adjust their body density in order to maintain their position in the water; therefore, they must constantly swim or they will sink. The bony fish, on the other hand, have a skeleton composed of bone and most also have a swim bladder. These fish can regulate the amount of gas in the bladder to control their density. Thus, the fish can remain at a given level in the water without expending large amounts of energy.

Scientists plan to study nuclear fusion involving extremely dense plasmas, up to six times denser than the center of the sun. The proposed National Ignition Facility (NIF) is expected to be funded and completed by 2010. This research facility will use high-power lasers to achieve self-sustaining fusion reactions, with densities reaching up to 1,000 g/cm³. Existing research facilities today are capable of creating plasma densities up to about 50 g/cm³. Compare these densities with what you might find at the center of the sun, about 160 g/cm³.

Finally, have you ever noticed the different floating characteristics of cans of the normal version of a carbonated cola beverage and a diet version? The surprising result is that the normal version usually sinks and the diet version usually floats. This has nothing to do with the amount of carbon dioxide in the two drinks. It is a result of the increase in density from the sugar added to the normal version, while the diet version has much less of an artificial sweetener that is much sweeter than sugar.

As with other ratios, density is obtained by dividing one number and unit by another number and unit. Thus, the density of an object with a volume of 5 cm³ and a mass of 10 g is

$$\text{density} = \frac{10\ \text{g}}{5\ \text{cm}^3} = 2\ \frac{\text{g}}{\text{cm}^3}$$

The density in this example is the ratio of 10 g to 5 cm³, or 10 g/5 cm³, or 2 g to 1 cm³. Thus, the density of the example object is the mass of *one* volume (a unit volume), or 2 g *for each* cm³.

Any unit of mass and any unit of volume may be used to express density. The densities of solids, liquids, and gases are usually expressed in grams per cubic centimeter (g/cm³), but the densities of liquids are sometimes expressed in grams per milliliter (g/mL). Using SI standard units, densities are expressed as kg/m³. Densities of some common substances are shown in table 1.4.

If matter is distributed the same throughout a volume, the *ratio* of mass to volume will remain the same no matter what mass and volume are being measured. Thus, a teaspoonful, a cup, or a lake full of freshwater at the same temperature will all have a density of about 1 g/cm³ or 1 kg/L.

CONCEPTS APPLIED
Density Examples

1. What is the mass density of this book? Measure the length, width, and height of this book in cm, then multiply to find the volume in cm³. Use a balance to find the mass of this book in grams. Compute the density of the book by dividing the mass by the volume. Compare the density in g/cm³ with other substances listed in table 1.4.

2. Compare the densities of some common liquids. Pour a cup of vinegar in a large bottle. Carefully add a cup of corn syrup, then a cup of cooking oil. Drop a coin, tightly folded pieces of aluminum foil, and toothpicks into the bottle. Explain what you observe in terms of density. Take care not to confuse the property of *density*, which describes the compactness of matter, with *viscosity*, which describes how much fluid resists flowing under normal conditions. (Corn syrup has a greater viscosity than water—is this true of density, too?)

Table 1.4 Densities of Some Common Substances

Substance	Mass Density (ρ) (g/cm³)
Aluminum	2.70
Copper	8.96
Iron	7.87
Lead	11.4
Water	1.00
Seawater	1.03
Mercury	13.6
Gasoline	0.680

SYMBOLS AND EQUATIONS

In the previous section, the relationship of density, mass, and volume was written with symbols. Mass density was represented by ρ, the lowercase letter rho in the Greek alphabet, mass was represented by m, and volume by V. The use of such symbols is established and accepted by convention, and these symbols are like the vocabulary of a foreign language. You learn what the symbols mean by use and practice, with the understanding that *each symbol stands for a very specific property or concept.* The symbols actually represent **quantities,** or *measured properties.* The symbol m thus represents a quantity of mass that is specified by a number and a unit, for example, 16 g. The symbol V represents a quantity of volume that is specified by a number and a unit, such as 17 cm³.

Symbols usually provide a clue about which quantity they represent, such as m for mass and V for volume. However, in some cases two quantities start with the same letter, such as volume and velocity, so the uppercase letter is used for one (V for volume) and the lowercase letter is used for the other (v for velocity). There are more quantities than upper- and lowercase letters, however, so letters from the Greek alphabet are also used, for example, ρ for mass density. Sometimes a subscript is used to identify a quantity in a particular situation, such as v_i for initial, or beginning, velocity and v_f for final velocity. Some symbols are also used to carry messages; for example, the Greek letter delta (Δ) is a message that means "the change in" a value. Other message symbols are the symbol $\therefore$, which means "therefore," and the symbol $\propto$, which means "is proportional to."

Symbols are used in an **equation,** a statement that describes a relationship where *the quantities on one side of the equal sign are identical to the quantities on the other side.* Identical refers to both the numbers and the units. Thus, in the equation describing the property of density, $\rho = m/V$, the numbers on both sides of the equal sign are identical (e.g., $5 = 10/2$). The units on both sides of the equal sign are also identical (e.g., $g/cm^3 = g/cm^3$).

Equations are used to (1) *describe a property,* (2) *define a concept,* or (3) *describe how quantities change relative to each other.* Understanding how equations are used in these three classes is basic to comprehension of physical science. Each class of uses is considered separately in the following discussion.

Describing a property. You have already learned that the compactness of matter is described by the property called density. Density is a ratio of mass to a unit volume, or $\rho = m/V$. The key to understanding this property is to understand the meaning of a ratio and what "per" or "for each" means. Other examples of properties that will be defined by ratios are how fast something is moving (speed) and how rapidly a speed is changing (acceleration).

Defining a concept. A physical science concept is sometimes defined by specifying a measurement procedure. This is called an *operational definition* because a procedure is established that defines a concept as well as telling you how to measure it. Concepts of what is meant by force, mechanical work, and mechanical power and concepts involved in electrical and magnetic interactions can be defined by measurement procedures.

Describing how quantities change relative to each other. Nature is full of situations where one or more quantities change in value, or vary in size, in response to changes in other quantities. Changing quantities are called **variables.** Your weight, for example, is a variable that changes in size in response to changes in another variable, for example, the amount of food you eat. You already know about the pattern, or relationship, between these two variables. With all other factors being equal, an increase in the amount of food you eat results in an increase in your weight. When two variables increase (or decrease) together in the same ratio, they are said to be in *direct proportion.* When two variables are in direct proportion, *an increase or decrease in one variable results in the same relative increase or decrease in a second variable.*

Figure 1.13

The volume of fuel you have added to the fuel tank is directly proportional to the amount of time that the fuel pump has been running. This relationship can be described with an equation by using a proportionality constant.

Variables do not always increase or decrease in direct proportion. Sometimes one variable *increases* while a second variable *decreases* in the same ratio. This is an *inverse proportion* relationship. Other common relationships include one variable increasing in proportion to the *square* or to the *inverse square* of a second variable. Here are the forms of these four different types of proportional relationships:

Direct	$a \propto b$
Inverse	$a \propto 1/b$
Square	$a \propto b^2$
Inverse square	$a \propto 1/b^2$

Proportionality statements describe in general how two variables change relative to each other, but a proportionality statement is *not* an equation. For example, consider the last time you filled your fuel tank at a service station (figure 1.13). You could say that the volume of gasoline in an empty tank you are filling is directly proportional to the amount of time that the fuel pump was running, or

$$\text{volume} \propto \text{time}$$

This is not an equation because the numbers and units are not identical on both sides. Considering the units, for example, it should be clear that minutes do not equal gallons; they are two different quantities. To make a statement of proportionality into an equation, you need to apply a *proportionality constant,* which is sometimes given the symbol k. For the fuel pump example the equation is

$$\text{volume} = (\text{time})(\text{constant})$$

or

$$V = tk$$

In the example, the constant is the flow of gasoline from the pump in gal/min (a ratio). Assume the rate of flow is 10 gal/min. In units, you can see why the statement is now an equality.

$$gal = (min)\left(\frac{gal}{min}\right)$$

$$gal = \frac{\cancel{min} \times gal}{\cancel{min}}$$

$$gal = gal$$

Here are some tips to consider if your instructor uses a problem-solving approach. You could think of an equation as a *set of instructions.* The density equation, for example, is $\rho = m/V$. The equation tells you that mass density is a ratio of mass to volume, and you can find the density by dividing the mass by the volume. If you have difficulty, you either do not know the instructions or do not know how to follow the instructions (algebraic rules). Appendix D has worked examples that can help you with both the instructions and how to follow them. See also appendix E, which deals with problem solving.

THE NATURE OF SCIENCE

Most humans are curious, at least when they are young, and are motivated to understand their surroundings. These traits have existed since antiquity and have proven to be a powerful motivation. In recent times the need to find out has motivated the launching of space probes to learn what is "out there," and humans have visited the moon to satisfy their curiosity. Curiosity and the motivation to understand nature were no less powerful in the past than today. Over two thousand years ago the ancient Greeks lacked the tools and technology of today and could only make conjectures about the workings of nature. These early seekers of understanding are known as *natural philosophers,* and they observed, thought, and wrote about the workings of all of nature. They are called philosophers because their understandings come from reasoning only, without experimental evidence. Nonetheless, some of their ideas were essentially correct and are still in use today. For example, the idea of matter being composed of *atoms* was first reasoned by certain ancient Greeks in the fifth century B.C. The idea of *elements,* basic components that make up matter, was developed much earlier but refined by the ancient Greeks in the fourth century B.C. The concept of what the elements are and the concept of the nature of atoms have changed over time, but the idea first came from ancient natural philosophers.

The Scientific Method

Some historians identify the time of Galileo and Newton, approximately three hundred years ago, as the beginning of modern science. Like the ancient Greeks, Galileo and Newton were interested in studying all of nature. Since the time of Galileo and Newton, the content of physical science has increased in scope and specialization, but the basic means of acquiring understanding, the scientific investigation, has changed little. A *scientific investigation* provides understanding through *experimental evidence,* as opposed to the conjectures based on thinking only of the ancient natural philosophers. In chapter 2, for example, you will learn how certain ancient Greeks described

how objects fall toward the earth with a thought-out, or reasoned, explanation. Galileo, on the other hand, changed how people thought of falling objects by developing explanations from both creative thinking and precise measurement of physical quantities, providing experimental evidence for his explanations. Experimental evidence provides explanations today, much as it did for Galileo, as relationships are found from precise measurements of physical quantities. Thus, scientific knowledge about nature has grown as measurements and investigations have led to understandings that lead to further measurements and investigations.

What is a scientific investigation and what methods are used to conduct one? Attempts have been made to describe scientific methods in a series of steps (define problem, gather data, make hypothesis, test, make conclusion), but no single description has ever been satisfactory to all concerned. Scientists do similar things in investigations but there are different approaches and different ways to evaluate what they find. Overall, the similar things might look like this:

1. Observe some aspect of nature.
2. Propose an explanation for something observed.
3. Use the explanation to make predictions.
4. Test predictions by doing an experiment or by making more observations.
5. Modify explanation as needed.
6. Return to step 3.

The exact approach a scientist uses depends on the individual doing the investigation as well as the particular field of science being studied.

Another way to describe what goes on during a scientific investigation is to consider what can be generalized. There are at least three separate activities that seem to be common to scientists in different fields as they conduct scientific investigations, and these generalized activities are:

- Collecting observations.
- Developing explanations.
- Testing explanations.

No particular order or routine can be generalized about these common elements. In fact, individual scientists might not even be involved in all three activities. Some, for example, might spend all of their time out in nature, "in the field" collecting data and generalizing about their findings. This is an acceptable means of scientific investigation in some fields of science. Yet, other scientists might spend all of their time indoors, at computer terminals, developing theoretical equations that offer explanations for generalizations made by others. Again, the work at a computer terminal is an acceptable means of scientific investigation. Thus, there is not an order of five steps that are followed, particularly by today's specialized scientists. This is one reason why many philosophers of science argue there is no such thing as *the* scientific method. There are common activities of observing, explaining, and testing in scientific investigations in different fields, and these activities will be discussed next.

Explanations and Investigations

Explanations in the natural sciences are concerned with things or events observed, and there can be several different means of developing or creating explanations. In general, explanations can come from the results of experiments, from an educated guess, or just from imaginative thinking. In fact, there are several examples in the history of science of valid explanations being developed even from dreams. Explanations go by various names, each depending on intended use or stage of development. For example, an explanation in an early stage of development is sometimes called a *hypothesis*. A **hypothesis** is a tentative thought- or experiment-derived explanation. It must be compatible with all observations and provide understanding of some aspect of nature, but the key word here is "tentative." A hypothesis is tested by experiment and is rejected, or modified, if a single observation or test does not fit. The successful testing of a hypothesis may lead to the design of experiments, or it could lead to the development of another hypothesis, which could, in turn, lead to the design of yet more experiments, which could lead to. . . . As you can see, this is a branching, ongoing process that is very difficult to describe in specific terms. In addition, it can be difficult to identify a conclusion, an endpoint in the process. The search for new concepts to explain experimental evidence may lead from a hypothesis to a new theory, which results in more new hypotheses. This is why one of the best ways to understand scientific methods is to study the history of science. Or, you can conduct a scientific investigation yourself.

In some cases a hypothesis may be tested by simply making additional observations. For example, if you hypothesize that a certain species of bird uses cavities in trees as places to build nests, you could observe several birds of the species and record the kinds of nests they build and where they are built.

Another common method for testing a hypothesis involves devising an experiment. An experiment is a re-creation of an event or occurrence in a way that enables a scientist to support or disprove a hypothesis. This can be difficult since a particular event may be influenced by a great many separate things. For example, the production of a song by a bird involves many activities of the bird's nervous and muscular systems and is influenced by a wide variety of environmental factors. It might seem that developing an understanding of the factors involved in birdsong production is an impossible task. To help unclutter such situations, scientists have devised what is known as a controlled experiment. A **controlled experiment** compares two situations that have all the influencing factors identical except one. The situation used as the basis of comparison is called the *control group* and the other is called the *experimental group*. The single influencing factor that is allowed to be different in the experimental group is called the *experimental variable*.

The situation involving birdsong production would have to be broken down into a large number of simple questions, as previously mentioned. Each question would provide the basis on which experimentation would occur. Each experiment would provide information about a small part of the total process of birdsong production. For example, in order to test the hypothesis that male sex hormones are involved in stimulating male birds to sing, an experiment could be performed in which one group of male birds had their testes removed (the experimental group), while the control group was allowed to develop normally. After the experiment, the new data (facts) gathered would be analyzed. If there were no differences between the two groups, scientists could conclude that the variable evidently did not have a cause-and-effect relationship (i.e., was not responsible for the event). However, if there were a difference, it would be likely that the variable was responsible for the difference between the control and experimental groups. In the case of songbirds, removal of the testes does change their singing behavior.

Scientists almost never accept the results of a single experiment, since a proposed hypothesis has to explain all experimental results. Otherwise, the hypothesis needs revision. For example, the operation necessary to remove the testes of male birds might cause illness or discomfort in some birds, resulting in less singing. A way to overcome this difficulty would be to subject all birds to the same surgery but to remove the testes of only half of them. (The control birds would still have their testes.) The results of the experiment are considered convincing only when there is one variable, many replicates (copies) of the same experiment have been conducted, and the results are consistent.

Furthermore, scientists often apply statistical tests to the results to help decide in an impartial manner if the results obtained are *valid* (meaningful; fit with other knowledge), *reliable* (give the same results repeatedly), and show cause-and-effect, or if they are just the result of random events.

During experimentation, scientists learn new information and formulate new questions that can lead to yet more experiments. One good experiment can result in a hundred new questions and experiments. The discovery of the structure of the DNA molecule by Watson and Crick resulted in thousands of experiments and stimulated the development of the entire field of molecular biology (figure 1.14). Similarly, the discovery of molecules that regulate the growth of plants resulted in much research about how the molecules work and which molecules might be used for agricultural purposes.

If the processes of questioning and experimentation continue, and evidence continually and consistently supports the original hypothesis and other closely related hypotheses, the scientific community will begin to see how these hypotheses and facts fit together into a broad pattern.

Scientific Laws

Sometimes you can observe a series of relationships that seem to happen over and over again. There is a popular saying, for example, that "if anything can go wrong, it will." This is called Murphy's law. It is called a *law* because it describes a relationship between events that seems to happen time after time. If you drop a slice of buttered bread, for example, it can land two ways, butter side up or butter side down. According to Murphy's law, it will land butter side down. With this example, you know at least one way of testing the validity of Murphy's law.

A

B

C

Figure 1.14

James D. Watson and Francis W. Crick are scientists who, in 1953, determined the structure of the DNA molecule, which contains the genetic information of a cell. (*A*) This photograph shows the model of DNA they constructed. The discovery of the structure of the DNA molecule was followed by much research into how the molecule codes information, how it makes copies of itself, and how the information is put into action. Ultimately, these lines of research have led to altering the DNA of bacteria so that they produce useful materials such as vitamins, proteins, and antibiotics. (*B*) Genetically altered bacteria are grown in special vats and the useful materials are harvested. (*C*) The bacterium *E. coli* is commonly found in the human intestine.

Another "popular saying" type of relationship seems to exist between the cost of a houseplant and how long it lives. You could call it the "law of houseplant longevity." The relationship is that the life of a houseplant is inversely proportional to its purchase price. This "law" predicts that a $10 houseplant will wilt and die within a month, but a 50¢ houseplant will live for years. The inverse relationship is between the variables of (1) cost and (2) life span, meaning the more you pay for a plant the shorter the time it will live. This would also mean that inexpensive plants will live for a long time. Since the relationship seems to occur time after time, it is called a law.

A **scientific law** describes an important relationship that is observed in nature to occur consistently time after time. The law is often identified with the name of a person associated with the formulation of the law. For example, with all other factors being equal, an increase in the temperature of the air in a balloon results in an increase in its volume. Likewise, a decrease in the temperature results in a decrease in the total volume of the balloon. The volume of the balloon varies directly with the temperature of the air in the balloon and this can be observed to occur consistently time after time. This relationship was first discovered in the later part of the eighteenth century by two French scientists, A. C. Charles and Joseph Gay-Lussac. Today, the relationship is sometimes called *Charles' law*. When you read about a scientific

law, you should remember that a law is a statement that means something about a relationship that you can observe time after time in nature.

Have you ever heard someone state that something behaved a certain way *because* of a scientific law? For example, a big truck accelerated slowly *because* of Newton's laws of motion. Perhaps this person misunderstands the nature of scientific laws. Scientific laws do not dictate the behavior of objects, they simply describe it. They do not say how things ought to act but rather how things *do* act. A scientific law is *descriptive;* it describes how things act.

Models and Theories

Often the part of nature being considered is too small or too large to be visible to the human eye and the use of a *model* is needed. A **model** (figure 1.15) is a description of a theory or idea that accounts for all known properties. The description can come in many forms, such as an actual physical model, a computer model, a sketch, an analogy, or an equation. No one has ever seen the whole solar system, for example, and all you can see in the real world is the movement of the sun, moon, and planets against a background of stars. A physical model or sketch of the solar system, however, will give you a pretty good idea of

A

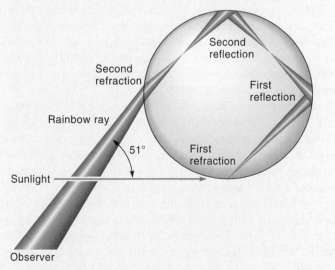

B

Figure 1.15

A model helps you visualize something that cannot be observed. You cannot observe what is making a double rainbow, for example, but models of light entering the upper and lower surface of a raindrop help you visualize what is happening. The drawings in (*B*) serve as a model that explains how a double rainbow is produced. (Also, see "The Rainbow" in chapter 7.)

what the solar system might look like. The physical model and the sketch are both models since they give you a mental picture of the solar system.

At the other end of the size scale, models of atoms and molecules are often used to help us understand what is happening in this otherwise invisible world. Also, a container of small, bouncing rubber balls can be used as a model to explain the relationships of Charles' law. This model helps you see what happens to invisible particles of air as the temperature, volume, and pressure of the gas change. Some models are better than others, and models constantly change along with our understanding about nature. Early twentieth-century models of atoms, for example, were based on a "planetary model," which had electrons in the role of planets moving around the nucleus, which played the role of the sun. Today, the model has changed as our understandings about the nature of the atom have changed. Electrons are now pictured as vibrating with certain wavelengths, which can make standing waves only at certain distances from the nucleus. Thus the model of the atom changed from one with electrons viewed as solid particles to one that views them as vibrations on a string.

The most recently developed scientific theory was refined and expanded during the 1970s. This theory concerns the surface of the earth, and it has changed our model of what the earth is like. At first, however, the basic idea of today's accepted theory was pure and simple conjecture. The term "conjecture" usually means an explanation or idea based on speculation, or one based on trivial grounds without any real evidence. Scientists would look at a map of Africa and South America, for example, and mull over how the two continents seem to be as pieces of a picture puzzle that had moved apart (figure 1.16). Any talk of moving continents was considered conjecture because it was not based on anything acceptable as real evidence.

Many years after the early musings about moving continents, evidence was collected from deep-sea drilling rigs that the ocean floor becomes progressively older toward the African and South American continents. This was good enough evidence to establish the "seafloor spreading hypothesis" that described the two continents moving apart.

If a hypothesis survives much experimental testing and leads, in turn, to the design of new experiments with the generation of new hypotheses that can be tested, you now have a working *theory*. A **theory** is defined as a broad, working hypothesis that is based on extensive experimental evidence. For example, the "seafloor spreading hypothesis" did survive requisite experimental testing, and together with other working hypotheses is today found as part of the *plate tectonic theory*. The plate tectonic theory describes how the continents have moved apart, just like pieces of a picture puzzle. Is this the same idea that was once considered conjecture? Sort of, but this time it is supported by experimental evidence.

The term *scientific theory* is reserved for historic schemes of thought that have survived the test of detailed examination for long periods of time. The *atomic theory*, for example, was developed in the late 1800s and has been the subject of extensive investigation and experimentation over the last century. The atomic theory and other scientific theories form the framework of scientific thought and experimentation today. Scientific the-

A

B

Figure 1.16

(*A*) Normal position of the continents on a world map. (*B*) A sketch of South America and Africa, suggesting that they once might have been joined together and subsequently separated by a continental drift.

ories point to new ideas about the behavior of nature and these ideas result in more experiments, more data to collect, and more explanations to develop. All of this may lead to a slight modification of an existing theory, a major modification, or perhaps the creation of an entirely new one. These activities continue in an ongoing attempt to satisfy the curiosity of people by understanding nature.

SCIENCE, NONSCIENCE, AND PSEUDOSCIENCE

As you can see from the discussion of the nature of science, a scientific approach to the world requires a certain way of thinking. There is an insistence on ample supporting evidence by numerous studies rather than easy acceptance of strongly stated opinions. Scientists must separate opinions from statements of fact. A scientist is a healthy skeptic.

Careful attention to detail is also important. Since scientists publish their findings and their colleagues examine their work, there is a strong desire to produce careful work that can be easily defended. This does not mean that scientists do not speculate and state opinions. When they do, however, they take great care to clearly distinguish fact from opinion.

There is also a strong ethic of honesty. Scientists are not saints, but the fact that science is conducted out in the open in front of one's peers tends to reduce the incidence of dishonesty. In addition, the scientific community strongly condemns and severely penalizes those who steal the ideas of others, perform shoddy science, or falsify data. Any of these infractions could lead to the loss of one's job and reputation.

From Experimentation to Application

The scientific method has helped us to understand and control many aspects of our natural world. Some information is extremely important in understanding the structure and functioning of things in our world but at first glance appears to have little practical value. For example, understanding the life cycle of a star may be important for people who are trying to answer questions about how the universe is changing, but it seems of little practical value to the average citizen. However, as our knowledge has increased, the time between first discovery to practical application has decreased significantly.

For example, scientists known as *genetic engineers* have altered the chemical code system of small organisms (microorganisms) so that they may produce many new drugs such as antibiotics, hormones, and enzymes. The ease with which these complex chemicals are produced would not have been possible had it not been for the information gained from the basic, theoretical sciences of microbiology, molecular biology, and genetics (see figure 1.14). Our understanding of how organisms genetically control the manufacture of proteins has led to the large-scale production of enzymes. Some of these chemicals can remove stains from clothing, deodorize, clean contact lenses, remove damaged skin from burn patients, and "stone wash" denim for clothing.

Another example is Louis Pasteur, a French chemist and microbiologist. Pasteur was interested in the theoretical problem of whether life could be generated from nonliving material. Much of his theoretical work led to practical applications in disease control. His theory that there are microorganisms that cause diseases and decay led to the development of vaccinations against rabies and the development of pasteurization for the preservation of foods (figure 1.17).

Science and Nonscience

The differences between science and nonscience are often based on the assumptions and methods used to gather and organize information and, most important, the testing of these assumptions. The difference between a scientist and a nonscientist is that a scientist continually challenges and tests principles and assumptions to determine a cause-and-effect relationship, whereas a nonscientist may not feel that this is important.

Figure 1.17

We can enjoy milk and other products because of pasteurization. This food preservation method was a direct result of experiments performed by Louis Pasteur (1822–1895).

Once you understand the nature of science, you will not have any trouble identifying astronomy, chemistry, physics, and biology as sciences. But what about economics, sociology, anthropology, history, philosophy, and literature? All of these fields may make use of certain central ideas that are derived in a logical way, but they are also nonscientific in some ways. Some things cannot be approached using the scientific method. Art, literature, theology, and philosophy are rarely thought of as sciences. They are concerned with beauty, human emotion, and speculative thought rather than with facts and verifiable laws. On the other hand, physics, chemistry, geology, and biology are always considered sciences.

Music is an area of study in a middle ground where scientific approaches may be used to some extent. "Good" music is certainly unrelated to science, but the study of how the human larynx generates the sound of a song is based on scientific principles. Any serious student of music will study the physics of sound and how the vocal cords vibrate to generate sound waves. Similarly, economists use mathematical models and established economic laws to make predictions about future economic conditions. However, the regular occurrence of unpredicted economic changes indicates that economics is far from scientific, since the reliability of predictions is a central criterion of science. Anthropology and sociology are also scientific in nature in many respects, but they cannot be considered true sciences because many of the generalizations they have developed cannot be tested by repeated experimentation. They also do not show a significantly high degree of cause-and-effect, or they have poor predictive value.

Pseudoscience

Pseudoscience ("pseudo" means "false") takes on the flavor of science but is not supportable as valid or reliable. Often, the purpose of pseudoscience is to confuse or mislead. The area of nutrition is flooded with pseudoscience. We all know that we must obtain certain nutrients like amino acids, vitamins, and minerals from the food that we eat or we may become ill. Many scientific experiments reliably demonstrate the validity of this information. However, in most cases, it has not been proven that the nutritional supplements so vigorously promoted are as useful or desirable as advertised. Rather, selected bits of scientific information (amino acids, vitamins, and minerals are essential to good health) have been used to create the feeling that additional amounts of these nutritional supplements are necessary or that they can improve your health. In reality, the average person eating a varied diet will obtain all of these nutrients in adequate amounts, and nutritional supplements are not required.

In addition, many of these products are labeled as organic or natural, with the implication that they have greater nutritive value because they are organically grown (grown without pesticides or synthetic fertilizers) or because they come from nature. The poisons curare, strychnine, and nicotine are all organic molecules that are produced in nature by plants that could be grown organically, but we would not want to include them in our diet.

Limitations of Science

By definition, science is a way of thinking and seeking information to solve problems. Therefore, scientific methods can be applied only to questions that have a factual basis. Questions concerning morals, value judgments, social issues, and attitudes cannot be answered using the scientific methods. What makes a painting great? What is the best type of music? Which wine is best? What color should I paint my car? These questions are related to values, beliefs, and tastes; therefore, scientific methods cannot be used to answer them.

Science is also limited by the ability of people to pry understanding from the natural world. People are fallible and do not always come to the right conclusions, because information is lacking or misinterpreted, but science is self-correcting. As new information is gathered, old incorrect ways of thinking must be changed or discarded. For example, at one time people were sure that the sun went around the earth. They observed that the sun rose in the east and traveled across the sky to set in the west. Since they could not feel the earth moving, it seemed perfectly logical that the sun traveled around the earth. Once they understood that the earth rotated on its axis, people began to understand that the rising and setting of the sun could be explained in other ways. A completely new concept of the relationship between the sun and the earth developed.

Although this kind of study seems rather primitive to us today, this change in thinking about the sun and the earth was a very important step in understanding the universe and how the various parts are related to one another. This background information was built upon by many generations of astronomers and space scientists, and finally led to space exploration.

People also need to understand that science cannot answer all the problems of our time. Although science is a powerful tool, there are many questions it cannot answer and many problems it cannot solve. The behavior and desires of people generate most of the problems societies face. Famine, drug abuse, and pollution are human-caused and must be resolved by humans. Science may provide some tools for social planners, politicians, and ethical thinkers, but science does not have, nor does it attempt to provide, all the answers to the problems of the human race. Science is merely one of the tools at our disposal.

SUMMARY

Science is a search for order in our surroundings. People have *concepts,* or mental images, about material *objects* and intangible *events* in their surroundings. Concepts are used for thinking and communicating. Concepts are based on *properties,* or attributes that describe a thing or event. Every property implies a *referent* that describes the property. Referents are not always explicit, and most communications require assumptions.

Measurement is a process that uses a well-defined and agreed-upon *referent* to describe a *standard unit.* The unit is compared to the property being defined by an *operation* that determines the *value* of the unit by *counting.* Measurements are always reported with a *number,* or value, and a *name* for the unit.

The two major *systems* of standard units are the *English system* and the *metric system.* The English system uses standard units that were originally based on human body parts, and the metric system uses standard units based on referents found in nature. The metric system also uses a system of prefixes to express larger or smaller amounts of units. The metric standard units for length, mass, and time are the *meter, kilogram,* and *second.*

Measurement information used to describe something is called *data.* One way to extract meanings and generalizations from data is to use a *ratio,* a simplified relationship between two numbers. Density is a ratio of mass to volume, or $\rho = m/V.$

Symbols are used to represent *quantities,* or measured properties. Symbols are used in *equations,* which are shorthand statements that describe a relationship where the quantities (both number values and units) are identical on both sides of the equal sign. Equations are used to (1) *describe* a property, (2) *define* a concept, or (3) *describe* how *quantities change* together.

Quantities that can have different values at different times are called *variables.* Variables that increase or decrease together in the same ratio are said to be in *direct proportion.* If one variable increases while the other decreases in the same ratio, the variables are in *inverse proportion.* Proportionality statements are not necessarily equations. A *proportionality constant* can be used to make such a statement into an equation. Proportionality constants might have numerical value only, without units, or they might have both value and units.

Modern science began about three hundred years ago during the time of Galileo and Newton. Since that time, *scientific investigation* has been used to provide *experimental evidence* about nature. The investigations provide *accurate, specific,* and *reliable* data that are used to develop and test *explanations.* A *hypothesis* is a tentative explanation that is accepted or rejected from experimental data. An accepted hypothesis may result in a *scientific law,* an explanation concerned with important phenomena. Laws are sometimes identified with the name of a scientist and can be expressed verbally, with an equation, or with a graph.

A *model* is used to help understand something that cannot be observed directly, explaining the unknown in terms of things already understood. Physical models, mental models, and equations are all examples of models that explain how nature behaves. A *theory* is a broad, detailed explanation that guides development and interpretations of experiments in a field of study.

Science and *nonscience* can be distinguished by the kinds of laws and rules that are constructed to unify the body of knowledge. Science involves the continuous *testing* of rules and principles by the collection of new facts. If the rules are not testable, or if no rules are used, it is not science. *Pseudoscience* uses scientific appearances to mislead.

Summary of Equations

1.1

$$\text{mass density} = \frac{\text{mass}}{\text{volume}}$$

$$\rho = \frac{m}{V}$$

KEY TERMS

controlled experiment (p. **13**)
data (p. **8**)
density (p. **9**)
equation (p. **11**)
hypothesis (p. **13**)
measurement (p. **4**)
model (p. **14**)
properties (p. **3**)

quantities (p. **10**)
referent (p. **3**)
scientific law (p. **14**)
standard unit (p. **4**)
theory (p. **16**)
unit (p. **4**)
variables (p. **11**)

APPLYING THE CONCEPTS

1. The process of comparing a property of an object to a well-defined and agreed-upon referent is called the process of
   (a) generalizing.
   (b) measurement.
   (c) graphing.
   (d) scientific investigation.
2. The height of an average person is closest to
   (a) 1.0 m.
   (b) 1.5 m.
   (c) 2.5 m.
   (d) 3.5 m.
3. Which of the following standard units is defined in terms of an object as opposed to an event?
   (a) kilogram
   (b) meter
   (c) second
   (d) none of the above
4. One-half liter of water has a mass of
   (a) 0.5 g.
   (b) 5 g.
   (c) 50 g.
   (d) 500 g.

5. A cubic centimeter (cm^3) of water has a mass of about 1
   (a) mL.
   (b) kg.
   (c) g.
   (d) dm.
6. Measurement information that is used to describe something is called
   (a) referents.
   (b) properties.
   (c) data.
   (d) a scientific investigation.
7. The property of volume is a measure of
   (a) how much matter an object contains.
   (b) how much space an object occupies.
   (c) the compactness of matter in a certain size.
   (d) the area on the outside surface.
8. As the volume of a cube becomes larger and larger, the surface-area-to-volume ratio
   (a) increases.
   (b) decreases.
   (c) remains the same.
   (d) sometimes increases and sometimes decreases.
9. If you consider a very small portion of a material that is the same throughout, the density of the small sample will be
   (a) much less.
   (b) slightly less.
   (c) the same.
   (d) greater.
10. A scientific law can be expressed as
    (a) a written concept.
    (b) an equation.
    (c) a graph.
    (d) any of the above.

Answers
1. b 2. b 3. a 4. d 5. c 6. c 7. b 8. b 9. c 10. d

QUESTIONS FOR THOUGHT

1. What is a concept?
2. What two things does a measurement statement always contain? What do the two things tell you?
3. Other than familiarity, what are the advantages of the English system of measurement?
4. Describe the metric standard units for length, mass, and time.
5. Does the density of a liquid change with the shape of a container? Explain.
6. Does a flattened pancake of clay have the same density as the same clay rolled into a ball? Explain.
7. Compare and contrast a scientific hypothesis and a scientific theory.
8. What is a model? How are models used?
9. Are all theories always completely accepted or completely rejected? Explain.
10. What is pseudoscience and how can you always recognize it?

PARALLEL EXERCISES

The exercises in groups A and B cover the same concepts. Solutions to group A exercises are located in appendix D.

Group A

Note: You will need to refer to table 1.4 to complete some of the following exercises.

1. What is your height in meters? In centimeters?
2. What is the mass density of mercury if 20.0 cm^3 has a mass of 272 g?
3. What is the mass of a 10.0 cm^3 cube of lead?
4. What is the volume of a rock with a mass density of 3.00 g/cm^3 and a mass of 600 g?
5. If you have 34.0 g of a 50.0 cm^3 volume of one of the substances listed in table 1.4, which one is it?
6. What is the mass of water in a 40 L aquarium?
7. A 2.1 kg pile of aluminum cans is melted, then cooled into a solid cube. What is the volume of the cube?
8. A cubic box contains 1,000 g of water. What is the length of one side of the box in meters? Explain your reasoning.
9. A loaf of bread (volume 3,000 cm^3) with a density of 0.2 g/cm^3 is crushed in the bottom of a grocery bag into a volume of 1,500 cm^3. What is the density of the mashed bread?
10. According to table 1.4, what volume of copper would be needed to balance a 1.00 cm^3 sample of lead on a two-pan laboratory balance?

Group B

Note: You will need to refer to table 1.4 to complete some of the following exercises.

1. What is your mass in kilograms? In grams?
2. What is the mass density of iron if 5.0 cm^3 has a mass of 39.5 g?
3. What is the mass of a 10.0 cm^3 cube of copper?
4. If ice has a mass density of 0.92 g/cm^3, what is the volume of 5,000 g of ice?
5. If you have 51.5 g of a 50.0 cm^3 volume of one of the substances listed in table 1.4, which one is it?
6. What is the mass of gasoline ($\rho = 0.680$ g/cm^3) in a 94.6 L gasoline tank?
7. What is the volume of a 2.00 kg pile of iron cans that are melted, then cooled into a solid cube?
8. A cubic tank holds 1,000.0 kg of water. What are the dimensions of the tank in meters? Explain your reasoning.
9. A hot dog bun (volume 240 cm^3) with a density of 0.15 g/cm^3 is crushed in a picnic cooler into a volume of 195 cm^3. What is the new density of the bun?
10. According to table 1.4, what volume of iron would be needed to balance a 1.00 cm^3 sample of lead on a two-pan laboratory balance?

Motion

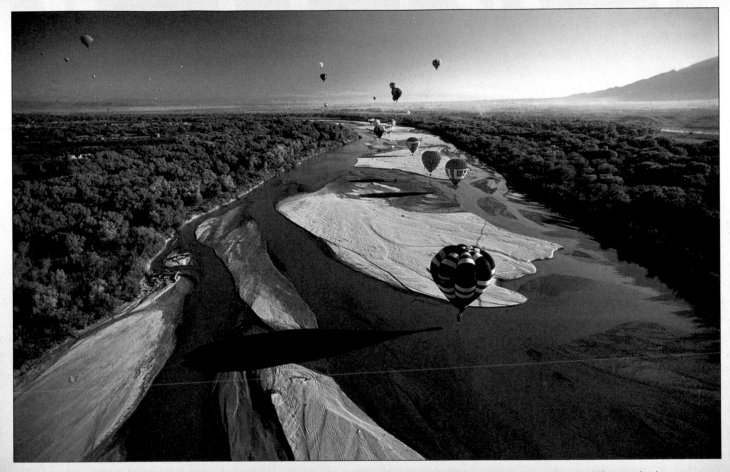

Information about the mass of a hot air balloon and forces on the balloon will enable you to predict if it is going to move up, down, or drift across the river. This chapter is about such relationships between force, mass, and changes in motion.

O U T L I N E

In chapter 1, you learned some "tools and rules" and some techniques for finding order in your surroundings. Order is often found in the form of patterns, or relationships between quantities that are expressed as equations. Equations can be used to (1) describe properties, (2) define concepts, and (3) describe how quantities change relative to one another. In all three uses, patterns are quantified, conceptualized, and used to gain a general understanding about what is happening in nature.

In the study of science, certain parts of nature are often considered and studied together for convenience. One of the more obvious groupings involves *movement*. Most objects around you appear to spend a great deal of time sitting quietly without motion.

Buildings, rocks, utility poles, and trees rarely, if ever, move from one place to another. Even things that do move from time to time sit still for a great deal of time. This includes you, automobiles, and bicycles (figure 2.1). On the other hand, the sun, the moon, and starry heavens always seem to move, never standing still. Why do things stand still? Why do things move?

Questions about motion have captured the attention of people for thousands of years. But the ancient people answered questions about motion with stories of mysticism and spirits that lived in objects. It was during the classic Greek culture, between 600 B.C. and 300 B.C., that people began to look beyond magic and spirits. One particular Greek philosopher, Aristotle, wrote a theory about the universe

that offered not only explanations about things such as motion but also offered a sense of beauty, order, and perfection. The theory seemed to fit with other ideas that people had and was held to be correct for nearly two thousand years after it was written. It was not until the work of Galileo and Newton during the 1600s that a new, correct understanding about motion was developed. The development of ideas about motion is an amazing and absorbing story. You will learn in this chapter how to describe and use some properties of motion. This will provide some basic understandings about motion, and will be very helpful in understanding some important aspects of astronomy and the earth sciences, as well as the movement of living things.

DESCRIBING MOTION

Motion is one of the more common events in your surroundings. You can see motion in natural events such as clouds moving, rain and snow falling, and streams of water, all moving in a never-ending cycle. Motion can also be seen in the activities of people who walk, jog, or drive various machines from place to place. Motion is so common that you would think everyone would intuitively understand the concepts of motion, but history indicates that it was only during the past three hundred years or so that people began to understand motion correctly. Perhaps the correct concepts are subtle and contrary to common sense, requiring a search for simple, clear concepts in an otherwise complex situation. The process of finding such order in a multitude of sensory impressions by taking measurable data, and then inventing a concept to describe what is happening, is the activity called *science*. We will now apply this process to motion.

What is motion? Consider a ball that you notice one morning in the middle of a lawn. Later in the afternoon, you notice that the ball is at the edge of the lawn, against a fence, and you wonder if the wind or some person moved the ball. You do not know if the wind blew it at a steady rate, if many gusts of wind moved it, or even if some children kicked it all over the yard. All you know for sure is that the ball has been moved because it is in a different position after some time passed. These are the two important aspects of motion: (1) a change of position and (2) the passage of time.

If you did happen to see the ball rolling across the lawn in the wind, you would see more than the ball at just two locations. You would see the ball moving continuously. You could consider, however, the ball in continuous motion to be a series of individual locations with very small time intervals. Moving involves a change of position during some time period. Motion is the act or process of something changing position.

The motion of an object is usually described with respect to something else that is considered to be not moving. (Such a sta-

tionary object is said to be "at rest.") Imagine that you are traveling in an automobile with another person. You know that you are moving across the land outside the car since your location on the highway changes from one moment to another. Observing your fellow passenger, however, reveals no change of position. You are in motion relative to the highway outside the car. You are not in motion relative to your fellow passenger. Your motion, and the motion of any other object or body, is the process of a change in position *relative* to some reference object or location. Thus *motion* can be defined as the act or process of changing position relative to some reference during a period of time.

MEASURING MOTION

You have learned that objects can be described by measuring certain fundamental properties such as mass and length. Since motion involves (1) a change of position and (2) the passage of *time*, the motion of objects can be described by using combinations of the fundamental properties of length and time. Combinations of these measurements describe three properties of motion: *speed*, *velocity*, and *acceleration*.

Speed

Suppose you are in a car that is moving over a straight road. How could you describe your motion? You need at least two measurements, (1) the distance you have traveled, and (2) the time that has elapsed while you covered this distance. Such a distance and time can be expressed as a ratio that describes your motion. This ratio is a property of motion called **speed**, which is a measure of how fast you are moving. Speed is defined as distance per unit of time, or

$$\text{speed} = \frac{\text{distance}}{\text{time}}$$

The units used to describe speed are usually miles/hour (mi/h), kilometers/hour (km/h), or meters/second (m/s).

Figure 2.1
The motion of this windsurfer, and of other moving objects, can be described in terms of the distance covered during a certain time period.

Let's go back to your car that is moving over a straight highway and imagine you are driving to cover equal distances in equal periods of time (figure 2.2). If you use a stopwatch to measure the time required to cover the distance between highway mile markers (those little signs with numbers along major highways), the time intervals will all be equal. You might find, for example, that one minute lapses between each mile marker. Such a uniform straight-line motion that covers equal distances in equal periods of time is the simplest kind of motion.

If your car were moving over equal distances in equal periods of time, it would have a *constant speed*. This means that the car is neither speeding up nor slowing down. It is usually difficult to maintain a constant speed. Other cars and distractions such as interesting scenery cause you to reduce your speed. At other times you increase your speed. If you calculate your speed over an entire trip, you are considering a large distance between two places and the total time that elapsed. The increases and decreases in speed would be averaged. Therefore, most speed calculations are for an *average speed*. The speed at any specific instant is called the *instantaneous speed*. To calculate the instantaneous speed, you would need to consider a very short time interval—one that approaches zero. An easier way would be to use the speedometer, which shows the speed at any instant.

It is easier to study the relationships between quantities if you use symbols instead of writing out the whole word. The letter v can be used to stand for speed when dealing with straight-line motion, which is the only kind of motion that will be considered in the problems in this text. A bar over the v ($\bar{v}$) is a symbol that means average (it is read "v-bar" or "v-average"). The letter d can be used to stand for distance and the letter t to stand for time. The relationship between average speed, distance, and time is therefore

$$\bar{v} = \frac{d}{t}$$

equation 2.1

This is one of the three types of equations that were discussed earlier, and in this case the equation defines a motion property (figure 2.3).

Constant, instantaneous, or average speeds can be measured with any distance and time units. Common units in the

Figure 2.2
This car is moving in a straight line over a distance of 1 mi each minute. Therefore, the car moves 60 mi in 60 min and has a speed of 60 mi/h.

Figure 2.3

Speed is distance per unit of time, which can be calculated from the equation or by finding the slope of a distance-versus-time graph.

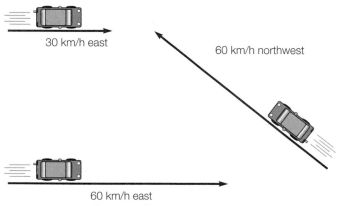

Figure 2.4

Velocity can be presented graphically with arrows. Here are three different velocities represented by three different arrows. The length of each arrow is proportional to the speed, and the arrowhead shows the direction of travel.

English system are miles/hour and feet/second. Metric units for speed are commonly kilometers/hour and meters/second. The ratio of any distance/time is usually read as distance per time, such as miles per hour.

Velocity

The word "velocity" is sometimes used interchangeably with the word "speed," but there is a difference. **Velocity** describes the *speed and direction* of a moving object. For example, a speed might be described as 60 km/h. A velocity might be described as 60 km/h to the west. To produce a change in velocity, either the speed or the direction is changed (or both are changed). A satellite moving with a constant speed in a circular orbit around the earth does not have a constant velocity since its direction of movement is constantly changing. Velocities can be represented graphically with arrows. The lengths of the arrows are proportional to the speed, and the arrowheads indicate the direction (figure 2.4).

Acceleration

Motion can be changed in three different ways: (1) by changing the speed, (2) by changing the direction of travel, or (3) by changing both the speed and direction of travel. Since velocity describes both the speed and the direction of travel, any of these three changes will result in a change of velocity. You need at least one additional measurement to describe a change of motion, which is how much time elapsed while the change was taking place. The change of velocity and time can be combined to define the *rate* at which the motion was changed. This rate is called **acceleration.** Acceleration is defined as a change of velocity per unit time, or

$$\text{acceleration} = \frac{\text{change of velocity}}{\text{time elapsed}}$$

Another way of saying "change in velocity" is the final velocity minus the initial velocity, so the relationship can also be written as

$$\text{acceleration} = \frac{\text{final velocity} - \text{initial velocity}}{\text{time elapsed}}$$

Acceleration due to a change in speed only can be calculated as follows. Consider a car that is moving with a constant, straight-line velocity of 60 km/h when the driver accelerates to 80 km/h. Suppose it takes 4 s to increase the velocity of 60 km/h to 80 km/h. The change in velocity is therefore 80 km/h minus 60 km/h, or 20 km/h. The acceleration was

$$\text{acceleration} = \frac{80\frac{km}{h} - 60\frac{km}{h}}{4s}$$

$$= \frac{20\frac{km}{h}}{4s}$$

$$= 5\frac{km/h}{s} \text{ or,}$$

$$= 5 \text{ km/h/s}$$

The average acceleration of the car was 5 km/h for each ("per") second. This is another way of saying that the velocity increases an average of 5 km/h in each second. The velocity of the car was 60 km/h when the acceleration began (initial velocity). At the end of 1 s, the velocity was 65 km/h. At the end of 2 s, it was 70 km/h; at the end of 3 s, 75 km/h; and at the end of 4 s (total time elapsed), the velocity was 80 km/h (final velocity). Note how fast the velocity is changing with time. In summary,

start (initial velocity)	60 km/h
first second	65 km/h
second second	70 km/h
third second	75 km/h
fourth second (final velocity)	80 km/h

As you can see, acceleration is really a description of how fast the speed is changing (figure 2.5); in this case, it is increasing 5 km/h each second.

$$a = \frac{v_f - v_i}{t} = \frac{80 \text{ km/h} - 60 \text{ km/h}}{4 \text{ s}} = 5\frac{km/h}{s}$$

$$\text{Slope} = \frac{\Delta y}{\Delta x} = \frac{10 \text{ km/h}}{2 \text{ s}}$$

$$= \frac{5 \text{ km/h}}{s}$$

Figure 2.5
This graph shows the speed increasing from 60 km/h to 80 km/h when moving in a straight line for 4 s. The acceleration, or change of velocity per unit of time, can be calculated from either the equation for acceleration or by calculating the slope of the straight-line graph. Both will tell you how fast the motion is changing with time.

Usually, you would want all the units to be the same, so you would convert km/h to m/s. A change in velocity of 5.0 km/h converts to 1.4 m/s and the acceleration would be 1.4 m/s/s. The units m/s per s mean what change of velocity (1.4 m/s) is occurring every second. The combination m/s/s is rather cumbersome, so it is typically treated mathematically to simplify the expression (to simplify a fraction, invert the divisor and multiply, or m/s × 1/s = m/s²). Remember that the expression 1.4 m/s² means the same as 1.4 m/s per s, a change of velocity in a given time period.

The relationship among the quantities involved in acceleration can be represented with the symbols a for average acceleration, v_f for final velocity, v_i for initial velocity, and t for time. The relationship is

$$a = \frac{v_f - v_i}{t}$$

equation 2.2

There are also other changes in the motion of an object that are associated with acceleration. One of the more obvious is a change that results in a decreased velocity. Your car's brakes, for example, can slow your car or bring it to a complete stop. This is *negative acceleration,* which is sometimes called *deceleration.* Another change in the motion of an object is a change of direction. Velocity encompasses both the rate of motion as well as direction, so a change of direction is an acceleration. The satellite moving with a constant speed in a circular orbit around the earth is constantly changing its direction of movement. It is therefore constantly accelerating because of this constant change in its motion. Your automobile has three devices that could change the state of its motion. Your automobile therefore has three accelerators—the gas pedal (which can increase magnitude of velocity), the brakes (which can decrease magnitude of velocity), and the steering wheel (which can change direction of velocity). (See figure 2.6.) The important thing to remember is that acceleration results from any *change* in the motion of an object.

Figure 2.6
Four different ways (*A–D*) to accelerate a car.

Travel

The super-speed magnetic levitation (maglev) train is a completely new technology based on magnetically suspending a train 3 to 10 cm (about 1 to 4 in) above a monorail, then moving it along with a magnetic field that travels along the monorail guides. The maglev train does not have friction between wheels and the rails since it does not have wheels. This lack of resistance and the easily manipulated magnetic fields makes very short acceleration distances possible. For example, a German maglev train can accelerate from 0 to 300 km/h (about 185 mi/h) over a distance of just 5 km (about 3 mi). A conventional train with wheels requires about 30 km (about 19 mi) in order to reach the same speed from a standing start. The maglev is attractive for short runs because of its superior acceleration. It is also attractive for longer runs because of its high top speed—up to about 500 km/h (about 310 mi/h). Today, only an aircraft can match such a speed.

CONCEPTS APPLIED
Acceleration Patterns

Suppose the radiator in your car has a leak and drops fall constantly, one every second. What pattern would the drops make on the pavement when you accelerate the car from a stoplight? What pattern would they make when you drive a constant speed? What pattern would you observe as the car comes to a stop? Use a marker to make dots on a sheet of paper that illustrate (1) acceleration, (2) constant speed, and (3) negative acceleration. Use words to describe the acceleration in each situation.

FORCES

The Greek philosopher Aristotle considered some of the first ideas about the causes of motion back in the fourth century B.C. However, he had it all wrong when he reportedly stated that a dropped object falls at a constant speed that is determined by its weight. He also incorrectly thought that an object moving across the earth's surface requires a continuously applied force in order to continue moving. These ideas were based on observing and thinking, not measurement, and no one checked to see if they were correct. It took about two thousands years before people began to correctly understand motion.

Aristotle did recognize an association between force and motion, and this much was acceptable. It is partly correct because a force is closely associated with *any* change of motion, as you will see. This section introduces the concept of a force, which will be developed more fully when the relationship between forces and motion is considered later.

A **force** is a push or a pull that is capable of changing the state of motion of an object. Consider, for example, the move-

---

Weather

Classification schemes are imaginative mental constructions used to show similarities and differences in objects or events. For example, the following describes two schemes used to help us classify storms that are not associated with weather fronts.

What is the difference between a tropical depression, tropical storm, and a hurricane? They are classified according to the *speed* of the maximum sustained surface winds. In the United States the maximum sustained surface wind is measured by averaging the wind speed over a 1-minute period. Here is the classification scheme:

Tropical Depression. This is a center of low pressure around which the winds are generally moving 55 km/h (about 35 mi/h) or less. The tropical depression might dissolve into nothing or it might develop into a more intense disturbance.

Tropical Storm. This is a more intense, highly organized center of low pressure with winds between 56 and 120 km/h (about 35 to 75 mi/h).

Hurricane. This is an intense low-pressure center with winds greater than 120 km/h (about 75 mi/h). A strong storm of this type is called a "hurricane" if it occurs over the Atlantic Ocean or the Pacific Ocean east of the international date line. It is called a "typhoon" if it occurs over the North Pacific Ocean west of the international date line. Hurricanes are further classified according to category and damage to be expected. Here is the classification scheme:

Category	Damage	Winds
1	minimal	120–153 km/h (75–95 mi/h)
2	moderate	154–177 km/h (96–110 mi/h)
3	extensive	178–210 km/h (111–130 mi/h)
4	extreme	211–250 km/h (131–155 mi/h)
5	catastrophic	>250 km/h (>155 mi/h)

---

ment of a ship from the pushing of two tugboats (figure 2.7). Tugboats can vary the strength of the force exerted on a ship, but they can also push in different directions. What effect does direction have on two forces acting on an object? If the tugboats were side by side, pushing in the same direction, the overall force is the sum of the two forces. If they act in exactly opposite directions, one pushing on each side of the ship, the overall force is the difference between the strength of the two forces. If they have the same strength, the overall effect is to cancel each other without producing any motion. The **net force** is the sum of all the forces acting on an object. Net force means "final," after the forces are added (figure 2.8).

Figure 2.7

The rate of movement and the direction of movement of this ship are determined by a combination of direction and size of force from each of the tugboats. Which direction are the two tugboats pushing? What evidence would indicate that one tugboat is pushing with a greater force? If the tugboat by the numbers is pushing with a greater force and the back tugboat is keeping the ship from moving, what will happen?

Figure 2.8

(A) When two parallel forces are acting on the ship in the same direction, the net force is the two forces added together. (B) When two forces are opposite and of equal size, the net force is zero. (C) When two parallel forces are not of equal size, the net force is the difference in the direction of the larger force.

When two parallel forces act in the same direction, they can be simply added. In this case, there is a net force that is equivalent to the sum of the two forces. When two parallel forces act in opposite directions, the net force is the difference between the two forces and is in the direction of the larger force. When two forces act neither in a way that is exactly together nor exactly opposite each other, the result will be like a new, different force having a new direction and strength.

Forces have a strength and direction that can be represented by force arrows. The tail of the arrow is placed on the object that feels the force, and the arrowhead points in the direction in which the force is exerted. The length of the arrow is proportional to the strength of the force. The use of force arrows helps you visualize and understand all the forces and how they contribute to the net force.

HORIZONTAL MOTION ON LAND

Everyday experience seems to indicate that Aristotle's idea about horizontal motion on the earth's surface is correct. After all, moving objects that are not pushed or pulled do come to rest in a short period of time. It would seem that an object keeps moving only if a force continues to push it. A moving automobile will slow and come to rest if you turn off the ignition. Likewise, a ball that you roll along the floor will slow until it comes to rest. Is the natural state of an object to be at rest, and is a force necessary to keep an object in motion? This is exactly what people thought until Galileo (figure 2.9) published his book *Two New Sciences* in 1638, which described his findings about motion.

The book had three parts that dealt with uniform motion, accelerated motion, and projectile motion. Galileo described details of simple experiments, measurements, calculations, and thought experiments as he developed definitions and concepts of motion. In one of his thought experiments, Galileo presented an argument against Aristotle's view that a force is needed to keep an object in motion. Galileo imagined an object (such as a ball) moving over a horizontal surface without the force of friction. He concluded that the object would move forever with a constant velocity as long as there was no unbalanced force acting to change the motion.

Why does a rolling ball slow to a stop? You know that a ball will roll farther across a smooth, waxed floor such as a bowling lane than it will across floor covered with carpet. The rough carpet offers more resistance to the rolling ball. The resistance of the floor friction is shown by a force arrow, F_{floor}, in figure 2.10. This force, along with the force arrow for air resistance, F_{air}, oppose the forward movement of the ball. Notice the dashed line arrow in part A of figure 2.10. There is no other force applied to the ball, so the rolling speed decreases until the ball finally comes to a complete stop. Now imagine what force you would need to exert by pushing with your hand, moving along with the ball to keep it rolling at a uniform rate. An examination of the forces in part B of figure 2.10 can help you determine the amount of force. The force you apply, $F_{applied}$, must counteract the resistance forces. It opposes the forces that are slowing down the ball as illustrated by the direction of the arrows. To determine how much force you should apply, look at the arrow equation. $F_{applied}$ has

Figure 2.9
Galileo (left) challenged the Aristotelian view of motion and focused attention on the concepts of distance, time, velocity, and acceleration.

the same length as the sum of the two resistance forces, but it is in the opposite direction of the resistance forces. Therefore, the overall force, F_{net}, is zero. The ball continues to roll at a uniform rate when you *balance* the force opposing its motion. It is reasonable, then, that if there were no opposing forces, you would not need to apply a force to keep it rolling. This was the kind of reasoning that Galileo did when he discredited the Aristotelian view that a force was necessary to keep an object moving. Galileo concluded that a moving object would continue moving with a constant velocity if no unbalanced forces were applied, that is, if the net force were zero.

It could be argued that the difference in Aristotle's and Galileo's views of forced motion is really a degree of analysis. After all, moving objects on the earth do come to rest unless continuously pushed or pulled. But Galileo's conclusion describes *why* they must be pushed or pulled and reveals the true nature of the motion of objects. Aristotle argued that the natural state of objects is to be at rest and attempted to explain why objects move. Galileo, on the other hand, argued that it is just as natural for an object to be moving and attempted to explain why they come to rest. Galileo called the behavior of matter to persist in

A No force—constant speed in straight line

B Force applied against direction of motion

C Force applied in same direction as motion

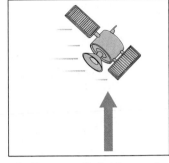

D Force applied sideways to direction of motion

Figure 2.10

(*A*) This ball is rolling to your left with no forces in the direction of motion. The sum of the force of floor friction (F_{floor}) and the force of air friction (F_{air}) result in a net force opposing the motion, so the ball slows to a stop. (*B*) A force is applied to the moving ball, perhaps by a hand that moves along with the ball. The force applied ($F_{applied}$) equals the sum of the forces opposing the motion, so the ball continues to move with a constant velocity.

Figure 2.11

Explain how the combination of drawings (*A–D*) illustrates inertia.

its state of motion **inertia.** Inertia is the *tendency of an object to remain in unchanging motion or at rest when the net force is zero.* The development of this concept changed the way people viewed the natural state of an object and opened the way for further understandings about motion. Today, it is understood that a satellite moving through free space will continue to do so with no unbalanced forces acting on it (figure 2.11A). An unbalanced force is needed to slow the satellite (figure 2.11B), increase its speed (figure 2.11C), or change its direction of travel (figure 2.11D).

FALLING OBJECTS

Did you ever wonder what happens to a falling rock during its fall? Aristotle reportedly thought that a rock falls at a uniform speed that is proportional to its weight. Thus, a heavy rock would fall at a faster uniform speed than a lighter rock. As stated in a popular story, Galileo discredited Aristotle's conclusion by dropping a solid iron ball and a solid wooden ball simultaneously from the top of the Leaning Tower of Pisa (figure 2.12). Both balls, according to the story, hit the ground nearly at the same time. To do this, they would have to fall with the same velocity. In other words, the velocity of a falling object does not depend on its weight. Any difference in freely falling bodies is explainable by air resistance. Soon after the time of Galileo the air pump was invented. The air pump could be used to remove the air from a glass tube. The effect of air resistance on falling objects could then be demonstrated by comparing how objects fall in the air with how they fall in an evacuated glass tube. You know that a coin falls faster when dropped with

Figure 2.12

According to a widespread story, Galileo dropped two objects with different weights from the Leaning Tower of Pisa. They reportedly hit the ground at about the same time, discrediting Aristotle's view that the speed during the fall is proportional to weight.

a feather in the air. A feather and heavy coin will fall together in the near vacuum of an evacuated glass tube because the effect of air resistance on the feather has been removed. When objects fall toward the earth without considering air resistance, they are said to be in **free fall.** Free fall considers only gravity and neglects air resistance.

Galileo concluded that light and heavy objects fall together in free fall, but he also wanted to know the details of what was going on while they fell. He now knew that the velocity of an object in free fall was *not* proportional to the weight of the object. He observed that the velocity of an object in free fall *increased* as the object fell and reasoned from this that the velocity

A Closer Look   A Bicycle Racer's Edge

Galileo was one of the first to recognize the role of friction in opposing motion. As shown in figure 2.10, friction with the surface and air friction combine to produce a net force that works against anything that is moving on the surface. This article is about air friction and some techniques that bike riders use to reduce that opposing force—perhaps giving them an edge in a close race.

The bike riders in box figure 2.1 are forming a single-file line, called a *paceline,* because the slipstream reduces the air resistance for a closely trailing rider. Cyclists say that riding in the slipstream of another cyclist will save much of their energy. In fact, the cyclists will be able to move 5 mi/h faster than they would expending the same energy while riding alone.

In a sense, riding in a slipstream means that you do not have to push as much air out of your way. It has been estimated that at 20 mi/h a cyclist must move a little less than half a ton of air out of the way every minute. One of the earliest demonstrations of how a slipstream can help a cyclist was done back about the turn of the century. Charles Murphy had a special bicycle trail built down the middle of a railroad track. Riding very close behind a special train caboose, Murphy

Box Figure 2.1
The object of the race is to be in the front, to finish first. If this is true, why are these racers forming a single-file line?

was able to reach a speed of over 60 mi/h for a one-mile course. More recently, cyclists have reached over 125 mi/h by following close, in the slipstream of a race car.

Along with the problem of moving about a half-ton of air out of the way every minute, there are two basic factors related to air resistance. These are (1) a turbulent versus a smooth flow of air, and (2) the problem of frictional drag. A turbulent flow of air contributes to air resistance because it causes the air to separate slightly on the back side, which increases the pressure on the

front of the moving object. This is why racing cars, airplanes, boats, and other racing vehicles are streamlined to a teardrop-like shape. This shape is not as likely to have the lower-pressure-producing air turbulence behind (and resulting greater pressure in front) because it smoothes or streamlines the air flow.

The frictional drag of air is similar to the frictional drag that occurs when you push a book across a rough tabletop. You know that smoothing the rough tabletop will reduce the frictional drag on the book. Likewise, the smoothing of a surface exposed to moving air will reduce air friction. Cyclists accomplish this "smoothing" by wearing smooth lycra clothing, and by shaving hair from arm and leg surfaces that are exposed to moving air. Each hair contributes to the overall frictional drag, and removal of the arm and leg hair can thus result in seconds saved. This might provide enough of an edge to win a close race. Shaving legs and arms, together with the wearing of lycra or some other tight, smooth-fitting garments, are just a few of the things a cyclist can do to gain an edge. Perhaps you will be able to think of more ways to reduce the forces that oppose motion.

Connections . . .

Sports

There are two different meanings for the term "free fall." In physics, "free fall" means the unconstrained motion of a body in a gravitational field, without considering air resistance. Without air resistance all objects are assumed to accelerate toward the surface at 9.8 m/s^2.

In the sport of skydiving, "free fall" means falling within the atmosphere without a drag-producing device such as a parachute. Air provides a resisting force that opposes the motion of a falling object, and the net force is the difference between the downward force (weight) and the upward force of air resistance. The weight of the falling object depends on the mass and acceleration from gravity, and this is the force downward. The resisting force is determined by at least two variables, (1) the area of the object exposed to the airstream, and (2) the speed of the falling object. Other variables such as streamlining, air temperature, and turbulence play a role, but the greatest effect seems to be from exposed area and the increased resistance as speed increases.

A skydiver's weight is constant, so the downward force is constant. Modern skydivers typically free-fall from about 3,650 m (about 12,000 ft) above the ground until about 750 m (about 2,500 ft), where they open their parachutes. After jumping from the plane, the diver at first accelerates toward the surface, reaching speeds up to about 185–210 km/h (about 115–130 mi/h). The air resistance increases with increased speed and the net force becomes less and less. Eventually, the downward weight force will be balanced by the upward air resistance force, and the net force becomes zero. The person now falls at a constant speed and we say the terminal velocity has been reached. It is possible to change your body position to vary your rate of fall up or down by 32 km/h (about 20 mi/h). However, by diving or "standing up" in free fall, experienced skydivers can learn to reach speeds of up to 290 km/h (about 180 mi/h). The record free fall speed, done without any special equipment, is 517 km/h (about 321 mi/h). Once the parachute opens, a descent rate of about 16 km/h (about 10 mi/h) is typical.

of the falling object would have to be (1) somehow proportional to the *time* of fall and (2) somehow proportional to the *distance* the object fell. If the time and distance were both related to the velocity of a falling object, how were they related to one another?

Galileo reasoned that a freely falling object should cover a distance *proportional to the square of the time of the fall* ($d \propto t^2$). In other words, the object should fall 4 times as far in 2 s as in 1 s ($2^2 = 4$), 9 times as far in 3 s ($3^2 = 9$), and so on. Galileo checked this calculation by rolling balls on an inclined board with a smooth groove in it. He used the inclined board to slow the motion of descent in order to measure the distance and time relationships, a necessary requirement since he lacked the accurate timing devices that exist today. He found, as predicted, that the falling balls moved through a distance proportional to the square of the time of falling. This also means that the *velocity of the falling object increased at a constant rate.* Recall that a change of velocity during some time period is called *acceleration*. In other words, a falling object *accelerates* toward the surface of the earth.

Since the velocity of a falling object increases at a constant rate, this must mean that falling objects are *uniformly accelerated* by the force of gravity. *All objects in free fall experience a constant acceleration.* During each second of fall, the object gains 9.8 m/s (32 ft/s) in velocity. This gain is the acceleration of the falling object, 9.8 m/s^2 (32 ft/s^2).

The acceleration of objects falling toward the earth varies slightly from place to place on the earth's surface because of the earth's shape and spin. The acceleration of falling objects decreases from the poles to the equator and also varies from place to place because the earth's mass is not distributed equally. The value of 9.8 m/s^2 (32 ft/s^2) is an approximation that is fairly close to, but not exactly, the acceleration due to gravity in any particular location. The acceleration due to gravity is important in a number of situations, so the acceleration from this force is given a special symbol, *g*.

CONCEPTS APPLIED
Falling Bodies

Galileo concluded that all objects fall together, with the same acceleration, when the upward force of air resistance is removed. It would be most difficult to remove air from the room, but it is possible to do some experiments that provide some evidence of how air influences falling objects.

1. Take a sheet of paper and your textbook and drop them side by side from the same height. Note the result.
2. Place the sheet of paper on top of the book and drop them at the same time. Do they fall together?
3. Crumple the sheet of paper into a loose ball and drop the ball and book side by side from the same height.
4. Crumple a sheet of paper into a very tight ball and again drop the ball and book side by side from the same height.

Explain any evidence you found concerning how objects fall.

COMPOUND MOTION

So far we have considered two types of motion: (1) the horizontal, straight-line motion of objects moving on the surface of the earth and (2) the vertical motion of dropped objects that accelerate toward the surface of the earth. A third type of motion occurs when an object is thrown, or projected, into the air. Essentially, such a projectile (rock, football, bullet, golf ball, or whatever) could be directed straight upward as a vertical projection, directed straight out as a horizontal projection, or directed at some angle between the vertical and the horizontal. Basic to understanding such compound motion is the observation that (1) gravity acts on objects *at all times,* no matter where they are, and (2) the acceleration due to gravity (*g*) is *independent of any motion* that an object may have.

Vertical Projectiles

Consider first a ball that you throw straight upward, a vertical projection. The ball has an initial velocity but then reaches a maximum height, stops for an instant, then accelerates back toward the earth. Gravity is acting on the ball throughout its climb, stop, and fall. As it is climbing, the force of gravity is accelerating it back to the earth. The overall effect during the climb is deceleration, which continues to slow the ball until the instantaneous stop. The ball then accelerates back to the surface just like a ball that has been dropped. If it were not for air resistance, the ball would return with the same velocity that it had initially. The velocity arrows for a ball thrown straight up are shown in figure 2.13.

Horizontal Projectiles

Horizontal projections are easier to understand if you split the complete motion into vertical and horizontal parts. Consider, for example, an arrow shot horizontally from a bow. The force of gravity accelerates the arrow downward, giving it an increasing downward velocity as it moves through the air. This increasing downward velocity is shown in figure 2.14 as increasingly longer velocity arrows (v_v). There are no forces in the horizontal direction if you ignore air resistance, so the horizontal velocity of the arrow remains the same as shown by the v_h velocity arrows. The combination of the increasing vertical (v_v) motion and the unchanging horizontal (v_h) motion causes the arrow to follow a curved path until it hits the ground.

An interesting prediction that can be made from the shot-arrow analysis is that an arrow shot horizontally from a bow will hit the ground at the same time as a second arrow that is simply dropped from the same height (figure 2.14). Would this be true of a bullet dropped at the same time as one fired horizontally from a rifle? The answer is yes; both bullets would hit the ground at the same time. Indeed, if you ignore air resistance, all the bullets and arrows should hit the ground at the same time if dropped or shot from the same height.

Golf balls, footballs, and baseballs are usually projected upward at some angle to the horizon. The horizontal motion of these projectiles is constant as before because there are no

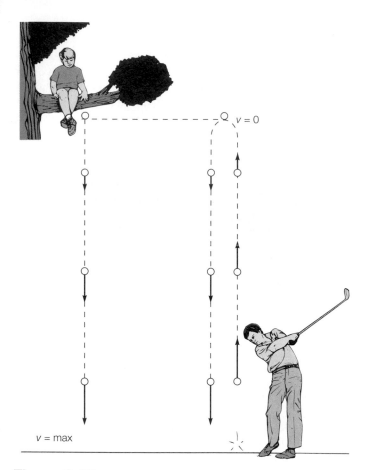

$v = 0$

$v = \text{max}$

Figure 2.13

On its way up, a vertical projectile such as this misdirected golf ball is slowed by the force of gravity until an instantaneous stop; then it accelerates back to the surface, just as another golf ball does when dropped from the same height. The straight up and down moving golf ball has been moved to the side in the sketch so we can see more clearly what is happening.

horizontal forces involved. The vertical motion is the same as that of a ball projected directly upward. The combination of these two motions causes the projectile to follow a curved path called a *parabola*, as shown in figure 2.15. The next time you have the opportunity, observe the path of a ball that has been projected at some angle (figure 2.16). Note that the second half of the path is almost a reverse copy of the first half. If it were not for air resistance, the two values of the path would be exactly the same. Also note the distance that the ball travels as compared to the angle of projection. An angle of projection of 45° results in the maximum distance of travel.

LAWS OF MOTION

Isaac Newton (1643–1727) was a quiet farm boy who seemed more interested in mathematics and tinkering than farming. He entered Trinity College of Cambridge University at the age of eighteen, where he enrolled in mathematics. He graduated four years later, the same year that the university was closed because

the bubonic plague, or Black Death, was ravaging Europe. During this time, Newton returned to his boyhood home, where he thought out most of the ideas that would later make him famous. Here, between the ages of twenty-three and twenty-four, he invented the field of mathematics called *calculus* and clarified his ideas on motion and gravitation. After the plague he returned to Cambridge, where he was appointed professor of mathematics at the age of twenty-six. He lectured and presented papers on optics. One paper on his theory about light and colors caused such a controversy that Newton resolved never to publish another line. Newton was a shy, introspective person who was too absorbed in his work for such controversy. In 1684, Edmund Halley (of Halley's comet fame) asked Newton to resolve a dispute involving planetary motions. Newton had already worked out the solution to this problem, in addition to other problems on gravity and motion. Halley persuaded the reluctant Newton to publish the material. Two years later, in 1687, Newton published *Principia,* which was paid for by Halley. Although he feared controversy, the book was accepted almost at once and established Newton as one of the greatest thinkers who ever lived.

Newton built his theory of motion on the previous work of Galileo and others. In fact, Newton's first law is similar to the concept of inertia described earlier by Galileo. Newton acknowledged the contribution of Galileo and others to his work, stating that if he had seen further than others "it was by standing upon the shoulders of giants."

Newton's First Law of Motion

Newton's first law of motion is also known as the *law of inertia* and is very similar to one of Galileo's findings about motion. Recall that Galileo used the term *inertia* to describe the tendency of an object to resist changes in motion. Newton's first law describes this tendency more directly. In modern terms (not Newton's words), the **first law of motion** is as follows:

> Every object retains its state of rest or its state of uniform straight-line motion unless acted upon by an unbalanced force.

This means that an object at rest will remain at rest unless it is put into motion by an unbalanced force; that is, the net force must be greater than zero. Likewise, an object moving with uniform straight-line motion will retain that motion unless a net force causes it to speed up, slow down, or change its direction of travel. Thus, Newton's first law describes the tendency of an object to resist *any* change in its state of motion.

Think of Newton's first law of motion when you ride standing in the aisle of a bus. The bus begins to move, and you, being an independent mass, tend to remain at rest. You take a few steps back as you tend to maintain your position relative to the ground outside. You reach for a seat back or some part of the bus. Once you have a hold on some part of the bus it supplies the forces needed to give you the same motion as the bus and you no longer find it necessary to step backward. You now have the same motion as the bus, and no forces are involved, at least until the bus goes around a curve.

Figure 2.14

A horizontal projectile has the same horizontal velocity throughout the fall as it accelerates toward the surface, with the combined effect resulting in a curved path. Neglecting air resistance, an arrow shot horizontally will strike the ground at the same time as one dropped from the same height above the ground, as shown here by the increasing vertical velocity arrows.

Figure 2.15

A football is thrown at some angle to the horizon when it is passed downfield. Neglecting air resistance, the horizontal velocity is a constant, and the vertical velocity decreases, then increases, just as in the case of a vertical projectile. The combined motion produces a parabolic path. Contrary to statements by sportscasters about the abilities of certain professional quarterbacks, it is impossible to throw a football with a "flat trajectory" because it begins to accelerate toward the surface as soon as it leaves the quarterback's hand.

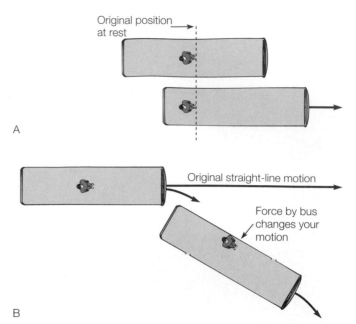

Figure 2.17

Top view of a person standing in the aisle of a bus. (*A*) The bus is at rest, and then starts to move forward. Inertia causes the person to remain in the original position, appearing to fall backward. (*B*) The bus turns to the right, but inertia causes the person to retain the original straight-line motion until forced in a new direction by the side of the bus.

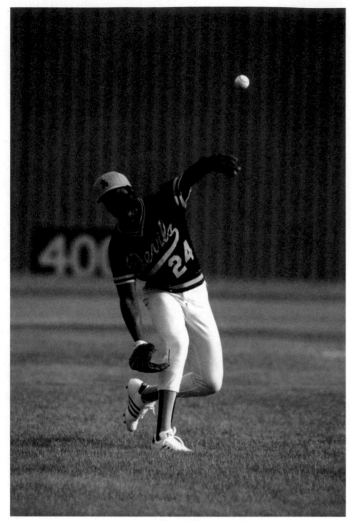

Figure 2.16

Without a doubt, this baseball player is aware of the relationship between the projection angle and the maximum distance acquired for a given projection velocity.

C O N C E P T S   A P P L I E D

First Law Experiment

Place a small ball on a flat part of the floor in a car, SUV, or pickup truck. First, predict what will happen to the ball in each of the following situations: (1) The vehicle moves forward from a stopped position. (2) The vehicle is moving at a constant speed. (3) The vehicle is moving at a constant speed, then turns to the right. (4) The vehicle is moving at a constant speed, then comes to a stop. Now, test your predictions, and then explain each finding in terms of Newton's first law of motion.

You now feel a tendency to move to the side of the bus. The bus has changed its straight-line motion, but you, again being an independent mass, tend to move straight ahead. The side of the seat forces you into following the curved motion of the bus. The forces you feel when the bus starts moving or turning are a result of your tendency to remain at rest or follow a straight path until forces correct your motion so that it is the same as that of the bus (figure 2.17).

Newton's Second Law of Motion

Newton had successfully used Galileo's ideas to describe the nature of motion. Newton's first law of motion explains that any object, once started in motion, will continue with a constant velocity in a straight line unless a force acts on the moving object. This law not only describes motion but establishes the role of a force as well. A change of motion is therefore *evidence* of the action of a net force. The association of forces and a change of motion is common in your everyday experience. You have felt forces on your back in an accelerating automobile, and you have felt other forces as the automobile turns or stops. You have also learned about gravitational forces that accelerate objects toward the surface of the earth. Unbalanced forces and acceleration are involved in any change of motion. Newton's second law of motion is a relationship between *net force, acceleration,* and *mass* that describes the cause of a change of motion (figure 2.18).

Consider the motion of you and a bicycle you are riding. Suppose you are riding your bicycle over level ground in a straight line at 10 miles per hour. Newton's first law tells you that you will continue with a constant velocity in a straight line as long as no external, unbalanced force acts on you and the bicycle. The force that you *are* exerting on the pedals seems to equal some external force that moves you and the bicycle along (more on this later). The force exerted as you move along is needed to *balance* the resisting forces of tire friction and air resistance. If

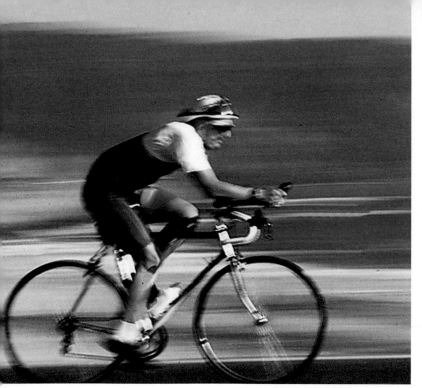

Figure 2.18
This bicycle rider knows about the relationship between force, acceleration, and mass.

net force is greater than zero. Likewise, you will slow down if you apply a force to the brakes, another kind of resisting friction. A third way to change your velocity is to apply a force on the handlebars, changing the direction of your velocity. Thus, *unbalanced forces* on you and your bicycle produce an *acceleration*.

Starting a bicycle from rest suggests a relationship between force and acceleration. You observe that the harder you push on the pedals, the greater your acceleration. If you double the net force, then you will also double the acceleration, reaching the same velocity in half the time. Likewise, if you triple the force you will increase the acceleration threefold. Recall that when quantities increase or decrease together in the same ratio, they are said to be *directly proportional*. The acceleration is therefore directly proportional to the force.

Suppose that your bicycle has two seats, and you have a friend who will ride with you. Suppose also that the addition of your friend on the bicycle will double the mass of the bike and riders. If you use the same net force as before, the bicycle will undergo a much smaller acceleration. In fact, with all other factors equal, doubling the mass and applying the same extra force will produce an acceleration of only half as much (figure 2.20). An even more massive friend would reduce the acceleration even more. If you triple the mass and apply the same extra force, the acceleration will be one-third as much. Recall that when a

these resisting forces were removed, you would not need to exert any force at all to continue moving at a constant velocity. The net force is thus the force you are applying minus the forces from tire friction and air resistance. The *net force* is therefore zero when you move at a constant speed in a straight line (figure 2.19).

If you now apply a greater force on the pedals the *extra* force you apply is unbalanced by friction and air resistance. Hence there will be a net force greater than zero, and you will accelerate. You will accelerate during, and *only* during, the time that the

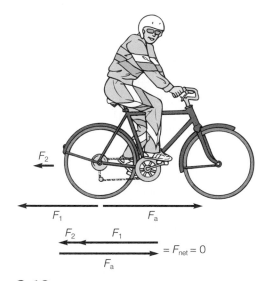

Figure 2.19
At a constant velocity the force of tire friction (F_1) and the force of air resistance (F_2) have a sum that equals the force applied (F_a). The net force is therefore zero.

Figure 2.20
More mass results in less acceleration when the same force is applied. With the same force applied, the riders and bike with twice the mass will have half the acceleration, with all other factors constant. Note that the second rider is not pedaling.

relationship between two quantities shows that one quantity increases as another decreases, in the same ratio, the quantities are said to be *inversely proportional*.

The acceleration of an object depends on *both* the *net force applied* and the *mass* of the object. The **second law of motion** is as follows:

The acceleration of an object is directly proportional to the net force acting on it and inversely proportional to the mass of the object.

If we express force in appropriate units we can write this statement as an equation,

$$a = \frac{F}{m}$$

By solving for F we rearrange the equation into the form it is most often expressed,

$$F = ma$$

equation 2.3

In the metric system you can see that the units for force will be the units for mass (m) times acceleration (a). The unit for mass is kg and the unit for acceleration is m/s². The combination of these units, (kg) (m/s²) is a unit of force called the **newton** (N), in honor of Isaac Newton. So,

$$1 \text{ newton} = 1 \text{ N} = 1 \frac{\text{kg} \cdot \text{m}}{\text{s}^2}$$

Newton's second law of motion is the essential idea of his work on motion. According to this law there is always a relationship between the acceleration, a net force, and the mass of an object. Implicit in this statement are two understandings, (1) that we are talking about the net force, meaning total external force acting on an object, and (2) that the motion statement is concerned with acceleration, not velocity.

Weight and Mass

What is the meaning of weight—is it the same concept as mass? Weight is a familiar concept to most people, and in everyday language the word is often used as having the same meaning as mass. In physics, however, there is a basic difference between weight and mass and this difference is very important in Newton's explanation of motion and the causes of motion.

Mass is defined as the property that determines how much an object resists a change in its motion. The greater the mass the greater the *inertia*, or resistance to change in motion. Consider, for example, that it is easier to push a small car into motion than to push a large truck into motion. The truck has more mass, and therefore more inertia. Newton originally defined mass as the "quantity of matter" in an object, and this definition is intuitively appealing. However, Newton needed to measure inertia because of its obvious role in motion and redefined mass as a measure of inertia.

You could use Newton's second law to measure a mass by exerting a force on the mass and measuring the resulting acceleration. This is not very convenient, so masses are usually mea-

CONCEPTS APPLIED

Second Law Experiment

Tie one end of a string to a book and the other end to a large rubber band. With your index finger curled in the loop of the rubber band, pull the book across a smooth tabletop. How much the rubber band stretches will provide a rough estimate of the force you are applying. (1) Pull the book with a constant velocity across the tabletop. Compare the force required for different constant velocities. (2) Accelerate the book at different rates. Compare the force required to maintain the different accelerations. (3) Use a different book with a greater mass and again accelerate the book at different rates. How does more mass change the results?

Based on your observations, can you infer a relationship between force, acceleration, and mass?

sured on a balance by comparing the force of gravity acting on a standard mass compared to the force of gravity acting on the unknown mass.

The force of gravity acting on a mass is the *weight* of an object. Weight is a force and has different units (N) from those of mass (kg). Since weight is a measure of the force of gravity acting on an object, the force can be calculated from Newton's second law of motion,

$$F = ma$$

or

downward force = (mass)(acceleration due to gravity)

or

$$\text{weight} = (\text{mass})(g)$$

$$\text{or} \quad w = mg$$

equation 2.4

You learned previously that g is the symbol used to represent acceleration due to gravity. Near the earth's surface, g has an approximate value of 9.8 m/s². To understand how g is applied to an object not moving, consider a ball you are holding in your hand. By supporting the weight of the ball you hold it stationary, so the upward force of your hand and the downward force of the ball (its weight) must add to a net force of zero. When you let go of the ball the gravitational force is the only force acting on the ball. The ball's weight is then the net force that accelerates it at g, the acceleration due to gravity. Thus, $F_{net} = w = ma = mg$. The weight of the ball never changes in a given location, so its weight is always equal to $w = mg$, even if the ball is not accelerating.

An important thing to remember is that *pounds* and *newtons* are units of *force* (table 2.1). A *kilogram*, on the other hand, is a measure of *mass*. Thus the English unit of 1.0 lb is comparable to the metric unit of 4.5 N (or 0.22 lb is equivalent to 1.0 N). Conversion tables sometimes show how to convert from pounds (a unit of weight) to kilograms (a unit of mass). This is possible because weight and mass are proportional in a given location on the surface of the earth. Using conversion factors from inside the front cover of this book, see if you can express your weight in pounds and newtons and your mass in kilograms.

Figure 2.21
Forces occur in matched pairs that are equal in strength and opposite in direction.

Table 2.1 Units of Mass and Weight in the Metric and English Systems of Measurement

	Mass	$\times$	Acceleration	$=$	Force
Metric system	kg	$\times$	$\dfrac{m}{s^2}$	$=$	$\dfrac{kg \cdot m}{s^2}$ (newton)
English system	$\left(\dfrac{lb}{ft/s^2}\right)$	$\times$	$\dfrac{ft}{s^2}$	$=$	lb (pound)

Newton's Third Law of Motion

Newton's first law of motion states that an object retains its state of motion when the net force is zero. The second law states what happens when the net force is *not* zero, describing how an object with a known mass moves when a given force is applied. The two laws give one aspect of the concept of a force; that is, if you observe that an object starts moving, speeds up, slows down, or changes its direction of travel, you can conclude that an unbalanced force is acting on the object. Thus, any change in the state of motion of an object is *evidence* that an unbalanced force has been applied.

Newton's third law of motion is also concerned with forces and considers how a force is produced. First, consider where a force comes from. A force is always produced by the interaction of two or more objects. There is always a second object pushing or pulling on the first object to produce a force. To simplify the many interactions that occur on the earth, consider a satellite freely floating in space. According to Newton's second law ($F = ma$), a force must be applied to change the state of motion of the satellite. What is a possible source of such a force? Perhaps an astronaut pushes on the satellite for 1 second. The satellite would accelerate *during* the application of the force, then move away from the original position at some constant velocity. The astronaut would also move away from the original position, but in the opposite direction (figure 2.21). A *single* force *does not exist* by itself. There is always a matched and opposite force that occurs at the same time. Thus, the astronaut exerted a momentary

force on the satellite, but the satellite evidently exerted a momentary force back on the astronaut as well, for the astronaut moved away from the original position in the opposite direction. Newton did not have astronauts and satellites to think about, but this is the kind of reasoning he did when he concluded that forces always occur in matched pairs that are equal and opposite. Thus the **third law of motion** is as follows:

> Whenever two objects interact, the force exerted on one object is equal in strength and opposite in direction to the force exerted on the other object.

The third law states that forces always occur in matched pairs that act in opposite directions and on two *different* bodies. Sometimes the third law of motion is expressed as follows: "For every action there is an equal and opposite reaction," but this can be misleading. Neither force is the cause of the other. The forces are at every instant the cause of each other and they appear and disappear at the same time. If you are going to describe the force exerted on a satellite by an astronaut, then you must realize that there is a simultaneous force exerted on the astronaut by the satellite. The forces (astronaut on satellite and satellite on astronaut) are equal in magnitude but opposite in direction.

Perhaps it would be more common to move a satellite with a small rocket. A satellite is maneuvered in space by firing a rocket in the direction opposite to the direction someone wants to move the satellite. Exhaust gases (or compressed gases) are accelerated in one direction, and exert an equal but opposite force on the satellite that accelerates it in the opposite direction. This is another example of the third law.

Consider how the pairs of forces work on the earth's surface. You walk by pushing your feet against the ground (figure 2.22). Of course you could not do this if it were not for friction. You would slide as on slippery ice without friction. But since friction does exist, you exert a backward horizontal force on the ground, and, as the third law explains, the ground exerts an equal and opposite force on you. You accelerate forward from the net force as explained by the second law. If the earth had the same mass as you, however, it would accelerate backward at the and the bike before you can apply $F = ma$. The only forces that will affect the forward motion of the bike system are the force of the ground pushing it forward and the frictional forces that oppose the forward motion. This is another example of the third law.

MOMENTUM

Sportscasters often refer to the *momentum* of a team, and newscasters sometimes refer to an election where one of the candidates has *momentum*. Both situations describe a competition where one side is moving toward victory and it is difficult to stop. It seems appropriate to borrow this term from the physical sciences because momentum is a property of movement. It takes a longer time to stop something from moving when it has a lot of momentum. The physical science concept of momentum is closely related to Newton's laws of motion. **Momentum** (*p*) is defined as the product of the mass (*m*) of an object and its velocity (*v*),

$$\text{momentum} = \text{mass} \times \text{velocity}$$

or

$$p = mv$$

equation 2.5

The astronaut in figure 2.23 has a mass of 60.0 kg and a velocity of 0.750 m/s as a result of the interaction with the satellite. The resulting momentum is therefore (60.0 kg) (0.750 m/s), or 45.0 kg·m/s. As you can see, the momentum would be greater if the astronaut had acquired a greater velocity or if the astronaut had a greater mass and acquired the same velocity. Momentum involves both the inertia and the velocity of a moving object.

Figure 2.22
The football player's foot is pushing against the ground, but it is the ground pushing against the foot that accelerates the player forward to catch a pass.

same rate that you were accelerated forward. The earth is much more massive than you, however, so any acceleration of the earth is a vanishingly small amount. The overall effect is that you are accelerated forward by the force the ground exerts on you.

Return now to the example of riding a bicycle that was discussed previously. What is the source of the *external* force that accelerates you and the bike? Pushing against the pedals is not external to you and the bike, so that force will *not* accelerate you and the bicycle forward. This force is transmitted through the bike mechanism to the rear tire, which pushes against the ground. It is the ground exerting an equal and opposite force against the system of you and the bike that accelerates you forward. You must consider the forces that act on the system of you

$$
\begin{array}{ll}
m = 60.0\ \text{kg} & m = 120.0\ \text{kg} \\
v = 0.750\ \text{m/s} & v = 0.375\ \text{m/s} \\
p = mv & p = mv \\
\quad = (60.0\ \text{kg})\,(0.750\ \tfrac{\text{m}}{\text{s}}) & \quad = (120.0\ \text{kg})\,(0.375\ \tfrac{\text{m}}{\text{s}}) \\
\quad = \boxed{45.0\ \tfrac{\text{kg·m}}{\text{s}}} & \quad = \boxed{45.0\ \tfrac{\text{kg·m}}{\text{s}}}
\end{array}
$$

Figure 2.23
Both the astronaut and the satellite received a force of 30.0 N for 1.50 s when they pushed on each other. Both then have a momentum of 45.0 kg·m/s in the opposite direction. This is an example of the law of conservation of momentum.

Swimming Scallop

Newton's laws of motion apply to animal motion as well as that of satellites and automobiles. Consider, for example, the dilemma of a growing scallop. A scallop is the shell often seen as a logo for a certain petroleum company, a fan-shaped shell with a radiating fluted pattern (box figure 2.2). The scallop is a marine mollusk that is most unusual since it is the only clamlike mollusk that is capable of swimming. By opening and closing its shell it is able to jet propel itself in a way by forcing water from the interior of the shell in a jetlike action. The popular seafood called "scallops" is the edible muscle that the scallop uses to close its shell.

A scallop is able to swim by orienting its shell at a proper angle and maintaining a minimum acceleration to prevent sinking. For example, investigations have found that one particular species of scallop must force enough water backward to move about 6 body lengths per second with a 10 degree angle of attack to maintain level swimming. Such a swimming effort can be maintained for up to about 20 seconds, enabling the scallop to escape predation or some other disturbing condition.

A more massive body limits the swimming ability of the scallop, as a greater force is needed to give a greater mass the same acceleration (as you

Box Figure 2.2
A scallop shell.

would expect from Newton's second law of motion). This problem becomes worse as the scallop grows larger and larger without developing a greater and greater jet force.

Conservation of Momentum

Notice that the momentum acquired by the satellite in figure 2.23 is *also* 45.0 kg·m/s. The astronaut gained a certain momentum in one direction, and the satellite gained the *very same momentum in the opposite direction.* Newton originally defined the second law in terms of a change of momentum being proportional to the net force acting on an object. Since the third law explains that the forces exerted on both the astronaut and satellite were equal and opposite, you would expect both objects to acquire equal momentum in the opposite direction. This result is observed any time objects in a system interact and the only forces involved are those between the interacting objects (figure 2.23). This statement leads to a particular kind of relationship called a *law of conservation.* In this case, the law applies to momentum and is called the **law of conservation of momentum:**

> **The total momentum of a group of interacting objects remains the same in the absence of external forces.**

Conservation of momentum, energy, and charge are among examples of conservation laws that apply to everyday situations. These situations always illustrate two understandings, that (1) each conservation law is an expression of symmetry that describes a physical principle that can be observed; and (2) each law holds regardless of the details of an interaction or how it took place. Since the conservation laws express symmetry that always occurs, they tell us what might be expected to happen, and what might be expected not to happen in a given situation. The symmetry also allows unknown quantities to be found by

analysis. The law of conservation of momentum, for example, is useful in analyzing motion in simple systems of collisions such as those of billiard balls, automobiles, or railroad cars. It is also useful in measuring action and reaction interactions, as in rocket propulsion, where the backward momentum of the exhaust gases equals the momentum given to the rocket in the opposite direction (figure 2.24). When this is done, momentum is always found to be conserved.

Impulse

Have you ever heard that you should "follow through" when hitting a ball? When you follow through, the bat is in contact with the ball for a longer period of time. The force of the hit is important, of course, but both the force and how long the force is applied determine the result. The product of the force and the time of application is called **impulse.** This quantity can be expressed as

$$\text{impulse} = Ft$$

where F is the force applied during the time of contact (t). The impulse you give the ball determines how fast the ball will move, and thus how far it will travel.

Impulse is related to the change of motion of a ball of a given mass, so the change of momentum (mv) is brought about by the impulse. This can be expressed as

$$\text{change of momentum} = (\text{applied force}) (\text{time of contact})$$

$$\Delta p = Ft$$

equation 2.6

Figure 2.24
According to the law of conservation of momentum, the momentum of the expelled gases in one direction equals the momentum of the rocket in the other direction in the absence of external forces.

the force applied by increasing the contact time, and hopefully you reduced the force sufficiently so the egg does not break.

Contact time is also important in safety. Automobile airbags, the padding in elbow and knee pads, and the plastic barrels off the highway in front of overpass supports are examples of designs intended to increase the contact time. Again, increasing the contact time reduces the force, since $\Delta p = Ft$. The impact force is reduced and so are the injuries. Think about this the next time you see a car that was crumpled and bent by a collision. The driver and passengers were probably saved from injuries that are more serious since more time was involved in stopping the car that crumpled. A car that crumples is a safer car in a collision.

FORCES AND CIRCULAR MOTION

Consider a communications satellite that is moving at a uniform speed around the earth in a circular orbit. According to the first law of motion there *must be* forces acting on the satellite, since it does *not* move off in a straight line. The second law of motion also indicates forces, since an unbalanced force is required to change the motion of an object.

Recall that acceleration is defined as a change in velocity, and that velocity has both strength and direction. Velocity is changed by a change in speed, direction, or both speed and direction. The satellite in a circular orbit is continuously being accelerated. This means that there is a continuously acting unbalanced force on the satellite that pulls it out of a straight-line path.

The force that pulls an object out of its straight-line path and into a circular path is called a **centripetal** (center-seeking) **force.** Perhaps you have swung a ball on the end of a string in a horizontal circle over your head. Once you have the ball moving, the only unbalanced force (other than gravity) acting on the ball is the centripetal force your hand exerts on the ball through the

where Δp is a change of momentum. You "follow through" while hitting a ball in order to increase the contact time. If the same force is used, a longer contact time will result in a greater impulse. A greater impulse means a greater change of momentum, and since the mass of the ball does not change, the overall result is a moving ball with a greater velocity. This means following through will result in more distance from hitting the ball with the same force. That's why it is important to follow through when you hit the ball.

Now consider bringing a moving object to a stop by catching it. In this case the mass and the velocity of the object are fixed at the time you catch it, and there is nothing you can do about these quantities. The change of momentum is equal to the impulse, and the force and time of force application *can* be manipulated. For example, consider how you would catch a raw egg that is tossed to you. You would probably move your hands with the egg as you catch it, increasing the contact time. Increasing the contact time has the effect of reducing the force, since $\Delta p = Ft$. You changed

Amusement park rides are designed to accelerate your body, sometimes producing changes in the acceleration (jerk) as well. This is done by changes in speed, changes in the direction of travel, or changes in both direction and speed. Many rides move in a circular path, since such movement is a constant acceleration.

Why do people enjoy amusement park rides? It is not the high speed, since your body is not very sensitive to moving at a constant speed. Moving at a steady 600 mi/h in an airplane, for example, provides little sensation when you are seated in an aisle seat in the central cabin.

Your body is not sensitive to high-speed traveling, but it is sensitive to acceleration and changes of acceleration. Acceleration affects the fluid in your inner ear, which controls your sense of balance. In most people, acceleration also produces a reaction that results in the release of the hormones epinephrine and norepinephrine from the adrenal medulla, located near the kidney. The heart rate increases, blood pressure rises, blood is shunted to muscles, and the breathing rate increases. You have probably experienced this reaction many times in your life, as when you nearly have an automobile accident or slip and nearly fall. In the case of an amusement park ride, your body adapts and you believe you enjoy the experience.

string. This centripetal force pulls the ball from its natural straight-line path into a circular path. There are no outward forces acting on the ball. The force that you feel on the string is a consequence of the third law; the ball exerts an equal and opposite force on your hand. If you were to release the string, the ball would move away from the circular path in a *straight line* that has a right angle to the radius at the point of release (figure 2.25). When you release the string, the centripetal force ceases, and the ball then follows its natural straight-line motion. If other forces were involved, it would follow some other path. Nonetheless, the apparent outward force has been given a name just as if it were a real force. The outward tug is called a *centrifugal force*.

The magnitude of the centripetal force required to keep an object in a circular path depends on the inertia, or mass, of the object and the acceleration of the object, just as you learned in the second law of motion. The acceleration of an object moving in a circle can be shown to be directly proportional to the square of the speed around the circle (v^2) and inversely proportional to the radius of the circle (r). (A smaller radius requires a greater acceleration.) Therefore, the acceleration of an object moving in uniform circular motion (a_c) is

$$a_c = \frac{v^2}{r}$$

equation 2.7

The magnitude of the centripetal force of an object with a mass (m) that is moving with a velocity (v) in a circular orbit of a radius (r) can be found by substituting equation 2.7 in $F = ma$, or

$$F = \frac{mv^2}{r}$$

equation 2.8

NEWTON'S LAW OF GRAVITATION

You know that if you drop an object, it always falls to the floor. You define *down* as the direction of the object's movement and *up* as the opposite direction. Objects fall because of the force of gravity, which accelerates objects at $g = 9.8$ m/s^2 (32 ft/s^2) and gives them weight, $w = mg$.

Gravity is an attractive force, a pull that exists between all objects in the universe. It is a mutual force that, just like all other forces, comes in matched pairs. Since the earth attracts you with a certain force, you must attract the earth with an exact opposite force. The magnitude of this force of mutual attraction depends on several variables. These variables were first described by Newton in *Principia,* his famous book on motion that was printed in 1687. Newton had, however, worked out his ideas much earlier, by the age of twenty-four, along with ideas about his laws of motion and the formula for centripetal acceleration. In a biography written by a friend in 1752, Newton stated that the notion of gravitation came to mind during a time of thinking that "was occasioned by the fall of an apple." He was thinking about why the moon stays in orbit around the earth rather

Figure 2.25

Centripetal force on the ball causes it to change direction continuously, or accelerate into a circular path. Without the force acting on it, the ball would continue in a straight line.

When do astronauts experience weightlessness, or "zero gravity"? Theoretically, the gravitational field of the earth extends to the whole universe. You know that it extends to the moon, and indeed, even to the sun some 93 million miles away. There is a distance, however, at which the gravitational force must become immeasurably small. But even at an altitude of 20,000 miles above the surface of the earth, gravity is measurable. At 20,000 miles, the value of g is about 1 ft/s² (0.3 m/s²) compared to 32 ft/s² (9.8 m/s²) on the surface. Since gravity does exist at these distances, how can an astronaut experience "zero gravity"?

Gravity does act on astronauts in spacecraft that are in orbit around the earth. The spacecraft stays in orbit, in fact, because of the gravitational attraction and because it has the correct tangential speed. If the tangential speed were less than 5 mi/s, the spacecraft would return to the earth. Astronauts fire their retro-rockets, which slow the tangential speed, causing the spacecraft to fall down to the earth. If the tangential speed were more than 7 mi/s, the spacecraft would fly off into space. The spacecraft stays in orbit be-

cause it has the right tangential speed to continuously "fall" around and around the earth. Gravity provides the necessary centripetal force that causes the spacecraft to fall out of its natural straight-line motion.

Since gravity is acting on the astronaut and spacecraft, the term *zero gravity* is not an accurate description of what is happening. The astronaut, spacecraft, and everything in it are experiencing *apparent weightlessness* because they are continuously falling toward the earth. Everything seems to float because everything is falling together. But, strictly speaking, everything still has weight, because weight is defined as a gravitational force acting on an object ($w = mg$).

Whether weightlessness is apparent or real, however, the effects on people are the same. Long-term orbital flights have provided evidence that the human body changes from the effect of weightlessness. Bones lose calcium and other minerals, the heart shrinks to a much smaller size, and leg muscles shrink so much on prolonged flights that astronauts cannot walk when they return to the earth. These changes occur because on the earth humans are constantly sub-

jected to the force of gravity. The nature of the skeleton and the strength of the muscles are determined by how the body reacts to this force. Metabolic pathways and physiological processes that maintain strong bones and muscles evolved having to cope with a specific gravitational force. When we are suddenly subjected to a place where gravity is significantly different, these processes result in weakened systems. If we had evolved on a planet with a different gravitational force, we would have muscles and bones that were adapted to the gravity on that planet. All organisms have evolved in a world with gravity. Many kinds of organisms have been used in experiments in space to try to develop a better understanding of how their systems work.

The problems related to prolonged weightlessness must be worked out before long-term weightless flights can take place. One solution to these problems might be a large, uniformly spinning spacecraft. The astronauts tend to move in a straight line, and the side of the turning spacecraft (now the "floor") exerts a force on them to make them go in a curved path. This force would act as an artificial gravity.

---

than moving off in a straight line as would be predicted by the first law of motion. Perhaps the same force that attracts the moon toward the earth, he thought, attracts the apple to the earth. Newton developed a theoretical equation for gravitational force that explained not only the motion of the moon but the motion of the whole solar system. Today, this relationship is known as the **universal law of gravitation:**

> **Every object in the universe is attracted to every other object with a force that is directly proportional to the product of their masses and inversely proportional to the square of the distances between them.**

In symbols, m_1 and m_2 can be used to represent the masses of two objects, d the distance between their centers, and G a constant of proportionality. The equation for the law of universal gravitation is therefore

$$F = G\frac{m_1 m_2}{d^2}$$

equation 2.9

This equation gives the magnitude of the attractive force that each object exerts on the other. The two forces are oppositely directed. The constant G is a universal constant, since the law applies to all objects in the universe. It was first measured experimentally by Henry Cavendish in 1798. The accepted value today is $G = 6.67 \times 10^{-11}$ N·m²/kg². Do not confuse G, the universal constant, with g, the acceleration due to gravity on the surface of the earth.

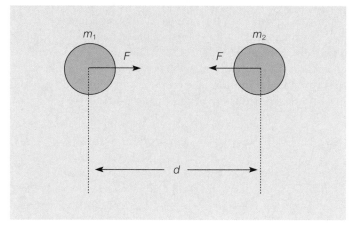

Figure 2.26

The variables involved in gravitational attraction. The force of attraction (F) is proportional to the product of the masses (m_1, m_2) and inversely proportional to the square of the distance (d) between the centers of the two masses.

Thus, the magnitude of the force of gravitational attraction is determined by the mass of the two objects and the distance between them (figure 2.26). The law also states that *every* object is attracted to every other object. You are attracted to all the objects around you—chairs, tables, other people, and so forth. Why don't you notice the forces between you and other objects? One or both of the interacting objects must be quite massive before a noticeable force results from the interaction. That is why you

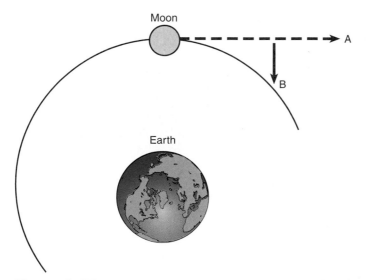

Figure 2.27

Gravitational attraction acts as a centripetal force that keeps the moon from following the straight-line path shown by the dashed line to position A. It was pulled to position B by gravity (0.0027 m/s²) and thus "fell" toward the earth the distance from the dashed line to B, resulting in a somewhat circular path.

do not notice the force of gravitational attraction between you and objects that are not very massive compared to the earth. The attraction between you and the earth overwhelmingly predominates, and that is all you notice.

The acceleration due to gravity, g, is about 9.8 m/s² and is practically a constant for relatively short distances above the surface. Notice, however, that Newton's law of gravitation is an inverse square law. This means if you double the distance, the force is $1/(2)^2$ or 1/4 as great. If you triple the distance, the force is $1/(3)^2$ or 1/9 as great. In other words, the force of gravitational attraction and g decrease inversely with the square of the distance from the earth's center.

Newton was able to calculate the acceleration of the moon toward the earth, about 0.0027 m/s². The moon "falls" toward the earth because it is accelerated by the force of gravitational attraction. This attraction acts as a *centripetal force* that keeps the moon from following a straight-line path as would be predicted from the first law. Thus, the acceleration of the moon keeps it in a somewhat circular orbit around the earth. Figure 2.27 shows that the moon would be in position A if it followed a straight-line path instead of "falling" to position B as it does. The moon thus "falls" around the earth. Newton was able to analyze the motion of the moon quantitatively as evidence that it is gravitational force that keeps the moon in its orbit. The law of gravitation was extended to the sun, other planets, and eventually the universe. The quantitative predictions of observed relationships among the planets were strong evidence that all objects obey the same law of gravitation. In addition, the law provided a means to calculate the mass of the earth, the moon, the planets, and the sun. Newton's law of gravitation, laws of motion, and work with mathematics formed the basis of most physics and technology for the next two centuries, as well as accurately describing the world of everyday experience.

SUMMARY

Motion can be measured by speed, velocity, and acceleration. *Speed* is a measure of how fast something is moving. It is a ratio of the distance covered between two locations to the time that elapsed while moving between the two locations. The *average speed* considers the distance covered during some period of time, while the *instantaneous speed* is the speed at some specific instant. *Velocity* is a measure of the speed and direction of a moving object. *Acceleration* is a change of velocity per unit of time.

A *force* is a push or a pull that can change the motion of an object. The *net force* is the sum of all the forces acting on an object.

Galileo determined that a continuously applied force is not necessary for motion and defined the concept of *inertia:* an object remains in unchanging motion in the absence of a net force. Galileo also determined that falling objects accelerate toward the earth's surface independent of the weight of the object. He found the acceleration due to gravity, g, to be 9.8 m/s² (32 ft/s²), and the distance an object falls is proportional to the square of the time of free fall ($d \propto t^2$).

Compound motion occurs when an object is projected into the air. Compound motion can be described by splitting the motion into vertical and horizontal parts. The acceleration due to gravity, g, is a constant that is acting at all times and acts independently of any motion that an object has. The path of an object that is projected at some angle to the horizon is therefore a parabola.

Newton's *first law of motion* is concerned with the motion of an object and the lack of a net force. Also known as the *law of inertia*, the first law states that an object will retain its state of straight-line motion (or state of rest) unless a net force acts on it.

The *second law of motion* describes a relationship between net force, mass, and acceleration. A *newton* of force is the force needed to give a 1.0 kg mass an acceleration of 1.0 m/s².

Weight is the downward force that results from the earth's gravity acting on the mass of an object. Weight is measured in *newtons* in the metric system and *pounds* in the English system.

Newton's *third law of motion* states that forces are produced by the interaction of *two different* objects. These forces always occur in matched pairs that are equal in size and opposite in direction.

Momentum is the product of the mass of an object and its velocity. In the absence of external forces, the momentum of a group of interacting objects always remains the same. This relationship is the *law of conservation of momentum. Impulse* is a change of momentum equal to a force times the time of application.

An object moving in a circular path must have a force acting on it, since it does not move in a straight line. The force that pulls an object out of its straight-line path is called a *centripetal force.* The centripetal force needed to keep an object in a circular path depends on the mass of the object, its velocity, and the radius of the circle.

The *universal law of gravitation* is a relationship between the masses of two objects, the distance between the objects, and a proportionality constant. Newton was able to use this relationship to show that gravitational attraction provides the centripetal force that keeps the moon in its orbit.

Summary of Equations

2.1
$$\text{average speed} = \frac{\text{distance}}{\text{time}}$$
$$\bar{v} = \frac{d}{t}$$

2.2
$$\text{acceleration} = \frac{\text{change of velocity}}{\text{time}}$$
$$= \frac{\text{final velocity} - \text{initial velocity}}{\text{time}}$$
$$a = \frac{v_f - v_i}{t}$$

2.3
$$\text{force} = \text{mass} \times \text{acceleration}$$
$$F = ma$$

2.4
$$\text{weight} = \text{mass} \times \text{acceleration due to gravity}$$
$$w = mg$$

2.5
$$\text{momentum} = \text{mass} \times \text{velocity}$$
$$p = mv$$

2.6
$$\text{change of momentum} = \text{force} \times \text{time}$$
$$\Delta p = Ft$$

2.7
$$\text{centripetal acceleration} = \frac{\text{velocity squared}}{\text{radius of circle}}$$
$$a_c = \frac{v^2}{r}$$

2.8
$$\text{centripetal force} = \frac{\text{mass} \times \text{velocity squared}}{\text{radius of circle}}$$
$$F = \frac{mv^2}{r}$$

2.9
$$\text{gravitational force} = \text{constant} \times \frac{\text{one mass} \times \text{another mass}}{\text{distance squared}}$$
$$F = G\frac{m_1 m_2}{d^2}$$

KEY TERMS

acceleration (p. **24**)
centripetal force (p. **40**)
first law of motion (p. **32**)
force (p. **26**)
free fall (p. **29**)
g (p. **31**)
impulse (p. **39**)
inertia (p. **29**)
law of conservation of
   momentum (p. **39**)

mass (p. **36**)
momentum (p. **38**)
net force (p. **26**)
newton (p. **36**)
second law of motion (p. **36**)
speed (p. **22**)
third law of motion (p. **37**)
universal law of gravitation (p. **42**)
velocity (p. **24**)

APPLYING THE CONCEPTS

1. A quantity of 5 m/s² is a measure of
   (a) metric area.
   (b) acceleration.
   (c) speed.
   (d) velocity.

2. An automobile has how many different devices that can cause it to undergo acceleration?
   (a) none
   (b) one
   (c) two
   (d) three or more

3. Ignoring air resistance, an object falling toward the surface of the earth has a *velocity* that is
   (a) constant.
   (b) increasing.
   (c) decreasing.
   (d) acquired instantaneously, but dependent on the weight of the object.

4. Ignoring air resistance, an object falling near the surface of the earth has an *acceleration* that is
   (a) constant.
   (b) increasing.
   (c) decreasing.
   (d) dependent on the weight of the object.

5. Two objects are released from the same height at the same time, and one has twice the weight of the other. Ignoring air resistance,
   (a) the heavier object hits the ground first.
   (b) the lighter object hits the ground first.
   (c) they both hit at the same time.
   (d) whichever hits first depends on the distance dropped.

6. A ball rolling across the floor slows to a stop because
   (a) there is a net force acting on it.
   (b) the force that started it moving wears out.
   (c) the forces are balanced.
   (d) the net force equals zero.

7. Considering the forces on the system of you and a bicycle as you pedal the bike at a constant velocity in a horizontal straight line,
   (a) the force you are exerting on the pedal is greater than the resisting forces.
   (b) all forces are in balance, with the net force equal to zero.
   (c) the resisting forces of air and tire friction are less than the force you are exerting.
   (d) the resisting forces are greater than the force you are exerting.
8. If you double the unbalanced force on an object of a given mass, the acceleration will be
   (a) doubled.
   (b) increased fourfold.
   (c) increased by one-half.
   (d) increased by one-fourth.
9. If you double the mass of a cart while it is undergoing a constant unbalanced force, the acceleration will be
   (a) doubled.
   (b) increased fourfold.
   (c) half as much.
   (d) one-fourth as much.
10. Doubling the distance between the center of an orbiting satellite and the center of the earth will result in what change in the gravitational attraction of the earth for the satellite?
    (a) one-half as much
    (b) one-fourth as much
    (c) twice as much
    (d) four times as much
11. If a ball swinging in a circle on a string is moved twice as fast, the force on the string will be
    (a) twice as great.
    (b) four times as great.
    (c) one-half as much.
    (d) one-fourth as much.
12. A ball is swinging in a circle on a string when the string length is doubled. At the same velocity, the force on the string will be
    (a) twice as great.
    (b) four times as great.
    (c) one-half as much.
    (d) one-fourth as much.

Answers
1. b 2. d 3. b 4. a 5. c 6. a 7. b 8. a 9. c 10. b 11. b 12. c

QUESTIONS FOR THOUGHT

1. An insect inside a bus flies from the back toward the front at 5.0 mi/h. The bus is moving in a straight line at 50.0 mi/h. What is the speed of the insect?
2. Disregarding air friction, describe all the forces acting on a bullet shot from a rifle into the air.
3. Can gravity act in a vacuum? Explain.
4. Is it possible for a small car to have the same momentum as a large truck? Explain.
5. What net force is needed to maintain the constant velocity of a car moving in a straight line? Explain.
6. How can an unbalanced force exist if every action has an equal and opposite reaction?
7. Why should you bend your knees as you hit the ground after jumping from a roof?
8. Is it possible for your weight to change as your mass remains constant? Explain.
9. What maintains the speed of the earth as it moves in its orbit around the sun?
10. Suppose you are standing on the ice of a frozen lake and there is no friction whatsoever. How can you get off the ice? (*Hint:* Friction is necessary to crawl or walk, so that will not get you off the ice.)
11. A rocket blasts off from a platform on a space station. An identical rocket blasts off from free space. Considering everything else to be equal, will the two rockets have the same acceleration? Explain.
12. An astronaut leaves a spaceship that is moving through free space to adjust an antenna. Will the spaceship move off and leave the astronaut behind? Explain.

PARALLEL EXERCISES

The exercises in groups A and B cover the same concepts. Solutions to group A exercises are located in appendix D.
Note: Neglect all frictional forces in all exercises.

Group A

1. How far away was a lightning strike if thunder is heard 5.00 seconds after seeing the flash? Assume that sound traveled at 350.0 m/s during the storm.
2. What is the acceleration of a car that moves from rest to 15.0 m/s in 10.0 s?
3. What is the average speed of a truck that makes a 285-mile trip in 5.0 hours?
4. What force will give a 40.0 kg grocery cart an acceleration of 2.4 m/s²?
5. An unbalanced force of 18 N will give an object an acceleration of 3 m/s². What force will give this very same object an acceleration of 10 m/s²?
6. A rocket pack with a thrust of 100 N accelerates a weightless astronaut at 0.5 m/s² through free space. What is the mass of the astronaut and equipment?
7. What is the momentum of a 100 kg football player who is moving at 6 m/s?
8. A car weighing 13,720 N is speeding down a highway with a velocity of 91 km/h. What is the momentum of this car?
9. A 15 g bullet is fired with a velocity of 200 m/s from a 6 kg rifle. What is the recoil velocity of the rifle?
10. A net force of 5,000.0 N accelerates a car from rest to 90.0 km/h in 5.0 s. (a) What is the mass of the car? (b) What is the weight of the car?
11. How much centripetal force is needed to keep a 0.20 kg ball on a 1.50 m string moving in a circular path with a speed of 3.0 m/s?
12. On the earth, an astronaut and equipment weigh 1,960.0 N. While weightless in space, the astronaut fires a 100 N rocket backpack for 2.0 s. What is the resulting velocity of the astronaut and equipment?

Group B

1. How many meters away is a cliff if an echo is heard one-half second after the original sound? Assume that sound traveled at 343 m/s on that day.
2. What is the acceleration of a car that moves from a speed of 5.0 m/s to a speed of 15 m/s during a time of 6.0 s?
3. What is the average speed of a car that travels 270.0 miles in 4.50 hours?
4. What force would an asphalt road have to give a 6,000 kg truck in order to accelerate it at 2.2 m/s² over a level road?
5. If a space probe weighs 39,200 N on the surface of the earth, what will be the mass of the probe on the surface of Mars?
6. On the earth, an astronaut and equipment weigh 1,960 N. Weightless in space, the motionless astronaut and equipment are accelerated by a rocket pack with a 100 N thruster that fires for 2 s. What is the resulting final velocity?
7. What is the momentum of a 30.0 kg shell fired from a cannon with a velocity of 500 m/s?
8. What is the momentum of a 39.2 N bowling ball with a velocity of 7.00 m/s?
9. A 30.0 kg shell fired from a 2,000 kg cannon will have a velocity of 500 m/s. What is the resulting velocity of the cannon?
10. A net force of 3,000.0 N accelerates a car from rest to 36.0 km/h in 5.00 s. (a) What is the mass of the car? (b) What is the weight of the car?
11. What tension must a 50.0 cm length of string support in order to whirl an attached 1,000.0 g stone in a circular path at 5.00 m/s?
12. A 200.0 kg astronaut and equipment move with a velocity of 2.00 m/s toward an orbiting spacecraft. How long will the astronaut need to fire a 100.0 N rocket backpack to stop the motion relative to the spacecraft?

Energy

The wind can be used as a source of energy. All you need is a way to capture the energy—such as these wind turbines in California—
and to live someplace where the wind blows enough to make it worthwhile.

O U T L I N E

The term *energy* is closely associated with the concepts of force and motion. Naturally moving matter, such as the wind or moving water, exerts forces. You have felt these forces if you have ever tried to walk against a strong wind or stand in one place in a stream of rapidly moving water. The motion and forces of moving air and moving water are used as *energy sources* (figure 3.1). The wind is an energy source as it moves the blades of a windmill performing useful work. Moving water is an energy source as it forces the blades of a water turbine to spin, turning an electric generator. Thus, moving matter exerts a force on objects in its path, and objects moved by the force can also be used as an energy source.

Matter does not have to be moving to supply energy; matter *contains* energy. Food supplied the energy for the muscular exertion of the humans and animals that accomplished most of the work before this century. Today, machines do the work that was formerly accomplished by muscular exertion. Machines also use the energy contained in matter. They use gasoline, for example, as they supply the forces and motion to accomplish work.

Moving matter and matter that contains energy can be used as energy sources to perform work. The concepts of work and energy and the relationship to matter are the topics of this chapter. You will learn how energy flows in and out of your surroundings as well as a broad, conceptual view of energy.

WORK

You learned earlier that the term *force* has a special meaning in science that is different from your everyday concept of force. In everyday use, you use the term in a variety of associations such as police force, economic force, or the force of an argument. Earlier, force was discussed in a general way as a push or pull. Then a more precise scientific definition of force was developed from Newton's laws of motion—a force is a result of an interaction that is capable of changing the state of motion of an object.

The word *work* represents another one of those concepts that has a special meaning in science that is different from your everyday concept. In everyday use, work is associated with a task to be accomplished or the time spent in performing the task. You might work at understanding science, for example, or you might tell someone that science is a lot of work. You also probably associate physical work, such as lifting or moving boxes, with how tired you become from the effort. The scientific definition of work is not concerned with tasks, time, or how tired you become from doing a task. It is concerned with the application of a force to an object and the distance the object moves as a result of the force. The **work** done on the object is defined as *the product of the applied force and the parallel distance through which the force acts:*

$$\text{work} = \text{force} \times \text{distance}$$
$$W = Fd$$

equation 3.1

Work, in the scientific sense, is the product of a force and the distance an object moves as a result of the force. There are two important considerations to remember about this definition: (1) something *must move* whenever work is done, and (2) the movement must be in the *same direction* as the direction of the force. When you move a book to a higher shelf in a bookcase you are doing work on the book. You apply a vertically upward force equal to the weight of the book as you move it in the same direction as the direction of the applied force. The work done on the book can therefore be calculated by multiplying the weight of the book by the distance it was moved (figure 3.2).

If you simply stand there holding the book, however, you are doing no work on the book. Your arm may become tired from holding the book, since you must apply a vertically upward

Figure 3.1

This is Glen Canyon Dam on the Colorado River between Utah and Arizona. The dam is 216 m (710 ft) tall, dropping 940 cubic meters of water per second (33,000 ft³/s) through 8 generating units. At capacity, it generates about 1.3 megawatts.

force equal to the weight of the book. But this force is not acting through a distance, since the book is not moving. According to equation 3.1, a distance of zero results in zero work. Only a force that results in motion in the same direction results in work.

Units of Work

The units of work can be obtained from the definition of work, $W = Fd$. In the metric system a force is measured in newtons (N), and distance is measured in meters (m), so the unit of work is

$$W = Fd$$
$$W = (\text{newton})(\text{meter})$$
$$W = (N)(m)$$

Figure 3.2

The force on the book moves it through a vertical distance from the second shelf to the fifth shelf, and work is done, $W = Fd$.

The newton-meter is therefore the unit of work. This derived unit has a name. The newton-meter is called a **joule** (J) (pronounced "jool").

$$1 \text{ joule} = 1 \text{ newton-meter}$$

The units for a newton are kg·m/s^2, and the unit for a meter is m. It therefore follows that the units for a joule are $\text{kg·m}^2/\text{s}^2$.

In the English system, the force is measured in pounds (lb), and the distance is measured in feet (ft). The unit of work in the English system is therefore the ft·lb. The ft·lb does not have a name of its own as the N·m does (figure 3.3). (Note that although the equation is $W = Fd$, and this means = (pounds) (feet), the unit is called the ft·lb.)

$W = Fd$
= (pounds) (feet)
= ft·lb

$W = Fd$
= (newtons) (meters)
= N·m
= joule

Figure 3.3

Work is done against gravity when lifting an object. Work is measured in joules or in foot-pounds.

Power

You are doing work when you walk up a stairway, since you are lifting yourself through a distance. You are lifting your weight (force exerted) the *vertical* height of the stairs (distance through which the force is exerted). Consider a person who weighs 120 lb and climbs a stairway with a vertical distance of 10 ft. This person will do (120 lb)(10 ft) or 1,200 ft·lb of work. Will the amount of work change if the person were to run up the stairs? The answer is no, the same amount of work is accomplished. Running up the stairs, however, is more tiring than walking up the stairs. You use the same amount of energy but at a greater *rate* when running. The rate at which energy is transformed or the rate at which work is done is called **power** (figure 3.4). Power is defined as work per unit of time,

$$\text{power} = \frac{\text{work}}{\text{time}}$$

$$P = \frac{W}{t}$$

equation 3.2

Considering just the work and time factors, the 120 lb person who ran up the 10 ft height of stairs in 4 seconds would have a power rating of

$$P = \frac{W}{t} = \frac{(120 \text{ lb}) (10 \text{ ft})}{4 \text{ s}} = 300 \frac{\text{ft · lb}}{\text{s}}$$

If the person had a time of 3 seconds on the same stairs, the power rating would be greater, 400 ft·lb/s. This is a greater *rate* of energy use, or greater power.

When the steam engine was first invented, there was a need to describe the rate at which the engine could do work. Since people at this time were familiar with using horses to do their work, the steam engines were compared to horses. James Watt, who designed a workable steam engine, defined **horsepower** as a power rating of 550 ft·lb/s (figure 3.5A). To convert a power rating in the English units of ft·lb/s to horsepower, divide the power rating by 550 ft·lb/s/hp. For example, the 120 lb person who had a power rating of 400 ft·lb/s had a horsepower of 400 ft·lb/s ÷ 550 ft·lb/s/hp, or 0.7 hp.

In the metric system, power is measured in joules per second. The unit J/s, however, has a name. A J/s is called a **watt** (W). The watt (figure 3.5B) is used with metric prefixes for large

$W = mgh$
$= (120 \text{ lb}) (10.0 \text{ ft})$
$= 1{,}200 \text{ ft·lb}$

Force = $w = mg$

120 lb

Height 10 ft

A

Force = $w = mg$

120 lb

$P = \dfrac{mgh}{t}$
$= \dfrac{(120 \text{ lb}) (10.0 \text{ ft})}{4 \text{ s}}$
$= 300 \dfrac{\text{ft·lb}}{\text{s}}$

Height 10 ft

B

Figure 3.4

(*A*) The work accomplished in climbing a stairway is the person's weight times the vertical distance. (*B*) The power level is the work accomplished per unit of time.

Time = 1.0 s

Force = 550 lb

Distance = 1.0 ft

A

Time = 1.0 s

Force = 1 newton (about 0.22 lb)

Distance = 1.0 m

B

Figure 3.5

(*A*) A horsepower is defined as a power rating of 550 ft·lb/s. (*B*) A watt is defined as a newton-meter per second, or joule per second.

numbers: 1,000 W = 1 kilowatt (kW) and 1,000,000 W = 1 megawatt (MW). It takes 746 W to equal 1 horsepower. One kilowatt is equal to about 1 1/3 horsepower. The electric utility company charges you for how much electrical energy you have used. Electrical energy is measured by power (kW) times the time of use (h). Thus, electrical energy is measured in kWh. We will return to kilowatts and kilowatt-hours later when we discuss electricity.

MOTION, POSITION, AND ENERGY

Closely related to the concept of work is the concept of **energy.** Energy can be defined as the *ability to do work.* This definition of energy seems consistent with everyday ideas about energy and physical work. After all, it takes a lot of energy to do a lot of work. In fact, one way of measuring the energy of something is to see how much work it can do. Likewise, when work is done *on* something, a change occurs in its energy level. The following examples will help clarify this close relationship between work and energy.

Potential Energy

Consider a book on the floor next to a bookcase. You can do work on the book by vertically raising it to a shelf. You can measure this work by multiplying the vertical upward force applied times the distance that the book is moved. You might find, for example, that you did an amount of work equal to 10 J on the book.

Suppose that the book has a string attached to it, as shown in figure 3.6. The string is threaded over a frictionless pulley and

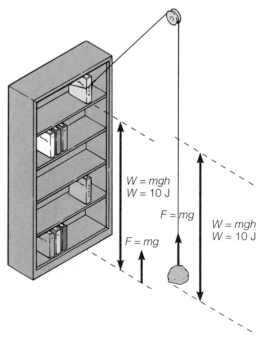

$W = mgh$
$W = 10 \text{ J}$

$F = mg$

$F = mg$

$W = mgh$
$W = 10 \text{ J}$

Figure 3.6

If moving a book from the floor to a high shelf requires 10 J of work, then the book will do 10 J of work on an object of the same mass when the book falls from the shelf.

attached to an object on the floor. If the book is caused to fall from the shelf, the object on the floor will be vertically lifted through some distance by the string. The falling book exerts a force on the object through the string, and the object is moved through a distance. In other words, the *book* did work on the object through the string, $W = Fd$.

The book can do more work on the object if it falls from a higher shelf, since it will move the object a greater distance. The higher the shelf, the greater the *potential* for the book to do work. The ability to do work is defined as energy. The energy that an object has because of its position is called **potential energy** (*PE*). Potential energy is defined as *energy due to position.* This type of potential energy is called *gravitational potential energy,* since it is a result of gravitational attraction. There are other types of potential energy, such as that in a compressed or stretched spring.

The potential energy of an object can be calculated from the work done *on* the object to change its position. You exert a force equal to its weight as you lift it some height above the floor, and the work you do is the product of the weight and height. Likewise, the amount of work the object *could* do because of its position is the product of its weight and height. For the metric unit of mass, weight is the product of the mass of an object times *g*, the acceleration due to gravity, so

$$\text{potential energy} = \text{weight} \times \text{height}$$
$$PE = mgh$$

equation 3.3

For English units, the pound *is* the gravitational unit of force, or weight, so equation 3.3 becomes $PE = (w)(h)$.

Under what conditions does an object have zero potential energy? Considering the book in the bookcase, you could say that the book has zero potential energy when it is flat on the floor. It can do no work when it is on the floor. But what if that floor happens to be the third floor of a building? You could, after all, drop the book out of a window. The answer is that it makes no difference. The same results would be obtained in either case since it is the *change of position* that is important in potential energy. The zero reference position for potential energy is therefore arbitrary. A zero reference point is chosen as a matter of convenience. Note that if the third floor of a building is chosen as the zero reference position, a book on ground level would have negative potential energy. This means that you would have to do work on the book to bring it back to the zero potential energy position (figure 3.7). You will learn more about negative energy levels later in the chapters on chemistry.

Kinetic Energy

Moving objects have the ability to do work on other objects because of their motion. A rolling bowling ball exerts a force on the bowling pins and moves them through a distance, but the ball loses speed as a result of the interaction (figure 3.8). A moving car has the ability to exert a force on a tree and knock it down, again with a corresponding loss of speed. Objects in motion have the ability to do work, so they have energy. The energy of motion is known as **kinetic energy.** Kinetic energy can be

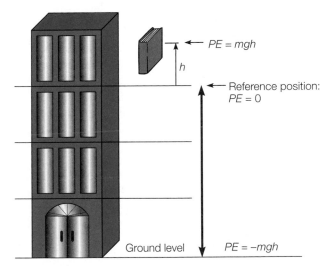

Figure 3.7

The zero reference level for potential energy is chosen for convenience. Here the reference position chosen is the third floor, so the book will have a negative potential energy at ground level.

measured (1) in terms of the work done to put the object in motion or (2) in terms of the work the moving object will do in coming to rest. Consider objects that you put into motion by throwing. You exert a force on a football as you accelerate it through a distance before it leaves your hand. The kinetic energy that the ball now has is equal to the work (force times distance) that you did on the ball. You exert a force on a baseball through a distance as the ball increases its speed before it leaves your hand. The kinetic energy that the ball now has is equal to the work that you did on the ball. The ball exerts a force on the

Figure 3.8

(A) Work is done on the bowling ball as a force (F_B) moves it through a distance. (B) This gives the ball a kinetic energy equal in amount to the work done on it. (C) The ball does work on the pins and has enough remaining energy to crash into the wall behind the pins.

hand of the person catching the ball and moves it through a distance. The net work done on the hand is equal to the kinetic energy that the ball had.

A baseball and a bowling ball moving with the same velocity do not have the same kinetic energy. You cannot knock down many bowling pins with a slowly rolling baseball. Obviously the more massive bowling ball can do much more work than a less massive baseball with the same velocity. Is it possible for the bowling ball and the baseball to have the same kinetic energy? The answer is yes, if you can give the baseball sufficient velocity. This might require shooting the baseball from a cannon, however. Kinetic energy is proportional to the mass of a moving object, but velocity has a greater influence. Consider two balls of the same mass, but one is moving twice as fast as the other. The ball with twice the velocity will do *four* times as much work as the slower ball. A ball with three times the velocity will do *nine* times as much work as the slower ball. Kinetic energy is proportional to the square of the velocity ($2^2 = 4$; $3^2 = 9$). The kinetic energy (KE) of an object is

$$\text{kinetic energy} = \frac{1}{2}(\text{mass})(\text{velocity})^2$$

$$KE = \frac{1}{2}mv^2$$

equation 3.4

Kinetic energy is measured in joules, as is work ($F \times d$ or N·m) and potential energy (mgh or N·m).

ENERGY FLOW

The key to understanding the individual concepts of work and energy is to understand the close relationship between the two. When you do work on something you give it energy of position (potential energy) or you give it energy of motion (kinetic energy). In turn, objects that have kinetic or potential energy can now do work on something else as the transfer of energy continues. Where does all this energy come from and where does it go? The answer to this question is the subject of this section on energy flow.

Energy Forms

Energy comes in various forms, and different terms are used to distinguish one form from another. Although energy comes in various *forms*, this does not mean that there are different *kinds* of energy. The forms are the result of the more common fundamental forces—gravitational, electromagnetic, and nuclear—and objects that are interacting. Energy can be categorized into five forms: (1) *mechanical*, (2) *chemical*, (3) *radiant*, (4) *electrical*, and (5) *nuclear*. The following is a brief discussion of each of the five forms of energy.

Mechanical energy is the form of energy of familiar objects and machines (figure 3.9). A car moving on a highway has kinetic mechanical energy. Water behind a dam has potential mechanical energy. The spinning blades of a steam turbine have kinetic mechanical energy. The form of mechanical energy is usually associated with the kinetic energy of everyday-sized objects and the potential energy that results from gravity. There are other possibilities (e.g., sound), but this description will serve the need for now.

Chemical energy is the form of energy involved in chemical reactions (figure 3.10). Chemical energy is released in the chemical reaction known as *oxidation*. The fire of burning wood is an example of rapid oxidation. A slower oxidation releases energy from food units in your body. As you will learn in the chemistry unit, chemical energy involves electromagnetic forces between the parts of atoms. Until then, consider the following comparison. Photosynthesis is carried on in green plants. The plants use the energy of sunlight to rearrange carbon dioxide and water into plant materials and oxygen. Leaving out many steps and generalizing, this could be represented by the following word equation:

energy + carbon dioxide + water = wood + oxygen

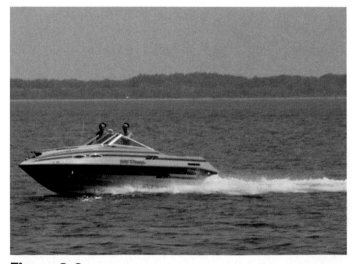

Figure 3.9

Mechanical energy is the energy of motion, or the energy of position, of many familiar objects. This boat has energy of motion.

A

B

Figure 3.10

Chemical energy is a form of potential energy that is released during a chemical reaction. Both (*A*) wood and (*B*) coal have chemical energy that has been stored through the process of photosynthesis. The pile of wood might provide fuel for a small fireplace for several days. The pile of coal might provide fuel for a power plant for a hundred days.

The plant took energy and two substances and made two different substances. This is similar to raising a book to a higher shelf in a bookcase. That is, the new substances have more energy than the original ones did. Consider a word equation for the burning of wood:

$$\text{wood} + \text{oxygen} = \text{carbon dioxide} + \text{water} + \text{energy}$$

Notice that this equation is exactly the reverse of photosynthesis. In other words, the energy used in photosynthesis was released during oxidation. Chemical energy is a kind of potential energy that is stored and later released during a chemical reaction.

Radiant energy is energy that travels through space (figure 3.11). Most people think of light or sunlight when considering this form of energy. Visible light, however, occupies only a small part of the complete electromagnetic spectrum, as shown in figure 3.12. Radiant energy includes light and all other parts of the

Figure 3.11

This demonstration solar cell array converts radiant energy from the sun to electrical energy, producing an average of 200,000 watts of electric power (after conversion).

Figure 3.12

The electromagnetic spectrum includes many categories of radiant energy. Note that visible light occupies only a tiny part of the entire spectrum.

spectrum. Infrared radiation is sometimes called "heat radiation" because of the association with heating when this type of radiation is absorbed. For example, you feel the interaction of infrared radiation when you hold your hand near a warm range element. However, infrared radiation is another type of radiant

Figure 3.13
The blades of a steam turbine. In a power plant, chemical or nuclear energy is used to heat water to steam, which is directed against the turbine blades. The mechanical energy of the turbine turns an electric generator. Thus, a power plant converts chemical or nuclear energy to mechanical energy, which is then converted to electrical energy.

energy. In fact, some snakes, such as the sidewinder, have pits between their eyes and nostrils that can detect infrared radiation emitted from warm animals. Microwaves are another type of radiant energy that is used in cooking. As with other forms of energy, light, infrared, and microwaves will be considered in more detail later. For now, consider all types of radiant energy to be forms of energy that travel through space.

Electrical energy is another form of energy from electromagnetic interactions that will be considered in detail later. You are familiar with electrical energy that travels through wires to your home from a power plant (figure 3.13), electrical energy that is generated by chemical cells in a flashlight, and electrical energy that can be "stored" in a car battery.

Nuclear energy is a form of energy often discussed because of its use as an energy source in power plants. Nuclear energy is another form of energy from the atom, but this time the energy involves the nucleus, the innermost part of an atom, and nuclear interactions.

Energy Conversion

Potential energy can be converted to kinetic energy and vice versa. The simple pendulum offers a good example of this conversion. A simple pendulum is an object, called a bob, suspended by a string or wire from a support. If the bob is moved to one side and then released, it will swing back and forth in an arc. At the moment that the bob reaches the top of its swing, it stops for an instant, then begins another swing. At the instant of stopping, the bob has 100 percent potential energy and no kinetic energy. As the bob starts back down through the swing, it is gaining kinetic energy and losing potential energy. At the instant the bob

Figure 3.14
This pendulum bob loses potential energy (*PE*) and gains an equal amount of kinetic energy (*KE*) as it falls through a distance *h*. The process reverses as the bob moves up the other side of its swing.

is at the bottom of the swing, it has 100 percent kinetic energy and no potential energy. As the bob now climbs through the other half of the arc, it is gaining potential energy and losing kinetic energy until it again reaches an instantaneous stop at the top, and the process starts over. The kinetic energy of the bob at the bottom of the arc is equal to the potential energy it had at the top of the arc (figure 3.14). Disregarding friction, the sum of the potential energy and the kinetic energy remains constant throughout the swing.

Any *form* of energy can be converted to another form. In fact, most technological devices that you use are nothing more than *energy-form converters* (figure 3.15). A lightbulb, for example, converts electrical energy to radiant energy. A car converts chemical energy to mechanical energy. A solar cell converts radiant energy to electrical energy, and an electric motor converts

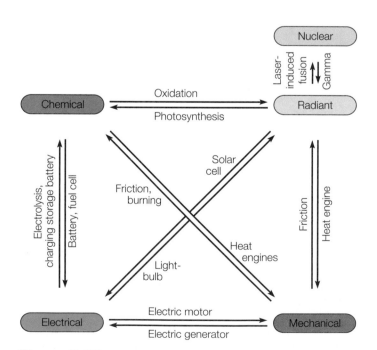

Figure 3.15
The energy forms and some conversion pathways.

Energy Converter Coaster

A roller coaster is an energy converter that swaps kinetic and potential energy back and forth. An outside energy source is used only to move the cars to the top of the first hill. The first hill is always the highest hill above the ground, and here the cars will have the most potential energy they will have for the entire ride.

When the coaster reaches the top of the first hill, it begins to move down a sloping track with increasing speed as potential energy is converted to kinetic energy. At the bottom of this hill the track then starts up a second hill, and the cars this time convert kinetic energy back to potential energy. Ideally, all the potential energy is converted to kinetic energy as the cars move down a hill, and all the kinetic energy is converted to potential energy as the cars move up a hill. There is no perfect energy conversion, and some energy is lost to air resistance and friction. The roller coaster design allows for these losses and leaves room for a slight surplus of kinetic energy. You know this is true because the operator must apply brakes at the end of the ride.

Is the speed of the moving roller coaster at the bottom of a hill directly proportional to the height of the hill? Ignoring friction, the speed of a coaster at the bottom of a hill is proportional to the *square root of the height* of the hill. This means that to double the speed of the coaster at the bottom of the hill you would need to build the hill more than four times higher. It would need to be more than four times higher because the coaster does not drop straight down. Doubling the height increases the theoretical speed only 40 percent.

What is the relationship between the weight of the people on a roller coaster and the speed achieved at the bottom of the hill? The answer is that the weight of the people does not matter. Ignoring air resistance and friction, a heavy roller coaster full of people and a lighter one with just a few people will both have the same speed at the bottom of the hill.

Draw a profile of a roller coaster ride to find out about potential and kinetic energy exchanges and the height of each hill. The profile should show the relative differences in height between the crest of each hill and the bottom of each upcoming dip. From such a profile you could find where in a ride the maximum speed would occur, as well as what speed to expect. As you study a profile keep in mind that the roller coaster is designed to produce many changes of speed—accelerations—rather than high speed alone.

---

electrical energy to mechanical energy. Each technological device converts some form of energy (usually chemical or electrical) to another form that you desire (usually mechanical or radiant).

It is interesting to trace the *flow of energy* that takes place in your surroundings. Suppose, for example, that you are riding a bicycle. The bicycle has kinetic mechanical energy as it moves along. Where did the bicycle get this energy? It came from you, as you use the chemical energy of food units to contract your muscles and move the bicycle along. But where did your chemical energy come from? It came from your food, which consists of plants, animals who eat plants, or both plants and animals. In any case, plants are at the bottom of your food chain. Plants convert radiant energy from the sun into chemical energy. Radiant energy comes to the plants from the sun because of the nuclear reactions that took place in the core of the sun. Your bicycle is therefore powered by nuclear energy that has undergone a number of form conversions!

Energy Conservation

Energy can be transferred from one object to another, and it can be converted from one form to another form. If you make a detailed accounting of all forms of energy before and after a transfer or conversion, the total energy will be *constant*. Consider your bicycle coasting along over level ground when you apply the brakes. What happened to the kinetic mechanical energy of the bicycle? It went into heating the rim and brakes of your bicycle, then eventually radiated to space as infrared radiation. All radiant energy that reaches the earth is eventually radiated back to space (figure 3.16). Thus, throughout all the form conversions and energy transfers that take place, the total sum of energy remains constant.

The total energy is constant in every situation that has been measured. This consistency leads to another one of the conservation laws of science, the **law of conservation of energy:**

> **Energy is never created or destroyed. Energy can be converted from one form to another but the total energy remains constant.**

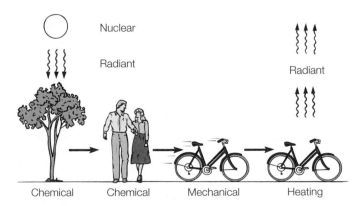

Figure 3.16

Energy arrives from the sun, goes through a number of conversions, then radiates back into space. The total sum leaving eventually equals the original amount that arrived.

Connections . . .

Bungee Jumping Energy

A group of college students in England started the sport of "bungee jumping" by jumping from bridges during the summer of 1979. They jumped with a rubber cord, and a common British word for rubber items is "bungee." Thus jumping from a high place with a long rubber cord became known as "bungee jumping." During the early 1990s towers, cranes, and hot air balloons were soon serving as bungee jumping platforms across the United States.

Today, commercial bungee jumping companies have specially made bungee cords, harnesses, and connectors. Let us consider what happens to a 645 N (mass = 65.8 kg) bungee jumper by first considering some hypothetical measurements. Our jumper is on a crane platform 27 m (90 ft) above the ground, in a body harness attached to a 9 m long (30 ft) bungee cord, which is attached to a 2 m length of cable from the crane. During the jump the bungee cord will stretch to a length 18 m, doubling in length.

Now, let us follow a hypothetical jump. After falling or jumping from the platform our jumper accelerates toward the surface at 9.8 m/s² in a free fall that lasts somewhere between one and two seconds. The bungee cord will now straighten to its full length, then stretch to double its length. At this instant of maximum stretch the velocity of the jumper and the cord is zero. For our hypothetical jumper the stretched cord results in a net upward force of three times the jumper's weight as it snaps back. From Newton's second law of motion, we find that the net upward acceleration would be 29.4 m/s². This acceleration is three times the normal *g*, meaning our hypothetical jumper was accelerated upward at about 3 *g*.

What is an acceleration of 3 *g*? One *g* is the normal acceleration due to gravity, which is 9.8 m/s². A 3 *g* acceleration of 29.4 m/s² will increase the velocity from zero to about 106 km/h (66 mi/h) in one second. In other words, our hypothetical bungee jumper was whisked back away from the ground at an acceleration equivalent to a car accelerating to a speed of 106 km/h (66 mi/h) in a second.

You may be wondering about the source of nuclear energy. Does a nuclear reaction create energy? Albert Einstein answered this question back in the early 1900s when he formulated his now-famous relationship between mass and energy, $E = mc^2$. This relationship will be discussed in detail in chapter 11. Basically, the relationship states that mass *is* a form of energy, and this has been experimentally verified many times.

Energy Transfer

Earlier it was stated that when you do work on something, you give it energy. The result of work could be increased kinetic mechanical energy, increased gravitational potential energy, or an increase in the temperature of an object. You could summarize this by stating that either *working* or *heating* is always involved any time energy is transformed. This is not unlike your financial situation. In order to increase or decrease your financial status, you need some mode of transfer, such as cash or checks, as a means of conveying assets. Just as with cash flow from one individual to another, energy flow from one object to another requires a mode of transfer. In energy matters the mode of transfer is working or heating. Any time you see working or heating occurring you know that an energy transfer is taking place. The next time you see heating, think about what energy form is being converted to what new energy form. (The final form is usually radiant energy.) Heating is the topic of chapter 4, where you will consider the role of heat in energy matters.

ENERGY SOURCES TODAY

Prometheus, according to ancient Greek mythology, stole fire from heaven and gave it to humankind. Fire has propelled human advancement ever since. All that was needed was something to burn—fuel for Prometheus's fire.

Any substance that burns can be used to fuel a fire, and various fuels have been used over the centuries as humans advanced. First, wood was used as a primary source for heating. Then coal fueled the industrial revolution. Eventually, humankind roared into the twentieth century burning petroleum. Today, petroleum is the most widely used source of energy (figure 3.17). It provides about 40 percent of the total energy used by the United States, but this dependence has been dropping since the 1970s. Natural gas contributes about 23 percent of the total energy used today. The use of coal provides about 21 percent of the total. Biomass—any material formed by photosynthesis—contributes about 3 percent of the total energy used today. Note that petroleum, coal, biomass, and natural gas are all chemical sources of energy, sources that are mostly burned for their energy. These chemical sources supply about 87 percent of the total energy consumed. About a third of this is burned for heating, and the rest is burned to drive engines or generators.

Nuclear energy and hydropower are the nonchemical sources of energy. These sources are used to generate electrical energy. The alternative sources of energy, such as solar and geothermal, provide about 0.5 percent of the total energy consumed in the United States today.

The energy-source mix has changed from past years, and it will change in the future. Wood supplied 90 percent of the energy until the 1850s, when the use of coal increased. Then, by 1910, coal was supplying about 75 percent of the total energy needs. Then petroleum began making increased contributions to the energy supply. Now increased economic and environmental constraints and a decreasing supply of petroleum are producing another supply shift. The present petroleum-based energy era is about to shift to a new energy era.

About 92 percent of the total energy consumed today is provided by four sources: (1) petroleum (including natural gas),

A

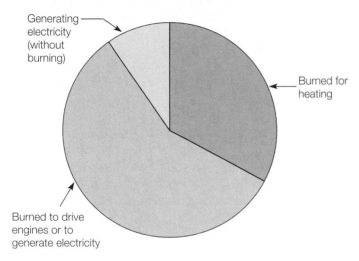

B

Figure 3.17

(*A*) The sources* of energy 1999 and (*B*) uses of this energy.

*Source: Energy Information Administration (www.eia.doe.gov/emeu/sep/us/frame.html).

(2) coal, (3) moving water, and (4) nuclear. The following is a brief introduction to these four sources.

Petroleum

The word *petroleum* is derived from the word "petra," meaning rock, and the word "oleum," meaning oil. Petroleum is oil that comes from oil-bearing rock. Natural gas is universally associated with petroleum and has similar origins. Both petroleum and natural gas form from organic sediments, materials that have settled out of bodies of water. Sometimes a local condition permits the accumulation of sediments that are exceptionally rich in organic material. This could occur under special conditions in a freshwater lake, or it could occur on shallow ocean basins. In either case, most of the organic material is from plankton—tiny free-floating animals and plants such as algae. It is from such accumulations of buried organic material that

petroleum and natural gas are formed. The exact process by which these materials become petroleum and gas is not understood. It is believed that bacteria, pressure, appropriate temperatures, and time are all important. Natural gas is formed at higher temperatures than is petroleum. Varying temperatures over time may produce a mixture of petroleum and gas or natural gas alone.

Petroleum forms a thin film around the grains of the rock where it formed. Pressure from the overlying rock and water move the petroleum and gas through the rock until it reaches a rock type or structure that stops it. If natural gas is present it occupies space above the accumulating petroleum. Such accumulations of petroleum and natural gas are the sources of supply for these energy sources.

Discussions about the petroleum supply and the cost of petroleum usually refer to a "barrel of oil." The *barrel* is an accounting device of 42 U.S. gallons. Such a 42-gallon barrel does not exist. When or if oil is shipped in barrels, each drum holds 55 U.S. gallons.

The supply of petroleum and natural gas is limited. Most of the continental drilling prospects appear to be exhausted, and the search for new petroleum supplies is now offshore. In general, over 25 percent of the United States' petroleum is estimated to come from offshore wells. The amount of imported petroleum has ranged from 30 to 50 percent over the years, with most imported oil coming from Canada, Saudi Arabia, Venezuela, Mexico, Iraq, and Nigeria.

Petroleum is used for gasoline (about 45 percent), diesel (about 40 percent), and heating oil (about 15 percent). Petroleum is also used in making medicine, clothing fabrics, plastics, and ink.

Coal

Petroleum and natural gas formed from the remains of tiny organisms that lived millions of years ago. Coal, on the other hand, formed from an accumulation of plant materials that collected under special conditions millions of years ago. Thus, petroleum, natural gas, and coal are called **fossil fuels.** Fossil fuels contain the stored radiant energy of organisms that lived millions of years ago.

The first thing to happen in the formation of coal was for plants in swamps to die and sink. Stagnant swamp water protected the plants and plant materials from consumption by animals and decomposition by microorganisms. Over time, chemically altered plant materials collected at the bottom of pools of water in the swamp. This carbon-rich material is *peat* (not to be confused with peat moss). Peat is used as a fuel in many places in the world. The flavor of Scotch (whisky) is the result of the peat fires used to brew the liquor. Peat is still being produced naturally in swampy areas today. Under pressure and at high temperatures peat will eventually be converted to coal. There are several stages, or *ranks* in the formation of coal. The lowest rank is lignite (brown coal), then subbituminous, then bituminous (soft coal), and the highest rank is anthracite (hard coal).

Each rank of coal has different burning properties and a different energy content. Coal also contains impurities of clay,

A Closer Look   Solar Technologies

An alternative source of energy is one that is different from the typical sources used today. The sources used today are the fossil fuels (coal, petroleum, and natural gas), nuclear, and falling water. Alternative sources could be solar, geothermal, hydrogen gas, fusion, or any other energy source that a new technology could utilize.

The term *solar energy* is used to describe a number of technologies that directly or indirectly utilize sunlight as an alternative energy source (box figure 3.1). There are eight main categories of these solar technologies:

1. **Solar cells.** A solar cell is a thin crystal of silicon, gallium, or some polycrystalline compound that generates electricity when exposed to light. Also called *photovoltaic devices,* solar cells have no moving parts and produce electricity directly, without the need for hot fluids or intermediate conversion states. Solar cells have been used extensively in space vehicles and satellites. Here on the earth, however, use has been limited to demonstration projects, remote site applications, and consumer specialty items such as solar-powered watches and calculators. The problem with solar cells today is that the manufacturing cost is too high (they are

Box Figure 3.1

Wind is another form of solar energy. This wind turbine generates electrical energy for this sailboat, charging batteries for backup power when the wind is not blowing. In case you are wondering, the turbine cannot be used to make a wind to push the boat. In accord with Newton's laws of motion, this would not produce a net force on the boat.

essentially handmade). Research is continuing on the development of highly efficient, affordable solar cells that could someday produce electricity for the home.

2. **Power tower.** This is another solar technology designed to generate electricity. One type of planned power tower will have a 171 m (560 ft) tower surrounded by

---

silt, iron oxide, and sulfur. The mineral impurities leave an ash when the coal is burned, and the sulfur produces sulfur dioxide, a pollutant.

Most of the coal mined today is burned by utilities to generate electricity (about 80 percent). The coal is ground to a face-powder consistency and blown into furnaces. This greatly increases efficiency but produces *fly ash,* ash that "flies" up the chimney. Industries and utilities are required by the federal Clean Air Act to remove sulfur dioxide and fly ash from plant emissions. About 20 percent of the cost of a new coal-fired power plant goes into air pollution control equipment. Coal is an abundant but dirty energy source.

Moving Water

Moving water has been used as a source of energy for thousands of years. It is considered a renewable energy source, inexhaustible as long as the rain falls. Today, hydroelectric plants generate about 3 percent of the United States' *total* energy consumption at about 2,400 power-generating dams across the nation. Hydropower furnished about 40 percent of the United States' electric power in 1940. Today, dams furnish 9 percent of the electric power. Energy consumption has increased, but hy-

dropower production has not kept pace because geography limits the number of sites that can be built.

Water from a reservoir is conducted through large pipes called penstocks to a powerhouse, where it is directed against turbine blades that turn a shaft on an electric generator.

Nuclear

Nuclear power plants use nuclear energy to produce electricity. Energy is released as the nuclei of uranium and plutonium atoms split, or undergo fission. The fissioning takes place in a large steel vessel called a *reactor.* Water is pumped through the reactor to produce steam, which is used to produce electrical energy, just as in the fossil fuel power plants. The nuclear processes are described in detail in chapter 11, and the process of producing electrical energy is described in detail in chapter 6. Nuclear power plants use nuclear energy to produce electricity, but there are opponents to this process. The electric utility companies view nuclear energy as *one* energy source used to produce electricity. They state that they have no allegiance to any one energy source but are seeking to utilize the most reliable and dependable of several energy sources. Petroleum, coal, and hydropower are also presently utilized as energy sources for some 9,000 special mirrors called heliostats. The heliostats will focus sunlight on a boiler at the top of the tower where salt (a mixture of sodium nitrate and potassium nitrate) will be heated to about 566°C (about 1,050°F). This molten salt will be pumped to a steam generator, and the steam will be used to drive a generator, just like other power plants. Water could be heated directly in the power tower boiler. Molten salt is used because it can be stored in an insulated storage tank for use when the sun is not shining, perhaps for up to twenty hours.

3. **Passive application.** In passive applications energy flows by natural means, without mechanical devices such as motors, pumps, and so forth. A passive solar house would include considerations like orientation of a house to the sun, the size and positioning of windows, and a roof overhang that lets sunlight in during the winter but keeps it out during the summer. There are different design plans to capture, store, and distribute solar energy throughout a house.

4. **Active application.** An active solar application requires a solar collector in which sunlight heats air, water, or some liquid. The liquid or air is pumped through pipes in a house to generate electricity, or it is used directly for hot water. Solar water heating makes more economic sense today than the other applications.

5. **Wind energy.** The wind has been used for centuries to move ships, grind grain into flour, and pump water. The wind blows, however, because radiant energy from the sun heats some parts of the earth's surface more than other parts. This differential heating results in pressure differences and the horizontal movement of air, which is called *wind.* Thus, wind is another form of solar energy. Wind turbines are used to generate electrical energy or mechanical energy. The biggest problem with wind energy is the inconsistency of the wind. Sometimes the wind speed is too great, and other times it is not great enough. Several methods of solving this problem are being researched.

6. **Biomass.** Biomass is any material formed by photosynthesis, including plants, trees, and crops, and any garbage, crop residue, or animal waste. Biomass can be burned directly as a fuel, converted into a gas fuel (methane), or converted into liquid fuels such as alcohol. The problems with using biomass include the energy expended in gathering the biomass, as well as the energy used to convert it to a gaseous or liquid fuel.

7. **Agriculture and industrial heating.** This is a technology that simply uses sunlight to dry grains, cure paint, or do anything that can be done with sunlight rather than using traditional energy sources.

8. **Ocean thermal energy conversion (OTEC).** This is an electric generating plant that would take advantage of the approximately 22°C (about 40°F) temperature difference between the surface and the depths of tropical, subtropical, and equatorial ocean waters.

Basically, warm water is drawn into the system to vaporize a fluid, which expands through a turbine generator. Cold water from the depths condenses the vapor back to a liquid form, which is then cycled back to the warm-water side. The concept has been tested several times and was found to be technically successful. The greatest interest seems to be islands that have warm surface waters (and cold depths) such as Hawaii, Puerto Rico, Guam, the Virgin Islands, and others.

electric power production. The electric utility companies are concerned that petroleum and natural gas are becoming increasingly expensive, and there are questions about long-term supplies. Hydropower has limited potential for growth, and solar energy is prohibitively expensive today. Utility companies see two major energy sources that are available for growth: coal and nuclear. There are problems and advantages to each, but the utility companies feel they must use coal and nuclear power until the new technologies, such as solar power, are economically feasible.

SUMMARY

Work is defined as the product of an applied force and the distance through which the force acts. Work is measured in newton-meters, a metric unit called a *joule. Power* is work per unit of time. Power is measured in *watts.* One watt is 1 joule per second. Power is also measured in *horsepower.* One horsepower is 550 ft·lb/s.

Energy is defined as the ability to do work. An object that is elevated against gravity has a potential to do work. The object is said to have *potential energy,* or *energy of position.* Moving objects have the ability to do work on other objects because of their motion. The *energy of motion* is called *kinetic energy.*

Work is usually done *against inertia, fundamental forces, friction, shape,* or *combinations of these.* As a result, there is a gain of *kinetic energy, potential energy, an increased temperature,* or *any combination of these.* Energy comes in the *forms of mechanical, chemical, radiant, electrical,* or *nuclear.* Potential energy can be *converted* to kinetic and kinetic can be *converted* to potential. Any form of energy can be *converted* to any other form. Most technological devices are *energy-form converters* that do work for you. Energy flows into and out of the surroundings, but the amount of energy is always constant. The *law of conservation of energy* states that *energy is never created or destroyed.* Energy conversion always takes place through *heating* or *working.*

The basic energy sources today are the chemical *fossil fuels* (petroleum, natural gas, and coal), *nuclear energy,* and *hydropower. Petroleum* and *natural gas* were formed from organic material of plankton, tiny free-floating plants and animals. A barrel of petroleum is 42 U.S. gallons, but such a container does not actually exist. *Coal* formed from plants that were protected from consumption by falling into a swamp. The decayed plant material, *peat,* was changed into the various *ranks* of coal by pressure and heating over some period of time. Coal is a dirty fuel that contains impurities and sulfur. Controlling air pollution from burning coal is costly. Moving water and nuclear energy are used for the generation of electricity.

Summary of Equations

3.1
$$\text{work} = \text{force} \times \text{distance}$$
$$W = Fd$$

3.2
$$\text{power} = \frac{\text{work}}{\text{time}}$$
$$P = \frac{W}{t}$$

3.3
$$\text{potential energy} = \text{weight} \times \text{height}$$
$$PE = mgh$$

3.4
$$\text{kinetic energy} = \frac{1}{2}(\text{mass})(\text{velocity})^2$$
$$KE = \frac{1}{2}mv^2$$

KEY TERMS

energy (p. **50**)
fossil fuels (p. **57**)
horsepower (p. **49**)
joule (p. **49**)
kinetic energy (p. **51**)

law of conservation of energy (p. **55**)
potential energy (p. **51**)
power (p. **49**)
watt (p. **49**)
work (p. **48**)

APPLYING THE CONCEPTS

1. The metric unit of a joule (J) is a unit of
   (a) potential energy.
   (b) work.
   (c) kinetic energy.
   (d) any of the above.
2. Power is
   (a) the rate at which work is done.
   (b) the rate at which energy is expended.
   (c) work per unit time.
   (d) any of the above.
3. Which of the following is a combination of units called a watt?
   (a) N·m/s
   (b) kg·m²/s²/s
   (c) J/s
   (d) all of the above
4. About how many watts are equivalent to 1 horsepower?
   (a) 7.5
   (b) 75
   (c) 750
   (d) 7,500
5. A kilowatt-hour is a unit of
   (a) power.
   (b) work.
   (c) time.
   (d) electrical charge.

6. The potential energy of a box on a shelf, relative to the floor, is a measure of
   (a) the work that was required to put the box on the shelf from the floor.
   (b) the weight of the box times the distance above the floor.
   (c) the energy the box has because of its position above the floor.
   (d) all of the above.
7. A rock on the ground is considered to have zero potential energy. In the bottom of a well, then, the rock would be considered to have
   (a) zero potential energy, as before.
   (b) negative potential energy.
   (c) positive potential energy.
   (d) zero potential energy, but will require work to bring it back to ground level.
8. Which quantity has the greatest influence on the amount of kinetic energy that a large truck has while moving down the highway?
   (a) mass
   (b) weight
   (c) velocity
   (d) size
9. Most energy comes to and leaves the earth in the form of
   (a) nuclear energy.
   (b) chemical energy.
   (c) radiant energy.
   (d) kinetic energy.
10. The law of conservation of energy is basically that
    (a) energy must not be used up faster than it is created or the supply will run out.
    (b) energy should be saved because it is easily destroyed.
    (c) energy is never created or destroyed.
    (d) you are breaking a law if you needlessly destroy energy.
11. The most widely used source of energy today is
    (a) coal.
    (b) petroleum.
    (c) nuclear.
    (d) moving water.
12. The accounting device of a "barrel of oil" is defined to hold how many U.S. gallons of petroleum?
    (a) 24
    (b) 42
    (c) 55
    (d) 100

Answers
1. d **2.** d **3.** d **4.** c **5.** b **6.** d **7.** b **8.** c **9.** c **10.** c **11.** b **12.** b

QUESTIONS FOR THOUGHT

1. How is work related to energy?
2. What is the relationship between the work done while moving a book to a higher bookshelf and the potential energy that the book has on the higher shelf?

3. Does a person standing motionless in the aisle of a moving bus have kinetic energy? Explain.

4. A lamp bulb is rated at 100 W. Why is a time factor not included in the rating?

5. Is a kWh a unit of work, energy, power, or more than one of these? Explain.

6. If energy cannot be destroyed, why do some people worry about the energy supplies?

7. A spring clamp exerts a force on a stack of papers it is holding together. Is the spring clamp doing work on the papers? Explain.

8. Why are petroleum, natural gas, and coal called *fossil fuels?*

9. From time to time, people claim to have invented a machine that will run forever without energy input and develops more energy than it uses (perpetual motion). Why would you have reason to question such a machine?

10. Define a joule. What is the difference between a joule of work and a joule of energy?

11. Compare the energy needed to raise a mass 10 m on the earth to the energy needed to raise the same mass 10 m on the moon. Explain the difference, if any.

12. What happens to the kinetic energy of a falling book when the book hits the floor?

PARALLEL EXERCISES

The exercises in groups A and B cover the same concepts. Solutions to group A exercises are located in appendix D.
Note: Neglect all frictional forces in all exercises.

Group A

1. A force of 200 N is needed to push a table across a level classroom floor for a distance of 3 m. How much work was done on the table?

2. An 880 N box is pushed across a level floor for a distance of 5.0 m with a force of 440 N. How much work was done on the box?

3. How much work is done in raising a 10.0 kg backpack from the floor to a shelf 1.5 m above the floor?

4. If 5,000 J of work is used to raise a 102 kg crate to a shelf in a warehouse, how high was the crate raised?

5. A 60.0 kg student runs up a 5.00 m high stairway in a time of 3.92 s. (a) How many watts of power did she develop? (b) How many horsepower is this?

6. What is the kinetic energy of a 2,000 kg car moving at 72 km/h?

7. How much work is needed to stop a 1,000.0 kg car that is moving straight down the highway at 54.0 km/h?

8. A 1,000 kg car stops on top of a 51.02 m hill. (a) How much energy was used in climbing the hill? (b) How much potential energy does the car have?

9. A horizontal force of 10.0 lb is needed to push a bookcase 5 ft across the floor. (a) How much work was done on the bookcase? (b) How much did the gravitational potential energy change as a result?

10. (a) How much work is done in moving a 2.0 kg book to a shelf 2.00 m high? (b) What is the potential energy of the book as a result? (c) How much kinetic energy will the book have as it hits the ground as it falls?

11. A 60.0 kg jogger moving at 2.0 m/s decides to double the jogging speed. How did this change in speed change the kinetic energy?

12. A 170.0 lb student runs up a stairway to a classroom 25.0 ft above ground level in 10.0 s. (a) How much work did the student do? (b) What was the average power output in horsepower?

Group B

1. Work of 1,200 J is done while pushing a crate across a level floor for a distance of 1.5 m. What force was used to move the crate?

2. How much work is done by a hammer that exerts a 980.0 N force on a nail, driving it 1.50 cm into a board?

3. A 5.0 kg textbook is raised a distance of 30.0 cm as a student prepares to leave for school. How much work did the student do on the book?

4. An electric hoist does 196,000 J of work in raising a 250.0 kg load. How high was the load lifted?

5. What is the horsepower of a 1,500.0 kg car that can go to the top of a 360.0 m high hill in exactly one minute?

6. What is the kinetic energy of a 30.0 g bullet that is traveling at 200.0 m/s?

7. How much work will be done by a 30.0 g bullet traveling at 200 m/s?

8. A 10.0 kg box is lifted 15 m above the ground by a construction crane. (a) How much work did the crane do on the box? (b) How much potential energy does the box have relative to the ground?

9. A force of 50.0 lb is used to push a box 10.0 ft across a level floor. (a) How much work was done on the box? (b) What is the change of potential energy as a result of this move?

10. (a) How much work is done in raising a 50.0 kg crate a distance of 1.5 m above a storeroom floor? (b) What is the change of potential energy as a result of this move? (c) How much kinetic energy will the crate have as it falls and hits the floor?

11. The driver of an 800.0 kg car decides to double the speed from 20.0 m/s to 40.0 m/s. What effect would this have on the amount of work required to stop the car, that is, on the kinetic energy of the car?

12. A 70.0 kg student runs up the stairs of a football stadium to a height of 10.0 m above the ground in 10.0 s. (a) What is the power of the student in kilowatts? (b) in horsepower?

Heat and Temperature

Sparks fly from a plate of steel as it is cut by an infrared laser. Today, lasers are commonly used to cut as well as weld metals, so the cutting and welding is done by light, not by a flame or electric current.

O U T L I N E

eat has been closely associated with the comfort and support of people throughout history. You can imagine the appreciation when your earliest ancestors first discovered fire and learned to keep themselves warm and cook their food. You can also imagine the wonder and excitement about 3000 B.C., when people put certain earthlike substances on the hot, glowing coals of a fire and later found metallic copper, lead, or iron. The use of these metals for simple tools followed soon afterwards. Today, metals are used to produce complicated engines that use heat for transportation and that do the work of moving soil and rock, construction, and agriculture. Devices made of heat-extracted metals are also used to control the temperature of structures, heating or cooling the air as necessary. Thus, the production and control of heat gradually built the basis of civilization today (figure 4.1).

The sources of heat are the energy forms that you learned about in chapter 3. The fossil fuels are *chemical* sources of heat. Heat is released when oxygen is combined with these fuels. Heat also results when *mechanical* energy does work against friction, such as in the brakes of a car coming to a stop. Heat also appears when *radiant* energy is absorbed. This is apparent when solar energy heats water in a solar collector or when sunlight melts snow. The transformation of *electrical* energy to heat is apparent in toasters, heaters, and ranges. *Nuclear* energy provides the heat to make steam in a nuclear power plant. Thus, all the energy forms can be converted to heat.

The relationship between energy forms and heat appears to give an order to nature, revealing patterns that you will want to understand. All that you need is some kind of explanation for the relationships—a model or theory that helps make sense of it all. This chapter is concerned with heat and temperature and their relationship to energy. It begins with a simple theory about the structure of matter, and then uses the theory to explain the concepts of heat, energy, and temperature changes.

THE KINETIC MOLECULAR THEORY

The idea that substances are composed of very small particles can be traced back to certain early Greek philosophers. The earliest record of this idea was written by Democritus during the fifth century B.C. He wrote that matter was empty space filled with tremendous numbers of tiny, indivisible particles called *atoms*. This idea, however, was not acceptable to most of the ancient Greeks, because matter seemed continuous, and empty space was simply not believable. The idea of atoms was rejected by Aristotle as he formalized his belief in continuous matter composed of the earth, air, fire, and water elements. Aristotle's belief about matter, like his beliefs about motion, predominated through the 1600s. Some people, such as Galileo and Newton, believed the ideas about matter being composed of tiny particles, or atoms, since this theory seemed to explain the behavior of matter. Widespread acceptance of the particle model did not occur, however, until strong evidence was developed through chemistry in the late 1700s and early 1800s. The experiments finally led to a collection of assumptions about the small particles of matter and the space around them. Collectively, the assumptions could be called the **kinetic molecular theory.** The following is a general description of some of these assumptions.

Figure 4.1
Heat and modern technology are inseparable. These glowing steel slabs, at over 1,100°C (about 2,000°F), are cut by an automatic flame torch. The slab caster converts 300 tons of molten steel into slabs in about 45 minutes. The slabs are converted to sheet steel for use in the automotive, appliance, and building industries.

Molecules

The basic assumption of the kinetic molecular theory is that all matter is made up of tiny, basic units of structure called *atoms.* Atoms are neither divided, created, nor destroyed during any type of chemical or physical change. There are similar groups of atoms that make up the pure substances known as chemical *elements.* Each element has its own kind of atom, which is different from the atoms of other elements. For example, hydrogen, oxygen, carbon, iron, and gold are chemical elements, and each has its own kind of atom.

In addition to the chemical elements, there are pure substances called *compounds* that have more complex units of structure (figure 4.2). Pure substances, such as water, sugar, and alcohol, are composed of atoms of two or more elements that join together in definite proportions. Water, for example, has structural units that are made up of two atoms of hydrogen tightly bound to one atom of oxygen (H_2O). These units are not easily broken apart and stay together as small physical particles of which water is composed. Each is the smallest particle of water that can exist, a molecule of water. A *molecule* is generally defined as a tightly bound group of atoms in which the atoms maintain their identity. How atoms become bound together to form molecules is discussed in chapter 9.

Figure 4.2

Metal atoms appear in the micrograph of a crystal of titanium niobium oxide, magnified 7,800,000 times by an electron microscope.

Some elements exist as gases at ordinary temperatures, and all elements are gases at sufficiently high temperatures. At ordinary temperatures, the atoms of oxygen, nitrogen, and other gases are paired in groups of two to form *diatomic molecules*. Other gases, such as helium, exist as single, unpaired atoms at ordinary temperatures. At sufficiently high temperatures, iron, gold, and other metals vaporize to form gaseous, single, unpaired atoms. In the kinetic molecular theory the term *molecule* has the additional meaning of the smallest, ultimate particle of matter that can exist. Thus, the ultimate particle of a gas, whether it is made up of two or more atoms bound together or of a single atom, is conceived of as a molecule. A single atom of helium, for example, is known as a *monatomic molecule*. For now, a **molecule** is defined as the smallest particle of a compound, or a gaseous element, that can exist and still retain the characteristic properties of that substance.

Molecules Interact

Some molecules of solids and liquids interact, strongly attracting and clinging to each other. When this attractive force is between the same kind of molecules, it is called *cohesion*. It is a stronger cohesion that makes solids and liquids different from gases, and without cohesion all matter would be in the form of gases. Sometimes one kind of molecule attracts and clings to a different kind of molecule. The attractive force between unlike molecules is called *adhesion*. Water wets your skin because the adhesion of water molecules and skin is stronger than the cohesion of water molecules. Some substances, such as glue, have a strong force of adhesion when they harden from a liquid state, and they are called adhesives.

Phases of Matter

Three phases of matter are common on the earth under conditions of ordinary temperature and pressure. These phases—or forms of existence—are solid, liquid, and gas. Each of these has a different molecular arrangement (figure 4.3). The different characteristics of each phase can be attributed to the molecular arrangements and the strength of attraction between the molecules (table 4.1).

Solids have definite shapes and volumes because they have molecules that are fixed distances apart and bound by relatively strong cohesive forces. Each molecule is a nearly fixed distance from the next, but it does vibrate and move around an equilibrium position. The masses of these molecules and the spacing between them determine the density of the solid. The hardness of a solid is the resistance of a solid to forces that tend to push its molecules further apart.

Liquids have molecules that are not confined to an equilibrium position as in a solid. The molecules of a liquid are close

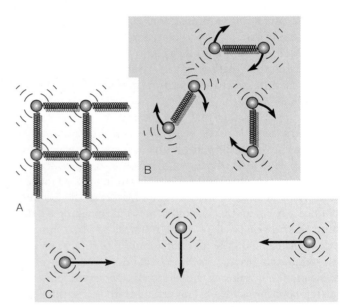

Figure 4.3

(*A*) In a solid, molecules vibrate around a fixed equilibrium position and are held in place by strong molecular forces. (*B*) In a liquid, molecules can rotate and roll over each other because the molecular forces are not as strong. (*C*) In a gas, molecules move rapidly in random, free paths.

Table 4.1	The Shape and Volume Characteristics of Solids, Liquids, and Gases Are Reflections of Their Molecular Arrangements*		
	Solids	**Liquids**	**Gases**
Shape	Fixed	Variable	Variable
Volume	Fixed	Fixed	Variable

*These characteristics are what would be expected under ordinary temperature and pressure conditions on the surface of the earth.

together and bound by cohesive forces that are not as strong as in a solid. This permits the molecules to move from place to place within the liquid. The molecular forces are strong enough to give the liquid a definite volume but not strong enough to give it a definite shape. Thus, a pint of milk is always a pint of milk (unless it is under tremendous pressure) and takes the shape of the container holding it. Because the forces between the molecules of a liquid are weaker than the forces between the molecules of a solid, a liquid cannot support the stress of a rock placed on it as a solid does. The liquid molecules *flow,* rolling over each other as the rock pushes its way between the molecules. Yet, the molecular forces are strong enough to hold the liquid together, so it keeps the same volume.

Gases are composed of molecules with weak cohesive forces acting between them. The gas molecules are relatively far apart and move freely in a constant, random motion that is changed often by collisions with other molecules. Gases therefore have neither fixed shapes nor fixed volumes.

There are other distinctions between the phases of matter. The term *vapor* is sometimes used to describe a gas that is usually in the liquid phase. Water vapor, for example, is the gaseous form of liquid water. Liquids and gases are collectively called *fluids* because of their ability to flow, a property that is lacking in most solids. A not-so-ordinary phase of matter on the earth is called *plasma.* Plasmas occur at extremely high temperatures and are made up of charged parts of atoms.

Molecules Move

Suppose you are in an evenly heated room with no air currents. If you open a bottle of ammonia, the odor of ammonia is soon noticeable everywhere in the room. According to the kinetic molecular theory, molecules of ammonia leave the bottle and bounce around among the other molecules making up the air until they are everywhere in the room, slowly becoming more evenly distributed. The ammonia molecules *diffuse,* or spread, throughout the room. The ammonia odor diffuses throughout the room faster if the air temperature is higher and slower if the air temperature is lower. This would imply a relationship between the temperature and the speed at which molecules move about.

The relationship between the temperature of a gas and the motion of molecules was formulated in 1857 by Rudolf Clausius. He showed that the temperature of a gas is proportional to the average kinetic energy of the gas molecules. This means that ammonia molecules have a greater average velocity at a higher temperature and a slower average velocity at a lower temperature. This explains why gases diffuse at a greater rate at higher temperatures. Recall, however, that kinetic energy involves the mass of the molecules as well as their velocity ($KE = 1/2\ mv^2$). It is the *average kinetic energy* that is proportional to the temperature, which involves the molecular mass as well as the molecular velocity. Whether the kinetic energy is jiggling, vibrating, rotating, or moving from place to place, the **temperature** of a substance is *a measure of the average kinetic energy of the molecules making up the substance* (figure 4.4).

Figure 4.4

The number of oxygen molecules with certain velocities that you might find in a sample of air at room temperature. Notice that a few are barely moving and some have velocities over 1,000 m/s at a given time, but the *average* velocity is somewhere around 500 m/s.

TEMPERATURE

If you ask people about the temperature, they usually respond with a referent ("hotter than the summer of '89") or a number ("68°F or 20°C"). Your response, or feeling, about the referent or number depends on a number of factors, including a *relative* comparison. A temperature of 20°C (68°F), for example, might seem cold during the month of July but warm during the month of January. The 20°C temperature is compared to what is expected at the time, even though 20°C is 20°C, no matter what month it is.

When people ask about the temperature, they are really asking *how hot or how cold something is.* Without a thermometer, however, most people can do no better than *hot* or *cold,* or perhaps *warm* or *cool,* in describing a relative temperature. Even then, there are other factors that confuse people about temperature. Your body judges temperature on the basis of the net *direction* of energy flow. You sense situations in which heat is flowing into your body as *warm* and situations in which heat is flowing from your body as *cool*. Perhaps you have experienced having your hands in snow for some time, then washing your hands in cold water. The cold water feels warm. Your hands are colder than the water, energy flows into your hands, and they communicate "warm."

Thermometers

The human body is a poor sensor of temperature, so a device called a *thermometer* is used to measure the hotness or coldness of something. Most thermometers are based on the relationship between some property of matter and changes in temperature. Almost all materials expand with increasing temperatures. A strip of metal is slightly longer when hotter and slightly shorter when cooler, but the change of length is too small to be useful in a thermometer. A more useful, larger change is obtained when two metals that have different expansion rates are bonded together in a strip. The bimetallic ("bi" = two; "metallic" = metal) strip will bend toward the metal with less expansion when the strip is heated (figure 4.5). Such a bimetallic strip is formed into a coil and used in thermostats and dial thermometers (figure 4.6).

The common glass thermometer is a glass tube with a bulb containing a liquid, usually mercury or colored alcohol, that expands up the tube with increases in temperature and contracts back toward the bulb with decreases in temperature. The height of this liquid column is used with a referent scale to measure temperature. Some thermometers, such as a fever thermometer, have a small constriction in the tube so the liquid cannot normally return to the bulb. Thus, the thermometer shows the highest reading, even if the temperature it measures has fluctuated up and down during the reading. The liquid must be forced back into the bulb by a small swinging motion, bulb-end down, then sharply stopping the swing with a snap of the wrist. The inertia of the mercury in the bore forces it past the constriction and into the bulb. The fever thermometer is then ready to use again.

Thermometer Scales

There are several referent scales used to define numerical values for measuring temperatures (figure 4.7). The **Fahrenheit scale**

Room temperature

Iron

Brass

A

When heated

Iron

Brass

B

Figure 4.5

(*A*) A bimetallic strip is two different metals, such as iron and brass, bonded together as a single unit, shown here at room temperature. (*B*) Since one metal expands more than the other, the strip will bend when it is heated. In this example, the brass expands more than the iron, so the bimetallic strip bends away from the brass.

Mercury vial

Bimetallic strip

Figure 4.6

This thermostat has a coiled bimetallic strip that expands and contracts with changes in the room temperature. The attached vial of mercury is tilted one way or the other, and the mercury completes or breaks an electric circuit that turns the heating or cooling system on or off.

was developed by the German physicist Gabriel D. Fahrenheit in about 1715. Fahrenheit invented a mercury-in-glass thermometer with a scale based on two arbitrarily chosen reference points. The original Fahrenheit scale was based on the temperature of

Connections . . .

Today, scientists have developed a different type of thermometer and a way around the problems of using a glass mercury fever thermometer. This new approach measures the internal core temperature by quickly reading infrared radiation from the eardrum. All bodies with a temperature above absolute zero emit radiation, including your body. The intensity of the radiation is a sensitive function of body temperature, so reading the radiation emitted will tell you about the temperature of that body.

The human eardrum is close to the hypothalamus, the body's thermostat, so a temperature reading taken here must be close to the temperature of the internal core. You cannot use a mercury thermometer in the ear because of the very real danger of puncturing the eardrum, along with obtaining doubtful readings from a mercury bulb. You can use a pyroelectric material to measure the radiation coming from the entrance to the ear canal, however, to quickly obtain a temperature reading. A pyroelectric ma-

terial is a polarized crystal that generates an electric charge in proportion to a temperature change.

The infrared fever thermometer has a short barrel, which is inserted in the ear canal opening. A button opens a shutter inside the battery-powered device, admitting infrared radiation for about 300 milliseconds. Infrared radiation from the ear canal increases the temperature of a thin pyroelectric crystal, which develops an electric charge. A current spike from the pyroelectric sensor now moves through some filters and converters, and into a microprocessor chip. The chip is programmed with the relationship between the body temperature and the infrared radiation emitted. Using this information, it calculates the temperature by measuring the current spike produced by the infrared radiation falling on the pyroelectric crystal. The microprocessor now sends the temperature reading to an LCD display on the outside of the device, where it can be read almost immediately.

Figure 4.7
The Fahrenheit, Celsius, and Kelvin temperature scales.

an ice and salt mixture for the lower reference point (0°) and the temperature of the human body as the upper reference point (about 100°). Thus, the original Fahrenheit scale was a centigrade scale with 100 divisions between the high and the low reference points. The distance between the two reference points was then divided into equal intervals called *degrees.* There were problems with identifying a "normal" human body temperature as a reference point, since body temperature naturally changes during a given day and from day to day. However, some people "normally" have a higher body temperature than others. Some may have a normal body temperature of 99.1°F while others have a temperature of 97°F. The average for a large population is 98.6°F. The only consistent thing about the human body tem-

perature is constant change. The standards for the Fahrenheit scale were eventually changed to something more consistent, the freezing point and the boiling point of water at normal atmospheric pressure. The original scale was retained with the new reference points, however, so the "odd" numbers of 32°F (freezing point of water) and 212°F (boiling point of water under normal pressure) came to be the reference points. There are 180 equal intervals, or degrees, between the freezing and boiling points on the Fahrenheit scale.

The **Celsius scale** was invented by Anders C. Celsius, a Swedish astronomer, in about 1735. The Celsius scale uses the freezing point and the boiling point of water at normal atmospheric pressure, but it has different arbitrarily assigned values. The Celsius scale identifies the freezing point of water as 0°C and the boiling point as 100°C. There are 100 equal intervals, or degrees, between these two reference points, so the Celsius scale is sometimes called the *centigrade* scale.

There is nothing special about either the Celsius scale or the Fahrenheit scale. Both have arbitrarily assigned numbers, and one is no more accurate than the other. The Celsius scale is more convenient because it is a decimal scale and because it has a direct relationship with a third scale to be described shortly, the Kelvin scale. Both scales have arbitrarily assigned reference points and an arbitrary number line that indicates *relative* temperature changes. Zero is simply one of the points on each number line and does *not* mean that there is no temperature. Likewise, since the numbers are relative measures of temperature change, 2° is not twice as hot as a temperature of 1° and 10° is not twice as hot as a temperature of 5°. The numbers simply mean some measure of temperature *relative to* the freezing and boiling points of water under normal conditions.

You can convert from one temperature to the other by considering two differences in the scales: (1) the difference in the degree size between the freezing and boiling points on the

Goosebumps and Shivering

For an average age and minimal level of activity, many people feel comfortable when the environmental temperature is about 25°C (77°F). Comfort at this temperature probably comes from the fact that the body does not have to make an effort to conserve or get rid of heat.

Changes in the body that conserve heat occur when the temperature of the air and clothing directly next to a person becomes less than 20°C, or if the body senses rapid heat loss. First, blood vessels in the skin are constricted. This slows the flow of blood near the surface, which reduces heat loss by conduction. Constriction of skin blood vessels reduces body heat loss, but may also cause the skin and limbs to become significantly cooler than the body core temperature (producing cold feet, for example).

Sudden heat loss, or a chill, often initiates another heat-saving action by the body. Skin hair is pulled upright, erected to slow heat loss to cold air moving across the skin. Contraction of a tiny muscle attached to the base of the hair shaft makes a tiny knot, or bump on the skin. These are sometimes called "goose bumps" or "chill bumps." Although "goose bumps" do not significantly increase insulation in humans, the equivalent response in birds and many mammals elevates hairs or feathers and greatly enhances insulation.

Further cooling after the blood vessels in the skin have been constricted results in the body taking yet another action. The body now begins to produce *more* heat, making up for heat loss through involuntary muscle contractions called "shivering." The greater the need for more body heat the greater the activity of shivering.

If the environmental temperatures rise above about 25°C (77°F), the body triggers responses that causes it to *lose* heat. One response is to make blood vessels in the skin larger, which increases blood flow in the skin. This brings more heat from the core to be conducted through the skin, then radiated away. It also causes some people to have a red blush from the increased blood flow in the skin. This action increases conduction through the skin, but radiation alone provides insufficient cooling at environmental temperatures above about 29°C (84°F). At about this temperature sweating begins and perspiration pours onto the skin to provide cooling through evaporation. The warmer the environmental temperature, the greater the rate of sweating and cooling through evaporation.

The actual responses to a cool, cold, warm, or hot environment will be influenced by a person's level of activity, age, and gender, and environmental factors such as the relative humidity, air movement, and combinations of these factors. Temperature is the single most important comfort factor. However, when the temperature is high enough to require perspiration for cooling, humidity also becomes an important factor in human comfort.

---

two scales, and (2) the difference in the values of the lower reference points.

The Fahrenheit scale has 180° between the boiling and freezing points (212°F − 32°F) and the Celsius scale has 100° between the same two points. Therefore each Celsius degree is 180/100 or 9/5 as large as a Fahrenheit degree. Each Fahrenheit degree is 100/180 or 5/9 of a Celsius degree. In addition, considering the difference in the values of the lower reference points (0°C and 32°F) gives the equations for temperature conversion.

$$T_F = \frac{9}{5} T_C + 32°$$

equation 4.1

$$T_C = \frac{5}{9}(T_F - 32°)$$

equation 4.2

There is a temperature scale that does not have arbitrarily assigned reference points and zero *does* mean nothing. This is not a relative scale but an absolute temperature scale called the **absolute scale,** or **Kelvin scale.** The zero point on the absolute scale is thought to be the lowest limit of temperature. **Absolute zero** is the *lowest temperature possible,* occurring when all random motion of molecules has ceased. Absolute zero is written as 0 K. A degree symbol is not used, and the K stands for the SI standard scale unit, Kelvin. The absolute scale uses the same degree size as the Celsius scale and −273°C = 0 K. Note in figure 4.7 that 273 K is the freezing point of water, and 373 K is the boiling point. You could think of the absolute scale as a Celsius scale with the zero point shifted by 273°. Thus, the relationship between the absolute and Celsius scales is

$$T_K = T_C + 273$$

equation 4.3

A temperature of absolute zero has never been reached, but scientists have cooled a sample of sodium to 700 nanokelvins, or 700 billionths of a kelvin above absolute zero.

HEAT

Suppose you have a bowl of hot soup or a cup of hot coffee that is too hot. What can you do to cool it? You can blow across the surface, which speeds evaporation and therefore results in cooling, but this is a slow process. If you were in a hurry, you would probably add something cooler, such as ice. Adding a cooler substance will cool the hot liquid.

You know what happens when you mix fluids or objects with a higher temperature with fluids or objects with a lower temperature. The warmer temperature object becomes cooler and the cooler temperature object becomes warmer. Eventually, both will have a temperature somewhere between the warmer and the cooler. This might suggest that something is moving between the warmer and cooler objects, changing the temperature. What is doing the moving?

The term **heat** is used to describe the "something" that moves between objects when two objects of different temperature are brought together. As you will learn in the next section, heat flow represents a form of energy transfer that take place between objects. For now, we will continue to think of heat as "something" that moves between objects of different temperatures, such as your bowl of hot soup and a cold ice cube.

The relationship that exists between energy and temperature will help explain the concept of heat, so we will consider it first. If you rub your hands together a few times they will feel a little warmer. If you rub them together vigorously for a while they will feel a lot warmer, maybe hot. A temperature increase takes place anytime mechanical energy causes one surface to rub against another (figure 4.8). The two surfaces could be solids, such as the two blocks, but they can also be the surface of a solid and a fluid, such as air. A solid object moving through the air encounters fluid friction, which results in a higher temperature of the surface. A high-velocity meteor enters the earth's atmosphere and is heated so much from the friction that it begins to glow, resulting in the fireball and smoke trail of a "falling star."

To distinguish between the energy of the object and the energy of its molecules, we use the terms "external" and "internal" energy. **External energy** is the total potential and kinetic energy of an everyday-sized object. All the kinetic and potential energy considerations discussed in previous chapters were about the external energy of an object.

Internal energy is the total kinetic and potential energy of the *molecules* of an object. The kinetic energy of a molecule can be much more complicated than straight-line velocity might suggest, however, as a molecule can have many different types of motion at the same time (pulsing, twisting, turning, etc.). Overall, internal energy is characterized by properties such as temperature, density, heat, volume, pressure of a gas, and so forth.

When you push a table across the floor, the observable *external* kinetic energy of the table is transferred to the *internal* ki-

Figure 4.9

External energy is the kinetic and potential energy that you can see. *Internal energy* is the total kinetic and potential energy of molecules. When you push a table across the floor, you do work against friction. Some of the external mechanical energy goes into internal kinetic and potential energy, and the bottom surface of the legs becomes warmer.

netic energy of the molecules between the table legs and the floor, resulting in a temperature increase (figure 4.9). The relationship between external and internal kinetic energy explains why the heating is proportional to the amount of mechanical energy used.

Heat as Energy Transfer

Temperature is a measure of the degree of hotness or coldness of a body, a measure that is based on the average molecular kinetic energy. Heat, on the other hand, is based on the *total internal energy* of the molecules of a body. You can see one difference in heat and temperature by considering a cup of water and a pitcher of water. If both the small and the large amount of water have the same temperature, both must have the same average molecular kinetic energy. Now, suppose you wish to cool both by, say, 20°. The pitcher of water would take much longer to cool, so it must be that the large amount of water has more internal energy (figure 4.10). Heat is a measure based on the *total* internal energy of the molecules of a body, and there is more total energy in a pitcher of water than in a cup of water at the same temperature.

How can we measure heat? Since it is difficult to see molecules, internal energy is difficult to measure directly. Thus heat is nearly always measured during the process of a body gaining or losing energy. This measurement procedure will also give us a working definition of heat:

> **Heat is a measure of the internal energy that has been absorbed or transferred from one body to another.**

The *process* of increasing the internal energy is called "heating" and the *process* of decreasing internal energy is called "cooling." The word "process" is italicized to emphasize that heat is energy in transit, not a material thing you can add or take away. Heat is understood to be a measure of internal energy that can be measured as energy flows in or out of an object.

There are two general ways that heating can occur. These are (1) from a temperature difference, with energy moving from

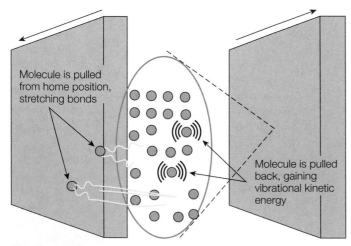

Figure 4.8

Here is how friction results in increased temperatures: Molecules on one moving surface will catch on another surface, stretching the molecular forces that are holding it. They are pulled back to their home position with a snap, resulting in a gain of vibrational kinetic energy.

Molecule is pulled from home position, stretching bonds

Molecule is pulled back, gaining vibrational kinetic energy

One liter
of water
at 90° C

250 milliliter
of water
at 90° C

Figure 4.10

Heat and temperature are different concepts, as shown by a liter of water (1,000 mL) and a 250 mL cup of water, both at the same temperature. You know the liter of water contains more internal energy since it will require more ice cubes to cool it to, say, 25°C than will be required for the cup of water. In fact, you will have to remove 48,750 *additional* calories to cool the liter of water.

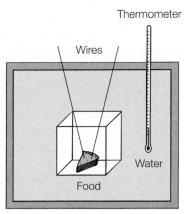

Figure 4.11

The Calorie value of food is determined by measuring the heat released from burning the food. If there is 10.0 kg of water and the temperature increased from 10°C to 20°C, the food contained 100 Calories (100,000 calories). The food illustrated here would release much more energy than this.

the region of higher temperature, and (2) from an object gaining energy by way of an energy-form conversion.

When a *temperature difference* occurs, energy is transferred from a region of higher temperature to a region of lower temperature. Energy flows from a hot range element, for example, to a pot of cold water on a range. It is a natural process for energy to flow from a region of a higher temperature to a region of a lower temperature just as it is natural for a ball to roll downhill. The temperature of an object and the temperature of the surroundings determines if heat will be transferred to or from an object. The terms "heating" and "cooling" describe the direction of energy flow, naturally moving from a region of higher energy to one of lower energy.

Measures of Heat

Since heating is a method of energy transfer, a quantity of heat can be measured just like any quantity of energy. The metric unit for measuring work, energy, or heat is the *joule*. However, the separate historical development of the concepts of heat and the concepts of motion resulted in separate units, some based on temperature differences.

The metric unit of heat is called the **calorie** (cal). A calorie is defined as the *amount of energy (or heat) needed to increase the temperature of 1 gram of water 1 degree Celsius*. A more precise definition specifies the degree interval from 14.5°C to 15.5°C because the energy required varies slightly at different temperatures. This precise definition is not needed for a general discussion. A **kilocalorie** (kcal) is the *amount of energy (or heat) needed to increase the temperature of 1 kilogram of water 1 degree Celsius*. The measure of the energy released by the oxidation of food is the kilocalorie, but it is called the Calorie (with a capital C) by nutritionists (figure 4.11). Confusion, can be avoided by making sure that the scientific calorie is never capitalized (cal) and the dieter's Calorie is always capitalized. The best solution would be to call the Calorie what it is, a kilocalorie (kcal).

The English system's measure of heating is called the **British thermal unit** (Btu). A Btu is *the amount of energy (or heat) needed to increase the temperature of 1 pound of water 1 degree Fahrenheit*. The Btu is commonly used to measure the heating or cooling rates of furnaces, air conditioners, water heaters, and so forth. The rate is usually expressed or understood to be in Btu per hour. A much larger unit is sometimes mentioned in news reports and articles about the national energy consumption. This unit is the *quad*, which is 1 quadrillion Btu (a million billion or 10^{15} Btu).

In metric units, the mechanical equivalent of heat is

$$4.184 \text{ J} = 1 \text{ cal}$$

or

$$4,184 \text{ J} = 1 \text{ kcal}$$

The establishment of this precise proportionality means that, fundamentally, mechanical energy and heat are different forms of the same thing.

Specific Heat

You can observe a relationship between heat and different substances by doing an experiment in "kitchen physics." Imagine that you have a large pot of liquid to boil in preparing a meal. There are three variables influencing how much heat you need:

1. The initial temperature of the liquid;
2. How much liquid is in the pot; and,
3. The nature of the liquid (water or soup, for example).

What this means specifically, is

1. **Temperature change.** The amount of heat needed is proportional to the temperature change. It takes more heat to raise the temperature of cool water, so this relationship could be written as $Q \propto \Delta T$.

Connections . . .

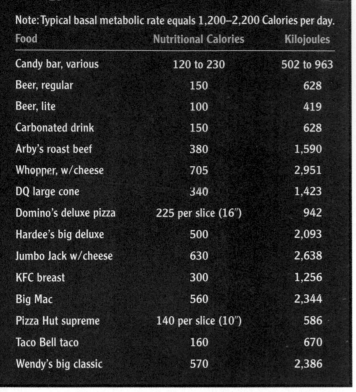

2. **Mass.** The amount of heat needed is also proportional to the amount of the substance being heated. A larger mass requires more heat to go through the same temperature change than a smaller mass. In symbols, $Q \propto m$.

3. **Substance.** Different materials require different amounts of heat to go through the same temperature range when their masses are equal (figure 4.12). This property is called the **specific heat** of a material, which is defined as the amount of heat needed to increase the temperature of 1 gram of a substance 1 degree Celsius.

Considering all the variables involved in our kitchen physics cooking experience, we find the heat (Q) need is described by the relationship of

$$Q = mc\Delta T$$

equation 4.4

where c is the symbol for specific heat. Specific heat is related to the internal structure of a substance; some of the energy goes into the internal potential energy of the molecules and some goes into the internal kinetic energy of the molecules. The different values for the specific heat of different substances are related to the number of molecules in the 1 gram sample and to the way they form a molecular structure.

Specific heat is responsible for the larger variation of temperature observed over land than near a large body of water. Table 4.2 gives the specific heat of soil as 0.200 cal/gC° and the specific heat of water as 1.00 cal/gC°. Since specific heat is defined as the amount of heat needed to increase the temperature

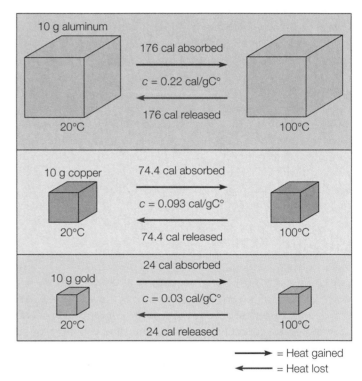

Figure 4.12

Of these three metals, aluminum needs the most heat per gram per degree when warmed, and releases the most heat when cooled. Why are the cubes different sizes?

of 1 gram of a substance 1 degree, this means 1 gram of water exposed to 1 calorie of sunlight will warm 1°C. One gram of soil exposed to 1 calorie of sunlight, on the other hand, will be warmed to 5°C since it only takes 0.2 calories to warm the soil 1°C. Thus the temperature is more even near large bodies of water since it is harder to change the temperature of the water.

Table 4.2 The Specific Heat of Selected Substances

Substance	Specific Heat (cal/gC° or kcal/kgC°)
Air	0.17
Aluminum	0.22
Concrete	0.16
Copper	0.093
Glass (average)	0.160
Gold	0.03
Ice	0.500
Iron	0.11
Lead	0.0305
Mercury	0.033
Seawater	0.93
Silver	0.056
Soil (average)	0.200
Steam	0.480
Water	1.00

Note: To convert to specific heat in J/kgC°, multiply each value by 4,184. Also note that 1 cal/gC° = 1 kcal/kgC°.

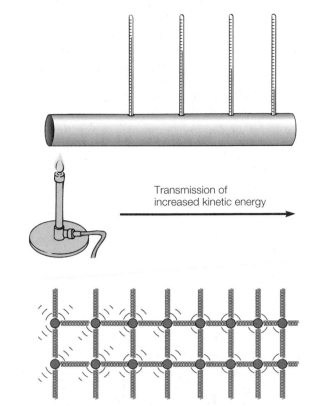

Figure 4.13
Thermometers placed in holes drilled in a metal rod will show that heat is conducted from a region of higher temperature to a region of lower temperature. The increased molecular activity is passed from molecule to molecule in the process of conduction.

Heat Flow

In a previous section you learned that heating is a transfer of energy that involves (1) a temperature difference or (2) energy-form conversions. Heat transfer that takes place because of a temperature difference takes place in three different ways: by conduction, convection, or radiation.

Conduction

Anytime there is a temperature difference, there is a natural transfer of heat from the region of higher temperature to the region of lower temperature. In solids, this transfer takes place as heat is *conducted* from a warmer place to a cooler one. Recall that the molecules in a solid vibrate in a fixed equilibrium position and that molecules in a higher temperature region have more kinetic energy, on the average, than those in a lower temperature region. When a solid, such as a metal rod, is held in a flame the molecules in the warmed end vibrate violently. Through molecular interaction, this increased energy of vibration is passed on to the adjacent, slower-moving molecules, which also begin to vibrate more violently. They, in turn, pass on more vibrational energy to the molecules next to them. The increase in activity thus moves from molecule to molecule, causing the region of increased activity to extend along the rod. This is called **conduction,** the transfer of energy from molecule to molecule (figure 4.13).

Most insulating materials are good insulators because they contain many small air spaces (figure 4.14). The small air spaces

Figure 4.14
Fiberglass insulation is rated in terms of R-value, a ratio of the conductivity of the material to its thickness.

are poor conductors because the molecules of air are far apart, compared to a solid, making it more difficult to pass the increased vibrating motion from molecule to molecule. Styrofoam, glass wool, and wool cloth are good insulators because they have many small air spaces, not because of the material they are made of. The best insulator is a vacuum, since there are no molecules to pass on the vibrating motion.

Wooden and metal parts of your desk have the same temperature, but the metal parts will feel cooler if you touch them. Metal is a better conductor of heat than wood and feels cooler because it conducts heat from your finger faster. This is the same reason that a wood or tile floor feels cold to your bare feet. You use an insulating rug to slow the conduction of heat from your feet.

Convection

Convection is the transfer of heat by a large-scale displacement of groups of molecules with relatively higher kinetic energy. In conduction, increased kinetic energy is passed from molecule to molecule. In convection, molecules with higher kinetic energy are moved from one place to another place. Conduction happens primarily in solids, but convection happens only in liquids and gases, where fluid motion can carry molecules with higher kinetic energy over a distance. When molecules gain energy, they move more rapidly and push more vigorously against their surroundings. The result is an expansion as the region of heated molecules pushes outward and increases the volume. Since the same amount of matter now occupies a larger volume, the overall density has been decreased.

In fluids, expansion sets the stage for convection. Warm, less dense fluid is pushed upward by the cooler, more dense fluid around it. In general, cooler air is more dense; it sinks and flows downhill. Cold air, being more dense, flows out near the bottom of an open refrigerator. You can feel the cold, dense air pouring from the bottom of an open refrigerator to your toes on the floor. On the other hand, you hold your hands *over* a heater because the warm, less dense air is pushed upward. In a room, warm air is pushed upward from a heater. The warm air spreads outward along the ceiling and is slowly displaced as newly warmed air is pushed upward to the ceiling. As the air cools it sinks over another part of the room, setting up a circulation pattern known as a *convection current* (figure 4.15). Convection currents can also be observed in a large pot of liquid that is heating on a range. You can see the warmer liquid being forced upward over the warmer parts of the range element, then sink over the cooler parts. Overall, convection currents give the pot of liquid an appearance of turning over as it warms.

Radiation

The third way that heat transfer takes place because of a temperature difference is called **radiation.** Radiation involves the form of energy called *radiant energy,* energy that moves through space. As you will learn in chapter 7, radiant energy includes visible light and many other forms as well. All objects with a temperature above absolute zero give off radiant energy. The absolute temperature of the object determines the rate, intensity, and kinds of radiant energy emitted. You know that visible

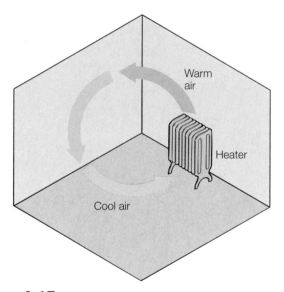

Figure 4.15

Convection currents move warm air throughout a room as the air over the heater becomes warmed, expands, and is moved upward by cooler air.

light is emitted if an object is heated to a certain temperature. A heating element on an electric range, for example, will glow with a reddish-orange light when at the highest setting, but it produces no visible light at lower temperatures, although you feel warmth in your hand when you hold it near the element. Your hand absorbs the nonvisible radiant energy being emitted from the element. The radiant energy does work on the molecules of your hand, giving them more kinetic energy. You sense this as an increase in temperature, that is, warmth.

All objects above absolute zero emit radiant energy, but all objects also absorb radiant energy. A hot object, however, emits more radiant energy than a cold object. The hot object will emit more energy than it absorbs from the colder object, and the

A Closer Look Passive Solar Design

Passive solar application is an economically justifiable use of solar energy today. Passive solar design uses a structure's construction to heat a living space with solar energy. There are few electric fans, motors, or other energy sources used. The passive solar design takes advantage of free solar energy; it stores and then distributes this energy through natural conduction, convection, and radiation.

Sunlight that reaches the earth's surface is mostly absorbed. Buildings, the ground, and objects become warmer as the radiant energy is absorbed. Nearly all materials, however, reradiate the absorbed energy at longer wavelengths, wavelengths too long to be visible to the human eye. The short wavelengths of sunlight pass readily through ordinary window glass, but the longer, reemitted wavelengths cannot. Therefore, sunlight passes through a window and warms objects inside a house. The reradiated longer wavelengths cannot pass readily back through the glass but are absorbed by certain molecules in the air. The temperature of the air is thus increased. This is called the "greenhouse effect." Perhaps you have experienced the effect when you left your car windows closed on a sunny, summer day.

In general, a passive solar home makes use of the materials from which it is constructed to capture, store, and distribute solar energy to its occupants. Sunlight enters the house through large windows facing south and warms a thick layer of concrete, brick, or stone. This energy "storage mass" then releases energy during the day, and, more important, during the night. This release of energy can be by direct radiation to occupants, by conduction to adjacent air, or by convection of air across the surface of the storage mass. The living space is thus heated without special plumbing or forced air circulation. As you can imagine, the key to a successful passive solar home is to consider every detail of natural energy flow, including the materials of which floors and walls are constructed, convective air circulation patterns, and the size and placement of windows. In addition, a passive solar home requires a different lifestyle and living patterns. Carpets, for example, would defeat the purpose of a storage-mass floor, since it would insulate the storage mass from sunlight. Glass is not a good insulator, so windows must have curtains or movable insulation panels to slow energy loss at night. This requires the daily activity of closing curtains or moving insulation panels at night and then open-

Box Figure 4.1
The direct solar gain design collects and stores solar energy in the living space.

ing curtains and moving panels in the morning. Passive solar homes, therefore, require a high level of personal involvement by the occupants.

There are three basic categories of passive solar design: (1) direct solar gain, (2) indirect solar gain, and (3) isolated solar gain.

A *direct solar gain* home is one in which solar energy is collected in the actual living space of the home (box figure 4.1). The advantage of this design is the large, open window space with a calculated overhang, which admits maximum solar energy in the winter but prevents solar gain in the summer. The disadvantage is that the occupants are living in the collection and storage components of the design and can place nothing (such as carpets and furniture) that would interfere with warming the storage mass in the floors and walls.

An *indirect solar gain* home uses a massive wall inside a window that serves as a storage mass. Such a wall, called a *Trombe wall,* is shown in box figure 4.2. The Trombe wall collects and stores solar energy, then warms the living space with radiant energy and convection currents. The disadvantage to the indirect solar gain design is that large windows are blocked by the Trombe wall. The advantage is that the occupants are not in direct contact with the solar collection and storage area, so they can place carpets and furniture as they wish. Controls to prevent energy loss at night are still necessary with this design.

An *isolated solar gain* home uses a structure that is separated from the living space to collect and store solar energy. Examples of an isolated gain design are an attached greenhouse or sun porch (box figure 4.3). Energy flow between the attached structure and the living space can be by conduction, convection, and radiation, which can be controlled by opening or closing off

Box Figure 4.2
The indirect solar gain design uses a Trombe wall to collect, store, and distribute solar energy.

Box Figure 4.3
The isolated solar gain design uses a separate structure to collect and store solar energy.

the attached structure. This design provides the best controls, since it can be completely isolated, opened to the living space as needed, or directly used as living space when the conditions are right. Additional insulation is needed for the glass at night, however, and for sunless winter days.

It has been estimated that building a passive solar home would cost about 10 percent more than building a traditional home of the same size. Considering the possible energy savings, you might believe that most homes would now have a passive solar design. They do not, however, as most new buildings require technology and large amounts of energy to maximize comfort. Yet, it would not require too much effort to consider where to place windows in relation to the directional and seasonal intensity of the sun and where to plant trees. Perhaps in the future you will have an opportunity to consider using the environment to your benefit through the natural processes of conduction, convection, and radiation.

colder object will absorb more energy from the hot object than it emits. There is, therefore, a net energy transfer that will take place by radiation as long as there is a temperature difference between the two objects.

ENERGY, HEAT, AND MOLECULAR THEORY

The kinetic molecular theory of matter is based on evidence from different fields of science, not just one subject area. Chemists and physicists developed some convincing conclusions about the structure of matter over the past 150 years, using carefully designed experiments and mathematical calculations that explained observable facts about matter. Step by step, the detailed structure of this submicroscopic, invisible world of particles became firmly established. Today, an understanding of this particle structure is basic to physics, chemistry, biology, geology, and practically every other science subject. This understanding has also resulted in present-day technology.

Phase Change

Solids, liquids, and gases are the three common phases of matter, and each phase is characterized by different molecular arrangements. The motion of the molecules in any of the three common phases can be increased by (1) adding heat through a temperature difference or (2) the absorption of one of the five forms of energy, which results in heating. In either case the temperature of the solid, liquid, or gas increases according to the specific heat of the substance, and more heating generally means higher temperatures.

More heating, however, does not always result in increased temperatures. When a solid, liquid, or gas changes from one phase to another, the transition is called a **phase change.** A phase change always absorbs or releases energy, *a quantity of heat that is not associated with a temperature change.* Since the quantity of heat associated with a phase change is not associated with a temperature change, it is called *latent heat.* Latent heat refers to the "hidden" energy of phase changes, which is energy (heat) that goes into or comes out of *internal potential energy* (figure 4.16).

There are three kinds of major phase changes that can occur: (1) *solid-liquid,* (2) *liquid-gas,* and (3) *solid-gas.* In each case the phase change can go in either direction. For example, the solid-liquid phase change occurs when a solid melts to a liquid or when a liquid freezes to a solid. Ice melting to water and water freezing to ice are common examples of this phase change and its two directions. Both occur at a temperature called the *freezing point* or the *melting point,* depending on the direction of the phase change. In either case, however, the freezing and melting points are the same temperature.

The liquid-gas phase change also occurs in two different directions. The temperature at which a liquid boils and changes to a gas (or vapor) is called the *boiling point.* The temperature at which a gas or vapor changes back to a liquid is called the *condensation point.* The boiling and condensation points are the same temperature. There are conditions other than boiling un-

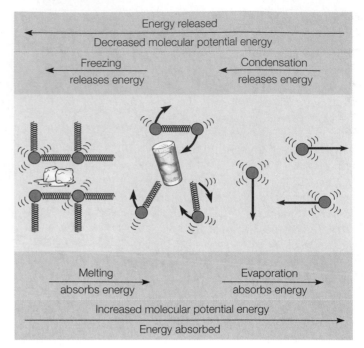

Figure 4.16

Each phase change absorbs or releases a quantity of latent heat, which goes into or is released from molecular potential energy.

der which liquids may undergo liquid-gas phase changes, and these conditions are discussed in the next section.

You probably are not as familiar with solid-gas phase changes, but they are common. A phase change that takes a solid directly to a gas or vapor is called *sublimation.* Mothballs and dry ice (solid CO_2) are common examples of materials that undergo sublimation, but frozen water, meaning common ice, also sublimates under certain conditions. Perhaps you have noticed ice cubes in a freezer become smaller with time as a result of sublimation. The frost that forms in a freezer, on the other hand, is an example of a solid-gas phase change that takes place in the other direction. In this case, water vapor forms the frost without going through the liquid state, a solid-gas phase change that takes place in an opposite direction to sublimation.

For a specific example, consider the changes that occur when ice is subjected to a constant source of heat (figure 4.17). Starting at the left side of the graph, you can see that the temperature of the ice increases from the constant input of heat. The ice warms according to $Q = mc\Delta T$, where c is the specific heat of ice. When the temperature reaches the melting point (0°C), it stops increasing as the ice begins to melt. More and more liquid water appears as the ice melts, but the temperature *remains* at 0°C even though heat is still being added at a constant rate. It takes a certain amount of heat to melt all of the ice. Finally, when all the ice is completely melted, the temperature again increases at a constant rate between the melting and boiling points. Then, at constant temperature the addition of heat produces another phase change, from liquid to gas. The quantity of heat involved in this phase change is used in doing the work of breaking the molecule-to-molecule bonds in the solid,

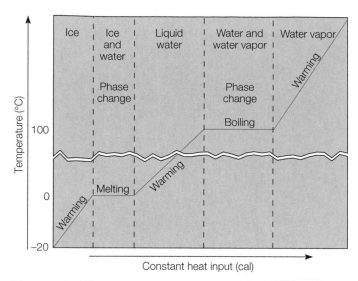

Figure 4.17

This graph shows three warming sequences and two phase changes with a constant input of heat. The ice warms to the melting point, then absorbs heat during the phase change as the temperature remains constant. When all the ice has melted, the now-liquid water warms to the boiling point, where the temperature again remains constant as heat is absorbed during this second phase change from liquid to gas. After all the liquid has changed to gas, continued warming increases the temperature of the water vapor.

making a liquid with molecules that are now free to move about and roll over one another. Since the quantity of heat (Q) is absorbed without a temperature change, it is called the **latent heat of fusion** (L_f). The latent heat of fusion is *the heat involved in a solid-liquid phase change in melting or freezing.* For water, the latent heat of fusion is 80.0 cal/g (144.0 Btu/lb). This means that every gram of ice that melts *absorbs* 80.0 cal of heat. Every gram of water that freezes *releases* 80.0 cal. The total heat involved in a solid-liquid phase change depends on the mass of the substance involved, so

$$Q = mL_f$$

equation 4.5

where L_f is the latent heat of fusion for the substance involved.

Refer again to figure 4.17. After the solid-liquid phase change is complete, the constant supply of heat increases the temperature of the water according to $Q = mc\Delta T$, where c is now the specific heat of liquid water. When the water reaches the boiling point, the temperature again remains constant even though heat is still being supplied at a constant rate. The quantity of heat involved in the liquid-gas phase change again goes into doing the work of overcoming the attractive molecular forces. This time the molecules escape from the liquid state to become single, independent molecules of gas. The quantity of heat (Q) absorbed or released during this phase change is called the **latent heat of vaporization** (L_v). The latent heat of vaporization is *the heat involved in a liquid-gas phase change where there is evaporation or condensation.* For water, the latent heat of vaporization is 540.0 cal/g (970.0 Btu/lb). This means that every

gram of water vapor that condenses on your bathroom mirror releases 540.0 cal, which warms the bathroom. The total heating depends on how much water vapor condensed, so

$$Q = mL_v$$

equation 4.6

where L_v is the latent heat of vaporization for the substance involved. The relationships between the quantity of heat absorbed during warming and phase changes are shown in figure 4.18. Some physical constants for water and heat are summarized in table 4.3.

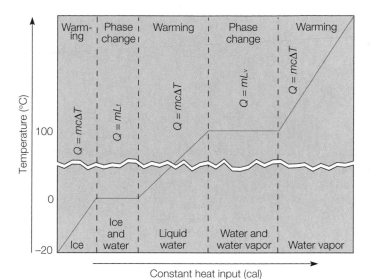

Figure 4.18

Compare this graph to the one in figure 4.17. This graph shows the relationships between the quantity of heat absorbed during warming and phase changes as water is warmed from ice at −20°C to water vapor at some temperature above 100°C. Note that the specific heat for ice, liquid water, and water vapor (steam) have different values.

Table 4.3 Some Physical Constants for Water and Heat		
Specific Heat (c)		
Water	$c = 1.00$ cal/gC°	
Ice	$c = 0.500$ cal/gC°	
Steam	$c = 0.480$ cal/gC°	
Latent Heat of Fusion		
L_f (water)	$L_f = 80.0$ cal/g	
Latent Heat of Vaporization		
L_v (water)	$L_v = 540.0$ cal/g	
Mechanical Equivalent of Heat		
1 kcal	4,184 J	

Evaporation and Condensation

Liquids do not have to be at the boiling point to change to a gas and, in fact, tend to undergo a phase change at any temperature when left in the open. The phase change occurs at any temperature but does occur more rapidly at higher temperatures. The temperature of the water is associated with the *average* kinetic energy of the water molecules. The word *average* implies that some of the molecules have a greater energy and some have less. If a molecule of water that has an exceptionally high energy is near the surface, and is headed in the right direction, it may overcome the attractive forces of the other water molecules and escape the liquid to become a gas. This is the process of *evaporation*. Evaporation reduces a volume of liquid water as water molecules leave the liquid state to become water vapor in the atmosphere (figure 4.19).

Water molecules that evaporate move about in all directions, and some will return, striking the liquid surface. The same forces that they escaped from earlier capture the molecules, returning them to the liquid state. This is called the process of condensation. Condensation is the opposite of evaporation. In *evaporation*, more molecules are leaving the liquid state than are returning. In *condensation*, more molecules are returning to the liquid state than are leaving. This is a dynamic, ongoing process with molecules leaving and returning continuously. The net number leaving or returning determines if evaporation or condensation is taking place (figure 4.20).

When the condensation rate *equals* the evaporation rate, the air above the liquid is said to be *saturated*. The air immediately next to a surface may be saturated, but the condensation of water molecules is easily moved away with air movement. There is no net energy flow when the air is saturated, since the heat car-

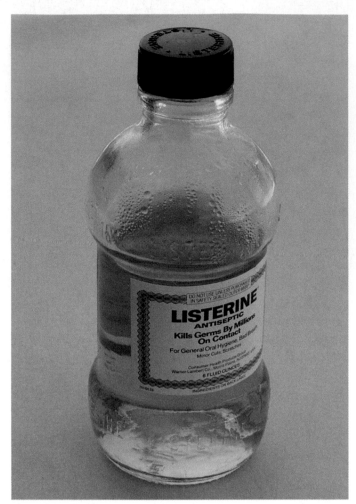

Figure 4.20

The inside of this closed bottle is isolated from the environment, so the space above the liquid becomes saturated. While it is saturated, the evaporation rate equals the condensation rate. When the bottle is cooled, condensation exceeds evaporation and droplets of liquid form on the inside surfaces.

ried away by evaporation is returned by condensation. This is why you fan your face when you are hot. The moving air from the fanning action pushes away water molecules from the air near your skin, preventing the adjacent air from becoming saturated, thus increasing the rate of evaporation. Think about this process the next time you see someone fanning his or her face.

There are four ways to increase the rate of evaporation. (1) An increase in the temperature of the liquid will increase the average kinetic energy of the molecules and thus increase the number of high-energy molecules able to escape from the liquid state. (2) Increasing the surface area of the liquid will also increase the likelihood of molecular escape to the air. This is why you spread out wet clothing to dry or spread out a puddle you want to evaporate. (3) Removal of water vapor from near the surface of the liquid will prevent the return of the vapor molecules to the liquid state and thus increase the net rate of evaporation. This is why things dry more rapidly on a windy day. (4) *Pressure* is defined as *force per unit area*, which can be measured in lb/in^2 or N/m^2. Gases exert a pressure, which is inter-

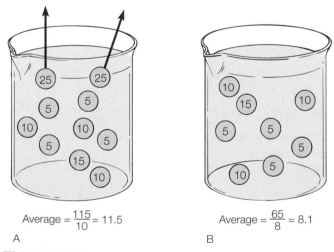

Average $= \frac{115}{10} = 11.5$ Average $= \frac{65}{8} = 8.1$

A   B

Figure 4.19

Temperature is associated with the average energy of the molecules of a substance. These numbered circles represent arbitrary levels of molecular kinetic energy that, in turn, represent temperature. The two molecules with the higher kinetic energy values [25 in (A)] escape, which lowers the average values from 11.5 to 8.1 (B). Thus evaporation of water molecules with more kinetic energy contributes to the cooling effect of evaporation, in addition to the absorption of latent heat.

Evaporative Coolers

Evaporative cooling is one of the earliest forms of mechanical air conditioning. An evaporative cooler (sometimes called a "swamp cooler") works by cooling outside air by evaporation, then blowing the cooler, but now more humid air through a house. Usually, an evaporative cooler moves a sufficient amount of air through a house to completely change the air every two or three minutes.

An evaporative cooler is a metal or fiberglass box with louvers. Inside the box is a fan and motor, and there might be a small pump to recycle water. Behind the louvers are loose pads made of wood shavings (excelsior) or some other porous material. Water is pumped from the bottom of the cooler and trickles down through the pads, thoroughly wetting them. The fan forces air into the house and dry, warm outside air moves through the wet pads into the house in a steady flow. Windows or a door in the house must be partly open or the air will not be able to flow through the house.

The water in the pads evaporates, robbing heat from the air moving through them. Each liter of water that evaporates could take over 540 kcal of heat from the air. The actual amount removed depends on the temperature of the water, the efficiency of the cooler, and the relative humidity of the outside air.

An electric fan cools you by evaporation, but an evaporative cooler cools the air that blows around you. Relative humidity is the main variable that determines how much an evaporative cooler will cool the air. The following data show the cooling power of a typical evaporative cooler at various relative humidities when the outside air is 38°C (about 100°F).

At This Humidity:	Air Will Be Cooled to:
10%	22°C (72°F)
20%	24°C (76°F)
30%	27°C (80°F)
40%	29°C (84°F)
50%	31°C (88°F)
60%	33°C (91°F)
70%	34°C (93°F)
80%	35°C (95°F)
90%	37°C (98°F)

The advantage to an evaporative cooler is the low operating cost compared to refrigeration. The disadvantages are that it doesn't cool the air that much when the humidity is high, it adds even more humidity to the air, and outside air with its dust, pollen, and pollution is continually forced through the house. Another disadvantage is that mineral deposits left by evaporating hard water could require a frequent maintenance schedule.

CONCEPTS APPLIED
Why It Is Called a Pop Can!

Obtain two empty, clean pop (soft drink) cans, a container of ice water with ice cubes, and a container of boiling water. You might want to "dry run" this experiment to make sure of the procedure before actually doing it.

Place about 2 cm of water in a pop can and heat it on a stove until the water boils and you see evidence of steam coming from the opening. Using tongs, quickly invert the can half way into a container of ice water. Note how much water runs from the can as your remove it from the ice water.

Repeat this procedure, this time inverting the can half way into a container of boiling water. Note how much water runs from the can as you remove it from the boiling water.

Explain your observations in terms of the kinetic molecular theory, evaporation, and condensation. It is also important to explain any differences observed between what happened to the two pop cans.

---

preted in terms of the kinetic molecular theory. Atmospheric pressure is discussed in detail in chapter 16. For now, consider that the atmosphere exerts a pressure of about 10.0 N/cm² (14.7 lb/in²) at sea level. The atmospheric pressure, as well as the intermolecular forces, tend to hold water molecules in the liquid state. Thus, reducing the atmospheric pressure will reduce one of the forces holding molecules in a liquid state. Perhaps you have noticed that wet items dry more quickly at higher elevations, where the atmospheric pressure is less.

Relative Humidity

There is a relationship between evaporation-condensation and the air temperature. If the air temperature is decreased, the average kinetic energy of the molecules making up the air is decreased. Water vapor molecules condense from the air when they slow enough that molecular forces can pull them into the liquid state. Fast-moving water vapor molecules are less likely to be captured than slow-moving ones. Thus, as the air temperature increases, there is less tendency for water vapor molecules to return to the liquid state. Warm air can therefore hold more water vapor than cool air. In fact, air at 38°C (100°F) can hold five times as much water vapor as air at 10°C (50°F) (figure 4.21).

The ratio of how much water vapor *is* in the air to how much water vapor *could be* in the air at a certain temperature is called **relative humidity.** This ratio is usually expressed as a percentage, and

$$\text{relative humidity} = \frac{\text{water vapor in air}}{\text{capacity at present temperature}} \times 100\%$$

$$R.H. = \frac{\text{g/m}^3(\text{present})}{\text{g/m}^3(\text{max})} \times 100\%$$

equation 4.7

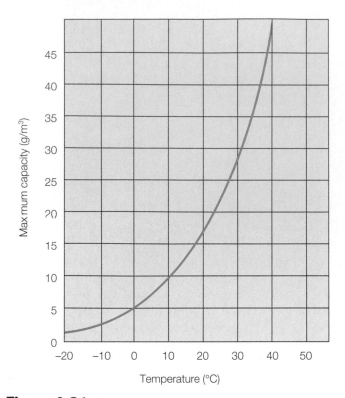

Figure 4.21

The curve shows the *maximum* amount of water vapor in g/m³ that can be in the air at various temperatures.

Figure 4.21 shows the maximum amount of water vapor that can be in the air at various temperatures. Suppose that the air contains 10 g/m³ of water vapor at 10°C (50°F). According to figure 4.21, the maximum amount of water vapor that *can be* in the air when the air temperature is 10°C (50°F) is 10 g/m³. Therefore the relative humidity is (10 g/m³) ÷ (10 g/m³) × 100%, or 100 percent. This air is therefore saturated. If the air held only 5 g/m³ of water vapor at 10°C, the relative humidity would be 50 percent, and 2 g/m³ of water vapor in the air at 10°C is 20 percent relative humidity.

As the air temperature increases, the capacity of the air to hold water vapor also increases. This means that if the *same amount* of water vapor is in the air during a temperature increase, the relative humidity will decrease. Thus, the relative humidity increases every night because the air temperature decreases, not because water vapor has been added to the air. The relative humidity is important because it is one of the things that controls the rate of evaporation, and the evaporation rate is one of the variables involved in how well you can cool yourself in hot weather.

Thermodynamics

The branch of physical science called *thermodynamics* is concerned with the study of heat and its relationship to mechanical energy, including the science of heat pumps, heat engines, and the transformation of energy in all its forms. The *laws of thermodynamics* describe the relationships concerning what hap-

pens as energy is transformed to work and the reverse, also serving as useful intellectual tools in meteorology, chemistry, and biology.

The **first law of thermodynamics** describes a relationship between changes in internal energy, work, and heat. It is a statement of the *law of conversation of energy,* that energy can neither be created nor destroyed; it can only be changed from one form into another. Energy can exist in several different forms, and it is possible to convert one form of energy into another. However, the total amount of energy remains constant.

From a human perspective, some forms of energy are more useful than others. We tend to make extensive use of electrical energy for a variety of purposes, but there is very little electrical energy available in nature. Therefore, we convert other forms of energy into electrical energy. When converting energy from one form to another, the energy is *degraded,* going from a more orderly form to a less orderly form. The meaning of "less orderly" is found in the example of dropping and breaking a drinking glass. The pieces of glass are less orderly, and they do not come together by themselves to fix the glass. The natural course of events is an increase of disorder. More ordered chemical energy eventually ends up as less ordered, low-level kinetic energy of the molecules. Disorder has increased.

One statement of the **second law of thermodynamics** is that the natural process proceeds toward a state of greater disorder. The measure of disorder is called **entropy.** Therefore, another way to state the second law of thermodynamics is to say that when energy is converted from one form to another, entropy increases. An alternative way to look at the idea of entropy is to say that entropy is a measure of disorder and that the amount of disorder (entropy) increases when energy conversions take place.

Thus the *natural process* is for a state of order to degrade into a state of disorder with a corresponding increase in entropy. This means that all the available energy of the universe is gradually diminishing, and over time, the universe should therefore approach a limit of maximum disorder called the *heat death* of the universe. The heat death of the universe is the theoretical limit of disorder, with all molecules spread far, far apart, vibrating slowly with a uniform low temperature.

CONCEPTS APPLIED

Thermodynamics in Action

The laws of thermodynamics are concerned with changes in energy and heat. This application explores some of these relationships.

Obtain an electric blender and a thermometer. Fill the blender half way with water, then let it remain undisturbed until the temperature is constant as shown by two consecutive temperature readings.

Run the blender at the highest setting for a short time, then stop and record the temperature of the water. Repeat this procedure several times.

Explain your observations in terms of thermodynamics. See if you can think of other experiments that show relationships between changes in energy and heat.

The heat death of the universe seems to be a logical consequence of the second law of thermodynamics, but scientists are not certain if the second law should apply to the whole universe. What do you think? Will the universe with all its complexities of organization end with the simplicity of spread-out and slowly vibrating molecules? As has been said, nature is full of symmetry—so why should the universe begin with a bang and end with a whisper?

SUMMARY

The kinetic theory of matter assumes that all matter is made up of tiny, ultimate particles of matter called *molecules*. A molecule is defined as the smallest particle of a compound, or a gaseous element, that can exist and still retain the characteristic properties of that substance. Molecules interact, attracting each other through a force of *cohesion*. Liquids, solids, and gases are the *phases of matter* that are explained by the molecular arrangements and forces of attraction between their molecules. A *solid* has a definite shape and volume because it has molecules that vibrate in a fixed equilibrium position with strong cohesive forces. A *liquid* has molecules that have cohesive forces strong enough to give it a definite volume but not strong enough to give it a definite shape. The molecules of a liquid can flow, rolling over each other. A *gas* is composed of molecules that are far apart, with weak cohesive forces. Gas molecules move freely in a constant, random motion.

The *temperature* of an object is related to the *average kinetic energy* of the molecules making up the object. A measure of temperature tells how hot or cold an object is on two arbitrary scales, the *Fahrenheit scale* and the *Celsius scale*. The *absolute scale*, or *Kelvin scale*, has the coldest temperature possible ($-273°C$) as zero (0 K).

The observable potential and kinetic energy of an object is the *external energy* of that object, while the potential and kinetic energy of the molecules making up the object is the *internal energy* of the object. Heat refers to the total internal energy and is a transfer of energy that takes place (1) because of a *temperature difference* between two objects, or (2) because of an *energy-form conversion*. An energy-form conversion is actually an energy conversion involving work at the molecular level, so all energy transfers involve *heating* and *working*.

A quantity of heat can be measured in *joules* (a unit of work or energy) or *calories* (a unit of heat). A *kilocalorie* is 1,000 calories, another unit of heat. A *Btu*, or *British thermal unit*, is the English system unit of heat. The *mechanical equivalent of heat* is 4,184 J = 1 kcal.

The *specific heat* of a substance is the amount of energy (or heat) needed to increase the temperature of 1 gram of a substance 1 degree Celsius. The specific heat of various substances is not the same because the molecular structure of each substance is different.

Energy transfer that takes place because of a temperature difference does so through conduction, convection, or radiation. *Conduction* is the transfer of increased kinetic energy from molecule to molecule. Substances vary in their ability to conduct heat, and those that are poor conductors are called *insulators*. Gases, such as air, are good insulators. The best insulator is a vacuum. *Convection* is the transfer of heat by the displacement of large groups of molecules with higher kinetic energy. Convection takes place in fluids, and the fluid movement that takes place because of density differences is called a *convection current*. *Radiation* is radiant energy that moves through space. All objects with an ab-

solute temperature above zero give off radiant energy, but all objects absorb it as well. Energy is transferred from a hot object to a cold one through radiation.

The transition from one phase of matter to another is called a *phase change*. A phase change always absorbs or releases a quantity of *latent heat* not associated with a temperature change. Latent heat is energy that goes into or comes out of *internal potential energy*. The *latent heat of fusion* is absorbed or released at a solid-liquid phase change. The latent heat of fusion for water is 80.0 cal/g (144.0 Btu/lb). The *latent heat of vaporization* is absorbed or released at a liquid-gas phase change. The latent heat of vaporization for water is 540.0 cal/g (970.0 Btu/lb).

Molecules of liquids sometimes have a high enough velocity to escape the surface through the process called *evaporation*. Evaporation is a cooling process, since the escaping molecules remove the latent heat of vaporization in addition to their high molecular energy. Vapor molecules return to the liquid state through the process called *condensation*. Condensation is the opposite of evaporation and is a warming process. When the condensation rate equals the evaporation rate, the air is said to be *saturated*. The rate of evaporation can be *increased* by (1) increased temperature, (2) increased surface area, (3) removal of evaporated molecules, and (4) reduced atmospheric pressure.

Warm air can hold more water vapor than cold air, and the ratio of how much water vapor is in the air to how much could be in the air at that temperature (saturation) is called *relative humidity*.

Thermodynamics is the study of heat and its relationship to mechanical energy, and the *laws of thermodynamics* describe these relationships: The *first law of thermodynamics* is a thermodynamic statement of the law of conservation of energy. The *second law of thermodynamics* states that heat flows from objects with a higher temperature to objects with a cooler temperature. The second law implies a *degradation of energy* as *high-quality* (more ordered) energy sources undergo *degradation* to *low-quality* (less ordered) sources. *Entropy* is a thermodynamic measure of disorder, and entropy is seen as continually increasing in the universe and may result in the maximum disorder called the *heat death* of the universe.

Summary of Equations

4.1
$$T_F = \frac{9}{5} T_C + 32°$$

4.2
$$T_C = \frac{5}{9}(T_F - 32°)$$

4.3
$$T_K = T_C + 273$$

4.4 Quantity of heat = (mass)(specific heat)(temperature change)
$$Q = mc\Delta T$$

4.5 Heat absorbed or released = (mass)(latent heat of fusion)
$$Q = mL_f$$

4.6 Heat absorbed or released = (mass)(latent heat of vaporization)
$$Q = mL_v$$

4.7 $\text{relative humidity} = \dfrac{\text{water vapor in air}}{\text{capacity at present temperature}} \times 100\%$

$$R.H. = \dfrac{\text{g/m}^3\text{(present)}}{\text{g/m}^3\text{(max)}} \times 100\%$$

KEY TERMS

absolute scale (p. **69**)
absolute zero (p. **69**)
British thermal unit (p. **71**)
calorie (p. **71**)
Celsius scale (p. **68**)
conduction (p. **73**)
convection (p. **74**)
entropy (p. **80**)
external energy (p. **70**)
Fahrenheit scale (p. **67**)
first law of thermodynamics (p. **80**)
heat (p. **70**)
internal energy (p. **70**)

Kelvin scale (p. **69**)
kilocalorie (p. **71**)
kinetic molecular theory (p. **64**)
latent heat of fusion (p. **77**)
latent heat of vaporization (p. **71**)
molecule (p. **65**)
phase change (p. **76**)
radiation (p. **74**)
relative humidity (p. **79**)
second law of thermodynamics (p. **80**)
specific heat (p. **72**)
temperature (p. **66**)

APPLYING THE CONCEPTS

1. The kinetic molecular theory explains the expansion of a solid material with increases of temperature as basically the result of
   (a) individual molecules expanding.
   (b) increased translational kinetic energy.
   (c) molecules moving a little farther apart.
   (d) heat taking up the spaces between molecules.
2. Using the absolute temperature scale, the freezing point of water is correctly written as
   (a) 0 K.
   (b) 0°K.
   (c) 273 K.
   (d) 273°K.
3. The metric unit of heat called a *calorie* is
   (a) the specific heat of water.
   (b) the energy needed to increase the temperature of 1 gram of water 1 degree Celsius.
   (c) equivalent to a little over 4 joules of mechanical work.
   (d) all of the above.
4. Table 4.2 lists the specific heat of soil as 0.20 kcal/kgC° and the specific heat of water as 1.00 kcal/kgC°. This means that if 1 kg of soil and 1 kg of water each receive 1 kcal of energy, ideally,
   (a) the water will be warmer than the soil by 0.8°C.
   (b) the soil will be 4°C warmer than the water.
   (c) the water will be 4°C warmer than the soil.
   (d) the water will warm by 1°C, and the soil will warm by 0.2°C.
5. The heat transfer that takes place by energy moving directly from molecule to molecule is called
   (a) conduction.
   (b) convection.
   (c) radiation.
   (d) none of the above.

6. The heat transfer that does not require matter is
   (a) conduction.
   (b) convection.
   (c) radiation.
   (d) impossible, for matter is always required.
7. Styrofoam is a good insulating material because
   (a) it is a plastic material that conducts heat poorly.
   (b) it contains many tiny pockets of air.
   (c) of the structure of the molecules making up the Styrofoam.
   (d) it is not very dense.
8. The transfer of heat that takes place because of density difference in fluids is
   (a) conduction.
   (b) convection.
   (c) radiation.
   (d) none of the above.
9. As a solid undergoes a phase change to a liquid state, it
   (a) releases heat while remaining at a constant temperature.
   (b) absorbs heat while remaining at a constant temperature.
   (c) releases heat as the temperature decreases.
   (d) absorbs heat as the temperature increases.
10. The condensation of water vapor actually
    (a) warms the surroundings.
    (b) cools the surroundings.
    (c) sometimes warms and sometimes cools the surroundings, depending on the relative humidity at the time.
    (d) neither warms nor cools the surroundings.
11. No water vapor is added to or removed from a sample of air that is cooling, so the relative humidity of this sample of air will
    (a) remain the same.
    (b) be lower.
    (c) be higher.
    (d) be higher or lower, depending on the extent of change.
12. Compared to cooler air, warm air can hold
    (a) more water vapor.
    (b) less water vapor.
    (c) the same amount of water vapor.
    (d) less water vapor, the amount depending on the humidity.

Answers
1. c **2.** c **3.** d **4.** b **5.** a **6.** c **7.** b **8.** b **9.** b **10.** a **11.** c **12.** a

QUESTIONS FOR THOUGHT

1. What is temperature? What is heat?
2. Explain why most materials become less dense as their temperature is increased.
3. Would the tight packing of more insulation, such as glass wool, in an enclosed space increase or decrease the insulation value? Explain.
4. A true vacuum bottle has a double-walled, silvered bottle with the air removed from the space between the walls. Describe how this design keeps food hot or cold by dealing with conduction, convection, and radiation.

5. Why is cooler air found in low valleys on calm nights?

6. Why is air a good insulator?

7. A piece of metal feels cooler than a piece of wood at the same temperature. Explain why.

8. What is condensation? Explain, on a molecular level, how the condensation of water vapor on a bathroom mirror warms the bathroom.

9. Which provides more cooling for a Styrofoam cooler, one with 10 lb of ice at 0°C or one with 10 lb of ice water at 0°C? Explain your reasoning.

10. Explain why a glass filled with a cold beverage seems to "sweat." Would you expect more sweating inside a house during the summer or during the winter? Explain.

11. Explain why a burn from 100°C steam is more severe than a burn from water at 100°C.

12. The relative humidity increases almost every evening after sunset. Explain how this is possible if no additional water vapor is added to or removed from the air.

PARALLEL EXERCISES

The exercises in groups A and B cover the same concepts. Solutions to group A exercises are located in appendix D.

Note: Neglect all frictional forces in all exercises.

Group A

1. The average human body temperature is 98.6°F. What is the equivalent temperature on the Celsius scale?

2. An electric current heats a 221 g copper wire from 20.0°C to 38.0°C. How much heat was generated by the current? ($c_{copper} = 0.093$ kcal/kgC°)

3. A bicycle and rider have a combined mass of 100.0 kg. How many calories of heat are generated in the brakes when the bicycle comes to a stop from a speed of 36.0 km/h?

4. A 15.53 kg loose bag of soil falls 5.50 m at a construction site. If all the energy is retained by the soil in the bag, how much will its temperature increase? ($c_{soil} = 0.200$ kcal/kgC°)

5. A 75.0 kg person consumes a small order of french fries (250.0 Cal) and wishes to "work off" the energy by climbing a 10.0 m stairway. How many vertical climbs are needed to use all the energy?

6. A 0.5 kg glass bowl ($c_{glass} = 0.2$ kcal/kgC°) and a 0.5 kg iron pan ($c_{iron} = 0.11$ kcal/kgC°) have a temperature of 68°F when placed in a freezer. How much heat will the freezer have to remove from each to cool them to 32°F?

7. A sample of silver at 20.0°C is warmed to 100.0°C when 896 cal is added. What is the mass of the silver? ($c_{silver} = 0.056$ kcal/kgC°)

8. A 300.0 W immersion heater is used to heat 250.0 g of water from 10.0°C to 70.0°C. About how many minutes did this take?

9. A 100.0 g sample of metal is warmed 20.0°C when 60.0 cal is added. What is the specific heat of this metal?

10. How much heat is needed to change 250.0 g of water at 80.0°C to steam at 100.0°C?

11. A 100.0 g sample of water at 20.0°C is heated to steam at 125.0°C. How much heat was absorbed?

12. In an electric freezer, 400.0 g of water at 18.0°C is cooled, frozen, and the ice is chilled to −5.00°C. How much total heat was removed from the water?

Group B

1. The Fahrenheit temperature reading is 98° on a hot summer day. What is this reading on the Kelvin scale?

2. A 0.25 kg length of aluminum wire is warmed 10.0°C by an electric current. How much heat was generated by the current? ($c_{aluminum} = 0.22$ kcal/kgC°)

3. A 1,000.0 kg car with a speed of 90.0 km/h brakes to a stop. How many calories of heat are generated by the brakes as a result?

4. A 1.0 kg metal head of a geology hammer strikes a solid rock with a velocity of 5.0 m/s. Assuming all the energy is retained by the hammer head, how much will its temperature increase? ($c_{head} = 0.11$ kcal/kgC°)

5. A 60.0 kg person will need to climb a 10.0 m stairway how many times to "work off" each excess Cal (kcal) consumed?

6. A 50.0 g silver spoon at 20.0°C is placed in a cup of coffee at 90.0°C. How much heat does the spoon absorb from the coffee to reach a temperature of 89.0°C?

7. If the silver spoon placed in the coffee in problem 6 causes it to cool 0.75°C, what is the mass of the coffee? (Assume $c_{coffee} = 1.0$ cal/gC°)

8. How many minutes would be required for a 300.0 W immersion heater to heat 250.0 g of water from 20.0°C to 100.0°C?

9. A 200.0 g china serving bowl is warmed 65.0°C when it absorbs 2.6 kcal of heat from a serving of hot food. What is the specific heat of the china dish?

10. A 500.0 g pot of water at room temperature (20.0°C) is placed on a stove. How much heat is required to change this water to steam at 100.0°C?

11. Spent steam from an electric generating plant leaves the turbines at 120.0°C and is cooled to 90.0°C liquid water by water from a cooling tower in a heat exchanger. How much heat is removed by the cooling tower water for each kilogram of spent steam?

12. Lead is a soft, dense metal with a specific heat of 0.028 kcal/kgC°, a melting point of 328.0°C, and a heat of fusion of 5.5 kcal/kg. How much heat must be provided to melt a 250.0 kg sample of lead with a temperature of 20.0°C?

Wave Motions and Sound

Compared to the sounds you hear on a calm day in the woods, the sounds from a waterfall can carry up to a million times more energy.

O U T L I N E

Sometimes you can feel the floor of a building shake for a moment when something heavy is dropped. You can also feel prolonged vibrations in the ground when a nearby train moves by. The floor of a building and the ground are solids that transmit vibrations from a disturbance. Vibrations are common in most solids because the solids are elastic, having a tendency to rebound, or snap back, after a force or an impact deforms them. Usually you cannot see the vibrations in a floor or the ground, but you sense they are there because you can feel them.

There are many examples of vibrations that you can see. You can see the rapid blur of a vibrating guitar string (figure 5.1). You can see the vibrating up-and-down movement of a bounced-upon diving board. Both the vibrating guitar string and the diving board set up a vibrating motion of air that you identify as a sound. You cannot see the vibrating motion of the air, but you sense it is there because you hear sounds.

There are many kinds of vibrations that you cannot see but can sense. Heat, as you have learned, is associated with molecular vibrations that are too rapid and too tiny for your senses to detect other than as an increase in temperature. Other invisible vibrations include electrons that vibrate, generating spreading electromagnetic radio waves or visible light. Thus vibrations take place as an observable motion of objects but are also involved in sound, heat, electricity, and light. The vibrations involved in all these phenomena are alike in many ways and all involve energy. Therefore, many topics of science are concerned with vibrational motion. In this chapter you will learn about the nature of vibrations and how they produce waves in general. These concepts will be applied to sound in this chapter and to electricity and electromagnetic radiation in later chapters.

FORCES AND ELASTIC MATERIALS

If you drop a rubber ball it bounces because it is capable of recovering its shape when it hits the floor. A ball of clay, on the other hand, does not recover its shape and remains a flattened blob on the floor. An *elastic* material is one that is capable of recovering its shape after a force deforms it. A rubber ball is elastic and a ball of clay is not elastic. You know a metal spring is elastic because you can stretch it or compress it and it always recovers its shape.

There is a direct relationship between the extent of stretching or compression of a spring and the amount of force applied to it. A large force stretches a spring a lot; a small force stretches it a little. As long as the applied force does not exceed the elastic limit of the spring, it will always return to its original shape when you remove the applied force. There are three important considerations about the applied force and the response of the spring:

1. The greater the applied force, the greater the compression or stretch of the spring from its original shape.
2. The spring appears to have an *internal restoring force*, which returns it to its original shape.
3. The farther the spring is pushed or pulled, the *stronger* the restoring force that returns the spring to its original shape.

Forces and Vibrations

A **vibration** is a back-and-forth motion that repeats itself. Such a motion is not restricted to any particular direction, and can be in many different directions at the same time. Almost any solid can be made to vibrate if it is elastic. To see how forces are involved in vibrations, consider the spring and mass in figure 5.2. The spring and mass are arranged so that the mass can freely move back and forth on a frictionless surface. When the mass has not been disturbed, it is at rest at an *equilibrium position* (figure 5.2A). At the equilibrium position the spring is not compressed or stretched, so it applies no force on the mass. If, however, the mass is pulled to the right (figure 5.2B) the spring

Figure 5.1
Vibrations are common in many elastic materials, and you can see and hear the results of many in your surroundings. Other vibrations in your surroundings, such as those involved in heat, electricity, and light, are invisible to the senses.

is stretched and applies a restoring force on the mass toward the left. The farther the mass is displaced, the greater the stretch of the spring and thus the greater the restoring force. The restoring force is proportional to the displacement and is in the opposite direction of the applied force.

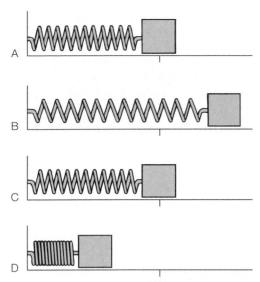

Figure 5.2

A mass on a frictionless surface is at rest at an equilibrium position (*A*) when undisturbed. When the spring is stretched (*B*) or compressed (*D*), then released (*C*), the mass vibrates back and forth because restoring forces pull opposite to and proportional to the displacement.

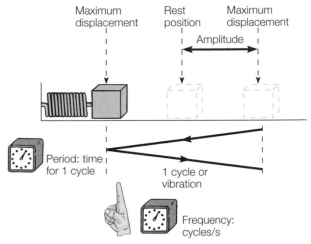

Figure 5.3

A vibrating mass attached to a spring is displaced from rest, or equilibrium position, and then released. The maximum displacement is called the *amplitude* of the vibration. A cycle is one complete vibration. The period is the time required for one complete cycle. The frequency is a count of how many cycles it completes in 1 s.

If the mass is now released, the restoring force is the only force acting (horizontally) on the mass, so it accelerates back toward the equilibrium position. This force will continuously decrease until the moving mass arrives back at the equilibrium position, where the force is zero. The mass will have a maximum velocity when it arrives, however, so it overshoots the equilibrium position and continues moving to the left (figure 5.2C). As it moves to the left of the equilibrium position, it compresses the spring, which exerts an increasing force on the mass. The moving mass comes to a temporary halt (figure 5.2D), but now the restoring force again starts it moving back toward the equilibrium position. The whole process repeats itself again and again as the mass moves back and forth over the same path.

The periodic vibration, or oscillation, of the mass is similar to many vibrational motions found in nature called *simple harmonic motion*. Simple harmonic motion is defined as the vibratory motion that occurs when there is a restoring force opposite to and proportional to a displacement.

The vibrating mass and spring system will continue to vibrate for a while, slowly decreasing with time until the vibrations stop completely. The slowing and stopping is due to air resistance and internal friction. If these could be eliminated or compensated for with additional energy, the mass would continue to vibrate with a repeating, or *periodic*, motion.

Describing Vibrations

A vibrating mass is described by measuring several variables (figure 5.3). The extent of displacement from the equilibrium position is called the **amplitude.** A vibration that has a mass displaced a greater distance from equilibrium thus has a greater amplitude than a vibration with less displacement.

A complete vibration is called a **cycle.** A cycle is the movement from some point, say the far left, all the way to the far right, and back to the same point again, the far left in this example. The **period** (*T*) is simply the time required to complete one cycle. For example, suppose 0.1 s is required for an object to move through one complete cycle, to complete the back-and-forth motion from one point, then back to that point. The period of this vibration is 0.1 s.

Sometimes it is useful to know how frequently a vibration completes a cycle every second. The number of cycles per second is called the **frequency** (*f*). For example, a vibrating object moves through 10 cycles in 1 s. The frequency of this vibration is 10 cycles per second. Frequency is measured in a unit called a **hertz** (Hz). The unit for a hertz is 1/s because a cycle does not have dimensions. Thus a frequency of 10 cycles per second is referred to as 10 hertz or 10 1/s.

The period and frequency are two ways of describing the time involved in a vibration. Since the period (*T*) is the total time involved in one cycle, and the frequency (*f*) is the number of cycles per second, the relationship is

$$T = \frac{1}{f}$$

equation 5.1

or

$$f = \frac{1}{T}$$

equation 5.2

You can obtain a graph of a vibrating object, which makes it easier to measure the amplitude, period, and frequency. If a pen is fixed to a vibrating mass and a paper is moved beneath it at a steady rate, it will draw a curve, as shown in figure 5.4. The

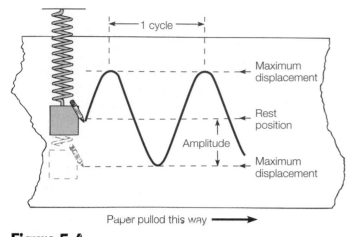

Figure 5.4
A graph of simple harmonic motion is described by a sinusoidal curve.

greater the amplitude of the vibrating mass, the greater the height of this curve. The greater the frequency, the closer together the peaks and valleys. Note the shape of this curve. This shape is characteristic of simple harmonic motion and is called a *sinusoidal,* or sine, graph. It is so named because it is the same shape as a graph of the sine function in trigonometry.

WAVES

A vibration can be a repeating, or *periodic,* type of motion that disturbs the surroundings. A *pulse* is a disturbance of a single event of short duration. Both pulses and periodic vibrations can create a physical *wave* in the surroundings. A wave is a disturbance that moves through a medium such as a solid or the air. A heavy object dropped on the floor, for example, makes a pulse that sends a mechanical wave that you feel. It might also make a sound wave in the air that you hear. In either case, the medium that transported a wave (solid floor or air) returns to its normal state after the wave has passed. The medium does not travel from place to place, the wave does. Two major considerations about a wave are that (1) a wave is a traveling disturbance, and (2) a wave transports energy.

You can observe waves when you drop a rock into a still pool of water. The rock pushes the water into a circular mound as it enters the water. Since it is forcing the water through a distance, it is doing work to make the mound. The mound starts to move out in all directions, in a circle, leaving a depression behind. Water moves into the depression, and a circular wave—mound and depression—moves from the place of disturbance outward. Any floating object in the path of the wave, such as a leaf, exhibits an up-and-down motion as the mound and depression of the wave pass. But the leaf merely bobs up and down and after the wave has passed, it is much in the same place as before the wave. Thus, it was the disturbance that traveled across the water, not the water itself. If the wave reaches a leaf floating near the edge of the water, it may push the leaf up and out of the water, doing work on the leaf. Thus, the wave is a moving disturbance that transfers energy from one place to another.

Kinds of Waves

If you could see the motion of an individual water molecule near the surface as a water wave passed, you would see it trace out a circular path as it moves up and over, down and back. This circular motion is characteristic of the motion of a particle reacting to a water wave disturbance. There are other kinds of waves, and each involves particles in a characteristic motion.

A **longitudinal wave** is a disturbance that causes particles to move closer together or farther apart in the same direction that the wave is moving. If you attach one end of a coiled spring to a wall and pull it tight, you will make longitudinal waves in the spring if you grasp the spring and then move your hand back and forth parallel to the spring. Each time you move your hand toward the length of the spring, a pulse of closer-together coils will move across the spring (figure 5.5A). Each time you pull your hand back, a pulse of farther-apart coils will move across the spring. The coils move back and forth in the same direction that the wave is moving, which is the characteristic movement in reaction to a longitudinal wave.

You will make a different kind of wave in the stretched spring if you now move your hand up and down perpendicular to the length of the spring. This creates a **transverse wave.** A transverse wave is a disturbance that causes motion perpendicular to the direction that the wave is moving. Particles responding to a transverse wave do not move closer together or farther apart in response to the disturbance; rather, they vibrate back and forth or up and down in a direction perpendicular to the direction of the wave motion (see figure 5.5B).

Whether you make mechanical longitudinal or transverse waves depends not only on the nature of the disturbance creating the waves, but also on the nature of the medium. Mechanical transverse waves can move through a material only if there is some interaction, or attachment, between the molecules making

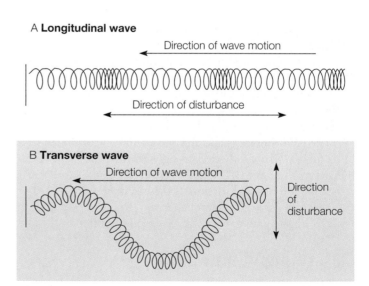

Figure 5.5
(*A*) Longitudinal waves are created in a spring when the free end is moved back and forth parallel to the spring. (*B*) Transverse waves are created in a spring when the free end is moved up and down.

up the medium. In a gas, for example, the molecules move about freely without attachments to one another. A pulse can cause these molecules to move closer together or farther apart, so a gas can carry a longitudinal wave. But if a gas molecule is caused to move up or down, there is no reason for other molecules to do the same, since they are not attached. Thus, a gas will carry mechanical longitudinal waves but not mechanical transverse waves. Likewise a liquid will carry mechanical longitudinal waves but not mechanical transverse waves since the liquid molecules simply slide past one another. The surface of a liquid, however, is another story because of surface tension. A surface water wave is, in fact, a combination of longitudinal and transverse wave patterns that produce the circular motion of a disturbed particle. Solids can and do carry both mechanical longitudinal and transverse waves because of the strong attachments between the molecules.

Waves in Air

Waves that move through the air are longitudinal, so sound waves must be longitudinal waves. A familiar situation will be used to describe the nature of a longitudinal wave moving through air before considering sound specifically. The situation involves a small room with no open windows and two doors that open into the room. When you open one door into the room the other door closes. Why does this happen? According to the kinetic molecular theory, the room contains many tiny, randomly moving gas molecules that make up the air. As you opened the door, it pushed on these gas molecules, creating a jammed-together zone of molecules immediately adjacent to the door. This jammed-together zone of air now has a greater density and pressure, which immediately spreads outward from the door as a pulse. The disturbance is rapidly passed from molecule to molecule, and the pulse of compression spreads through the room. The pulse of greater density and increased pressure of air reached the door at the other side of the room, and the composite effect of the molecules impacting the door, that is, the increased pressure, caused it to close.

If the door at the other side of the room does not latch, you can probably cause it to open again by pulling on the first door quickly. By so doing, you send a pulse of thinned-out molecules of lowered density and pressure. The door you pulled quickly pushed some of the molecules out of the room. Other molecules quickly move into the region of less pressure, then back to their normal positions. The overall effect is the movement of a thinned-out pulse that travels through the room. When the pulse of slightly reduced pressure reaches the other door, molecules exerting their normal pressure on the other side of the door cause it to move. After a pulse has passed a particular place, the molecules are very soon homogeneously distributed again due to their rapid, random movement.

If you were to swing a door back and forth it would be a vibrating object. As it vibrates back and forth it would have a certain frequency in terms of the number of vibrations per second. As the vibrating door moves toward the room, it creates a pulse of jammed-together molecules called a **condensation** (or compression) that quickly moves throughout the room. As the vibrating door moves away from the room, a pulse of thinned-out molecules called a **rarefaction** quickly moves throughout the room. The vibrating door sends repeating pulses of condensation (increased density and pressure) and rarefaction (decreased density and pressure) through the room as it moves back and forth (figure 5.6). You know that the pulses transmit energy because they produce movement, or do work on, the other door. Individual molecules execute a harmonic motion about their equilibrium position and can do work on a movable object. Energy is thus transferred by this example of longitudinal waves.

Hearing Waves in Air

You cannot hear a vibrating door because the human ear normally hears sounds originating from vibrating objects with a frequency between 20 and 20,000 Hz. Longitudinal waves with frequencies less than 20 Hz are called **infrasonic.** You usually *feel* sounds below 20 Hz rather than hearing them, particularly if you are listening to a good sound system. Longitudinal waves above 20,000 Hz are called **ultrasonic.** Although 20,000 Hz is usually considered the upper limit of hearing, the actual limit varies from person to person and becomes lower and lower with increasing age. Humans do not hear infrasonic nor ultrasonic sounds, but various animals have different limits. Dogs, cats,

Ultrasonic waves are mechanical waves that have frequencies above the normal limit of hearing of the human ear. The arbitrary upper limit is about 20,000 Hz, so an ultrasonic wave has a frequency of 20,000 Hz or greater. Intense ultrasonic waves are used in many ways in industry and medicine.

Industrial and commercial applications of ultrasound utilize lower-frequency ultrasonic waves in the 20,000 to 60,000 Hz range. Commercial devices that send ultrasound through the air include burglar alarms and rodent repellers. An ultrasonic burglar alarm sends ultrasonic spherical waves through the air of a room. The device is adjusted to ignore echoes from the contents of the room. The presence of a person provides a new source of echoes, which activates the alarm. Rodents emit ultrasonic frequencies up to 150,000 Hz that are used in rodent communication when they are disturbed or during aggressive behavior. The ultrasonic rodent repeller generates ultrasonic waves of similar frequency. Other commercial applications of ultrasound include sonar and depth measurements, cleaning and drilling, welding plastics and metals, and material flaw detection. Ultrasonic cleaning baths are used to remove dirt and foreign matter from solid surfaces, usually within a liquid solvent. The ultrasonic waves create vapor bubbles in the liquid,

which vibrate and emit audible and ultrasonic sound waves. The audible frequencies are often heard as a hissing or frying sound.

Medical applications of ultrasound use frequencies in the 1,000,000 to 20,000,000 Hz range. Ultrasound in this frequency range cannot move through the air because the required displacement amplitudes of the gas molecules in the air are less than the average distance between the molecules. Thus, a gas molecule that is set into motion in this frequency range cannot collide with other gas molecules to transmit the energy of the wave. Intense ultrasound is used for cleaning teeth and disrupting kidney stones. Less intense ultrasound is used for therapy (heating and reduction of pain). The least intense ultrasounds are used for diagnostic imaging. The largest source of exposure of humans to ultrasound is for the purpose of ultrasonic diagnostic imaging, particularly in fertility and pregnancy cases. In the United States, ultrasonic scans are routinely done when the baby is 12 weeks to confirm the due date, and around 20 weeks to check for structural abnormalities.

Originally developed in the late 1950s, ultrasonic medical machines have been improved and today make images of outstanding detail and clarity. One type of ultrasonic medical machine

uses a transducer probe, which emits an ultrasonic pulse that passes into the body. Echoes from the internal tissues and organs are reflected back to the transducer, which sends the signals to a computer. Another pulse of ultrasound is then sent out after the echoes from the first pulse have returned. The strength of and number of pulses per second vary with the application, ranging from hundreds to thousands of pulses per second. The computer constructs a picture from the returning echoes, showing an internal view without the use of more dangerous X rays.

Ultrasonic scanners have been refined to the point that the surface of the ovaries can now be viewed, showing the number and placement of developing eggs. The ovary scan is typically used in conjunction with fertility-stimulating drugs, where multiple births are possible, and to identify the exact time of ovulation. As early as four weeks after conception, the ultrasonic scanner is used to identify and monitor the fetus. By the thirteenth week, an ultrasonic scan can show the fetal heart movement, bone and skull size, and internal organs. The ultrasonic scanner is often used to show the position of the fetus and placenta for the purpose of amniocentesis, which involves withdrawing a sample of amniotic fluid from the uterus for testing.

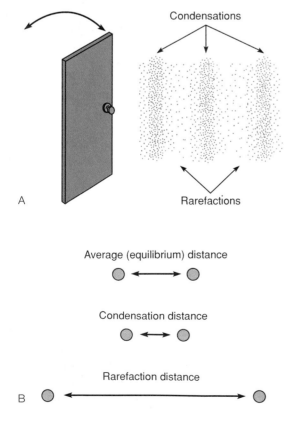

A

Condensations

Rarefactions

Average (equilibrium) distance

Condensation distance

Rarefaction distance

B

Figure 5.6

(*A*) Swinging the door inward produces pulses of increased density and pressure called *condensations*. Pulling the door outward produces pulses of decreased density and pressure called *rarefactions*. (*B*) In a condensation, the average distance between gas molecules is momentarily decreased as the pulse passes. In a rarefaction, the average distance is momentarily increased.

rats, and bats can hear higher frequencies than humans. Dogs can hear an ultrasonic whistle when a human hears nothing, for example. Some bats make and hear sounds of frequencies up to 100,000 Hz as they navigate and search for flying insects in total darkness. Scientists discovered recently that elephants communicate with extremely low-frequency sounds over distances of several kilometers. Humans cannot detect such low-frequency sounds. This raises the possibility of infrasonic waves that other animals can detect that we cannot.

A tuning fork that vibrates at 260 Hz makes longitudinal waves much like the swinging door, but these longitudinal waves are called *sound waves* because they are within the frequency range of human hearing. The prongs of a struck tuning fork vibrate, moving back and forth. This is more readily observed if the prongs of the fork are struck, then held against a sheet of paper or plunged into a beaker of water. In air, the vibrating prongs first

Connections . . .

The Human Ear

Box figure 5.1 shows the anatomy of the ear. The sound that arrives at the ear is first funneled by the external ear to the tympanum, also known as the eardrum. The cone-shaped nature of the external ear focuses sound on the tympanum and causes it to vibrate at the same frequency as the sound waves reaching it. Attached to the tympanum are three tiny bones known as the malleus (hammer), incus (anvil), and stapes (stirrup). The malleus is attached to the tympanum, the incus is attached to the malleus and stapes, and the stapes is attached to a small, membrane-covered opening called the oval window in a snail-shaped structure known as the cochlea. The vibration of the tympanum causes the tiny bones (malleus, incus, and stapes) to vibrate, and they, in turn, cause a corresponding vibration in the membrane of the oval window.

The cochlea of the ear is the structure that detects sound, and it consists of a snail-shaped set of fluid-filled tubes. When the oval window vibrates, the fluid in the cochlea begins to move, causing a membrane in the cochlea, called the basilar membrane, to vibrate. High-pitched, short-wavelength sounds cause the basilar membrane to vibrate at the base of the cochlea near the oval window. Low-pitched, long-wavelength sounds vibrate the basilar membrane far from the oval window. Loud sounds cause the basilar membrane to vibrate more vigorously than do faint sounds. Cells on this membrane depolarize when they are stimulated by its vibrations. Since they synapse with neurons, messages can be sent to the brain.

Because sounds of different wavelengths stimulate different portions of the cochlea, the brain is able to determine the pitch of a sound. Most sounds consist of a mixture of pitches that are heard. Louder sounds stimulate the membrane more forcefully, causing the sensory cells in the cochlea to send more nerve impulses per second. Thus, the brain is able to perceive the loudness of various sounds, as well as the pitch.

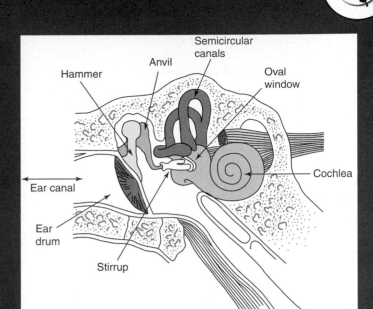

Box Figure 5.1
A schematic sketch of the human ear.

Associated with the cochlea are two fluid-filled chambers and a set of fluid-filled tubes called the semicircular canals. These structures are not involved in hearing but are involved in maintaining balance and posture. In the walls of these canals and chambers are cells similar to those found on the basilar membrane. These cells are stimulated by movements of the head and by the position of the head with respect to the force of gravity. The constantly changing position of the head results in sensory input that is important in maintaining balance.

---

move toward you, pushing the air molecules into a condensation of increased density and pressure. As the prongs then move back, a rarefaction of decreased density and pressure is produced. The alternation of increased and decreased pressure pulses moves from the vibrating tuning fork and spreads outward equally in all directions, much like the surface of a rapidly expanding balloon (figure 5.7). Your eardrum is forced in and out by the pulses it receives. It now vibrates with the same frequency as the tuning fork. The vibrations of the eardrum are transferred by three tiny bones to a fluid in a coiled chamber. Here, tiny hairs respond to the frequency and size of the disturbance, activating nerves that transmit the information to the brain. The brain interprets a frequency as a sound with a certain **pitch.** High-frequency sounds are interpreted as high-pitched musical notes, for example, and low-frequency sounds are interpreted as low-pitched musical notes. The brain then selects certain sounds from all you hear, and you "tune" to certain ones, enabling you to listen to whatever

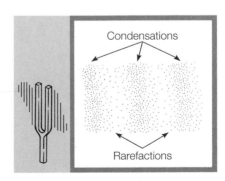

Figure 5.7
A vibrating tuning fork produces a series of condensations and rarefactions that move away from the tuning fork. The pulses of increased and decreased pressure reach your ear, vibrating the eardrum. The ear sends nerve signals to the brain about the vibrations, and the brain interprets the signals as sounds.

sounds you want while ignoring the background noise, which is made up of all the other sounds.

DESCRIBING WAVES

A tuning fork vibrates with a certain frequency and amplitude, producing a longitudinal wave of alternating pulses of increased-pressure condensations and reduced-pressure rarefactions. A graph of the frequency and amplitude of the vibrations is shown in figure 5.8A, and a representation of the condensations and rarefactions is shown in figure 5.8B. The wave pattern can also be represented by a graph of the changing air pressure of the traveling sound wave, as shown in figure 5.8C. This graph can be used to define some interesting concepts associated with sound waves. Note the correspondence between the (1) amplitude, or displacement, of the vibrating prong, (2) the pulses of condensations and rarefactions, and (3) the changing air pressure. Note also the correspondence between the frequency of the vibrating prong and the frequency of the wave cycles.

Figure 5.9 shows the terms commonly associated with waves from a continuously vibrating source. The wave *crest* is the maximum disturbance from the undisturbed (rest) position. For a sound wave, this would represent the maximum increase of air pressure. The wave *trough* is the maximum disturbance in the opposite direction from the rest position. For a sound wave, this would represent the maximum decrease of air pressure. The *amplitude* of a wave is the displacement from rest to the crest *or* from rest to the trough. The time required for a

Figure 5.9

Here are some terms associated with periodic waves. The *wavelength* is the distance from a part of one wave to the same part in the next wave, such as from one crest to the next. The *amplitude* is the displacement from the rest position. The *period* is the time required for a wave to repeat itself, that is, the time for one complete wavelength to move past a given location.

wave to repeat itself is the *period* (*T*). To repeat itself means the time required to move through one full wave, such as from the crest of one wave to the crest of the next wave. This length in which the wave repeats itself is called the **wavelength** (the symbol is λ, which is the Greek letter lambda). Wavelength is measured in centimeters or meters just like any other length.

There is a relationship between the wavelength, period, and speed of a wave. The relationship is

$$v = \lambda f$$

equation 5.3

which we will call the *wave equation*. This equation tells you that the velocity of a wave can be obtained from the product of the wavelength and the frequency. Note that it also tells you that the wavelength and frequency are inversely proportional at a given velocity.

SOUND WAVES

The transmission of a sound wave requires a medium, that is, a solid, liquid, or gas to carry the disturbance. Therefore, sound does not travel through the vacuum of outer space, since there is nothing to carry the vibrations from a source. The nature of the molecules making up a solid, liquid, or gas determines how well or how rapidly the substance will carry sound waves. The two variables are (1) the inertia of the molecules and (2) the strength of the interaction, if the molecules are attached to one another. Thus, hydrogen gas, with the least massive molecules with no interaction or attachments, will carry a sound wave at 1,284 m/s (4,213 ft/s) when the temperature is 0°C. More massive helium gas molecules have more inertia and carry a sound wave at only 965 m/s (3,166 ft/s) at the same temperature. A solid, however, has molecules that are strongly attached so vibrations are passed rapidly from molecule to molecule. Steel, for example, is highly elastic, and sound will move through a steel rail at 5,940 m/s (19,488 ft/s). Thus there is a reason for the old saying, "Keep your ear to the ground," because sounds move through solids more rapidly than through a gas (table 5.1).

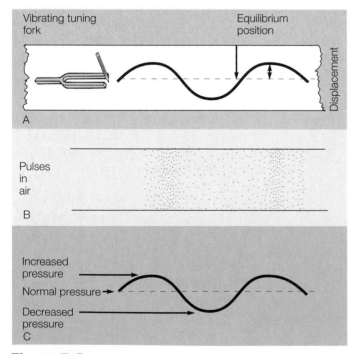

Figure 5.8

Compare the (*A*) back-and-forth vibrations of a tuning fork with (*B*) the resulting condensations and rarefactions that move through the air and (*C*) the resulting increases and decreases of air pressure on a surface that intercepts the condensations and rarefactions.

Table 5.1 Speed of Sound in Various Materials		
Medium	**m/s**	**ft/s**
Carbon dioxide (0°C)	259	850
Dry air (0°C)	331	1,087
Helium (0°C)	965	3,166
Hydrogen (0°C)	1,284	4,213
Water (25°C)	1,497	4,911
Seawater (25°C)	1,530	5,023
Lead	1,960	6,430
Glass	5,100	16,732
Steel	5,940	19,488

Velocity of Sound in Air

Most people have observed that sound takes some period of time to move through the air. If you watch a person hammering on a roof a block away, the sounds of the hammering are not in sync with what you see. Light travels so rapidly that you can consider what you see to be simultaneous with what is actually happening for all practical purposes. Sound, however, travels much more slowly and the sounds arrive late in comparison to what you are seeing. This is dramatically illustrated by seeing a flash of lightning, then hearing thunder seconds later. Perhaps you know of a way to estimate the distance to a lightning flash by timing the interval between the flash and boom. You will learn a precise way to measure this distance shortly.

The air temperature influences how rapidly sound moves through the air. The gas molecules in warmer air have a greater kinetic energy than those of cooler air. The molecules of warmer air therefore transmit an impulse from molecule to molecule more rapidly. More precisely, the speed of a sound wave increases 0.60 m/s (2.0 ft/s) for *each* Celsius degree increase in temperature. In *dry* air at sea-level density (normal pressure) and 0°C (32°F), the velocity of sound is about 331 m/s (1,087 ft/s).

Refraction and Reflection

When you drop a rock into a still pool of water, circular patterns of waves move out from the disturbance. These water waves are on a flat, two-dimensional surface. Sound waves, however, move in three-dimensional space like a rapidly expanding balloon. Sound waves are *spherical waves* that move outward from the source. Spherical waves of sound move as condensations and rarefactions from a continuously vibrating source at the center. If you identify the same part of each wave in the spherical waves, you have identified a *wave front*. For example, the crests of each condensation could be considered as a wave front. From one wave front to the next, therefore, identifies one complete wave or wavelength. At some distance from the source, a small part of a spherical wave front can be considered a *linear wave front* (figure 5.10).

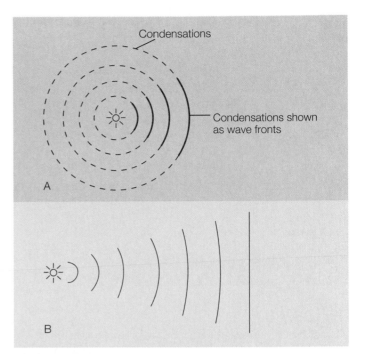

Figure 5.10

(*A*) Spherical waves move outward from a sounding source much as a rapidly expanding balloon. This two-dimensional sketch shows the repeating condensations as spherical wave fronts. (*B*) Some distance from the source, a spherical wave front is considered a linear, or plane, wave front.

Waves move within a homogeneous medium such as a gas or a solid at a fairly constant rate but gradually lose energy to friction. When a wave encounters a different condition (temperature, humidity, or nature of material), however, drastic changes may occur rapidly. The division between two physical conditions is called a *boundary*. Boundaries are usually encountered (1) between different materials or (2) between the same materials with different conditions. An example of a wave moving between different materials is a sound made in the next room that moves through the air to the wall and through the wall to the air in the room where you are. The boundaries are air-wall and wall-air. If you have ever been in a room with "thin walls," it is obvious that sound moved through the wall and air boundaries.

An example of sound waves moving through the same material with different conditions is found when a wave front moves through air of different temperatures. Since sound travels faster in warm air than in cold air, the wave front becomes bent. The bending of a wave front between boundaries is called *refraction*. Refraction changes the direction of travel of a wave front. Consider, for example, that on calm, clear nights the air near the earth's surface is cooler than air farther above the surface. Air at rooftop height above the surface might be four or five degrees warmer under such ideal conditions. Sound will travel faster in the higher, warmer air than it will in the lower, cooler air close to the surface. A wave front will therefore become bent, or refracted, toward the ground on a cool night and you will be able to hear sounds from farther away than on warm nights (figure 5.11A). The opposite process occurs during the day as the

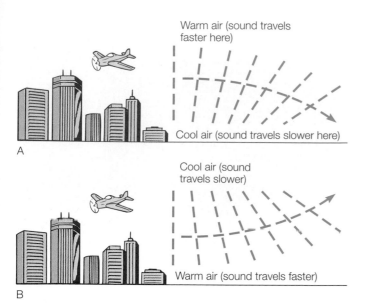

Warm air (sound travels faster here)

Cool air (sound travels slower here)

A

Cool air (sound travels slower)

Warm air (sound travels faster)

B

Figure 5.11

(*A*) Since sound travels faster in warmer air, a wave front becomes bent, or refracted, toward the earth's surface when the air is cooler near the surface. (*B*) When the air is warmer near the surface, a wave front is refracted upward, away from the surface.

earth's surface becomes warmer from sunlight (figure 5.11B). Wave fronts are refracted upward because part of the wave front travels faster in the warmer air near the surface. Thus, sound does not seem to carry as far in the summer as it does in the winter. What is actually happening is that during the summer the wave fronts are refracted away from the ground before they travel very far.

When a wave front strikes a boundary that is parallel to the front the wave may be absorbed, transmitted, or undergo *reflection,* depending on the nature of the boundary medium, or the wave may be partly absorbed, partly transmitted, partly reflected, or any combination thereof. Some materials, such as hard, smooth surfaces, reflect sound waves more than they absorb them. Other materials, such as soft, ruffly curtains, absorb sound waves more than they reflect them. If you have ever been in a room with smooth, hard walls and with no curtains, carpets, or furniture, you know that sound waves may be reflected several times before they are finally absorbed.

Do you sing in the shower? Many people do because the tone is more pleasing than singing elsewhere. The walls of a shower are usually hard and smooth, reflecting sounds back and forth several times before they are absorbed. The continuation of many reflections causes a tone to gain in volume. Such mixing of reflected sounds with the original is called **reverberation.** Reverberation adds to the volume of a tone, and it is one of the factors that determines the acoustical qualities of a room, lecture hall, or auditorium. An open-air concert sounds flat without the reverberation of an auditorium, and is usually enhanced electronically to make up for the lack of reflected sounds. Too much reverberation in a room or classroom is not good since the spoken word is not as sharp. Sound-absorbing materials are therefore used on the walls and floors where clear, distinct speech is

Figure 5.12

This closed-circuit TV control room is acoustically treated by covering the walls with sound-absorbing baffles.

important (figure 5.12). The carpet and drapes you see in a movie theater are not decorator items, but are there to absorb sounds.

If a reflected sound arrives after 0.10 s, the human ear can distinguish the reflected sound from the original sound. A reflected sound that can be distinguished from the original is called an **echo.** Thus, a reflected sound that arrives before 0.10 s is perceived as an increase in volume and is called a reverberation, but a sound that arrives after 0.10 s is perceived as an echo.

Sound wave echoes are measured to determine the depth of water or to locate underwater objects by a *sonar* device. The word *sonar* is taken from *so*und *na*vigation *r*anging. The device generates an underwater ultrasonic sound pulse, then measures the elapsed time for the returning echo. Sound waves travel at about 1,531 m/s (5,023 ft/s) in seawater at 25°C (77°F). A 1 s lapse between the ping of the generated sound and the echo return would mean that the sound traveled 5,023 ft for the round trip. The bottom would be half this distance below the surface (figure 5.13).

Interference

Waves interact with a boundary much as a particle would, reflecting or refracting because of the boundary. A moving ball, for example, will bounce from a surface at the same angle it strikes the surface, just as a wave does. A particle or a ball, however, can be in only one place at a time, but waves can be spread over a distance at the same time. You know this since many different people in different places can hear the same sound at the same time.

Another difference between waves and particles is that two or more waves can exist in the same place at the same time. When two patterns of waves meet, they pass through each other without refracting or reflecting. However, at the place where they meet the waves interfere with each other, producing a *new* disturbance. This new disturbance has a different amplitude, which is the algebraic sum of the amplitudes of the two separate wave patterns. If the wave crests or wave troughs ar-

A Echo

$t = 0.10$ s
$v = 343$ m/s
$d = 34$ m
$\frac{d}{2} = 17$ m

$\frac{1}{2} \cdot 34$ m

B Sonar

$t = 1.0$ s
$v = 5{,}023$ ft/s
$d = 5{,}023$ ft
depth $= \frac{1}{2} \cdot d$

$\frac{1}{2} \cdot 5{,}023$ ft

Figure 5.13

(*A*) At room temperature, sound travels at 343 m/s. In 0.10 s, sound would travel 34 m. Since the sound must travel to a surface and back in order for you to hear an echo, the distance to the surface is one-half the total distance. (*B*) Sonar measures a depth by measuring the elapsed time between an ultrasonic sound pulse and the echo. The depth is one-half the round trip.

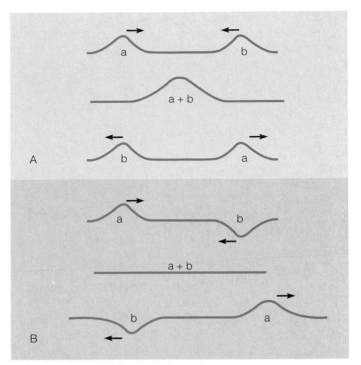

Figure 5.14

(*A*) Constructive interference occurs when two equal, in-phase waves meet. (*B*) Destructive interference occurs when two equal, out-of-phase waves meet. In both cases, the wave displacements are superimposed when they meet, but they then pass through one another and return to their original amplitudes.

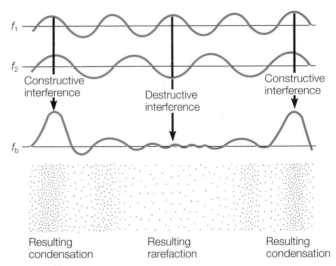

Figure 5.15

Two waves of equal amplitude but slightly different frequencies interfere destructively and constructively. The result is an alternation of loudness called a *beat*.

rive at the same place at the same time, the two waves are said to be *in phase.* The result of two waves arriving in phase is a new disturbance with a crest and trough that has greater displacement than either of the two separate waves. This is called *constructive interference* (figure 5.14A). If the trough of one wave arrives at the same place and time as the crest of another wave, the waves are completely *out of phase.* When two waves are completely out of phase, the crest of one wave (positive displacement) will cancel the trough of the other wave (negative displacement), and the result is zero total disturbance, or no wave. This is called *destructive interference* (figure 5.14B). If the two sets of wave patterns do not have the exact same amplitudes or wavelengths, they will be neither completely in phase nor completely out of phase. The result will be partly constructive or destructive interference, depending on the exact nature of the two wave patterns.

Suppose that two vibrating sources produce sounds that are in phase, equal in amplitude, and equal in frequency. The resulting sound will be increased in volume because of constructive interference. But suppose the two sources are slightly different in frequency, for example, 350 Hz and 352 Hz. You will hear a regularly spaced increase and decrease of sound known as **beats.** Beats occur because the two sound waves experience al-

ternating constructive and destructive interferences (figure 5.15). The phase relationship changes because of the difference in frequency, as you can see in the illustration. These alternating constructive and destructive interference zones are moving from the source to the receiver, and the receiver hears the results as a rapidly rising and falling sound level. The beat frequency is the difference between the frequencies of the two sources. A 352 Hz source and 350 Hz source sounded together would result in a beat frequency of 2 Hz. Thus, as two frequencies are closer and closer together, fewer beats will be heard per second. You may be familiar with the phenomenon of beats if you have ever flown in an airplane with two engines. If one engine is running slightly faster than the other, you hear a slow beat. The beat frequency (f_b) is equal to the absolute difference in frequency of two interfering waves with slightly different frequencies, or

$$f_b = f_2 - f_1$$

equation 5.4

ENERGY AND SOUND

All waves involve the transportation of energy, including sound waves. The vibrating mass and spring in figure 5.2 vibrate with an amplitude that depends on how much work you did on the mass in moving it from its equilibrium position. More work on the mass results in a greater displacement and a greater amplitude of vibration. A vibrating object that is producing sound waves will produce more intense condensations and rarefactions if it has a greater amplitude. The intensity of a sound wave is a measure of the energy the sound wave is carrying (figure 5.16). **Intensity** is defined as the power (in watts) transmitted by a wave to a unit area (in square meters) that is perpendicular to the waves.

Loudness

The *loudness* of a sound is a subjective interpretation that varies from person to person. Loudness is also related to (1) the energy of a vibrating object, (2) the condition of the air the sound wave travels through, and (3) the distance between you and the vibrating source. Furthermore, doubling the amplitude of the vibrating source will quadruple the *intensity* of the resulting sound wave, but the sound will not be perceived as four times as loud. The relationship between perceived loudness and the intensity of a sound wave is not a linear relationship. In fact, a sound that is perceived as twice as loud requires ten times the intensity, and quadrupling the loudness requires a one-hundred-fold increase in intensity.

The human ear is very sensitive, capable of hearing sounds with intensities as low as 10^{-12} W/m^2 and is not made uncomfortable by sound until the intensity reaches about 1 W/m^2. The second intensity is a million million (10^{12}) times greater than the first. Within this range, the subjective interpretation of intensity seems to vary by powers of ten. This observation led to the development of the **decibel scale** to measure the intensity level. The scale is a ratio of the intensity level of a given sound to the threshold of hearing, which is defined as 10^{-12} W/m^2 at 1,000 Hz. In keeping with the power-of-ten subjective interpretations of intensity, a logarithmic scale is used rather than a linear scale. Originally the scale was the logarithm of the ratio of the intensity level of a sound to the threshold of hearing. This definition set the zero point at the threshold of human hearing. The unit was named the *bel* in honor of Alexander Graham Bell. This unit was too large to be practical, so it was reduced by one-tenth and called a *decibel*. The intensity level of a sound is therefore measured in decibels (table 5.2). Compare the decibel noise level of familiar sounds listed in table 5.2, and note that each increase of 10 on the decibel scale is matched by a *multiple* of 10 on the intensity level. For example, moving from a decibel level of 10 to a decibel level of 20 requires *ten times* more intensity. Likewise, moving from a decibel level of 20 to 40 requires a one-hundred-fold increase in the intensity level. As you can see, the decibel scale is not a simple linear scale.

Resonance

You know that sound waves transmit energy when you hear a thunderclap rattle the windows. In fact, the sharp sounds from an explosion have been known not only to rattle but also break windows. The source of the energy is obvious when thunderclaps or explosions are involved. But sometimes energy transfer occurs through sound waves when it is not clear what is hap-

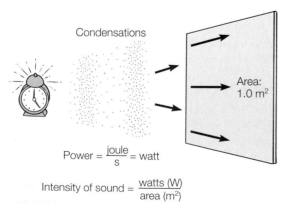

Power = $\dfrac{\text{joule}}{\text{s}}$ = watt

Intensity of sound = $\dfrac{\text{watts (W)}}{\text{area (m}^2)}$

Figure 5.16

The intensity of a sound wave is the rate of energy transferred to an area perpendicular to the waves. Intensity is measured in watts per square meter, W/m^2.

Table 5.2	Comparison of Noise Levels in Decibels with Intensity		
Example	**Response**	**Decibels**	**Intensity W/m^2**
Least needed for hearing	Barely perceived	0	1×10^{-12}
Calm day in woods	Very, very quiet	10	1×10^{-11}
Whisper (15 ft)	Very quiet	20	1×10^{-10}
Library	Quiet	40	1×10^{-8}
Talking	Easy to hear	65	3×10^{-6}
Heavy street traffic	Conversation difficult	70	1×10^{-5}
Pneumatic drill (50 ft)	Very loud	95	3×10^{-3}
Jet plane (200 ft)	Discomfort	120	1

Noise Pollution Solution?

Bells, jet planes, sirens, motorcycles, jackhammers, and construction noises all contribute to the constant racket that has become commonplace in many areas. Such noise pollution is everywhere, and it is a rare place where you can escape the ongoing din. Earplugs help reduce the noise level, but they also block sounds that you want to hear, such as music and human voices. In addition, earplugs are more effective against high-frequency sounds than they are against lower ones like aircraft engines or wind noise.

A better solution for noise pollution might be the relatively new "antinoise" technology that cancels sound waves before they reach your ears. A microphone detects background noise and transmits information to microprocessors about the noise waveform. The microprocessors then generate an "antinoise" signal that is 180 degrees out of phase with the background noise. When the noise and antinoise meet, they undergo destructive interference and significantly reduce the final sound loudness.

The noise-canceling technology is available today as a consumer portable electronic device with a set of headphones, looking much like a portable tape player. The headphones can be used to combat steady, ongoing background noise such as you might hear from a spinning computer drive, a vacuum cleaner, or while inside the cabin of a jet plane. Some vendors claim their device cancels up to 40 percent of whirring air conditioner noise, 80 percent of ongoing car noise, or up to 95 percent of constant airplane cabin noise. You hear all the other sounds—people talking, warning sounds, and music—as the microprocessors are not yet fast enough to match anything but a constant sound source.

Noise-canceling microphones can also be used to limit background noise that muddles the human voice in telephone booths, in teleconferencing, and during cellular phone use from a number of high-noise places. With improved microprocessor and new digital applications, noise-canceling technology will help bring higher sound quality to voice-driven applications. It may also find extended practical use in helping to turn down the volume on our noisy world.

---

pening. A truck drives down the street, for example, and one window rattles but the others do not. A singer shatters a crystal water glass by singing a single note, but other objects remain undisturbed. A closer look at the nature of vibrating objects and the transfer of energy will explain these phenomena.

Almost any elastic object can be made to vibrate and will vibrate freely at a constant frequency after being sufficiently disturbed. Entertainers sometimes discover this fact and appear on late-night talk shows playing saws, wrenches, and other odd objects as musical instruments. All material objects have a **natural frequency** of vibration determined by the materials and shape of the objects. The natural frequencies of different wrenches enable an entertainer to use the suspended tools as if they were the bars of a xylophone.

If you have ever pumped a swing, you know that small forces can be applied at any frequency. If the frequency of the applied forces matches the natural frequency of the moving swing, there is a dramatic increase in amplitude. When the two frequencies match, energy is transferred very efficiently. This condition, when the frequency of an external force matches the natural frequency, is called **resonance.** The natural frequency of an object is thus referred to as the *resonant frequency,* that is, the frequency at which resonance occurs.

A silent tuning fork will resonate if a second tuning fork with the same frequency is struck and vibrates nearby (figure 5.17). You will hear the previously silent tuning fork sounding if you stop the vibrations of the struck fork by touching it. The waves of condensations and rarefactions produced by the struck tuning fork produce a regular series of impulses that match the natural frequency of the silent tuning fork. This illustrates that at resonance, relatively little energy is required to start vibrations.

Struck tuning fork

Not struck, but vibrating, tuning fork

Figure 5.17

When the frequency of an applied force, including the force of a sound wave, matches the natural frequency of an object, energy is transferred very efficiently. The condition is called *resonance.*

A truck causing vibrations as it is driven past a building may cause one window to rattle while others do not. Vibrations caused by the truck have matched the natural frequency of this window but not the others. The window is undergoing resonance from the sound wave impulses that matched its natural frequency. It is also resonance that enables a singer to break a water glass. If the tone is at the resonant frequency of the glass, the resulting vibrations may be large enough to shatter it.

Resonance considerations are important in engineering. A large water pump, for example, was designed for a nuclear power plant. Vibrations from the electric motor matched the resonant frequency of the impeller blades, and they shattered after a short period of time. The blades were redesigned to have a different natural frequency when the problem was discovered. Resonance vibrations are particularly important in the design of buildings.

A

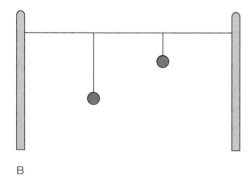

B

Figure 5.18
The "Concepts Applied" section demonstrates that one of these pendulum arrangements will show resonance, and the other will not. Can you predict which one will show resonance?

SOURCES OF SOUNDS

All sounds have a vibrating object as their source. The vibrations of the object send pulses or waves of condensations and rarefactions through the air. These sound waves have physical properties that can be measured, such as frequency and intensity. Subjectively, your response to frequency is to identify a certain pitch. A high-frequency sound is interpreted as a high-pitched sound, and a low-frequency sound is interpreted as a low-pitched sound. Likewise, a greater intensity is interpreted as increased loudness, but there is not a direct relationship between intensity and loudness as there is between frequency and pitch.

There are other subjective interpretations about sounds. Some sounds are bothersome and irritating to some people but go unnoticed by others. In general, sounds made by brief, irregular vibrations such as those made by a slamming door, dropped book, or sliding chair are called *noise*. Noise is characterized by sound waves with mixed frequencies and jumbled intensities (figure 5.19). On the other hand, there are sounds made by very regular, repeating vibrations such as those made by a tuning fork. A tuning fork produces a *pure tone* with a sinusoidal curved pressure variation and regular frequency. Yet a tuning fork produces a tone that most people interpret as bland. You would not call a tuning fork sound a musical note! Musical sounds from instruments have a certain frequency and loudness, as do noise and pure tones, but you can readily identify the source of the very same musical note made by two different instruments. You recognize it as a musical note, not noise and not a pure tone. You also recognize if the note was produced by a violin or a guitar. The difference is in the wave form of the sounds made by the two instruments, and the difference is called the *sound quality*. How does a musical instrument produce a sound of a characteristic quality? The answer may be found by looking at instruments that make use of vibrating strings.

Vibrating Strings

A stringed musical instrument, such as a guitar, has strings that are stretched between two fixed ends. When a string is plucked, waves of many different frequencies travel back and forth on the string, reflecting from the fixed ends. Many of these waves

Figure 5.19
Different sounds that you hear include (*A*) noise, (*B*) pure tones, and (*C*) musical notes.

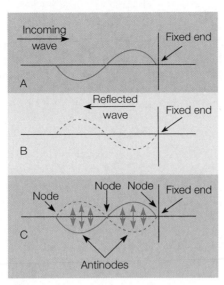

Figure 5.20

An incoming wave on a cord with a fixed end (*A*) meets a reflected wave (*B*) with the same amplitude and frequency, producing a standing wave (*C*). Note that a standing wave of one wavelength has three nodes and two antinodes.

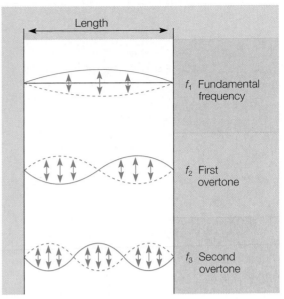

Figure 5.21

A stretched string of a given length has a number of possible resonant frequencies. The lowest frequency is the fundamental, f_1; the next higher frequencies, or overtones, shown are f_2 and f_3.

quickly fade away but certain frequencies resonate, setting up patterns of waves. Before considering these resonant patterns in detail, keep in mind that (1) two or more waves can be in the same place at the same time, traveling through one another from opposite directions; (2) a confined wave will be reflected at a boundary, and the reflected wave will be inverted (a crest becomes a trough); and (3) reflected waves interfere with incoming waves of the same frequency to produce **standing waves.** Figure 5.20 is a graphic "snapshot" of what happens when reflected wave patterns meet incoming wave patterns. The incoming wave is shown as a solid line, and the reflected wave is shown as a dotted line. The result is (1) places of destructive interference, called *nodes,* which show no disturbance, and (2) loops of constructive interference, called *antinodes,* which take place where the crests and troughs of the two wave patterns produce a disturbance that rapidly alternates upward and downward. This pattern of alternating nodes and antinodes does not move along the string and is thus called a *standing wave.* Note that the standing wave for *one wavelength* will have a node at both ends and in the center and also two antinodes. Standing waves occur at the natural, or resonant, frequencies of the string, which are a consequence of the nature of the string, the string length, and the tension in the string. Since the standing waves are resonant vibrations, they continue as all other waves quickly fade away.

Since the two ends of the string are not free to move, the ends of the string will have nodes. The *longest* wave that can make a standing wave on such a string has a wavelength (λ) that is twice the length (*L*) of the string. Since frequency (*f*) is inversely proportional to wavelength ($f = v/\lambda$ from equation 5.3), this longest wavelength has the lowest frequency possible, called the **fundamental frequency.** The fundamental frequency has one antinode, which means that the length of the string has one-half a wavelength. The fundamental frequency (f_1) determines the pitch of the *basic* musical note being sounded and is called the first har-

monic. Other resonant frequencies occur at the same time, however, since other standing waves can also fit onto the string. A higher frequency of vibration (f_2) could fit two half-wavelengths between the two fixed nodes. An even higher frequency (f_3) could fit three half-wavelengths between the two fixed nodes (figure 5.21). Any whole number of halves of the wavelength will permit a standing wave to form. The frequencies (f_2, f_3, etc.) of these wavelengths are called the *overtones,* or harmonics. It is the presence and strength of various overtones that give a musical note from a certain instrument its characteristic quality. The fundamental and the overtones add together to produce the characteristic *sound quality,* which is different for the same-pitched note produced by a violin and by a guitar (figure 5.22).

Figure 5.22

A combination of the fundamental and overtone frequencies produces a composite waveform with a characteristic sound quality.

How Animals Make Sounds

There are several ways that animals intentionally make sounds for the purpose of communication. Many animals force air through openings or over surfaces to cause sounds. In the simplest case the air may simply produce a whistling or hissing sound as the air is forced through a small opening. If the opening is larger, the sound may be a snort or booming sound. In many vertebrate animals air is forced past vocal cords that are caused to vibrate in a manner similar to a violin string. The tension on the vocal cords can be changed to vary the pitch of the sound that can then be used to encode different messages.

Many insects such as crickets and katydids rub parts of their bodies (wings) together and generate vibrations that cause sounds. Different species produce different frequencies or durations of sound.

Other insects, such as cicadas, use muscles to deform a portion of their external skeleton. When the portion of the skeleton pops back to its normal shape, a sound is produced. By repeating this process very rapidly, cicadas produce the high-pitched buzzing sound you can hear high in trees in late summer.

Some animals, like woodpeckers, actually select specific trees that have good sound-producing properties and hammer at them to produce sounds that advertise their presence. Some have even been known to select aluminum flagpoles.

Some birds purposely make sounds with their wings. Ruffed grouse rapidly move their wings forward and stop them abruptly. The first few beats are slow and gradually speed up causing a blur of sound at the end of the episode. This action creates a booming sound that can be heard for several hundred meters.

Since nodes must be located at the ends, only half-wavelengths ($1/2\lambda$) can fit on a string of a given length (L), so the fundamental frequency of a string is $1/2\lambda = L$, or $\lambda = 2L$. Substituting this value in the wave equation (solved for frequency, f) will give the relationship for finding the fundamental frequency and the overtones when the string length and velocity of waves on the string are known. The relationship is

$$f_n = \frac{nv}{2L}$$

equation 5.5

where $n = 1, 2, 3, 4 \ldots$, and $n = 1$ is the fundamental frequency and $n = 2$, $n = 3$, and so forth are the overtones.

The vibrating string produces a waveform with overtones, so instruments that have vibrating strings are called *harmonic instruments*. Instruments that use an air column as a sound maker are also harmonic instruments. These include all the wind instruments such as the clarinet, flute, trombone, trumpet, pipe organ, and many others. The various wind instruments have different ways of making a column of air vibrate. In the flute, air vibrates as it moves over a sharp edge, while in the clarinet, saxophone, and other reed instruments, it vibrates through fluttering thin reeds. The air column in brass instruments, on the other hand, is vibrated by the tightly fluttering lips of the player.

The length of the air column determines the frequency, and woodwind instruments have holes in the side of a tube that are opened or closed to change the length of the air column. The resulting tone depends on the length of the air column and the resonate overtones.

Sounds from Moving Sources

When the source of a sound is stationary, equally spaced sound waves expand from a source in all directions. But, if the sounding source starts moving, then successive sound waves become displaced in the direction of movement and this changes the pitch. For example, the siren of an approaching ambulance seems to change pitch when the ambulance passes you. The sound wave is "squashed" as the ambulance approaches you and you hear a higher-frequency siren than the people inside the ambulance. When the ambulance passes you, the sound waves are "stretched" and you hear a lower-frequency siren. The overall effect of a higher pitch as a source approaches, and then a lower pitch as it moves away is called the **Doppler effect.** The Doppler effect is evident if you stand by a street and an approaching car sounds its horn as it drives by you. You will hear a higher-pitched horn as the car approaches, which shifts to a lower-pitched horn as the waves go by you. The driver of the car, however, will hear the continual, true pitch of the horn since the driver is moving with the source (figure 5.23).

A Doppler shift is also noted if the observer is moving and the source of sound is stationary. When the observer moves toward the source, the wave fronts are encountered more frequently than if the observer were standing still. As the observer moves away from the source, the wave fronts are encountered less frequently than if the observer were not moving. An observer on a moving train approaching a crossing with a sounding bell thus hears a high-pitched bell that shifts to a lower-pitched bell as the train whizzes by the crossing.

When an object moves through the air at the speed of sound, it keeps up with its own sound waves. All the successive wave fronts pile up on one another, creating a large wave disturbance called a *shock wave*. The shock wave from a supersonic airplane is a cone-shaped shock wave of intense condensations trailing backward at an angle dependent on the speed of the aircraft. Wherever this cone of superimposed crests passes, a **sonic boom** occurs. The many crests have been added together, each contributing to the pressure increase. The human ear cannot differentiate between such a pressure wave created by a supersonic aircraft and a pressure wave created by an explosion.

Connections . . .

The Doppler effect was named after the Austrian scientist Christian Doppler, who first demonstrated the effect using sound waves back in 1842. The same principle applies to electromagnetic radiation as well as sound, but now the shifts are in frequency of the radiation. A lower frequency is observed when a source of light is moving away, and this is called a "red shift." Also, a "blue shift" toward a higher frequency occurs when a source of light is moving toward an observer. Radio waves will also experience such shifts of frequency, and weather radar that measures frequency changes as a result of motion is called *Doppler radar.*

Weather radar broadcasts short radio waves from an antenna. When directed at a storm, the waves are reflected back to the antenna by rain, snow, and hail. Reflected radar waves are electronically converted and dis-

played on a monitor, showing the location and intensity of precipitation. A Doppler radar also measures frequency shifts in the reflected radio waves. Waves from objects moving toward the antenna show a higher frequency, and waves from objects moving away from the antenna show a lower frequency. These shifts of frequency are measured, then displayed as the speed and direction of winds moving raindrops and other objects in the storm.

Weather forecasters can direct a Doppler radar machine to measure different elevations of a storm system. This shows highly accurate information that can be used to identify, for example, where and when a tornado might form, the intensity of storm winds in a given area, and even provide an estimate of how much precipitation fell from the storm.

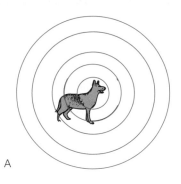

A

Motion of source →

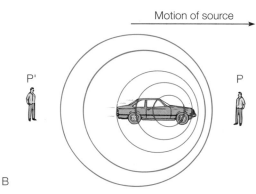

P'

P

B

Figure 5.23

(*A*) Spherical sound waves from a stationary source spread out evenly in all directions. (*B*) If the source is moving, an observer at position P will experience more wave crests per second than an observer at position P′. The observer at P interprets this as a higher pitch, and the phenomenon is called the *Doppler effect.*

Does a sonic boom occur only when an airplane breaks the sound barrier? The answer is no, an airplane traveling at or faster than the speed of sound produces the shock wave continuously, and a sonic boom will be heard everywhere the plane drags its cone-shaped shock wave.

The Austrian physicist Ernst Mach published a paper in 1877 laying out the principles of supersonics. He also came up

with the idea of using a ratio of the velocity of an object to the velocity of sound. Today, this ratio is called the Mach number. A plane traveling at the speed of sound has a Mach number of 1, a plane traveling at twice the speed of sound has a Mach number of 2, and so on. Ernest Mach was also the first to describe what is happening to produce a sonic boom, and he observed the existence of a conical shock wave formed by a projectile as it approached the speed of sound.

SUMMARY

Elastic objects *vibrate,* or move back and forth, in a repeating motion when disturbed by some external force. They are able to do this because they have an *internal restoring force* that returns them to their original positions after being deformed by some external force. If the internal restoring force is opposite to and proportional to the deforming displacement, the vibration is called a *simple harmonic motion.* The extent of displacement is called the *amplitude,* and one complete back-and-forth motion is one *cycle.* The time required for one cycle is a *period.* The *frequency* is the number of cycles per second, and the unit of frequency is the *hertz.* A *graph* of the displacement as a function of time for a simple harmonic motion produces a *sinusoidal* graph.

Periodic, or repeating, vibrations or the *pulse* of a single disturbance can create *waves,* disturbances that carry energy through a medium. A wave that disturbs particles in a back-and-forth motion in the direction of the wave travel is called a *longitudinal wave.* A wave that disturbs particles in a motion perpendicular to the direction of wave travel is called a *transverse wave.* The nature of the medium and the nature of the disturbance determine the type of wave created.

Waves that move through the air are longitudinal and cause a back-and-forth motion of the molecules making up the air. A zone of molecules forced closer together produces a *condensation,* a pulse of increased density and pressure. A zone of reduced density and pressure is a *rarefaction.* A vibrating object produces condensations and rarefactions that expand outward from the source. If the frequency is between 20 Hz and 20,000 Hz, the human ear perceives the waves as sound of a certain *pitch*. High frequency is interpreted as high-pitched sound, and low frequency as low-pitched sound.

A graph of pressure changes produced by condensations and rarefactions can be used to describe sound waves. The condensations produce *crests,* and the rarefactions produce *troughs.* The *amplitude* is the maximum change of pressure from the normal. The *wavelength* is the distance between any two successive places on a wave train, such as the distance from one crest to the next crest. The *period* is the time required for a wave to repeat itself. The *velocity* of a wave is how quickly a wavelength passes. The *frequency* can be calculated from the *wave equation,* $v = \lambda f$.

Sound waves can move through any medium but not a vacuum. The velocity of sound in a medium depends on the molecular inertia and strength of interactions. Sound, therefore, travels most rapidly through a solid, then a liquid, then a gas. In air, sound has a greater velocity in warmer air than in cooler air because the molecules of air are moving about more rapidly, therefore transmitting a pulse more rapidly.

Sound waves are *reflected* or *refracted* from a *boundary,* which means a change in the transmitting medium. Reflected waves that are *in phase* with incoming waves undergo *constructive interference* and waves that are *out of phase* undergo *destructive interference.* Two waves that are otherwise alike but with slightly different frequencies produce an alternating increasing and decreasing of loudness called *beats.*

The *energy* of a sound wave is called the wave *intensity,* which is measured in watts per square meter. The intensity of sound is expressed on the *decibel scale,* which relates it to changes in loudness as perceived by the human ear.

All elastic objects have *natural frequencies* of vibration that are determined by the materials they are made of and their shapes. When energy is transferred at the natural frequencies, there is a dramatic increase of amplitude called *resonance.* The natural frequencies are also called *resonant frequencies.*

Sounds are compared by pitch, loudness, and *quality.* The quality is determined by the instrument sounding the note. Each instrument has its own characteristic quality because of the resonant frequencies that it produces. The basic, or *fundamental, frequency* is the longest standing wave that it can make. The fundamental frequency determines the basic note being sounded, and other resonant frequencies, or standing waves called *overtones* or *harmonics,* combine with the fundamental to give the instrument its characteristic quality.

A moving source of sound or a moving observer experiences an apparent shift of frequency called the *Doppler effect.* If the source is moving as fast or faster than the speed of sound, the sound waves pile up into a *shock wave* called a *sonic boom.* A sonic boom sounds very much like the pressure wave from an explosion.

Summary of Equations

5.1
$$\text{period} = \frac{1}{\text{frequency}}$$
$$T = \frac{1}{f}$$

5.2
$$\text{frequency} = \frac{1}{\text{period}}$$
$$f = \frac{1}{T}$$

5.3
$$\text{velocity} = (\text{wavelength})(\text{frequency})$$
$$v = \lambda f$$

5.4
$$\text{beat frequency} = \text{one frequency} - \text{other frequency}$$
$$f_b = f_2 - f_1$$

5.5
$$\text{resonant frequency} = \frac{\text{number} \times \text{velocity on string}}{2 \times \text{length of string}}$$

where number 1 = fundamental frequency, and numbers 2, 3, 4, and so on = overtones.

$$f_n = \frac{nv}{2L}$$

KEY TERMS

amplitude (p. **87**)
beats (p. **95**)
condensation (p. **89**)
cycle (p. **87**)
decibel scale (p. **96**)
Doppler effect (p. **100**)
echo (p. **94**)
frequency (p. **87**)
fundamental frequency (p. **99**)
hertz (p. **87**)
infrasonic (p. **89**)
intensity (p. **96**)
longitudinal wave (p. **88**)

natural frequency (p. **97**)
period (p. **87**)
pitch (p. **91**)
rarefaction (p. **89**)
resonance (p. **97**)
reverberation (p. **94**)
sonic boom (p. **100**)
standing waves (p. **99**)
transverse wave (p. **88**)
ultrasonic (p. **89**)
vibration (p. **86**)
wavelength (p. **92**)

APPLYING THE CONCEPTS

1. The time required for a vibrating object to complete one full cycle is the
   (a) frequency.
   (b) amplitude.
   (c) period.
   (d) hertz.
2. The unit of cycles per second is called a
   (a) hertz.
   (b) lambda.
   (c) wave.
   (d) watt.
3. The period of a vibrating object is related to the frequency, since they are
   (a) directly proportional.
   (b) inversely proportional.
   (c) frequently proportional.
   (d) not proportional.
4. A longitudinal mechanical wave causes particles of a material to move
   (a) back and forth in the same direction the wave is moving.
   (b) perpendicular to the direction the wave is moving.
   (c) in a circular motion in the direction the wave is moving.
   (d) in a circular motion opposite the direction the wave is moving.

5. Transverse mechanical waves will move only through
   (a) solids.
   (b) liquids.
   (c) gases.
   (d) all of the above.
6. Longitudinal mechanical waves will move only through
   (a) solids.
   (b) liquids.
   (c) gases.
   (d) all of the above.
7. The characteristic of a wave that is responsible for what you interpret as pitch is the wave
   (a) amplitude.
   (b) shape.
   (c) frequency.
   (d) height.
8. The number of cycles that a vibrating tuning fork experiences each second is related to the resulting sound wave characteristic of
   (a) frequency.
   (b) amplitude.
   (c) wave height.
   (d) quality.
9. Sound waves travel faster in
   (a) solids as compared to liquids.
   (b) liquids as compared to gases.
   (c) warm air as compared to cooler air.
   (d) all of the above.
10. Sound interference is necessary to produce the phenomenon known as
   (a) resonance.
   (b) decibels.
   (c) beats.
   (d) reverberation.
11. The efficient transfer of energy that takes place at a natural frequency is known as
   (a) resonance.
   (b) beats.
   (c) the Doppler effect.
   (d) reverberation.
12. An observer on the ground will hear a sonic boom from an airplane traveling faster than the speed of sound
   (a) only when the plane breaks the sound barrier.
   (b) as the plane is approaching.
   (c) when the plane is directly overhead.
   (d) after the plane has passed by.

Answers
1. c **2.** a **3.** b **4.** a **5.** a **6.** d **7.** c **8.** a **9.** d **10.** c **11.** a **12.** d

QUESTIONS FOR THOUGHT

1. What is a wave?
2. Is it possible for a transverse wave to move through air? Explain.
3. A piano tuner hears three beats per second when a tuning fork and a note are sounded together and six beats per second after the string is tightened. What should the tuner do next, tighten or loosen the string? Explain.
4. Why do astronauts on the moon have to communicate by radio even when close to one another?
5. What is resonance?
6. Explain why sounds travel faster in warm air than in cool air.
7. Do all frequencies of sound travel with the same velocity? Explain your answer by using the wave equation.
8. What eventually happens to a sound wave traveling through the air?
9. What gives a musical note its characteristic quality?
10. Does a supersonic aircraft make a sonic boom only when it cracks the sound barrier? Explain.
11. What is an echo?
12. Why are fundamental frequencies and overtones also called resonant frequencies?

PARALLEL EXERCISES

The exercises in groups A and B cover the same concepts. Solutions to group A exercises are located in appendix D.

Group A

1. A vibrating object produces periodic waves with a wavelength of 50 cm and a frequency of 10 Hz. How fast do these waves move away from the object?

2. The distance between the center of a condensation and the center of an adjacent rarefaction is 1.50 m. If the frequency is 112.0 Hz, what is the speed of the wave front?

3. Water waves are observed to pass under a bridge at a rate of one complete wave every 4.0 s. (a) What is the period of these waves? (b) What is the frequency?

4. A sound wave with a frequency of 260 Hz moves with a velocity of 330 m/s. What is the distance from one condensation to the next?

5. The following sound waves have what velocity?
   (a) Middle C, or 256 Hz and 1.34 m λ.
   (b) Note A, or 440.0 Hz and 78.0 cm λ.
   (c) A siren at 750.0 Hz and λ of 45.7 cm.
   (d) Note from a stereo at 2,500.0 Hz and λ of 13.72 cm.

6. You hear an echo from a cliff 4.80 s after you shout, "Hello." How many feet away is the cliff if the speed of sound is 1,100 ft/s?

7. During a thunderstorm, thunder was timed 4.63 s after lightning was seen. How many feet away was the lightning strike if the speed of sound is 1,140 ft/s?

8. If the velocity of a 440 Hz sound is 1,125 ft/s in the air and 5,020 ft/s in seawater, find the wavelength of this sound (a) in air, (b) in seawater.

Group B

1. A tuning fork vibrates 440.0 times a second, producing sound waves with a wavelength of 78.0 cm. What is the velocity of these waves?

2. The distance between the center of a condensation and the center of an adjacent rarefaction is 65.23 cm. If the frequency is 256.0 Hz, how fast are these waves moving?

3. A warning buoy is observed to rise every 5.0 s as crests of waves pass by it. (a) What is the period of these waves? (b) What is the frequency?

4. Sound from the siren of an emergency vehicle has a frequency of 750.0 Hz and moves with a velocity of 343.0 m/s. What is the distance from one condensation to the next?

5. The following sound waves have what velocity?
   (a) 20.0 Hz, λ of 17.15 m
   (b) 200.0 Hz, λ of 1.72 m
   (c) 2,000.0 Hz, λ of 17.15 cm
   (d) 20,000.0 Hz, λ of 1.72 cm

6. A ship at sea sounds a whistle blast, and an echo returns from the coastal land 10.0 s later. How many kilometers is it to the coastal land if sound travels at 337 m/s?

7. How many seconds will elapse between seeing lightning and hearing the thunder if the lightning strikes 1 mile (5,280 ft) away and sound travels at 1,151 ft/s?

8. A 600.0 Hz sound has a velocity of 1,087.0 ft/s in the air and a velocity of 4,920.0 ft/s in water. Find the wavelength of this sound (a) in the air, (b) in the water.

Electricity

A thunderstorm produces an interesting display of electrical discharge. Each bolt can carry over 150,000 amperes of current with a voltage of 100 million volts.

O U T L I N E

The previous chapters have been concerned with *mechanical* concepts, explanations of the motion of objects that exert forces on one another. These concepts were used to explain straight-line motion, the motion of free fall, and the circular motion of objects on the earth as well as the circular motion of planets and satellites. The mechanical concepts were based on Newton's laws of motion and are sometimes referred to as Newtonian physics. The mechanical explanations were then extended into the submicroscopic world of matter through the kinetic molecular theory. The objects of motion were now particles, molecules that exert force on one another, and concepts associated with heat were interpreted as the motion of these particles. In a further extension of Newtonian concepts, mechanical explanations were given for concepts associated with sound, a mechanical disturbance that follows the laws of motion as it moves through the molecules of matter.

You might wonder, as did the scientists of the 1800s, if mechanical interpretations would also explain other natural phenomena such as electricity, chemical reactions, and light. A mechanical model would be very attractive since it already explained so many other facts of nature, and scientists have always looked for basic, unifying theories. Mechanical interpretations were tried, as electricity was considered as a moving fluid, and light was considered as a mechanical wave moving through a material fluid. There were many unsolved puzzles with such a model and gradually it was recognized that electricity, light, and chemical reactions could not be explained by mechanical interpretations. Gradually, the point of view changed from a study of particles to a study of the properties of the *space* around the particles. In this chapter you will learn about electric charge in terms of the space around particles. This model of electric charge, called the *field model,* will be used to develop concepts about electric current, the electric circuit, and electrical work and power. A relationship between electricity and the fascinating topic of magnetism is discussed next, including what magnetism is and how it is produced. The relationship is then used to explain the mechanical production of electricity, (figure 6.1) how electricity is measured, and how electricity is used in everyday technological applications.

ELECTRIC CHARGE

It was a big mystery for thousands of years. No one could figure out why a rubbed piece of amber, which is fossilized tree resin, would attract small pieces of paper, thread, and hair. This unexplained attraction was called the "amber effect." Then about one hundred years ago, Joseph J. Thomson found the answer while experimenting with electric currents. From these experiments, Thomson was able to reach the conclusion that negatively charged particles were present in all matter, and in fact might be the stuff of which matter is made. The amber effect was traced to the movement of these particles, so they were called *electrons* after the Greek word for amber. The word *electricity* is also based on the Greek word for amber.

Today, we understand that the basic unit of matter is the *atom,* which is made up of electrons and other particles such as *protons* and *neutrons.* The atom is considered to have a dense center part called a *nucleus* that contains the closely situated protons and neutrons. The electrons move around the nucleus at some relatively greater distance (figure 6.2). For understanding electricity, you need only to consider the protons in the nucleus, the electrons that move around the nucleus, and the fact that electrons can be moved from an atom and caused to move to or from an object. Details on the nature of protons, neutrons, electrons, and models on how the atom is constructed will be considered in chapter 8.

Electrons and protons have a property called **electric charge.** Electrons have a *negative electric charge* and protons have a *positive electric charge.* The negative and positive description simply means that these two properties are opposite; it does not mean that one charge is better than the other. Charge is as fundamental to these subatomic particles as gravitational attraction is fundamental to masses. This means that you cannot separate gravity from a mass, and you cannot separate charge from an electron or a proton.

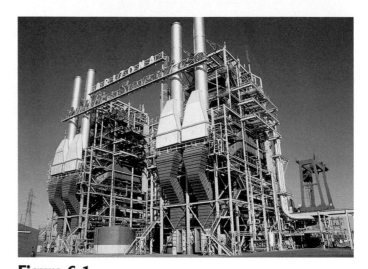

Figure 6.1

The importance of electrical power seems obvious in a modern industrial society. What is not so obvious is the role of electricity in magnetism, light, chemical change, and as the very basis for the structure of matter. All matter, in fact, is electrical in nature, as you will see.

Electric charges interact to produce what is called the **electrical force.** Like charges produce a repulsive electrical force as positive repels positive, and negative repels negative. Unlike charges produce an attractive electrical force as positive and negative charges attract each other. You can remember how this happens with the simple rule of "*like charges repel and unlike charges attract.*"

Ordinary atoms are usually neutral because there is a balance between the number of positively charged protons and the number of negatively charged electrons. A number of different physical and chemical interactions can result in an atom gaining or losing electrons. In either case, the atom is said to be *ionized*

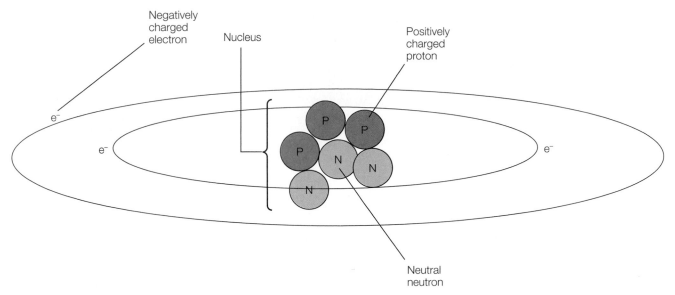

Figure 6.2

A very highly simplified model of an atom has most of the mass in a small, dense center called the *nucleus*. The nucleus has positively charged protons and neutral neutrons. Negatively charged electrons move around the nucleus at a much greater distance than is suggested by this simplified model. Ordinary atoms are neutral because there is balance between the number of positively charged protons and negatively charged electrons.

and *ions* are produced as a result. An atom that is ionized by losing electrons results in a *positive ion* since it has a net positive charge. An atom that is ionized by gaining electrons results in a *negative ion* since it has a net negative charge (figure 6.3).

Electrons can be moved from atom to atom to create ions. They can also be moved from one object to another by friction. Since electrons are negatively charged, an object that acquires an excess of electrons becomes a negatively charged body. The loss of electrons by another body results in a deficiency of electrons, which results in a positively charged object. Thus, *electric charges on objects result from the gain or loss of electrons.* Because the electric charge is confined to an object and is not moving, it is called an **electrostatic charge.** You probably call this charge *static electricity.* Static electricity is an accumulated electric charge at rest, that is, one that is not moving. When you comb your hair with a hard rubber comb, the comb becomes negatively charged because electrons are transferred from your hair to the comb, and your hair acquires a positive charge (figure 6.4). Both the negative charge on the comb from an excess of electrons and the positive charge on your hair from a deficiency of electrons are charges that are momentarily at rest, so they are electrostatic charges.

Once charged by friction, objects such as the rubber comb soon return to a neutral, or balanced, state by the movement of electrons. This happens more quickly on a humid day because water vapor assists with the movement of electrons to or from charged objects. Thus, static electricity is more noticeable on dry days than on humid ones.

A charged object can also exert a force of attraction on a second object that does not have a net charge. For example, you can give your hard rubber comb a negative charge by combing your hair. Your charged comb will attract tiny pieces

Figure 6.3

(*A*) A neutral atom has no net charge because the numbers of electrons and protons are balanced. (*B*) Removing an electron produces a net positive charge; the charged atom is called a positive ion. (*C*) The addition of an electron produces a net negative charge and a negative ion.

Figure 6.5

Charging by induction. The comb has become charged by friction, acquiring an excess of electrons. The paper (*A*) normally has a random distribution of (+) and (−) charges. (*B*) When the charged comb is held close to the paper, there is a reorientation of charges because of the repulsion of like charges. This leaves a net positive charge on the side close to the comb, and since unlike charges attract, the paper is attracted to the comb.

Figure 6.4

Arbitrary numbers of protons (+) and electrons (−) on a comb and in hair (*A*) before and (*B*) after combing. Combing transfers electrons from the hair to the comb by friction, resulting in a negative charge on the comb and a positive charge on the hair.

of paper that do not have a charge, pulling them to the comb. This happens because the negative charge on the comb repels electrons on the tiny bits of paper, giving the paper a positive charge on the side nearest the comb (figure 6.5). These unlike charges attract and the paper is pulled to the comb.

Measuring Electrical Charge

As you might have experienced, sometimes you receive a slight shock after walking across a carpet, and sometimes you are really zapped. You receive a greater shock when you have accumulated a greater electric charge. Since there is less electric charge at one time and more at another, it should be evident that charge occurs in different amounts, and these amounts can be measured. The magnitude of an electric charge is identified with the number of electrons that have been transferred onto or away from an object. The quantity of such a charge (q) is measured in a unit called a **coulomb** (C). The coulomb is a fundamental metric unit of measure like the meter, kilogram, and second.

Every electron has a charge of -1.60×10^{-19} C and every proton has a charge of $+1.60 \times 10^{-19}$ C. To accumulate a negative charge of 1 C you would need to accumulate more than 6 billion billion electrons.

The charge on an electron (or proton), 1.60×10^{-19} C, is the smallest common charge known (more exactly $1.6021892 \times 10^{-19}$ C). It is the **fundamental charge** of the electron ($e^{-} = 1.60$

$\times 10^{-19}$ C) and the proton ($p^{+} = 1.60 \times 10^{-19}$ C). All charged objects have multiples of this fundamental charge.

Measuring Electrical Force

Recall that two objects with like charges, (−) and (−) or (+) and (+), produce a repulsive force, and two objects with unlike charges, (−) and (+), produce an attractive force. The size of either force depends on the amount of charge of each object and on the distance between the objects. The relationship is known as **Coulomb's law,** which is

$$F = k \frac{q_1 q_2}{d^2}$$

equation 6.1

where k has the value of 9.00×10^9 newton·meters²/coulomb² (9.00×10^9 N·m²/C²).

The force between the two charged objects is repulsive if q_1 and q_2 are the same charge and attractive if they are different (like charges repel, unlike charges attract). Whether the force is attractive or repulsive, you know that both objects feel the same force, as described by Newton's third law of motion. In addition, the strength of this force decreases if the distance between the objects increases. (A doubling of the distance reduces the force to 1/4 of the original value.)

The model of a *field* is useful in understanding how a charge can attract or repel another charge some distance away. This model does not consider the force that one object exerts on another one through space. Instead, it considers the condition of space around a charge. The condition of space around an electric charge is changed by the presence of the charge. The charge produces a *force field* in the space around it. Since this force field is produced by an electrical charge, it is called an **electric field.** Imagine a second electric charge, called a "test charge," that is far enough away from the electric charge that no forces are experienced. As you move the test charge closer and closer, it will experience an increasing force as it enters the electric field. The test charge can be used to identify the electric field that spreads out and around the space of an electric charge.

An electric field can be visualized by making a map of the field. The field is represented by electric field lines that indicate the strength and direction of the force the field would exert on the field of another charge. The direction of the field lines always point outward around a positively charged particle and point inward around a negatively charged particle. The spacing of the field lines shows the strength of the field. The field is stronger where the lines are closer together and weaker where they are farther apart (figure 6.6).

A

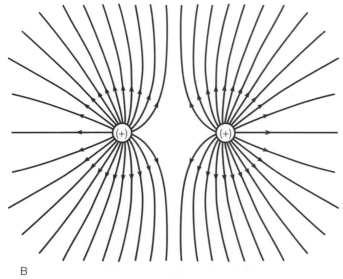

B

Figure 6.6

Lines of force diagrams for (A) a negative charge and (B) a positive charge when the charges have the same strength as the test charge.

ELECTRIC CURRENT

Electric current means a flow of charge in the same way that "water current" means a flow of water. An electric current is flow of charge and this flow can be either negative or positive. An electric wire in a car is usually a copper wire covered with a plastic insulation. When the radio is operated by using such a wire, electrons are moved from one terminal of the car battery through the radio, then back toward the battery. In a computer monitor, electrons are accelerated inside the picture tube to the back of the screen, producing an image as directed by the processor. Finally, a computer printer is plugged into a household current, which has an insulated copper wire with electrons moving back and forth. In all of these examples there was movement of charge, also known as an electric current.

There are many examples of different types of current, but all can be compared by the amount of electric charge that flows per second. The relationship is

$$\text{electric current} = \frac{\text{quantity of charge}}{\text{time}}$$

$$I = \frac{q}{t}$$

equation 6.2

The unit for current is coulomb/second, which is called an **ampere** (A or **amp** for short). A current of 1 ampere in a wire is 1 coulomb of charge flowing through the wire every second. A 2 amp current is 2 coulombs of charge flowing through the wire every second, and so on (figure 6.7).

Figure 6.7

A simple electric circuit carrying a current of 1.00 coulomb per second through a cross section of a conductor has a current of 1.00 amp.

Charge can flow easily through some materials, such as metals because they have many loosely attached electrons that can be easily moved from atom to atom. A substance that allows charges to flow easily is called a *conductor*. Materials such as plastic, wood, and rubber hold tightly to their electrons and do not readily allow for the flow of charge. A substance that does not allow charges to flow is called an *insulator* (table 6.1). Thus, metal wires are used to conduct an electric current from one place to another and rubber, glass, and plastics are used as insulators to keep the current from going elsewhere.

There is a third class of materials, such as silicon and germanium, that sometimes conduct and sometimes insulate, depending on the conditions and how pure they are. These materials are called *semiconductors,* and their special properties make possible a number of technological devices such as computers, electrostatic copying machines, laser printers, solar cells, and so forth.

Table 6.1 Electrical Conductors and Insulators

Conductors	Insulators
Silver	Rubber
Copper	Glass
Gold	Carbon (diamond)
Aluminum	Plastics
Carbon (graphite)	Wood
Tungsten	
Iron	
Lead	
Nichrome	

Resistance

Insulators have a property of opposing, or reducing a current, and this property is called *electrical resistance*. A good electrical conductor has a very low electrical resistance, and a good electrical insulator has a very high electrical resistance. The actual magnitude of electrical resistance of a metal wire conductor depends on four variables (figure 6.8):

1. **Material.** Different materials have different resistances, as shown by the list of conductors in table 6.1. Silver, for example, is at the top of the list because it offers the least resistance, followed by copper, gold, then aluminum. Of the materials listed in the table, nichrome is the conductor with the greatest resistance. By definition, conductors have less electrical resistance than insulators, which have a very large electrical resistance.

2. **Length.** The resistance of a conductor varies directly with the length; that is, a longer wire has more resistance and a shorter wire has less resistance. The longer the wire is, the greater the resistance.

3. **Diameter.** The resistance varies inversely with the cross-sectional area of a conductor. A thick wire has a greater cross-sectional area and therefore has less resistance than a thin wire. The thinner the wire is, the greater the resistance.

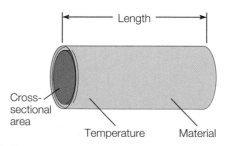

Figure 6.8

The four factors that influence the resistance of an electrical conductor are the material the conductor is made of, the length of the conductor, the cross-sectional area of the conductor, and the temperature of the conductor.

There is more than one way to power an electric vehicle, and a rechargeable lead-acid battery is not the best answer. There is a newly developing technology that uses a hydrogen-powered fuel cell to power an electric vehicle. Storage batteries, dry cells, and fuel cells all operate by using a chemical reaction to produce electricity. The difference is that the fuel cell does not need charging like a storage battery or need replacing like a dry cell. A fuel cell can use a fuel to run continuously, and that is why it is called a "fuel cell."

Fuel cells are able to generate electricity directly onboard an electric vehicle, so heavy batteries are not needed. A short driving range is not a problem, and time lost charging the batteries is not a problem. Instead, electricity is produced electrochemically, in a device without any moving parts. As you probably know, energy is required to separate water into its component gases of hydrogen and oxygen. Thus, as you might expect, energy is released when hydrogen and oxygen combine to form water. It is this energy that a fuel cell uses to produce an electric current.

One design of a fuel cell has two electrodes, one positive and one negative, where the reactions that produce electricity take place. A proton exchange membrane (PEM) separates the electrodes, which also use catalysts to speed chemical reactions. In general terms, hydrogen atoms enter the fuel cell at the positive terminal where a chemical reaction strips them of their electrons (box figure 6.1). The hydrogen atoms are now positive ions, which pass through the membrane and leave their electrons behind. Once through the membrane the hydrogen ions combine with oxygen, forming water, which drains from the cell. The electrons on the original side of the membrane increase in number, building up an electrical potential difference. Once a wire connects the two sides of the membrane the electrons are able to make a current, providing electricity. As long as such a fuel cell is supplied with hydrogen and oxygen, it will generate electricity to run a car.

The silent-running, nonpolluting fuel cell with no moving parts sounds too good to be true,

Box Figure 6.1
A schematic of a PEM fuel cell.

but the technology works. The technology has been too expensive for everyday use until recently, but is now more affordable. Fuel-cell-powered vehicles can operate directly on compressed hydrogen gas or liquid hydrogen and when they do, the only emission is water vapor. Direct use of hydrogen is also very efficient, with a 50 to 60 percent efficiency compared to the typical 15 to 20 percent efficiency for automobiles that run on petroleum in an internal combustion engine. Other fuels can also be used by running them through an onboard reformer, which transforms the fuel to hydrogen. Methanol or natural gas, for example, can be used with significantly less CO_2, CO, HC, and NO_x emissions than produced by an internal combustion engine.

An added advantage to the use of methanol is that the existing petroleum fuel distribution system (tanks, pumps, etc.) can be used to distribute this liquid fuel. Liquid or compressed hydrogen, on the other hand, require a completely new type of distribution system.

A fuel-cell powered electric vehicle gives the emission benefits of a battery-powered vehicle without the problems of constantly recharging the batteries. A 1999 DaimlerChrysler test car using hydrogen gas to power the fuel cells in a converted Mercedes-Benz reached top speeds of 145 km/h (90 mi/h) and traveled 450 km (280 mi) before refueling. Before long, you may see a fuel-cell vehicle in your neighborhood. It is the car of the future, which is needed now for the environment.

4. **Temperature.** For most materials, the resistance increases with increases in temperature. This is a consequence of the increased motion of electrons and ions at higher temperatures, which increases the number of collisions. At very low temperatures, the resistance of some materials approaches zero, and the materials are said to be *superconductors*.

AC and DC

Another aspect of the nature of an electric current is the direction the charge is flowing. The wires in a car have currents that always move in one direction, and this is called a **direct current** (DC). Chemical batteries, fuel cells, and solar cells produce a direct current, and direct currents are utilized in electronic devices. Electric utilities and most of the electrical industry, on the other hand, use an **alternating current** (AC). An alternating current, as the name implies, moves the electrons alternately one way, then the other way. Since household electric circuits use alternating current, there is no movement of electrons from the electrical outlets through the wires. Instead an electric *field* moves back and forth through a wire at nearly the speed of light, causing electrons to move back and forth. This movement constitutes a current that flows one way, then the other with the changing field. The current changes like this 120 times a second in a 60 hertz alternating current.

THE ELECTRIC CIRCUIT

An electric current is established in a conductor when an electric field exerts a force on charges in the conductor. A car battery, for example, is able to light a bulb because the battery produces an electric field that forces electrons to move through the light-bulb filament. An **electric circuit** contains some device, such as a battery or electric generator, that acts as a source of energy as it forces charges to move out one terminal, through the wires of the circuit, and then back in the other terminal (figure 6.9). The charges do work in another part of the circuit as they light bulbs, run motors, or provide heat. The charges flow through connecting wires to make a continuous path, and the number of coulombs of charge that leaves one terminal is the same as the number of coulombs of charge that enters the other terminal. The electric field moves through the circuit at nearly the speed of light, forcing the electrons to move along. The electrons, however, actually move through the circuit very slowly (figure 6.10).

Work is done by the device that creates the electric field (battery, for example) as it exerts a force on the electrons and moves them through a distance in the circuit. Disregarding any losses due to resistance, the work done in some device (lamp, for example) is equal to the work done by the battery. The amount of work can be quantified by considering the work done and the size of the charge moved, and this ratio is used to define *voltage*. A **volt** (V) is the ratio of work to charge moved,

$$\text{voltage} = \frac{\text{work}}{\text{charge moved}}$$

$$V = \frac{W}{q}$$

equation 6.3

In units,

$$1 \text{ volt (V)} = \frac{1 \text{ joule (J)}}{1 \text{ coulomb (C)}}$$

The voltage of any electric charge is the energy transfer per coulomb. The energy transfer can be measured by the work that is done to move the charge or by the work that the charge can do because of its position in the field. This is perfectly analogous to the work that must be done to give an object gravitational potential energy or to the work that the object can potentially do because of its new position. Thus, when a 12-volt battery is charged, 12 joules of work are done to transfer 1 coulomb of charge from an outside source against the electric field of the battery terminal. When the 12-volt battery is used,

Figure 6.9

A simple electric circuit has an energy source (such as a generator or battery), some device (such as a lamp or motor) where work is done, and continuous pathways for the current to follow.

Figure 6.10

Electrons move very slowly in a direct current circuit. With a drift velocity of 0.01 cm/s, more than 5 hours would be required for an electron to travel 200 cm from a car battery to the brake light. It is the electric field, not the electrons, that moves at near the speed of light in an electric circuit.

it does 12 joules of work for each coulomb of charge transferred from one terminal of the battery through the electrical system and back to the other terminal. Household circuits usually have a difference of potential of 120 or 240 volts. A voltage of 120 means that each coulomb of charge that moves through the circuit can do 120 joules of work in some electrical device.

The current in a circuit depends on the resistance as well as the voltage that is causing the current. If a conductor offers a small resistance, less voltage would be required to push an amp of current through the circuit. If a conductor offers more resistance, then more voltage will be required to push the same amp of current through the circuit. Resistance (R) is therefore a ratio between the voltage (V) and the resulting current (I). This ratio is

$$R = \frac{V}{I}$$

In units, this ratio is

$$1 \text{ ohm } (\Omega) = \frac{1 \text{ volt (V)}}{1 \text{ amp (A)}}$$

The ratio of volts/amps is the unit of resistance called an **ohm** (Ω) after the German physicist who discovered the relationship.

Another way to show the relationship between the voltage, current, and resistance is

$$V = IR$$

equation 6.4

which is known as **Ohm's law.** This is one of three ways to show the relationship, but this way (solved for V) is convenient for easily solving the equation for other unknowns.

ELECTRICAL POWER AND WORK

All electric circuits have three parts in common:

1. A voltage source, such as a battery or electric generator that uses some nonelectric source of energy to do work on electrons.
2. An electric device, such as a lightbulb or electric motor where work is done by the electric field.
3. Conducting wires that maintain the current between the electric device and voltage source.

The work done by a voltage source (battery, electric generator) is equal to the work done by the electric field in an electric device (lightbulb, electric motor) plus the energy lost to resistance. Resistance is analogous to friction in a mechanical device, so low-resistance conducting wires are used to reduce this loss. Disregarding losses to resistance, electrical work can therefore be measured where the voltage source does work in moving charges.

If you include a time factor with the work done in moving charges you will be considering the *power output* of the voltage source. The power is determined by the voltage and current in the following relationship:

$$\text{power} = \text{voltage} \times \text{current}$$
$$P = VI$$

equation 6.5

In units, you can see that multiplying the current ($I = C/s$) by the voltage ($V = J/C$) yields

$$\frac{\text{coulomb}}{\text{second}} \times \frac{\text{joule}}{\text{coulomb}} = \frac{\text{joule}}{\text{second}}$$

A joule/second is a unit of power called the **watt.** Therefore, electrical power is measured in units of watts. Note that the relationship between power output and the current is directly proportional. Therefore, the greater the current supplied the greater the power output.

Household electric devices are designed to operate on a particular voltage, usually 120 or 240. They therefore draw a certain current to produce the designed power. Information about these requirements is usually found somewhere on the device. A lightbulb, for example, is usually stamped with the designed power, such as 100 watts. Other electric devices may be stamped with amp and volt requirements. You can determine the power produced in these devices by using equation 6.5; that is, amps $\times$ volts = watts (figure 6.11). Another handy conversion factor to remember is that 746 watts are equivalent to 1 horsepower.

A

B

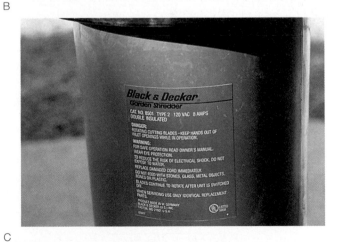

C

Figure 6.11

What do you suppose it would cost to run each of these appliances for one hour? (*A*) This lightbulb is designed to operate on a potential difference of 120 volts and will do work at the rate of 100 W. (*B*) The finishing sander does work at the rate of 1.6 amp $\times$ 120 volts, or 192 W. (*C*) The garden shredder does work at the rate of 8 amps $\times$ 120 volts, or 960 W.

The example of a household circuit in box figure 6.2B shows a light and four wall outlets in a parallel circuit. The use of the term "parallel" means that a current can flow through any of the separate branches, without having to first go through any of the other devices. It does not imply that the branches are necessarily lined up with each other. Lights and outlets in this circuit all have at least two wires, one that carries the electrical load and one that maintains a potential difference by serving as a system ground. The load-carrying wire is usually black (or red) and the system ground is usually white. A third wire, usually bare or green, is used as a ground for an appliance.

Too many appliances running in a circuit—or a short circuit—can result in a very large current, perhaps great enough to cause strong heating and possibly a fire. A fuse or circuit breaker prevents this by disconnecting the circuit when it reaches a preset value, usually 15 or 20 amps. A fuse contains a short piece of metal that melts when heated by too large a current, creating a gap when the current through the circuit reaches the preset rating. The gap disconnects the circuit, just as cutting the wire or using a switch. The circuit breaker has the same purpose, but uses the proportional relationship between the magnitude of a current and the strength of the magnetic field that forms around the conductor. When the current reaches a preset level, the magnetic field is strong enough to open a spring-loaded switch. The circuit breaker is reset by flipping the switch back to its original position.

Besides a fuse or circuit breaker, a modern household electric circuit has three-pronged plugs, polarized plugs, and ground fault interrupters (GFI) to help protect people and property from electrical damage. A *three-pronged plug* provides a grounding wire through a (usually round) prong on the plug. The grounding wire connects the housing of an appliance directly to the ground. If there is a short circuit, the current will

Box Figure 6.2

(*A*) One circuit breaker in this photo of a circuit breaker panel has tripped, indicating an overload or short circuit. (*B*) This could be circuit number 3 that is protected by a simple circuit breaker.

A

The electric utility charge for the electrical work done is at a rate of cents per kilowatt-hour. The rate varies from place to place across the country, depending on the cost of producing the power. You can predict the cost of running a particular electric appliance with the following equation,

$$\text{cost} = \frac{(\text{watts})(\text{time})(\text{rate})}{1{,}000 \frac{\text{watt}}{\text{kilowatt}}}$$

equation 6.6

If the watt power rating is not given it can be obtained by multiplying amps times volts. Also note that since the time unit is in hours, if you want to know the cost of running an appliance for a number of minutes, the time must be converted to the decimal equivalent of an hour (x min ÷ 60 min).

As you may have noticed, the electric cord to certain appliances such as hair dryers becomes warm when the appliance is used. Heating occurs in any conductor that has any resistance, and it represents a loss of useable energy. The amount of heating depends on the resistance and the size of the current. Increasing the current or increasing any of the four resistance factors discussed earlier will cause an increase in heating. A large current in a wire of high resistance can become very hot, hot enough to melt insulation and ignite materials. This is why all circuits have fuses or circuit breakers, to "break" the circuit if the current exceeds a preset safe limit (see "A Closer Look: Household Circuits and Safety").

take the path of least resistance—through the grounding wire—rather than through a person.

A *polarized plug* has one of the two flat prongs larger than the other. An alternating current moves back and forth with a frequency of 60 Hz, and polarized in this case has nothing to do with positive or negative. A polarized plug in an AC circuit means that one prong always carries the load. The smaller plug is connected to the load-carrying wire and the larger one is connected to the neutral wire. An ordinary, nonpolarized plug can fit into an outlet either way, which means there is a 50–50 chance that one of the wires will be the one that carries the load. The polarized plug always has the load-carrying wire on the same side of the circuit, so the switch can be wired in so it is always on the load-carrying side. The switch will function the same on either wire, but when it is on the ground wire the appliance contains a load-carrying wire, just waiting for a short circuit. When the switch is on the load-carrying side the appliance does not have this potential safety hazard.

Yet another safety device called a *ground-fault interrupter* (GFI) offers a different kind of protection. Normally, the current in the load-carrying and system ground wire is the same. If a short circuit occurs, some of the current might be diverted directly to the ground or to the appliance ground. A GFI device monitors the load-carrying and system ground wires and if any difference is detected, it trips, opening the circuit within a fraction of a second. This is much quicker than the regular fuse or general circuit breaker can react, and the difference might be enough to prevent a fatal shock. The GFI device is usually placed in an outside or bathroom circuit, places where currents might be diverted through people with wet feet. Note the GFI can also be tripped by a line surge that might occur during an electrical thunderstorm.

Shocking Costs

You can predict the cost of running an electric appliance with just a few calculations. For example, suppose you want to know the cost of using a 1,300-watt hair dryer for 20 minutes if the utility charges 10 cents per kilowatt-hour. With these figures, equation 6.6 would look like this:

$$\text{cost} = \frac{(1{,}300\ \text{W})(0.33\ \text{h})(\$0.10/\text{kWh})}{1{,}000\ \dfrac{\text{W}}{\text{kW}}}$$

Find answers to one or more of the following questions about the cost of running an electric appliance:

- What is your monthly electrical cost for watching television?
- What is the cost of drying your hair with a blow dryer?
- How much would you save by hanging out your clothes to dry rather than using an electric dryer?
- Compare the cost of using the following appliances: coffeemaker, toaster, can opener, vegetable steamer, microwave oven, and blender.
- How much does the electricity cost per month for use of your desk lamp?
- Of all the electric devices in a typical household, which three have the greatest monthly electrical cost?

Inside a Dry Cell

The common dry cell used in a flashlight produces electrical energy from a chemical reaction between ammonium chloride and the zinc can (box figure 6.3). The reaction leaves a negative charge on the zinc and a positive charge on the carbon rod. Manganese dioxide takes care of hydrogen gas, which is a by-product of the reaction. Dry cells always produce 1.5 volts, regardless of their size. Larger voltages are produced by combinations of smaller cells (making a true "battery").

Metal cap

Zinc can
(negative
electrode)

Manganese dioxide
and graphite
mixture (depolarizer)

Porous
barrier

Ammonium chloride
and inert fillers
(electrolyte)

Carbon rod

(Steel jacket
not shown)

Box Figure 6.3
A schematic sketch of a dry cell.

MAGNETISM

The ability of a certain naturally occurring rock to attract iron has been known since at least 600 B.C. This rock was called "Magnesian stone" since it was discovered near the ancient city of Magnesia in Turkey. Knowledge about the iron-attracting properties of the Magnesian stone grew slowly. About A.D. 100, the Chinese learned to magnetize a piece of iron with the stone, and sometime before A.D. 1000, they learned to use the magnetized iron or stone as a direction finder (compass). The rock that attracts iron is known today as the mineral named *magnetite.*

Magnetite is a natural magnet that strongly attracts iron and steel but also attracts cobalt and nickel. Such substances that are attracted to magnets are said to have *ferromagnetic* properties, or simply magnetic properties. Iron, cobalt, and nickel are considered to have magnetic properties, and most other common materials are considered not to have magnetic properties. Most of these nonmagnetic materials, however, are slightly attracted or slightly repelled by a strong magnet. In addition, certain rare earth elements as well as certain metal oxides exhibit strong magnetic properties.

Every magnet has two **magnetic poles,** or ends, about which the force of attraction seems to be concentrated. Iron filings or other small pieces of iron are attracted to the poles of

Figure 6.12
Every magnet has ends, or poles, about which the magnetic properties seem to be concentrated. As this photo shows, more iron filings are attracted to the poles, revealing their location.

a magnet, for example, revealing their location (figure 6.12). A magnet suspended by a string will turn, aligning itself in a north-south direction. The north-seeking pole is called the *north pole* of the magnet. The south-seeking pole is likewise named the *south pole* of the magnet. All magnets have both a north pole and a south pole and neither pole can exist by itself. You cannot separate a north pole from a south pole. If a magnet is broken into pieces, each new piece will have its own north and south pole (figure 6.13).

You are probably familiar with the fact that two magnets exert forces on each other. For example, if you move the north pole of one magnet near the north pole of a second magnet, each will experience a repelling force. A repelling force also occurs if two south poles are moved close together. But if the north pole of one magnet is brought near the south pole of a second magnet, an attractive force occurs. The rule is *"like magnetic poles repel and unlike magnetic poles attract."*

A similar rule of "likes charges repel and unlike charges attract" was used for electrostatic charges, so you might wonder if there is some similarity between charges and poles. The answer is they are not related. A magnet has no effect on a charged glass rod and the charged glass rod has no effect on either pole of a magnet.

Figure 6.13
A bar magnet cut into halves always makes new, complete magnets with both a north and a south pole. The poles always come in pairs, and the separation of a pair into single poles, called monopoles, has never been accomplished.

Connections . . .

Do animals use the earth's magnetic fields to navigate? Since animals move from place to place to meet their needs it is useful to be able to return to a nest, water hole, den, or favorite feeding spot. This requires some sort of memory of their surroundings (a mental map) and a way of determining direction. Often it is valuable to have information about distance as well. Direction can be determined by such things as magnetic fields, identified landmarks, scent trails, or reference to the sun or stars. If the sun or stars are used for navigation, then some sort of time sense is also needed since these bodies move in the sky.

Instinctive behaviors are automatic, preprogrammed, and genetically determined. Such behaviors are found in a wide range of organisms from simple one-celled protozoans to complex vertebrates. These behaviors are performed correctly the first time without previous experience when the proper stimulus is given. A stimulus is some change in the internal or external environment of the organism that causes it to react. The reaction of the organism to the stimulus is called a response.

An organism can respond only to stimuli it can recognize. For example, it is difficult for us as humans to appreciate what the world seems like to a bloodhound. The bloodhound is able to identify individuals by smell, whereas we have great difficulty detecting, let alone distinguishing, many odors. Some animals, like dogs, deer, and mice, are color-blind and are able to see only shades of gray. Others, such as honeybees, can see ultraviolet light, which is invisible to us. Some birds and other animals are able to detect the magnetic field of the earth.

There is evidence that some birds navigate by compass direction, that is, they fly as if they had a compass in their heads. They seem to be able to sense magnetic north. Their ability to sense magnetic fields has been proven at the U.S. Navy's test facility in Wisconsin. The weak magnetism radiated from this test site has changed the flight pattern of migrating birds, but it is yet to be proven that birds use the magnetism of the earth to guide their migration. Homing pigeons are famous for their ability to find their way home. They make use of a wide variety of clues, but it has been shown that one of the clues they use involves magnetism. Birds with tiny magnets glued to the sides of their heads were very poor navigators, while others with nonmagnetic objects attached to the sides of their heads did not lose their ability to navigate.

A magnet moved into the space near a second magnet experiences a magnetic force as it enters the **magnetic field** of the second magnet. A magnetic field can be represented by *magnetic field lines.* By convention, magnetic field lines are drawn to indicate how the *north pole* of a tiny imaginary magnet would point when in various places in the magnetic field. Arrowheads indicate the direction that the north pole would point, thus defining the direction of the magnetic field. The strength of the magnetic field is greater where the lines are closer together and weaker where they are farther apart. Figure 6.14 shows the magnetic field lines around the familiar bar magnet.

The north end of a magnetic compass needle points north because the earth has a magnetic field. The earth's magnetic field is shaped and oriented as if there were a huge bar magnet inside the earth (figure 6.15). The geographic North Pole is the axis of the earth's rotation, and this pole is used to determine the direction of true north on maps. A magnetic compass does not point to true north because the north magnetic pole and the geographic North Pole are in two different places. The difference is called *magnetic declination.* The map in figure 6.16 shows approximately how many degrees east or west of true north a compass needle will point in different locations. Magnetic declination must be considered when navigating with a compass.

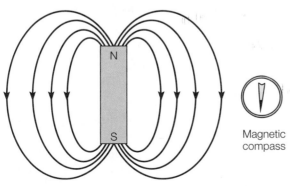

Figure 6.14
These lines are a map of the magnetic field around a bar magnet. The needle of a magnetic compass will follow the lines, with the north end showing the direction of the field.

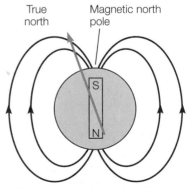

Figure 6.15
The earth's magnetic field. Note that the magnetic north pole and the geographic North Pole are not in the same place. Note also that the magnetic north pole acts as if the south pole of a huge bar magnet were inside the earth. You know that it must be a magnetic south pole, since the north end of a magnetic compass is attracted to it, and opposite poles attract.

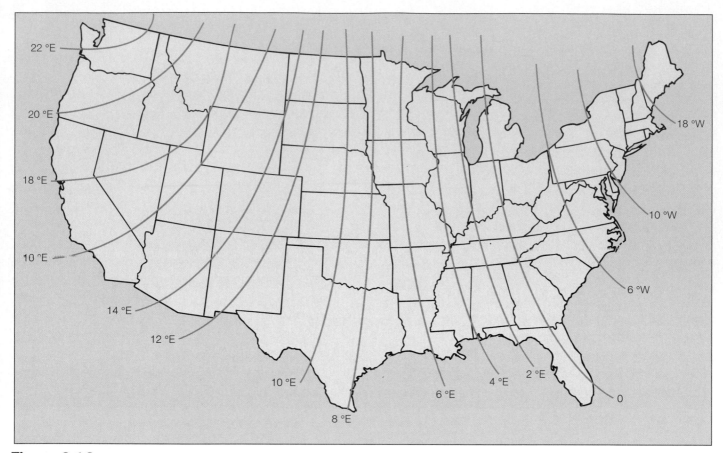

Figure 6.16

This magnetic declination map shows the approximate number of degrees east or west of the true geographic north that a magnetic compass will point in various locations.

Note in figure 6.15 that the earth's magnetic field acts as if there is a huge bar magnet inside the earth with a south pole near the earth's geographic North Pole. This is not an error. The north pole of a magnet is attracted to the south pole of a second magnet, and the north pole of a compass needle points to the north. Therefore, the bar magnet must be arranged as shown. This apparent contradiction is a result of naming the magnetic poles after their "seeking" direction.

Moving Charges and Magnetic Fields

Recall that every electric charge is surrounded by an electric field. If the charge is moving, it is surrounded by an electric field *and* a magnetic field. The magnetic field is formed in the shape of circles around the moving charge. A DC current is many charges moving through a conductor in response to an electric field, and these moving charges also produce a magnetic field. The field is in the shape of circles around the length of a current-carrying wire (figure 6.17). This relationship suggests that electricity and magnetism are two different manifestations of charges in motion. The electric field of a charge, for example, is fixed according to the fundamental charge of the particle. The magnetic field, however, changes with the velocity of the moving charge. The magnetic field does

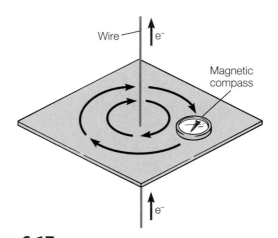

Figure 6.17

A magnetic compass shows the presence and direction of the magnetic field around a straight length of current-carrying wire.

not exist at all if the charge is not moving, and the strength of the magnetic field increases with increases in velocity. It seems clear that magnetic fields are produced by the motion of charges, or electric currents. Thus, a magnetic field is a property of the space around a moving charge.

You can see the shape of the magnetic field established by the current by running a straight wire vertically through a sheet of paper. The wire is connected to a battery, and iron filings are sprinkled on the paper. The filings will become aligned as each tiny piece of iron is moved parallel to the field. Overall, filings near the wire form a pattern of circles with the wire in the center.

A current-carrying wire that is formed into a loop has circular magnetic field lines that pass through the inside of the loop in the same direction. This has the effect of concentrating the field lines, which increases the magnetic field intensity. Since the field lines all pass through the loop in the same direction, one side of the loop will have a north pole and the other side a south pole (figure 6.18).

Many loops of wire formed into a cylindrical coil are called a **solenoid.** When a current passes through the loops of wire in a solenoid, each loop contributes field lines along the length of the cylinder (figure 6.19). The overall effect is a magnetic field around the solenoid that acts just like the magnetic field of a bar magnet. This **electromagnet** can be turned on or off by turning the current on or off. In addition, the strength of the electromagnet depends on the size of the current and the number of loops. The strength of the electromagnet can also be increased by placing a piece of soft iron in the coil.

A solenoid can serve as an electrical switch or valve. It can be used as a water valve by placing a spring-loaded movable piece of iron inside the wire coil. When a current flows in such a coil the iron is pulled into the coil by the magnetic field, turning the hot or cold water on in a washing machine or dishwasher, for example. Solenoids are also used as mechanical switches on VCRs, automobile starters, and signaling devices such as door bells and buzzers.

Electrons in atoms are moving around the nucleus, so they produce a magnetic field. Electrons also have a magnetic field associated with their spin. In most materials these magnetic fields cancel one another and neutralize the overall magnetic effect. In other materials, such as an iron magnet, the electrons are

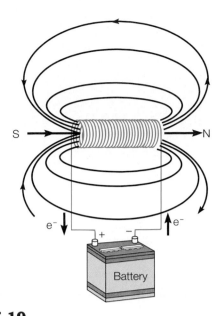

Figure 6.19
When a current is run through a cylindrical coil of wire, a solenoid, it produces a magnetic field like the magnetic field of a bar magnet.

arranged and oriented in such a way that individual magnetic fields are aligned. Each field becomes essentially a tiny magnet with a north and south pole. In an unmagnetized piece of iron the fields are oriented in all possible directions and effectively cancel any overall magnetic effect. The net magnetism is therefore zero or near zero.

Magnetic Fields Interact

An electric charge at rest does not have a magnetic field, so it does not interact with a magnetic field from a magnet. A moving electric charge does produce a magnetic field and this field will interact with another magnetic field, producing a force. A DC current is many charges moving through a wire, so it follows that a current-carrying wire will also interact with another magnetic field, producing a force. This force varies, depending on the size and direction of the current and the orientation of the external field. With a constant current, the force is at a maximum when the direction of the current is at a right angle to the other magnetic field.

Since you cannot measure electricity directly, it must be measured indirectly through one of the effects that it produces. Recall that the strength of the magnetic field around a current-carrying wire is proportional to the size of the current. Thus, one way to measure a current is to measure the magnetic field that it produces. A device that measures the size of a current from the size of its magnetic field is called a **galvanometer** (figure 6.20). A galvanometer has a coil of wire that can rotate on pivots in the magnetic field of a permanent magnet. When there is a current in the coil, the magnetic field produced is attracted and repelled by the field of the permanent magnet. The larger the current the greater the force and the more the coil will rotate. The amount of movement of the coil is proportional to the current in the coil, and the pointer shows this on a scale. With

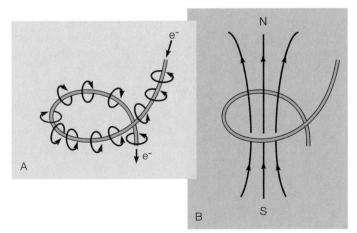

Figure 6.18
(*A*) Forming a wire into a loop causes the magnetic field to pass through the loop in the same direction. (*B*) This gives one side of the loop a north pole and the other side a south pole.

Figure 6.20

A galvanometer measures the direction and relative strength of an electric current from the magnetic field it produces. A coil of wire wrapped around an iron core becomes an electromagnet that rotates in the field of a permanent magnet. The rotation moves a pointer on a scale.

certain modifications and applications, the device can be used to measure current (ammeter), voltage (voltmeter), and resistance (ohmmeter).

An electric motor is able to convert electrical energy to mechanical energy by using the force produced by interacting magnetic fields. Basically, a motor has two working parts, a stationary electromagnet and an electromagnet that moves. The movable electromagnet rotates in the magnetic field of the stationary one. It turns fan blades, compressors, drills, pulleys, or other devices that do mechanical work.

Different designs of electric motors are used for various applications, but the simple demonstration motor shown in figure 6.21 can be used as an example of the basic operating principle.

Figure 6.21

A schematic of a simple electric motor.

Both the stationary and the movable electromagnets are connected to an electric current. When the current is turned on, the unlike poles of the two fields attract, rotating the movable electromagnet for a half turn. The motor has a simple device that now reverses the direction of the current. This switches the magnetic poles on the movable electromagnet, which is now repelled for another half turn. The device again reverses the polarity, and the motion continues in one direction. An actual motor has many wire loops in both electromagnets to obtain a useful force, and change the direction of the current often. This gives the motor a smoother operation with a greater turning force.

A Moving Magnet Produces an Electric Field

So far, you have learned what happens when we move a charge: (1) a moving charge and a current-carrying wire produce a magnetic field and (2) a second magnetic field exerts a force on a moving charge and exerts a force on a current-carrying wire as their magnetic fields interact. Since a moving charge produces a magnetic field, what should we expect if we move a magnet? The answer is that a *moving magnet* produces an *electric field* in the shape of circles around the path of the magnet. Here we see some similarity: A moving charge produces a magnetic field and a moving magnet produces an electric field. Thus we observe that electricity and magnetism interact, but only when there is motion.

A moving magnet produces an electric field that is circular around the path of the magnet. If you place a coil of wire near the moving magnet, the created electric field will interact with charges in the wire, forcing them to move as a current. The process of creating an electric current with a moving magnetic field is called **electromagnetic induction.** One way to produce electromagnetic induction is to move a bar magnet into or out of a coil of wire (figure 6.22). A galvanometer shows that the induced current flows one way when the bar magnet is moved toward the coil and flows the other way when the bar magnet is moved away from the coil. The same effect occurs if you move the coil back and forth over a stationary magnet. Furthermore, no current is detected when the magnetic field and the coil of wire are not moving. Thus, electromagnetic induction depends on the relative motion of the magnetic field and the coil of wire.

Figure 6.22

A current is induced in a coil of wire moved through a magnetic field. The direction of the current depends on the direction of motion.

It does not matter which one moves or changes, but one must move or change relative to the other for electromagnetic induction to occur.

Electromagnetic induction occurs when the loop of wire moves across magnetic field lines or when magnetic field lines moves across the loop. The size of the induced voltage is proportional to (1) the number of wire loops passing through the magnetic field lines, (2) the strength of the magnetic field, and (3) the rate at which magnetic field lines pass through the wire.

An **electric generator** is a device that converts mechanical energy into electrical energy. A simple generator is built much like an electric motor, with an axle with many wire loops in a coil that rotates. The coil is turned by some form of mechanical energy, such as a water turbine or a steam turbine, which uses steam generated from fossil fuels or nuclear energy. As the coil rotates in a magnetic field, a current is induced in the coil.

The use of the electric generator began in the 1890s, when George Westinghouse designed and built a waterwheel-powered 75-kilowatt AC generator near Ouray, Colorado (figure 6.23). Thomas Edison built a DC power plant in New York City about the same time (figure 6.24). In 1896 a 3.7 megawatt power plant at Niagara Falls sent electrical power to factories in Buffalo, New York.

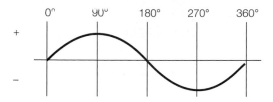

Figure 6.23

(*A*) Schematic of a simple alternator (AC generator) with one output loop. (*B*) Output of the single loop turning in a constant magnetic field, which alternates the induced current each half-cycle.

Figure 6.24

(*A*) Schematic of a simple DC generator with one output loop. (*B*) Output of the single loop turning in a constant magnetic field. The split ring (commutator) reverses the sign of the output when the voltage starts to reverse, so the induced current has half-cycle voltages of a constant sign, which is the definition of direct current.

C O N C E P T S   A P P L I E D

Simple Generator

1. Make a coil of insulated bell wire (#18 copper wire) by wrapping fifty windings around a cardboard tube from a roll of paper. Tape the coil at several places so it does not come apart and discard the cardboard tube.

2. Make a current-detecting instrument from a magnetic compass and some thin insulated wire (the thinner the better). Wrap the thin insulated wire in parallel windings around the compass. Make as many parallel windings as you can, but leave enough room to see both ends of the compass needle. Connect the wire ends to the coil you made in step 1.

3. Orient the compass so the needle is parallel to the wire around the compass. When a current passes through the coil of wire the magnetic field produced will cause the needle to move, showing the presence of a current.

4. First, move a bar magnet into and out of the stationary coil of wire and observe the compass needle. Second, move the coil of wire back and forth over a stationary bar magnet and observe the compass needle.

5. Experiment with a larger coil of wire, bar magnets of greater or weaker strengths, and by moving the coil at varying speeds. See how many generalizations you can make concerning electromagnetic induction.

Current from a power plant goes to a transformer to step up the voltage. A **transformer** is a device that steps up or steps down the AC voltage. It has two basic parts: (1) a *primary* or "input" coil, and (2) a *secondary* or "output" coil, which is close by. Both coils are often wound on a single iron core but are always fully insulated from each other. When an alternating current flows through the primary coil, a magnetic field grows around the coil to a maximum size, collapses to zero, then grows to a maximum size with an opposite polarity. This happens 120 times a second as the alternating current oscillates at 60 hertz. The growing and collapsing magnetic field moves across the wires in the secondary coil, inducing a voltage in the secondary coil.

The size of the induced voltage in the secondary coil is proportional to the number of wire loops in the two coils. Each turn or loop of the output coil has the same voltage induced in it, so more loops in an output coil means higher voltage output. If the output coil has the same number of loops as the input coil, the induced voltage in the secondary coil will be the same as the voltage in the primary coil. If the secondary coil has one-tenth as many loops as the primary coil, then the induced voltage in the secondary coil will be one-tenth the voltage in the primary coil. This is called a **step-down transformer** since the voltage was stepped down in the secondary coil. On the other hand, if the secondary coil has ten times more loops than the primary coil, then the voltage will be *increased* by a factor of 10. This is a **step-up transformer.** How much the voltage is stepped up or stepped down depends on the ratio of wire loops in the primary and secondary coils (figure 6.25). Note that the volts per wire loop are the same in each coil. The relationship is

$$\frac{\text{volts}_{\text{primary}}}{(\text{number of loops})_{\text{primary}}} = \frac{\text{volts}_{\text{secondary}}}{(\text{number of loops})_{\text{secondary}}}$$

or

$$\frac{V_{\text{p}}}{N_{\text{p}}} = \frac{V_{\text{s}}}{N_{\text{s}}}$$

equation 6.7

A step-up or step-down transformer steps up or steps down the voltage of an alternating current according to the ratio of wire loops in the primary and secondary coils. Assuming no losses in the transformer, the *power input* on the primary coil equals the *power output* on the secondary coil. Since $P = IV$, you can see that when the voltage is stepped up the current is correspondingly decreased, as

$$\text{power input} = \text{power output}$$
$$\text{watts input} = \text{watts output}$$
$$(\text{amps} \times \text{volts})_{\text{in}} = (\text{amps} \times \text{volts})_{\text{out}}$$

or

$$V_{\text{p}}I_{\text{p}} = V_{\text{s}}I_{\text{s}}$$

equation 6.8

Energy losses in transmission are reduced by stepping up the voltage. Recall that electrical resistance results in an energy loss and a corresponding absolute temperature increase in the conducting

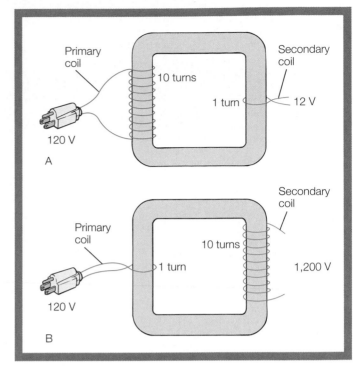

Figure 6.25

(A) This step-down transformer has ten turns on the primary for each turn on the secondary and reduces the voltage from 120 V to 12 V. (B) This step-up transformer increases the voltage from 120 V to 1,200 V, since there are ten turns on the secondary to each turn on the primary.

A Closer Look   Solar Cells

You may be familiar with many solid-state devices such as calculators, computers, word processors, digital watches, VCRs, digital stereos, and camcorders. All of these are called solid-state devices because they use a solid material, such as the semiconductor silicon, in an electric circuit in place of vacuum tubes. Solid-state technology developed from breakthroughs in the use of semiconductors during the 1950s, and the use of thin pieces of silicon crystal is common in many electric circuits today.

A related technology also uses thin pieces of a semiconductor such as silicon but not as a replacement for a vacuum tube. This technology is concerned with photovoltaic devices, also called *solar cells,* that generate electricity when exposed to light (box figure 6.5). A solar cell is unique in generating electricity since it produces electricity directly, without moving parts or chemical reactions, and potentially has a very

long lifetime. This reading is concerned with how a solar cell generates electricity.

Light is an electromagnetic wave in a certain wavelength range. This wave is a pulse of electric and magnetic fields that are locked together, moving through space as they exchange energy back and forth, regenerating each other in an endless cycle until the wave is absorbed, giving up its energy. Light also has the ability to give its energy to an electron, knocking it out of a piece of metal. This phenomenon, called the *photoelectric effect,* is described in chapter 7.

Solid-state technology is mostly concerned with crystals and electrons in crystals. A *crystal* is a solid with an ordered, three-dimensional arrangement of atoms. A crystal of common table salt, sodium chloride, has an ordered arrangement that produces a cubic structure. You can see this cubic structure of table salt if you look closely at a few grains. The regular geometric arrangement of

atoms (or ions) in a crystal is called the *lattice,* or framework, of the crystal structure.

A highly simplified picture of a crystal of silicon has each silicon atom bonded with four other silicon atoms, with each pair of atoms sharing two electrons. Normally, this ties up all the electrons and none are free to move and produce an electric current. This happens in the dark in silicon crystals, and the silicon crystal is an insulator. Light, however, can break electrons free in a silicon crystal, so that they can move in a current. Silicon is therefore a semiconductor.

The conducting properties of silicon can be changed by *doping,* that is, artificially forcing atoms of other elements into the crystal lattice. Phosphorus, for example, has five electrons in its outermost shell compared to the four in a silicon atom. When phosphorus atoms replace silicon atoms in the crystal lattice, there are extra electrons not tied up in the two electron bonds. The

—Continued top of next page

Box Figure 6.6

The cell barrier forms at the p-n junction between the n-silicon and the p-silicon. The barrier creates a "one-way" door that accumulates negative charges in the n-silicon.

Box Figure 6.5

Solar cells are economical in remote uses such as (*A*) navigational aids and (*B*) communications. The solar panels in both of these examples are oriented toward the south.

extra electrons move easily through the crystal lattice, carrying a charge. Since the phosphorus-doped silicon carries a negative charge it is called an *n-type* semiconductor. The n means negative charge carrier.

A silicon crystal doped with boron will have atoms in the lattice with only three electrons in the outermost shell. This results in a deficiency, that is, electron "holes" that act as positive charges. A hole can move as an electron is attracted to it, but it leaves another hole elsewhere, from where it moved. Thus, a flow of electrons in one direction is equivalent to a flow of holes in the opposite direction. A hole, therefore, behaves as a positive charge. Since the boron-doped silicon carries a positive charge it is called a *p-type* semiconductor. The p means positive charge carrier.

The basic operating part of a silicon solar cell is typically an 8 cm wide and 3×10^{-1} mm (about one-hundredth of an inch) thick wafer cut from a silicon crystal. One side of the wafer is doped with boron to make p-silicon, and the other side is doped with phosphorus to make n-silicon. The place of contact between the two is called the p-n junction, which creates a *cell barrier*. The cell barrier forms as electrons are attracted from the n-silicon to the holes in the p-silicon. This creates a very thin zone of negatively charged p-silicon and positively charged n-silicon (box figure 6.6). Thus, an internal electric field is established at the p-n junction, and the field is the cell barrier. The barrier is about 3×10^{-5} mm (about one-millionth of an inch) thick.

A metal base plate is attached to the p-silicon side of the wafer, and a grid of metal contacts to the n-silicon side for electrical contacts. The grid is necessary to allow light into the cell. The entire cell is then coated with a transparent plastic covering.

The cell is thin, and light can penetrate through the p-n junction. Light impacts the p-silicon, freeing electrons. Low-energy free electrons might combine with a hole, but high-energy electrons cross the cell barrier into the n-silicon. The electron loses some of its energy, and the barrier prevents it from returning, creating an excess negative charge in the n-silicon and a positive charge in the p-silicon. This establishes a potential of about 0.5 volt, which will drive a 2 amp current (8 cm cell) through a circuit connected to the electrical contacts.

Solar cells are connected different ways for specific electrical energy requirements in about a 1 meter arrangement called a *module.* Several modules are used to make a solar *panel,* and panels are the units used to design a solar cell *array.*

Today, solar cells are essentially handmade and are economical only in remote power uses (navigational aids, communications, or irrigation pumps) and in consumer specialty items (solar-powered watches and calculators). Research continues on finding methods of producing highly efficient, highly reliable solar cells that are affordably priced.

wire. If the current is large, there are many collisions between the moving electrons and positive ions of the wire, resulting in a large energy loss. Each collision takes energy from the electric field, diverting it into increased kinetic energy of the positive ions and thus increased temperature of the conductor. The energy lost to resistance is therefore reduced by lowering the current, which is what a transformer does by increasing the voltage. Hence, electric power companies step up the voltage of generated power for economical transmission. A step-up transformer at a power plant, for example, might step up the voltage from 22,000 volts to 500,000 volts for transmission across the country to a city (figure 6.26A). This step up in voltage correspondingly reduces the current, lowering the resistance losses to a more acceptable 4 or 5 percent over long distances. A step-down transformer at a substation near the city reduces the voltage to several thousand volts for transmission around the city. Additional step-down transformers reduce this voltage to 120 volts for transmission to three or four houses (figure 6.26B).

A

Figure 6.26

Energy losses in transmission are reduced by increasing the voltage, so the voltage of generated power is stepped up at the power plant. (*A*) These transformers, for example, might step up the voltage from tens to hundreds of thousands of volts. After a step-down transformer reduces the voltage at a substation, still another transformer (*B*) reduces the voltage to 120 volts for transmission to three or four houses.

B

SUMMARY

The basic unit of matter is the *atom*, which is made up of *protons*, *electrons*, and other particles. Protons and electrons have a property called *electric charge*; electrons have a *negative electric charge* and protons have a *positive electric charge*. The charges interact and *like charges repel* and *unlike charges attract*.

Electrons can be moved and *electrostatic charge*, or *static electricity*, results from a surplus or deficiency of electrons.

A *quantity of charge* (*q*) is measured in units of *coulombs* (C), the charge equivalent to the transfer of 6.24×10^{18} charged particles such as the electron. The *fundamental charge* of an electron or proton is 1.60×10^{-19} coulomb. The *electrical forces* between two charged objects can be calculated from the relationship between the quantity of charge and the distance between two charged objects. The relationship is known as *Coulomb's law*.

A flow of electric charge is called an *electric current* (*I*). Current (*I*) is measured as the *rate* of flow of charge, the quantity of charge (*q*) through a conductor in a period of time (*t*). The unit of current in coulomb/second is called an *ampere* or *amp* for short (A).

An *electric circuit* has some device that does work in moving charges through wires to do work in another part of the circuit. The *work done* and the *size of the charge* moved defines *voltage*. A *volt* (V) is the ratio of work to charge moved, $V = W/q$. The ratio of volts/amps in a circuit is the *unit of resistance* called an *ohm*. *Ohm's law* is $V = IR$.

Disregarding the energy lost to resistance, the *work* done by a voltage source is equal to the work accomplished in electrical devices in a circuit. The *rate of doing work is power*, or *work per unit time*, $P = W/t$. *Electrical power* can be calculated from the relationship of $P = IV$, which gives the power unit of *watts*.

Magnets have two poles about which their attraction is concentrated. When free to turn, one pole moves to the north and the other to the south. The north-seeking pole is called the *north pole* and the south-seeking pole is called the *south pole*. *Like poles repel* one another and *unlike poles attract*.

A *current-carrying wire* has magnetic field lines of closed, *concentric circles* that are at right angles to the length of wire. The *direction* of the magnetic field depends on the direction of the current. A coil of many loops is called a *solenoid* or *electromagnet*. The electromagnet is the working part in electric meters, electromagnetic switches, and the electric motor.

When a loop of wire is moved in a magnetic field, or if a magnetic field is moved past a wire loop, a voltage is induced in the wire loop. The interaction is called *electromagnetic induction*. An electric generator is a rotating coil of wire in a magnetic field. The coil is rotated by mechanical energy, and electromagnetic induction induces a voltage, thus converting mechanical energy to electrical energy. A *transformer* steps up or steps down the voltage of an alternating current. The ratio of input and output voltage is determined by the number of loops in the primary and secondary coils. Increasing the voltage decreases the current, which makes long-distance transmission of electrical energy economically feasible.

Summary of Equations

6.1 electrical force $=$ (constant) $\times \dfrac{\text{charge on one object} \times \text{charge on second object}}{\text{distance between objects squared}}$

$$F = k\frac{q_1 q_2}{d^2}$$

where $k = 9.00 \times 10^9$ newton·meters²/coulomb²

6.2 $\quad$ electrical current $= \dfrac{\text{quantity of charge}}{\text{time}}$

$$I = \frac{q}{t}$$

6.3 $\quad$ voltage $= \dfrac{\text{work}}{\text{charge moved}}$

$$V = \frac{W}{q}$$

6.4 $\quad$ voltage = current $\times$ resistance

$$V = IR$$

6.5 $\quad$ power = voltage $\times$ current

$$P = VI$$

6.6 $\quad$ cost $= \dfrac{(\text{work})(\text{time})(\text{rate})}{1,000 \frac{\text{watt}}{\text{kilowatt}}}$

6.7 $\quad \dfrac{\text{volts}_{\text{primary}}}{(\text{number of loops})_{\text{primary}}} = \dfrac{\text{volts}_{\text{secondary}}}{(\text{number of loops})_{\text{secondary}}}$

$$\frac{V_p}{N_p} = \frac{V_s}{N_s}$$

6.8 $\quad$ (amps $\times$ volts)$_{\text{in}}$ = (amps $\times$ volts)$_{\text{out}}$

$$V_p I_p = V_s I_s$$

KEY TERMS

alternating current (p. **112**)
amp (p. **109**)
ampere (p. **109**)
coulomb (p. **108**)
Coulomb's law (p. **109**)
direct current (p. **111**)
electric charge (p. **106**)
electric circuit (p. **112**)
electric current (p. **109**)
electric field (p. **109**)
electrical force (p. **106**)
electromagnet (p. **119**)
electromagnetic induction (p. **121**)
electrostatic charge (p. **107**)
fundamental charge (p. **108**)
galvanometer (p. **119**)
magnetic field (p. **117**)
magnetic poles (p. **116**)
ohm (p. **113**)
Ohm's law (p. **113**)
solenoid (p. **119**)
step-down transformer (p. **122**)
step-up transformer (p. **122**)
transformer (p. **122**)
volt (p. **112**)
watt (p. **113**)

APPLYING THE CONCEPTS

1. An object that acquires an excess of electrons becomes a (an)
   (a) ion.
   (b) negatively charged object.
   (c) positively charged object.
   (d) electrical conductor.
2. Which of the following is most likely to acquire an electrostatic charge?
   (a) electrical conductor
   (b) electrical nonconductor
   (c) Both are equally likely.
   (d) None of the above are correct.
3. A quantity of electric charge is measured in a unit called a (an)
   (a) coulomb.
   (b) volt.
   (c) amp.
   (d) watt.
4. The unit that describes the potential difference that occurs when a certain amount of work is used to move a certain quantity of charge is called the
   (a) ohm.
   (b) volt.
   (c) amp.
   (d) watt.
5. An electric current is measured in units of
   (a) coulomb.
   (b) volt.
   (c) amp.
   (d) watt.
6. In an electric current the electrons are moving
   (a) at a very slow rate.
   (b) at the speed of light.
   (c) faster than the speed of light.
   (d) at a speed described as "Warp 8."
7. If you multiply amps × volts, the answer will be in units of
   (a) resistance.
   (b) work.
   (c) current.
   (d) power.
8. The unit of resistance is the
   (a) watt.
   (b) ohm.
   (c) amp.
   (d) volt.
9. Compared to a thick wire, a thin wire of the same length, material, and temperature has
   (a) less electrical resistance.
   (b) more electrical resistance.
   (c) the same electrical resistance.
   (d) None of the above are correct.
10. A permanent magnet has magnetic properties because
    (a) the magnetic fields of its electrons are balanced.
    (b) of an accumulation of monopoles in the ends.
    (c) the magnetic domains are aligned.
    (d) All of the above are correct.
11. A current-carrying wire has a magnetic field around it because
    (a) a moving charge produces a magnetic field of its own.
    (b) the current aligns the magnetic domains in the metal of the wire.
    (c) the metal was magnetic before the current was established and the current enhanced the magnetic effect.
    (d) None of the above are correct.
12. A step-up transformer steps up (the)
    (a) power.
    (b) current.
    (c) voltage.
    (d) All of the above are correct.

Answers

1. b **2.** b **3.** a **4.** b **5.** c **6.** a **7.** d **8.** b **9.** b **10.** c **11.** a **12.** c

QUESTIONS FOR THOUGHT

1. Explain why a balloon that has been rubbed sticks to a wall for a while.
2. Explain what is happening when you walk across a carpet and receive a shock when you touch a metal object.
3. Why does a positively or negatively charged object have multiples of the fundamental charge?
4. Explain how you know that it is an electric field, not electrons, that moves rapidly through a circuit.
5. Is a kWh a unit of power or a unit of work? Explain.
6. What is the difference between AC and DC?
7. What is a magnetic pole? How are magnetic poles named?
8. How is an unmagnetized piece of iron different from the same piece of iron when it is magnetized?
9. Explain why the electric utility company increases the voltage of electricity for long-distance transmission.
10. Describe how an electric generator is able to generate an electric current.
11. Why does the north pole of a magnet point to the geographic North Pole if like poles repel?
12. Explain what causes an electron to move toward one end of a wire when the wire is moved across a magnetic field.

PARALLEL EXERCISES

The exercises in groups A and B cover the same concepts. Solutions to group A exercises are located in appendix D.

Group A

1. A rubber balloon has become negatively charged from being rubbed with a wool cloth, and the charge is measured as 1.00×10^{-14} C. According to this charge, the balloon contains an excess of how many electrons?

2. An electric current through a wire is 6.00 C every 2.00 s. What is the magnitude of this current?

3. A current of 4.00 A flows through a toaster connected to a 120.0 V circuit. What is the resistance of the toaster?

4. What is the current in a 60.0 Ω resistor when the potential difference across it is 120.0 V?

5. A lightbulb with a resistance of 10.0 Ω allows a 1.20 A current to flow when connected to a battery. (a) What is the voltage of the battery? (b) What is the power of the lightbulb?

6. A small radio operates on 3.00 V and has a resistance of 15.0 Ω. At what rate does the radio use electric energy?

7. A 1,200 W hair dryer is operated on a 120 V circuit for 15 min. If electricity costs $0.10/kWh, what was the cost of using the blow dryer?

8. An automobile starter rated at 2.00 hp draws how many amps from a 12.0 V battery?

9. An average-sized home refrigeration unit has a 1/3 hp fan motor for blowing air over the inside cooling coils, a 1/3 hp fan motor for blowing air over the outside condenser coils, and a 3.70 hp compressor motor. (a) All three motors use electric energy at what rate? (b) If electricity costs $0.10/kWh, what is the cost of running the unit per hour? (c) What is the cost for running the unit 12 hours a day for a 30-day month?

10. A 15 ohm toaster is turned on in a circuit that already has a 0.20 hp motor, three 100 W lightbulbs, and a 600 W electric iron that are on. Will this trip a 15 A circuit breaker? Explain.

11. A power plant generator produces a 1,200 V, 40 A alternating current that is fed to a step-up transformer before transmission over the high lines. The transformer has a ratio of 200 to 1 wire loops. (a) What is the voltage of the transmitted power? (b) What is the current?

12. A step-down transformer has an output of 12 V and 0.5 A when connected to a 120 V line. Assuming no losses: (a) What is the ratio of primary to secondary loops? (b) What current does the transformer draw from the line? (c) What is the power output of the transformer?

Group B

1. An inflated rubber balloon is rubbed with a wool cloth until an excess of a billion electrons is on the balloon. What is the magnitude of the charge on the balloon?

2. A wire carries a current of 2.0 A. At what rate is the charge flowing?

3. A current of 0.83 A flows through a lightbulb in a 120 V circuit. What is the resistance of this lightbulb?

4. What is the voltage across a 60.0 Ω resistor with a current of 3⅓ amp?

5. A 10.0 Ω lightbulb is connected to a 12.0 V battery. (a) What current flows through the bulb? (b) What is the power of the bulb?

6. A lightbulb designed to operate in a 120.0 V circuit has a resistance of 192 Ω. At what rate does the bulb use electric energy?

7. What is the monthly energy cost of leaving a 60 W bulb on continuously if electricity costs 10¢ per kWh?

8. An electric motor draws a current of 11.5 A in a 240 V circuit. (a) What is the power of this motor in W? (b) How many horsepower is this?

9. A swimming pool requiring a 2.0 hp motor to filter and circulate the water runs for 18 hours a day. What is the monthly electrical cost for running this pool pump if electricity costs 10¢ per kWh?

10. Is it possible for two people to simultaneously operate 1,300 W hair dryers on the same 120 V circuit without tripping a 15 A circuit breaker? Explain.

11. A step-up transformer has a primary coil with 100 loops and a secondary coil with 1,500 loops. If the primary coil is supplied with a household current of 120 V and 15 A, (a) what voltage is produced in the secondary circuit? (b) What current flows in the secondary circuit?

12. The step-down transformer in a local neighborhood reduces the voltage from a 7,200 V line to 120 V. (a) If there are 125 loops on the secondary, how many are on the primary coil? (b) What current does the transformer draw from the line if the current in the secondary is 36 A? (c) What are the power input and output?

Light

This fiber optics bundle carries pulses of light from an infrared laser to carry much more information than could be carried by electrons moving through wires. This is part of a dramatic change underway, a change that will first find a hybrid "optoelectronics" replacing the more familiar "electronics" of electrons and wires.

O U T L I N E

You use light and your eyes more than any other sense to learn about your surroundings. All of your other senses—touch, taste, sound, and smell—involve matter, but the most information is provided by light. Yet, light seems more mysterious than matter. You can study matter directly, measuring its dimensions, taking it apart, and putting it together to learn about it. Light, on the other hand, can only be studied indirectly in terms of how it behaves (figure 7.1). Once you understand its behavior, you know everything there is to know about light. Anything else is thinking about what the behavior means.

The behavior of light has stimulated thinking, scientific investigations, and debate for hundreds of years. The investigations and debate have occurred because light cannot be directly observed, which makes the exact nature of light very difficult to pin down. For example, you know that light moves energy from one place to another place. You can feel energy from the sun as sunlight warms you, and you know that light has carried this energy across millions of miles of empty space. The ability of light to move energy like this could be explained (1) as energy transported by waves, just as sound waves carry energy from a source, or (2) as the kinetic energy of a stream of moving particles, which give up their energy when they strike a surface. The movement of energy from place to place could be explained equally well by a wave model of light or by a particle model of light. When two possibilities exist like this in science, experiments are designed and measurements are made to support one model and reject the other. Light, however, presents a baffling dilemma. Some experiments provide evidence that light consists of waves and not a stream of moving particles. Yet other experiments provide evidence of just the opposite, that light is a stream of particles and not a wave. Evidence for accepting a wave or particle model seems to depend on which experiments are considered.

The purpose of using a model is to make new things understandable in terms of what is already known. When these new things concern light, three models are useful in visualizing separate behaviors. Thus, the electromagnetic wave model will be used to describe how light is created at a source. Another model, a model of light as a ray, a small beam of light, will be used to discuss some common properties of light such as reflection and the refraction, or bending, of light. Finally, properties of light that provide evidence for a particle model will be discussed before ending with a discussion of the present understanding of light.

SOURCES OF LIGHT

The sun, and other stars, lightbulbs, and burning materials all give off light. When something produces light it is said to be **luminous.** The sun is a luminous object that provides almost all of the *natural* light on the earth. A small amount of light does reach the earth from the stars but not really enough to see by on a moonless night. The moon and planets shine by reflected light and do not produce their own light, so they are not luminous.

Burning has been used as a source of *artificial* light for thousands of years. A wood fire and a candle flame are luminous because of their high temperatures. When visible light is given off as a result of high temperatures, the light source is said to be **incandescent.** A flame from any burning source, an ordinary lightbulb, and the sun are all incandescent sources because of high temperatures.

How do incandescent objects produce light? One explanation is given by the electromagnetic wave model. This model describes a relationship between electricity, magnetism, and light. The model pictures an electromagnetic wave as forming whenever an electric charge is *accelerated* by some external force. The acceleration produces a wave consisting of electrical and magnetic fields that become isolated from the accelerated charge, moving off into space. As the wave moves through space, the two fields exchange energy back and forth, continuing on until they are absorbed by matter and give up their energy.

The frequency of an electromagnetic wave depends on the acceleration of the charge; the greater the acceleration, the higher the frequency of the wave that is produced. The complete range of frequencies is called the *electromagnetic spectrum* (figure 7.2). The spectrum ranges from radio waves at the low-frequency end of the spectrum to gamma rays at the high-frequency end. Visible light occupies only a small part of the middle portion of the complete spectrum.

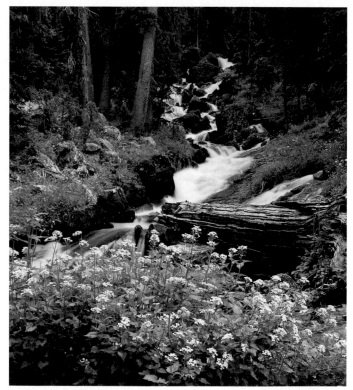

Figure 7.1

Light, sounds, and odors can identify the pleasing environment of this garden, but light provides the most information. Sounds and odors can be identified and studied directly, but light can only be studied indirectly, that is, in terms of how it behaves. As a result, the behavior of light has stimulated thinking, scientific investigations, and debate for hundreds of years. Perhaps you have wondered about light and its behaviors. What is light?

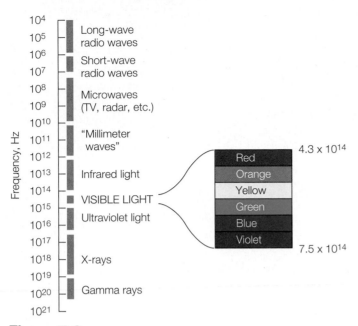

Figure 7.2

The electromagnetic spectrum. All electromagnetic waves have the same fundamental character and the same speed in a vacuum, but many aspects of their behavior depend on their frequency.

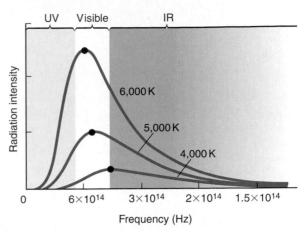

Figure 7.3

Blackbody spectra for several different temperatures. The frequency of the peak of the curve (shown by dot) shifts to higher frequency at higher temperatures.

Visible light is emitted from incandescent sources at high temperatures, but actually electromagnetic radiation is given off from matter at *any* temperature. This radiation is called **blackbody radiation,** which refers to an idealized material (the *blackbody*) that perfectly absorbs and perfectly emits electromagnetic radiation. From the electromagnetic wave model, the radiation originates from the acceleration of charged particles near the surface of an object. The frequency of the blackbody radiation is determined by the energy available for accelerating charged particles, that is, the temperature of the object. Near absolute zero, there is little energy available and no radiation is given off. As the temperature of an object is increased, more energy is available, and this energy is distributed over a range of values, so more than one frequency of radiation is emitted. A graph of the frequencies emitted from the range of available energy is thus somewhat bell-shaped. The steepness of the curve and the position of the peak depend on the temperature (figure 7.3). As the temperature of an object increases, there is an increase in the *amount* of radiation given off, and the peak radiation emitted progressively *shifts* toward higher and higher frequencies.

At room temperature the radiation given off from an object is in the infrared region, invisible to the human eye. When the temperature of the object reaches about 700°C (about 1,300°F), the peak radiation is still in the infrared region, but the peak has shifted enough toward the higher frequencies that a little visible light is emitted as a dull red glow. As the temperature of the object continues to increase, the amount of radiation increases, and the peak continues to shift toward shorter wavelengths. Thus, the object begins to glow brighter, and the color changes from red, to orange, to yellow, and eventually to white. The association of this color change with temperature is noted in the referent description of an object being "red hot," "white hot," and so forth.

The incandescent flame of a candle or fire results from the blackbody radiation of carbon particles in the flame. At a blackbody temperature of 1,500°C (about 2,700°F), the carbon particles emit visible light in the red to yellow frequency range. The tungsten filament of an incandescent lightbulb is heated to about 2,200°C (about 4,000°F) by an electric current. At this temperature, the visible light emitted is in the reddish, yellow-white range.

The radiation from the sun, or sunlight, comes from the sun's surface, which has a temperature of about 5,700°C (about 10,000°F). As shown in figure 7.4, the sun's radiation has a broad spectrum centered near the yellow-green frequency. Your eye is most sensitive to this frequency of sunlight. The spectrum of sunlight before it travels through the earth's atmosphere is infrared (about 51 percent), visible light (about 40 percent), and

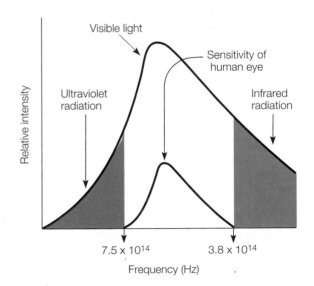

Figure 7.4

Sunlight is about 9 percent ultraviolet radiation, 40 percent visible light, and 51 percent infrared radiation before it travels through the earth's atmosphere.

Ultraviolet Light and Life

Ultraviolet light is a specific portion of the electromagnetic spectrum that has a significant impact on living things. UV light has been subdivided into different categories based on wavelength—the shorter the wavelength the higher the energy level. Ultraviolet light causes mutations when absorbed by DNA.

Class	Wavelength (nanometers)	Characteristics	Biological Effects
UV-A	320–400	Most passes through the ozone layer; can pass through window glass	Causes skin damage and promotes skin cancer
UV-B	280–320	90% blocked by ozone layer; amount is highest at midday	Causes sunburn, skin damage, and skin cancer
UV-C	100–280	Nearly all blocked by ozone layer	Kills bacteria and viruses

- Ultraviolet light is important in forming Vitamin D in the skin. When cholesterol in the skin is struck by ultraviolet light, some of the cholesterol is converted to Vitamin D. Therefore some exposure to ultraviolet light is good. However, since large amounts of ultraviolet light damage the skin, the normal tanning reaction seen in the skin of light-skinned people is a protective reaction to the damaging effects of ultraviolet light. The darker skin protects the deeper layers of the skin from damage.

- Tanning booths have mostly UV-A with a small amount of UV-B. Dosages in tanning booths are usually higher than in sunlight.

- Since UV-C disrupts DNA it can be used to sterilize medical instruments, drinking water, and wastewater. A standard mercury vapor light emits much of its light in the UV-C range, but the lamp is coated to prevent it from leaving the lamp. Specially manufactured lamps without the coating and ultraviolet transparent glass are used to provide UV-C for germicidal purposes.

- Some animals are able to see light in the ultraviolet spectrum. In particular, bees do not see the color red but are able to see ultraviolet. When flowers are photographed with filters that allow ultraviolet light to be recorded on the film, flowers often have a different appearance from what we see. The pattern assists the bees in finding the source of nectar within the flower.

---

ultraviolet (about 9 percent). Sunlight originated as energy released from nuclear reactions in the sun's core. This energy requires about a million years to work its way up to the surface. At the surface, the energy from the core accelerates charged particles, which then emit light like tiny antennas. The sunlight requires about eight minutes to travel the distance from the sun's surface to the earth.

PROPERTIES OF LIGHT

You can see luminous objects from the light they emit, and you can see nonluminous objects from the light they reflect, but you cannot see the path of the light itself. For example, you cannot see a flashlight beam unless you fill the air with chalk dust or smoke. The dust or smoke particles reflect light, revealing the path of the beam. This simple observation must be unknown to the makers of science fiction movies, since they always show visible laser beams zapping through the vacuum of space.

Some way to represent the invisible travels of light is needed in order to discuss some of its properties. Throughout history a **light ray model** has been used to describe the travels of light. The meaning of this model has changed over time, but it has always been used to suggest that "something" travels in *straight-line paths.* The light ray is a line that is drawn to represent the straight-line travel of light. A line is drawn to represent this imaginary beam to illustrate the law of reflection (as from a mirror) and the law of refraction (as through a lens). There are limits to using a light ray for explaining some properties of light, but it works very well in explaining mirrors, prisms, and lenses.

Light Interacts with Matter

A ray of light travels in a straight line from a source until it encounters some object or particles of matter (figure 7.5). What happens next depends on several factors, including (1) the smoothness of the surface, (2) the nature of the material, and (3) the angle at which the light ray strikes the surface.

The *smoothness* of the surface of an object can range from perfectly smooth to extremely rough. If the surface is perfectly smooth, rays of light undergo *reflection,* leaving the surface parallel to each other. A mirror is a good example of a very smooth surface that reflects light this way (figure 7.6A). If a surface is not

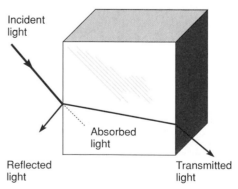

Figure 7.5

Light that interacts with matter can be reflected, absorbed, or transmitted through transparent materials. Any combination of these interactions can take place, but a particular substance is usually characterized by what it mostly does to light.

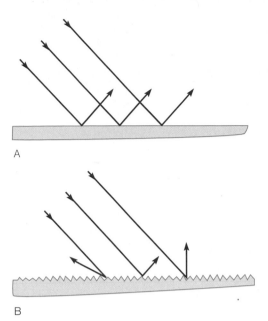

Figure 7.6
(*A*) Rays reflected from a perfectly smooth surface are parallel to each other.
(*B*) Diffuse reflection from a rough surface causes rays to travel in many random directions.

Figure 7.7
Light travels in a straight line, and the color of an object depends on which wavelengths of light the object reflects. Each of these flowers absorbs the colors of white light and reflects the color that you see.

smooth, the light rays are reflected in many random directions as *diffuse reflection* takes place (figure 7.6B). Rough and irregular surfaces and dust in the air make diffuse reflections. It is diffuse reflection that provides light in places not in direct lighting, such as under a table or under a tree. Such shaded areas would be very dark without diffuse reflection of light.

Some *materials* allow much of the light that falls on them to move through the material without being reflected. Materials that allow transmission of light through them are called *transparent*. Glass and clear water are examples of transparent materials. Many materials do not allow transmission of any light and are called *opaque*. Opaque materials reflect light, absorb light, or some combination of partly absorbing and partly reflecting light (figure 7.7). The light that is reflected varies with wavelength and gives rise to the perception of color, which will be discussed shortly. Absorbed light gives up its energy to the material and may be reemitted at a different wavelength, or it may simply show up as a temperature increase.

The *angle* of the light ray to the surface and the nature of the material determine if the light is absorbed, transmitted through a transparent material, or reflected. Vertical rays of light, for example, are mostly transmitted through a transparent material with some reflection and some absorption. If the rays strike the surface at some angle, however, much more of the light is reflected, bouncing off the surface. Thus, the glare of reflected sunlight is much greater around a body of water in the late afternoon than when the sun is directly overhead.

Light that interacts with matter is reflected, transmitted, or absorbed, and all combinations of these interactions are possible. Materials are usually characterized by which of these interactions they *mostly* do, but this does not mean that other interactions are not occurring too. For example, a window glass is usually characterized as a transmitter of light. Yet, the glass always *reflects* about 4 percent of the light that strikes it. The reflected light usually goes unnoticed during the day because of the bright light that is transmitted from the outside. When it is dark outside you notice the reflected light as the window glass now appears to act much like a mirror. A one-way mirror is another example of both reflection and transmission occurring (figure 7.8). A mirror is usually characterized as a reflector of light. A one-way mirror, however, has a very thin silvering that reflects most of the light but still transmits a little. In a lighted room, a one-way mirror appears to reflect light just as any other mirror does. But a person behind the mirror in a dark room can see into the lighted room by means of the transmitted light. Thus, you know that this mirror transmits as well as reflects light. One-way mirrors are used to unobtrusively watch for shoplifters in many businesses.

Reflection

Most of the objects that you see are visible from diffuse reflection. For example, consider some object such as a tree that you see during a bright day. Each *point* on the tree must reflect light in all directions, since you can see any part of the tree from any angle (figure 7.9). As a model, think of bundles of light rays entering your eye, which enable you to see the tree. This means that you can see any part of the tree from any angle because different bundles of reflected rays will enter your eye from different parts of the tree.

Light rays that are diffusely reflected move in all possible directions, but rays that are reflected from a smooth surface, such as a mirror, leave the mirror in a definite direction. Suppose you look at a tree in a mirror. There is only one place on the mirror where you look to see any one part of the tree. Light is reflecting

A

B

Figure 7.8

(*A*) The one-way mirror around the top of the wall in this store reflects most of the light that strikes it. (*B*) It also transmits some light to a person behind the mirror in a darkened room.

Figure 7.9

Bundles of light rays are reflected diffusely in all directions from every point on an object. Only a few light rays are shown from only one point on the tree in this illustration. The light rays that move to your eyes enable you to see the particular point from which they were reflected.

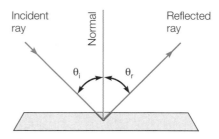

Figure 7.10

The law of reflection states that the angle of incidence (θ_i) is equal to the angle of reflection (θ_r). Both angles are measured from the *normal*, a reference line drawn perpendicular to the surface at the point of reflection.

off the mirror from all parts of the tree, but the only rays that reach your eye are the rays that are reflected at a certain angle from the place where you look. The relationship between the light rays moving from the tree and the direction in which they are reflected from the mirror to reach your eyes can be understood by drawing three lines: (1) a line representing an original ray from the tree, called the *incident ray,* (2) a line representing a reflected ray, called the *reflected ray,* and (3) a reference line that is perpendicular to the reflecting surface and is located at the point where the incident ray struck the surface. This line is called the *normal.* The angle between the incident ray and the normal is called the *angle of incidence,* θ_i, and the angle between the reflected ray and the normal is called the *angle of reflection,* θ_r (figure 7.10). The *law of reflection,* which was known to the ancient Greeks, is that the *angle of incidence equals the angle of reflection,* or

$$\theta_i = \theta_r$$

equation 7.1

Figure 7.11 shows how the law of reflection works when you look at a flat mirror. Light is reflected from all points on the box, and of course only the rays that reach your eyes are detected. These rays are reflected according to the law of reflection, with the angle of reflection equaling the angle of incidence. If you move your head slightly, then a different bundle of rays reaches your eyes. Of all the bundles of rays that reach your eyes, only two rays from a point are shown in the illustration. After these two rays are reflected, they continue to spread apart at the same rate that they were spreading before reflection. Your eyes and brain do not know that the rays have been reflected, and the diverging rays appear to come from behind the mirror, as the dashed lines show. The image, therefore, appears to be the same distance *behind* the mirror as the box is from the front of the mirror. Thus, a mirror image is formed where the rays of light *appear* to originate. This is called a **virtual image.** A virtual image is the result of your eyes' and brain's interpretations of light

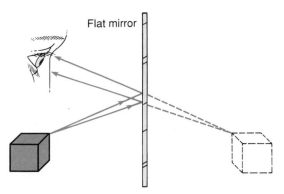

Figure 7.11

Light rays leaving a point on the block are reflected according to the law of reflection, and those reaching your eye are seen. After reflecting, the rays continue to spread apart at the same rate. You interpret this to be a block the same distance behind the mirror. You see a virtual image of the block, because light rays do not actually move from the image.

rays, not actual light rays originating from an image. Light rays that do originate from the other kind of image are called a **real image.** A real image is like the one displayed on a movie screen, with light originating from the image. A virtual image cannot be displayed on a screen, since it results from an interpretation.

Curved mirrors are either *concave*, with the center part curved inward, or *convex*, with the center part bulging outward. A concave mirror can be used to form an enlarged virtual image, such as a shaving or makeup mirror, or it can be used to form a real image, as in a reflecting telescope. Convex mirrors, for example, the mirrors on the sides of trucks and vans, are often

used to increase the field of vision. Convex mirrors are also used above an aisle in a store to show a wide area.

Refraction

You may have observed that an object that is partly in the air and partly in water appears to be broken, or bent, where the air and water meet. When a light ray moves from one transparent material to another, such as from water through air, the ray undergoes a change in the direction of travel at the boundary between the two materials. This change of direction of a light ray at the boundary is called **refraction.** The amount of change can be measured as an angle from the normal, just as it was for the angle of reflection. The incoming ray is called the *incident ray* as before, and the new direction of travel is called the *refracted ray*. The angles of both rays are measured from the normal (figure 7.12).

Refraction results from a *change in speed* when light passes from one transparent material into another. The speed of light in a vacuum is 3.00×10^8 m/s, but it is slower when moving through a transparent material. In water, for example, the speed of light is reduced to about 2.30×10^8 m/s. The speed of light has a magnitude that is specific for various transparent materials.

When light moves from one transparent material to another transparent material with a *slower* speed of light, the ray is refracted *toward* the normal (figure 7.13A). For example, light travels through air faster than through water. Light traveling from air into water is therefore refracted toward the normal as it enters the water. On the other hand, if light has a *faster* speed in the new material, it is refracted *away* from the normal. Thus, light traveling from water into the air is refracted away from the normal as it enters the air (figure 7.13B).

The magnitude of refraction depends on (1) the angle at which light strikes the surface and (2) the ratio of the speed of light in the two transparent materials. An incident ray that is perpendicular (90°) to the surface is not refracted at all. As the angle of incidence is increased, the angle of refraction is also

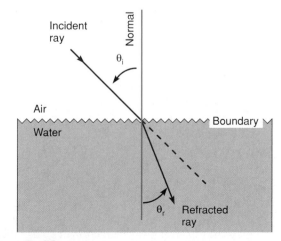

Figure 7.12

A ray diagram shows refraction at the boundary as a ray moves from air through water. Note that θ_i does not equal θ_r in refraction.

A

B

Figure 7.13

(*A*) A light ray moving to a new material with a slower speed of light is refracted toward the normal ($\theta_i > \theta_r$). (*B*) A light ray moving to a new material with a faster speed is refracted away from the normal ($\theta_i < \theta_r$).

increased. There is a limit, however, that occurs when the angle of refraction reaches 90°, or along the water surface. Figure 7.14 shows rays of light traveling from water to air at various angles. When the incident ray is about 49°, the angle of refraction that results is 90°, along the water surface. This limit to the angle of incidence that results in an angle of refraction of 90° is called the *critical angle* for a water-to-air surface (figure 7.14). At any incident angle greater than the critical angle, the light ray does not move from the water to the air but is *reflected* back from the sur-

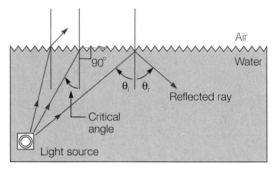

Figure 7.14

When the angle of incidence results in an angle of refraction of 90°, the refracted light ray is refracted along the water surface. The angle of incidence for a material that results in an angle of refraction of 90° is called the *critical angle*. When the incident ray is at this critical angle or greater, the ray is reflected internally. The critical angle for water is about 49°, and for a diamond it is about 25°.

face as if it were a mirror. This is called **total internal reflection** and implies that the light is trapped inside if it arrived at the critical angle or beyond. Faceted transparent gemstones such as the diamond are brilliant because they have a small critical angle and thus reflect much light internally. Total internal reflection is also important in fiber optics.

As was stated earlier, refraction results from a change in speed when light passes from one transparent material into another. The ratio of the speeds of light in the two materials determines the magnitude of refraction at any given angle of incidence. The greatest speed of light possible, according to current theory, occurs when light is moving through a vacuum. The speed of light in a vacuum is accurately known to nine decimals but is usually rounded to 3.00×10^8 m/s for general discussion. The speed of light in a vacuum is a very important constant in physical science, so it is given a symbol of its own, *c*. The ratio of

Historians tell us there are many early stories and legends about the development of ancient optical devices. The first glass vessels were made about 1500 B.C., so it is possible that samples of clear, transparent glass were available soon after. One legend claimed that the ancient Chinese invented eyeglasses as early as 500 B.C. A burning glass (lens) was mentioned in an ancient Greek play written about 424 B.C. Several writers described how Archimedes saved his hometown of Syracuse with a burning glass in about 214 B.C. Syracuse was besieged by Roman ships when Archimedes supposedly used the burning glass to focus sunlight on the ships, setting them on fire. It is not known if this story is true or not, but it is known that the Romans indeed did have burning glasses. Glass spheres, which were probably used to start fires, have been found in Roman ruins, including a convex lens recovered from the ashes of Pompeii.

Today, lenses are no longer needed to start fires, but they are common in cameras, scanners, optical microscopes, eyeglasses, lasers, binoculars, and many other optical devices. Lenses are no longer just made from glass, and today many are made from a transparent, hard plastic that is shaped into a lens.

The primary function of a lens is to form an image of a real object by refracting incoming parallel light rays. Lenses have two basic shapes, with the center of a surface either bulging in or bulging out. The outward bulging shape is thicker at the center than around the outside edge and this is called a *convex lens* (box figure 7.1A). The other basic lens shape is just the opposite, thicker around the outside edge than at the center and is called a *concave lens* (box figure 7.1B).

Convex lenses are used to form images in magnifiers, cameras, eyeglasses, projectors, telescopes, and microscopes (box figure 7.2). Concave lenses are used in some eyeglasses and in combinations with the convex lens to correct for defects. The convex lens is the most commonly used lens shape.

Your eyes are optical devices with convex lenses. Box figure 7.3 shows the basic structure. First, a transparent hole called the *pupil* allows light to enter the eye. The size of the pupil is controlled by the *iris,* the colored part that is a muscular diaphragm. The *lens* focuses a sharp image on the back surface of the eye, the *retina.* The retina is made up of millions of light-sensitive structures, and nerves carry signals (impulses) from the retina through the optic nerve to the brain.

Box Figure 7.1

(*A*) Convex lenses are called converging lenses since they bring together, or converge, parallel rays of light.
(*B*) Concave lenses are called diverging lenses since they spread apart, or diverge, parallel rays of light.

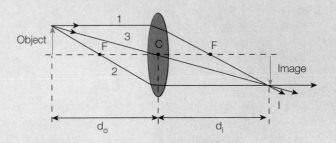

Box Figure 7.2

A convex lens forms an inverted image from refracted light rays of an object outside the focal point. Convex lenses are mostly used to form images in cameras, film or overhead projectors, magnifying glasses, and eyeglasses.

The lens is a convex, pliable material held in place and changed in shape by the attached *ciliary muscle.* When the eye is focused on a distant object the ciliary muscle is completely relaxed. Looking at a closer object requires the contraction of the ciliary muscles to change the curvature of the lens. This adjustment of focus by action of the ciliary muscle is called *accommodation.* The closest distance an object can be seen without a blurred image is called the *near point,* and this is the limit to accommodation.

The near point moves outward with age as the lens becomes less pliable. By middle age the near point may be twice this distance or greater, creating the condition known as farsightedness. The condition of farsightedness, or *hyperopia,* is a problem associated with aging (called presbyopia). Hyperopia can be caused at an early age by an eye that is too short or by problems with the cornea or lens that focuses the image behind the retina. Farsightedness can be corrected with a convex lens as shown in box figure 7.4A.

Nearsightedness, or *myopia,* is a problem caused by an eye that is too long or problems with the cornea or lens that focus the image in front of the retina. Nearsightedness can be corrected with a concave lens as shown in box figure 7.4B.

The microscope is an optical device used to make things look larger. It is essentially a system of two lenses, one to produce an image of the object being studied, and the other to act as a magnifying glass and enlarge that image. The power of the microscope is basically determined by the *objective lens,* which is placed close to the specimen on the stage of the microscope. Light is

—Continued top of next page

Continued—

Box Figure 7.3
Light rays from a distant object are focused by the lens onto the retina, a small area on the back of the eye.

A

B

Box Figure 7.4
Eyeglasses can help (*A*) a person with presbyopia (farsighted) or (*B*) a person with myopia (nearsighted).

Box Figure 7.5
A simple microscope uses a system of two lenses, which are an objective lens that makes an enlarged image of the specimen, and an eyepiece lens that makes an enlarged image of that image.

Box Figure 7.6
This illustrates how the path of light moves through a simple reflecting astronomical telescope. Several different designs and placement of mirrors are possible.

projected up through the specimen, and the objective lens makes an enlarged image of the specimen inside the tube between the two lenses. The *eyepiece lens* is positioned so that it makes a sharp enlarged image of the image produced by the objective lens (box figure 7.5).

Telescopes are optical instruments used to provide enlarged images of near and distant objects. There are two major types of telescopes, *refracting* telescopes that use two lenses, and *reflecting* telescopes that use combinations of mirrors, or a mirror and a lens. The refracting telescope has two lenses, with the objective lens forming a reduced image, which is viewed with an eyepiece lens to enlarge that image. In reflecting telescopes, mirrors are used instead of lenses to collect the light (box figure 7.6).

Finally, the *digital camera* is one of the more recently developed light-gathering and photograph-taking optical instruments. This camera has a group of small photocells, with perhaps thousands lined up on the focal plane behind a converging lens. An image falls on the array, and each photocell stores a charge that is

proportional to the amount of light falling on the cell. A microprocessor measures the amount of charge registered by each photocell and considers it as a pixel, a small bit of the overall image. A shade of gray or a color is assigned to each pixel and the image is ready to be enhanced, transmitted to a screen, printed, or magnetically stored for later use. Modern digital cameras have an advantage since the picture is available instantly, without the use of darkrooms, chemicals, or light-sensitive paper. A photo can be immediately downloaded from the camera into a computer, and then perhaps loaded onto your website, ready for the whole world to view.

---

c to the speed of light in some transparent material, *v*, is called the **index of refraction,** *n*, of that material or

$$n = \frac{c}{v}$$

equation 7.2

The indexes of refraction for some substances are listed in table 7.1. The values listed are constant physical properties and can be used to identify a specific substance. Note that a larger value means a greater refraction at a given angle. Of the materials listed, diamond refracts light the most and air the least. The index for air is nearly 1, which means that light is slowed only slightly in air.

Note that table 7.1 shows that colder air at 0°C (32°F) has a higher index of refraction than warmer air at 30°C (86°F), which means that light travels faster in warmer air. This difference explains the "wet" highway that you sometimes see at a distance in the summer. The air near the road is hotter on a clear, calm day. Light rays traveling toward you in this hotter air are refracted upward as they enter the cooler air. Your brain interprets this refracted light as *reflected* light. Light traveling downward from other cars is also refracted upward toward you, and you think you are seeing cars "reflected" from the wet highway (figure 7.15). When you reach the place where the "water" seemed to be, it disappears, only to appear again farther down the road.

Connections . . .

Diamond Quality

The beauty of a diamond depends largely on optical properties, including degree of refraction, but how well the cutter shapes the diamond determines the brilliance and reflection of light from a given stone. The "finish" describes the precision of facet placement on a cut stone, but overall quality involves much more. There are four factors that are generally used to determine the quality of a cut diamond. These are sometimes called the four C's of Carat, Color, Clarity, and Cut.

Carat is a measure of the weight of a diamond. When measuring diamonds (as opposed to gold), one metric carat is equivalent to 0.2 g, so a 5 carat diamond would have a mass of 1 g. Diamonds are weighed in to the nearest hundredth of a carat, and 100 points equal 1 carat. Thus a 3 point diamond is 3/100 of a carat, with a mass of about 0.006 g.

Color is a measure of how much (or how little) color a diamond has. The color grade is based on the color of the body of a diamond, not the color of light dispersion from the stone. The color grades range from colorless, near colorless, faint yellow, very light yellow, and down to light yellow. It is measured by comparing a diamond to a standard of colors, but the color of a diamond weighing one-half carat or less is often difficult to determine.

Clarity is a measure of inclusions (grains or crystals), cleavages, or other blemishes. Such inclusions and flaws may affect transparency, brilliance, and even the durability of a diamond. The clarity grade ranges from flawless down to industrial grade stones, which may appear milky from too many tiny inclusions.

Cut involves the relationships between the sizes of a stone's major physical features and their various angles. They determine, above all else, the limit to which a diamond will accomplish its optical potential. There are various types of faceted cuts, including the round brilliant cut, with a flat top, a flat bottom, and about fifty-eight facets. Facet shapes of square, triangular, diamond-shaped, and trapezoidal cuts might be used.

The details of a given diamond's four C's determine how expensive the diamond is going to be, and a top grade in all factors would be very expensive. No two stones are alike, and it is possible to find a brilliant diamond with a few lower grades in some of the factors that is a good value.

Table 7.1 Index of Refraction	
Substance	$n = c/v$
Glass	1.50
Diamond	2.42
Ice	1.31
Water	1.33
Benzene	1.50
Carbon tetrachloride	1.46
Ethyl alcohol	1.36
Air (0°C)	1.00029
Air (30°C)	1.00026

Sometimes convection currents produce a mixing of warmer air near the road with the cooler air just above. This mixing refracts light one way, then the other, as the warmer and cooler air mix. This produces a shimmering or quivering that some people call "seeing heat." They are actually seeing changing refraction, which is a *result* of heating and convection. In addition to causing distant objects to quiver, the same effect causes the point source of light from stars to appear to twinkle. The light from closer planets does not twinkle because the many light rays from the disklike sources are not refracted together as easily as the fewer rays from the point sources of stars. The light from planets will appear to quiver, however, if the atmospheric turbulence is great.

Figure 7.15

Mirages are caused by hot air near the ground refracting, or bending light rays upward into the eyes of a distant observer. The observer believes he is seeing an upside down image reflected from water on the highway.

CONCEPTS APPLIED

Internal Reflection

Seal a flashlight in a clear plastic bag to waterproof it, then investigate the critical angle and total internal reflection in a swimming pool, play pool, or large tub of water. In a darkened room or at night, shine the flashlight straight up from beneath the water, then at different angles until it shines almost horizontally beneath the surface. Report your observation of the critical angle for the water used.

Dispersion and Color

Electromagnetic waves travel with the speed of light with a whole spectrum of waves of various frequencies and wavelengths. The speed of electromagnetic waves (c) is related to the wavelength (λ) and the frequency (f) by a form of the wave equation, or

$$c = \lambda f$$

equation 7.3

Figure 7.16

The flowers appear to be red because they reflect light in the 7.9×10^{-7} m to 6.2×10^{-7} m range of wavelengths.

Visible light is the part of the electromagnetic spectrum that your eyes can detect, a narrow range of wavelength from about 7.90×10^{-7} m to 3.90×10^{-7} m. In general, this range of visible light can be subdivided into ranges of wavelengths that you perceive as colors (figure 7.16). These are the colors of the rainbow, and there are six distinct colors that blend one into another. These colors are *red, orange, yellow, green, blue,* and *violet.* The corresponding ranges of wavelengths and frequencies of these colors are given in table 7.2.

In general, light is interpreted to be white if it has the same mixture of colors as the solar spectrum. That sunlight is made up of component colors was first investigated in detail by Isaac Newton. While a college student, Newton became interested in grinding lenses, light, and color. At the age of twenty-three, Newton visited a local fair and bought several triangular glass prisms and proceeded to conduct a series of experiments with a beam of sunlight in his room. In 1672, he reported the results of his experiments with prisms and color, concluding that white light is a mixture of all the independent colors. Newton found that a beam of sunlight falling on a glass prism in a darkened room produced a band of colors he called a *spectrum.* Further, he found that a second glass prism would not subdivide each separate color but would combine all the colors back into white sunlight. Newton concluded that sunlight consists of a mixture of the six colors.

A glass prism separates sunlight into a spectrum of colors because the index of refraction is different for different wavelengths of light. The same processes that slow the speed of light in a transparent substance have a greater effect on short wavelengths than they do on longer wavelengths. As a result, violet light is refracted most, red light is refracted least, and the other colors are refracted between these extremes. This results in a beam of white light being separated, or dispersed, into a spectrum when it is refracted. Any transparent material in which the index of refraction varies with wavelength has the property of *dispersion.* The dispersion of light by ice crystals sometimes produces a colored halo around the sun and the moon.

EVIDENCE FOR WAVES

The nature of light became a topic of debate toward the end of the 1600s as Isaac Newton published his *particle theory* of light. He believed that the straight-line travel of light could be better explained as small particles of matter that traveled at great speed from a source of light. Particles, reasoned Newton, should follow a straight line according to the laws of motion. Waves, on the other hand, should bend as they move, much as water waves on a pond bend into circular shapes as they move away from a disturbance. About the same time that Newton developed his particle theory of light, Christian Huygens (pronounced "hi-ganz") was concluding that light is not a stream of particles but rather a longitudinal wave.

Both theories had advocates during the 1700s, but the majority favored Newton's particle theory. By the beginning of the 1800s, new evidence was found that favored the wave theory, evidence that could not be explained in terms of anything but waves.

Interference

In 1801, Thomas Young published evidence of a behavior of light that could only be explained in terms of a wave model of light. Young's experiment is illustrated in figure 7.17A. Light from a single source is used to produce two beams of light that are in phase, that is, having their crests and troughs together as they move away from the source. This light falls on a card with

Table 7.2	Range of Wavelengths and Frequencies of the Colors of Visible Light	
Color	**Wavelength (in meters)**	**Frequency (in hertz)**
Red	7.9×10^{-7} to 6.2×10^{-7}	3.8×10^{14} to 4.8×10^{14}
Orange	6.2×10^{-7} to 6.0×10^{-7}	4.8×10^{14} to 5.0×10^{14}
Yellow	6.0×10^{-7} to 5.8×10^{-7}	5.0×10^{14} to 5.2×10^{14}
Green	5.8×10^{-7} to 4.9×10^{-7}	5.2×10^{14} to 6.1×10^{14}
Blue	4.9×10^{-7} to 4.6×10^{-7}	6.1×10^{14} to 6.6×10^{14}
Violet	4.6×10^{-7} to 3.9×10^{-7}	6.6×10^{14} to 7.7×10^{14}

A rainbow is a spectacular, natural display of color that is supposed to have a pot of gold under one end. Understanding the why and how of a rainbow requires information about water droplets and knowledge of how light is reflected and refracted. This information will also explain why the rainbow seems to move when you move—making it impossible to reach the end to obtain that mythical pot of gold.

First, note the pattern of conditions that occurs when you see a rainbow. It usually appears when the sun is shining low in one part of the sky and rain is falling in the opposite part. With your back to the sun, you are looking at a zone of raindrops that are all showing red light, another zone that are all showing violet light, with zones of the other colors between (ROYGBV). For a rainbow to form like this requires a surface that refracts and reflects the sunlight, a condition met by spherical raindrops.

Water molecules are put together in such a way that they have a positive side and a negative side, and this results in strong molecular attractions. It is the strong attraction of water molecules for one another that results in the phenomenon of surface tension. Surface tension is the name given to the surface of water acting as if it is covered by an ultra-thin elastic membrane that is contracting. It is surface tension that pulls raindrops into a spherical shape as they fall through the air.

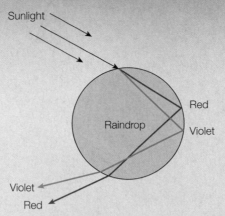

Box Figure 7.7

Light is refracted when it enters a raindrop and when it leaves. The part that leaves the front surface of the raindrop is the source of the light in thousands upon thousands of raindrops from which you see zones of color—a rainbow.

Box figure 7.7 shows one thing that can happen when a ray of sunlight strikes a single spherical raindrop near the top of the drop. At this point some of the sunlight is reflected, and some is refracted into the raindrop. The refraction disperses the light into its spectrum colors, with the violet light being refracted most and red the least. The refracted light travels through the drop to the opposite side, where some of it might be reflected back into the drop. The reflected part

travels back through the drop again, leaving the front surface of the raindrop. As it leaves, the light is refracted for a second time. The combined refraction, reflection, and second refraction is the source of the zones of colors you see in a rainbow. This also explains why you see a rainbow in the part of the sky opposite from the sun.

The light from any one raindrop is one color, and that color comes from all drops on the arc of a circle that is a certain angle between the incoming sunlight and the refracted light. Thus the raindrops in the red region refract red light toward your eyes at an angle of 42°, and all other colors are refracted over your head by these drops. Raindrops in the violet region refract violet light toward your eyes at an angle of 40° and the red and other colors toward your feet. Thus the light from any one drop is seen as one color, and all drops showing this color are on the arc of a circle. An arc is formed because the angle between the sunlight and the refracted light of a color is the same for each of the spherical drops.

There is sometimes a fainter secondary rainbow, with colors reversed, that forms from sunlight entering the bottom of the drop, reflecting twice, and then refracting out the top. The double reflection reverses the colors, and the angles are 50° for the red and 54° for the violet (see figure 1.15 on page 15).

Figure 7.17

(A) The arrangement for Young's double-slit experiment. Sunlight passing through the first slit is coherent and falls on two slits close to each other. Light passing beyond the two slits produces an interference pattern on a screen. (B) The double-slit pattern of a small-diameter beam of light from a helium-neon laser.

Lasers

A laser is a device that produces a coherent beam of single-frequency, in-phase light. The beam comes from atoms that have been stimulated by electricity. Most ordinary light sources produce incoherent light; light that is emitted randomly and at different frequencies. The coherent light from a laser has the same frequency, phase, and direction, so it does not tend to spread out and can be very intense. This has made possible a number of specialized applications and the list of uses continues to grow.

There are different kinds of lasers in use and new ones are under development. One common type of laser is a gas-filled tube with mirrors at both ends. The mirror at one end is only partly silvered, which allows light to escape as the laser beam. The distance between the mirrors matches the resonate frequency of the light produced, so the trapped light will set up an optical standing wave. An electric discharge produces fast electrons that raise the energy level of the electrons of the specific gas atoms in the tube.

The electrons of the energized gas atoms emit a particular frequency of light as they drop back to the original level, and this emitted light sets up the standing wave. The standing wave stimulates other atoms of the gas, resulting in the emission of more light at the same frequency and phase.

Lasers are everywhere today and have connections with a wide variety of technologies. At the supermarket a laser and detector unit reads the bar code on each grocery item. The laser sends the pattern to a computer, which sends a price to the register as well as tracking the store inventory. A low-powered laser and detector also reads your CD music disk or MP3 disk and can be used to make a three-dimensional image. Most laser printers use a laser, and a laser makes the operational part of a fiber optics communication system. Stronger lasers are used for cutting, drilling, and welding. Lasers are used extensively in many different medical procedures, from welding a detached retina to bloodless surgery.

---

two slits, each less than a millimeter in width. The light is diffracted from each slit, moving out from each as an expanding arc. Beyond the card the light from one slit crosses over the light from the other slit to produce a series of bright lines on a screen. Young had produced a phenomenon of light called **interference,** and interference can only be explained by waves!

The pattern of bright lines and dark zones is called an *interference pattern* (figure 7.17B). The light moved from each slit in phase, crest to crest and trough to trough. Light from both slits traveled the same distance directly across to the screen, so they arrived in phase. The crests from the two slits are superimposed here, and constructive interference produces a bright line in the center of the pattern. But for positions above and below the center, the light from the two slits must travel different distances to the screen. At a certain distance above and below the bright center line, light from one slit had to travel a greater distance and arrives one-half wavelength after light from the other slit. Destructive interference produces a zone of darkness at these positions. Continuing up and down the screen, a bright line of constructive interference will occur at each position where the distance traveled by light from the two slits differs by any whole number of wavelengths. A dark zone of destructive interference will occur at each position where the distance traveled by light from the two slits differs by any half-wavelength. Thus, bright lines occur above and below the center bright line at positions representing differences in paths of 1, 2, 3, 4, and so on wavelengths. Similarly, zones of darkness occur above and below the center bright line at positions representing differences in paths of ½, 1½, 2½, 3½, and so on wavelengths. Young found all of the experimental data such as this in full agreement with predictions from a wave theory of light. About fifteen years later, A. J. Fresnel (pronounced "fray-nel") demonstrated mathematically that diffraction as well as other behaviors of light could be fully explained with the wave theory. In 1821 Fresnel determined that the wavelength of red light was about 8×10^{-7} m

and of violet light about 4×10^{-7} m, with other colors in between these two extremes. The work of Young and Fresnel seemed to resolve the issue of considering light to be a stream of particles or a wave, and it was generally agreed that light must be waves.

Polarization

Huygens' wave theory and Newton's particle theory could explain some behaviors of light satisfactorily, but there were some behaviors that neither (original) theory could explain. Both theories failed to explain some behaviors of light, such as light moving through certain transparent crystals. For example, a slice of the mineral tourmaline transmits what appears to be a low-intensity greenish light. But if a second slice of tourmaline is placed on the first and rotated, the transmitted light passing through both slices begins to dim. The transmitted light is practically zero when the second slice is rotated 90°. Newton suggested that this behavior had something to do with "sides" or "poles" and introduced the concept of what is now called the *polarization* of light.

The waves of Huygens' wave theory were longitudinal, moving like sound waves, with wave fronts moving in the direction of travel. A longitudinal wave could not explain the polarization behavior of light. In 1817, Young modified Huygens' theory by describing the waves as *transverse*, vibrating at right angles to the direction of travel. This modification helped explain the polarization behavior of light transmitted through the two crystals and provided firm evidence that light is a transverse wave. As shown in figure 7.18A, **unpolarized light** is assumed to consist of transverse waves vibrating in all conceivable random directions. Polarizing materials, such as the tourmaline crystal, transmit light that is vibrating in one direction only, such as the vertical direction in figure 7.18B. Such a wave is said to be **polarized,** or *plane-polarized*, since it vibrates only

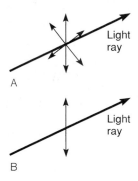

Figure 7.18

(*A*) Unpolarized light has transverse waves vibrating in all possible directions perpendicular to the direction of travel. (*B*) Polarized light vibrates only in one plane. In this illustration, the wave is vibrating in a vertical direction only.

in one plane. The single crystal polarized light by transmitting only waves that vibrate parallel to a certain direction while selectively absorbing waves that vibrate in all other directions. Your eyes cannot tell the difference between unpolarized and polarized light, so the light transmitted through a single crystal looks just like any other light. When a second crystal is placed on the first, the amount of light transmitted depends on the alignment of the two crystals (figure 7.19). When the two crystals are *aligned,* the polarized light from the first crystal passes through the second with little absorption. When the crystals are *crossed* at 90°, the light transmitted by the first is vibrating in a plane that is absorbed by the second crystal, and practically all the light is absorbed. At some other angle, only a fraction of the polarized light from the first crystal is transmitted by the second.

You can verify whether or not a pair of sunglasses is made of polarizing material by rotating a lens of one pair over a lens of a second pair. Light is transmitted when the lenses are aligned but mostly absorbed at 90° when the lenses are crossed.

Light is completely polarized when all the waves are removed except those vibrating in a single direction. Light is partially polarized when some of the waves are in a particular orientation, and any amount of polarization is possible. There are several means of producing partially or completely polarized light, including (1) selective absorption, (2) reflection, and (3) scattering.

Selective absorption is the process that takes place in certain crystals, such as tourmaline, where light in one plane is transmitted and all the other planes are absorbed. A method of manufacturing a polarizing film was developed in the 1930s by Edwin H. Land. The film is called *Polaroid.* Today Polaroid is made of long chains of hydrocarbon molecules that are aligned in a film. The long-chain molecules ideally absorb all light waves that are parallel to their lengths and transmit light that is perpendicular to their lengths. The direction that is *perpendicular* to the oriented molecular chains is thus called the polarization direction or the *transmission axis.*

Reflected light with an angle of incidence between 1° and 89° is partially polarized as the waves parallel to the reflecting surface are reflected more than other waves. Complete polarization, with all waves parallel to the surface, occurs at a particular angle of incidence. This angle depends on a number of variables, including the nature of the reflecting material. Figure 7.20 illustrates polarization by reflection. Polarizing sunglasses reduce the glare of reflected light because they have vertically oriented transmission axes. This absorbs the horizontally oriented reflected light. If you turn your head from side to side so as to rotate your sunglasses while looking at a reflected glare, you will see the intensity of the reflected light change. This means that the reflected light is partially polarized.

The phenomenon called *scattering* occurs when light is absorbed and reradiated by particles about the size of gas molecules that make up the air. Sunlight is initially unpolarized. When it strikes a molecule, electrons are accelerated and vibrate horizontally and vertically. The vibrating charges reradiate polarized light. Thus, if you look at the blue sky with a pair of polarizing sunglasses and rotate them, you will observe that light from the sky is polarized. Bees are believed to be able to detect polarized skylight and use it to orient the direction of their flights. Violet and blue light have the shortest wavelengths of visible light, and red and orange light have the longest. The violet

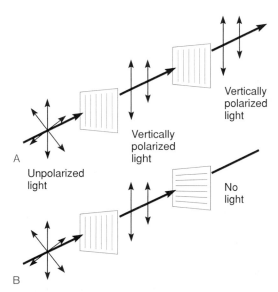

Figure 7.19

(*A*) Two crystals that are aligned both transmit vertically polarized light that looks like any other light. (*B*) When the crystals are crossed, no light is transmitted.

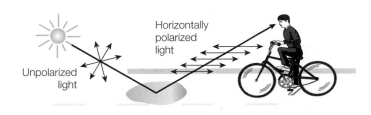

Figure 7.20

Light that is reflected becomes partially or fully polarized in a horizontal direction, depending on the incident angle and other variables.

Figure 7.21
A setup for observing the photoelectric effect. Light strikes the negatively charged plate, and electrons are ejected. The ejected electrons move to the positively charged plate and can be measured as a current in the circuit.

and blue rays of sunlight are scattered the most. At sunset the path of sunlight through the atmosphere is much longer than when the sun is more directly overhead. Much of the blue and violet have been scattered away as a result of the longer path through the atmosphere at sunset. The remaining light that comes through is mostly red and orange, so these are the colors you see at sunset.

EVIDENCE FOR PARTICLES

In 1850, J. L. Foucault was able to prove that light travels much slower in transparent materials than it does in air. This was in complete agreement with the wave theory and completely opposed to the particle theory. By the end of the 1800s, Maxwell's theoretical concept of electric and magnetic fields changed the concept of light from mechanical waves to waves of changing electric and magnetic fields. Further evidence removed the necessity for ether, the material supposedly needed for waves to move through. Light was now seen as electromagnetic waves that could move through empty space. By this time it was possible to explain all behaviors of light moving through empty space or through matter with a wave theory. Yet, there were nagging problems that the wave theory could not explain. In general, these problems concerned light that is absorbed by or emitted from matter.

Photoelectric Effect

Light is a form of energy, and it gives its energy to matter when it is absorbed. Usually the energy of absorbed light results in a temperature increase, such as the warmth you feel from absorbed sunlight. Sometimes, however, the energy from absorbed light results in other effects. In some materials, the energy is acquired by electrons, and some of the electrons acquire sufficient energy to jump out of the material. The movement of electrons as a result of energy acquired from light is known as the **photo-**

electric effect. The photoelectric effect is put to a practical use in a solar cell, which transforms the energy of light into an electric current (figure 7.21).

The energy of light can be measured with great accuracy. The kinetic energy of electrons after they absorb light can also be measured with great accuracy. When these measurements were made of the light and electrons involved in the photoelectric effect, some unexpected results were observed. Monochromatic light, that is, light of a single, fixed frequency, was used to produce the photoelectric effect. First, a low-intensity, or dim, light was used, and the numbers and energy of the ejected electrons were measured. Then a high-intensity light was used, and the numbers and energy of the ejected electrons were again measured. Measurement showed that (1) low-intensity light caused fewer electrons to be ejected, and high-intensity light caused many to be ejected, and (2) all electrons ejected from low- or high-intensity light ideally had the *same* kinetic energy. Surprisingly, the kinetic energy of the ejected electrons was found to be *independent* of the light intensity. This was contrary to what the wave theory of light would predict, since a stronger light should mean that waves with more energy have more energy to give to the electrons. Here is a behavior involving light that the wave theory could not explain.

Quantization of Energy

In addition to the problem of the photoelectric effect, there were problems with blackbody radiation, light emitted from hot objects. The experimental measurements of light emitted through blackbody radiation did not match predictions made from theory. In 1900, Max Planck (pronounced "plonk"), a German physicist, found that he could fit the experimental measure-

A compact disc (CD) is a laser-read (also called *optically read*) data storage device on which music, video, or any type of computer data can be stored. Three types in popular use today are the CD audio, DVD (or MP3), and the CD-ROM used to store any type of computer data. All three kinds of players rotate between 200 to 500 revolutions per minute, but the drive changes speed to move the head at a constant linear velocity over the recording track, faster near the inner hub and slower near the outer edge of the disc. Furthermore, the CD drive reads from the inside out, so the disc will slow as it is played.

The CD itself is a 12 cm diameter, 1.3 mm thick sandwich of a hard plastic core, a mirrorlike layer of metallic aluminum, and a tough, clear plastic overcoating that protects the thin layer of aluminum. The CD records digitized data; music, video, or computer data that have been converted into a string of binary numbers. First, a master disc is made. The binary numbers are translated into a series of pulses that are fed to a laser. The laser is focused onto a photosensitive material on a spinning master disc. Whenever there is a pulse in the signal, the laser burns a small oval pit into the surface, making a pattern of pits and bumps on the track of the master disc. The laser beam is incredibly small, making marks about a micron or so in diameter. A micron is one-millionth of a meter, so you can fit a tremendous number of data tracks onto the disc, which has each track spaced 1.6 microns apart. Next, the CD audio or CD-ROM discs are made by using the master disc as a mold. Soft plastic is pressed against the master disc in a vacuum-forming machine so the small physical marks—the pits and bumps made by the laser—are pressed into the plastic. This makes a record of the strings of binary numbers that were etched into the master disc by the strong but tiny laser beam. During playback, a low-powered laser beam is reflected off the track to read the binary marks on it. The optical sensor head contains a tiny diode laser, a lens, mirrors, and tracking devices that can move the head in three directions. The head moves side to side to keep the head over a single track (within 1.6 micron), it moves up and down to keep the laser beam in focus, and it moves forward and backward as a fine adjustment to maintain a constant linear velocity.

The advantages of the CD audio and video discs over conventional records or tapes include more uniform and accurate frequency response, a complete absence of background noise, and absence of wear—since nothing mechanical touches the surface of the disc when it is played. The advantages of the CD-ROM over the traditional magnetic floppy disks or magnetic hard disks include storage capacity, long-term reliability, and the impossibility of a head crash with a resulting loss of data.

The disadvantage of the CD audio and CD-ROM discs is the lack of ability to do writing or rewriting. Rewritable optical media are available, and these are called CD-R and CD-RW.

A CD-R records data to a disc by using a laser to burn spots into an organic dye. Such a "burned" spot reflects less light than an area that was not heated by the laser. This is designed to mimic the way light reflects off of pits and bumps of a normal CD, except this time the string of binary numbers are burned (nonreflective) and not burned areas (reflective). Since this is similar to how data on a normal CD are represented, a CD-R disc can generally be used in a CD audio or CD-ROM player as if it were a pressed CD. The dyes in a CD-R disc are photosensitive organic compounds that are similar to those used in making photographs. The color of a CD-R disc is a result of the kind of dye that was used in the recording layer combined with the type of reflective coating used. Some of these dye and reflective coating combinations appear green, some appear blue, and others appear to be gold. Once a CD-R disc is burned it cannot be rewritten or changed.

The CD-RW is designed to have the ability to do writing or rewriting. It uses a different technology, but again mimics the way light reflects off the pits and bumps of a pressed CD. Instead of a dye-based recording layer, the CD-RW uses a compound made from silver, indium, antimony, and tellurium. This layer has a property that permits rewriting the information on a disc. This property is that when it is heated to a certain temperature and cooled it becomes crystalline. However, when it is heated to a higher temperature and cooled it becomes noncrystalline. A crystalline surface reflects a laser beam, while a noncrystalline surface absorbs the laser beam. The CD-RW is again designed to mimic the way light reflects off of pits and bumps of a normal CD, except this time the string of binary numbers are noncrystalline (nonreflective) and crystalline areas (reflective). In order to write, erase, and read, the CD-RW recorder must have three different laser powers. It must have (1) a high power to heat spots to about 600°C, which cool rapidly and make noncrystalline spots that are less reflective. It must have (2) a medium power to erase data by heating the media to about 200°C, which allows the media to crystallize and have a uniform reflectivity. Finally, it must have (3) a low setting that is used for finding and reading nonreflective and the more reflective areas of a disc. The writing and rewriting of a CD-RW can be repeated hundreds of times. A problem is that not all CD players can read the less reflective data spots of a CD-RW unless it has a dual-wavelength head that meets multiread specifications.

---

ments and theory together by assuming that the vibrating molecules that emitted the light could only have a *discrete amount* of energy. Instead of energy existing through a continuous range of amounts, Planck found that the vibrating molecules could only have energy in multiples of energy in certain amounts, or **quanta** (meaning "discrete amounts"; *quantum* is singular, and *quanta* plural).

Planck's discovery of quantized energy states was a radical, revolutionary development, and most scientists, including Planck, did not believe it at the time. Planck, in fact, spent considerable time and effort trying to disprove his own discovery. It was, however, the beginning of the quantum theory, which was eventually to revolutionize physics.

Five years later, in 1905, Albert Einstein applied Planck's quantum concept to the problem of the photoelectric effect. Einstein described the energy in a light wave as quanta of energy called **photons.** Each photon has an energy E that is related to the frequency f of the light through Planck's constant h, or

$$E = hf$$

equation 7.4

The value of Planck's constant is 6.63×10^{-34} J·s. This relationship says that higher-frequency light (e.g., blue light at 6.50×10^{14} Hz) has more energy than lower-frequency light (e.g., red light at 4.00×10^{14} Hz). The energy of such high- and low-frequency light can be verified by experiment.

The photon theory also explained the photoelectric effect. According to this theory, light is a stream of moving photons. It is the number of photons in this stream that determines if the light is dim or intense. A high-intensity light has many, many photons, and a low-intensity light has only a few photons. At any particular fixed frequency, all the photons would have the same energy, the product of the frequency and Planck's constant (hf). When a photon interacts with matter, it is absorbed and gives up all of its energy. In the photoelectric effect, this interaction takes place between photons and electrons. When an intense light is used, there are more photons to interact with the electrons, so more electrons are ejected. The energy given up by each photon is a function of the frequency of the light, so at a fixed frequency, the energy of each photon, hf, is the same, and the acquired kinetic energy of each ejected electron is the same. Thus, the photon theory explains the measured experimental results of the photoelectric effect.

The photoelectric effect is explained by considering light to be photons with quanta of energy, not a wave of continuous energy. This is not the only evidence about the quantum nature of light, and more will be presented in chapter 8. But, as you can see, there is a dilemma. The electromagnetic wave theory and the photon theory seem incompatible. Some experiments cannot be explained by the wave theory and seem to support the photon theory. Other experiments are contradictions, providing seemingly equal evidence to reject the photon theory in support of the wave theory.

THE PRESENT THEORY

Today, light is considered to have a dual nature, sometimes acting like a wave and sometimes acting like a particle. A wave model is useful in explaining how light travels through space and how it exhibits such behaviors as refraction, interference, and diffraction. A particle model is useful in explaining how light is emitted from and absorbed by matter, exhibiting such behaviors as blackbody radiation and the photoelectric effect. Together, both of these models are part of a single theory of light, a theory that pictures light as having both particle and wave properties. Some properties are more useful when explaining some observed behaviors, and other properties are more useful when explaining other behaviors.

Frequency is a property of a wave, and the energy of a photon is a property of a particle. Both frequency and the energy of a photon are related in equation 7.4, $E = hf$. It is thus possible to describe light in terms of a frequency (or wavelength) or in terms of a quantity of energy. Any part of the electromagnetic spectrum can thus be described by units of frequency, wavelength, or energy, which are alternative means of describing light. The radio radiation parts of the spectrum are low-frequency, low-energy, and long-wavelength radiations. Radio radiations have more wave properties and practically no particle properties, since the energy levels are low. Gamma radiation, on the other hand, is high-frequency, high-energy, and short-wavelength radiation. Gamma radiation has more particle properties, since the extremely short wavelengths have very high energy levels. The more familiar part of the spectrum, visible light, is between these two extremes and exhibits both wave and particle properties, but it never exhibits both properties at the same time in the same experiment.

Part of the problem in forming a concept or mental image of the exact nature of light is understanding this nature in terms of what is already known. The things you already know about are observed to be particles, or objects, or they are observed to be waves. You can see objects that move through the air, such as baseballs or footballs, and you can see waves on water or in a field of grass. There is nothing that acts like a moving object in some situations but acts like a wave in other situations. Objects are objects, and waves are waves, but objects do not become waves, and waves do not become objects. If this dual nature did exist it would seem very strange. Imagine, for example, holding a book at a certain height above a lake (figure 7.22). You can make measurements and calculate the kinetic energy the book will have when dropped into the lake. When it hits the water, the book disappears, and water waves move away from the point of impact in a circular pattern that moves across the water. When the waves reach another person across the lake, a book identical to the one you dropped pops up out of the water as the waves disappear. As it leaves the water across the lake, the book has the same kinetic energy that your book had when it hit the water in front of you. You and the other person could measure things about either book, and you could measure things about the waves, but you could not measure both at the same time. You might say that this behavior is not only strange but impossible. Yet, it is an analogy to the observed behavior of light.

As stated, light has a dual nature, sometimes exhibiting the properties of a wave and sometimes exhibiting the properties of moving particles but never exhibiting both properties at the same time. Both the wave and the particle nature are accepted as being part of one model today, with the understanding that the exact nature of light is not describable in terms of anything that is known to exist in the everyday-sized world. Light is an extremely small-scale phenomenon that must be different, without a sharp distinction between a particle and a wave. Evidence about this strange nature of an extremely small-scale phenomenon will be considered again in chapter 8 as a basis for introducing the quantum theory of matter.

Figure 7.22

It would seem very strange if there were not a sharp distinction between objects and waves in our everyday world. Yet this appears to be the nature of light.

SUMMARY

Electromagnetic radiation is emitted from all matter with a temperature above absolute zero, and as the temperature increases, more radiation and shorter wavelengths are emitted. Visible light is emitted from matter hotter than about 700°C, and this matter is said to be *incandescent.* The sun, a fire, and the ordinary lightbulb are incandescent sources of light.

The behavior of light is shown by a light ray model that uses straight lines to show the straight-line path of light. Light that interacts with matter is *reflected* with parallel rays, moves in random directions by *diffuse reflection* from points, or is *absorbed,* resulting in a temperature increase. Matter is *opaque,* reflecting light, or *transparent,* transmitting light.

In reflection, the incoming light, or *incident ray,* has the same angle as the *reflected ray* when measured from a perpendicular from the point of reflection, called the *normal.* That the two angles are equal is called the *law of reflection.* The law of reflection explains how a flat mirror forms a *virtual image,* one from which light rays do not originate. Light rays do originate from the other kind of image, a *real image.*

Light rays are bent, or *refracted,* at the boundary when passing from one transparent media to another. The amount of refraction depends on the *incident angle* and the *index of refraction,* a ratio of the speed of light in a vacuum to the speed of light in the media. When the refracted angle is 90°, *total internal reflection* takes place. This limit to the angle of incidence is called the *critical angle,* and all light rays with an incident angle at or beyond this angle are reflected internally.

Each color of light has a range of wavelengths that forms the *spectrum* from red to violet. A glass prism has the property of *dispersion,* separating a beam of white light into a spectrum. Dispersion occurs because the index of refraction is different for each range of colors, with short wavelengths refracted more than larger ones.

A wave model of light can be used to explain diffraction, interference, and polarization, all of which provide strong evidence for the wavelike nature of light. *Interference* occurs when light passes through two small slits or holes and produces an *interference pattern* of bright lines and dark zones. *Polarized light* vibrates in one direction only, in a plane. Light can be polarized by certain materials, by reflection, or by scattering. Polarization can only be explained by a transverse wave model.

A wave model fails to explain observations of light behaviors in the *photoelectric effect* and *blackbody radiation.* Max Planck found that he could modify the wave theory to explain blackbody radiation by assuming that vibrating molecules could only have discrete amounts, or *quanta,* of energy and found that the quantized energy is related to the frequency and a constant known today as *Planck's constant.* Albert Einstein applied Planck's quantum concept to the photoelectric effect and described a light wave in terms of quanta of energy called *photons.* Each photon has an energy that is related to the frequency and Planck's constant.

Today, the properties of light are explained by a model that incorporates both the wave and the particle nature of light. Light is considered to have both wave and particle properties and is not describable in terms of anything known in the everyday-sized world.

Summary of Equations

7.1   angle of incidence = angle of reflection

$$\theta_i = \theta_r$$

7.2   index of refraction $= \dfrac{\text{speed of light in vacuum}}{\text{speed of light in material}}$

$$n = \frac{c}{v}$$

7.3   speed of light in vacuum = (wavelength)(frequency)

$$c = \lambda f$$

7.4 $\begin{matrix}\text{energy of}\\\text{photon}\end{matrix} = \left(\begin{matrix}\text{Planck's}\\\text{constant}\end{matrix}\right)$ (frequency)

$$E = hf$$

KEY TERMS

blackbody radiation (p. **131**)
incandescent (p. **130**)
index of refraction (p. **138**)
interference (p. **142**)
light ray model (p. **132**)
luminous (p. **130**)
photoelectric effect (p. **144**)
photons (p. **145**)

polarized (p. **142**)
quanta (p. **145**)
real image (p. **135**)
refraction (p. **135**)
total internal reflection (p. **136**)
unpolarized light (p. **142**)
virtual image (p. **134**)

APPLYING THE CONCEPTS

1. An object is hot enough to emit a dull red glow. When this object is heated even more, it will emit
   (a) shorter-wavelength, higher-frequency radiation.
   (b) longer-wavelength, lower-frequency radiation.
   (c) the same wavelengths as before, but with more energy.
   (d) more of the same wavelengths with more energy.
2. The difference in the light emitted from a candle, an incandescent lightbulb, and the sun is basically from differences in
   (a) energy sources.
   (b) materials.
   (c) temperatures.
   (d) phases of matter.
3. Before it travels through the earth's atmosphere, sunlight is mostly
   (a) infrared radiation.
   (b) visible light.
   (c) ultraviolet radiation.
   (d) blue light.

4. You are able to see in shaded areas, such as under a tree, because light has undergone
   (a) refraction.
   (b) incident bending.
   (c) a change in speed.
   (d) diffuse reflection.

5. The ratio of the speed of light in a vacuum to the speed of light in some transparent materials is called
   (a) the critical angle.
   (b) total internal reflection.
   (c) the law of reflection.
   (d) the index of refraction.

6. Any part of the electromagnetic spectrum, including the colors of visible light, can be measured in units of
   (a) wavelength.
   (b) frequency.
   (c) energy.
   (d) any of the above.

7. A prism separates the colors of sunlight into a spectrum because
   (a) each wavelength of light has its own index of refraction.
   (b) longer wavelengths are refracted more than shorter wavelengths.
   (c) red light is refracted the most, violet the least.
   (d) all of the above.

8. Which of the following can only be explained by a wave model of light?
   (a) reflection
   (b) refraction
   (c) interference
   (d) photoelectric effect

9. The polarization behavior of light is best explained by considering light to be
   (a) longitudinal waves.
   (b) transverse waves.
   (c) particles.
   (d) particles with ends, or poles.

10. Max Planck made the revolutionary discovery that the energy of vibrating molecules involved in blackbody radiation existed only in
    (a) multiples of certain fixed amounts.
    (b) amounts that smoothly graded one into the next.
    (c) the same, constant amount of energy in all situations.
    (d) amounts that were never consistent from one experiment to the next.

11. Einstein applied Planck's quantum discovery to light and found
    (a) a direct relationship between the energy and frequency of light.
    (b) that the energy of a photon divided by the frequency of the photon always equaled a constant known as Planck's constant.
    (c) that the energy of a photon divided by Planck's constant always equaled the frequency.
    (d) all of the above.

12. Today, light is considered to be
    (a) tiny particles of matter that move through space, having no wave properties.
    (b) electromagnetic waves only, with no properties of particles.
    (c) a small-scale phenomenon without a sharp distinction between particle and wave properties.
    (d) something that is completely unknown.

Answers

1. a 2. c 3. a 4. d 5. d 6. d 7. a 8. c 9. b 10. a 11. d 12. c

QUESTIONS FOR THOUGHT

1. What determines if an electromagnetic wave emitted from an object is a visible light wave or a wave of infrared radiation?

2. What model of light does the polarization of light support? Explain.

3. Which carries more energy, red light or blue light? Should this mean anything about the preferred color of warning and stop lights? Explain.

4. What model of light is supported by the photoelectric effect? Explain.

5. What happens to light that is absorbed by matter?

6. One star is reddish, and another is bluish. Do you know anything about the relative temperatures of the two stars? Explain.

7. When does total internal reflection occur? Why does this occur in the diamond more than other gemstones?

8. Why does a highway sometimes appear wet on a hot summer day when it is not wet?

9. How can you tell if a pair of sunglasses is polarizing or not?

10. Explain why the intensity of reflected light appears to change if you tilt your head from side to side while wearing polarizing sunglasses.

11. What was so unusual about Planck's findings about blackbody radiation? Why was this considered revolutionary?

12. Why are both the photon model and the electromagnetic wave model accepted today as a single theory? Why was this so difficult for people to accept at first?

PARALLEL EXERCISES

The exercises in groups A and B cover the same concepts. Solutions to group A exercises are located in appendix D.

Group A

1. What is the speed of light while traveling through (a) water and (b) ice?
2. How many minutes are required for sunlight to reach the earth if the sun is 1.50×10^8 km from the earth?
3. How many hours are required before a radio signal from a space probe near the planet Pluto reaches Earth, 6.00×10^9 km away?
4. A light ray is reflected from a mirror with an angle 10° to the normal. What was the angle of incidence?
5. Light travels through a transparent substance at 2.20×10^8 m/s. What is the substance?
6. The wavelength of a monochromatic light source is measured to be 6.00×10^{-7} m in an experiment. (a) What is the frequency? (b) What is the energy of a photon of this light?
7. At a particular location and time, sunlight is measured on a 1 square meter solar collector with an intensity of 1,000.0 W. If the peak intensity of this sunlight has a wavelength of 5.60×10^{-7} m, how many photons are arriving each second?
8. A light wave has a frequency of 4.90×10^{14} cycles per second. (a) What is the wavelength? (b) What color would you observe (see table 7.2)?
9. What is the energy of a gamma photon of frequency 5.00×10^{20} Hz?
10. What is the energy of a microwave photon of wavelength 1.00 mm?

Group B

1. (a) What is the speed of light while traveling through a vacuum? (b) While traveling through air at 30°C? (c) While traveling through air at 0°C?
2. How much time is required for reflected sunlight to travel from the moon to the earth if the distance between the earth and the moon is 3.85×10^5 km?
3. How many minutes are required for a radio signal to travel from Earth to a space station on Mars if the planet Mars is 7.83×10^7 km from Earth?
4. An incident light ray strikes a mirror with an angle of 30° to the surface of the mirror. What is the angle of the reflected ray?
5. The speed of light through a transparent substance is 2.00×10^8 m/s. What is the substance?
6. A monochromatic light source used in an experiment has a wavelength of 4.60×10^{-7} m. What is the energy of a photon of this light?
7. In black-and-white photography, a photon energy of about 4.00×10^{-19} J is needed to bring about the changes in the silver compounds used in the film. Explain why a red light used in a darkroom does not affect the film during developing.
8. The wavelength of light from a monochromatic source is measured to be 6.80×10^{-7} m. (a) What is the frequency of this light? (b) What color would you observe?
9. How much greater is the energy of a photon of ultraviolet radiation ($\lambda = 3.00 \times 10^{-7}$ m) than the energy of an average photon of sunlight ($\lambda = 5.60 \times 10^{-7}$ m)?
10. At what rate must electrons in a wire vibrate to emit microwaves with a wavelength of 1.00 mm?

CHAPTER EIGHT

Atoms and Periodic Properties

This is a picture of pure zinc, one of the 89 naturally occuring elements found on the earth.

OUTLINE

The development of the modern atomic model illustrates how modern scientific understanding comes from many different fields of study. For example, you will learn how studies of electricity led to the discovery that atoms have subatomic parts called *electrons*. The discovery of radioactivity led to the discovery of more parts, a central nucleus that contains protons and neutrons. Information from the absorption and emission of light was used to construct a model of how these parts are put together, a model resembling a miniature solar system with electrons circling the nucleus. The solar system model had initial, but limited, success and was inconsistent with other understandings about matter and energy. Modifications of this model were attempted, but none solved the problems. Then the discovery of wave properties of matter led to an entirely new model of the atom (figure 8.1).

The atomic model will be put to use in later chapters to explain the countless varieties of matter and the changes that matter undergoes. In addition, you will learn how these changes can be manipulated to make new materials, from drugs to ceramics. In short, you will learn how understanding the atom and all the changes it undergoes not only touches your life directly but shapes and affects all parts of civilization.

ATOMIC STRUCTURE DISCOVERED

Did you ever wonder how scientists could know about something so tiny that you cannot see it, even with the most powerful optical microscope? The atom is a tiny unit of matter, so small that one gram of hydrogen contains about 600,000,000,000,000,000,000,000 (six-hundred-thousand-billion-billion or 6×10^{23}) atoms. Even more unbelievable is that atoms are not individual units, but are made up of even smaller particles. How is it possible that scientists are able to tell you about the parts of something so small that it cannot be seen? The answer is that these things cannot be observed directly, but their existence can be inferred from experimental evidence. The following story describes the evidence and how scientists learned about the parts—electrons, the nucleus, protons, neutrons, and others—and how they are all arranged in the atom.

The atomic concept is very old, dating back to ancient Greek philosophers some 2,500 years ago. The ancient Greeks also reasoned about the way that pure substances are put together. A glass of water, for example, appears to be the same throughout. Is it the same? Two plausible, but conflicting, ideas were possible as an intellectual exercise. The water could have a *continuous* structure; that is, it could be completely homogeneous throughout. The other idea was that the water only appears to be continuous but is actually *discontinuous*. This means that if you continue to divide the water into smaller and smaller volumes, you would eventually reach a limit to this dividing, a particle that could not be further subdivided. The Greek philosopher Democritus (460–362 B.C.) developed this model in the fourth century B.C., and he called the indivisible particle an *atom*, from a Greek word meaning "uncuttable." However, neither Plato nor Aristotle accepted the atomic theory of matter and it was not until about two thousand years later that the atomic concept of matter was reintroduced. In the early 1800s, the English chemist John Dalton brought back the ancient Greek idea of hard, indivisible atoms to explain chemical reactions. Five statements will summarize his theory. As you will soon see, today we know that statement 2 is not strictly correct:

1. Indivisible minute particles called atoms make up all matter.
2. All the atoms of an element are exactly alike in shape and mass.

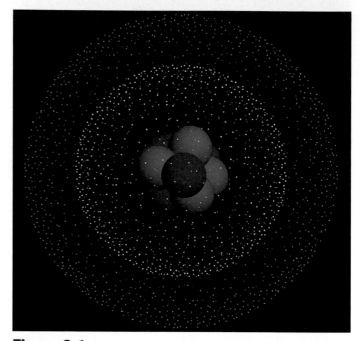

Figure 8.1

This is a computer-generated model of a beryllium atom, showing the nucleus and 1s, 2s electron orbitals. This configuration can also be predicted from information on a periodic table.

3. The atoms of different elements differ from one another in their masses.
4. Atoms chemically combine in definite whole-number ratios to form chemical compounds.
5. Atoms are neither created nor destroyed in chemical reactions.

During the 1800s, Dalton's concept of hard, indivisible atoms was familiar to most scientists. Yet, the existence of atoms was not generally accepted by all scientists. There was skepticism about something that could not be observed directly. Strangely, full acceptance of the atom came in the early 1900s with the discovery that the atom was not indivisible after all. The atom has parts that give it an internal structure. The first part to be discovered was the *electron*, a part that was discovered through studies of electricity.

Discovery of the Electron

Scientists of the late 1800s were interested in understanding the nature of the recently discovered electric current. To observe a current directly they tried to produce one by itself, away from wires, by removing the air from a tube and then running it through the vacuum. When metal plates inside a tube were connected to the negative and positive terminals of a high-voltage source (figure 8.2) a greenish beam was observed that seemed to move from the cathode (negative terminal) through the empty tube and collect at the anode (positive terminal). Since this mysterious beam seemed to come out of the cathode it was said to be a *cathode ray*.

The English physicist J. J. Thomson figured out what the cathode ray was in 1897. He placed charged metal plates on each side of the beam (figure 8.3) and found that the beam was deflected away from the negative plate. Since it was known that like charges repel, this meant that the beam was composed of negatively charged particles.

The cathode ray was also deflected when caused to pass between the poles of a magnet. By balancing the deflections made by the magnet with the deflections made by the electric field, Thomson could determine the ratio of the charge to mass for an individual particle. Today, the charge-to-mass ratio is considered to be 1.7584×10^{11} coulomb/kilogram. A significant part of Thomson's experiments was that he found the charge-to-mass ratio was the same no matter what gas was in the tube and of what materials the electrodes were made. Thomson had discovered the **electron,** a fundamental particle of matter.

A method for measuring the charge and mass of the electron was worked out by an American physicist, Robert A. Millikan, around 1906. Millikan used an apparatus like the one illustrated in figure 8.4 to measure the charge on tiny droplets of oil. Millikan found that none of the droplets had a charge less

Figure 8.3

A cathode ray passed between two charged plates is deflected toward the positively charged plate. The ray is also deflected by a magnetic field. By measuring the deflection by both, J. J. Thomson was able to calculate the ratio of charge to mass. He was able to measure the deflection because the detecting screen was coated with zinc sulfide, a substance that produces a visible light when struck by a charged particle.

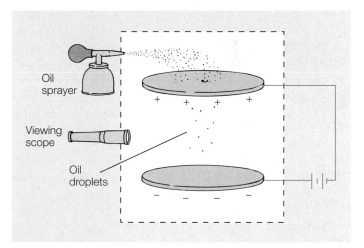

Figure 8.4

Millikan measured the charge of an electron by balancing the pull of gravity on oil droplets with an upward electrical force. Knowing the charge-to-mass ratio that Thomson had calculated, Millikan was able to calculate the charge on each droplet. He found that all the droplets had a charge of 1.60×10^{-19} coulomb or multiples of that charge. The conclusion was that this had to be the charge of an electron.

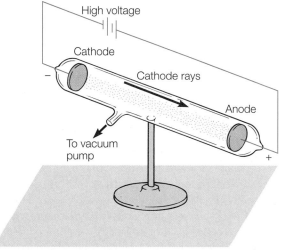

Figure 8.2

A vacuum tube with metal plates attached to a high-voltage source produces a greenish beam called *cathode rays*. These rays move from the cathode (negative charge) to the anode (positive charge).

than one particular value (1.60×10^{-19} coulomb) and that larger charges on various droplets were always multiples of this unit of charge. Since all of the droplets carried the single unit of charge or multiples of the single unit, the unit of charge was understood to be the charge of a single electron.

Knowing the charge of a single electron and knowing the charge-to-mass ratio that Thomson had measured now made it possible to calculate the mass of a single electron. The mass of an electron was thus determined to be about 9.11×10^{-31} kg, or about 1/1,840 of the mass of the lightest atom, hydrogen.

Thomson had discovered the negatively charged electron, and Millikan had measured the charge and mass of the electron. But atoms themselves are electrically neutral. If an electron is part of an atom, there must be something else that is positively charged, canceling the negative charge of the electron. The next step in the sequence of understanding atomic structure would be to find what is neutralizing the negative charge and to figure out how all the parts are put together.

Thomson had proposed a model for what was known about the atom at the time. He suggested that an atom could be a blob of positively charged matter in which electrons were stuck like "raisins in plum pudding." If the mass of a hydrogen atom is due to the electrons embedded in a positively charged matrix, then 1,840 electrons would be needed together with sufficient positive matter to make the atom electrically neutral.

The Nucleus

The nature of radioactivity and matter were the research interests of a British physicist, Ernest Rutherford. In 1907, Rutherford was studying the scattering of alpha particles directed toward a thin sheet of metal. As shown in figure 8.5, alpha particles from a radioactive source were allowed to move through a small opening in a lead container, so only a narrow beam of the massive, fast-moving particles would penetrate a very thin sheet of gold. The alpha particles were detected by plates, which produced a small flash of light when struck by alpha particles.

Rutherford found that most of the alpha particles went straight through the foil. However, he was astounded to find that some alpha particles were deflected at very large angles and some were even reflected backward. He could account for this only by assuming that the massive, positively charged alpha particles were repelled by a massive positive charge concentrated in a small region of the atoms in the gold foil (figure 8.6). He concluded that an atom must have a tiny, massive, and positively charged **nucleus** surrounded by electrons.

From measurements of the scattering, Rutherford estimated electrons must be moving around the nucleus at a dis-

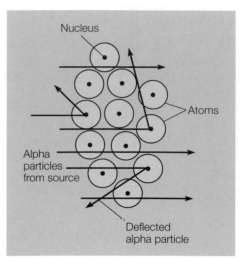

Figure 8.6

Rutherford's nuclear model of the atom explained the alpha-scattering results as positive alpha particles experiencing a repulsive force from the positive nucleus. Measurements of the percent of alpha particles passing straight through, and of the various angles of scattering of those coming close to the nuclei, gave Rutherford a means of estimating the size of the nucleus.

tance 100,000 times the radius of the nucleus. This means the volume of an atom is mostly empty space. A few years later Rutherford was able to identify the discrete unit of positive charge, which we now call a **proton.** Rutherford also speculated about the existence of a neutral particle in the nucleus, a neutron. The **neutron** was eventually identified in 1932 by James Chadwick.

Today, the number of protons in the nucleus of an atom is called the **atomic number.** All the atoms of a particular element have the same number of protons in their nucleus, so all atoms of an element have the same atomic number. Hydrogen has an atomic number of 1, so any atom that has one proton in its nucleus is an atom of the element hydrogen. Today, scientists have identified 113 different kinds of elements, each with a different number of protons.

The neutrons of the nucleus, along with the protons, contribute to the mass of an atom. Although all the atoms of an element must have the same number of protons in their nuclei, the number of neutrons may vary. Atoms of an element that have different numbers of neutrons are called **isotopes.** There are three isotopes of hydrogen illustrated in figure 8.7. All three isotopes have the same number of protons and electrons, but one isotope has no neutrons, one isotope has 1 neutron (deuterium), and one isotope has 2 neutrons (tritium).

An atom is very tiny, and it is impossible to find the mass of a given atom. It is possible, however, to compare the mass of one atom to another. The mass of any atom is compared to the mass of an atom of a particular isotope of carbon. This particular carbon isotope is assigned a mass of exactly 12.00 units called **atomic mass units** (u). This isotope, called *carbon-12,* provides the standard to which the masses of all other isotopes are compared. The relative mass of any isotope is based on the mass of a carbon-12 isotope.

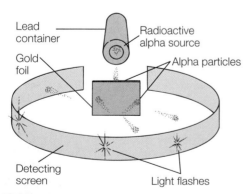

Figure 8.5

Rutherford and his co-workers studied alpha particle scattering from a thin metal foil. The alpha particles struck the detecting screen, producing a flash of visible light. Measurements of the angles between the flashes, the metal foil, and the source of the alpha particles showed that the particles were scattered in all directions, including straight back toward the source.

Figure 8.7

The three isotopes of hydrogen have the same number of protons but different numbers of neutrons. Hydrogen-1 is the most common isotope. Hydrogen-2, with an additional neutron, is named *deuterium,* and hydrogen-3 is called *tritium.*

Table 8.1 Selected Atomic Weights Calculated from Mass and Abundance of Isotopes

Stable Isotopes	Mass of Isotope Compared to C-12	Abundance	Atomic Weight
$^{1}_{1}H$	1.007	99.985%	
$^{2}_{1}H$	2.0141	0.015%	1.0079
$^{9}_{4}Be$	9.01218	100%	9.01218
$^{14}_{7}N$	14.00307	99.63%	
$^{15}_{7}N$	15.00011	0.37%	14.0067
$^{16}_{8}O$	15.99491	99.759%	
$^{17}_{8}O$	16.99914	0.037%	
$^{18}_{8}O$	17.00016	0.204%	15.9994
$^{19}_{9}F$	18.9984	100%	18.9984
$^{20}_{10}Ne$	19.99244	90.92%	
$^{21}_{10}Ne$	20.99395	0.257%	
$^{22}_{10}Ne$	21.99138	8.82%	20.179
$^{27}_{13}Al$	26.9815	100%	26.9815

The relative mass of the hydrogen isotope without a neutron is 1.007 when compared to carbon-12. The relative mass of the hydrogen isotope with 1 neutron is 2.0141 when compared to carbon-12. Elements occur in nature as a mixture of isotopes, and the contribution of each is calculated in the atomic weight. **Atomic weight** is a weighted average of the isotopes based on their mass compared to carbon-12, and their relative abundance in nature. Of all the hydrogen isotopes, for example, 99.985 percent occurs as the isotope without a neutron and 0.015 percent is the isotope with 1 neutron (the other isotope is not considered because it is radioactive). The fractional part of occurrence is multiplied by the relative atomic mass for each isotope and the results are summed to obtain the atomic weight. Table 8.1 gives the atomic weight of hydrogen as 1.0079 as a result of this calculation.

The sum of the number of protons and neutrons in a nucleus of an atom is called the **mass number** of that atom. Mass numbers are used to identify isotopes. A hydrogen atom with 1 proton and 1 neutron has a mass number of 1 + 1, or 2, and is referred to as hydrogen-2. A hydrogen atom with 1 proton and 2 neutrons has a mass number of 1 + 2, or 3, and is referred to as hydrogen-3. Using symbols, hydrogen-3 is written as

$$^{3}_{1}H$$

where H is the chemical symbol for hydrogen, the subscript to the bottom left is the atomic number, and the superscript to the top left is the mass number.

How are the electrons moving around the nucleus? It might occur to you, as it did to Rutherford and others, that an atom might be similar to a miniature solar system. In this analogy, the nucleus is in the role of the sun, electrons in the role of moving planets in their orbits, and electrical attractions between the nucleus and electrons in the role of gravitational attraction. There are, however, big problems with this idea. If electrons were mov-

ing in circular orbits, they would continually change their direction of travel and would therefore be accelerating. According to the Maxwell model of electromagnetic radiation, an accelerating electric charge emits electromagnetic radiation such as light. If an electron gave off light, it would lose energy. The energy loss would mean that the electron could not maintain its orbit, and it would be pulled into the oppositely charged nucleus. The atom would collapse as electrons spiraled into the nucleus. Since atoms do not collapse like this, there is a significant problem with the solar system model of the atom.

CONCEPTS APPLIED

Atomic Parts

Identify the number of protons, neutrons, and electrons in an atom of $^{16}_{8}O$. Write your answer before reading the solution in the next paragraph.

The subscript to the bottom left is the atomic number. Atomic number is defined as the number of protons in the nucleus, so this number identifies the number of protons as 8. Any atom with 8 protons is an atom of oxygen, which is identified with the symbol O. The superscript to the top left identifies the mass number of this isotope of oxygen, which is 16. The mass number is defined as the sum of the number of protons and the number of neutrons in the nucleus. Since you already know the number of protons is 8 (from the atomic number), then the number of neutrons is 16 minus 8, or 8 neutrons. Since a neutral atom has the same number of electrons as protons, an atom of this oxygen isotope has 8 protons, 8 neutrons, and 8 electrons.

Now, can you describe how many protons, neutrons, and electrons are found in an atom of $^{17}_{8}O$? Compare your answer with a classmate's to check.

THE BOHR MODEL

Niels Bohr was a young Danish physicist who visited Rutherford's laboratory in 1912 and became very interested in questions about the solar system model of the atom. He wondered what determined the size of the electron orbits and the energies of the electrons. He wanted to know why orbiting electrons did not give off electromagnetic radiation. Seeking answers to questions such as these led Bohr to incorporate the *quantum concept* of Planck and Einstein with Rutherford's model to describe the electrons in the outer part of the atom. This quantum concept will be briefly reviewed before proceeding with the development of Bohr's model of the hydrogen atom.

The Quantum Concept

In the year 1900, Max Planck introduced the idea that matter emits and absorbs energy in discrete units that he called **quanta.** Planck had been trying to match data from spectroscopy experiments with data that could be predicted from the theory of electromagnetic radiation. In order to match the experimental findings with the theory, he had to assume that specific, discrete amounts of energy were associated with different frequencies of radiation. In 1905, Albert Einstein extended the quantum concept to light, stating that light consists of discrete units of energy that are now called **photons.** The energy of a photon is directly proportional to the frequency of vibration, and the higher the frequency of light, the greater the energy of the individual photons. In addition, the interaction of a photon with matter is an "all-or-none" affair, that is, matter absorbs an entire photon or none of it. The relationship between frequency (f) and energy (E) is

$$E = hf$$

equation 8.1

where h is the proportionality constant known as *Planck's constant* (6.63×10^{-34} J·s). This relationship means that higher-frequency light, such as ultraviolet, has more energy than lower-frequency light, such as red light.

Atomic Spectra

Planck was concerned with hot solids that emit electromagnetic radiation. The nature of this radiation, called *blackbody radiation,* depends on the temperature of the source. When this light is passed through a prism, it is dispersed into a *continuous spectrum,* with one color gradually blending into the next as in a rainbow. Today, it is understood that a continuous spectrum comes from solids, liquids, and dense gases because the atoms interact, and all frequencies within a temperature-determined range are emitted. Light from an incandescent gas, on the other hand, is dispersed into a **line spectrum,** narrow lines of colors with no light between the lines (figure 8.8). The atoms in the incandescent gas are able to emit certain characteristic frequencies, and each frequency is a line of color that represents a definite value of energy. The line spectra are specific for a substance, and increased or decreased temperature changes only the intensity of the lines of colors. Thus hydrogen always produces the same colors of lines in the same position. Helium has its own specific set of lines, as do other substances. Line spectra are a kind of fingerprint that can be used to identify a gas. A line spectrum might also extend beyond visible light into ultraviolet, infrared, and other electromagnetic regions.

In 1885, a Swiss mathematics teacher named J. J. Balmer was studying the regularity of spacing of the hydrogen line spectra. Balmer was able to develop an equation that fit all the visible lines. These four lines became known as the *Balmer series.* Other series were found later, outside the visible part of the spectrum (figure 8.9).

Such regularity of observable spectral lines must reflect some unseen regularity in the atom. At this time it was known that hydrogen had only one electron. How could one electron produce series of spectral lines with such regularity?

Bohr's Theory

An acceptable model of the hydrogen atom would have to explain the characteristic line spectra and their regularity as de-

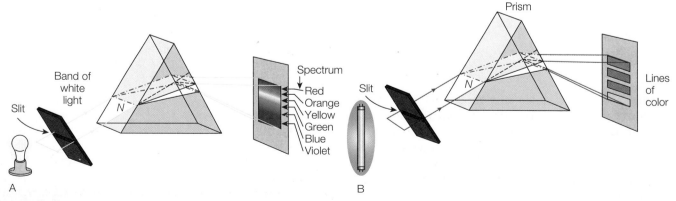

Figure 8.8

(*A*) Light from incandescent solids, liquids, or dense gases produces a continuous spectrum as atoms interact to emit all frequencies of visible light. (*B*) Light from an incandescent gas produces a line spectrum as atoms emit certain frequencies that are characteristic of each element.

Figure 8.9

Atomic hydrogen produces a series of characteristic line spectra in the ultra-violet, visible, and infrared parts of the total spectrum. The visible light spectra always consist of two violet lines, a blue-green line, and a bright red line.

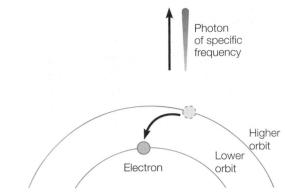

Figure 8.10

Each time an electron makes a "quantum leap," moving from a higher-energy orbit to a lower-energy orbit, it emits a photon of a specific frequency and energy value.

scribed by Balmer. In fact, a successful model should be able to predict the occurrence of each color line as well as account for its origin. By 1913, Bohr was able to do this by applying the quantum concept to a solar system model of the atom. He began by considering the single hydrogen electron to be a single "planet" revolving in a circular orbit around the nucleus. There were three sets of rules that described this electron:

1. **Allowed orbits.** An electron can revolve around an atom only in specific allowed orbits. Bohr considered the electron to be a particle with a known mass in motion around the nucleus, and used Newtonian mechanics to calculate the distances of the allowed orbits. According to the Bohr model, electrons can exist only in one of these allowed orbits and nowhere else.

2. **Radiationless orbits.** An electron in an allowed orbit does not emit radiant energy as long as it remains in the orbit. According to Maxwell's theory of electromagnetic radiation, an accelerating electron should emit an electromagnetic wave, such as light, which would move off into space from the electron. Bohr recognized that electrons moving in a circular orbit are accelerating, since they are changing direction continuously. Yet, hydrogen atoms did not emit light in their normal state. Bohr decided that the situation must be different for orbiting electrons and that electrons could stay in their allowed orbits and not give off light. He postulated this rule as a way to make his theory consistent with other scientific theories.

3. **Quantum leaps.** An electron gains or loses energy only by moving from one allowed orbit to another (figure 8.10). In the Bohr model, the energy an electron has depends on which allowable orbit it occupies. The only way that an electron can change its energy is to jump from one allowed orbit to another in quantum "leaps." An electron must acquire energy to jump from a lower orbit to a higher one. Likewise, an electron gives up energy when jumping from a higher orbit to a lower one. Such jumps must be all at once, not part way and not gradual. An electron acquires energy from high temperatures or from electrical discharges to jump to a higher orbit. An electron jumping from a higher to a lower orbit gives up energy in the form of light. A single photon is emitted when a downward jump occurs,

and the energy of the photon is *exactly* equal to the difference in the energy level of the two orbits.

The energy level diagram in figure 8.11 shows the energy states for the orbits of a hydrogen atom. The lowest energy state is the **ground state** (or normal state). The higher states are the **excited states.** The electron in a hydrogen atom would normally occupy the ground state, but high temperatures or electric discharge can give the electron sufficient energy to jump to one of the excited states. Once in an excited state, the electron immediately jumps back to a lower state, as shown by the arrows in the figure. The length of the arrow represents the frequency of the photon that the electron emits in the process. A hydrogen atom can give off only one photon at a time, and the many lines of a

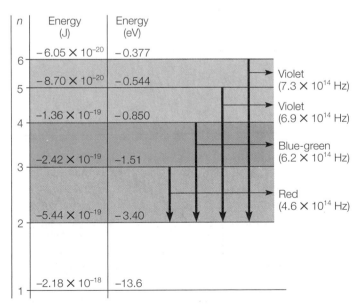

Figure 8.11

An energy level diagram for a hydrogen atom, not drawn to scale. The energy levels (*n*) are listed on the left side, followed by the energies of each level in J and eV. The color and frequency of the visible light photons emitted are listed on the right side, with the arrow showing the orbit moved from and to.

NSLS: The National Synchrotron Light Source

The National Synchrotron Light Source (NSLS) is a research facility that shows close connections between the concepts presented in this chapter and the growing nature of interdisciplinary research. Scientists from all over the country do research at NSLS, but all projects are reviewed and approved based on scientific importance. In return for using the facility for free, the researchers agree to publish their results in open literature where others can read about their experiments and use the results in their own work.

A synchrotron is a machine that uses powerful magnets to guide electrons through a circular tube, a ring with almost all the air pumped out of it. An electron gun sends bunches of electrons through the vacuum and, as they round each bend in the ring, they give off energy in the form of light. This is called "synchrotron light."

*Information from the Internet: www.nsls.bnl/gov/bnl.

More than two thousand researchers use the NSLS each year. Many of these researchers are working in the field of materials sciences, a field interested in the development of transistors and chip technology for computers and processors. A fast-growing field is life sciences, particularly in the area of structural biology. Many major pharmaceutical companies are using synchrotron light to solve the structures of certain biological molecules, and then to design drugs to inhibit the molecules' actions. Medical centers and laboratories are using the NSLS in their research on antidepressants, joint diseases such as arthritis, and HIV. The Protein Data Bank is an archive of experimentally determined three-dimensional structures of biological macromolecules, serving a global community of researchers, educators, and students.*

---

hydrogen line spectrum come from many atoms giving off many photons at the same time.

As you can see, the energy level diagram in figure 8.11 shows how the change of known energy levels from known orbits results in the exact frequencies of the color lines in the Balmer series. Bohr's theory did explain the lines in the hydrogen spectrum with a remarkable degree of accuracy. However, the model did not have much success with larger atoms. The Bohr model could not explain the spectra line of larger atoms with its single quantum number. A German physicist, A. Sommerfield, tried to modify Bohr's model by adding elliptical orbits in addition to Bohr's circular orbits. It soon became apparent that the "patched up" model, too, was not adequate. Bohr had made the rule that there were radiationless orbits without an explanation, and he did not have an explanation for the quantized orbits. There was something fundamentally incomplete about the model.

QUANTUM MECHANICS

The Bohr model of the atom successfully accounted for the line spectrum of hydrogen and provided an understandable mechanism for the emission of photons by atoms. However, the model did not predict the spectra of any atom larger than hydrogen, and there were other limitations. A new, better theory was needed. The roots of a new theory would again come from experiments with light. Experiments with light had established that sometimes light behaves like a stream of particles, and at other times it behaves like a wave. Eventually scientists began to accept that light has both wave properties and particle properties, which is now referred to as the *wave-particle duality of light*. In 1923 Louis de Broglie, a French physicist, reasoned that symmetry is usually found in nature, so if a particle of light has a dual nature, then particles such as electrons should too.

Recall that waves confined on a fixed string establish resonant modes of vibration called *standing waves* (see chapter 5). Only certain fundamental frequencies and harmonics can exist

on a string, and the combination of the fundamental and overtones gives the stringed instrument its particular quality. The same result of resonant modes of vibrations is observed in *any* situation where waves are confined to a fixed space. Characteristic standing wave patterns depend on the wavelength and wave velocity for waves formed on strings, in enclosed columns of air, or for any kind of wave in a confined space. Electrons are confined to the space near a nucleus, and electrons have wave properties, so an electron in an atom must be a confined wave. Does an electron form a characteristic wave pattern? This was the question being asked in about 1925 when Heisenberg, Schrödinger, Dirac, and others applied the wave nature of the electron to develop a new model of the atom based on the mechanics of electron waves. The new theory is now called *wave mechanics*, or quantum mechanics.

Erwin Schrödinger, an Austrian physicist, treated the atom as a three-dimensional system of waves to derive the *Schrödinger equation*. Instead of the simple circular planetary orbits of the Bohr model, solving the Schrödinger equation results in a description of three-dimensional shapes of the patterns that develop when electron waves are confined by a nucleus. Schrödinger first considered the hydrogen atom, calculating the states of vibration that would be possible for an electron wave confined by a nucleus. He found that the frequency of these vibrations, when multiplied by Planck's constant, matched exactly, to the last decimal point, the observed energies of the quantum states of the hydrogen atom. The conclusion is that the wave nature of the electron is the important property to consider for a successful model of the atom.

The quantum mechanics theory of the atom proved to be very successful; it confirmed all the known experimental facts and predicted new discoveries.

Quantum mechanics does share at least one idea with the Bohr model. This is that an electron emits a photon when jumping from a higher state to a lower one. The Bohr model, however, considered the *particle* nature of an electron moving in a circular orbit with a definitely assigned position at a given time. Quantum mechanics considers the *wave* nature, with the electron as a confined wave with well-defined shapes and frequencies. A wave is not localized like a particle and is spread out in space. The quantum mechanics model is, therefore, a series of orbitlike smears, or fuzzy statistical representations, of where the electron might be found.

The quantum mechanics model is a highly mathematical treatment of the mechanics of matter waves. In addition, the wave properties are considered as three-dimensional problems, and three quantum numbers are needed to describe the fuzzy electron cloud. The mathematical detail will not be presented here. The following is a qualitative description of the main ideas in the quantum mechanics model. It will describe the results of the mathematics and will provide a mental visualization of what it all means.

First, understand that the quantum mechanical theory is not an extension or refinement of the Bohr model. The Bohr model considered electrons as particles in circular orbits that could be only certain distances from the nucleus. The quantum mechanical model, on the other hand, considers the electron as a wave and considers the energy of its harmonics, or modes, of standing waves. In the Bohr model, the location of an electron was certain—in an orbit. In the quantum mechanical model, the electron is a spread-out wave.

Quantum mechanics describes the energy state of an electron wave with four *quantum numbers:*

1. **Distance from the nucleus.** The *principal quantum number* describes the *main energy level* of an electron in terms of its most probable distance from the nucleus. The lowest energy state possible is closest to the nucleus and is assigned the principal quantum number of 1 ($n = 1$). Higher states are assigned progressively higher positive whole numbers of $n = 2$, $n = 3$, $n = 4$, and so on. Electrons with higher principal quantum numbers have higher energies and are located farther from the nucleus.

2. **Energy sublevel.** The *angular momentum quantum number* defines energy sublevels within the main energy levels. Each sublevel is identified with a letter. The first four of these letters, in order of increasing energy, are s, p, d, and f. The letter s represents the lowest sublevel, and the letter f represents the highest sublevel. A principal quantum number and a letter indicating the angular momentum quantum number are combined to identify the main energy state and energy sublevel of an electron. For an electron in the lowest main energy level, $n = 1$, and in the lowest sublevel, s, the number and letter are 1s (read as "one-s"). Thus 1s indicates an electron that is as close to the nucleus as possible in the lowest energy sublevel possible.

   There are limits to how many sublevels can occupy each of the main energy levels. Basically, the lowest main energy level can have only the lowest sublevel, and another sublevel is added as you move up through the main energy levels. Thus, the lowest main energy level, $n = 1$, can have only the s sublevel. The $n = 2$ can have s and p sublevels. The $n = 3$ main energy level can have the s, p, and d sublevels. Finally, the $n = 4$ main energy level can have all four sublevels, with s, p, d, and f.

The Bohr model considered the location of an electron as certain, like a tiny shrunken marble in an orbit. The quantum mechanics model considers the electron as a wave, and knowledge of its location is very uncertain. The **Heisenberg uncertainty principle** states that you cannot measure the exact position of a wave because a wave is spread out. One cannot specify the position and the momentum of a spread-out electron. The location of the electron can be described only in terms of *probabilities* of where it might be at a given instant. The probability of location is described by a fuzzy region of space called an **orbital.** An orbital defines the space where an electron is likely to be found. Orbitals have characteristic three-dimensional shapes and sizes and are identified with electrons of characteristic energy levels (figure 8.12). An orbital shape represents where an electron could probably be located at any particular instant. This "probability cloud" could likewise have any particular orientation in space, and the direction of this orientation is uncertain.

3. **Orientation in space.** An external magnetic field applied to an atom produces different energy levels that are related to the orientation of the orbital to the magnetic field. The orientation of an orbital in space is described by the *magnetic quantum number.* This number is related to the energies of orbitals as they are oriented in space relative to an external magnetic field, a kind of energy sub-sublevel. In general, the lowest-energy sublevel (s) has only one orbital orientation. The next higher-energy sublevel (p) can have three orbital orientations (figure 8.13). The d sublevel can have five orbital orientations, and the highest

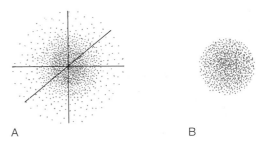

A   B

Figure 8.12

(*A*) An electron distribution sketch representing probability regions where an electron is most likely to be found. (*B*) A boundary surface, or contour, that encloses about 90 percent of the electron distribution shown in (*A*). This three-dimensional space around the nucleus, where there is the greatest probability of finding an electron, is called an *orbital.*

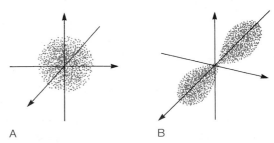

A   B

Figure 8.13

(*A*) A contour representation of an s orbital. (*B*) A contour representation of a p orbital.

Table 8.2 Quantum Numbers and Electron Distribution to $n = 4$

Main Energy Level	Energy Sublevels	Maximum Number of Electrons	Maximum Number of Electrons per Main Energy Level
$n = 1$	s	2	2
$n = 2$	s	2	
	p	6	8
$n = 3$	s	2	
	p	6	
	d	10	18
$n = 4$	s	2	
	p	6	
	d	10	
	f	14	32

sublevel, f, can have a total of seven different orientations (table 8.2).

4. **Direction of spin.** Detailed studies have shown that an electron spinning one way (say clockwise) in an external magnetic field would have a different energy than one spinning the other way (say counterclockwise). The *spin quantum number* describes these two spin orientations (figure 8.14).

Electron spin is an important property that helps determine the electronic structure of an atom. As it turns out, two electrons spinning in opposite directions produce unlike magnetic fields that are attractive, balancing some of the normal repulsion from two like charges. Two electrons of opposite spin, called an electron pair, can thus occupy the same orbital. Wolfgang Pauli, a German physicist, summarized this idea in 1924. His summary, now known as the **Pauli exclusion principle,** states that *no two electrons in an atom can have the same four quantum numbers.* This principle provides the key for understanding the electron structure of atoms.

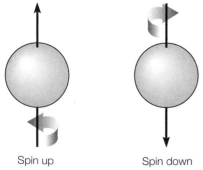

Spin up   Spin down

Figure 8.14

Experimental evidence supports the concept that electrons can be considered to spin one way or the other as they move about an orbital under an external magnetic field.

ELECTRON CONFIGURATION

The arrangement of electrons in orbitals is called the *electron configuration.* Before you can describe the electron arrangement, you need to know how many electrons are present in an atom. An atom is electrically neutral, so the number of protons (positive charge) must equal the number of electrons (negative charge). The atomic number therefore identifies the number of electrons as well as the number of protons. Now that you have a means of finding the number of electrons, consider the various energy levels to see how the electron configuration is determined.

According to the Pauli exclusion principle, no two electrons in an atom can have all four quantum numbers the same. As it works out, this means there can only be *a maximum of two electrons in any given orbital.* There are four things to consider: (1) the main energy level, (2) the energy sublevel, (3) the number of orbital orientations, and (4) the electron spin. Recall that the lowest-energy level is $n = 1$, and successive numbers identify progressively higher-energy levels. Recall also that the energy sublevels, in order of increasing energy, are s, p, d, and f. This electron configuration is written in shorthand, with 1s standing for the lowest-energy sublevel of the first energy level. A superscript gives the number of electrons present in a sublevel. Thus, the electron configuration for a helium atom, which has two electrons, is written as $1s^2$. This combination of symbols has the following meaning: The symbols mean an atom with two electrons in the s sublevel of the first main energy level.

Table 8.3 gives the electron configurations for the first twenty elements. The configurations of the p energy sublevel have been condensed in this table. There are three possible orientations of the p orbital, each with two electrons (figure 8.15). This is shown as p^6, which designates the number of electrons in all of the three possible p orientations. Note that the sum of the electrons in all the orbitals equals the atomic number. Note also that as you proceed from a lower atomic number to a higher one, the higher element has the same configuration as the element before it with the addition of one more electron. In general, it is then possible to begin with the simplest atom, hydrogen, and add one electron at a time to the order of energy sublevels and obtain the electron configuration for all the elements. The exclusion principle limits the number of electrons in any orbital, and allowances will need to be made for the more complex behavior of atoms with many electrons.

The energies of the orbital are different for each element, and there are several factors that influence their energies. The first orbitals are filled in a straightforward 1s, 2s, 2p, 3s, then 3p order. Then the order becomes contrary to what you might expect. One useful way of figuring out the order in which orbitals are filled is illustrated in figure 8.16. Each row of this matrix represents a principal energy level with possible energy sublevels increasing from left to right. The order of filling is indicated by the diagonal arrows. There are exceptions to the order of filling shown by the matrix, but it works for most of the elements.

Table 8.3 Electron Configuration for the First Twenty Elements

Atomic Number	Element	Electron Configuration
1	Hydrogen	$1s^1$
2	Helium	$1s^2$
3	Lithium	$1s^2 2s^1$
4	Beryllium	$1s^2 2s^2$
5	Boron	$1s^2 2s^2 2p^1$
6	Carbon	$1s^2 2s^2 2p^2$
7	Nitrogen	$1s^2 2s^2 2p^3$
8	Oxygen	$1s^2 2s^2 2p^4$
9	Fluorine	$1s^2 2s^2 2p^5$
10	Neon	$1s^2 2s^2 2p^6$
11	Sodium	$1s^2 2s^2 2p^6 3s^1$
12	Magnesium	$1s^2 2s^2 2p^6 3s^2$
13	Aluminum	$1s^2 2s^2 2p^6 3s^2 3p^1$
14	Silicon	$1s^2 2s^2 2p^6 3s^2 3p^2$
15	Phosphorus	$1s^2 2s^2 2p^6 3s^2 3p^3$
16	Sulfur	$1s^2 2s^2 2p^6 3s^2 3p^4$
17	Chlorine	$1s^2 2s^2 2p^6 3s^2 3p^5$
18	Argon	$1s^2 2s^2 2p^6 3s^2 3p^6$
19	Potassium	$1s^2 2s^2 2p^6 3s^2 3p^6 4s^1$
20	Calcium	$1s^2 2s^2 2p^6 3s^2 3p^6 4s^2$

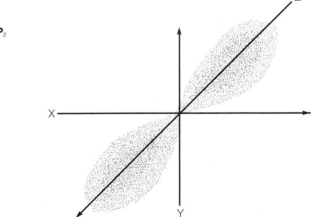

Figure 8.15

There are three possible orientations of the p orbital, and these are called p_x, p_y, and p_z. Each orbital can hold two electrons, so a total of six electrons are possible in the three orientations; thus the notation p^6.

CONCEPTS APPLIED
Firework Configuration

Certain strontium (atomic number 38) chemicals are used to add the pure red color to flares and fireworks. Write the electron configuration of strontium and do this before looking at the solution below.

First, note that an atomic number of 38 means a total of thirty-eight electrons. Second, refer to the order of filling the matrix in figure 8.16. Remember that only two electrons can occupy an orbital, but there are three orientations of the p orbital, for a total of six electrons. There are likewise five possible orientations of the d orbital, for a total of ten electrons. Starting at the lowest energy level, two electrons go in 1s, making $1s^2$; then two go in 2s, making $2s^2$. That is a total of four electrons so far. Next $2p^6$ and $3s^2$ use eight more electrons, for a total of twelve so far. The $3p^6$, $4s^2$, $3d^{10}$, and $4p^6$ use up twenty-four more electrons, for a total of thirty-six. The remaining two go into the next sublevel, $5s^2$, and the complete answer is

Strontium: $1s^2\ 2s^2\ 2p^6\ 3s^2\ 3p^6\ 4s^2\ 3d^{10}\ 4p^6\ 5s^2$

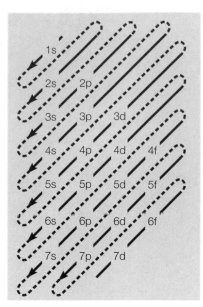

Figure 8.16
A matrix showing the order in which the orbitals are filled. Start at the top left, then move from the head of each arrow to the tail of the one immediately below it. This sequence moves from the lowest-energy level to the next higher level for each orbital.

THE PERIODIC TABLE

The periodic table is made up of rows and columns of cells, with each element having its own cell in a specific location. The cells are not arranged symmetrically. The arrangement has a meaning, both about atomic structure and about chemical behaviors. The key to meaningful, satisfying use of the table is to understand the code of this structure. The following explains some of what the code means. It will facilitate your understanding of the code if you refer frequently to a periodic table during the following discussion (figure 8.17).

An element is identified in each cell with its chemical symbol. The number above the symbol is the atomic number of the element, and the number below the symbol is the rounded atomic weight of the element. Horizontal rows of elements run from left to right with increasing atomic numbers. Each row is called a *period*. The periods are numbered from 1 to 7 on the left side. A vertical column of elements is called a *family* (or group) of elements. Elements in families have similar properties, but this is more true of some families than others.

As shown in table 8.4, all of the elements in the first column have an outside electron configuration of one electron. With the exception of hydrogen, the elements of the first column are shiny, low-density metals that are so soft you can cut them easily with a knife. These metals are called the *alkali metals* because they react violently with water to form an alkaline solution. The alkali metals do not occur in nature as free elements because they are so reactive. Hydrogen is a unique element in the periodic table. It is not an alkali metal, but it is placed in the group because it seems to fit there because it has one electron in its outer orbit.

CONCEPTS APPLIED
Spiral Origins

Make a periodic table. List the elements in boxes on a roll of adding machine tape. Have someone hold the tape up by the hydrogen end while you make a spiral with the noble gases appearing one below the other. Use cellophane tape to hold the noble gases family together, and then cut the tape just to the right of this group. Spread this cut spiral flat and a long form of the periodic table will be the result. Moving the inner transition elements shown in figure 8.17 (58 to 71 and 90 to 103) will produce the familiar short form of the periodic table.

CONCEPTS APPLIED
Periodic Practice

Identify the periodic table period and family of the element silicon. Write your answer before reading the solution in the next paragraph.

According to the list of elements on the inside back cover of this text, silicon has the symbol Si and an atomic number of 14. The square with the symbol Si and the atomic number 14 is located in the third period (third row) and in the column identified as IVA (14).

Now, can you identify the period and family of the element iron? Compare your answer with a classmate's to check.

Table 8.4 Electron Structures of the Alkali Metal Family

Element	Electron Configuration	Number of Electrons in Main Energy Level						
		1ST	2ND	3RD	4TH	5TH	6TH	7TH
Lithium (Li)	[He] $2s^1$	2	1	—	—	—	—	—
Sodium (Na)	[Ne] $3s^1$	2	8	1	—	—	—	—
Potassium (K)	[Ar] $4s^1$	2	8	8	1	—	—	—
Rubidium (Rb)	[Kr] $5s^1$	2	8	18	8	1	—	—
Cesium (Cs)	[Xe] $6s^1$	2	8	18	18	8	1	—
Francium (Fr)	[Rn] $7s^1$	2	8	18	32	18	8	1

The elements in the second column all have an outside configuration of two electrons and are called the *alkaline earth metals*. The alkaline earth metals are soft, reactive metals but not as reactive or soft as the alkali metals. Calcium and magnesium, in the form of numerous compounds, are familiar examples of this group.

The elements in group VIIA (17) all have an outside configuration of seven electrons, needing only one more electron to completely fill the outer orbitals. These elements are called the *halogens*. The halogens are very reactive nonmetals. The halogens fluorine and chlorine are greenish-colored gases.

Periodic Table of the Elements

Transition Elements

Inner Transition Elements

Group	Metals / Semiconductors / Nonmetals
Metals	
Semiconductors	
Nonmetals	

Alkali Metals — IA (1)
Alkaline Earth Metals — IIA (2)
Halogens — VIIA (17)
Noble gases — VIIIA (18)

Period 1
Hydrogen 1 H 1.008
Helium 2 He 4.003

Period 2
Lithium 3 Li 6.941
Beryllium 4 Be 9.012
Boron 5 B 10.81
Carbon 6 C 12.01
Nitrogen 7 N 14.01
Oxygen 8 O 16.00
Fluorine 9 F 19.00
Neon 10 Ne 20.18

Period 3
Sodium 11 Na 22.99
Magnesium 12 Mg 24.31
Aluminum 13 Al 26.98
Silicon 14 Si 28.09
Phosphorus 15 P 30.97
Sulfur 16 S 32.07
Chlorine 17 Cl 35.45
Argon 18 Ar 39.95

Period 4
Potassium 19 K 39.10
Calcium 20 Ca 40.08
Scanium 21 Sc 44.96
Titanium 22 Ti 47.88
Vanadium 23 V 50.94
Chromium 24 Cr 52.00
Maganese 25 Mn 54.94
Iron 26 Fe 55.85
Cobalt 27 Co 58.93
Nickel 28 Ni 58.69
Copper 29 Cu 63.55
Zinc 30 Zn 65.39
Gallium 31 Ga 69.72
Germanium 32 Ge 72.61
Arsenic 33 As 74.92
Selenium 34 Se 78.96
Bromine 35 Br 79.90
Krypton 36 Kr 83.80

Period 5
Rubidium 37 Rb 85.47
Strontium 38 Sr 87.62
Yttrium 39 Y 88.91
Zirconium 40 Zr 91.22
Niobium 41 Nb 92.91
Molybdenum 42 Mo 95.94
Technetium 43 Tc (98)
Ruthenium 44 Ru 101.1
Rhodium 45 Rh 102.9
Palladium 46 Pd 106.4
Silver 47 Ag 107.9
Cadmium 48 Cd 112.4
Indium 49 In 114.8
Tin 50 Sn 118.7
Antimony 51 Sb 121.8
Tellurium 52 Te 127.6
Iodine 53 I 126.9
Xenon 54 Xe 131.3

Period 6
Cesium 55 Cs 132.9
Barium 56 Ba 137.3
Lanthanum 57 La 138.9
Hafnium 72 Hf 178.5
Tantalum 73 Ta 180.9
Tungsten 74 W 183.8
Rhenium 75 Re 186.2
Osmium 76 Os 190.2
Iridium 77 Ir 192.2
Platinum 78 Pt 195.1
Gold 79 Au 197.0
Mercury 80 Hg 200.6
Thallium 81 Tl 204.4
Lead 82 Pb 207.2
Bismuth 83 Bi 209.0
Polonium 84 Po (209)
Astatine 85 At (210)
Radon 86 Rn (222)

Period 7
Francium 87 Fr (223)
Radium 88 Ra (226)
Actinium 89 Ac (227)
Rutherfordium 104 Rf (261)
Dubnium 105 Db (262)
Seaborgium 106 Sg (266)
Bohrium 107 Bh (264)
Hassium 108 Hs (277)
Meitnerium 109 Mt (268)
Ununnillium 110 Uun (281)
Unununium 111 Uuu (272)
Ununbium 112 Uub (285)
Ununquadium 114 Uuq (289)

Transition element group labels:
IIIB (3), IVB (4), VB (5), VIB (6), VIIB (7), VIIIB (8), VIIIB (9), VIIIB (10), IB (11), IIB (12), IIIA (13), IVA (14), VA (15), VIA (16)

†Lanthanides 6:
Cerium 58 Ce 140.1
Praseodymium 59 Pr 140.9
Neodymium 60 Nd 144.2
Promethium 61 Pm (145)
Samarium 62 Sm 150.4
Europium 63 Eu 152.0
Gadolinium 64 Gd 157.3
Terbium 65 Tb 158.9
Dysprosium 66 Dy 162.5
Holmium 67 Ho 164.9
Erbium 68 Er 167.3
Thulium 69 Tm 168.9
Ytterbium 70 Yb 173.0
Lutetium 71 Lu 175.0

‡Actinides 7:
Thorium 90 Th 232.0
Protactinium 91 Pa 231.0
Uranium 92 U 238.0
Neptunium 93 Np (237)
Plutonium 94 Pu (244)
Americium 95 Am (243)
Curium 96 Cm (247)
Berkelium 97 Bk (247)
Californium 98 Cf (251)
Einsteinium 99 Es (252)
Fermium 100 Fm (257)
Mendelevium 101 Md (258)
Nobelium 102 No (259)
Lawrencium 103 Lr (262)

Key
element name — Hydrogen
atomic number — 1
symbol of element — H
atomic weight — 1.008

Values in parentheses are the mass numbers of the most stable or best-known isotopes.

Names and symbols for elements 110–114 are under review.

Figure 8.17

The periodic table of the elements.

Compounds of the rare earths were first identified when they were isolated from uncommon minerals in the late 1700s. The elements are very reactive and have similar chemical properties, so they were not recognized as elements until some fifty years later. Thus, they were first recognized as earths, that is, nonmetal substances, when in fact they are metallic elements. They were also considered to be rare since, at that time, they were known to occur only in uncommon minerals. Today, these metallic elements are known to be more abundant in the earth than gold, silver, mercury, or tungsten. The rarest of the rare earths, thulium, is twice as abundant as silver. The rare earth elements are neither rare nor earths, and they are important materials in glass, electronic, and metallurgical industries.

You can identify the rare earths in the two lowermost rows of the periodic table. These rows contain two series of elements that actually belong in periods 6 and 7, but they are moved below so that the entire table is not so wide. Together, the two series are called the inner transition elements. The top series is fourteen elements wide from elements 58 through 71. Since this series belongs next to element 57, lanthanum, it is sometimes called the *lanthanide series*. This series is also known as the rare earths. The second series of fourteen elements is called the *actinide series*. These are mostly the artificially prepared elements that do not occur naturally.

You may never have heard of the rare earth elements, but they are key materials in many advanced or high-technology products. Lanthanum, for example, gives glass special refractive properties and is used in optic fibers and expensive camera lenses. Samarium, neodymium, and dysprosium are used to manufacture crystals used in lasers. Samarium, ytterbium, and terbium have special magnetic properties that have made possible new electric motor designs, magnetic-optical devices in computers, and the creation of a ceramic superconductor. Other rare earth metals are also being researched for use in possible high-temperature superconductivity materials. Many rare earths are also used in metal alloys; for example, an alloy of cerium is used to make heat-resistant jet-engine parts. Erbium is also used in high-performance metal alloys. Dysprosium and holmium have neutron-absorbing properties and are used in control rods to control nuclear fission. Europium should be mentioned because of its role in making the red color of color television screens. The rare earths are relatively abundant metallic elements that play a key role in many common and high-technology applications. They may also play a key role in superconductivity research.

Table 8.5 Electron Structures of the Noble Gas Family

Element	Electron Configuration	Number of Electrons in Main Energy Level						
		1ST	2ND	3RD	4TH	5TH	6TH	7TH
Helium (He)	$1s^2$	2	—	—	—	—	—	—
Neon (Ne)	[He] $2s^2 2p^6$	2	8	—	—	—	—	—
Argon (Ar)	[Ne] $3s^2 3p^6$	2	8	8	—	—	—	—
Krypton (Kr)	[Ar] $4s^2 3d^{10} 4p^6$	2	8	18	8	—	—	—
Xenon (Xe)	[Kr] $5s^2 4d^{10} 5p^6$	2	8	18	18	8	—	—
Radon (Rn)	[Xe] $6s^2 4f^{14} 5d^{10} 6p^6$	2	8	18	32	18	8	—

Bromine is a reddish brown liquid and iodine is a dark purple solid. Halogens are used as disinfectants, bleaches, and combined with a metal as a source of light in halogen lights. Halogens react with metals to form a group of chemicals called *salts*, such as sodium chloride. In fact, the word *halogen* is Greek, meaning "salt former."

As shown in table 8.5, the elements in group VIIIA (18) have orbitals that are filled to capacity. These elements are colorless, odorless gases that almost never react with other elements to form compounds. Sometimes they are called the *noble gases* because they are chemically inert, perhaps indicating they are above the other elements. They have also been called the *rare gases* because of their scarcity and *inert gases* because they are mostly chemically inert, not forming compounds. The noble gases are inert because they have filled outer electron configurations, a particularly stable condition.

Each period *begins* with a single electron in a new orbital. Second, each period *ends* with the filling of an orbital, completing the maximum number of electrons that can occupy that main energy level. Since the first A family is identified as IA, this means that all the atoms of elements in this family have one electron in their outer orbitals. All the atoms of elements in family IIA have two electrons in their outer orbitals. This pattern continues on to family VIIIA, in which all the atoms of elements have eight electrons in their outer orbitals except helium. Thus, the number identifying the A families *also identifies the number of electrons in the outer orbitals,* with the exception of helium. Helium is nonetheless similar to the other elements in this family, since all have filled outer orbitals. The electron theory of chemical bonding, which is discussed in chapter 9, states that only the electrons in the outermost orbitals of an atom are involved in chemical reactions. Thus, *the outer orbital electrons are mostly responsible for the chemical properties of an element.* Since the members of a family all have similar outer configurations, you would expect them to have similar chemical behaviors, and they do.

The members of the A-group families are called the *main-group* or **representative elements.** The members of the B-group families are called the **transition elements** (or metals).

Metals, Nonmetals, and Semiconductors

As indicated earlier, chemical behavior is mostly concerned with the outer shell electrons. The outer shell electrons, that is, the highest energy level electrons, are conveniently represented with an **electron dot notation,** made by writing the chemical symbol with dots around it indicating the number of outer shell electrons. Electron dot notations are shown for the representative elements in figure 8.18. Again, note the pattern in figure 8.18—all the noble gases are in group VIIIA, and all (except helium) have eight outer electrons. All the group IA elements (alkali metals) have one dot, all the IIA elements have two dots, and so on. This pattern will explain the difference in metals, nonmetals, and a third group of in-between elements called semiconductors.

There are several ways to classify matter according to properties. One example is to group substances according to the physical properties of metals and nonmetals—luster, conductivity, malleability, and ductility. Metals and nonmetals also have certain chemical properties that are related to their positions in the periodic table. Figure 8.19 shows where the *metals, nonmetals,* and *semiconductors* are located. Note that about 80 percent of all the elements are metals.

The noble gases have completely filled outer orbitals in their highest energy levels, and this is a particularly stable arrangement. Other elements react chemically, either *gaining or losing electrons to attain a filled outermost energy level like the noble gases.* When an atom loses or gains electrons, it acquires an

Figure 8.18

Electron dot notation for the representative elements.

unbalanced electron charge and is called an **ion.** An atom of lithium, for example, has three protons (plus charges) and three electrons (negative charges). If it loses the outermost electron, it now has an outer filled orbital structure like helium, a noble gas. It is also now an ion, since it has three protons (3+) and two electrons (2−), for a net charge of 1+. A lithium ion thus has a 1+ charge.

Elements with one, two, or three outer electrons tend to lose electrons to form positive ions. The metals lose electrons like this, and the *metals are elements that lose electrons to form positive ions* (figure 8.20). Nonmetals, on the other hand, are elements with five to seven outer electrons that tend to acquire electrons to fill their outer orbitals. *Nonmetals are elements that gain electrons to form negative ions.* In general, elements located in the left two-thirds or so of the periodic table are metals. The nonmetals are on the right side of the table.

The dividing line between the metals and nonmetals is a steplike line from the left top of group IIIA down to the bottom left of group VIIA. This is not a line of sharp separation between

Figure 8.19

The location of metals, nonmetals, and semiconductors in the periodic table.

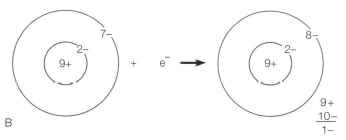

Figure 8.20

(*A*) Metals lose their outer electrons to acquire a noble gas structure and become positive ions. Lithium becomes a 1+ ion as it loses its one outer electron. (*B*) Nonmetals gain electrons to acquire an outer noble gas structure and become negative ions. Fluorine gains a single electron to become a 1− ion.

the metals and nonmetals, and elements *along* this line sometimes act like metals, sometimes like nonmetals, and sometimes like both. These hard-to-classify elements are called **semiconductors** (or *metalloids*). Silicon, germanium, and arsenic have physical properties of nonmetals; for example, they are brittle materials that cannot be hammered into a new shape. Yet these elements conduct electric currents under certain conditions. The ability to conduct an electric current is a property of a metal, and nonmalleability is a property of nonmetals, so as you can see, these semiconductors have the properties of both metals and nonmetals.

The transition elements, which are all metals, are located in the B-group families. Unlike the representative elements, which form vertical families of similar properties, the transition elements tend to form horizontal groups of elements with similar properties. Iron (Fe), cobalt (Co), and nickel (Ni) in

Connections . . .

Nutrient Elements

All molecules required to support living things are called nutrients. Some nutrients are inorganic molecules such as calcium, iron, or potassium, and others are organic molecules such as carbohydrates, proteins, fats, and vitamins. All minerals are elements and cannot be synthesized by the body. Because they are elements, they cannot be broken down or destroyed by metabolism or cooking. They commonly occur in many foods and in water. Minerals retain their characteristics whether they are in foods or in the body, and each plays a different role in metabolism.

Minerals can function as regulators, activators, transmitters, and controllers of various enzymatic reactions. For example, sodium ions (Na^+) and potassium ions (K^+) are important in the transmission of nerve impulses, while magnesium ions (Mg^{2+}) facilitate energy release during reactions involving ATP. Without iron, not enough hemoglobin would be formed to transport oxygen, a condition called anemia, and a lack of calcium may re-

sult in osteoporosis. Osteoporosis is a condition that results from calcium loss leading to painful, weakened bones. There are many minerals that are important in your diet. In addition to those just mentioned, you need chlorine, cobalt, copper, iodine, phosphorus, potassium, sulfur, and zinc to remain healthy. With few exceptions, adequate amounts of minerals are obtained in a normal diet. Calcium and iron supplements may be necessary, particularly in women.

Osteoporosis is a nutritional deficiency disease that results in a change in the density of the bones as a result of the loss of bone mass. Bones that have undergone this change look lacy or like Swiss cheese, with larger than normal holes. A few risk factors found to be associated with this disease are being female and fair skinned; having a sedentary lifestyle; using alcohol, caffeine, and tobacco; being anorexic; and having reached menopause.

group VIIIB, for example, are three horizontally arranged metallic elements that show magnetic properties.

A family of representative elements all form ions with the same charge. Alkali metals, for example, all lose an electron to form a 1+ ion. The transition elements have *variable charges.* Some transition elements, for example, lose their one outer electron to form 1+ ions (copper, silver). Copper, because of its special configuration, can also lose an additional electron to form a 2+ ion. Thus, copper can form either a 1+ ion or a 2+ ion. Most transition elements have two outer s orbital electrons and lose them both to form 2+ ions (iron, cobalt, nickel), but some of these elements also have special configurations that permit them to lose more of their electrons. Thus, iron and cobalt, for example, can form either a 2+ ion or a 3+ ion. Much more can be interpreted from the periodic table, and more generalizations will be made as the table is used in the following chapters.

SUMMARY

Attempts at understanding matter date back to ancient Greek philosophers, who viewed matter as being composed of *elements,* or simpler substances. Two models were developed that considered matter to be (1) *continuous,* or infinitely divisible, or (2) *discontinuous,* made up of particles called *atoms.*

In the early 1800s, Dalton published an *atomic theory,* reasoning that matter was composed of hard, indivisible atoms that were joined together or dissociated during chemical change.

When a good air pump to provide a vacuum was invented in 1885, *cathode rays* were observed to move from the negative terminal in an evacuated glass tube. The nature of cathode rays was a mystery. The mystery was solved in 1887 when Thomson discovered they were negatively charged particles now known as *electrons.* Thomson had discovered the first elementary particle of which atoms are made and measured their charge-to-mass ratio.

Rutherford developed a solar system model based on experiments with alpha particles scattered from a thin sheet of metal. This model had a small, massive, and positively charged *nucleus* surrounded by moving electrons. These electrons were calculated to be at a distance from the nucleus of 100,000 times the radius of the nucleus, so the volume of an atom is mostly empty space. Later, Rutherford proposed that the nucleus contained two elementary particles: *protons* with a positive charge and *neutrons* with no charge. The *atomic number* is the number of protons in an atom. Atoms of elements with different numbers of neutrons are called *isotopes.* The mass of each isotope is compared to the mass of carbon-12, which is assigned a mass of exactly 12.00 *atomic mass units.* The mass contribution of the isotopes of an element according to their abundance is called the *atomic weight* of an element. Isotopes are identified by their *mass number,* which is the sum of the number of protons and neutrons in the nucleus. Isotopes are identified by their chemical symbol, with the atomic number as a subscript and the mass number as a superscript.

Bohr developed a model of the hydrogen atom to explain the characteristic *line spectra* emitted by hydrogen. His model specified that (1) electrons can move only in allowed orbits, (2) electrons do not emit radiant energy when they remain in an orbit, and (3) electrons move from one allowed orbit to another when they gain or lose energy. When an electron jumps from a higher orbit to a lower one, it gives up energy in the form of a single photon. The energy of the photon corresponds to the difference in energy between the two levels. The Bohr model worked well for hydrogen but not for other atoms.

Schrödinger and others used the wave nature of the electron to develop a new model of the atom called *wave mechanics,* or *quantum mechanics.* This model was found to confirm exactly all the experimental data as well as predict new data. The quantum mechanical model describes the energy state of the electron in terms of quantum numbers based on the wave nature of the electron. The quantum numbers defined the *probability* of the location of an electron in terms of fuzzy regions of space called *orbitals.*

The *periodic table* has horizontal rows of elements called *periods* and vertical columns of elements called *families.* Families have the same outer orbital electron configurations, and it is the electron configuration that is mostly responsible for the chemical properties of an element.

Summary of Equations

8.1   energy = (Planck's constant) (frequency)

$$E = hf$$

where $h = 6.63 \times 10^{-34}$ J·s

KEY TERMS

atomic mass units (p. **154**)   mass number (p. **155**)
atomic number (p. **154**)   neutron (p. **154**)
atomic weight (p. **155**)   nucleus (p. **154**)
electron (p. **153**)   orbital (p. **159**)
electron dot notation   Pauli exclusion principle
    (p. **165**)   (p. **160**)
excited states (p. **157**)   photons (p. **156**)
ground state (p. **157**)   proton (p. **154**)
Heisenberg uncertainty   quanta (p. **156**)
    principle (p. **159**)   representative elements
ion (p. **165**)   (p. **165**)
isotopes (p. **154**)   semiconductors (p. **166**)
line spectrum (p. **156**)   transition elements (p. **165**)

APPLYING THE CONCEPTS

1. The electron was discovered through experiments with
   (a) radioactivity.
   (b) light.
   (c) matter waves.
   (d) electricity.

2. Thomson was convinced that he had discovered a subatomic particle, the electron, from the evidence that
   (a) the charge-to-mass ratio was the same for all materials.
   (b) cathode rays could move through a vacuum.
   (c) electrons were attracted toward a negatively charged plate.
   (d) the charge was always 1.60×10^{-19} coulomb.

3. The existence of a tiny, massive, and positively charged nucleus was deduced from the observation that
   (a) fast, massive, and positively charged alpha particles all move straight through metal foil.
   (b) alpha particles were deflected by a magnetic field.
   (c) some alpha particles were deflected by metal foil.
   (d) None of the above are correct.

4. According to Rutherford's calculations, the volume of an atom is mostly
   (a) occupied by protons and neutrons.
   (b) filled with electrons.
   (c) occupied by tightly bound protons, electrons, and neutrons.
   (d) empty space.

5. Hydrogen, with its one electron, produces a line spectrum in the visible light range with
   (a) one color line.
   (b) two color lines.
   (c) three color lines.
   (d) four color lines.

6. According to the Bohr model, an electron gains or loses energy only by
   (a) moving faster or slower in an allowed orbit.
   (b) jumping from one allowed orbit to another.
   (c) being completely removed from an atom.
   (d) jumping from one atom to another atom.

7. When an electron in a hydrogen atom jumps from an orbit farther from the nucleus to an orbit closer to the nucleus, it
   (a) emits a single photon with an energy equal to the energy difference of the two orbits.
   (b) emits four photons, one for each of the color lines observed in the line spectrum of hydrogen.
   (c) emits a number of photons dependent on the number of orbit levels jumped over.
   (d) None of the above are correct.

8. The quantum mechanics model of the atom is based on
   (a) the quanta, or measured amounts of energy of a moving particle.
   (b) the energy of a standing electron wave that can fit into an orbit.
   (c) calculations of the energy of the three-dimensional shape of a circular orbit of an electron particle.
   (d) Newton's laws of motion, but scaled down to the size of electron particles.

9. The Bohr model of the atom described the energy state of electrons with one quantum number. The quantum mechanics model uses how many quantum numbers to describe the energy state of an electron?
   (a) one
   (b) two
   (c) four
   (d) ten

10. An electron in the second main energy level and the second sublevel is described by the symbols
    (a) 1s.
    (b) 2s.
    (c) 1p.
    (d) 2p.

11. The space in which it is probable that an electron will be found is described by a(an)
    (a) circular orbit.
    (b) elliptical orbit.
    (c) orbital.
    (d) geocentric orbit.

12. Two electrons can occupy the same orbital because they have different
    (a) principal quantum numbers.
    (b) angular momentum quantum numbers.
    (c) magnetic quantum numbers.
    (d) spin quantum numbers.

Answers

1. d 2. a 3. c 4. d 5. d 6. b 7. a 8. b 9. c 10. d 11. c 12. d

QUESTIONS FOR THOUGHT

1. What was the experimental evidence that Thomson had discovered the existence of a subatomic particle when working with cathode rays?

2. Describe the experimental evidence that led Rutherford to the concept of a nucleus in an atom.

3. What is the main problem with a solar system model of the atom?

4. Compare the size of an atom to the size of its nucleus.

5. What does *atomic number* mean? How does the atomic number identify the atoms of a particular element? How is the atomic number related to the number of electrons in an atom?

6. An atom has 11 protons in the nucleus. What is the atomic number? What is the name of this element? What is the electron configuration of this atom?

7. How is the atomic weight of an element determined?

8. Describe the three main points in the Bohr model of the atom.

9. Why do the energies of electrons in an atom have negative values? (*Hint:* It is *not* because of the charge of the electron.)

10. Which has the lowest energy, an electron in the first energy level ($n = 1$) or an electron in the third energy level ($n = 3$)? Explain.

11. What is similar about the Bohr model of the atom and the quantum mechanical model? What are the fundamental differences?

12. What is the difference between a hydrogen atom in the ground state and one in the excited state?

PARALLEL EXERCISES

The exercises in groups A and B cover the same concepts. Solutions to group A exercises are located in appendix D.

Group A

1. How much energy is needed to move an electron in a hydrogen atom from $n = 2$ to $n = 6$? (See figure 8.11 for needed values.)
2. What frequency of light is emitted when an electron in a hydrogen atom jumps from $n = 6$ to $n = 2$? What color would you see?
3. How much energy is needed to completely remove the electron from a hydrogen atom in the ground state?
4. Thomson determined the charge-to-mass ratio of the electron to be -1.76×10^{11} coulomb/kilogram. Millikan determined the charge on the electron to be -1.60×10^{-19} coulomb. According to these findings, what is the mass of an electron?
5. Using any reference you wish, write the complete electron configurations for (a) boron, (b) aluminum, and (c) potassium.
6. Explain how you know that you have the correct *total* number of electrons in your answers for 5a, 5b, and 5c.
7. Refer to figure 8.17 *only,* and write the complete electron configurations for (a) argon, (b) zinc, and (c) bromine.

Group B

1. How much energy is needed to move an electron in a hydrogen atom from the ground state ($n = 1$) to $n = 3$?
2. What frequency of light is emitted when an electron in a hydrogen atom jumps from $n = 2$ to the ground state ($n = 1$)?
3. How much energy is needed to completely remove an electron from $n = 2$ in a hydrogen atom?
4. If the charge-to-mass ratio of a proton is 9.58×10^7 coulomb/kilogram and the charge is 1.60×10^{-19} coulomb, what is the mass of the proton?
5. Using any reference you wish, write the complete electron configurations for (a) nitrogen, (b) phosphorus, and (c) chlorine.
6. Explain how you know that you have the correct *total* number of electrons in your answers for 5a, 5b, and 5c.
7. Referring to figure 8.17 *only,* write the complete electron configuration for (a) neon, (b) sulfur, and (c) calcium.

Chemical Reactions

A chemical change occurs when iron rusts, and rust is a different substance with physical and chemical properties different from iron.
This rusted anchor makes a colorful display on the bow of a grain ship.

O U T L I N E

In chapter 8, you learned how the modern atomic theory is used to describe the structures of atoms of different elements. The electron structures of different atoms successfully account for the position of elements in the periodic table as well as for groups of elements with similar properties. On a large scale, all metals were found to have a similarity in electron structure, as were nonmetals. On a smaller scale, chemical families such as the alkali metals were found to have the same outer electron configurations. Thus, the modern atomic theory accounts for observed similarities between elements in terms of atomic structure.

So far, only individual, isolated atoms have been discussed; we have not considered how atoms of elements join together to produce compounds. There is a relationship between the electron structure of atoms and the reactions they undergo to produce specific compounds. Understanding this relationship will explain the changes that matter itself undergoes. For example, hydrogen is a highly flammable, gaseous element that burns with an explosive reaction. Oxygen, on the other hand, is a gaseous element that supports burning. As you know, hydrogen and oxygen combine to form water. Water is a liquid that neither burns nor supports burning. What happens when atoms of elements such as hydrogen and oxygen join to form molecules such as water? Why do such atoms join and why do they stay together? Why does water have properties different from the elements that combine to produce it? And finally, why is water H_2O and not H_3O or H_4O?

Answers to questions about why and how atoms join together in certain numbers are provided by considering the electronic structures of the atoms. Chemical substances are formed from the interactions of electrons as their structures merge, forming new patterns that result in molecules with new properties. It is the new electron pattern of the water molecule that gives water properties that are different from the oxygen or hydrogen from which it formed (figure 9.1). Understanding how electron structures of atoms merge to form new patterns is understanding the changes that matter itself undergoes, the topic of this chapter.

COMPOUNDS

Matter that you see around you seems to come in a wide variety of sizes, shapes, forms, and kinds. Are there any patterns in the apparent randomness that will help us comprehend matter? Yes, there are many patterns and one of the more obvious is that all matter occurs as either a mixture or as a pure substance (figure 9.2). A *mixture* has unlike parts and a composition that varies from sample to sample. For example, sand from a beach is a variable mixture of things such as bits of rocks, minerals, and sea shells.

Mixtures can be separated into their component parts by physical means. For example, you can physically separate the parts making up a sand mixture by using a magnifying glass and tweezers to move and isolate each part. A solution of salt in water is a mixture since the amount of salt dissolved in water can have a variable composition. But how do you separate the parts of a solution? One way is to evaporate the water, leaving the salt behind. There are many methods for separating mixtures, but all involve a **physical change.** A physical change does not alter the *identity* of matter. When water is evaporated from the salt solution, for example, it changes to a different state of matter (water vapor) but is still recognized as water. Physical changes involve physical properties only; no new substances are formed. Examples of physical changes include evaporation, condensation, melting, freezing, and dissolving, as well as reshaping processes such as crushing or bending.

Mixtures can be physically separated into *pure substances,* materials that are the same throughout and have a fixed, definite composition. If you closely examine a sample of table salt, you will see that it is made up of hundreds of tiny cubes. Any one of these cubes will have the same properties as any other cube, including a salty taste. Sugar, like table salt, has all of its parts alike. Unlike salt, sugar grains have no special shape or form, but each grain has the same sweet taste and other properties as any other sugar grain.

If you heat salt and sugar in separate containers, you will find very different results. Salt, like some other pure substances,

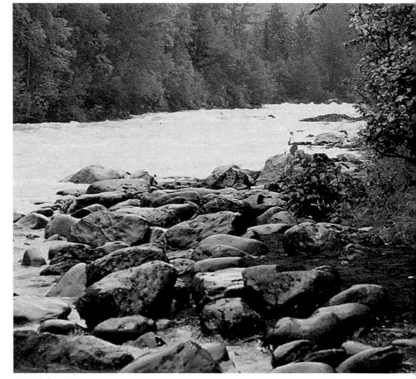

Figure 9.1

Water is the most abundant liquid on the earth and is necessary for all life. Because of water's great dissolving properties, any sample is a solution containing solids, other liquids, and gases from the environment. This stream also carries suspended, ground-up rocks, called *rock flour,* from a nearby glacier.

undergoes a physical change and melts, changing back to a solid upon cooling with the same properties that it had originally. Sugar, however, changes to a black material upon heating, while it gives off water vapor. The black material does not change back to sugar upon cooling. The sugar has *decomposed* to a new substance, while the salt did not. The new substance is carbon, and

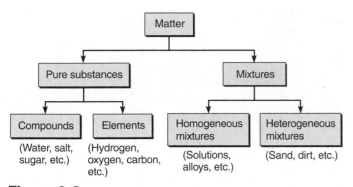

Figure 9.2

A classification scheme for matter.

it has properties completely different from sugar. The original substance (sugar) and its properties have changed to a new substance (carbon) with new properties. The sugar has gone through a **chemical change.** A chemical change alters the identity of matter, producing new substances with different properties. In this case, the chemical change was one of decomposition. Heat produced a chemical change by decomposing the sugar into carbon and water vapor.

The decomposition of sugar always produces the same mass ratio of carbon to water, so sugar has a fixed, definite composition. A **compound** is a pure substance that can be decomposed by a chemical change into simpler substances with a fixed mass ratio. This means that sugar is a compound (figure 9.3).

A pure substance that cannot be broken down into anything simpler by chemical or physical means is an **element.** Sugar is decomposed by heating into carbon and water vapor. Carbon cannot be broken down further, so carbon is an ele-

ment. It has been known since about 1800 that water is a compound that can be broken down by electrolysis into hydrogen and oxygen, two gases that cannot be broken down to anything simpler. So sugar is a compound made from the elements of carbon, hydrogen, and oxygen.

But what about the table salt? Is table salt a compound? Table salt is a stable compound that is not decomposed by heating. It melts at a temperature of about 800°C, and then returns to the solid form with the same salty properties upon cooling. Electrolysis—the splitting of a compound by means of electricity—can be used to decompose table salt into the elements sodium and chlorine, positively proving that it is a compound.

Pure substances are either compounds or elements. Decomposition through heating and decomposition through electrolysis are two means of distinguishing between compounds and elements. If a substance can be decomposed into something simpler, you know for sure that it is a compound. If the substance cannot be decomposed, it might be an element, or it might be a stable compound that resists decomposition. More testing would be necessary before you can be confident that you have identified an element. Most pure substances are compounds. There are millions of different compounds but only 113 known elements at the present time. These elements are the fundamental materials of which all matter is made.

ELEMENTS

All the natural elements were discovered by 1940. The naturally occurring elements generally range from hydrogen (atomic number 1) to uranium (atomic number 92) with a few exceptions as follows:

- Technetium (43) and promethium (61), which are not found anywhere on earth unless they are artificially produced.
- Francium (87) and astatine (85), which exist only a short time before undergoing radioactive decay, so they do not exist in any significant quantities. Half a sample of francium decays in 21 minutes. The most stable form of astatine decays in 7 1/2 hours.
- Plutonium (94), which has been found occurring naturally in small amounts.

The other elements that do not occur naturally were artificially made during nuclear physics experiments and are not found on the earth. Thus, there are 89 naturally occurring elements that occur in significant quantities and 24 short-lived and artificially prepared elements, for a total of 113 that are known at the present time.

Elements are not equally abundant and only a few are common. In table 9.1, for example, you can see that only eight elements make up about 99 percent of the solid surface of the earth. Oxygen is most abundant, making up about 50 percent of the weight of the earth's crust. Silicon makes up more than 25 percent, so these two nonmetals alone make up about 75 percent of the earth's solid surface. Almost all the rest is made up of just six metals, as shown in the table.

The number of common elements is limited elsewhere, too. Only two elements make up about 99 percent of the atmospheric air around the earth. Air is mostly nitrogen (about

A   B

Figure 9.3

Sugar (*A*) is a compound that can be easily decomposed to simpler substances by heating. (*B*) One of the simpler substances is the black element carbon, which cannot be further decomposed by chemical or physical means.

Table 9.1 Elements Making up 99 Percent of Earth's Crust

Element (Symbol)	Percent by Weight
Oxygen (O)	46.6
Silicon (Si)	27.7
Aluminum (Al)	8.1
Iron (Fe)	5.0
Calcium (Ca)	3.6
Sodium (Na)	2.8
Potassium (K)	2.6
Magnesium (Mg)	2.1

78 percent) and oxygen (about 21 percent), with traces of five other elements and compounds. Water on the earth is hydrogen and oxygen, of course, but seawater also contains elements in solution. These elements are chlorine (55 percent), sodium (31 percent), sulfur (8 percent), and magnesium (4 percent). Only

Figure 9.4

The elements of aluminum, iron, oxygen, and silicon make up about 88 percent of the earth's solid surface. Water on the surface and in the air as clouds and fog is made up of hydrogen and oxygen. The air is 99 percent nitrogen and oxygen. Hydrogen, oxygen, and carbon make up 97 percent of a person. Thus, almost everything you see in this picture is made up of just six elements.

three elements make up about 97 percent of your body. These elements are hydrogen (60 percent), oxygen (26 percent), and carbon (11 percent). Generally, all of this means that the elements are not equally distributed or equally abundant in nature (figure 9.4).

CHEMICAL CHANGE

The air you breathe, the liquids you drink, and all the things around you are elements, compounds, or mixtures. Most are compounds, however, and very few are pure elements. Water, sugar, gasoline, and chalk are examples of compounds. Each can be broken down into the elements that make it up. Recall that elements are basic substances that cannot be broken down into simpler substances. Examples of elements are hydrogen, carbon, and calcium. Why and how these elements join together in different ways to form different compounds is the subject of this chapter.

You have already learned that elements are made up of atoms that can be described by the modern atomic theory. You can also consider an **atom** to be *the smallest unit of an element that can exist alone or in combination with other elements.* Compounds are formed when atoms are held together by an attractive force called a *chemical bond.* The chemical bond binds individual atoms together in a compound. A molecule is generally thought of as a tightly bound group of atoms that maintains its identity. More specifically, a **molecule** is defined as *the smallest particle of a compound, or a gaseous element, that can exist and still retain the characteristic chemical properties of a substance.* Compounds with one type of chemical bond, as you will see, have molecules that are electrically neutral groups of atoms held together strongly enough to be considered independent units. For example, water is a compound. The smallest unit of water that can exist alone is an electrically neutral unit made up of two hydrogen atoms and one oxygen atom held together by chemical bonds. The concept of a molecule will be expanded as chemical bonds are discussed.

Compounds occur naturally as gases, liquids, and solids. Many common gases occur naturally as molecules made up of two or more atoms. For example, at ordinary temperatures hydrogen gas occurs as molecules of two hydrogen atoms bound together. Oxygen gas also usually occurs as molecules of two oxygen atoms bound together. Both hydrogen and oxygen occur naturally as *diatomic molecules* ("di-" means "two"). Oxygen sometimes occurs as molecules of three oxygen atoms bound together. These *triatomic molecules* ("tri-" means "three") are called *ozone*.

When molecules of any size are formed or broken down into simpler substances, new materials with new properties are produced. This kind of a change in matter is called a chemical change, and the process is called a chemical reaction. A **chemical reaction** is defined as

> **a change in matter in which different chemical substances are created by forming or breaking chemical bonds.**

In general, chemical bonds are formed when atoms of elements are bound together to form compounds. Chemical bonds are broken when a compound is decomposed into simpler substances. Chemical bonds are electrical in nature, formed by electrical attractions, as discussed in chapter 6.

Chemical reactions happen all the time, all around you. A growing plant, burning fuels, and your body's utilization of food all involve chemical reactions. These reactions produce different chemical substances with greater or smaller amounts of internal potential energy (see chapter 4 for a discussion of internal potential energy). Energy is *absorbed* to produce new chemical substances with more internal potential energy. Energy is *released* when new chemical substances are produced with less internal potential energy (figure 9.5). In general, changes in internal potential energy are called **chemical energy.** For example, new chemical substances are produced in green plants through the process called *photosynthesis*. A green plant uses radiant energy (sunlight), carbon dioxide, and water to produce new chemical materials and oxygen. These new chemical materials, the stuff that leaves, roots, and wood are made of, contain more chemical energy than the carbon dioxide and water they were made from.

A **chemical equation** is a way of describing what happens in a chemical reaction. The chemical reaction of photosynthesis can be described by using words in an equation:

$$\text{energy (sunlight)} + \frac{\text{carbon}}{\text{dioxide molecules}} + \frac{\text{water}}{\text{molecules}} \rightarrow \frac{\text{plant}}{\text{material molecules}} + \frac{\text{oxygen}}{\text{molecules}}$$

The substances that are changed are on the left side of the word equation and are called *reactants*. The reactants are carbon dioxide molecules and water molecules. The equation also indicates that energy is absorbed, since the term *energy* appears on the left side. The arrow means *yields*. The new chemical substances are on the right side of the word equation and are called

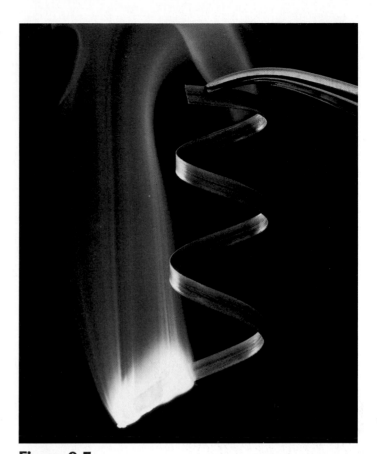

Figure 9.5
Magnesium is an alkaline earth metal that burns brightly in air, releasing heat and light. As chemical energy is released, a new chemical substance is formed. The new chemical material is magnesium oxide, a soft powdery material that forms an alkaline solution in water (called *milk of magnesia*).

products. Reading the photosynthesis reaction as a sentence you would say, "Carbon dioxide and water use energy to react, yielding plant materials and oxygen."

The plant materials produced by the reaction have more internal potential energy, also known as *chemical energy,* than the reactants. You know this from the equation because the term *energy* appears on the left side but not the right. This means that the energy on the left went into internal potential energy on the right. You also know this because the reaction can be reversed to release the stored energy (figure 9.6). When plant materials (such as wood) are burned, the materials react with oxygen, and chemical energy is released in the form of radiant energy (light) and high kinetic energy of the newly formed gases and vapors. In words,

$$\frac{\text{plant}}{\text{material molecules}} + \frac{\text{oxygen}}{\text{molecules}} \rightarrow \frac{\text{carbon}}{\text{dioxide molecules}} + \frac{\text{water}}{\text{molecules}} + \text{energy}$$

If you compare the two equations, you will see that burning is the opposite of the process of photosynthesis! The energy released in burning is exactly the same amount of solar energy that was stored as internal potential energy by the plant. Such chemical changes, in which chemical energy is stored in one reaction and released by another reaction, are the result of the making,

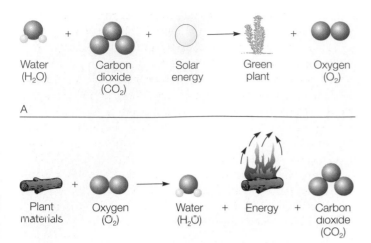

Water
(H_2O)
+
Carbon
dioxide
(CO_2)
+
Solar
energy
→
Green
plant
+
Oxygen
(O_2)

A

Plant
materials
+
Oxygen
(O_2)
→
Water
(H_2O)
+
Energy
+
Carbon
dioxide
(CO_2)

B

Figure 9.6

(*A*) New chemical bonds are formed as a green plant makes new materials and stores solar energy through the photosynthesis process. (*B*) The chemical bonds are later broken, and the same amount of energy and the same original materials are released. The same energy and the same materials are released rapidly when the plant materials burn, and they are released slowly when the plant decomposes.

then the breaking, of chemical bonds. Chemical bonds were formed by utilizing energy to produce new chemical substances. Energy was released when these bonds were broken to produce the original substances. In this example, chemical reactions and energy flow can be explained by the making and breaking of chemical bonds. Chemical bonds can be explained in terms of changes in the electron structures of atoms. Thus, the place to start in seeking understanding about chemical reactions is the electron structure of the atoms themselves.

VALENCE ELECTRONS AND IONS

As discussed in chapter 8, it is the number of electrons in the outermost orbital that usually determines the chemical properties of an atom. These outer electrons are called **valence electrons,** and it is the valence electrons that participate in chemical bonding. The inner electrons are in stable, fully occupied orbitals and do not participate in chemical bonds. The representative elements (the A-group families) have valence electrons in the outermost orbitals, which contain from one to eight valence electrons. Recall that you can easily find the number of valence electrons by referring to a periodic table. The number at the top of each representative family is the same as the number of outer orbital electrons (with the exception of helium).

The noble gases have filled outer orbitals and do not normally form compounds. Apparently, half-filled and filled orbitals are particularly stable arrangements. Atoms have a tendency to seek such a stable, filled outer orbital arrangement such as the one found in the noble gases. For the representative elements, this tendency is called the **octet rule.** The octet rule states that *atoms attempt to acquire an outer orbital with eight electrons* through chemical reactions. This rule is a generaliza-

tion, and a few elements do not meet the requirement of eight electrons but do seek the same general trend of stability. There are a few other exceptions, and the octet rule should be considered a generalization that helps keep track of the valence electrons in most representative elements.

The family number of the representative element in the periodic table tells you the number of valence electrons and what the atom must do to reach the stability suggested by the octet rule. For example, consider sodium (Na). Sodium is in family IA, so it has one valence electron. If the sodium atom can get rid of this outer valence electron through a chemical reaction, it will have the same outer electron configuration as an atom of the noble gas neon (Ne) (compare figure 9.7B and 9.7C).

When a sodium atom (Na) loses an electron to form a sodium ion (Na^+), it has the same, stable outer electron configuration as a neon atom (Ne). The sodium ion (Na^+) is still a form of sodium since it still has eleven protons. But it is now a sodium *ion*, not a sodium *atom*, since it has eleven protons (eleven positive charges) and now has ten electrons (ten negative charges) for a total of

$$11 + \text{(protons)}$$
$$\underline{10 - \text{(electrons)}}$$
$$1 + \text{(net charge on sodium ion)}$$

This charge is shown on the chemical symbol of Na^+ for the *sodium ion.* Note that the sodium nucleus and the inner orbitals

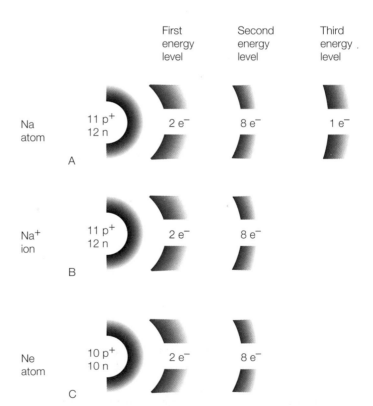

| | First energy level | Second energy level | Third energy level |

Na atom — 11 p⁺ 12 n — 2 e⁻ — 8 e⁻ — 1 e⁻

A

Na^+ ion — 11 p⁺ 12 n — 2 e⁻ — 8 e⁻

B

Ne atom — 10 p⁺ 10 n — 2 e⁻ — 8 e⁻

C

Figure 9.7

(*A*) A sodium atom has two electrons in the first energy level, eight in the second energy level, and one in the third level. (*B*) When it loses its one outer, or valence, electron, it becomes a positively charged sodium ion with the same electron structure as an atom of neon (*C*).

do not change when the sodium atom is ionized. The sodium ion is formed when a sodium atom loses its valence electron, and the process can be described by

$$energy \;+\; Na \cdot \longrightarrow Na^{+} \;+\; e^{-}$$

where Na · is the electron dot symbol for sodium, and the e^{-} is the electron that has been pulled off the sodium atom.

CHEMICAL BONDS

Atoms gain or lose electrons through a chemical reaction to achieve a state of lower energy, the stable electron arrangement of the noble gas atoms. Such a reaction results in a **chemical bond,** an *attractive force that holds atoms together in a compound.* There are three general classes of chemical bonds: (1) ionic bonds, (2) covalent bonds, and (3) metallic bonds.

Ionic bonds are formed when atoms *transfer* electrons to achieve the noble gas electron arrangement. Electrons are given up or acquired in the transfer, forming positive and negative ions. The electrostatic attraction between oppositely charged ions forms ionic bonds, and ionic compounds are the result. In general, ionic compounds are formed when a metal from the left side of the periodic table reacts with a nonmetal from the right side.

Covalent bonds result when atoms achieve the noble gas electron structure by *sharing* electrons. Covalent bonds are generally formed between the nonmetallic elements on the right side of the periodic table.

Metallic bonds are formed in solid metals such as iron, copper, and the other metallic elements that make up about 80 percent of all the elements. The atoms of metals are closely packed and share many electrons in a "sea" that is free to move throughout the metal, from one metal atom to the next. Metallic bonding accounts for metallic properties such as high electrical conductivity.

Ionic, covalent, and metallic bonds are attractive forces that hold atoms or ions together in molecules and crystals. There are two ways to describe what happens to the electrons when one of these bonds is formed, by considering (1) the new patterns formed when atomic orbitals overlap to form a combined orbital called a *molecular orbital* or (2) the atoms in a molecule as *isolated atoms* with changes in their outer shell arrangements. The molecular orbital description considers that the electrons belong to the whole molecule and form a molecular orbital with its own shape, orientation, and energy levels. The isolated atom description considers the electron energy levels as if the atoms in the molecule were alone, isolated from the molecule. The isolated atom description is less accurate than the molecular orbital description, but it is less complex and more easily understood. Thus, the following details about chemical bonding will mostly consider individual atoms and ions in compounds.

Ionic Bonds

An **ionic bond** is defined as the *chemical bond of electrostatic attraction* between negative and positive ions. Ionic bonding occurs when an atom of a metal reacts with an atom of a nonmetal. The reaction results in a transfer of one or more valence electrons from the metal atom to the valence shell of the nonmetal atom. The atom that loses electrons becomes a positive ion, and the atom that gains electrons becomes a negative ion. Oppositely charged ions attract one another, and when pulled together, they form an ionic solid with the ions arranged in an orderly geometric structure (figure 9.8). This results in a crystalline solid that is typical of salts such as sodium chloride (figure 9.9).

As an example of ionic bonding, consider the reaction of sodium (a soft reactive metal) with chlorine (a pale yellow-green gas). When an atom of sodium and an atom of chlorine collide, they react violently as the valence electron is transferred from the sodium to the chlorine atom. This produces a sodium ion

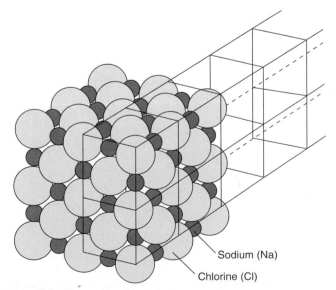

Sodium (Na)

Chlorine (Cl)

Figure 9.8

Sodium chloride crystals are composed of sodium and chlorine ions held together by electrostatic attraction. A crystal builds up, giving the sodium chloride crystal a cubic structure.

Figure 9.9

You can clearly see the cubic structure of these ordinary table salt crystals because they have been magnified about ten times.

and a chlorine ion. The reaction can be illustrated with electron dot symbols as follows:

$$Na \cdot \ + \ \cdot \overset{..}{\underset{..}{Cl}} : \ \longrightarrow \ Na^+ \ (: \overset{..}{\underset{..}{Cl}} :)^-$$

As you can see, the sodium ion transferred its valence electron, and the resulting ion now has a stable electron configuration. The chlorine atom accepted the electron in its outer orbital to acquire a stable electron configuration. Thus, a stable positive ion and a stable negative ion are formed. Because of opposite electrical charges, the ions attract each other to produce an ionic bond. When many ions are involved, each Na^+ ion is surrounded by six Cl^- ions, and each Cl^- ion is surrounded by six Na^+ ions. This gives the resulting solid NaCl its crystalline cubic structure, as shown in figure 9.9. In the solid state, all the sodium ions and all the chlorine ions are bound together in one giant unit. Thus, the term *molecule* is not really appropriate for ionic solids such as sodium chloride. But the term is sometimes used anyway, since any given sample will have the same number of Na^+ ions as Cl^- ions.

The sodium-chlorine reaction can be represented with electron dot notation as occurring in three steps:

1. $energy \ + \ Na \cdot \ \longrightarrow \ Na^+ \ + \ e^-$

2. $\cdot \overset{..}{\underset{..}{Cl}} : \ + \ e^- \ \longrightarrow \ (: \overset{..}{\underset{..}{Cl}} :)^- \ + \ energy$

3. $Na^+ \ + \ (: \overset{..}{\underset{..}{Cl}} :)^- \ \longrightarrow \ Na^+ \ (: \overset{..}{\underset{..}{Cl}} :)^- + \ energy$

The energy released in steps 2 and 3 is greater than the energy absorbed in step 1, and an ionic bond is formed. The energy released is called the **heat of formation.** It is also the amount of energy required to decompose the compound (sodium chloride) into its elements. The reaction does not take place in

steps as described, however, but occurs all at once. Note again, as in the photosynthesis-burning reactions described earlier, that the total amount of chemical energy is conserved. The energy released by the formation of the sodium chloride compound is the *same* amount of energy needed to decompose the compound.

Ionic bonds are formed by electron transfer, and electrons are conserved in the process. This means that electrons are not created or destroyed in a chemical reaction. The same total number of electrons exists after a reaction that existed before the reaction. There are two rules you can use for keeping track of electrons in ionic bonding reactions:

1. Ions are formed as atoms gain or lose valence electrons to achieve the stable noble gas structure.
2. There must be a balance between the number of electrons lost and the number of electrons gained by atoms in the reaction.

The sodium-chlorine reaction follows these two rules. The loss of one valence electron from a sodium atom formed a stable sodium ion. The gain of one valence electron by the chlorine atom formed a stable chlorine ion. Thus, both ions have noble gas configurations (rule 1), and one electron was lost and one was gained, so there is a balance in the number of electrons lost and the number gained (rule 2).

The **formula** of a compound *describes what elements are in the compound and in what proportions.* Sodium chloride contains one positive sodium ion for each negative chlorine ion. The formula of the compound sodium chloride is NaCl. If there are no subscripts at the lower right part of each symbol, it is understood that the symbol has a number "1." Thus, NaCl indicates a compound made up of the elements sodium and chlorine, and there is one sodium atom for each chlorine atom.

Calcium (Ca) is an alkaline metal in family IIA, and fluorine (F) is a halogen in family VIIA. Since calcium is a metal and fluorine is a nonmetal, you would expect calcium and fluorine atoms to react, forming a compound with ionic bonds. Calcium must lose two valence electrons to acquire a noble gas configuration. Fluorine needs one valence electron to acquire a noble gas configuration. So calcium needs to lose two electrons and fluorine needs to gain one electron to achieve a stable configuration (rule 1). Two fluorine atoms, each acquiring one electron, are needed to balance the number of electrons lost and the number of electrons gained. The compound formed from the reaction, calcium fluoride, will therefore have a calcium ion with a charge of plus two for every fluorine ion with a charge of minus one. Recalling that electron dot symbols show only the outer valence electrons, you can see that the reaction is

which shows that a calcium atom transfers two electrons, one each to two fluorine atoms. Now showing the results of the reaction, a calcium ion is formed from the loss of two electrons (charge 2+) and two fluorine ions are formed by gaining one electron each (charge 1−):

The formula of the compound is therefore CaF_2, with the subscript 2 for fluorine and the understood subscript 1 for calcium. This means that there are two fluorine atoms for each calcium atom in the compound.

Sodium chloride (NaCl) and magnesium fluoride (MgF_2) are examples of compounds held together by ionic bonds. Such compounds are called **ionic compounds.** Ionic compounds of the representative elements are generally white, crystalline solids that form colorless solutions. Sodium chloride, the most common example, is common table salt. Many of the transition elements form colored compounds that make colored solutions. Ionic compounds dissolve in water, producing a solution of ions that can conduct an electric current.

In general, the elements in families IA and IIA of the periodic table tend to form positive ions by losing electrons. The ion charge for these elements equals the family number of these elements. The elements in families VIA and VIIA tend to form negative ions by gaining electrons. The ion charge for these elements equals their family number minus 8. The elements in families IIIA and VA have less of a tendency to form ionic compounds, except for those in higher periods. Common ions of representative elements are given in table 9.2. The

Table 9.2 Common Ions of Some Representative Elements

Element	Symbol	Ion
Lithium	Li	1+
Sodium	Na	1+
Potassium	K	1+
Magnesium	Mg	2+
Calcium	Ca	2+
Barium	Ba	2+
Aluminum	Al	3+
Oxygen	O	2−
Sulfur	S	2−
Hydrogen	H	1+, 1−
Fluorine	F	1−
Chlorine	Cl	1−
Bromine	Br	1−
Iodine	I	1−

Table 9.3 Common Ions of Some Transition Elements

Single-Charge Ions

Element	Symbol	Charge
Zinc	Zn	2+
Tungsten	W	6+
Silver	Ag	1+
Cadmium	Cd	2+

Variable-Charge Ions

Element	Symbol	Charge
Chromium	Cr	2+, 3+, 6+
Manganese	Mn	2+, 4+, 7+
Iron	Fe	2+, 3+
Cobalt	Co	2+, 3+
Nickel	Ni	2+, 3+
Copper	Cu	1+, 2+
Tin	Sn	2+, 4+
Gold	Au	1+, 3+
Mercury	Hg	1+, 2+
Lead	Pb	2+, 4+

transition elements form positive ions of several different charges. Some common ions of the transition elements are listed in table 9.3.

The single-charge representative elements and the variable-charge transition elements form single, monatomic negative ions. There are also many polyatomic (*poly* means "many") negative ions, charged groups of atoms that act like a single unit in ionic compounds. Polyatomic ions are held together by covalent bonds, which are discussed in the next section.

Covalent Bonds

Most substances do not have the properties of ionic compounds since they are not composed of ions. Most substances are molecular, composed of electrically neutral groups of atoms that are tightly bound together. As noted earlier, many gases are diatomic, occurring naturally as two atoms bound together as an electrically neutral molecule. Hydrogen, for example, occurs as molecules of H_2 and no ions are involved. The hydrogen atoms are held together by a covalent bond. A **covalent bond** is a *chemical bond formed by the sharing of a pair of electrons.* In the diatomic hydrogen molecule each hydrogen atom contributes a single electron to the shared pair. Both hydrogen atoms count the shared pair of electrons in achieving their noble gas configuration. Hydrogen atoms both share one pair of electrons, but other elements might share more than one pair to achieve a noble gas structure.

Aluminium Fluoride

Use electron dot notation to predict the formula of a compound formed when aluminum (Al) combines with fluorine (F).

Aluminum, atomic number 13, is in family IIIA so it has three valence electrons and an electron dot notation of

$$\overset{\bullet}{\underset{\bullet}{Al}}\bullet$$

According to the octet rule, the aluminum atom would need to lose three electrons to acquire the stable noble gas configuration. Fluorine, atomic number 9, is in family VIIA so it has seven valence electrons and an electron dot notation of

$$\bullet\overset{\bullet\bullet}{\underset{\bullet\bullet}{F}}\bullet$$

Fluorine would acquire a noble gas configuration by accepting one electron. Three fluorine atoms, each acquiring one electron, are needed to balance the three electrons lost by aluminum. The reaction can be represented as

The ratio of aluminum atoms to fluorine atoms in the compound is 1:3. The formula for aluminum fluoride is therefore AlF_3.

Predict the formula of the compound formed between aluminum and oxygen using electron dot notation. (Answer: Al_2O_3).

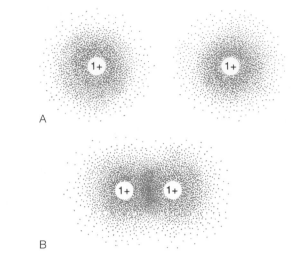

Figure 9.10

(A) Two hydrogen atoms, each with its own probability distribution of electrons about the nucleus. (B) When the hydrogen atoms bond, a new electron distribution pattern forms around the entire molecule, and both atoms share the two electrons.

helium. A dashed circle around each symbol shows that both atoms have two electrons:

$$H\bullet \;+\; H\bullet \;\longrightarrow\; \left(H\overset{\bullet}{\underset{\bullet}{:}}H\right)$$

Hydrogen and fluorine react to form a covalent molecule (how this is known will be discussed shortly), and this bond can be represented with electron dots. Fluorine is in the VIIA family, so you know an atom of fluorine has seven valence electrons in the outermost energy level. The reaction is

$$H\bullet \;+\; \bullet\overset{\bullet\bullet}{\underset{\bullet\bullet}{F}}{:} \;\longrightarrow\; \left(H\overset{\bullet\bullet}{\underset{\bullet\bullet}{:}F{:}}\right)$$

Each atom shares a pair of electrons to achieve a noble gas configuration. Hydrogen achieves the helium configuration, and fluorine achieves the neon configuration. All the halogens have seven valence electrons and all need to gain one electron (ionic bond) or share an electron pair (covalent bond) to achieve a noble gas configuration. This also explains why the halogen gases occur as diatomic molecules. Two fluorine atoms can achieve a noble gas configuration by sharing a pair of electrons:

$$\bullet\overset{\bullet\bullet}{\underset{\bullet\bullet}{F}}{:} \;+\; \bullet\overset{\bullet\bullet}{\underset{\bullet\bullet}{F}}{:} \;\longrightarrow\; \left({:}\overset{\bullet\bullet}{\underset{\bullet\bullet}{F}}\overset{\bullet\bullet}{\underset{\bullet\bullet}{F}}{:}\right)$$

The diatomic hydrogen (H_2) and fluorine (F_2), hydrogen fluoride (HF), and water (H_2O) are examples of compounds held together by covalent bonds. A compound held together by covalent bonds is called a **covalent compound.** In general, covalent compounds form from nonmetallic elements on the right side of the periodic table. For elements in families IVA through VIIA, the number of unpaired electrons (and thus the number of covalent bonds formed) is eight minus the family number.

Consider how the covalent bond forms between two hydrogen atoms by imagining two hydrogen atoms moving toward one another. Each atom has a single electron. As the atoms move closer and closer together, their orbitals begin to overlap. Each electron is attracted to the oppositely charged nucleus of the other atom and the overlap tightens. Then the repulsive forces from the like-charged nuclei will halt the merger. A state of stability is reached between the two nuclei and two electrons, and an H_2 molecule has been formed. The two electrons are now shared by both atoms, and the attraction of one nucleus for the other electron and vice versa holds the atoms together (figure 9.10).

Electron dot notation can be used to represent the formation of covalent bonds. For example, the joining of two hydrogen atoms to form an H_2 molecule can be represented as

$$H\bullet \;+\; H\bullet \;\longrightarrow\; H\overset{\bullet}{\underset{\bullet}{:}}H$$

Since an electron pair is *shared* in a covalent bond, the two electrons move throughout the entire molecular orbital. Since each hydrogen atom now has both electrons on an equal basis, each can be considered now to have the noble gas configuration of

Table 9.4 Structures and Compounds of Nonmetallic Elements Combined with Hydrogen

Nonmetallic Elements	Element (E represents any element of family)	Compound
Family IVA: C, Si, Ge	$\cdot\overset{\cdot}{\underset{\cdot}{E}}\cdot$	H $H\!:\!\overset{H}{\underset{H}{E}}\!:\!H$
Family VA: N, P, As, Sb	$\cdot\overset{\cdot}{\underset{\cdot}{E}}\cdot$	$H\!:\!\overset{H}{\underset{H}{E}}\!:\!H$
Family VIA: O, S, Se, Te	$\cdot\overset{\cdot\cdot}{\underset{\cdot}{E}}\cdot$	$H\!:\!\overset{\cdot\cdot}{\underset{\cdot\cdot}{E}}\!:\!H$
Family VIIA: F, Cl, Br, I	$\cdot\overset{\cdot\cdot}{\underset{\cdot\cdot}{E}}\!:$	$H\!:\!\overset{\cdot\cdot}{\underset{\cdot\cdot}{E}}\!:$

You can get a lot of information from the periodic table from generalizations like this one. For another generalization, compare table 9.4 with the periodic table. The table gives the structures of nonmetals combined with hydrogen and the resulting compounds.

Two dots can represent a lone pair of valence electrons, or they can represent a bonding pair, a single pair of electrons being shared by two atoms. Bonding pairs of electrons are often represented by a simple line between two atoms. For example,

$$H : H \qquad \text{is shown as} \qquad H - H$$

and

$$:\overset{\cdot\cdot}{\underset{H\ \ H}{O}}: \qquad \text{is shown as} \qquad \overset{O}{\underset{H\quad H}{\diagup\ \diagdown}}$$

Note that the line between the two hydrogen atoms represents an electron pair, so each hydrogen atom has two electrons in the outer shell, as does helium. In the water molecule, each hydrogen atom has two electrons as before. The oxygen atom has two lone pairs (a total of four electrons) and two bonding pairs (a total of four electrons) for a total of eight electrons. Thus, oxygen has acquired a stable octet of electrons.

A covalent bond in which a single pair of electrons is shared by two atoms is called a *single covalent bond* or simply a **single bond.** Some atoms have two unpaired electrons and can share more than one electron pair. A **double bond** is a covalent bond formed when *two pairs* of electrons are shared by two atoms. This happens mostly in compounds involving atoms of the elements C, N, O, and S. Ethylene, for example, is a gas given off from ripening fruit. The electron dot formula for ethylene is

$$\overset{H}{\underset{H}{:}}\overset{\cdot\ \ \ \cdot}{C} :: \overset{\cdot\ \ \ \cdot}{C} \overset{H}{\underset{H}{:}} \qquad \text{or} \qquad \overset{H}{\underset{H}{\diagdown}}C = C\overset{H}{\underset{H}{\diagup}}$$

The ethylene molecule has a double bond between two carbon atoms. Since each line represents two electrons, you can simply count the lines around each symbol to see if the octet rule has been satisfied. Each H has one line, so each H atom is sharing two electrons. Each C has four lines so each C atom has eight electrons, satisfying the octet rule.

A **triple bond** is a covalent bond formed when *three pairs* of electrons are shared by two atoms. Triple bonds occur mostly in compounds with atoms of the elements C and N. Acetylene, for example, is a gas often used in welding torches (figure 9.11). The electron dot formula for acetylene is

$$H : C ::: C : H \qquad \text{or} \qquad H - C \equiv C - H$$

The acetylene molecule has a triple bond between two carbon atoms. Again, note that each line represents two electrons. Each C atom has four lines, so the octet rule is satisfied.

Figure 9.11
Acetylene is a hydrocarbon consisting of two carbon atoms and two hydrogen atoms held together by a triple covalent bond between the two carbon atoms. When mixed with oxygen gas (the tank to the right), the resulting flame is hot enough to cut through most metals.

Ionic Compound Names

Ionic compounds formed by metal ions are named by stating the name of the metal (positive ion), then the name of the nonmetal (negative ion). Ionic compounds formed by variable-charge ions have an additional rule to identify which variable-charge ion is involved. There was an old way of identifying the charge on the ion by adding either "-ic" or "-ous" to the name of the metal. The suffix "-ic" meant the higher of two possible charges, and the suffix "-ous" meant the lower of two possible charges. For example, iron has two possible charges, 2+ or 3+. The old system used the Latin name for the root. The Latin name for iron is ferrum, so a higher charged iron ion (3+) was named a ferric ion. The lower charged iron ion (2+) was called a ferrous ion. You still hear the old names sometimes, but chemists now have a better way to identify the variable-charge ion. The newer system uses the English name of the metal with Roman numerals in parentheses to indicate the charge number. Thus, an iron ion with a charge of 2+ is called an iron(II) ion and an iron ion with a charge of 3+ is an iron(III) ion. Box table 9.1 gives some of the modern names for variable-charge ions. These names are used with the name of a nonmetal ending in "-ide," just like the single-charge ions in ionic compounds made up of two different elements.

Some ionic compounds contain three or more elements, and so are more complex than a combination of a metal ion and a nonmetal ion.

Box Table 9.1

Modern Names of Some Variable-Charge Ions

Ion	Name of Ion
Fe^{2+}	Iron(II) ion
Fe^{3+}	Iron(III) ion
Cu^+	Copper(I) ion
Cu^{2+}	Copper(II) ion
Pb^{2+}	Lead(II) ion
Pb^{4+}	Lead(IV) ion
Sn^{2+}	Tin(II) ion
Sn^{4+}	Tin(IV) ion
Cr^{2+}	Chromium(II) ion
Cr^{3+}	Chromium(III) ion
Cr^{6+}	Chromium(VI) ion

This is possible because they have **polyatomic ions,** groups of two or more atoms that are bound together tightly and behave very much like a single monatomic ion (box table 9.2). For example, the OH^- ion is an oxygen atom bound to a hydrogen atom with a net charge of 1-. This polyatomic ion is called a *hydroxide ion.* The hydroxide compounds make up one of the main groups of ionic compounds, the *metal hydroxides.* A metal hydroxide is an ionic compound consisting of a metal with the hydroxide ion. Another main group consists of the salts with polyatomic ions.

Box Table 9.2

Some Common Polyatomic Ions

Ion Name	Formula
Acetate	$(C_2H_3O_2)^-$
Ammonium	$(NH_4)^+$
Borate	$(BO_3)^{3-}$
Carbonate	$(CO_3)^{2-}$
Chlorate	$(ClO_3)^-$
Chromate	$(CrO_4)^{2-}$
Cyanide	$(CN)^-$
Dichromate	$(Cr_2O_7)^{2-}$
Hydrogen carbonate (or bicarbonate)	$(HCO_3)^-$
Hydrogen sulfate (or bisulfate)	$(HSO_4)^-$
Hydroxide	$(OH)^-$
Hypochlorite	$(ClO)^-$
Nitrate	$(NO_3)^-$
Nitrite	$(NO_2)^-$
Perchlorate	$(ClO_4)^-$
Permanganate	$(MnO_4)^-$
Phosphate	$(PO_4)^{3-}$
Phosphite	$(PO_3)^{3-}$
Sulfate	$(SO_4)^{2-}$
Sulfite	$(SO_3)^{2-}$

COMPOSITION OF COMPOUNDS

As you can imagine, there are literally millions of different chemical compounds from all the possible combinations of over ninety natural elements held together by ionic or covalent bonds. Each of these compounds has its own name, so there are millions of names and formulas for all the compounds. In the early days, compounds were given *common names* according to how they were used, where they came from, or some other means of identifying them. Thus, sodium carbonate was called soda, and closely associated compounds were called baking soda (sodium bicarbonate), washing soda (sodium carbonate), caustic soda (sodium hydroxide), and the bubbly drink made by reacting soda with acid was called soda water, later called soda pop (figure 9.12). Potassium carbonate was extracted from charcoal by soaking in water and came to be called potash. Such common names are colorful, and some are descriptive, but it was impossible to keep up with the names as the number of known compounds grew. So a systematic set of rules was

Figure 9.12

These substances are made up of sodium and some form of a carbonate ion. All have common names with the term "soda" for this reason. Soda water (or "soda pop") was first made by reacting soda (sodium carbonate) with an acid, so it was called "soda water."

The metal hydroxides are named by identifying the metal first and the term *hydroxide* second. Thus, NaOH is named sodium hydroxide and KOH is potassium hydroxide. The salts are similarly named, with the metal (or ammonium ion) identified first, then the name of the polyatomic ion. Thus, $NaNO_3$ is named sodium nitrate and $NaNO_2$ is sodium nitrite. Note that the suffix "-ate" means the polyatomic ion with one more oxygen atom than the "-ite" ion. For example, the chlor*ate* ion is $(ClO_3)^-$, and the chlor*ite* ion is $(ClO_2)^-$. Sometimes more than two possibilities exist, and more oxygen atoms are identified with the prefix "per-", and less with the prefix "hypo-". Thus, the *perchlorate* ion is $(ClO_4)^-$ and the *hypochlorite* ion is $(ClO)^-$.

Covalent Compound Names

Covalent compounds are molecular, and the molecules are composed of two *nonmetals,* as opposed to the metal and nonmetal elements that make up ionic compounds. The combinations of nonmetals do not present simple names as the ionic compounds did, so a different set of rules for naming and formula writing is needed.

Ionic compounds were named by stating the name of the positive metal ion, then the name of the negative nonmetal ion with an "-ide" ending. This system is not adequate for naming the covalent compounds. To begin, covalent compounds are composed of two or more nonmetal atoms that form a molecule. It is possible for some atoms to form single, double, or even triple bonds with other atoms, including atoms of the same element. The net result is that the same two elements can form more than one kind of covalent compound. Carbon and oxygen, for example, can combine to form the gas released from burning and respiration, carbon dioxide (CO_2). Under certain conditions the very same elements combine to produce a different gas, the poisonous carbon monoxide (CO). Similarly, sulfur and oxygen can combine differently to produce two different covalent compounds. A successful system for naming covalent compounds must therefore provide a means of identifying different compounds made of the same elements. This is accomplished by using a system of Greek prefixes (see box table 9.3). The rules are as follows:

1. The first element in the formula is named first with a prefix indicating the number of atoms if the number is greater than one.
2. The stem name of the second element in the formula is next. A prefix is used with the stem if two elements form more than one compound. The suffix "-ide" is again used to indicate a compound of only two elements. For example, CO is carbon monoxide and CO_2 is carbon dioxide. The compound BF_3 is boron trifluoride and N_2O_4 is dinitrogen tetroxide. Knowing the formula and the prefix and stem information in box table 9.3, you can write the name of any covalent compound made up

of two elements by ending it with "-ide." Conversely, the name will tell you the formula. However, there are a few polyatomic ions with "-ide" endings that are compounds made up of more than just two elements (hydroxide and cyanide). Compounds formed with the ammonium ion will also have an "-ide" ending, and these are also made up of more than two elements.

Box Table 9.3

Prefixes and Element Stem Names

Prefixes		Stem Names	
PREFIX	MEANING	ELEMENT	STEM
Mono-	1	Hydrogen	Hydr-
Di-	2	Carbon	Carb-
Tri-	3	Nitrogen	Nitr-
Tetra-	4	Oxygen	Ox-
Penta-	5	Fluorine	Fluor-
Hexa-	6	Phosphorus	Phosph-
Hepta-	7	Sulfur	Sulf-
Octa-	8	Chlorine	Chlor-
Nona-	9	Bromine	Brom-
Deca-	10	Iodine	Iod-

Note: the *a* or *o* ending on the prefix is often dropped if the stem name begins with a vowel, for example, "tetroxide," not "tetraoxide."

---

developed to determine the name and formula of each compound. Once you know the rules, you can write the formula when you hear the name. Conversely, seeing the formula will tell you the systematic name of the compound. This can be an interesting intellectual activity and can also be important when reading the list of ingredients to understand the composition of a product.

There is a different set of systematic rules to be used with ionic compounds and covalent compounds, but there are a few rules in common. For example, a compound made of only two different elements always ends with the suffix "-ide." So when you hear the name of a compound ending with "-ide" you automatically know that the compound is made up of only two elements. Sodium chlor*ide* is an ionic compound made up of sodium and chlorine ions. Carbon diox*ide* is a covalent compound with carbon and oxygen atoms. Thus, the systematic name tells you what elements are present in a compound with an "-ide" ending.

Ionic Compound Formulas

The formulas for ionic compounds are easy to write. There are two rules:

1. **The symbols.** Write the symbol for the positive element first, followed by the symbol for the negative element (same order as in the name).
2. **The subscripts.** Add subscripts to indicate the numbers of ions needed to produce an electrically neutral compound.

As an example, let us write the formula for the compound calcium chloride. The name tells you that this compound consists of positive calcium ions and negative chlorine ions. The suffix "-ide" tell you there are only two elements present. Following rule 1, the symbols would be CaCl.

For rule 2 note the calcium ion is Ca^{2+}, and the chlorine ion is Cl^-. You know the calcium is plus two and chlorine is negative one by applying the atomic theory, knowing their positions in the periodic table, or by using a table of ions and their charges. To be electrically neutral, the compound must have an equal number of pluses and minuses. Thus, you will need two negative chlorine ions for every calcium ion with its 2+ charge. Therefore, the formula is $CaCl_2$. The total charge of two chlorines is thus 2−, which balances the 2+ charge on the calcium ion.

One easy way to write a formula showing that a compound is electrically neutral is to cross over the absolute charge numbers (without plus or minus signs) and use them as subscripts. For example, the symbols for the calcium ion and the chlorine ion are

$$Ca^{2+}Cl^{1-}$$

Crossing the absolute numbers as subscripts, as follows

and then dropping the charge numbers gives

$$Ca_1 \, Cl_2$$

No subscript is written for 1; it is understood. The formula for calcium chloride is thus

$$CaCl_2$$

The crossover technique works because ionic bonding results from a transfer of electrons,

and the net charge is conserved. A calcium ion has a 2+ charge because the atom lost two electrons and two chlorine atoms gain one electron each, for a total of two electrons gained. Two electrons lost equals two electrons gained, and the net charge on calcium chloride is zero, as it has to be.

When using the crossover technique it is sometimes necessary to reduce the ratio to the lowest common multiple. Thus, Mg_2O_2 means an equal ratio of magnesium and oxygen ions, so the correct formula is MgO.

The formulas for variable-charge ions are easy to write, since the Roman numeral tells you the charge number. The formula for tin(II) fluoride is written by crossing over the charge numbers (Sn^{2+}, F^{1-}), and the formula is SnF_2.

The formulas for ionic compounds with polyatomic ions are written from combinations of positive metal ions or the ammonium ion with the polyatomic ions, as listed in box table 9.2. Since the polyatomic ion is a group of atoms that has a charge and stays together in a unit, it is sometimes necessary to indicate this with parentheses. For example, magnesium hydroxide is composed of Mg^{2+} ions and $(OH)^{1-}$ ions. Using the crossover technique to write the formula, you get

The parentheses are used and the subscript is written *outside* the parenthesis to show that the entire hydroxide unit is taken twice. The formula $Mg(OH)_2$ means

which shows that the pluses equal the minuses. Parentheses are not used, however, when only one polyatomic ion is present. Sodium hydroxide is NaOH, not $Na(OH)_1$.

Covalent Compound Formulas

The systematic name tells you the formula for a covalent compound. The gas that dentists use as an anesthetic, for example, is dinitrogen monoxide. This tells you there are two nitrogen atoms and one oxygen atom in the molecule, so the for-

mula is N_2O. A different molecule composed of the very same elements is nitrogen dioxide. Nitrogen dioxide is the pollutant responsible for the brownish haze of smog. The formula for nitrogen dioxide is NO_2. Other examples of formulas from systematic names are carbon dioxide (CO_2) and carbon tetrachloride (CCl_4).

Formulas of covalent compounds indicate a pattern of how many atoms of one element combine with atoms of another. Carbon, for example, combines with no more than two oxygen atoms to form carbon dioxide. Carbon combines with no more than four chlorine atoms to form carbon tetrachloride. Electron dot formulas show these two molecules as

$$\ddot{O} :: C :: \ddot{O} \qquad \begin{array}{c} :\ddot{Cl}: \\ :\ddot{Cl}:\ddot{C}:\ddot{Cl}: \\ :\ddot{Cl}: \end{array}$$

Using a dash to represent bonding pairs, we have

$$O = C = O \qquad \begin{array}{c} Cl \\ | \\ Cl - C - Cl \\ | \\ Cl \end{array}$$

In both of these compounds, the carbon atom forms four covalent bonds with another atom. The number of covalent bonds that an atom can form is called its *valence.* Carbon has a valence of four and can form single, double, or triple bonds. Here are the possibilities for a single carbon atom (combining elements not shown):

$$\begin{array}{cc} | & | \\ -C- & -C= \\ | & \\ =C= & -C\equiv \end{array}$$

Hydrogen has only one unshared electron, so the hydrogen atom has a valence of one. Oxygen has a valence of two and nitrogen has a valence of three. Here are the possibilities for hydrogen, oxygen, and nitrogen:

$$H- \qquad -\ddot{O}- \qquad :\ddot{O}=$$

$$-\ddot{N}- \qquad -\ddot{N}= \qquad :N\equiv$$

Figure 9.13
The charcoal used in a grill is basically carbon. The carbon reacts with oxygen to yield carbon dioxide. The chemical equation for this reaction, $C + O_2 \rightarrow CO_2$, contains the same information as the English sentence but has quantitative meaning as well.

CHEMICAL EQUATIONS

Chemical reactions occur when bonds between the outermost parts of atoms are formed or broken. Bonds are formed, for example, when a green plant uses sunlight—a form of energy—to create molecules of sugar, starch, and plant fibers. Bonds are broken and energy is released when you digest the sugars and starches or when plant fibers are burned. Chemical reactions thus involve changes in matter, the creation of new materials with new properties, and energy exchanges. So far you have considered chemical symbols as a concise way to represent elements, and formulas as a concise way to describe what a compound is made of. There is also a concise way to describe a chemical reaction, the *chemical equation.*

Word equations are useful in identifying what has happened before and after a chemical reaction. The substances that existed before a reaction are called *reactants,* and the substances that exist after the reaction are called the *products.* The equation has a general form of

$$\text{reactants} \rightarrow \text{products}$$

where the arrow signifies a separation in time; that is, it identifies what existed before the reaction and what exists after the reaction. For example, the charcoal used in a barbecue grill is carbon (figure 9.13). The carbon reacts with oxygen while burning, and the reaction (1) releases energy and (2) forms carbon dioxide. The reactants and products for this reaction can be described as

$$\text{carbon} + \text{oxygen} \rightarrow \text{carbon dioxide}$$

The arrow means *yields,* and the word equation is read as, "Carbon reacts with oxygen to yield carbon dioxide." This word equation describes what happens in the reaction but says nothing about the quantities of reactants or products.

Chemical symbols and formulas can be used in the place of words in an equation and the equation will have a whole new meaning. For example, the equation describing carbon reacting with oxygen to yield carbon dioxide becomes

$$C + O_2 \rightarrow CO_2$$

(balanced)

The new, added meaning is that one atom of carbon (C) reacts with one molecule of oxygen (O_2) to yield one molecule of carbon dioxide (CO_2). Note that the equation also shows one atom of carbon and two atoms of oxygen (recall that oxygen occurs as a diatomic molecule) as reactants on the left side and one atom of carbon and two atoms of oxygen as products on the right side. Since the same number of each kind of atom appears on both sides of the equation, the equation is said to be *balanced.*

You would not want to use a charcoal grill in a closed room because there might not be enough oxygen. An insufficient supply of oxygen produces a completely different product, the poisonous gas, carbon monoxide (CO). An equation for this reaction is

$$C + O_2 \rightarrow CO$$

(not balanced)

As it stands, this equation describes a reaction that violates the **law of conservation of mass,** that matter is neither created nor destroyed in a chemical reaction. From the point of view of an equation, this law states that

$$\text{mass of reactants} = \text{mass of products}$$

Mass of reactants here means all that you start with, including some that might not react. Thus elements are neither created nor destroyed, and this means the elements present and their mass. In any chemical reaction the kind and mass of the reactive elements are identical to the kind and mass of the product elements.

Natural gas is mostly methane, CH_4, which burns by reacting with oxygen (O_2) to produce carbon dioxide (CO_2) and water vapor (H_2O). Write a balanced chemical equation for this reaction by following a procedure of four steps.

Step 1. Write the correct formulas for the reactants and products in an unbalanced equation. For the burning of methane, the unbalanced, but otherwise correct, formula equation would be

$$CH_4 + O_2 \rightarrow CO_2 + H_2O$$
(not balanced)

Step 2. Inventory the number of each kind of atom on both sides of the unbalanced equation. In the example there are

Reactants	Products
1 C	1 C
4 H	2 H
2 O	3 O

This step shows that the H and O are unbalanced.

Step 3. Determine where to place coefficients in front of formulas to balance the equation. It is often best to focus on the simplest thing you can do with whole number ratios. The H and the O are unbalanced, for example, and there are 4 H atoms on the left and 2 H atoms on the right. Placing a coefficient 2 in front of H_2O will balance the H atoms:

$$CH_4 + O_2 \rightarrow CO_2 + 2\,H_2O$$
(not balanced)

Now take a second inventory:

Reactants	Products
1 C	1 C
4 H	4 H
2 O	4 O ($O_2 + 2$ O)

This shows the O atoms are still unbalanced with 2 on the left and 4 on the right. Placing a coefficient of 2 in front of O_2 will balance the O atoms.

$$CH_4 + 2\,O_2 \rightarrow CO_2 + 2\,H_2O$$
(balanced)

Step 4. Take another inventory to determine if the numbers of atoms on both sides are now equal. If they are, determine if the coefficients are in the lowest possible whole-number ratio. The inventory is now

Reactants	Products
1 C	1 C
4 H	4 H
4 O	4 O

The number of each kind of atom on each side of the equation is the same, and the ratio of $1:2 \rightarrow 1:2$ is the lowest possible whole-number ratio. The equation is balanced, which is illustrated with sketches of molecules in box figure 9.1.

Balancing chemical equations is mostly a trial-and-error procedure. But with practice, you will find there are a few generalized "role models" that can be useful in balancing equations for many simple reactions. The key to success at balancing equations is to think it out step-by-step while remembering the following:

1. Atoms are neither lost nor gained nor do they change their identity in a chemical reaction. The same kind and number of atoms in the reactants must appear in the products, meaning atoms are conserved.

2. A correct formula of a compound cannot be changed by altering the number or placement of subscripts. Changing subscripts changes the identity of a compound and the meaning of the entire equation.

3. A coefficient in front of a formula multiplies everything in the formula by that number.

Here are a few generalizations that can be helpful in balancing equations:

1. Look first to formulas of compounds with the most atoms and try to balance the atoms or compounds from which they were formed or to what they decomposed.

2. You should treat polyatomic ions that appear on both sides of the equation as independent units with a charge. That is, consider the polyatomic ion as a unit while taking an inventory rather than the individual atoms making up the polyatomic ion. This will save time and simplify the procedure.

3. Both the "crossover technique" and the use of "fractional coefficients" can be useful in finding the least common multiple to balance an equation.

4. The physical state of reactants and products in a reaction is often identified by the symbols (g) for gas, (l) for liquid, (s) for solid, and (aq) for an aqueous solution, which means water. If a gas escapes, this is identified with an arrow pointing up ($\uparrow$). A solid formed from a solution is identified with an arrow pointing down ($\downarrow$). The Greek symbol delta (Δ) is often used under or over the yield sign to indicate a change of temperature or other physical values.

Reaction:	Methane reacts with oxygen to yield carbon dioxide and water
Balanced equation:	$CH_4 + 2\,O_2 \longrightarrow CO_2 + 2\,H_2O$

| Sketches representing molecules: | |
| Meaning: | 1 molecule of methane + 2 molecules of oxygen → 1 molecule of carbon dioxide + 2 molecules of water |

Box Figure 9.1

Compare the numbers of each kind of atom in the balanced equation with the numbers of each kind of atom in the sketched representation. Both the equation and the sketch have the same number of atoms in the reactants and in the products.

From the point of view of atoms, the law of conservation of mass means that *atoms are neither created nor destroyed in the chemical reaction.* A chemical reaction is the making or breaking of chemical bonds between atoms or groups of atoms. Atoms are not lost or destroyed in the process, nor are they changed to a different kind. The equation for the formation of carbon monoxide has two oxygen atoms in the reactants (O_2) but only one in the product (CO). An atom of oxygen has disappeared somewhere, and that violates the law of conservation of mass. You cannot fix the equation by changing the CO to a CO_2, because this would change the identity of the compounds. Carbon monoxide is a poisonous gas that is different from carbon dioxide, a relatively harmless product of burning and respiration. *You cannot change the subscript in a formula* because that would change the formula. A different formula means a different composition and thus a different compound.

You cannot change the subscripts of a formula, but you can place a number called a *coefficient* in *front* of the formula. Changing a coefficient changes the *amount* of a substance, not the identity. Thus 2 CO means two molecules of carbon monoxide and 3 CO means three molecules of carbon monoxide. If there is no coefficient, 1 is understood as with subscripts. The meaning of coefficients and subscripts is illustrated in figure 9.14.

Placing a coefficient of 2 in front of the C and a coefficient of 2 in front of the CO in the equation will result in the same numbers of each kind of atom on both sides:

$$2\,C + O_2 \rightarrow 2\,CO$$

Reactants:	2 C	Products:	2 C
	2 O		2 O

The equation is now balanced.

Generalizing from groups of chemical reactions also makes it possible to predict what will happen in similar reactions. For example, the combustion of methane (CH_4), propane (C_3H_8), and octane (C_8H_{18}) all involve a *hydrocarbon,* a compound of the elements hydrogen and carbon. Each hydrocarbon reacts with O_2, yielding CO_2 and releasing the energy of combustion. Generalizing from these reactions, you could predict that the combustion of *any* hydrocarbon would involve the combination of atoms of the hydrocarbon molecule with O_2 to produce CO_2 and H_2O with the release of energy. Such reactions could be analyzed by chemical experiments, and the products could be identified by their physical and chemical properties. You would find your predictions based on similar reactions would be correct, thus justifying predictions from such generalizations. Butane, for example, is a hydrocarbon with the formula C_4H_{10}. The balanced equation for the combustion of butane is

$$2\,C_4H_{10(g)} + 13\,O_{2(g)} \rightarrow 8\,CO_{2(g)} + 10\,H_2O_{(g)}$$

You could extend the generalization further, noting that the combustion of compounds containing oxygen as well as carbon and hydrogen also produces CO_2 and H_2O (figure 9.15). These compounds are *carbohydrates,* composed of carbon and water. Glucose, for example, is a compound with the formula $C_6H_{12}O_6$. Glucose combines with oxygen to produce CO_2 and H_2O, and the balanced equation is

$$C_6H_{12}O_{6(s)} + 6\,O_{2(g)} \rightarrow 6\,CO_{2(g)} + 6\,H_2O_{(g)}$$

Note that three molecules of oxygen were not needed from the O_2 reactant since the other reactant, glucose, contains six oxygen atoms per molecule. An inventory of atoms will show that the equation is thus balanced.

C	means	●	One atom of carbon
O	means	○	One atom of oxygen
O_2	means	◐◐	One molecule of oxygen consisting of two atoms of oxygen
CO	means	●◐	One molecule of carbon monoxide consisting of one atom of carbon attached to one atom of oxygen
CO_2	means	◐●◐	One molecule of carbon dioxide consisting of one atom of carbon attached to two atoms of oxygen
$3\,CO_2$	means	◐●◐ ◐●◐ ◐●◐	Three molecules of carbon dioxide, each consisting of one atom of carbon attached to two atoms of oxygen

Figure 9.14

The meaning of subscripts and coefficients used with a chemical formula. The subscripts tell you how many atoms of a particular element are in a compound. The coefficient tells you about the quantity, or number, of molecules of the compound.

Figure 9.15

Hydrocarbons are composed of the elements hydrogen and carbon. Propane (C_3H_8) and gasoline, which contains a mixture of molecules such as octane (C_8H_{18}) are examples of hydrocarbons. *Carbohydrates* are composed of the elements of hydrogen, carbon, and oxygen. Table sugar, for example, is the carbohydrate $C_{12}H_{22}O_{11}$. Generalizing, all hydrocarbons and carbohydrates react completely with oxygen to yield CO_2 and H_2O.

Combustion is a rapid reaction with O_2 that releases energy, usually with a flame. A very similar, although much slower reaction takes place in plant and animal respiration. In respiration, carbohydrates combine with O_2 and release energy used for biological activities. This reaction is slow compared to combustion and requires enzymes to proceed at body temperature. Nonetheless, CO_2 and H_2O are the products.

TYPES OF CHEMICAL REACTIONS

The reactions involving hydrocarbons and carbohydrates with oxygen are examples of an important group of chemical reactions called *oxidation-reduction* reactions. When the term "oxidation" was first used, it specifically meant reactions involving the combination of oxygen with other atoms. But fluorine, chlorine, and other nonmetals were soon understood to have similar reactions as oxygen, so the definition was changed to one concerning the shifts of electrons in the reaction.

An **oxidation-reduction reaction** (or **redox reaction**) is broadly defined as a reaction in which electrons are transferred from one atom to another. As is implied by the name, such a reaction has two parts and each part tells you what happens to the electrons. *Oxidation* is the part of a redox reaction in which there is a loss of electrons by an atom. *Reduction* is the part of a redox reaction in which there is a gain of electrons by an atom. The name also implies that in any reaction in which oxidation occurs reduction must take place, too. One cannot take place without the other.

Substances that take electrons from other substances are called **oxidizing agents.** Oxidizing agents take electrons from the substances being oxidized. Oxygen is the most common oxidizing agent, and several examples have already been given about how it oxidizes foods and fuels. Chlorine is another commonly used oxidizing agent, often for the purposes of bleaching or killing bacteria (figure 9.16).

A **reducing agent** supplies electrons to the substance being reduced. Hydrogen and carbon are commonly used reducing agents. Carbon is commonly used as a reducing agent to

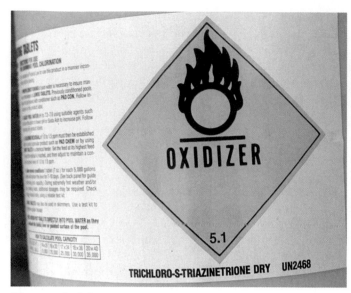

Figure 9.16

Oxidizing agents take electrons from other substances that are being oxidized. Oxygen and chlorine are commonly used, strong oxidizing agents.

extract metals from their ores. For example, carbon (from coke, which is coal that has been baked) reduces Fe_2O_3, an iron ore, in the reaction

$$2\ Fe_2O_{3(s)} + 3\ C_{(s)} \rightarrow 4\ Fe_{(s)} + 3\ CO_2 \uparrow$$

The Fe in the ore gained electrons from the carbon, the reducing agent in this reaction.

Many chemical reactions can be classified as redox or nonredox reactions. Another way to classify chemical reactions is to consider what is happening to the reactants and products. This type of classification scheme leads to four basic categories of chemical reactions, which are (1) *combination,* (2) *decomposition,* (3) *replacement,* and (4) *ion exchange* reactions. The first three categories are subclasses of redox reactions. It is in the ion exchange reactions that you will find the first example of a reaction that is not a redox reaction.

Combination Reactions

A **combination reaction** is a synthesis reaction in which two or more substances combine to form a single compound. The combining substances can be (1) elements, (2) compounds, or (3) combinations of elements and compounds. In generalized form, a combination reaction is

$$X + Y \rightarrow XY$$

Many redox reactions are combination reactions. For example, metals are oxidized when they burn in air, forming a metal oxide. Consider magnesium, which gives off a bright white light as it burns:

$$2\ Mg_{(s)} + O_{2(g)} \rightarrow 2\ MgO_{(s)}$$

The rusting of metals is oxidation that takes place at a slower pace than burning, but metals are nonetheless oxidized in

---

the process. Again noting the generalized form of a combination reaction, consider the rusting of iron:

$$4 \, Fe_{(s)} + 3 \, O_{2(g)} \rightarrow 2 \, Fe_2O_{3(s)}$$

Nonmetals are also oxidized by burning in air, for example, when carbon burns with a sufficient supply of O_2:

$$C_{(s)} + O_{2(g)} \rightarrow CO_{2(g)}$$

Note that all the combination reactions follow the generalized form of $X + Y \rightarrow XY$.

Decomposition Reactions

A **decomposition reaction,** as the term implies, is the opposite of a combination reaction. In decomposition reactions a compound is broken down (1) into the elements that make up the compound, (2) into simpler compounds, or (3) into elements and simpler compounds. Decomposition reactions have a generalized form of

$$XY \rightarrow X + Y$$

Decomposition reactions generally require some sort of energy, which is usually supplied in the form of heat or electrical energy. An electric current, for example, decomposes water into hydrogen and oxygen:

$$2 \, H_2O_{(1)} \overset{\text{electricity}}{\rightarrow} 2 \, H_{2(g)} + O_{2(g)}$$

Mercury(II) oxide is decomposed by heat, an observation that led to the discovery of oxygen:

$$2HgO_{(s)} \overset{\Delta}{\rightarrow} 2 \, Hg_{(s)} + O_2\uparrow$$

Note that all the decomposition reactions follow the generalized form of $XY \rightarrow X + Y$.

Replacement Reactions

In a **replacement reaction,** an atom or polyatomic ion is replaced in a compound by a different atom or polyatomic ion. The replaced part can be either the negative or positive part of the compound. In generalized form, a replacement reaction is

$$XY + Z \rightarrow XZ + Y$$
(negative part replaced)

or

$$XY + A \rightarrow AY + X$$
(positive part replaced)

Replacement reactions occur because some elements have a stronger electron-holding ability than other elements. Elements that have the least ability to hold on to their electrons are the most chemically active. Figure 9.17 shows a list of chemical activity of some metals, with the most chemically active at the top. Hydrogen is included because of its role in acids. Take a few minutes to look over the generalizations listed in figure 9.17. The generalizations apply to combination, decomposition, and replacement reactions.

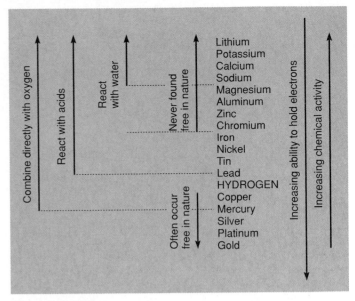

Figure 9.17

The activity series for common metals, together with some generalizations about the chemical activities of the metals. The series is used to predict which replacement reactions will take place and which reactions will not occur. (Note that hydrogen is not a metal and is placed in the series for reference to acid reactions.)

Replacement reactions take place as more active metals give up electrons to elements lower on the list with a greater electron-holding ability. For example, aluminum is higher on the activity series than copper. When aluminum foil is placed in a solution of copper(II) chloride, aluminum is oxidized, losing electrons to the copper. The loss of electrons from metallic aluminum forms aluminum ions in solution, and the copper comes out of solution as a solid metal (figure 9.18).

$$2 \, Al_{(s)} + 3 \, CuCl_{2(aq)} \rightarrow 2 \, AlCl_{3(aq)} + 3 \, Cu_{(s)}$$

A metal will replace any metal ion in solution that it is above in the activity series. If the metal is listed below the metal ion in solution, no reaction occurs. For example, $Ag_{(s)} + CuCl_{2(aq)} \rightarrow$ no reaction.

The very active metals (lithium, potassium, calcium, and sodium) react with water to yield metal hydroxides and hydrogen. For example,

$$2 \, Na_{(s)} + 2 \, H_2O_{(1)} \rightarrow 2 \, NaOH_{(aq)} + H_2 \uparrow$$

Acids yield hydrogen ions in solution, and metals above hydrogen in the activity series will replace hydrogen to form a metal salt. For example,

$$Zn_{(s)} + H_2SO_{4(aq)} \rightarrow ZnSO_{4(aq)} + H_2 \uparrow$$

In general, the energy involved in replacement reactions is less than the energy involved in combination or decomposition reactions.

Connections . . .

The Catalytic Converter

The modern automobile produces two troublesome products in the form of (1) nitrogen monoxide (NO) and (2) hydrocarbons from the incomplete combustion of gasoline. These products from the exhaust enter the air to react in sunlight, eventually producing an irritating haze known as *photochemical smog*. To reduce photochemical smog, modern automobiles are fitted with a catalytic converter as part of the automobile exhaust system (box figure 9.2). This reading is about how the catalytic converter combats smog-forming pollutants.

Chemical reactions proceed at a rate that is affected by (1) the concentration of the reactants, (2) the temperature at which a reaction occurs, and (3) the surface area of the reaction. In general, a higher concentration, higher temperatures, and greater surface area mean a faster reaction. The problem with nitrogen monoxide is that it is easily oxidized to nitrogen dioxide (NO_2), a reddish brown, damaging gas that also plays a key role in the formation of photochemical smog. Nitrogen dioxide and the hydrocarbons are oxidized slowly in the air when left to themselves. What is needed is a means to decompose nitrogen monoxide and uncombusted hydrocarbons rapidly before they are released into the air.

The rate at which a chemical reaction proceeds is affected by a *catalyst*, a material that speeds up a chemical reaction without being permanently changed by the reaction. Apparently, molecules require a certain amount of energy to change the chemical bonds that tend to keep them as they are, unreacted. This certain amount of energy is called the *activation energy*, and it represents an energy barrier that must be overcome before a chemical reaction can take place. This explains why chemical reactions proceed at a faster rate at higher temperatures. At higher temperatures, molecules have greater average kinetic energies; thus they already have part of the minimum energy needed for a reaction to take place.

A catalyst appears to speed a chemical reaction by lowering the activation energy. Molecules become temporarily attached to the surface of the catalyst, which weakens the chemical bonds holding the molecule together. Thus, the weakened molecule is easier to break apart and the activation energy is lowered. Some catalysts do this better with some specific compounds than others, and extensive chemical research programs are devoted to finding new and more effective catalysts.

Automobile catalytic converters use unreactive metals, such as platinum, and transition metal oxides such as copper(II) oxide and chromium(III) oxide. Catalytic reactions that occur in the converter include the following:

$$H_2O + CO \xrightarrow{\text{catalyst}} H_2 + CO_2$$

$$2\,NO + 2CO \xrightarrow{\text{catalyst}} N_2 + 2\,CO_2$$

$$2\,NO + 2\,H_2 \xrightarrow{\text{catalyst}} N_2 + 2\,H_2O$$

$$2\,O_2 + CH^*_4 \xrightarrow{\text{catalyst}} CO_2 + 2\,H_2O$$

*or other hydrocarbon

Thus nitrogen monoxide is reduced to nitrogen gas, and hydrocarbons are oxidized to CO_2 and H_2O.

A catalytic converter can reduce or oxidize about 90 percent of the hydrocarbons, 85 percent of the carbon monoxide, and 40 percent of the nitrogen monoxide from exhaust gases. Other controls, such as exhaust gas recirculation (EGR), are used to further reduce NO formation.

Box Figure 9.2
This silver-colored canister is the catalytic converter. The catalytic converter is located between the engine and the muffler, which is farther back toward the rear of the car.

Ion Exchange Reactions

An **ion exchange reaction** is a reaction that takes place when the ions of one compound interact with the ions of another compound, forming (1) a solid that comes out of solution (a precipitate), (2) a gas, or (3) water.

A water solution of dissolved ionic compounds is a solution of ions. For example, solid sodium chloride dissolves in water to become ions in solution,

$$NaCl_{(s)} \rightarrow Na^+_{(aq)} + Cl^-_{(aq)}$$

If a second ionic compound is dissolved with a solution of another, a mixture of ions results. The formation of a precipitate, a gas, or water, however, removes ions from the solution, and this must occur before you can say that an ionic exchange reaction has taken place. For example, water being treated for domestic use sometimes carries suspended matter that is removed by adding aluminum sulfate and calcium hydroxide to the water. The reaction is

$$3\,Ca(OH)_{2(aq)} + Al_2(SO_4)_{3(aq)} \rightarrow 3\,CaSO_{4(aq)} + 2\,Al(OH)_3\downarrow$$

Figure 9.18

This shows a reaction between metallic aluminum and the blue solution of copper(II) chloride. Aluminum is above copper in the activity series, and aluminum replaces the copper ions from the solution as copper is deposited as a metal. The aluminum loses electrons to the copper and forms aluminum ions in solution.

The aluminum hydroxide is a jellylike solid, which traps the suspended matter for sand filtration. The formation of the insoluble aluminum hydroxide removed the aluminum and hydroxide ions from the solution, so an ion exchange reaction took place.

In general, an ion exchange reaction has the form

$$AX + BY \rightarrow AY + BX$$

where one of the products removes ions from the solution. The calcium hydroxide and aluminum sulfate reaction took place as the aluminum and calcium ions traded places. A solubility table such as the one in appendix B will tell you if an ionic exchange reaction has taken place. Aluminum hydroxide is insoluble, according to the table, so the reaction did take place. No ionic exchange reaction occurred if the new products are both soluble.

Another way for an ion exchange reaction to occur is if a gas or water molecule forms to remove ions from the solution. When an acid reacts with a base (an alkaline compound), a salt and water are formed

$$HCl_{(aq)} + NaOH_{(aq)} \rightarrow NaCl_{(aq)} + H_2O_{(l)}$$

The reactions of acids and bases are discussed in chapter 10.

SUMMARY

Mixtures are made up of *unlike parts* with a *variable composition. Pure substances* are the *same throughout* and have a *definite composition.* Mixtures can be separated into their components by *physical changes,* changes that do not alter the identity of matter. Some pure substances can be broken down into simpler substances by a *chemical change,* a change that *alters the identity of matter as it produces new substances with different properties.* A pure substance that can be decomposed by chemical change into simpler substances with a definite composition is

a *compound.* A pure substance that cannot be broken down into anything simpler is an *element.*

A *chemical change* produces new substances by making or breaking chemical bonds. The process of chemical change is called a *chemical reaction.* During a chemical reaction different chemical substances with greater or lesser amounts of internal potential energy are produced. *Chemical energy* is the change of internal potential energy during a chemical reaction. A *chemical equation* is a shorthand way of describing a chemical reaction. An equation shows the substances that are changed, the *reactants,* on the left side, and the new substances produced, the *products,* on the right side.

Chemical reactions involve *valence electrons,* the electrons in the outermost energy level of an atom. Atoms tend to lose or acquire electrons to achieve the configuration of the noble gases with stable, filled outer orbitals. This tendency is generalized as the *octet rule,* that atoms lose or gain electrons to acquire the noble gas structure of eight electrons in the outer orbital. Atoms form negative or positive *ions* in the process.

A chemical bond is an attractive force that holds atoms together in a compound. Chemical bonds formed when atoms transfer electrons to become ions are *ionic bonds.* An ionic bond is an electrostatic attraction between oppositely charged ions. Chemical bonds formed when ions share electrons are *covalent bonds.*

Ionic bonds result in *ionic compounds* with a crystalline structure. The energy released when an ionic compound is formed is called the *heat of formation.* It is the same amount of energy that is required to decompose the compound into its elements.

A *formula* of a compound uses symbols to tell what elements are in a compound and in what proportions. Ions of representative elements have a single, fixed charge, but many transition elements have variable charges. Electrons are conserved when ionic compounds are formed, and the ionic compound is electrically neutral. A formula shows the overall balance of charges.

Covalent compounds are molecular, composed of electrically neutral groups of atoms bound together by *covalent bonds.* The sharing of a pair of electrons, with each atom contributing a single electron to the shared pair, forms a *single covalent bond.* Covalent bonds formed when two pairs of electrons are shared are called *double bonds* and a *triple bond* is the sharing of three pairs of electrons.

Compounds are named with different rules for ionic and covalent compounds. Both ionic and covalent compounds that are made up of only two different elements always end with an *-ide* suffix, but there are a few exceptions.

The rule for naming variable-charge ions states the English name and gives the charge with Roman numerals in parentheses. Ionic compounds are electrically neutral, and formulas must show a balance of charge. The *crossover technique* is an easy way to write formulas that show a balance of charge.

Covalent compounds are molecules of two or more nonmetal atoms held together by a covalent bond. The system for naming covalent compounds uses Greek prefixes to identify the numbers of atoms, since more than one compound can form from the same two elements (CO and CO_2, for example).

A concise way to describe a chemical reaction is to use formulas in a *chemical equation.* A chemical equation with the same number of each kind of atom on both sides is called a *balanced equation.* A balanced equation is in accord with the *law of conservation of mass,* which states that atoms are neither created nor destroyed in a chemical reaction. To balance a chemical equation, *coefficients* are placed in front of chemical formulas. Subscripts of formulas may not be changed since this would change the formula, meaning a different compound.

One important group of chemical reactions is called *oxidation-reduction reactions*, or *redox* reactions for short. Redox reactions are reactions where shifts of electrons occur. The process of losing electrons is called *oxidation*, and the substance doing the losing is said to be *oxidized*. The process of gaining electrons is called *reduction*, and the substance doing the gaining is said to be *reduced*. Substances that take electrons from other substances are called *oxidizing agents*. Substances that supply electrons are called *reducing agents*.

Chemical reactions can also be classified as (1) *combination*, (2) *decomposition*, (3) *replacement*, or (4) *ion exchange*. The first three of these are redox reactions, but ion exchange is not.

KEY TERMS

atom (p. **174**)
chemical bond (p. **177**)
chemical change (p. **173**)
chemical energy (p. **175**)
chemical equation (p. **175**)
chemical reaction (p. **175**)
combination reaction (p. **188**)
compound (p. **173**)
covalent bond (p. **179**)
covalent compound (p. **180**)
decomposition reaction (p. **189**)
double bond (p. **181**)
element (p. **173**)
formula (p. **178**)
heat of formation (p. **178**)
ion exchange reaction (p. **190**)
ionic bond (p. **177**)

ionic compounds (p. **179**)
law of conservation of mass (p. **185**)
molecule (p. **174**)
octet rule (p. **176**)
oxidation-reduction reaction (p. **188**)
oxidizing agents (p. **188**)
physical change (p. **172**)
polyatomic ion (p. **182**)
redox reaction (p. **188**)
reducing agent (p. **188**)
replacement reaction (p. **189**)
single bond (p. **181**)
triple bond (p. **181**)
valence electrons (p. **176**)

APPLYING THE CONCEPTS

1. Which of the following represents a chemical change?
   (a) heating a sample of ice until it melts
   (b) tearing a sheet of paper into tiny pieces
   (c) burning a sheet of paper
   (d) All of the above are correct.

2. A pure substance that cannot be decomposed into anything simpler by chemical or physical means is called a (an)
   (a) element.
   (b) compound.
   (c) mixture.
   (d) isotope.

3. The electrons that participate in chemical bonding are (the)
   (a) valence electrons.
   (b) electrons in fully occupied orbitals.
   (c) stable inner electrons.
   (d) all of the above.

4. Which type of chemical bond is formed between two atoms by the sharing of electrons?
   (a) ionic
   (b) covalent
   (c) metallic

5. Which combination of elements forms crystalline solids that will dissolve in water, producing a solution of ions that conduct an electric current?
   (a) metal and metal
   (b) metal and nonmetal
   (c) nonmetal and nonmetal
   (d) All of the above are correct.

6. An inorganic compound made of only two different elements has a name that always ends with the suffix
   (a) -ite.
   (b) -ate.
   (c) -ide.
   (d) -ous.

7. Dihydrogen monoxide is a compound with the common name of
   (a) laughing gas.
   (b) water.
   (c) smog.
   (d) rocket fuel.

8. A chemical equation is balanced by changing (the)
   (a) subscripts.
   (b) superscripts.
   (c) coefficients.
   (d) any of the above as necessary to achieve a balance.

9. Since wood is composed of carbohydrates, you should expect what gases to exhaust from a fireplace when complete combustion takes place?
   (a) carbon dioxide, carbon monoxide, and pollutants
   (b) carbon dioxide and water vapor
   (c) carbon monoxide and smoke
   (d) It depends on the type of wood being burned.

10. When carbon burns with an insufficient supply of oxygen, carbon monoxide is formed according to the following equation: $2 C + O_2 \rightarrow 2 CO$. What category of chemical reaction is this?
    (a) combination
    (b) ion exchange
    (c) replacement
    (d) None of the above are correct.

11. Of the elements listed below, the one with the greatest chemical activity is
    (a) aluminum.
    (b) zinc.
    (c) iron.
    (d) mercury.

12. A balanced chemical equation has
    (a) the same number of molecules on both sides of the equation.
    (b) the same kinds of molecules on both sides of the equation.
    (c) the same number of each kind of atom on both sides of the equation.
    (d) all of the above.

Answers
1. c 2. a 3. a 4. b 5. b 6. c 7. b 8. c 9. b 10. a 11. a 12. c

QUESTIONS FOR THOUGHT

1. What is the difference between a chemical change and a physical change? Give three examples of each.
2. Describe how the following are alike and how they are different: (a) a sodium atom and a sodium ion, and (b) a sodium ion and a neon atom.
3. What is the difference between an ionic and covalent bond? What do atoms forming the two bond types have in common?
4. What is the octet rule?
5. What is a polyatomic ion? Give the names and formulas for several common polyatomic ions.
6. Write the formula for magnesium hydroxide. Explain what the parentheses mean.
7. What is the basic difference between a single bond and a double bond?
8. What is the law of conservation of mass? How do you know if a chemical equation is in accord with this law?
9. Describe in your own words how a chemical equation is balanced.
10. How is the activity series for metals used to predict if a replacement reaction will occur or not?
11. What must occur in order for an ion exchange reaction to take place? What is the result if this does not happen?
12. Predict the products for the following reactions: (a) The combustion of ethyl alcohol (C_2H_5OH), and (b) the rusting of aluminum (Al).

PARALLEL EXERCISES

The exercises in groups A and B cover the same concepts. Solutions to group A exercises are located in appendix D.

Group A

1. How many outer orbital electrons are found in an atom of:
   (a) Li
   (b) N
   (c) F
   (d) Cl
   (e) Ra
   (f) Be
2. Write electron dot notations for the following elements:
   (a) Boron
   (b) Bromine
   (c) Calcium
   (d) Potassium
   (e) Oxygen
   (f) Sulfur
3. Identify the charge on the following ions:
   (a) Boron
   (b) Bromine
   (c) Calcium
   (d) Potassium
   (e) Oxygen
   (f) Nitrogen
4. Name the following polyatomic ions:
   (a) $(OH)^-$
   (b) $(SO_3)^{2-}$
   (c) $(ClO)^-$
   (d) $(NO_3)^-$
   (e) $(CO_3)^{2-}$
   (f) $(ClO_4)^-$

Group B

1. How many outer orbital electrons are found in an atom of:
   (a) Na
   (b) P
   (c) Br
   (d) I
   (e) Te
   (f) Sr
2. Write electron dot notations for the following elements:
   (a) Aluminum
   (b) Fluorine
   (c) Magnesium
   (d) Sodium
   (e) Carbon
   (f) Chlorine
3. Identify the charge on the following ions:
   (a) Aluminum
   (b) Chlorine
   (c) Magnesium
   (d) Sodium
   (e) Sulfur
   (f) Hydrogen
4. Name the following polyatomic ions:
   (a) $(C_2H_3O_2)^-$
   (b) $(HCO_3)^-$
   (c) $(SO_4)^{2-}$
   (d) $(NO_2)^-$
   (e) $(MnO_4)^-$
   (f) $(CO_3)^{2-}$

5. Use the crossover technique to write formulas for the following compounds:
   (a) Iron(III) hydroxide
   (b) Lead(II) phosphate
   (c) Zinc carbonate
   (d) Ammonium nitrate
   (e) Potassium hydrogen carbonate
   (f) Potassium sulfite

6. Write formulas for the following covalent compounds:
   (a) Carbon tetrachloride
   (b) Dihydrogen monoxide
   (c) Manganese dioxide
   (d) Sulfur trioxide
   (e) Dinitrogen pentoxide
   (f) Diarsenic pentasulfide

7. Name the following covalent compounds:
   (a) CO
   (b) CO_2
   (c) CS_2
   (d) N_2O
   (e) P_4S_3
   (f) N_2O_3

8. Write balanced chemical equations for each of the following unbalanced reactions:
   (a) $SO_2 + O_2 \rightarrow SO_3$
   (b) $P + O_2 \rightarrow P_2O_5$
   (c) $Al + HCl \rightarrow AlCl_3 + H_2$
   (d) $NaOH + H_2SO_4 \rightarrow Na_2SO_4 + H_2O$
   (e) $Fe_2O_3 + CO \rightarrow Fe + CO_2$
   (f) $Mg(OH)_2 + H_3PO_4 \rightarrow Mg_3(PO_4)_2 + H_2O$

9. Identify the following as combination, decomposition, replacement, or ion exchange reactions:
   (a) $NaCl_{(aq)} + AgNO_{3(aq)} \rightarrow NaNO_{3(aq)} + AgCl\downarrow$
   (b) $H_2O_{(l)} + CO_{2(g)} \rightarrow H_2CO_{3(l)}$
   (c) $2\,NaHCO_{3(s)} \rightarrow Na_2CO_{3(s)} + H_2O_{(g)} + CO_{2(g)}$
   (d) $2\,Na_{(s)} + Cl_{2(g)} \rightarrow 2\,NaCl_{(s)}$
   (e) $Cu_{(s)} + 2\,AgNO_{3(aq)} \rightarrow Cu(NO_3)_{2(aq)} + 2\,Ag_{(s)}$
   (f) $CaO_{(s)} + H_2O_{(l)} \rightarrow Ca(OH)_{2(aq)}$

10. Write complete, balanced equations for each of the following reactions:
   (a) $C_5H_{12(g)} + O_{2(g)} \rightarrow$
   (b) $HCl_{(aq)} + NaOH_{(aq)} \rightarrow$
   (c) $Al_{(s)} + Fe_2O_{3(s)} \rightarrow$
   (d) $Fe_{(s)} + CuSO_{4(aq)} \rightarrow$
   (e) $MgCl_{(aq)} + Fe(NO_3)_{2(aq)} \rightarrow$
   (f) $C_6H_{10}O_{5(s)} + O_{2(g)} \rightarrow$

5. Use the crossover technique to write formulas for the following compounds:
   (a) Aluminum hydroxide
   (b) Sodium phosphate
   (c) Copper(II) chloride
   (d) Ammonium sulfate
   (e) Sodium hydrogen carbonate
   (f) Cobalt(II) chloride

6. Write formulas for the following covalent compounds:
   (a) Silicon dioxide
   (b) Dihydrogen sulfide
   (c) Boron trifluoride
   (d) Dihydrogen dioxide
   (e) Carbon tetrafluoride
   (f) Nitrogen trihydride

7. Name the following covalent compounds:
   (a) N_2O
   (b) SO_2
   (c) SiC
   (d) PF_5
   (e) $SeCl_6$
   (f) N_2O_4

8. Write balanced chemical equations for each of the following unbalanced reactions:
   (a) $NO + O_2 \rightarrow NO_2$
   (b) $KClO_3 \rightarrow KCl + O_2$
   (c) $NH_4Cl + Ca(OH)_2 \rightarrow CaCl_2 + NH_3 + H_2O$
   (d) $NaNO_3 + H_2SO_4 \rightarrow Na_2SO_4 + HNO_3$
   (e) $PbS + H_2O_2 \rightarrow PbSO_4 + H_2O$
   (f) $Al_2(SO_4)_3 + BaCl_2 \rightarrow AlCl_3 + BaSO_4$

9. Identify the following as combination, decomposition, replacement, or ion exchange reactions:
   (a) $ZnCO_{3(s)} \rightarrow ZnO_{(s)} + CO_2\uparrow$
   (b) $2\,NaBr_{(aq)} + Cl_{2(g)} \rightarrow 2\,NaCl_{(aq)} + Br_{2(g)}$
   (c) $2\,Al_{(s)} + 3\,Cl_{2(g)} \rightarrow 2\,AlCl_{3(s)}$
   (d) $Ca(OH)_{2(aq)} + H_2SO_{4(aq)} \rightarrow CaSO_{4(aq)} + 2\,H_2O_{(l)}$
   (e) $Pb(NO_3)_{2(aq)} + H_2S_{(g)} \rightarrow 2\,HNO_{3(aq)} + PbS\downarrow$
   (f) $C_{(s)} + ZnO_{(s)} \rightarrow Zn_{(s)} + CO\uparrow$

10. Write complete, balanced equations for each of the following reactions:
   (a) $C_3H_{6(g)} + O_{2(g)} \rightarrow$
   (b) $H_2SO_{4(aq)} + KOH_{(aq)} \rightarrow$
   (c) $C_6H_{12}O_{6(s)} + O_{2(g)} \rightarrow$
   (d) $Na_3PO_{4(aq)} + AgNO_{3(aq)} \rightarrow$
   (e) $NaOH_{(aq)} + Al(NO_3)_{3(aq)} \rightarrow$
   (f) $Mg(OH)_{2(aq)} + H_3PO_{4(aq)} \rightarrow$

Water and Solutions

Water is often referred to as the *universal solvent* because it makes so many different kinds of solutions. Eventually, moving water can dissolve solid rock, carrying it away in solution.

O U T L I N E

What do you think about when you see a stream (figure 10.1)? Do you wonder about the water quality and what might be dissolved in the water? Do you wonder where the stream comes from and if it will ever run out of water?

Many people can look at a stream, but they might think about different things. A farmer might think about how the water could be diverted and used for his crops. A city planner might wonder if the water is safe for domestic use, and if not, what it would cost to treat the water. Others might wonder if the stream has large fish they could catch. Many large streams can provide water for crops, domestic use, and recreation, and still meet the requirements for a number of other uses.

It is the specific properties of water that make it important for agriculture, domestic use, and recreation. Living things evolved in a watery environment, so that water and its properties are essential to life on the earth. Some properties of water, such as the ability to dissolve almost anything, also make water very easy to pollute. This chapter is concerned with some of the unique properties of water, water solutions, and household use of water.

HOUSEHOLD WATER

Water is an essential resource, not only because it is required for life processes but also because of its role in a modern society. Water is used in the home for drinking and cooking (2 percent), cleaning dishes (6 percent), laundry (11 percent), bathing (23 percent), toilets (29 percent), and for maintaining lawns and gardens (29 percent).

The water supply is obtained from streams, lakes, and reservoirs on the surface, or from groundwater pumped from below the surface. Surface water contains more sediments, bacteria, and possible pollutants than water from a well because it is exposed to the atmosphere and water runs off the land into streams and rivers. Surface water requires filtering to remove suspended particles, treatment to kill bacteria, and sometimes processing to remove pollution. Well water is generally cleaner, but still might require treatment to kill bacteria and remove pollution that has seeped through the ground from waste dumps, agricultural activities, or from industrial sites.

Most pollutants are usually too dilute to be considered a significant health hazard, but there are exceptions. There are five types of contamination found in U.S. drinking water that are responsible for the most widespread danger, and these are listed in table 10.1. In spite of these general concerns and other occasional local problems, the U.S. water supply is considered to be among the cleanest in the world.

Demand for domestic water sometimes exceeds the immediate supply in some growing metropolitan areas. This is most common during the summer, when water demand is high and rainfall is often low. Communities in these areas often have public education campaigns designed to help reduce the demand for water. For example, did you know that taking a tub bath can use up to 135 liters (about 36 gal) of water compared to only 95 liters (about 25 gal) for a regular shower? Even more water is saved by a shower that does not run continuously—wetting down, soaping up, and rinsing off uses only 15 liters (about 4 gal) of water. You can also save about 35 liters (about 9 gal) of water by not letting the water run continuously while brushing your teeth.

It is often difficult to convince people to conserve water when it is viewed as an inexpensive, limitless supply. However, efforts to conserve water increase dramatically as the cost to the household consumer increases.

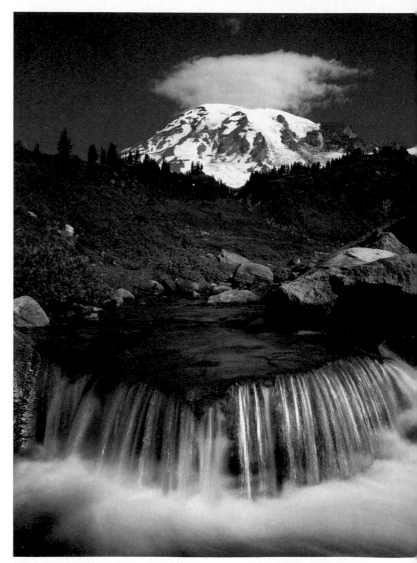

Figure 10.1
A freshwater stream has many potential uses.

The issues involved in maintaining a safe water supply are better understood by considering some of the properties of water and water solutions. These are the topics of the following sections.

Table 10.1 Possible Pollution Problems in the U.S. Water Supply

Pollutant	Source	Risk
Lead	Lead pipes in older homes; solder in copper pipes, brass fixtures	Nerve damage, miscarriage, birth defects, high blood pressure, hearing problems
Chlorinated solvents	Industrial pollution	Cancer
Trihalomethanes	Chlorine disinfectant reacting with other pollutants	Liver damage, kidney damage, possibly cancer
PCBs	Industrial waste, older transformers	Liver damage, possibly cancer
Bacteria and viruses	Septic tanks, outhouses, overflowing sewer lines	Gastrointestinal problems, serious diseases

PROPERTIES OF WATER

Water is essential for life since living organisms are made up of cells filled with water and a great variety of dissolved substances. Foods are mostly water, with fruits and vegetables containing up to 95 percent water and meat consisting of about 50 percent water. Your body is over 70 percent water by weight. Since water is such a large component of living things, understanding the properties of water is important to understanding life. One important property is water's unusual ability to act as a solvent. Water is called a "universal solvent" because of its ability to dissolve most molecules. In living things these dissolved molecules can be transported from one place to another by diffusion or by some kind of a circulatory system.

The usefulness of water does not end with its unique abilities as a solvent and transporter; it has many more properties that are useful, although unusual. For example, unlike other liquids, water in its liquid phase has a greater density than solid water (ice). This important property enables solid ice to float on the surface of liquid water, insulating the water below and permitting fish and other water organisms to survive the winter. If ice were denser than water, it would sink, freezing all lakes and rivers from the bottom up. Fish and most organisms that live in water would not be able to survive in a lake or river of solid ice.

As described in chapter 4, water is also unusual because it has a high specific heat. The same amount of sunlight falling on equal masses of soil and water will warm the soil 5°C for each 1°C increase in water temperature. Thus, it will take five times more sunlight to increase the temperature of the water to as much as the soil. This enables large bodies of water to moderate the temperature, making it more even.

A high latent heat of vaporization is yet another unusual property of water. This particular property enables people to dissipate large amounts of heat by evaporating a small amount of water. Since people carry this evaporative cooling system with them, they can survive some very warm desert temperatures, for example.

Finally, other properties of water are not crucial for life, but are interesting nonetheless. For example, why do all snowflakes have six sides? Is it true that no two snowflakes are alike? The unique structure of the water molecule will explain water's unique solvent abilities, why solid water is less dense than liquid water, its high specific heat, its high latent heat of vaporization, and perhaps why no two snowflakes seem to be alike.

Structure of the Water Molecule

In chapter 9 you learned that atoms combine two ways. Atoms from opposite sides of the periodic table form ionic bonds after transferring one or more electrons. Atoms from the right side of the periodic table form covalent bonds by sharing one or more pairs of electrons. This distinction is clear-cut in many compounds, but not in water. The way atoms share electrons in a water molecule is not exactly covalent, but it is not ionic either.

In a molecule of water an oxygen atom shares a pair of electrons with two hydrogen atoms. Oxygen has six outer electrons and needs two more to satisfy the octet rule, achieving the noble gas structure of eight. Each hydrogen atom needs one more electron to fill its outer orbital with two. Therefore, one oxygen atom bonds with two hydrogen atoms, forming H_2O. Both oxygen and hydrogen are more stable with the outer orbital configuration of the noble gases (neon and helium in this case).

Electrons are shared in a water molecule, but not equally. Oxygen, with its eight positive protons, has a greater attraction for the shared electrons than do either of the hydrogens with a single proton. Therefore, the shared electrons spend more time around the oxygen part of the molecule than they do around the hydrogen part. This results in the oxygen end of the molecule being more negative than the hydrogen end. When electrons in a covalent bond are not equally shared the molecule is said to be polar. A **polar molecule** has a *dipole* ("di" = two; "pole" = side or end), meaning it has a positive end and a negative end.

A water molecule has a negative center at the oxygen end and a positive center at the hydrogen end. The positive charges on the hydrogen end are separated, giving the molecule a bent arrangement rather than in a straight line. Figure 10.2A shows a model of a water molecule showing its polar nature.

It is the polar structure of the water molecule that is responsible for many of the unique properties of water. Polar molecules of any substances have attractions between the positive end of a molecule and the negative end of another molecule. When the polar molecule has hydrogen at one end and fluorine, oxygen, or nitrogen on the other part, the attractions are strong enough to make a type of bonding called **hydrogen bonding.** Hydrogen bonding is a strong bond that occurs between the

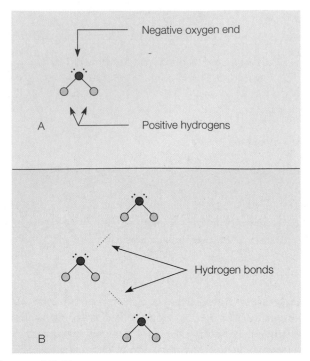

Figure 10.2

(*A*) The water molecule is polar, with centers of positive and negative charges. (*B*) Attractions between these positive and negative centers establish hydrogen bonds between adjacent molecules.

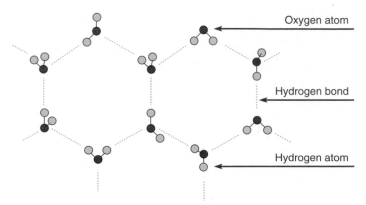

Figure 10.3

The hexagonal structure of ice. Hydrogen bonding between the oxygen atom and two hydrogen atoms of other water molecules results in an arrangement, which forms the open, hexagonal structure of ice.

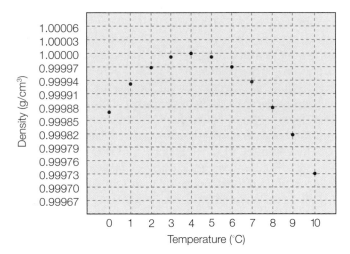

Figure 10.4

The density of water from 0°C to 10°C. The density of water is at a maximum at 4°C, becoming less dense as it is cooled or warmed from this temperature. Hydrogen bonding explains this unusual behavior.

hydrogen end of a molecule and the fluorine, oxygen or nitrogen end of other similar molecules. A better name for this would be a hydrogen-fluorine bond, a hydrogen-oxygen bond, or a hydrogen-nitrogen bond. However, for brevity the second part of the bond is not named and all the hydrogen-something bonds are simply known of as "hydrogen" bonds. The dotted line between the hydrogen and oxygen molecules in figure 10.2B represents a hydrogen bond. A dotted line is used to represent a strong bond that is not as strong as the bond represented by the solid line of a covalent compound.

Hydrogen bonding accounts for the physical properties of water, including its unusual density changes with changes in temperature. Figure 10.3 shows the hydrogen-bonded structure of ice. Water molecules form a six-sided hexagonal structure that extends out for billions of molecules. The large channels, or holes, in the structure result in ice being less dense than water. The shape of the hexagonal arrangement also suggests why snowflakes always have six sides. Why does it seem like no two snowflakes are alike? Perhaps the answer can be found in the almost infinite variety of shapes that can be built from billions and billions of tiny hexagons of ice crystals.

When ice is warmed, the increased vibrations of the molecules begin to expand and stretch the hydrogen bond structure. When ice melts, about 15 percent of the hydrogen bonds break and the open structure collapses into the more compact arrangement of liquid water. As the liquid water is warmed from 0°C still more hydrogen bonds break down, and the density of the water steadily increases. At 4°C the expansion of water from the increased molecular vibrations begins to predominate and the

density decreases steadily with further warming (figure 10.4). Thus, water has its greatest density at a temperature of 4°C.

The heat of fusion, specific heat, and heat of vaporization of water are unusually high when compared to other chemically similar substances. These high values are accounted for by the additional energy needed to break hydrogen bonds.

The Dissolving Process

A **solution** is a homogeneous mixture of ions or molecules of two or more substances. *Dissolving* is the process of making a solution. During dissolving, the different components that make up the solution become mixed. For example, when sugar dissolves in water, molecules of sugar become uniformly dispersed throughout the molecules of water. The uniform taste of sweetness of any part of the sugar solution is a result of this uniform mixing.

The general terms of *solvent* and *solute* identify the the components of a solution. The solvent is the component present in the larger amount. The solute is the component that dissolves in the solvent. Atmospheric air, for example, is about 78 percent nitrogen, so nitrogen is considered the solvent. Oxygen (about 21 percent), argon (about 0.9 percent), and other gases make up the solutes. If one of the components of a solution is a liquid, it is usually identified as the solvent. An *aqueous solution* is a solution of a solid, a liquid, or a gas in water.

A solution is formed when the molecules or ions of two or more substances become homogeneously mixed. But the process of dissolving must be more complicated than the simple mixing together of particles because (1) solutions become saturated, meaning there is a limit on solubility, and (2) some substances are *insoluble*, not dissolving at all or at least not noticeably. In general, the forces of attraction between molecules or ions of the solvent and solute determine if something will dissolve and if there are limits on the solubility. These forces of attraction and their role in the dissolving process will be considered in the following examples.

First, consider the dissolving process in gaseous and liquid solutions. In a gas, the intermolecular forces are small, so gases can mix in any proportion. Fluids that can mix in any proportion like this are called **miscible fluids.** Fluids that do not mix are called *immiscible fluids.* Air is a mixture of gases, so gases (including vapors) are miscible.

Liquid solutions can dissolve a gas, another liquid, or a solid. Gases are miscible in liquids, and a carbonated beverage (your favorite cola) is the common example, consisting of carbon dioxide dissolved in water. Whether or not two given liquids form solutions depends on some similarities in their molecular structures. The water molecule, for example, is a polar molecule with a negative end and a positive end. On the other hand, carbon tetrachloride (CCl_4) is a molecule with polar bonds that are symmetrically arranged. Because of the symmetry, CCl_4 has no negative or positive ends, so it is nonpolar. Thus some liquids have polar molecules, and some have nonpolar molecules. The general rule for forming solutions is *like dissolves like.* A nonpolar compound, such as carbon tetrachloride, will dissolve oils and greases because they are nonpolar compounds. Water, a polar compound, will not dissolve the nonpolar oils and greases. Carbon tetrachloride was at one time used as a cleaning solvent because of its oil and grease dissolving abilities. Its use is no longer recommended because it causes liver damage.

Some molecules, such as soap, have a part of the molecule that is polar and a part that is nonpolar. Washing with water alone will not dissolve oils because water and oil are immiscible. When soap is added to the water, however, the polar end of the soap molecule is attracted to the polar water molecules, and the nonpolar end is absorbed into the oil. A particle (larger than a molecule) is formed, and the oil is washed away with the water.

The "like dissolves like" rule applies to solids and liquid solvents as well as liquids and liquid solvents. Polar solids, such as salt, will readily dissolve in water, which has polar molecules, but do not dissolve readily in oil, grease, or other nonpolar solvents. Polar water readily dissolves salt because the charged polar water molecules are able to exert an attraction on the ions, pulling

them away from the crystal structure. Thus, ionic compounds dissolve in water.

Ionic compounds vary in their solubilities in water. This difference is explained by the existence of two different forces involved in an ongoing "tug of war." One force is the attraction between an ion on the surface of the crystal and a water molecule, an *ion-polar molecule force.* When solid sodium chloride and water are mixed together, the negative ends of the water molecules (the oxygen ends) become oriented toward the positive sodium ions on the crystal. Likewise, the positive ends of water molecules (the hydrogen ends) become oriented toward the negative chlorine ions. The attraction of water molecules for ions is called **hydration.** If the force of hydration is greater than the attraction between the ions in the solid, they are pulled away from the solid, and dissolving occurs (figure 10.5). Not considering the role of water in this dissolving process, the equation is

$$Na^+Cl^-_{(s)} \rightarrow Na^+_{(aq)} + Cl^-_{(aq)}$$

which shows that the ions were separated from the solid to become a solution of ions. In other compounds the attraction between the ions in the solid might be greater than the energy of hydration. In this case, the ions of the solid would win the "tug of war," and the ionic solid is insoluble.

The saturation of soluble compounds is explained in terms of hydration eventually occupying a large number of the polar water molecules. Fewer available water molecules means less attraction on the ionic solid, with more solute ions being pulled back to the surface of the solid. The tug of war continues back and forth as an equilibrium condition is established.

Concentration of Solutions

The relative amounts of solute and solvent are described by the concentration of a solution. In general, a solution with a large amount of solute is *concentrated,* and a solution with much less solute is *dilute.* The terms "dilute" and "concentrated" are somewhat arbitrary, and it is sometimes difficult to know the difference between a solution that is "weakly concentrated" and one that is "not very diluted." More meaningful information is provided by measurement of the *amount of solute in a solution.* There are different ways to express concentration measurements, each lending itself to a particular kind of solution or to how the information will be used. For example, you read about concentrations of parts per million in an article about pollution, but most of the concentration of solutions sold in stores are reported in percent by volume or percent by weight. Each of these concentrations is concerned with the amount of *solute* in the *solution.*

Concentration ratios that describe small concentrations of solute are sometimes reported as a ratio of *parts per million* (ppm) or *parts per billion* (ppb). This ratio could mean ppm by volume or ppm by weight, depending if the solution is a gas or a liquid. For example, a drinking water sample with 1 ppm Na^+ by weight has 1 weight measure of solute, sodium ions, *in* every 1,000,000 weight measures of the total solution. By way of analogy, 1 ppm expressed in money means 1 cent in every $10,000 (which is one million cents). A concentration of 1 ppb means

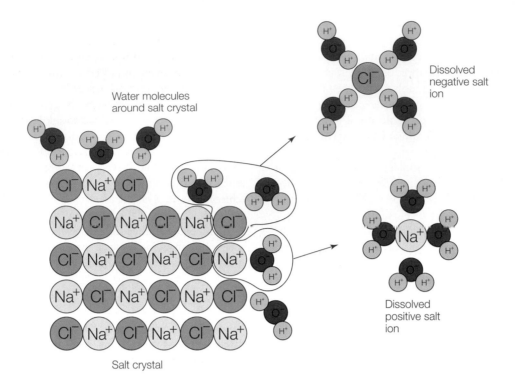

Water molecules around salt crystal

Dissolved negative salt ion

Dissolved positive salt ion

Salt crystal

Figure 10.5

An ionic solid dissolves in water because the number of water molecules around the surface is greater than the number of other ions of the solid. The attraction between polar water molecules and a charged ion enables the water molecules to pull ions away from the crystal so the salt crystals dissolve in the water.

1 cent in $10,000,000. Thus, the concentrations of very dilute solutions, such as certain salts in seawater, minerals in drinking water, and pollutants in water or in the atmosphere are often reported in ppm or ppb.

The concentration term of *percent by volume* is defined as the *volume of solute in 100 volumes of solution*. This concentration term is just like any other percentage ratio, that is, "part" divided by the "whole" times 100 percent. The distinction is that the part and the whole are concerned with a volume of solute and a volume of solution. Knowing the meaning of percent by volume can be useful in consumer decisions. Rubbing alcohol, for example, can be purchased at a wide range of prices. The various brands range from a concentration, according to the labels, of "12% by volume" to "70% by volume." If the volume unit is mL, a "12% by volume" concentration contains 12 mL of pure isopropyl (rubbing) alcohol in every 100 mL of solution. The "70% by volume" contains 70 mL of isopropyl alcohol in every 100 mL of solution. The relationship for percent by volume is

$$\frac{\text{volume solute}}{\text{volume solution}} \times 100\% \text{ solution} = \% \text{ solute}$$

The concentration term of *percent by weight* is defined as the *weight of solute in 100 weight units of solution*. This concentration term is just like any other percentage composition, the difference being that it is concerned with the weight of solute (the part) in a weight of solution (the whole). Hydrogen peroxide, for example, is usually sold in a concentration of "3% by weight." This means that 3 oz (or other weight units) of pure hydrogen peroxide are in 100 oz of solution. Since weight is proportional to mass in a given location, mass units such as grams are sometimes used to calculate a percent by weight. The relationship for percent by weight (using mass units) is

$$\frac{\text{mass of solute}}{\text{mass of solution}} \times 100\% \text{ solution} = \% \text{ solute}$$

Both percent by volume and percent by weight are defined as the volume or weight per 100 units of solution because per-

Connections . . .

Ppm or Ppb to Percent

Sometimes it is useful to know the conversion factors between ppm or ppb and the more familiar percent concentration by weight. These factors are ppm ÷ (1×10^4) = percent concentration and ppb ÷ (1×10^7) = percent concentration. For example, very hard water (water containing Ca^{2+} or Mg^{2+} ions), by definition, contains more than 300 ppm of the ions. This is a percent concentration of 300 ÷ 1×10^4, or 0.03 percent. To be suitable for agricultural purposes, irrigation water must not contain more than 700 ppm of total dissolved salts, which means a concentration no greater than 0.07 percent salts.

Figure 10.6

Salinity is a measure of the amount of salts dissolved in 1 kg of solution. If 1,000 g of seawater were evaporated, 35.0 g of salts would remain as 965.0 g of water leave.

1,000 g
seawater

35 g
of salts

+

965 g
of water

cent *means* parts per hundred. The measure of dissolved salts in seawater is called *salinity*. **Salinity** is defined as the mass of salts dissolved in 1,000 g of solution. As illustrated in figure 10.6, evaporation of 965 g of water from 1,000 g of seawater will leave an average of 35 g salts. Thus, the average salinity of the seawater is 35‰. Note the ‰, which means parts per thousand just as % means parts per hundred. Thus, the average salinity of seawater is 35‰, which means there are 35 g of salts dissolved in every 1,000 g of seawater. The equivalent percent measure for salinity is 3.5%, which equals 35‰.

A **mole** is a measure of amount used in chemistry. One mole is defined as the amount of a substance that contains the same number of elementary units as there are atoms in exactly 12 grams of the carbon-12 isotope. The number of units in this case is called *Avogadro's number,* which is 6.02×10^{23}—a very large number. This measure can be compared with identifying amounts in the grocery store by the dozen. You know that a dozen is 12 of something. Now you know that a *mole* is 6.02×10^{23} of what ever you are measuring.

Chemists use a measure of concentration that is convenient for considering chemical reactions of solutions. The measure is based on moles of solute, since a mole is a known number of particles (atoms, molecules, or ions). The concentration term of **molarity** (M) is defined as the number of moles of solute dissolved in one liter of solution. Thus,

$$\text{Molarity(M)} = \frac{\text{moles of solute}}{\text{liters of solution}}$$

An aqueous solution of NaCl that has a molarity of 1.0 contains one mole of NaCl per liter of solution. To make such a solution you would place 58.5 g (1.0 mole) NaCl in a beaker, then add water to make 1 liter of solution.

Solubility

Gases and liquids appear to be soluble in all proportions, but there is a limit to how much solid can be dissolved in a liquid. You may have noticed that a cup of hot tea will dissolve several teaspoons of sugar, but the limit of solubility is reached quickly in a glass of iced tea. The limit of how much sugar will dissolve seems to depend on the temperature of the tea. More sugar added to the cold tea after the limit is reached will not dissolve, and solid sugar granules begin to accumulate at the bottom of

the glass. At this limit the sugar and tea solution is said to be *saturated.* Dissolving does not actually stop when a solution becomes saturated and undissolved sugar continues to enter the solution. However, dissolved sugar is now returning to the undissolved state at the same rate as it is dissolving. The overall equilibrium condition of sugar dissolving as sugar is coming out of solution is called a **saturated solution.** A saturated solution is a *state of equilibrium that exists between dissolving solute and solute coming out of solution.* You actually cannot see the dissolving and coming out of solution that occurs in a saturated solution because the exchanges are taking place with particles the size of molecules or ions.

Not all compounds dissolve as sugar does, and more or less of a given compound may be required to produce a saturated solution at a particular temperature. In general, the difficulty of dissolving a given compound is referred to as *solubility.* More specifically, the **solubility** of a solute is defined as the *concentration that is reached in a saturated solution at a particular temperature.* Solubility varies with the temperature as the sodium and potassium salt examples show in figure 10.7. These solubility curves describe the amount of solute required to reach the saturation equilibrium at a particular temperature. In general, the solubilities of most ionic solids increase with temperature, but there are exceptions. In addition, some salts release heat when dissolved in water, and other salts absorb heat when dissolved. The "instant cold pack" used for first aid is a bag of water containing a second bag of ammonium nitrate (NH_4NO_3). When the bag of ammonium nitrate is broken, the compound dissolves and absorbs heat.

Figure 10.7

Approximate solubility curves for sodium nitrate, potassium nitrate, potassium chloride, and sodium chloride.

You can usually dissolve more of a solid, such as salt or sugar, as the temperature of the water is increased. Contrary to what you might expect, gases usually become *less* soluble in water as the temperature increases. As a glass of water warms, small bubbles collect on the sides of the glass as dissolved air comes out of solution. The first bubbles that appear when warming a pot of water to boiling are also bubbles of dissolved air coming out of solution. This is why water that has been boiled usually tastes "flat." The dissolved air has been removed by the heating. The "normal" taste of water can be restored by pouring the boiled water back and forth between two glasses. The water dissolves more air during this process, restoring the usual taste.

Changes in pressure have no effect on the solubility of solids in liquids but greatly affect the solubility of gases. The release of bubbles (fizzing) when a bottle or can of soda is opened, occurs because pressure is reduced on the beverage and dissolved carbon dioxide comes out of solution. In general, *gas solubility decreases with temperature and increases with pressure.* As usual, there are exceptions to this generalization.

PROPERTIES OF WATER SOLUTIONS

Pure solvents have characteristic physical and chemical properties that are changed by the addition of a solute. Following are some of the more interesting changes.

Electrolytes

Water solutions of ionic substances will conduct an electric current, so they are called **electrolytes.** Ions must be present and free to move in a solution to carry the charge, so electrolytes are solutions containing ions. Pure water will not conduct an electric current as it is a covalent compound, which ionizes only very slightly. Water solutions of sugar, alcohol, and most other covalent compounds are nonconductors, so they are called *nonelectrolytes.* Nonelectrolytes are covalent compounds that form molecular solutions, so they cannot conduct an electric current.

Some covalent compounds are nonelectrolytes as pure liquids but become electrolytes when dissolved in water. Pure hydrogen chloride (HCl), for example, does not conduct an electric current, so you can assume that it is a molecular substance. When dissolved in water, hydrogen chloride does conduct a current, so it must now contain ions. Evidently, the hydrogen chloride has become *ionized* by the water. The process of forming ions from molecules is called *ionization.* Hydrogen chloride, just as water, has polar molecules. The positive hydrogen atom on the HCl molecule is attracted to the negative oxygen end of a water molecule, and the force of attraction is strong enough to break the hydrogen-chlorine bond, forming charged particles (figure 10.8). The reaction is

$$H_2O_{(1)} + HCl_{(1)} \rightarrow H_3O^+{}_{(aq)} + Cl^-{}_{(aq)}$$

The H_3O^+ ion is called a **hydronium ion.** A hydronium ion is basically a molecule of water with an attached hydrogen ion. The presence of the hydronium ion gives the solution new

Figure 10.8

Three representations of water and hydrogen chloride in an ionizing reaction. (*A*) Sketches of molecules involved in the reaction. (*B*) Electron dot equation of the reaction. (*C*) The chemical equation for the reaction. Each of these representations shows the hydrogen being pulled away from the chlorine atom to form H_3O^+, the hydronium ion.

chemical properties; the solution is no longer hydrogen chloride but is *hydrochloric acid.* Hydrochloric acid, and other acids, will be discussed shortly.

Boiling Point

Boiling occurs when the pressure of the vapor escaping from a liquid (vapor pressure) is equal to the atmospheric pressure on the liquid. The *normal* boiling point is defined as the temperature at which the vapor pressure is equal to the average atmospheric pressure at sea level. For pure water, this temperature is 100°C (212°F). It is important to remember that boiling is a purely physical process. No bonds within water molecules are broken during boiling.

The vapor pressure over a solution is *less* than the vapor pressure over the pure solvent at the same temperature. Molecules of a liquid can escape into the air only at the surface of the liquid, and the presence of molecules of a solute means that fewer solvent molecules can be at the surface to escape. Thus, the vapor pressure over a solution is less than the vapor pressure over a pure solvent (figure 10.9).

Because the vapor pressure over a solution is less than that over the pure solvent, the solution boils at a higher temperature. A higher temperature is required to increase the vapor pressure to that of the atmospheric pressure. Some cooks have been observed to add a "pinch" of salt to a pot of water before boiling. Is this to increase the boiling point, and therefore cook the food more quickly? How much does a pinch of salt increase the boiling temperature? The answers are found in the relationship between the concentration of a solute and the boiling point of the solution.

It is the number of solute particles (ions or molecules) at the surface of a solution that increases the boiling point. A mole is a measure that can be defined as a number of particles called Avogadro's number. Since the number of particles at the surface is proportional to the ratio of particles in the solution, the con-

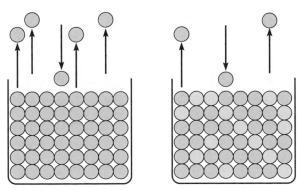

Figure 10.9

The rate of evaporation, and thus the vapor pressure, is less for a solution than for a solvent in the pure state. The greater the solute concentration, the less the vapor pressure.

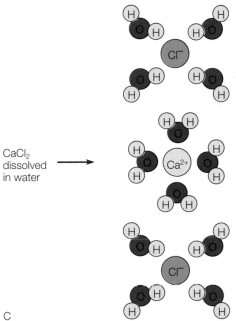

Figure 10.10

Since ionic compounds dissolve by separation of ions, they provide more particles in solution than molecular compounds. (*A*) A mole of sugar provides Avogadro's number of particles. (*B*) A mole of NaCl provides two times Avogadro's number of particles. (*C*) A mole of $CaCl_2$ provides three times Avogadro's number of particles.

centration of the solute will directly influence the increase in the boiling point. In other words, the boiling point of any dilute solution is increased proportional to the concentration of the solute. For water, the boiling point is increased 0.521°C for every mole of solute dissolved in 1,000 g of water. Thus, any water solution will boil at a higher temperature than pure water. Since it boils at a higher temperature, it also takes a longer time to reach the boiling point.

It makes no difference what substance is dissolved in the water; one mole of solute in 1,000 g of water will elevate the boiling point by 0.521°C. A mole contains Avogadro's number of particles, so a mole of any solute will lower the vapor pressure by the same amount. Sucrose, or table sugar, for example, is $C_{12}H_{22}O_{11}$ and has a gram-formula weight of 342 g. Thus 342 g of sugar in 1,000 g of water (about a liter) will increase the boiling point by 0.521°C. Therefore, if you measure the boiling point of a sugar solution you can determine the concentration of sugar in the solution. For example, pancake syrup that boils at 100.261°C (sea-level pressure) must contain 171 g of sugar dissolved in 1,000 g of water. You know this because the increase of 0.261°C over 100°C is one-half of 0.521°C. If the boiling point were increased by 0.521°C over 100°C, the syrup would have the full gram-formula weight (342 g) dissolved in a kg of water.

Since it is the number of particles of solute in a specific sample of water that elevates the boiling point, different effects are observed in dissolved covalent and dissolved ionic compounds (figure 10.10). Sugar is a covalent compound, and the solute is molecules of sugar moving between the water molecules. Sodium chloride, on the other hand, is an ionic compound and dissolves by the separation of ions, or

$$Na^+Cl^-_{(s)} \rightarrow Na^+_{(aq)} + Cl^-_{(aq)}$$

This equation tells you that one mole of NaCl separates into one mole of sodium ions and one mole of chlorine ions for a total of *two* moles of solute. The boiling point elevation of a solution made from one mole of NaCl (58.5 g) is therefore multiplied by two, or $2 \times 0.521°C = 1.04°C$. The boiling point of a solution made by adding 58.5 g of NaCl to 1,000 g of water is therefore 101.04°C at normal sea-level pressure.

Now back to the question of how much a pinch of salt increases the boiling point of a pot of water. Assuming the pot contains about a liter of water (about a quart), and assuming that a "pinch" of salt has a mass of about 0.2 gram, the boiling point will be increased by 0.0037°C. Thus, there must be some reason other than increasing the boiling point that a cook adds a pinch of salt to a pot of boiling water. Perhaps the salt is for seasoning?

Freezing Point

Freezing occurs when the kinetic energy of molecules has been reduced sufficiently so the molecules can come together, forming the crystal structure of the solid. Reduced kinetic energy of the molecules, that is, reduced temperature, results in a specific freezing point for each pure liquid. The *normal* freezing point for pure water, for example, is 0°C (32°F) under normal pressure. The presence of solute particles in a solution interferes with the water molecules as they attempt to form the six-sided hexagonal structure. The water molecules cannot get by the solute particles until the kinetic energy of the solute particles is reduced, that is, until the temperature is below the normal freezing point. Thus, the presence of solute particles lowers the freezing point, and solutions freeze at a lower temperature than the pure solvent.

The freezing-point depression of a solution has a number of interesting implications for solutions such as seawater. When seawater freezes, the water molecules must work their way around the salt particles as was described earlier. Thus, the solute particles are *not* normally included in the hexagonal structure of ice. Ice formed in seawater is practically pure water. Since the solute was *excluded* when the ice formed, the freezing of seawater increases the salinity. Increased salinity means increased concentration, so the freezing point of seawater is further depressed and more ice forms only at a lower temperature. When this additional ice forms, more pure water is removed and the process goes on. Thus, seawater does not have a fixed freezing point but has a lower and lower freezing point as more and more ice freezes.

The depression of the freezing point by a solute has a number of interesting applications in colder climates. Salt, for example, is spread on icy roads to lower the freezing point (and thus the melting point) of the ice. Calcium chloride, $CaCl_2$, is a salt that is often used for this purpose. Water in a car radiator would also freeze in colder climates if a solute, called antifreeze, were not added to the radiator water. Methyl alcohol has been used as an antifreeze because it is soluble in water and does not damage the cooling system. Methyl alcohol, however, has a low boiling point and tends to boil away. Ethylene glycol has a higher boiling point, so it is called a "permanent" antifreeze. Like other solutes, ethylene glycol also raises the boiling point, which is an added benefit for summer driving.

ACIDS, BASES, AND SALTS

The electrolytes known as *acids, bases,* and *salts* are evident in environmental quality, foods, and everyday living. Environmental quality includes the hardness of water, which is determined by the presence of certain salts, the acidity of soils, which determines how well plants grow, and acid rain, which is a by-product of industry and automobiles. Many concerns about air and water pollution are often related to the chemistry concepts of acids, bases, and salts.

Properties of Acids and Bases

Acids and bases are classes of chemical compounds that have certain characteristic properties. These properties can be used to identify if a substance is an acid or a base (tables 10.2 and 10.3). The following are the properties of *acids* dissolved in water:

Table 10.2 Some Common Acids

Name	Formula	Comment
Acetic acid	CH_3COOH	A weak acid found in vinegar
Boric acid	H_3BO_3	A weak acid used in eyedrops
Carbonic acid	H_2CO_3	The weak acid of carbonated beverages
Formic acid	HCOOH	Makes the sting of insects and certain plants
Hydrochloric acid	HCl	Also called muriatic acid; used in swimming pools, soil acidifiers, and stain removers
Lactic acid	$CH_3CHOHCOOH$	Found in sour milk, sauerkraut, and pickles; gives tart taste to yogurt
Nitric acid	HNO_3	A strong acid
Phosphoric acid	H_3PO_4	Used in cleaning solutions; added to carbonated beverages for tartness
Sulfuric acid	H_2SO_4	Also called oil of vitriol; used as battery acid and in swimming pools

Table 10.3 Some Common Bases

Name	Formula	Comment
Sodium hydroxide	NaOH	Also called lye or caustic soda; a strong base used in oven cleaners and drain cleaners
Potassium hydroxide	KOH	Also called caustic potash; a strong base used in drain cleaners
Ammonia	NH_3	A weak base used in household cleaning solutions
Calcium hydroxide	$Ca(OH)_2$	Also called slaked lime; used to make brick mortar
Magnesium hydroxide	$Mg(OH)_2$	Solution is called milk of magnesia; used as antacid and laxative

1. Acids have a sour taste such as the taste of citrus fruits.
2. Acids change the color of certain substances; for example, litmus changes from blue to red when placed in an acid solution (figure 10.11A).
3. Acids react with active metals, such as magnesium or zinc, releasing hydrogen gas.
4. Acids *neutralize* bases, forming water and salts from the reaction.

Likewise, *bases* have their own characteristic properties. Bases are also called alkaline substances, and the following are the properties of bases dissolved in water:

1. Bases have a bitter taste, for example, the taste of caffeine.
2. Bases reverse the color changes that were caused by acids. Red litmus is changed back to blue when placed in a solution containing a base (figure 10.11B).

A

B

Figure 10.11

(A) Acid solutions will change the color of blue litmus to red. (B) Solutions of bases will change the color of red litmus to blue.

caustic potash. These two bases are used in products known as drain cleaners. They open plugged drains because of their caustic action, turning grease, hair, and other organic "plugs" into soap and other soluble substances that are washed away. A weaker base is a solution of ammonia (NH_3), which is often used as a household cleaner. A solution of magnesium hydroxide, $Mg(OH)_2$, has a common name of milk of magnesia and is sold as an antacid and laxative.

Many natural substances change color when mixed with acids or bases. You may have noticed that tea changes color slightly, becoming lighter, when lemon juice (which contains citric acid) is added. Some plants have flowers of one color when grown in acidic soil and flowers of another color when grown in basic soil. A vegetable dye that changes color in the presence of acids or bases can be used as an *acid-base indicator*. An indicator is simply a vegetable dye that is used to distinguish between acid and base solutions by a color change. Litmus, for example, is an acid-base indicator made from a dye extracted from certain species of lichens. The dye is applied to paper strips, which turn red in acidic solutions and blue in basic solutions.

Explaining Acid-Base Properties

Comparing the lists in tables 10.2 and 10.3, you can see that acids and bases appear to be chemical opposites. Notice in table 10.2 that the acids all have an H, or hydrogen atom, in their formulas. In table 10.3, most of the bases have a hydroxide ion, OH^-, in their formulas. Could this be the key to acid-base properties?

The modern concept of an acid considers the properties of acids in terms of the hydronium ion, H_3O^+. As was mentioned earlier, the hydronium ion is a water molecule to which an H^+ ion is attached. Since a hydrogen ion is a hydrogen atom without its single electron, it could be considered as an ion consisting of a single proton. Thus the H^+ ion can be called a *proton*. An **acid** is defined as any substance that is a *proton donor* when dissolved in water, increasing the hydronium ion concentration. For example, hydrogen chloride dissolved in water has the following reaction:

3. Basic solutions feel slippery on the skin. They have a *caustic* action on plant and animal tissue, converting tissue into soluble materials. A strong base, for example, reacts with fat to make soap and glycerine. This accounts for the slippery feeling on the skin.

4. Bases *neutralize* acids, forming water and salts from the reaction.

Tasting an acid or base to see if it is sour or bitter can be hazardous, since some are highly corrosive or caustic. Many organic acids are not as corrosive and occur naturally in foods. Citrus fruit, for example, contains citric acid, vinegar is a solution of acetic acid, and sour milk contains lactic acid. The stings or bites of some insects (bees, wasps, and ants) and some plants (stinging nettles) are painful because an organic acid, formic acid, is injected by the insect or plant. Your stomach contains a solution of hydrochloric acid. In terms of relative strength, the hydrochloric acid in your stomach is about ten times stronger than the carbonic acid (H_2CO_3) of carbonated beverages.

Examples of bases include solutions of sodium hydroxide (NaOH), which has a common name of lye or caustic soda, and potassium hydroxide (KOH), which has a common name of

The dotted circle and arrow were added to show that the hydrogen chloride donated a proton to a water molecule. The resulting solution contains H_3O^+ ions and has acid properties, so the solution is called hydrochloric acid. It is the H_3O^+ ion that is responsible for the properties of an acid.

The bases listed in table 10.3 all appear to have a hydroxide ion, OH^-. Water solutions of these bases do contain OH^- ions, but the definition of a base is much broader. A **base** is defined as any substance that is a *proton acceptor* when dissolved in water, increasing the hydroxide ion concentration. For example, ammonia dissolved in water has the following reaction:

$$NH_{3(g)} + H_2O_{(l)} \rightarrow (NH_4)^+ + OH^-$$

The dotted circle and arrow show that the ammonia molecule accepted a proton from a water molecule, providing a hydroxide ion. The resulting solution contains OH^- ions and has basic properties, so a solution of ammonium hydroxide is a base.

Carbonates, such as sodium carbonate (Na_2CO_3), form basic solutions because the carbonate ion reacts with water to produce hydroxide ions.

$$(CO_3)^{2-}_{(aq)} + H_2O_{(l)} \rightarrow (HCO_3)^-_{(aq)} + OH^-_{(aq)}$$

Thus, sodium carbonate produces a basic solution.

Acids could be thought of as simply solutions of hydronium ions in water, and bases could be considered solutions of hydroxide ions in water. The proton donor and proton acceptor definition is much broader, and it does include the definition of acids and bases as hydronium and hydroxide compounds. The broader, more general definition covers a wider variety of reactions and is therefore more useful.

The modern concept of acids and bases explains why the properties of acids and bases are **neutralized,** or lost, when acids and bases are mixed together. For example, consider the hydronium ion produced in the hydrochloric acid solution and the hydroxide ion produced in the ammonia solution. When these solutions are mixed together, the hydronium ion reacts with the hydroxide ion, and

$$H_3O^+_{(aq)} + OH^+_{(aq)} \rightarrow H_2O_{(l)} + H_2O_{(l)}$$

Thus, a proton is transferred from the hydronium ion (an acid), and the proton is accepted by the hydroxide ion (a base). Water is produced, and both the acid and base properties disappear or are neutralized.

Strong and Weak Acids and Bases

Acids and bases are classified according to their degree of ionization when placed in water. *Strong acids* ionize completely in water, with all molecules dissociating into ions. Nitric acid, for example, reacts completely in the following equation:

$$HNO_{3(aq)} + H_2O_{(l)} \rightarrow H_3O^+_{(aq)} + (NO_3)^-_{(aq)}$$

Nitric acid, hydrochloric acid (figure 10.12), and sulfuric acid are common strong acids.

Figure 10.12

Hydrochloric acid (HCl) has the common name of *muriatic* acid. Hydrochloric acid is a strong acid used in swimming pools, soil acidifiers, and stain removers.

Acids that ionize only partially produce fewer hydronium ions, so they are *weak acids.* Vinegar, for example, contains acetic acid that reacts with water in the following reaction:

$$HC_2H_3O_2 + H_2O \rightarrow H_3O^+ + (C_2H_3O_2)^-$$

Only about 1 percent or less of the acetic acid molecules ionize, depending on the concentration.

Bases are also classified as strong or weak. A *strong base* is completely ionized in solution. Sodium hydroxide, or lye, is the most common example of a strong base. It dissolves in water to form a solution of sodium and hydroxide ions:

$$Na^+OH^-_{(s)} \rightarrow Na^+_{(aq)} + OH^-_{(aq)}$$

A *weak base* is only partially ionized. Ammonia, magnesium hydroxide, and calcium hydroxide are examples of weak bases. Magnesium and calcium hydroxide are only slightly soluble in water, and this reduces the *concentration* of hydroxide ions in a solution.

The pH Scale

The strength of an acid or a base is usually expressed in terms of a range of values called a **pH scale.** The pH scale is based on the concentration of the hydronium ion (in moles/L) in an acidic or a basic solution. To understand how the scale is able to express both acid and base strength in terms of the hydronium ion, first note that pure water is very slightly ionized in the reaction:

$$H_2O_{(1)} + H_2O_{(1)} \rightarrow H_3O^+{}_{(aq)}OH^-{}_{(aq)}$$

The amount of self-ionization by water has been determined through measurements. In pure water at 25°C or any neutral water solution at that temperature, the H_3O^+ concentration is 1×10^{-7} moles/L, and the OH^- concentration is also 1×10^{-7} moles/L. Since both ions are produced in equal numbers, then the H_3O^+ concentration equals the OH^- concentration, and pure water is neutral, neither acidic nor basic.

In general, adding an acid substance to pure water increases the H_3O^+ concentration. Adding a base substance to pure water increases the OH^- concentration. Adding a base also *reduces* the H_3O^+ concentration as the additional OH^- ions are able to combine with more of the hydronium ions to produce un-ionized water. Thus, at a given temperature, an increase in OH^- concentration is matched by a *decrease* in H_3O^+ concentration. The concentration of the hydronium ion can be used as a measure of acidic, neutral, and basic solutions. In general, (1) acidic solutions have H_3O^+ concentrations above 1×10^{-7} moles/L, (2) neutral solutions have H_3O^+ concentrations equal to 1×10^{-7} moles/L, and (3) basic solutions have H_3O^+ concentrations less than 1×10^{-7} moles/L. These three statements lead directly to the pH scale, which is named from the French *pouvoir hydrogene,* meaning "hydrogen power." Power refers to the exponent of the hydronium ion concentration, and the pH is a *power of ten notation that expresses the H_3O^+ concentration.*

A neutral solution has a pH of 7.0. Acidic solutions have pH values below 7, and smaller numbers mean greater acidic properties. Increasing the OH^- concentration decreases the H_3O^+ concentration, so the strength of a base is indicated on the same scale with values greater than 7. Note that the pH scale is logarithmic, so a pH of 2 is ten times as acidic as a pH of 3. Likewise, a pH of 10 is one hundred times as basic as a pH of 8. Figure 10.13 is a diagram of the pH scale, and table 10.4 compares the pH of some common substances (figure 10.14).

Properties of Salts

Salt is produced by a neutralization reaction between an acid and a base. A **salt** is defined as any ionic compound except those with hydroxide or oxide ions. Table salt, NaCl, is but one example of this large group of ionic compounds. As an example of a salt produced by a neutralization reaction, consider the reaction of HCl (an acid in solution) with $Ca(OH)_2$ (a base in solution). The reaction is

$$2HCl_{(aq)} + Ca(OH)_{2(aq)} \rightarrow CaCl_{2(aq)} + 2H_2O_{(1)}$$

This is an ionic exchange reaction that forms molecular water, leaving Ca^{2+} and Cl^- in solution. As the water is evaporated, these ions begin forming ionic crystal structures as the solution concentration increases. When the water is all evaporated, the white crystalline salt of $CaCl_2$ remains.

H_3O^+ Concentration (moles/liters)	pH	Meaning
1×10^{-0} (=1)	0	
1×10^{-1}	1	
1×10^{-2}	2	
1×10^{-3}	3	Increasing acidity
1×10^{-4}	4	
1×10^{-5}	5	
1×10^{-6}	6	
1×10^{-7}	7	Neutral
1×10^{-8}	8	
1×10^{-9}	9	
1×10^{-10}	10	Increasing basicity
1×10^{-11}	11	
1×10^{-12}	12	
1×10^{-13}	13	
1×10^{-14}	14	

Figure 10.13
The pH scale.

Table 10.4 The Approximate pH of Some Common Substances

Substance	pH (or pH Range)
Hydrochloric acid (4%)	0
Gastric (stomach) solution	1.6–1.8
Lemon juice	2.2–2.4
Vinegar	2.4–3.4
Carbonated soft drinks	2.0–4.0
Grapefruit	3.0–3.2
Oranges	3.2–3.6
Acid rain	4.0–5.5
Tomatoes	4.2–4.4
Potatoes	5.7–5.8
Natural rainwater	5.6–6.2
Milk	6.3–6.7
Pure water	7.0
Seawater	7.0–8.3
Blood	7.4
Sodium bicarbonate solution	8.4
Milk of magnesia	10.5
Ammonia cleaning solution	11.9
Sodium hydroxide solution	13.0

Figure 10.14

The pH increases as the acidic strength of these substances decreases from left to right. Did you know that lemon juice is more acidic than vinegar? That a soft drink is more acidic than orange juice or grapefruit juice?

If sodium hydroxide had been used as the base instead of calcium hydroxide, a different salt would have been produced:

$$HCL_{(aq)} + NaOH_{(aq)} \rightarrow NaCl_{(aq)} + H_2O_{(l)}$$

Salts are also produced when elements combine directly, when an acid reacts with a metal, and by other reactions.

Salts are essential in the diet both as electrolytes and as a source of certain elements, usually called *minerals* in this context. Plants must have certain elements that are derived from water-soluble salts. Potassium, nitrates, and phosphate salts are often used to supply the needed elements. There is no scientific evidence that plants prefer to obtain these elements from natural sources, as compost, or from chemical fertilizers. After all, a nitrate ion is a nitrate ion, no matter what its source. Table 10.5 lists some common salts and their uses.

Hard and Soft Water

Salts vary in their solubility in water, and a solubility chart appears in appendix B. Table 10.6 lists some generalizations concerning the various common salts. Some of the salts are dissolved by water that will eventually be used for domestic supply. When the salts are soluble calcium or magnesium compounds, the water will contain calcium or magnesium ions in solution. A solution of Ca^{2+} or Mg^{2+} ions is said to be *hard water* because it is hard to make soap lather in the water. "Soft" water, on the other hand, makes a soap lather easily. The difficulty occurs because soap is a sodium or potassium compound that is soluble in water. The calcium or magnesium ions, when present, replace the sodium or potassium ions in the soap compound, forming an insoluble compound. It is this insoluble compound that forms a "bathtub ring" and also collects on clothes being washed, preventing cleansing.

The key to "softening" hard water is to remove the troublesome calcium and magnesium ions (figure 10.15). If the hard-

Table 10.5 Some Common Salts and Their Uses

Common Name	Formula	Use
Alum	$KAl(SO_4)_2$	Medicine, canning, baking powder
Baking soda	$NaHCO_3$	Fire extinguisher, antacid, deodorizer, baking powder
Bleaching powder (chlorine tablets)	$CaOCl_2$	Bleaching, deodorizer, disinfectant in swimming pools
Borax	$Na_2B_4O_7$	Water softener
Chalk	$CaCO_3$	Antacid tablets, scouring powder
Chile saltpeter	$NaNO_3$	Fertilizer
Cobalt chloride	$CoCl_2$	Hygrometer (pink in damp weather, blue in dry weather)
Epsom salt	$MgSO_4 \cdot 7\,H_2O$	Laxative
Fluorspar	CaF_2	Metallurgy flux
Gypsum	$CaSO_4 \cdot 2\,H_2O$	Plaster of Paris, soil conditioner
Lunar caustic	$AgNO_3$	Germicide and cauterizing agent
Niter (or saltpeter)	KNO_3	Meat preservative, makes black gunpowder (75 parts KNO_3, 15 of carbon, 10 of sulfur)
Potash	K_2CO_3	Makes soap, glass
Rochelle salt	$KNaC_4H_4O_6$	Baking powder ingredient
TSP	Na_3PO_4	Water softener, fertilizer

Table 10.6 Generalizations About Salt Solubilities

Salts	Solubility	Exceptions
Sodium Potassium Ammonium	Soluble	None
Nitrate Acetate Chlorate	Soluble	None
Chlorides	Soluble	Ag and Hg (I) are insoluble
Sulfates	Soluble	Ba, Sr, and Pb are insoluble
Carbonates Phosphates Silicates	Insoluble	Na, K, and NH_4 are soluble
Sulfides	Insoluble	Na, K, and NH_4 are soluble: Mg, Ca, Sr, and Ba decompose

are accounted for by the modern concepts of each. *Acids* are *proton donors* that form *hydronium ions* (H_3O^+) in water solutions. *Bases* are *proton acceptors* that form *hydroxide ions* (OH^-) in water solutions. The strength of an acid or base is measured on the *pH scale,* a power of ten notation of the hydronium ion concentration. On the scale, numbers from 0 up to 7 are acids, 7 is neutral, and numbers above 7 and up to 14 are bases. Each unit represents a tenfold increase or decrease in acid or base properties.

A *salt* is any ionic compound except those with hydroxide or oxide ions. Salts provide plants and animals with essential elements. The solubility of salts varies with the ions that make up the compound. Solutions of magnesium or calcium produce *hard water,* water in which it is hard to make soap lather. Hard water is softened by removing the magnesium and calcium ions.

KEY TERMS

acid (p. **205**)
base (p. **206**)
electrolytes (p. **202**)
hydration (p. **199**)
hydrogen bonding (p. **197**)
hydronium ion (p. **202**)
miscible fluids (p. **199**)
molarity (p. **201**)
mole (p. **201**)

neutralized (p. **206**)
pH scale (p. **206**)
polar molecule (p. **197**)
salinity (p. **201**)
salt (p. **207**)
saturated solution (p. **201**)
solubility (p. **201**)
solution (p. **198**)

APPLYING THE CONCEPTS

1. Which of the following is *not* a solution?
   (a) seawater
   (b) carbonated water
   (c) sand
   (d) brass

2. Atmospheric air is a homogeneous mixture of gases that is mostly nitrogen gas. The nitrogen is therefore (the)
   (a) solvent.
   (b) solution.
   (c) solute.
   (d) none of the above.

3. A concentration of 500 ppm is reported in a news article. This is the same concentration as
   (a) 0.005%.
   (b) 0.05%.
   (c) 5%.
   (d) 50%.

4. According to the label, a bottle of vodka has a 40% by volume concentration. This means the vodka contains 40 mL of pure alcohol
   (a) in each 140 mL of vodka.
   (b) to every 100 mL of water.
   (c) to every 60 mL of vodka.
   (d) mixed with water to make 100 mL vodka.

5. A bottle of vinegar is 4% by weight, so you know that the solution contains 4 weight units of pure vinegar with
   (a) 96 weight units of water.
   (b) 100 weight units of water.
   (c) 104 weight units of water.

6. If a salt solution has a salinity of 40‰, what is the equivalent percentage measure?
   (a) 400%
   (b) 40%
   (c) 4%
   (d) 0.4%

7. Water has the greatest density at what temperature?
   (a) 100°C
   (b) 20°C
   (c) 4°C
   (d) 0°C

8. A solid salt is insoluble in water so the strongest force must be the
   (a) ion-water molecule force.
   (b) ion-ion force.
   (c) force of hydration.
   (d) polar molecule force.

9. The ice that forms in freezing seawater is
   (a) pure water.
   (b) the same salinity as liquid seawater.
   (c) more salty than liquid seawater.
   (d) more dense than liquid seawater.

10. Which of the following would have a pH of *less* than 7?
    (a) a solution of ammonia
    (b) a solution of sodium chloride
    (c) pure water
    (d) carbonic acid

11. Which of the following would have a pH of *more* than 7?
    (a) a solution of ammonia
    (b) a solution of sodium chloride
    (c) pure water
    (d) carbonic acid

12. Substance A has a pH of 2 and substance B has a pH of 3. This means that
    (a) substance A has more basic properties than substance B.
    (b) substance B has more acidic properties than substance A.
    (c) substance A is ten times more acidic than substance B.
    (d) substance B is ten times more acidic than substance A.

Answers
1. c 2. a 3. b 4. d 5. a 6. c 7. c 8. b 9. a 10. d 11. a 12. c

QUESTIONS FOR THOUGHT

1. How is a solution different from other mixtures?
2. Explain why some ionic compounds are soluble while others are insoluble in water.
3. Explain why adding salt to water increases the boiling point.
4. A deep lake in Minnesota is covered with ice. What is the water temperature at the bottom of the lake? Explain your reasoning.
5. Explain why water has a greater density at 4°C than at 0°C.
6. What is hard water? How is it softened?
7. According to the definition of an acid and the definition of a base, would the pH increase, decrease, or remain the same when NaCl is added to pure water? Explain.
8. What is a hydrogen bond? Explain how a hydrogen bond forms.
9. What feature of a soap molecule gives it cleaning ability?
10. What ion is responsible for (a) acidic properties? (b) for basic properties?

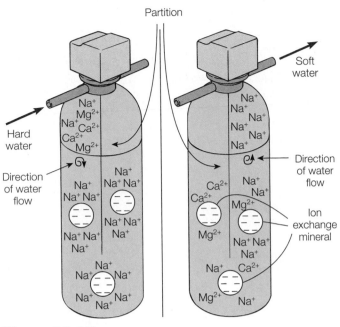

Figure 10.15
A water softener exchanges sodium ions for the calcium and magnesium ions of hard water. Thus, the water is now soft, but it contains sodium ions.

ness is caused by magnesium or calcium *bicarbonates,* the removal is accomplished by simply heating the water. Upon heating, they decompose, forming an insoluble compound that effectively removes the ions from solution. The decomposition reaction for calcium bicarbonate is

$$Ca^{2+}(HCO_3)_{2(aq)} \rightarrow CaCO_{3(s)} + H_2O_{(1)} + CO_2\uparrow$$

The reaction is the same for magnesium bicarbonate. As the solubility chart in appendix B shows, magnesium and calcium car-

bonates are insoluble, so the ions are removed from solution in the solid that is formed. Perhaps you have noticed such a white compound forming around faucets if you live where bicarbonates are a problem. Commercial products to remove such deposits usually contain an acid, which reacts with the carbonate to make a new, soluble salt that can be washed away.

Water hardness is also caused by magnesium or calcium *sulfate,* which requires a different removal method. Certain chemicals such as sodium carbonate (washing soda), trisodium phosphate (TSP), and borax will react with the troublesome ions, forming an insoluble solid that removes them from solution. For example, washing soda and calcium sulfate react as follows:

$$Na_2CO_{3(aq)} + CaSO_{4(aq)} \rightarrow Na_2SO_{4(aq)} + CaCO_3\downarrow$$

Calcium carbonate is insoluble; thus the calcium ions are removed from solution before they can react with the soap. Many laundry detergents have Na_2CO_3, TSP, or borax $(Na_2B_4O_7)$ added to soften the water. TSP causes other problems, however, as the additional phosphates in the waste water can act as a fertilizer, stimulating the growth of algae to such an extent that other organisms in the water die.

A water softener unit is an ion exchanger. The unit contains a mineral that exchanges sodium ions for calcium and magnesium ions as water is run through it. The softener is regenerated periodically by flushing with a concentrated sodium chloride solution. The sodium ions replace the calcium and magnesium ions, which are carried away in the rinse water. The softener is then ready for use again. The frequency of renewal cycles depends on the water hardness, and each cycle can consume from four to twenty pounds of sodium chloride per renewal cycle. In general, water with less than 75 ppm calcium and magnesium ions is called soft water; with greater concentrations, it is called hard water. The greater the concentration above 75 ppm, the harder the water.

A Closer Look   Acid Rain

Acid rain is a general term used to describe any acidic substances, wet or dry, that fall from the atmosphere. Wet acidic deposition could be in the form of rain, but snow, sleet, and fog could also be involved. Dry acidic deposition could include gases, dust, or any solid particles that settle out of the atmosphere to produce an acid condition.

Pure, unpolluted rain is naturally acidic. Carbon dioxide in the atmosphere is absorbed by rainfall, forming carbonic acid (H_2CO_3). Carbonic acid lowers the pH of pure rainfall to a range of 5.6 to 6.2. Decaying vegetation in local areas can provide more CO_2, making the pH even lower. A

pH range of 4.5 to 5.0, for example, has been measured in remote areas of the Amazon jungle. Human-produced exhaust emissions of sulfur and nitrogen oxides can lower the pH of rainfall even more, to a 4.0 to 5.5 range. This is the pH range of acid rain.

The sulfur and nitrogen oxides that produce acid rain come from exhaust emissions of industries and electric utilities that burn coal and from the exhaust of cars, trucks, and buses (box figure 10.1). The emissions are sometimes called "SO_x" and "NO_x," which is read "socks" and "knox." The x subscript implies the variable presence of any or

all of the oxides, for example, nitrogen monoxide (NO), nitrogen dioxide (NO_2), and dinitrogen tetroxide (N_2O_4) for NO_x.

SO_x and NO_x are the raw materials of acid rain and are not themselves acidic. They react with other atmospheric chemicals to form sulfates and nitrates, which combine with water vapor to form sulfuric acid (H_2SO_4) and nitric acid (HNO_3). These are the chemicals of concern in acid rain.

Many variables influence how much and how far SO_x and NO_x are carried in the atmosphere and if they are converted to acid rain or

—Continued top of next page

Continued—

Box Figure 10.1
Natural rainwater has a pH of 5.6 to 6.2. Exhaust emissions of sulfur and nitrogen oxides can lower the pH of rainfall to a range of 4.0 to 5.5. The exhaust emissions come from industries, electric utilities, and automobiles. Not all emissions are as visible as those pictured in this illustration.

simply return to the surface as a dry gas or particles. During the 1960s and 1970s, concerns about local levels of pollution led to the replacement of short smokestacks of about 60 m (about 200 ft) with taller smokestacks of about 200 m (about 650 ft). This reduced the local levels of pollution by dumping the exhaust higher in the atmosphere where winds could carry it away. It also set the stage for longer-range transport of SO$_x$ and NO$_x$ and their eventual conversion into acids.

There are two main reaction pathways by which SO$_x$ and NO$_x$ are converted to acids: (1) reactions in the gas phase and (2) reactions in the liquid phase, such as in water droplets in clouds and fog. In the gas phase, SO$_x$ and NO$_x$ are oxidized to acids, mainly by hydroxyl ions and ozone, and the acid is absorbed by cloud droplets and precipitated as rain or snow. Most of the nitric acid in acid rain and about one-fourth of the sulfuric acid is formed in gas-phase reactions. Most of the liquid-phase reactions that produce sulfuric acid involve the absorbed SO$_x$ and hydrogen peroxide (H$_2$O$_2$), ozone, oxygen, and particles of carbon, iron oxide, and manganese oxide particles. These particles also come from the exhaust of fossil fuel combustion.

Acid rain falls on land, bodies of water, forests, crops, buildings, and people. Concerns about acid rain center on its environmental impact on lakes, forests, crops, materials, and hu-

man health. Lakes in different parts of the world, for example, have been increasing in acidity over the past fifty years. Lakes in northern New England, the Adirondacks, and parts of Canada now have a pH of less than 5.0, and correlations have been established between lake acidity and decreased fish populations. Trees, mostly conifers, are dying at unusually rapid rates in the northeastern United States. Red spruce in Vermont's Green Mountains and the mountains of New York and New Hampshire have been affected by acid rain as have pines in New Jersey's Pine Barrens. It is believed that acid rain leaches essential nutrients, such as calcium, from the soil and also mobilizes aluminum ions. The aluminum ions disrupt the water equilibrium of fine root hairs, and when the root hairs die, so do the trees.

Human-produced emissions of sulfur and nitrogen oxides from burning fossil fuels are the cause of acid rain. The heavily industrialized northeastern part of the United States, from the Midwest through New England, releases sulfur and nitrogen emissions that result in a precipitation pH of 4.0 to 4.5. This region is the geographic center of the nation's acid rain problem. The solution to the problem is found in (1) using fuels other than fossil fuels and (2) reducing the thousands of tons of SO$_x$ and NO$_x$ that are dumped into the atmosphere per day when fossil fuels are used.

SUMMARY

A water molecule consists of two hydrogen atoms and an oxygen atom with bonding and electron pairs in a tetrahedral arrangement. This results in a *bent molecular arrangement,* with 105° between the hydrogen atoms. Electrons spend more time around the oxygen, producing a *polar molecule,* with centers of negative and positive charge. Polar water molecules interact. The force of attraction is called a *hydrogen bond.* The hydrogen bond accounts for the decreased density of ice, the high heat of fusion, and the high heat of vaporization of water. The hydrogen bond is also involved in the *dissolving* process.

A *solution* is a homogeneous mixture of ions or molecules of two or more substances. The substance present in the large amount is the *solvent,* and the *solute* is dissolved in the solvent. If one of the components is a liquid, however, it is called the solvent.

Fluids that mix in any proportion are called *miscible fluids,* and *immiscible fluids* do not mix. Polar substances dissolve in polar solvents, but not nonpolar solvents, and the general rule is *like dissolves like.* Thus oil, a nonpolar substance, is immiscible in water, a polar substance.

The relative amount of solute in a solvent is called the *concentration* of a solution. Concentrations are measured (1) in *parts per million*

(ppm) or *parts per billion* (ppb), (2) *percent by volume,* the volume of a solute per 100 volumes of solution, (3) *percent by weight,* the weight of solute per 100 weight units of solution, and (4) *salinity,* the mass of salts in 1 kg of solution.

A limit to dissolving solids in a liquid occurs when the solution is *saturated.* A *saturated solution* is one with equilibrium between solute dissolving and solute coming out of solution. The *solubility* of a solid is the concentration of a saturated solution at a particular temperature.

Water solutions that carry an electric current are called *electrolytes,* and nonconductors are called *nonelectrolytes.* In general, ionic substances make electrolyte solutions, and molecular substances make nonelectrolyte solutions. Polar molecular substances may be *ionized* by polar water molecules, however, making an electrolyte from a molecular solution.

The *boiling point of a solution* is greater than the boiling point of the pure solvent, and the increase depends only on the concentration of the solute (at a constant pressure). For water, the boiling point is increased 0.521°C for each mole of solute in each kg of water. The *freezing point of a solution* is lower than the freezing point of the pure solvent, and the depression also depends on the concentration of the solute.

Acids, bases, and salts are chemicals that form ionic solutions in water, and each can be identified by simple properties. These properties

CHAPTER ELEVEN

Nuclear Reactions

With the top half of the steel vessel and control rods removed, fuel rod bundles can be replaced in the water-flooded nuclear reactor.

OUTLINE

The ancient alchemist dreamed of changing one element into another, such as lead into gold. The alchemist was never successful, however, because such changes were attempted with chemical reactions. Chemical reactions are reactions that involve only the electrons of atoms. Electrons are shared or transferred in chemical reactions, and the internal nucleus of the atom is unchanged. Elements thus retain their identity during the sharing or transferring of electrons. This chapter is concerned with a different kind of reaction, one that involves the *nucleus* of the atom. In nuclear reactions, the nucleus of the atom is often altered, changing the identity of the elements involved. The ancient alchemist's dream of changing one element into another was actually a dream of achieving a nuclear change, that is, a nuclear reaction.

Understanding nuclear reactions is important because although fossil fuels are the major source of energy today, there are growing concerns about (1) air pollution from fossil fuel combustion, (2) increasing levels of CO_2 from fossil fuel combustion, which may be warming the earth (the greenhouse effect), and (3) the dwindling fossil fuel supply itself, which cannot last forever. Energy experts see nuclear energy as a means of meeting rising energy demands in an environmentally acceptable way. However, the topic of nuclear energy is controversial, and discussions of it often result in strong emotional responses. Decisions about the use of nuclear energy require some understandings about nuclear reactions and some facts about radioactivity and radioactive materials (figure 11.1). These understandings and facts are the topics of this chapter.

NATURAL RADIOACTIVITY

Natural **radioactivity** is the spontaneous emission of particles or energy from an atomic nucleus as it disintegrates. It was discovered in 1896 by Henri Becquerel, a French scientist who was very interested in the recent discovery of X rays. Becquerel was experimenting with fluorescent minerals, minerals that give off visible light after being exposed to sunlight. He wondered if fluorescent minerals emitted X rays in addition to visible light. From previous work with X rays, Becquerel knew that they would penetrate a wrapped, light-tight photographic plate, exposing it as visible light exposes an unprotected plate. Thus, Becquerel decided to place a fluorescent uranium mineral on a protected photographic plate while the mineral was exposed to sunlight. Sure enough, he found a silhouette of the mineral on the plate when it was developed. Believing the uranium mineral emitted X rays, he continued his studies until the weather turned cloudy. Storing a wrapped, protected photographic plate and the uranium mineral together during the cloudy weather, Becquerel returned to the materials later and developed the photographic plate to again find an image of the mineral (figure 11.2). He concluded that the mineral was emitting an "invisible radiation" that was not induced by sunlight. Becquerel named the emission of invisible radiation *radioactivity*. Materials that have the property of radioactivity are called *radioactive* materials.

Becquerel's discovery led to the beginnings of the modern atomic theory and to the discovery of new elements. Ernest Rutherford studied the nature of radioactivity and found that there are three kinds, which are today known by the first three letters of the Greek alphabet—alpha (α), beta (β), and gamma (γ). These Greek letters were used at first before the nature of the radiation was known. Today, an **alpha particle** (sometimes called an alpha ray) is known to be the nucleus of a helium atom, that is, two protons and two neutrons. A **beta particle** (or beta ray) is a high-energy electron. A **gamma ray** is electromagnetic radiation, as is light, but of very short wavelength (figure 11.3).

It was Rutherford's work with alpha particles that resulted in the discovery of the nucleus and the proton (see chapter 8). At Becquerel's suggestion, Madame Marie Curie searched for other radioactive materials and, in the process, discovered two

Figure 11.1

Decisions about nuclear energy require some understanding of nuclear reactions and the nature of radioactivity. This is one of the three units of the Palo Verde Nuclear Generating Station in Arizona. With all three units running, enough power is generated to meet the electrical needs of nearly 4 million people.

new elements, polonium and radium. More radioactive elements have been discovered since that time, and, in fact, all the isotopes of all the elements with an atomic number greater than 83 (bismuth) are radioactive. As a result of radioactive disintegration, the nucleus of an atom often undergoes a change of identity, becoming a simpler nucleus. The spontaneous disintegration of a given nucleus is a purely natural process and cannot be controlled or influenced. The natural spontaneous disinte-

A

B

Figure 11.2

Radioactivity was discovered by Henri Becquerel when he exposed a light-tight photographic plate to a radioactive mineral, then developed the plate. (*A*) A photographic film is exposed to a uraninite ore sample. (*B*) The film, developed normally after a four-day exposure to uranite. Becquerel found an image like this one and deduced that the mineral gave off invisible radiation that he called radioactivity.

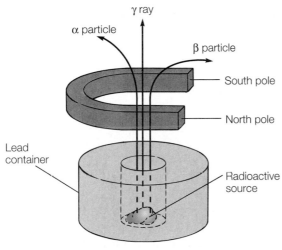

Figure 11.3

Radiation passing through a magnetic field shows that massive, positively charged alpha particles are deflected one way, and less massive beta particles with their negative charge are greatly deflected in the opposite direction. Gamma rays, like light, are not deflected.

gration or decomposition of a nucleus is also called **radioactive decay.** Although it is impossible to know *when* a given nucleus will undergo radioactive decay, as you will see later, it is possible to deal with the *rate* of decay for a given radioactive material with precision.

Nuclear Equations

There are two main subatomic particles in the nucleus, the proton and the neutron. The proton and neutron are called **nucleons.** Recall that the number of protons, the *atomic number,*

determines what element an atom is, and that all atoms of a given element have the same number of protons. The number of neutrons varies in *isotopes,* which are atoms with the same atomic number but different numbers of neutrons. The number of protons and neutrons together determines the *mass number,* so different isotopes of the same element are identified with their mass numbers. Thus, the two most common, naturally occurring isotopes of uranium are referred to as uranium-238 and uranium-235, and the 238 and 235 are the mass numbers of these isotopes. Isotopes are also represented by the following symbol:

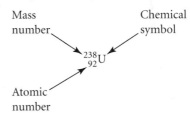

Subatomic particles involved in nuclear reactions are represented by symbols with the following form:

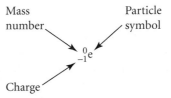

Symbols are used in an equation for a nuclear reaction that is written much like a chemical reaction with reactants and products. When a uranium-238 nucleus emits an alpha particle ($^4_2$He), for example, it loses two protons and two neutrons. The nuclear reaction is written in equation form as

$$^{238}_{92}\text{U} \rightarrow ^{234}_{90}\text{Th} + ^4_2\text{He}$$

The *products* of this nuclear reaction from the decay of a uranium-238 nucleus are (1) the alpha particle ($^4_2$He) given off and (2) the nucleus, which remains after the alpha particle leaves the original nucleus. What remains is easily determined since all nuclear equations must show conservation of charge and conservation of the total number of nucleons. Therefore, (1) the number of protons (positive charge) remains the same, and the sum of the subscripts (atomic number, or numbers of protons) in the reactants must equal the sum of the subscripts in the products; and (2) the total number of nucleons remains the same, and the sum of the superscripts (atomic mass, or number of protons plus neutrons) in the reactants must equal the sum of the superscripts in the products. The new nucleus remaining after the emission of an alpha particle, therefore, has an atomic number of 90 ($92 - 2 = 90$). According to the table of atomic numbers on the inside back cover of this text, this new nucleus is thorium (Th). The mass of the thorium isotope is 238 minus 4, or 234. The emission of an alpha particle thus decreases the number of protons by 2 and the mass number by 4. From the subscripts, you can see that the total charge is conserved ($92 = 90 + 2$). From the superscripts, you can see that the total number of nucleons is also conserved ($238 = 234 + 4$). The mass numbers (superscripts) and the atomic numbers (subscripts) are *balanced* in a correctly written nuclear equation. Such nuclear equations are considered to be independent of any chemical form or chemical reaction. Nuclear reactions are independent and separate from chemical reactions, whether or not the atom is in the pure element or in a compound. Each particle that is involved in nuclear reactions has its own symbol with a superscript indicating mass number and a subscript indicating the charge. These symbols, names, and numbers are given in table 11.1.

The Nature of the Nucleus

The modern atomic theory does not picture the nucleus as a group of stationary protons and neutrons clumped together by some "nuclear glue." The protons and neutrons are understood to be held together by a *nuclear force*, a strong fundamental force of attraction that is functional only at very short distances, on the order of 10^{-15} m or less. At distances greater than about 10^{-15} m the nuclear force is negligible, and the weaker *electromagnetic force*, the force of repulsion between like charges, is the operational force. Thus, like-charged protons experience a re-

CONCEPTS APPLIED
Plutonium Decay Equation

A plutonium-242 nucleus undergoes radioactive decay, emitting an alpha particle. Write the nuclear equation for this nuclear reaction.

Step 1: The table of atomic weights on the inside back cover gives the atomic number of plutonium as 94. Plutonium-242 therefore has a symbol of $^{242}_{94}$Pu. The symbol for an alpha particle is ($^4_2$He), so the nuclear equation so far is

$$^{242}_{94}\text{Pu} \rightarrow {}^4_2\text{He} + ?$$

Step 2: From the subscripts, you can see that $94 = 2 + 92$, so the new nucleus has an atomic number of 92. The table of atomic weights identifies element 92 as uranium with a symbol of U.

Step 3: From the superscripts, you can see that the mass number of the uranium isotope formed is $242 - 4 = 238$, so the product nucleus is $^{238}_{92}$U and the complete nuclear equation is

$$^{242}_{94}\text{Pu} \rightarrow {}^4_2\text{He} + {}^{238}_{92}\text{U}$$

Step 4: Checking the subscripts ($94 = 2 + 92$) and the superscripts ($242 = 4 + 238$), you can see that the nuclear equation is balanced.

What is the product nucleus formed when radium emits an alpha particle? (Answer: Radon-222, a chemically inert, radioactive gas.)

pulsive force when they are farther apart than about 10^{-15} m. When closer together than 10^{-15} m, the short-range, stronger nuclear force predominates, and the protons experience a strong attractive force. This explains why the like-charged protons of the nucleus are not repelled by their like electric charges.

Observations of radioactive decay reactants and products and experiments with nuclear stability have led to a *shell model of the nucleus.* This model considers the protons and neutrons moving in energy levels in the nucleus analogous to the orbital structure of electrons in the outermost part of the atom. As in the electron orbitals, there are certain configurations of nuclear shells that have a greater stability than others. Considering electrons, filled and half-filled orbitals are more stable than other arrangements, and maximum stability occurs with the noble gases and their 2, 10, 18, 36, 54, and 86 electrons. Considering the nucleus, atoms with 2, 8, 20, 28, 50, 82, or 126 protons or neutrons have a maximum nuclear stability. The stable numbers are not the same for electrons and nucleons because of differences in nuclear and electromagnetic forces.

Isotopes of uranium, radium, and plutonium, as well as other isotopes, emit an alpha particle during radioactive decay to a simpler nucleus. The alpha particle is a helium nucleus, ($^4_2$He). The alpha particle contains two protons as well as two neutrons, which is one of the nucleon numbers of stability, so you would expect the helium nucleus (or alpha particle) to have a stable nucleus, and it does. *Stable* means it does not undergo radioactive decay. Pairs of protons and pairs of

Table 11.1	Names, Symbols, and Properties of Particles in Nuclear Equations		
Name	**Symbol**	**Mass Number**	**Charge**
Proton	$^1_1$H (or $^1_1$p)	1	1+
Electron	$^0_{-1}$e (or $^0_{-1}\beta$)	0	1−
Neutron	$^1_0$n	1	0
Gamma photon	$^0_0\gamma$	0	0

neutrons have increased stability, just as pairs of electrons in a molecule do. As a result, nuclei with an *even number* of both protons and neutrons are, in general, more stable than nuclei with odd numbers of protons and neutrons. There are a little more than 150 stable isotopes with an even number of protons and an even number of neutrons, but there are only 5 stable isotopes with odd numbers of each. Just as in the case of electrons, there are other factors that come into play as the nucleus becomes larger and larger with increased numbers of nucleons.

The results of some of these factors are shown in figure 11.4, which is a graph of the number of neutrons versus the number of protons in nuclei. As the number of protons increases, the neutron-to-proton ratio of the *stable nuclei* also increases in a **band of stability.** Within the band the neutron-to-proton ratio increases from about 1:1 at the bottom left to about 1½:1 at the top right. The increased ratio of neutrons is needed to produce a stable nucleus as the number of protons increases. Neutrons provide additional attractive *nuclear* (not electrical) forces, which counter the increased electrical repulsion from a larger number of positively charged protons. Thus, more neutrons are required in larger nuclei to produce a stable nucleus. However, there is a limit to the additional attractive forces that can be provided by more and more neutrons, and all isotopes of all elements with more than 83 protons are unstable and thus undergo radioactive decay.

The generalizations about nuclear stability provide a means of predicting if a particular nucleus is radioactive. The generalizations are as follows:

1. All isotopes with an atomic number greater than 83 have an unstable nucleus.
2. Isotopes that contain 2, 8, 20, 28, 50, 82, or 126 protons or neutrons in their nucleus occur in more stable isotopes than those with other numbers of protons or neutrons.
3. Pairs of protons and pairs of neutrons have increased stability, so isotopes that have nuclei with even numbers of both protons and neutrons are generally more stable than nuclei with odd numbers of both protons and neutrons.
4. Isotopes with an atomic number less than 83 are stable when the ratio of neutrons to protons in the nucleus is about 1:1 in isotopes with up to 20 protons, but the ratio increases in larger nuclei in a band of stability (see figure 11.4). Isotopes with a ratio to the left or right of this band are unstable and thus will undergo radioactive decay.

Types of Radioactive Decay

Through the process of radioactive decay, an unstable nucleus becomes a more stable one with less energy. The three more familiar types of radiation emitted—alpha, beta, and gamma—were introduced earlier.

1. **Alpha emission.** Alpha (α) emission is the expulsion of an alpha particle (^4_2He) from an unstable, disintegrating nucleus. The alpha particle, a helium nucleus, travels from 2 to 12 cm through the air, depending on the energy of emission from the source. An alpha particle is easily stopped by a sheet of paper close to the nucleus. As an example of alpha emission, consider the decay of a radon-222 nucleus,

$$^{222}_{86}\text{Rn} \rightarrow {}^{218}_{84}\text{Po} + {}^4_2\text{He}$$

The spent alpha particle eventually acquires two electrons and becomes an ordinary helium atom.

2. **Beta emission.** Beta (β^-) emission is the expulsion of a different particle, a beta particle, from an unstable disintegrating nucleus. A beta particle is simply an electron ($^{\ 0}_{-1}\text{e}$) ejected from the nucleus at a high speed. The emission of a beta particle *increases the number of protons* in a nucleus. It is as if a neutron changed to a proton by emitting an electron, or

$$^1_0\text{n} \rightarrow {}^1_1\text{p} + {}^{\ 0}_{-1}\text{e}$$

Carbon-14 is a carbon isotope that decays by beta emission:

$$^{14}_6\text{C} \rightarrow {}^{14}_7\text{N} + {}^{\ 0}_{-1}\text{e}$$

Note that the number of protons increased from six to seven, but the mass number remained the same. The mass number is unchanged because the mass of the expelled electron (beta particle) is negligible.

Beta particles are more penetrating than alpha particles and may travel several hundred centimeters through the air. They can be stopped by a thin layer of metal close to the emitting nucleus, such as a 1 cm thick piece of aluminum. A spent beta particle may eventually join an ion to become part of an atom, or it may remain a free electron.

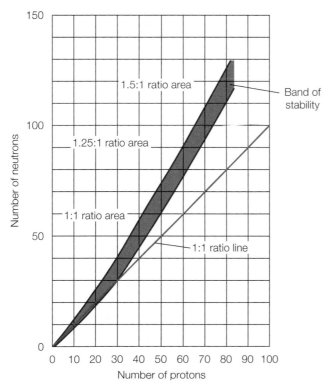

Figure 11.4

The shaded area indicates stable nuclei, which group in a band of stability according to their neutron-to-proton ratio. As the size of nuclei increases, so does the neutron-to-proton ratio that represents stability. Nuclei outside this band of stability are radioactive.

Table 11.5 Approximate Single Dose, Whole Body Effects of Radiation Exposure

Level	Comment
0.130 rem	Average annual exposure to natural background radiation
0.500 rem	Upper limit of annual exposure to general public
25.0 rem	Threshold for observable effects such as blood count changes
100.0 rem	Fatigue and other symptoms of radiation sickness
200.0 rem	Definite radiation sickness, bone marrow damage, possibility of developing leukemia
500.0 rem	Lethal dose for 50 percent of individuals
1,000.0 rem	Lethal dose for all

radiation striking living tissue is usually very small, but it can cause biological damage nonetheless because chemical bonds are broken and free polyatomic ions are produced by radiation.

The amount of radiation received by a human is expressed in terms of radiological dose. Radiation dose is usually written in units of a **rem,** which takes into account the possible biological damage produced by different types of radiation. Doses are usually expressed in terms of fractions of the rem, for example, a *millirem* (mrem). A millirem is 1/1,000 of a rem and is the unit of choice when low levels of radiation are discussed. The SI unit for radiation dose is the *millisievert* (mSv). Both the millirem and the millisievert relate ionizing radiation and biological effect to humans. The natural radiation that people receive from nature in one day is about 1 millirem (0.01 millisievert). A single dose of 100,000 to 200,000 millirem (1,000 to 2,000 millisievert) can cause radiation sickness in humans (table 11.5). A single dose of 500,000 millirem (5,000 millisievert) results in death about 50 percent of the time.

Another measure of radiation received by a material is the **rad.** The term *rad* is from *rad*iation *a*bsorbed *d*ose. The SI unit for radiation received by a material is the *gray* (Gy). One gray is equivalent to an exposure of 100 rad.

Radiation Exposure

Natural radioactivity is a part of your environment, and you receive between 100 and 500 millirems each year from natural sources. This radiation from natural sources is called **background radiation.** Background radiation comes from outer space in the form of cosmic rays and from unstable isotopes in the ground, building materials, and foods. Many activities and situations will increase your yearly exposure to radiation. For example, the atmosphere absorbs some of the cosmic rays from space, so the less atmosphere above you, the more radiation you will receive. You are exposed to one additional millirem per year for each 100 feet you live above sea level. You receive approximately 0.3 millirem for each hour spent on a jet flight. Airline crews receive an additional 300 to 400 millirems per year because they spend so much time high in the atmosphere. Additional radiation exposure comes from medical X rays, television

sets, and luminous objects such as watch and clock dials. In general, the background radiation exposure for the average person is about 130 millirems per year.

What are the consequences of radiation exposure? Radiation can be a hazard to living organisms because it produces ionization along its path of travel. This ionization can (1) disrupt chemical bonds in essential macromolecules such as DNA and (2) produce molecular fragments, which are free polyatomic ions that can interfere with enzyme action and other essential cell functions. Tissues with highly active cells are more vulnerable to radiation damage than others, such as blood-forming tissue. Thus, one of the symptoms of an excessive radiation exposure is an altered blood count. Table 11.5 compares the estimated results of various levels of acute radiation exposure.

Radiation is not a mysterious, unique health hazard. It is a hazard that should be understood and neither ignored nor exaggerated. Excessive radiation exposure should be avoided, just as you avoid excessive exposure to other hazards such as certain chemicals, electricity, or even sunlight. Everyone agrees that *excessive* radiation exposure should be avoided, but there is some controversy about long-term, low-level exposure and its possible role in cancer. Some claim that tolerable low-level exposure does not exist because that is not possible. Others point to many studies comparing high and low background radioactivity with cancer mortality data. For example, no cancer mortality differences could be found between people receiving 500 or more millirems a year and those receiving less than 100 millirems a year. The controversy continues, however, because of lack of knowledge about long-term exposure. Two models of long-term, low-level radiation exposure have been proposed: (1) a linear model and (2) a threshold model. The *linear model* proposes that any radiation exposure above zero is damaging and can produce cancer and genetic damage. The *threshold model* proposes that the human body can repair damage and get rid of damaging free polyatomic ions up to a certain exposure level called the threshold (figure 11.9). The controversy over long-term, low-level radiation exposure will probably continue until there is clear evidence about

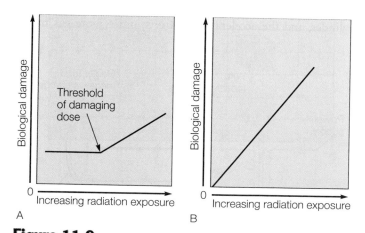

Figure 11.9

Graphic representation of the (*A*) threshold model and (*B*) linear model of low-level radiation exposure. The threshold model proposes that the human body can repair damage up to a threshold. The linear model proposes that any radiation exposure is damaging.

which model is correct. Whichever is correct will not lessen the need for rational risks versus cost-benefit analyses of all energy alternatives.

NUCLEAR ENERGY

As discussed, some nuclei are unstable because they are too large or because they have an unstable neutron-to-proton ratio. These unstable nuclei undergo radioactive decay, eventually forming products of greater stability. An example of this radioactive decay is the alpha emission reaction of uranium-238 to thorium-234,

$$^{238}_{92}U \rightarrow ^{234}_{90}Th + ^{4}_{2}He$$

$$238.0003 \text{ u} \rightarrow 233.9942 \text{ u} + 4.00150 \text{ u}$$

The numbers below the nuclear equation are the *nuclear* masses (u) of the reactant and products. As you can see, there seems to be a loss of mass in the reaction,

$$233.9942 + 4.00150 - 238.0003 = -0.0046 \text{ u}$$

This change in mass is related to the energy change according to the relationship that was formulated by Albert Einstein in 1905. The relationship is

$$E = mc^2$$

equation 11.1

where E is a quantity of energy, m is a quantity of mass, and c is a constant equal to the speed of light in a vacuum, 3.00×10^8 m/s. According to this relationship, matter and energy are the same thing, and energy can be changed to matter and vice versa.

The relationship between mass and energy explains why the mass of a nucleus is always *less* than the sum of the masses of the individual particles of which it is made.

The difference between the mass of the individual nucleons making up a nucleus and the actual mass of the nucleus is called the **mass defect** of the nucleus. The explanation for the mass defect is again found in $E = mc^2$. When nucleons join to make a nucleus, energy is released as the more stable nucleus is formed.

The energy equivalent released when a nucleus is formed is the same as the **binding energy,** the energy required to break the nucleus into individual protons and neutrons. The binding energy of the nucleus of any isotope can be calculated from the mass defect of the nucleus.

The ratio of binding energy to nucleon number is a reflection of the stability of a nucleus (figure 11.10). The greatest binding energy per nucleon occurs near mass number 56, then decreases for both more massive and less massive nuclei. This means that more massive nuclei can gain stability by splitting into smaller nuclei with the release of energy. It also means that less massive nuclei can gain stability by joining together with the release of energy. The slope also shows that more energy is released in the coming-together process than in the splitting process.

The nuclear reaction of splitting a massive nucleus into more stable, less massive nuclei with the release of energy is **nuclear fission** (figure 11.11). Nuclear fission occurs rapidly in an

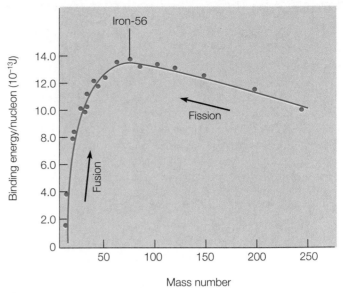

Figure 11.10

The maximum binding energy per nucleon occurs around mass number 56, then decreases in both directions. As a result, fission of massive nuclei and fusion of less massive nuclei both release energy.

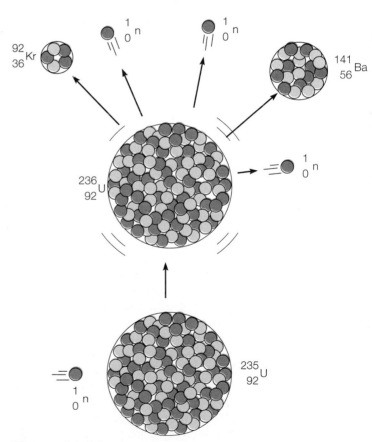

Figure 11.11

The fission reaction occurring when a neutron is absorbed by a uranium-235 nucleus. The deformed nucleus splits any number of ways into lighter nuclei, releasing neutrons in the process.

Nuclear medicine had its beginnings in 1946 when radioactive iodine was first successfully used to treat thyroid cancer patients. Then physicians learned that radioactive iodine could also be used as a diagnostic tool, providing a way to measure the function of the thyroid and to diagnose thyroid disease. More and more physicians began to use nuclear medicine to diagnose thyroid disease as well as to treat hyperthyroidism and other thyroid problems. Nuclear medicine is a branch of medicine using radiation or radioactive materials to diagnose as well as treat diseases.

The development of new nuclear medicine technologies, such as new cameras, detection instruments, and computers, has led to a remarkable increase in the use of nuclear medicine as a diagnostic tool. Today, there are nearly 100 different nuclear medicine imaging procedures. These provide unique, detailed information about virtually every major organ system within the body, information that was unknown just years ago. Treatment of disease with radioactive materials continues to be a valuable part of nuclear medicine, too. The material that follows will consider some techniques of using nuclear medicine as a diagnostic tool, followed by a short discussion of the use of radioactive materials in the treatment of disease.

Nuclear medicine provides diagnostic information about organ function, compared to conventional radiology, which provides images about the structure. For example, a conventional X-ray image will show if a bone is broken or not, while a bone imaging nuclear scan will show changes caused by tumors, hair-line fractures, or arthritis. There are procedures for making detailed structural X-ray pictures of internal organs such as the liver, kidney, or heart, but these images often cannot provide diagnostic information, showing only the structure. Nuclear medicine scans, on the other hand, can provide information about how much heart tissue is still alive after a heart attack or if a kidney is working, even when there are no detectable changes in organ appearance.

An X-ray image is produced when X rays pass through the body and expose photographic film on the other side. Some X-ray exams improve photographic contrast by introducing certain substances. A barium sulfate "milk shake," for example, can be swallowed to highlight the esophagus, stomach, and intestine. More information is provided if X rays are used in a CAT scan (CAT stands for "Computed Axial Tomography"). The CAT scan is a diagnostic test that combines the use of X rays with computer technology. The CAT scan shows organs of interest by making X-ray images from many different angles as the source of the X rays moves around the patient. Contrast-improving substances, such as barium sulfate, might also be used with a CAT scan. In any case, CAT scan images are assembled by a computer into a three-dimensional picture that can show organs, bones, and tissues in great detail.

The gamma camera is a key diagnostic imaging tool used in nuclear medicine. Its use requires a radioactive material, called a radiopharmaceutical, to be injected into or swallowed by the patient. A given radiopharmaceutical tends to go to a specific organ of the body, for example, radioactive iodine tends to go to the thyroid gland, and others go to other organs. Gamma emitting radiopharmaceuticals are used with the gamma camera, and the gamma camera collects and processes these gamma rays to produce images. These images provide a way of studying the structure as well as measuring the function of the selected organ. Together, the structure and function provide a way of identifying tumors, areas of infection, or other problems. The patient experiences little or no discomfort and the radiation dose is small.

A SPECT scan (Single Photon Emission Computerized Tomography) is an imaging technique employing a gamma camera that rotates around the patient, measuring gamma rays and computing their point of origin. Cross-sectional images of a three-dimensional internal organ can be obtained from such data, resulting in images that have higher resolution and thus more diagnostic information than a simple gamma camera image. A gallium radiopharmaceutical is often used in a scan to diagnose and follow the progression of tumors or infections. Gallium scans also can be used to evaluate the heart, lungs, or any other organ that may be involved with inflammatory disease.

Use of MRI (Magnetic Resonance Imaging) also produces images as an infinite number of projections through the body. Unlike CAT, gamma, or SPECT scans, MRI does not use any form of ionizing radiation. MRI uses magnetic

---

atomic bomb explosion and occurs relatively slowly in a nuclear reactor. The nuclear reaction of less massive nuclei, coming together to form more stable, and more massive, nuclei with the release of energy is **nuclear fusion.** Nuclear fusion occurs rapidly in a hydrogen bomb explosion and occurs continually in the sun, releasing the energy essential for the continuation of life on the earth.

Nuclear Fission

Nuclear fission was first accomplished in the late 1930s when researchers were attempting to produce isotopes by bombarding massive nuclei with neutrons. In 1938 two German scientists, Otto Hahn and Fritz Strassman, identified the element barium in a uranium sample that had been bombarded with neutrons. Where the barium came from was a puzzle at the time, but soon afterward Lise Meitner, an associate who had moved to Sweden, deduced that the uranium nuclei had split, producing barium. The reaction might have been

$$_{0}^{1}n + {}_{92}^{235}U \rightarrow {}_{56}^{141}Ba + {}_{36}^{92}Kr + 3{}_{0}^{1}n$$

The phrase "might have been" is used because a massive nucleus can split in many different ways, producing different products. About thirty-five different, less massive elements have been identified among the fission products of uranium-235. Some of these products are fission fragments, and some are produced by unstable fragments that undergo radioactive decay. Selected fission fragments are listed in table 11.6 on page 226, together with their major modes of radioactive decay and half-lives. Some of the isotopes are the focus of concern about nuclear wastes, the topic of the reading at the end of this chapter.

The fission of a uranium-235 nucleus produces two or three neutrons along with other products. These neutrons can each collide with other uranium-235 nuclei where they are ab-fields, radio waves, and a computer to produce detailed images. As the patient enters an MRI scanner, his body is surrounded by a large magnet. The technique requires a very strong magnetic field, a field so strong that it aligns the nuclei of the person's atoms. The scanner sends a strong radio signal, temporarily knocking the nuclei out of alignment. When the radio signal stops, the nuclei return to the aligned position, releasing their own faint radio frequencies. These radio signals are read by the scanner, which uses them in a computer program to produce very detailed images of the human anatomy (box figure 11.1).

The PET scan (Positron Emission Tomography) is the most recent technological development in nuclear medicine imaging, producing 3D images superior to gamma camera images. This technique is built around a radiopharmaceutical that emits positrons (like an electron with a positive charge). Positrons collide with electrons, releasing a burst of energy in the form of photons. Detectors track the emissions and feed the information into a computer. The computer has a program to plot the source of radiation and translates the data into an image. Positron-emitting radiopharmaceuticals used in a PET scan can be low atomic weight elements like carbon, nitrogen, and oxygen. This is important for certain purposes since these are the same elements found in many biological substances like sugar, urea, or carbon dioxide. Thus a PET scan can be used to study processes in organs such as the brain and heart where glucose is being broken down or oxygen is being consumed. This diagnos-

A   B

Box Figure 11.1
(A) MRI Scan of brain, (B) CAT Scan of brain.

tic method can be used to detect epilepsy or brain tumors, among other problems.

Radiopharmaceuticals used for diagnostic examinations are selected for their affinity for certain organs, if they emit sufficient radiation to be easily detectable in the body, and if they have a rather short half-life, preferably no longer than a few hours. Useful radioisotopes that meet these criteria for diagnostic purposes are technetium-99, gallium-67, indium-111, iodine-123, iodine-131, thallium-201, and krypton-81.

The goal of therapy in nuclear medicine is to use radiation to destroy diseased or cancerous tissue while sparing adjacent healthy tissue. Few

radioactive therapeutic agents are injected or swallowed, with the exception of radioactive iodine—mentioned earlier as a treatment for cancer of the thyroid. Useful radioisotopes for therapeutical purposes are iodine-131, phosphorus-32, iridium-192, and gold-198. The radioactive source placed in the body for local irradiation of a tumor is normally iridium-192. A nuclear pharmaceutical is a physiologically active carrier to which a radioisotope is attached. Today, it is possible to manufacture chemical or biological carriers that migrate to a particular part of the human body, and this is the subject of much on-going medical research.

sorbed, causing fission with the release of more neutrons, which collide with other uranium-235 nuclei to continue the process. A reaction where the products are able to produce more reactions in a self-sustaining series is called a **chain reaction.** A chain reaction is self-sustaining until all the uranium-235 nuclei have fissioned or until the neutrons fail to strike a uranium-235 nucleus (figure 11.12).

You might wonder why all the uranium in the universe does not fission in a chain reaction. Natural uranium is mostly uranium-238, an isotope that does not fission easily. Only about 0.7 percent of natural uranium is the highly fissionable uranium-235. This low ratio of readily fissionable uranium-235 nuclei makes it unlikely that a stray neutron would be able to achieve a chain reaction.

In order to achieve a chain reaction, there must be (1) a sufficient mass with (2) a sufficient concentration of fissionable nuclei. When the mass and concentration are sufficient to sus-

tain a chain reaction the amount is called a **critical mass.** Likewise, a mass too small to sustain a chain reaction is called a *subcritical mass.* A mass of sufficiently pure uranium-235 (or plutonium-239) that is large enough to produce a rapidly accelerating chain reaction is called a *supercritical mass.* An atomic bomb is simply a device that uses a small, conventional explosive to push subcritical masses of fissionable material into a supercritical mass. Fission occurs almost instantaneously in the supercritical mass, and tremendous energy is released in a violent explosion.

Nuclear Power Plants

The nuclear part of a nuclear power plant is the *nuclear reactor,* a steel vessel in which a controlled chain reaction of fissionable material releases energy (figure 11.13). In the most popular design, called a pressurized light-water reactor, the fissionable

Table 11.6 Fragments and Products from Nuclear Reactors Using Fission of Uranium-235

Isotope	Major Mode of Decay	Half-Life	Isotope	Major Mode of Decay	Half-Life
Tritium	Beta	12.26 years	Cerium-144	Beta, gamma	285 days
Carbon-14	Beta	5,930 years	Promethium-147	Beta	2.6 years
Argon-41	Beta, gamma	1.83 hours	Samarium-151	Beta	90 years
Iron-55	Electron capture	2.7 years	Europium-154	Beta, gamma	16 years
Cobalt-58	Beta, gamma	71 days	Lead-210	Beta	22 years
Cobalt-60	Beta, gamma	5.26 years	Radon-222	Alpha	3.8 days
Nickel-63	Beta	92 years	Radium-226	Alpha, gamma	1,620 years
Krypton-85	Beta, gamma	10.76 years	Thorium-229	Alpha	7,300 years
Strontium-89	Beta	5.4 days	Thorium-230	Alpha	26,000 years
Strontium-90	Beta	28 years	Uranium-234	Alpha	2.48×10^5 years
Yttrium-91	Beta	59 days	Uranium-235	Alpha, gamma	7.13×10^8 years
Zirconium-93	Beta	9.5×10^5 years	Uranium-238	Alpha	4.51×10^9 years
Zirconium-95	Beta, gamma	65 days	Neptunium-237	Alpha	2.14×10^6 years
Niobium-95	Beta, gamma	35 days	Plutonium-238	Alpha	89 years
Technetium-99	Beta	2.1×10^5 years	Plutonium-239	Alpha	24,360 years
Ruthenium-106	Beta	1 year	Plutonium-240	Alpha	6,760 years
Iodine-129	Beta	1.6×10^7 years	Plutonium-241	Beta	13 years
Iodine-131	Beta, gamma	8 days	Plutonium-242	Alpha	3.79×10^5 years
Xenon-133	Beta, gamma	5.27 days	Americium-241	Alpha	458 years
Cesium-134	Beta, gamma	2.1 years	Americium-243	Alpha	7,650 years
Cesium-135	Beta	2×10^6 years	Curium-242	Alpha	163 days
Cesium-137	Beta	30 years	Curium-244	Alpha	18 years
Cerium-141	Beta	32.5 days			

material is enriched 3 percent uranium-235 and 97 percent uranium-238 that has been fabricated in the form of small ceramic pellets (figure 11.14A). The pellets are encased in a long zirconium alloy tube called a *fuel rod*. The fuel rods are locked into a *fuel rod assembly* by locking collars, arranged to permit pressurized water to flow around each fuel rod (figure 11.14B) and to allow the insertion of *control rods* between the fuel rods. *Control rods* are constructed of materials, such as cadmium, that absorb neutrons. The lowering or raising of control rods within the fuel rod assemblies slows or increases the chain reaction by varying the amount of neutrons absorbed. When they are lowered completely into the assembly, enough neutrons are absorbed to stop the chain reaction.

It is physically impossible for the low-concentration fuel pellets to form a supercritical mass. A nuclear reactor in a power plant can only release energy at a comparatively slow rate, and it is impossible for a nuclear power plant to produce a nuclear explosion. In a pressurized water reactor the energy released is carried away from the reactor by pressurized water in a closed pipe called the *primary loop* (figure 11.15). The water is pressurized at about 150 atmospheres (about 2,200 lb/in^2) to keep it from boiling, since its temperature may be 350°C (about 660°F).

In the pressurized light-water (ordinary water) reactor the circulating pressurized water acts as a coolant, carrying heat away from the reactor. The water also acts as a *moderator*, a substance that slows neutrons so they are more readily absorbed by uranium-235 nuclei. Other reactor designs use heavy water (deuterium dioxide) or graphite as a moderator.

Water from the closed primary loop is circulated through a heat exchanger called a *steam generator* (figure 11.15). The pressurized high-temperature water from the reactor moves through hundreds of small tubes inside the generator as *feedwater* from the *secondary loop* flows over the tubes. The water in the primary loop heats feedwater in the steam generator and then returns to the nuclear reactor to become heated again. The feedwater is heated to steam at about 235°C (455°F) with a pressure of about 68 atmospheres (1,000 lb/in^2). This steam is piped to the turbines, which turn an electric generator (figure 11.16).

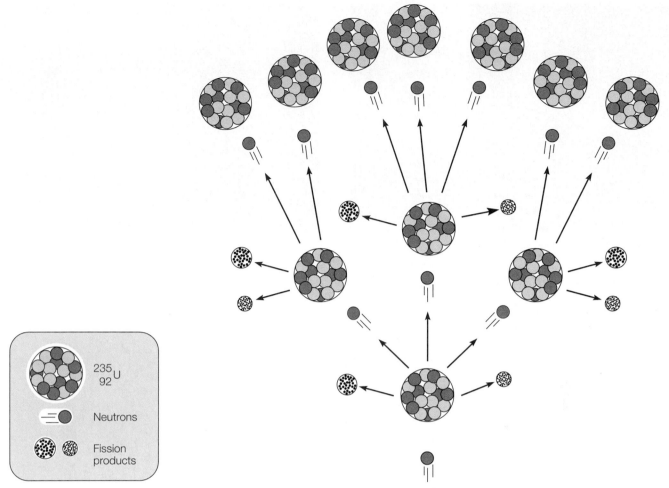

Figure 11.12

A schematic representation of a chain reaction. Each fissioned nucleus releases neutrons, which move out to fission other nuclei. The number of neutrons can increase quickly with each series.

Figure 11.13

A schematic representation of the basic parts of a nuclear reactor. The largest commercial nuclear power plant reactors are nine- to eleven-inch-thick steel vessels with a stainless steel liner, standing about 40 feet high with a diameter of 16 feet. Such a reactor has four pumps, which move 440,000 gallons of water per minute through the primary loop.

After leaving the turbines, the spent steam is condensed back to liquid water in a second heat exchanger receiving water from the cooling towers. Again, the cooling water does not mix with the closed secondary loop water. The cooling-tower water enters the condensing heat exchanger at about 32°C (90°F) and leaves at about 50°C (about 120°F) before returning to a cooling tower, where it is cooled by evaporation. The feedwater is preheated, then recirculated to the steam generator to start the cycle over again. The steam is condensed back to liquid water because of the difficulty of pumping and reheating steam.

After a period of time the production of fission products in the fuel rods begins to interfere with effective neutron transmission, so the reactor is shut down annually for refueling. During refueling about one-third of the fuel that had the longest exposure in the reactor is removed as "spent" fuel. New fuel rod assemblies are inserted to make up for the part removed (figure 11.17). However, only about 4 percent of the "spent" fuel is unusable waste, about 94 percent is uranium-238, 0.8 percent is uranium-235, and about 0.9 percent is plutonium (figure 11.18). Thus, "spent" fuel rods contain an appreciable amount of

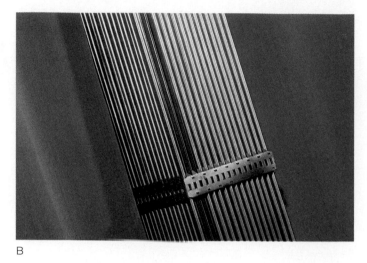

A

B

Figure 11.14

(*A*) These are uranium oxide fuel pellets that are stacked inside fuel rods, which are then locked together in a fuel rod assembly. (*B*) A fuel rod assembly. See also figure 11.17, which shows a fuel rod assembly being loaded into a reactor.

Figure 11.15

A schematic general system diagram of a pressurized water nuclear power plant, not to scale. The containment building is designed to withstand an internal temperature of 300°F at a pressure of 60 lbs/in^2 and still maintain its leak-tight integrity.

usable uranium and plutonium. For now, spent reactor fuel rods are mostly stored in cooling pools at the nuclear plant sites. In the future, a decision will be made either to reprocess the spent fuel, recovering the uranium and plutonium through chemical reprocessing, or put the fuel in terminal storage. Concerns about reprocessing are based on the fact that plutonium-239 and uranium-235 are fissionable and could possibly be used by terrorist groups to construct nuclear explosive devices. Six other

countries do have reprocessing plants, however, and the spent fuel rods represent a significant energy source. Some energy experts say that it would be inappropriate to dispose of such an energy source.

The technology to dispose of fuel rods exists if the decision is made to do so. The longer half-life waste products are mostly alpha emitters. These metals could be converted to oxides, mixed with powdered glass (or a ceramic), melted, and then

Figure 11.16

The turbine deck of a nuclear generating station. There is one large generator in line with four steam turbines in this non-nuclear part of the plant. The large silver tanks are separators that remove water from the steam after it has left the high-pressure turbine and before it is recycled back into the low-pressure turbines.

Figure 11.17

Spent fuel rod assemblies are removed and new ones are added to a reactor head during refueling. This shows an initial fuel load to a reactor, which has the upper part removed and set aside for the loading.

poured into stainless steel containers. The solidified canisters would then be buried in a stable geologic depository. The glass technology is used in France for disposal of high-level wastes. Buried at two-thousand- to three-thousand-foot depths in solid granite, the only significant means of the radioactive wastes reaching the surface would be through groundwater dissolving the stainless steel, glass, and waste products and then transporting them back to the surface. Many experts believe that if such groundwater dissolving were to take place it would require thousands of years. The radioactive isotopes would thus undergo natural radioactive decay by the time they could reach the surface. Nonetheless, research is continuing on nuclear waste and its disposal. In the meantime, the question of whether it is best to reprocess fuel rods or place them in permanent storage remains unanswered.

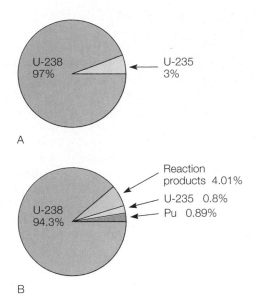

Figure 11.18

The composition of the nuclear fuel in a fuel rod (*A*) before and (*B*) after use over a three-year period in a nuclear reactor.

What is the volume of nuclear waste under question? If all the spent fuel rods from all the commercial nuclear plants accumulated up to the year 2000 were reprocessed, then mixed with glass, the total amount of glassified waste would make a pile on one football field an estimated 4 m (about 13 ft) high.

Nuclear Fusion

As the graph of nuclear binding energy versus mass numbers shows (see figure 11.10), nuclear energy is released when (1) massive nuclei such as uranium-235 undergo fission and (2) when less massive nuclei come together to form more massive nuclei through nuclear fusion. Nuclear fusion is responsible for the energy released by the sun and other stars. At the present halfway point in the sun's life—with about 5 billion years to go—the core is now 35 percent hydrogen and 65 percent helium. Through fusion, the sun converts about 650 million tons of hydrogen to 645 million tons of helium every second. The other roughly 5 million tons of matter are converted into energy. Even at this rate, the sun has enough hydrogen to continue the process for an estimated 5 billion years. There are several fusion reactions that take place between hydrogen and helium isotopes, including the following:

$$^1_1H + {}^1_1H \rightarrow {}^2_1H + {}^0_1e$$

$$^2_1H + {}^2_1H \rightarrow {}^3_2H + {}^1_0n$$

$$^3_2He + {}^3_2He \rightarrow {}^4_2He + 2{}^1_1H$$

The fusion process would seem to be a desirable energy source on Earth because (1) two isotopes of hydrogen, deuterium (^2_1H) and tritium (^3_1H), undergo fusion at a relatively low temperature; (2) the supply of deuterium is practically unlimited, with each gallon of seawater containing about a teaspoonful of

Three Mile Island (down-river from Harrisburg, Pennsylvania, U.S.A.) and Chernobyl (former U.S.S.R., now Ukraine) are two nuclear power plants that became famous because of accidents. Here is a brief accounting of what happened.

Three Mile Island

It was March 28, 1979, and the 880-megawatt Three Mile Island Nuclear Plant, operated by Metropolitan Edison Co., was going full blast. At 4:00 A.M. that morning the main feedwater pump that pumps water to the steam generator failed for some unexplained reason (follow this description in figure 11.15). Backup feedwater pumps kicked in, but the valves that should have been open were closed for maintenance, blocking the backup source of water for the steam generator. All of a sudden there was not a source of water for the steam generator that removes heat from the primary loop. Events began to happen quickly at that point.

The computer sensors registered that the steam generator was not receiving water, and the computer began to follow a shutdown procedure. First, the turbine was shut down as steam was vented from the steam line out through the turbine building, sounding much like a large jet plane. Within 6 seconds the reactor was shut down (scrammed) with the control rods dropping down between the fuel rods in the reactor vessel. Fissioning began to slow, but the reactor was still hot.

Between 3 and 6 seconds a significant event occurred when a pressure relief valve opened on the primary loop, releaving the excess pressure that was generated because feedwater was not entering the steam generator to remove heat from the primary loop. The valve should have closed when the excess pressure was released, but it did not close. It was stuck in an open position. Pressurized water and steam was pouring from the primary loop into the containment building. As water was lost from the primary loop, temperatures inside the reactor began to climb. The loss of pressure resulted in high-pressure water flashing into steam. If an operator had pressed the right button In the control room it would have closed the open valve, but this did not happen until 32 minutes later.

At this point the reactor could have recovered from the events. Two minutes after the initial shut down the computer sensors noted the loss of pressure from the open valve and kicked in the emergency core cooling system, which pumps more water into the reactor vessel. However, for some unknown reason, the control room operators shut down one pump 4 1/2 minutes after the initial event and the second pump 6 minutes later.

Water continued to move through the open pressure relief valve into the containment building. At 7 1/2 minutes after the start of the event the radioactive water on the floor was two feet deep and the sump pumps started pumping water into tanks in the auxiliary building. This water would become the source of the radioactivity that did escape the plant. It escaped because the pump seals leaked and spilled radioactive water. The filters had been removed from the auxiliary building air vents, allowing radioactive gases to escape.

Eleven minutes after the start of the event, an operator restarted the emergency core cooling system that had been turned off. With the cooling water flowing again, the pressure in the reactor stopped falling. The fuel rods, some 36,000 in this reactor, had not yet suffered any appreciable damage. This would be taken care of by the next incredible event.

What happened next was that operators began turning off the emergency cooling pumps, perhaps because they were vibrating too much. In any case, with the pumps off the water level in the reactor fell again, this time uncovering the fuel rods. Within minutes, the temperature was high enough to rupture fuel rods, dropping radioactive oxides into the bottom of the reactor vessel. The operators now had a general emergency.

Eleven hours later the operators decided to start the main reactor coolant pump. This pump had shut down at the beginning of the series of events. Water again covered the fuel rods, and pressure and temperature stabilized.

The consequences of this series of events were as follows:

1. Local residences did receive some radiation from the release of gases. They received 10 millirem (0.1 millisievert) in a low-exposure area and up to 25 millirem (0.25 millisievert) in a high-exposure area.
2. Cleaning up the damaged reactor vessel and core required more than ten years to

---

deuterium dioxide; and (3) enormous amounts of energy are released with no radioactive by-products.

The oceans contain enough deuterium to generate electricity for the entire world for millions of years, and tritium can be constantly produced by a fusion device. Researchers know what needs to be done to tap this tremendous energy source. The problem is *how* to do it in an economical, continuous energy-producing fusion reactor. The problem, one of the most difficult engineering tasks ever attempted, is meeting three basic fusion reaction requirements of (1) temperature, (2) density, and (3) time (figure 11.19):

1. **Temperature.** Nuclei contain protons and are positively charged, so they experience the electromagnetic repulsion of like charges. This force of repulsion can be overcome, moving the nuclei close enough to fuse together, by giving the nuclei sufficient kinetic energy. The fusion reaction of deuterium and tritium, which has the lowest temperature requirements of any fusion reaction

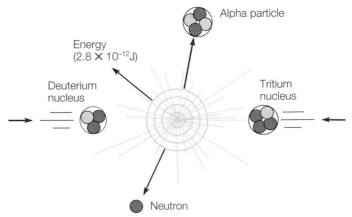

Figure 11.19

A fusion reaction between a tritium nucleus and a deuterium nucleus requires a certain temperature, density, and time of containment to take place.

cut it up, pack it into canisters, and ship everything to the Federal Nuclear Reservation at Idaho Falls.

3. The cost of the clean up was more than $1 billion.

4. Changes were implemented at other nuclear power plants as a consequence: Pressure relief valves have been removed, operators can no longer turn off the emergency cooling system, and operators must now spend about one-fourth of their time in training.

Chernobyl

The Soviet-designed Chernobyl reactor was a pressurized water reactor with individual fuel channels, which is very different from the pressurized water reactors used in the United States. The Chernobyl reactor was constructed with each fuel assembly in an individual pressure tube with circulating pressurized water. This heated water was circulated to a steam separator, and the steam was directed to turbines, which turned a generator to produce electricity. Graphite blocks surrounded the pressure tubes, serving as moderators to slow the neutrons involved in the chain reaction. The graphite was cooled by a mixture of helium and nitrogen. The reactor core was located in a concrete bunker that acted as a radiation shield. The top part was a steel cap and shield that supported the fuel assemblies. There were no containment buildings around the Soviet reactors as there are in the United States.

The Chernobyl accident was the result of a combination of a poorly engineered reactor design, poorly trained reactor operators, and serious mistakes made by the operators on the day of the accident.

The reactor design was flawed because at low power, steam tended to form pockets in the water-filled fuel channels, creating a condition of instability. Instability occurred because (1) steam is not as efficient at cooling as is liquid water, and (2) liquid water acts as a moderator and neutron absorber while steam does not. Excess steam therefore leads to overheating and increased power generation. Increased power can lead to increased steam generation, which leads to further increases in power. This coupled response is very difficult to control because it feeds itself.

On April 25, 1986, the operators of Chernobyl unit 4 started a test to find out how long the turbines would spin and supply power following the loss of electrical power. The operators disabled the automatic shutdown mechanisms, and then started the test early on April 26, 1986. The plan was to stabilize the reactor at 1,000 MW, but an error was made and the power fell to about 30 MW and pockets of steam became a problem. They tried to increase the power by removing all the control rods. At 1:00 A.M., they were able to stabilize the reactor at 200 MW. Then instability returned and the operators were making continuous adjustments to maintain a constant power. They reduced the feedwater to maintain steam pressure, and this created even more steam voids in the fuel channels. Power surged to a very high level and fuel elements ruptured. A steam explosion moved the reactor cap, exposing individual fuel channels and releasing fission products to the environment. A second explosion knocked a hole in the roof, exposed more of the reactor core, and the reactor graphite burst into flames. The graphite burned for nine days, releasing about 324 million Ci (12×10^{18} Bq) into the environment.

The fiery release of radioactivity was finally stopped by using a helicopter to drop sand, boron, lead, and other materials onto the burning graphite reactor. After the fire was out, the remains of the reactor were covered with a large concrete shelter.

In addition to destroying the reactor, the accident killed 30 people, and 28 of these were from radiation exposure. Another 134 people were treated for acute radiation poisoning and all recovered from the immediate effects.

Cleanup crews over the next year received about 10 rem (100 millisievert) to 25 rem (250 millisievert) and some received as much as 50 rem (500 millisievert). In addition to this direct exposure, large expanses of Belarus, Ukraine, and Russia were contaminated by radioactive fallout from the reactor fire. Hundreds of thousands of people have been resettled into less contaminated areas. The World Health Organization and other international agencies have studied the data to understand the impact of radiation-related disease. These studies do confirm a rising incidence of thyroid cancer, but no increases in leukemia so far.

known at the present time, requires temperatures on the order of 100 million degrees Celsius.

2. **Density.** There must be a sufficiently dense concentration of heavy hydrogen nuclei, on the order of $10^{14}/cm^3$, so many reactions occur in a short time.

3. **Time.** The nuclei must be confined at the appropriate density up to a second or longer at pressures of at least 10 atmospheres to permit a sufficient number of reactions to take place.

The temperature, density, and time requirements of a fusion reaction are interrelated. A short time of confinement, for example, requires an increased density, and a longer confinement time requires less density. The primary problems of fusion research are the high-temperature requirements and confinement. No material in the world can stand up to a temperature of 100 million degrees Celsius, and any material container would be instantly vaporized. Thus, research has centered on meeting the fusion reaction requirements without a material container. Two approaches are being tested, *magnetic confinement* and *inertial confinement.*

Magnetic confinement utilizes a *plasma,* a very hot gas consisting of atoms that have been stripped of their electrons because of the high kinetic energies. The resulting positively and negatively charged particles respond to electrical and magnetic forces, enabling researchers to develop a "magnetic bottle," that is, magnetic fields that confine the plasma and avoid the problems of material containers that would vaporize. A magnetically confined plasma is very unstable, however, and researchers have compared the problem to trying to carry a block of jello on a pair of rubber bands. Different magnetic field geometries and magnetic "mirrors" are the topics of research in attempts to stabilize the hot, wobbly plasma. Electric currents, injection of fast

There are two general categories of nuclear wastes: (1) low-level wastes and (2) high-level wastes. The *low-level wastes* are produced by the normal operation of a nuclear reactor. Radioactive isotopes sometimes escape from fuel rods in the reactor and in the spent fuel storage pools. These isotopes are removed from the water by ion-exchange resins and from the air by filters. The used resins and filters will contain the radioactive isotopes and will become low-level wastes. In addition, any contaminated protective clothing, tools, and discarded equipment also become low-level wastes.

Low-level liquid wastes are evaporated, mixed with cement, then poured into fifty-five-gallon steel drums. Solid wastes are compressed and placed in similar drums. The drums are currently disposed of by burial in government-licensed facilities. In general, low-level waste has an activity of less than 1.0 curie per cubic foot. Contact with the low-level waste could expose a person to up to 20 millirems per hour of contact.

High-level wastes from nuclear power plants are spent nuclear fuel rods. At the present time most of the commercial nuclear power plants have these rods in temporary storage at the plant site. These rods are "hot" in the radioac-

Box Figure 11.2

This is a standard warning sign for a possible radioactive hazard. Such warning signs would have to be maintained around a nuclear waste depository for thousands of years.

tive sense, producing about 100,000 curies per cubic foot. They are also hot in the thermal sense, continuing to generate heat for months after removal from the reactor. The rods are cooled by heat exchangers connected to storage pools; they

could otherwise achieve an internal temperature as high as 800°C for several decades. In the future, these spent fuel rods will be reprocessed or disposed of through terminal storage.

Agencies of the United States federal government have also accumulated millions of gallons of high-level wastes from the manufacture of nuclear weapons and nuclear research programs. These liquid wastes are stored in million-gallon stainless steel containers that are surrounded by concrete. The future of this large amount of high-level wastes may be evaporation to a solid form or mixture with a glass or ceramic matrix, which is melted and poured into stainless steel containers. These containers would be buried in solid granite rock in a stable geologic depository. Such high-level wastes must be contained for thousands of years as they undergo natural radioactive decay (box figure 11.2). Burial at a depth of two thousand to three thousand feet in solid granite would provide protection from exposure by explosives, meteorite impact, or erosion. One major concern about this plan is that a hundred generations later, people might lose track of what is buried in the nuclear garbage dump.

ions, and radio frequency (microwave) heating methods are also being studied.

Inertial confinement is an attempt to heat and compress small frozen pellets of deuterium and tritium with energetic laser beams or particle beams, producing fusion. The focus of this research is new and powerful lasers, light ion and heavy ion beams. If successful, magnetic or inertial confinement will provide a long-term solution for future energy requirements.

SUMMARY

Radioactivity is the spontaneous emission of particles or energy from an unstable atomic nucleus. The modern atomic theory pictures the nucleus as protons and neutrons held together by a short-range *nuclear force* that has moving *nucleons* (protons and neutrons) in *energy shells* analogous to the shell structure of electrons. A graph of the number of neutrons to the number of protons in a nucleus reveals that stable nuclei have a certain neutron-to-proton ratio in a *band of stability*. Nuclei that are above or below the band of stability, and nuclei that are beyond atomic number 83, are radioactive and undergo *radioactive decay*.

Three common examples of radioactive decay involve the emission of an *alpha particle*, a *beta particle*, and a *gamma ray*. An alpha particle is a helium nucleus, consisting of two protons and two neutrons. A beta particle is a high-speed electron that is ejected from the nucleus. A gamma ray is a short-wavelength electromagnetic radiation from an

excited nucleus. In general, nuclei with an atomic number of 83 or larger become more stable by alpha emission. Nuclei with a neutron-to-proton ratio that is too large become more stable by beta emission. Gamma ray emission occurs from a nucleus that was left in a high-energy state by the emission of an alpha or beta particle.

Each radioactive isotope has its own specific *radioactive decay rate*. This rate is usually described in terms of *half-life*, the time required for one-half the unstable nuclei to decay.

Radiation is measured by (1) its effects on photographic film, (2) the number of ions it produces, or (3) the flashes of light produced on a phosphor. It is measured at a source in units of a *curie*, defined as 3.70×10^{10} nuclear disintegrations per second. It is measured where received in units of a *rad*, defined as 1×10^{-5} J. A *rem* is a measure of radiation that takes into account the biological effectiveness of different types of radiation damage. In general, the natural environment exposes everyone to 100 to 500 millirems per year, an exposure called *background radiation*. Lifestyle and location influence the background radiation received, but the average is 130 millirems per year.

Energy and mass are related by Einstein's famous equation of $E = mc^2$, which means that *matter can be converted to energy and energy to matter*. The mass of a nucleus is always less than the sum of the masses of the individual particles of which it is made. This *mass defect* of a nucleus is equivalent to the energy released when the nucleus was formed according to $E = mc^2$. It is also the *binding energy*, the energy required to break the nucleus apart into nucleons.

When the binding energy is plotted against the mass number, the greatest binding energy per nucleon is seen to occur for an atomic number near that of iron. More massive nuclei therefore release energy by fission, or splitting to more stable nuclei. Less massive nuclei release energy by fusion, the joining of less massive nuclei to produce a more stable, more massive nucleus. Nuclear fission provides the energy for atomic explosions and nuclear power plants. Nuclear fusion is the energy source of the sun and other stars and also holds promise as a future energy source for humans.

Summary of Equations

11.1   energy = mass × the speed of light squared

$$E = mc^2$$

KEY TERMS

alpha particle (p. **214**)	half-life (p. **219**)
background radiation (p. **222**)	mass defect (p. **223**)
band of stability (p. **217**)	nuclear fission (p. **223**)
beta particle (p. **214**)	nuclear fusion (p. **224**)
binding energy (p. **223**)	nucleons (p. **215**)
chain reaction (p. **225**)	rad (p. **222**)
critical mass (p. **225**)	radioactive decay (p. **215**)
curie (p. **221**)	radioactivity (p. **214**)
gamma ray (p. **214**)	rem (p. **222**)

APPLYING THE CONCEPTS

1. A high-speed electron ejected from a nucleus during radioactive decay is called a (an)
   (a) alpha particle.
   (b) beta particle.
   (c) gamma ray.
   (d) none of the above.
2. The ejection of an alpha particle from a nucleus results in
   (a) an increase in the atomic number by one.
   (b) an increase in the atomic mass by four.
   (c) a decrease in the atomic number by two.
   (d) none of the above.
3. An atom of radon-222 loses an alpha particle to become a more stable atom of
   (a) radium.
   (b) bismuth.
   (c) polonium.
   (d) radon.
4. A sheet of paper will stop a (an)
   (a) alpha particle.
   (b) beta particle.
   (c) gamma ray.
   (d) none of the above.
5. The most penetrating of the three common types of nuclear radiation is the
   (a) alpha particle.
   (b) beta particle.
   (c) gamma ray.
   (d) All have equal penetrating ability.
6. An atom of an isotope with an atomic number greater than 83 will probably emit a (an)
   (a) alpha particle.
   (b) beta particle.
   (c) gamma ray.
   (d) none of the above.
7. An atom of an isotope with a large neutron-to-proton ratio will probably emit a (an)
   (a) alpha particle.
   (b) beta particle.
   (c) gamma ray.
   (d) none of the above.
8. The rate of radioactive decay can be increased by increasing the
   (a) temperature.
   (b) pressure.
   (c) size of the sample.
   (d) None of the above are correct.
9. Isotope A has a half-life of seconds, and isotope B has a half-life of millions of years. Which isotope is more radioactive?
   (a) It depends on the sample size.
   (b) isotope A
   (c) isotope B
   (d) Unknown, from the information given.
10. A measure of radioactivity at the *source* is (the)
    (a) curie.
    (b) rad.
    (c) rem.
    (d) any of the above.
11. A measure of radiation received that considers the biological effect resulting from the radiation is (the)
    (a) curie.
    (b) rad.
    (c) rem.
    (d) any of the above.
12. Used fuel rods from a nuclear reactor contain about
    (a) 96% usable uranium and plutonium.
    (b) 33% usable uranium and plutonium.
    (c) 4% usable uranium and plutonium.
    (d) 0% usable uranium and plutonium.

Answers

1. b 2. c 3. c 4. a 5. c 6. a 7. b 8. c 9. b 10. a 11. c 12. a

QUESTIONS FOR THOUGHT

1. How is a radioactive material different from a material that is not radioactive?
2. What is radioactive decay? Describe how the radioactive decay rate can be changed if this is possible.
3. Describe three kinds of radiation emitted by radioactive materials. Describe what eventually happens to each kind of radiation after it is emitted.
4. How are positively charged protons able to stay together in a nucleus since like charges repel?
5. What is half-life? Give an example of the half-life of an isotope, describing the amount remaining and the time elapsed after five half-life periods.

6. Would you expect an isotope with a long half-life to be more, the same, or less radioactive than an isotope with a short half-life? Explain.

7. What is (a) a curie? (b) a rad? (c) a rem?

8. What is meant by background radiation? What is the normal radiation dose for the average person from background radiation?

9. Why is there controversy about the effects of long-term, low levels of radiation exposure?

10. What is a mass defect? How is it related to the binding energy of a nucleus? How can both be calculated?

11. Compare and contrast nuclear fission and nuclear fusion.

PARALLEL EXERCISES

The exercises in groups A and B cover the same concepts. Solutions to group A exercises are located in appendix D.

Group A

Note: You will need the table of atomic weights inside the back cover of this text.

1. Give the number of protons and the number of neutrons in the nucleus of each of the following isotopes:
   (a) cobalt-60
   (b) potassium-40
   (c) neon-24
   (d) lead-208

2. Write the nuclear symbols for each of the nuclei in exercise 1.

3. Predict if the nuclei in exercise 1 are radioactive or stable, giving your reasoning behind each prediction.

4. Write a nuclear equation for the decay of the following nuclei as they give off a beta particle:
   (a) $^{56}_{26}\text{Fe}$
   (b) $^{7}_{4}\text{Be}$
   (c) $^{64}_{29}\text{Cu}$
   (d) $^{24}_{11}\text{Na}$
   (e) $^{214}_{82}\text{Pb}$
   (f) $^{32}_{15}\text{P}$

5. Write a nuclear equation for the decay of the following nuclei as they undergo alpha emission:
   (a) $^{235}_{92}\text{U}$
   (b) $^{226}_{88}\text{Ra}$
   (c) $^{239}_{94}\text{Pu}$
   (d) $^{214}_{83}\text{Bi}$
   (e) $^{230}_{90}\text{Th}$
   (f) $^{210}_{84}\text{Po}$

6. The half-life of iodine-131 is 8 days. How much of a 1.0 oz sample of iodine-131 will remain after 32 days?

7. How much energy must be supplied to break a single iron-56 nucleus into separate protons and neutrons? (The mass of an iron-56 nucleus is 55.9206 u, one proton is 1.00728 u, and one neutron is 1.00867 u.)

Group B

Note: You will need the table of atomic weights inside the back cover of this text.

1. Give the number of protons and the number of neutrons in the nucleus of each of the following isotopes:
   (a) aluminum-25
   (b) technetium-95
   (c) tin-120
   (d) mercury-200

2. Write the nuclear symbols for each of the nuclei in exercise 1.

3. Predict if the nuclei in exercise 1 are radioactive or stable, giving your reasoning behind each prediction.

4. Write a nuclear equation for the beta emission decay of each of the following:
   (a) $^{14}_{6}\text{C}$
   (b) $^{60}_{27}\text{Co}$
   (c) $^{24}_{11}\text{Na}$
   (d) $^{241}_{94}\text{Pu}$
   (e) $^{131}_{53}\text{I}$
   (f) $^{210}_{82}\text{Pb}$

5. Write a nuclear equation for each of the following alpha emission decay reactions:
   (a) $^{241}_{95}\text{Am}$
   (b) $^{232}_{90}\text{Th}$
   (c) $^{223}_{88}\text{Ra}$
   (d) $^{234}_{92}\text{U}$
   (e) $^{242}_{96}\text{Cm}$
   (f) $^{237}_{93}\text{Np}$

6. If the half-life of cesium-137 is 30 years, how much time will be required to reduce a 1.0 kg sample to 1.0 g?

7. How much energy is needed to separate the nucleons in a single lithium-7 nucleus? (The mass of a lithium-7 nucleus is 7.01435 u, one proton is 1.00728 u, and one neutron is 1.00867 u.)

The Universe

This is a planetary nebula in the constellation Aquarius. Planetary nebulae are clouds of ionized gases with no relationship to any planet. They were named long ago, when they appeared similar to the planets Neptune and Uranus when viewed through early telescopes.

O U T L I N E

Astronomy is an exciting field of science that has fascinated people since the beginnings of recorded history. Ancient civilizations searched the heavens in wonder, some recording on clay tablets what they observed. Many religious and philosophical beliefs were originally based on interpretations of these ancient observations. Today, we are still awed by space, but now we are fascinated with ideas of space travel, black holes, and the search for extraterrestrial life. Throughout history, people have speculated about the universe and their place in it and watched the sky and wondered (figure 12.1). What is out there and what does it all mean? Are there other people on other planets, looking at the star in their sky that is our sun, wondering if we exist?

Until about thirty years ago progress in astronomy was limited to what could be observed and photographed. Developments in technology then began to provide the details of what is happening in the larger expanses of space away from the earth. This included new data made available from the development of infrared, radio, and X-ray telescopes and the Hubble Space Telescope. New data and the discovery of pulsars, neutron stars, and black holes began to fit together like the pieces of a puzzle. Theoretical models emerged about how stars, galaxies, and the universe have evolved. This chapter is concerned with these topics and how the stars are arranged in space. The chapter concludes with theoretical models of how the universe began and what may happen to it in the future.

THE NIGHT SKY

Early civilizations had a much better view of the night sky before city lights, dust, and pollution obscured much of it. Today, you must travel far from cities, perhaps to a remote mountaintop, to see a clear night sky as early people observed it. Back then, people could clearly see the motion of the moon and stars night after night, observing recurring cycles of motion. These cycles became important as people associated them with certain events. Thus, watching the sun, moon, and star movements became a way to identify when to plant crops, when to harvest, and when it was time to plan for other events. Observing the sky was an important activity, and many early civilizations built observatories with sighting devices to track and record astronomical events. Stonehenge, for example, was an ancient observatory built in England around 2600 B.C. by Neolithic people (figure 12.2).

Light from the stars and planets must pass through the earth's atmosphere to reach you and this affects the light. Stars appear as point *sources* of light, and each star generates its own light. The stars seem to twinkle because density differences in the atmosphere refract the point of starlight one way, then the other, as the air moves. The result is the slight dancing about and change in intensity called twinkling. The points of starlight are much steadier when viewed on a calm night or when viewed from high in the mountains where there is less atmosphere for the starlight to pass through. Astronauts outside the atmosphere see no twinkling, and the stars appear as steady point sources of light.

Back at ground level, within the atmosphere, the *reflected* light from a planet does not seem to twinkle. A planet appears as a disk of light rather than a point source, so refraction from moving air of different densities does not affect the image as much. Sufficient air movement can cause planets to appear to shimmer, however, just as a road appears to shimmer on a hot summer day.

How far away is a star? When you look at the sky it appears that all the stars are at the same distance. It seems impossible to know anything about the actual distance to any given star. Standard referent units of length such as kilometers or miles have little meaning in astronomy since there are no referent points of

A

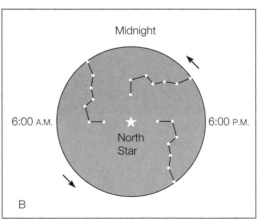

B

Figure 12.1

Ancient civilizations used celestial cycles of motion as clocks and calendars. (*A*) This photograph shows the path of stars around the North Star. (*B*) A "snapshot" of the position of the Big Dipper over a period of 24 hours as it turns around the North Star. This shows how the Big Dipper can be used to help you keep track of time.

Figure 12.2
The stone pillars of Stonehenge were positioned so that they could be used to follow the movement of the sun and moon with the seasons of the year.

comparison. Distance without a referent point can be measured in terms of angles or time. The unit of astronomical distance that uses time is the **light-year** (ly). A light-year is the distance that light travels in one year, about 9.5×10^{12} km (about 6×10^{12} mi).

If you could travel by spaceship a few hundred light-years from the earth, you would observe the sun shrink to a bright point of light among the billions and billions of other stars. The sun is just an ordinary star with an average brightness. Like the other stars, the sun is a massive, dense ball of gases with a surface heated to incandescence by energy released from fusion reactions deep within. Since the sun is an average star, it can be used as a reference for understanding all the other stars.

ORIGIN OF STARS

Theoretically, a star with the mass of the sun was born from swirling clouds of hydrogen gas in the deep space between other stars. Such interstellar (between stars) clouds are called **nebulae.** These clouds consist of random, swirling atoms of gases that have little gravitational attraction for one another because they have little mass. Complex motion of stars, however, can produce

a shock wave that causes particles to move closer together, making local compressions. Their mutual gravitational attraction then begins to pull them together into a cluster. The cluster grows as more atoms are pulled in, which increases the mass and thus the gravitational attraction, and still more atoms are pulled in from farther away. Theoretical calculations indicate that on the order of 1×10^{57} atoms are necessary, all within a distance of 3 trillion kilometers (about 1.9 trillion miles). When these conditions occur, the cloud of gas atoms begins to condense by gravitational attraction to a **protostar,** an accumulation of gases that will become a star.

Gravitational attraction pulls the average protostar from a cloud with a diameter of trillions of kilometers (trillions of miles) down to a dense sphere with a diameter of 2.5 million kilometers (1.6 million miles) or so. As gravitational attraction accelerates the atoms toward the center, they gain kinetic energy, and the interior temperature increases. Over a period of some 10 million years of contracting and heating, the temperature and density conditions at the center of the protostar are sufficient to start nuclear fusion reactions. Pressure from hot gases and energy from increasing fusion reactions begin to balance the gravitational attraction over the next 17 million years, and the newborn, average star begins its stable life, which will continue for the next 10 billion years.

The interior of an average star, such as the sun, is modeled after the theoretical pressure, temperature, and density conditions that would be necessary to produce the observed energy and light from the surface. This model describes the interior as a set of three shells: (1) the core, (2) a radiation zone, and (3) the convection zone (figure 12.3).

Our model describes the **core** as a dense, very hot region where nuclear fusion reactions release gamma and X-ray radiation. The density of the core is about twelve times that of solid

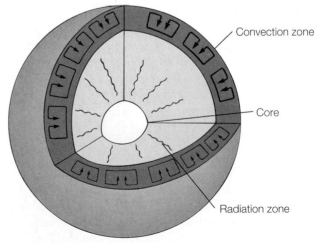

Figure 12.3
The structure of an average, mature star such as the sun consists of a core, a radiation zone, and a convection zone. Hydrogen fusion reactions occur in the core, releasing gamma and X-ray radiation. This radiation moves through the radiation zone from particle to particle, eventually heating gases at the base of the convection zone. Convection cells carry energy to the surface, where it is emitted to space as visible light, ultraviolet radiation, and infrared radiation.

lead. Because of the exceedingly hot conditions, however, the core remains in a gaseous state even at this density.

The model describes the **radiation zone** as less dense than the core, having a density about the same as the density of water. Energy in the form of gamma and X rays from the core is absorbed and reemitted by collisions with atoms in this zone. The radiation slowly diffuses outward because of the countless collisions over a distance comparable to the distance between the earth and the moon. It could take millions of years before this radiation finally escapes the radiation zone.

In the model the **convection zone** begins about seven-tenths of the way to the surface, where the density of the gases is about 1 percent the density of water. Gases at the bottom of this zone are heated by radiation from the radiation zone below, expand from the heating, and rise to the surface by convection. At the surface, the gases emit energy in the form of visible light, ultraviolet radiation, and infrared radiation, which moves out into space. As they lose energy, the gases contract in volume and sink back to the radiation zone to become heated again, continuously carrying energy from the radiation zone to the surface in convection cells. The surface is continuously heated by the convection cells as it gives off energy to space, maintaining a temperature of about 5,800 K (about 5,500°C).

As an average star, the sun converts about 1.4×10^{17} kg of matter to energy every year as hydrogen nuclei are fused to produce helium. The sun was born about 5 billion years ago and has sufficient hydrogen in the core to continue shining for another 4 or 5 billion years. Other stars, however, have masses that are much greater or much less than the mass of the sun so they have different life spans. More massive stars generate higher temperatures in the core because they have a greater gravitational contraction from their greater masses. Higher temperatures mean increased kinetic energy, which results in increased numbers of collisions between hydrogen nuclei with the end result an increased number of fusion reactions. Thus, a more massive star uses up its hydrogen more rapidly than a less massive star. On the other hand, stars that are less massive than the sun use their hydrogen at a slower rate so they have longer life spans. The life spans of the stars range from a few million years for large, massive stars, to 10 billion years for average stars like the sun, to trillions of years for small, less massive stars.

BRIGHTNESS OF STARS

Stars generate their own light, but some stars appear brighter than others in the night sky. As you can imagine, this difference in brightness could be related to (1) the amount of light produced by the stars, (2) the size of each star, or (3) the distance to a particular star. A combination of these factors is responsible for the brightness of a star as it appears to you in the night sky. A classification scheme for different levels of brightness that you see is called the **apparent magnitude** scale (table 12.1).

The apparent magnitude of a star depends on how far away stars are in addition to differences in the stars themselves. Stars at a farther distance will appear fainter, and those closer will appear brighter, just as any other source of light. To compensate for distance differences, astronomers calculate the brightness that

Table 12.1 The Apparent Magnitude Scale Comparing Some Familiar Objects and Some Observable Limits

stars would appear to have if they were all at a defined, standard distance. The brightness of a star at this distance is called the **absolute magnitude.** The sun, for example, is the closest star and has an apparent magnitude of −26.7 at an average distance from the earth. When viewed from the standard distance the sun would have an absolute magnitude of +4.8, which is about the brightness of a faint star.

STAR TEMPERATURE

If you observe the stars on a clear night you will notice that some are brighter than others, and you will also notice some color differences. Some stars have a reddish color, some have a bluish white color, and others have a yellowish color. This color difference is understood to be a result of the relationship that exists between the color and the temperature of a glowing object. The

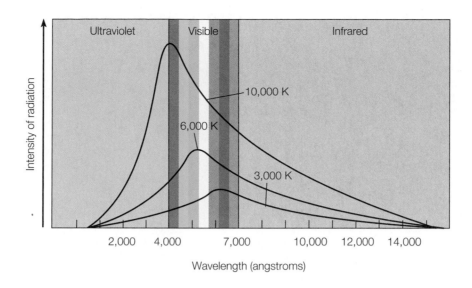

Figure 12.4

The distribution of radiant energy emitted is different for stars with different surface temperatures. Note that the peak radiation of a cooler star is more toward the red part of the spectrum, and the peak radiation of a hotter star is more toward the blue part of the spectrum.

colors of the various stars are a result of the temperatures of the stars (figure 12.4). You see a cooler star as reddish in color and comparatively hotter stars as bluish white. Stars with in-between temperatures, such as the sun, appear to have a yellowish color.

Astronomers use information about the star temperature and spectra as the basis for a star classification scheme. Originally, the classification scheme was based on sixteen categories according to the strength of the hydrogen line spectra (see line

Table 12.2 Major Stellar Spectral Types and Temperatures

Type	Color	Temperature (K)	Comment
O	Bluish	30,000–80,000	Spectrum with ionized helium and hydrogen but little else; short-lived and rare stars
B	Bluish	10,000–30,000	Spectrum with neutral helium, none ionized
A	Bluish	7,500–10,000	Spectrum with no helium, strongest hydrogen, some magnesium and calcium
F	White	6,000–7,500	Spectrum with ionized calcium, magnesium, neutral atoms of iron
G	Yellow	5,000–6,000	The spectral type of the sun. Spectrum shows sixty-seven elements in the sun
K	Orange-red	3,500–5,000	Spectrum packed with lines from neutral metals
M	Reddish	2,000–3,500	Band spectrum of molecules, e.g., titanium oxide; other related spectral types (R, N, and S) are based on other molecules present in each spectral type

spectra in chapter 8). The groups were identified alphabetically with A for the group with the strongest hydrogen line spectrum, B for slightly weaker lines, and on to the last group with the faintest lines. Later, astronomers realized that the star temperature was the important variable, so they rearranged the categories according to decreasing temperatures. The original letter categories were retained, however, resulting in classes of stars with the hottest temperature first and the coolest last with the sequence O B A F G K M. Table 12.2 compares the color, temperature ranges, and other features of the stellar spectra classification scheme.

STAR TYPES

In 1910, Henry Russell in the United States and Ejnar Hertzsprung in Denmark independently developed a scheme to classify stars with a temperature-luminosity graph. The graph is called the **Hertzsprung-Russell diagram,** or the *H-R diagram* for short. The diagram is a plot with temperature indicated by spectral types, and the true brightness indicated by absolute magnitude. The diagram, as shown in figure 12.5, plots temperature by spectral types sequenced O through M types, so the

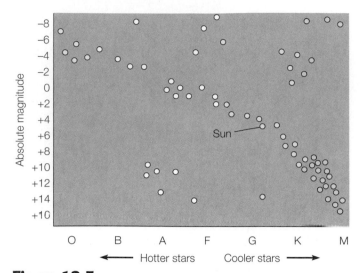

Figure 12.5

The Hertzsprung-Russell diagram.

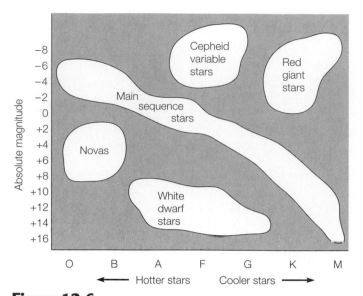

Figure 12.6

The regions of the Cepheid variable, red giant, main sequence, and white dwarf stars and novas on the H-R diagram.

temperature decreases from left to right. The hottest, brightest stars are thus located at the top left of the diagram, and the coolest, faintest stars are located at the bottom right.

Each dot is a data point representing the surface temperature and brightness of a particular star. The sun, for example, is a type G star with an absolute magnitude of about +5, which places the data point for the sun almost in the center of the diagram. This means that the sun is an ordinary, average star with respect to both surface temperature and true brightness.

Most of the stars plotted on an H-R diagram fall in or close to a narrow band that runs from the top left to the lower right. This band is made up of **main sequence stars.** Stars along the main sequence band are normal, mature stars that are using their nuclear fuel at a steady rate. Those stars on the upper left of the main sequence are the brightest, bluest, and most massive stars on the sequence. Those at the lower right are the faintest, reddest, and least massive of the stars on the main sequence. In general, most of the main sequence stars have masses that fall between a range from ten times greater than the mass of the sun (upper left) to one-tenth the mass of the sun (lower right). The extremes, or ends, of the main sequence range from about sixty times more massive than the sun to one-twenty-fifth of the sun's mass. It is the *mass* of a main sequence star that determines its brightness, its temperature, and its location on the H-R diagram. High-mass stars on the main sequence are brighter, hotter, and have shorter lives than low-mass stars. These relationships do not apply to the other types of stars in the H-R diagram.

As shown in figure 12.6, there are two groups of stars that have a different set of properties than the main sequence stars. The **red giant stars** are bright, but low-temperature, giants. These reddish stars are enormously bright for their temperature because they are very large, with an enormous surface area giving off light. A red giant might be one hundred times larger but have the same mass as the sun. These low-density red giants are located in the upper right part of the H-R diagram. The **white dwarf stars,** on the other hand, are located at the lower left because they are faint, white-hot stars. A white dwarf is faint be-

cause it is small, perhaps twice the size of the earth. It is also very dense, with a mass approximately equal to the sun's. During its lifetime a star will be found in different places on the H-R diagram as it undergoes changes. Red giants and white dwarfs are believed to be evolutionary stages that aging stars pass through, and the path a star takes across the diagram is called an evolutionary track. During the lifetime of the sun, it will be a main sequence star, a red giant, and then a white dwarf.

Stars such as the sun emit a steady light because the force of gravitational contraction is balanced by the outward flow of energy. *Variable stars,* on the other hand, are stars that change in brightness over a period of time. A **Cepheid variable** is a bright variable star that is used to measure distances. There is a general relationship between the period and the brightness: the longer the time needed for one pulse, the greater the apparent brightness of that star. The period-brightness relationship to distance was calibrated by comparing the apparent brightness with the absolute magnitude (true brightness) of a Cepheid at a known distance with a known period. This calibration allowed astronomers to calculate the distance to a star by using the inverse-square law of brightness.

Edwin Hubble used the Cepheid period-brightness relationship to find the distances to other galaxies and discovered yet another relationship, that the greater the distance to a galaxy, the greater a shift in spectral lines toward the red end of the spectrum (red shift). This relationship is called Hubble's law, and it forms the foundation for understanding our expanding universe. Measuring red shift provides another means of establishing distances to other, far-out galaxies.

THE LIFE OF A STAR

A star is born in a gigantic cloud of gas and dust in interstellar space, then spends billions of years of calmly shining while it fuses hydrogen nuclei in the core. How long a star shines and what happens to it when it uses up the hydrogen in the core depends on the mass of the star. Of course no one has observed a life cycle of over billions of years. The life cycle of a star is a theoretical outcome based on what is known about nuclear reactions. The predicted outcomes seem to agree with observations of stars today, with different groups of stars that can be plotted on the H-R diagram. Thus, the groups of stars on the diagram—main sequence, red giants, and white dwarfs, for example—are understood to be stars in various stages of their lives.

Protostar Stage. The first stage in the theoretical model of the life cycle of a star is the formation of the protostar. As gravity pulls the gas of a protostar together, the density, pressure, and temperature increase from the surface down to the center. Eventually, the conditions are right for nuclear fusion reactions to begin in the core, which requires a temperature of 10 million kelvins. The initial fusion reaction essentially combines four hydrogen nuclei to form a helium nucleus with the release of much energy. This energy heats the core beyond the temperature reached by gravitational contraction, eventually to 16 million kelvins. Since the star is a gas, the increased temperature expands the volume of the star. The outward pressure of expansion balances the inward pressure from gravitational collapse, and the star settles down to a balanced condition of calmly converting hydrogen to helium in the core, radiating the energy released into space (figure 12.7). The theoretical time elapsed from the initial formation and collapse of the protostar to the main sequence is about 50 million years for a star of a solar mass.

Main Sequence Stage. Where the star is located on the main sequence and what happens to it next depend only on how massive it is. The more massive stars have higher core temperatures and use up their hydrogen more rapidly as they shine at higher surface temperatures (O type stars). Less massive stars shine at lower surface temperatures (M type stars) as they use their fuel at a slower rate. The overall life span on the main sequence ranges from millions of years for O type stars to trillions of years for M type stars. An average one-solar-mass star will last about 10 billion years.

Red Giant Stage. The next stage in the theoretical life of a star begins when much of the hydrogen in the core has been fused into helium. With fewer hydrogen fusion reactions, less energy is released and less outward balancing pressure is produced, so the star begins to collapse. The collapse heats the core, which now is composed primarily of helium, and the surrounding shell where hydrogen still exists. The increased temperature causes the hydrogen in the shell to undergo fusion, and the increased release of energy causes the outer layers of the star to expand. With an increased surface area, the amount of radiation emitted per unit area is less, and the star acquires the properties of a brilliant red giant. Its position on the H-R diagram changes since it now has different brightness and temperature properties. (The star has not physically *moved.* The changing properties move its temperature brightness data point, not the star, to a new position.)

Back Toward Main Sequence. After about five hundred million years as a red giant, the star now has a surface temperature of about 4,000 kelvins compared to its main sequence surface temperature of 6,000 kelvins. The radius of the red giant is now a thousand times greater, a distance that will engulf the earth when the sun reaches this stage, assuming the earth is in the same position as it is today. Even though the surface temperature has decreased from the expansion, the helium core is continually heating and eventually reaches a temperature of 100 million kelvins, the critical temperature necessary for the helium nuclei to undergo fusion to produce carbon. The red giant now has helium fusion reactions in the core and hydrogen fusion reactions in a shell around the core. This changes the radius, the surface temperature, and the brightness with the overall result depending on the composition of the star. In general, the radius and brightness decrease when this stage is reached, moving the star back toward the main sequence (figure 12.8).

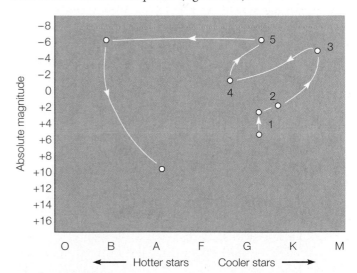

Figure 12.8
The evolution of a star of solar mass as it depletes hydrogen in the core (1), fuses hydrogen in the shell to become a red giant (2 to 3), becomes hot enough to produce helium fusion in the core (3 to 4), then expands to a red giant again as helium and hydrogen fusion reactions move out into the shells (4 to 5). It eventually becomes unstable and blows off the outer shells to become a white dwarf star.

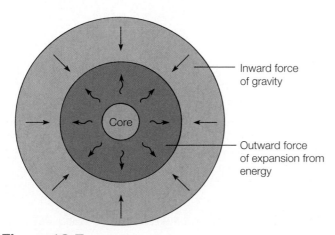

Figure 12.7
A star becomes stable when the outward forces of expansion from the energy released in nuclear fusion reactions balance the inward forces of gravity.

Figure 12.9
The blown-off outer layers of stars form ringlike structures called *planetary nebulae.*

Beginning of the End for Less Massive Stars. After millions of years of helium fusion reactions, the core is gradually converted to a carbon core and helium fusion begins in the shell surrounding the core. The core reactions decrease as the star now has a helium-fusing shell surrounded by a second, hydrogen-fusing shell. This releases additional energy, and the star again expands to a red giant for the second time. A star the size of the sun or less massive may cool enough at this point that nuclei at the surface become neutral atoms rather than a plasma. As neutral atoms they can absorb radiant energy coming from within the star, heating the outer layers. Changes in temperature produce changes in pressure, which change the balance between the temperature, pressure, and the internal energy generation rate. The star begins to expand outward from heating. The expanded gases are cooled by the expansion process, however, and are pulled back to the star by gravity, only to be heated and expand outward again. In other words, the outer layers of the star begin to pulsate in and out. Finally, a violent expansion blows off the outer layers of the star, leaving the hot core. Such blown-off outer layers of a star form circular nebulae called *planetary nebulae* (figure 12.9). The nebulae continue moving away from the core, eventually adding to the dust and gases between the stars. The remaining carbon core and helium-fusing shell begin gravitationally to contract to a small, dense *white dwarf* star. A star with the original mass of the sun, or less, slowly cools from white to red, then to a black lump of carbon in space (figure 12.10).

Beginning of the End for Massive Stars. A more massive star will have a different theoretical ending than the slow cooling of a white dwarf. A massive star will contract, just as the less massive stars, after blowing off its outer shells. In a more mas-

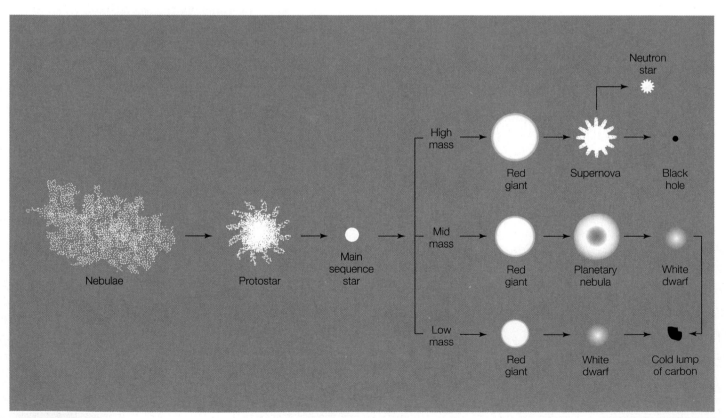

Figure 12.10
This flowchart shows some of the possible stages in the birth and aging of a star. The differences are determined by the mass of the star.

sive star, however, heat from the contraction may reach the critical temperature of 600 million kelvins to begin carbon fusion reactions. Thus, a more massive star may go through a carbon fusing stage and other fusion reaction stages that will continue to produce new elements until the element iron is reached. After iron, energy is no longer released by the fusion process, and the star has used up all of its energy sources. Lacking an energy source, the star is no longer able to maintain its internal temperature. The star loses the outward pressure of expansion from the high temperature, which had previously balanced the inward pressure from gravitational attraction. The star thus collapses, then rebounds like a compressed spring into a catastrophic explosion called a **supernova.** A supernova produces a brilliant light in the sky that may last for months before it begins to dim as the new elements that were created during the life of the star diffuse into space. These include all the elements up to iron that were produced by fusion reactions during the life of the star and heavier elements that were created during the instant of the explosion. All the elements heavier than iron were created as some less massive nuclei disintegrated in the explosion, joining with each other and with lighter nuclei to produce the nuclei of the elements from iron to uranium. As you will see in chapter 13, these newly produced, scattered elements will later become the building blocks for new stars and planets such as the sun and the earth.

The remains of the compressed core after the supernova have yet another fate if the core has a remaining mass greater than 1.4 solar masses. The gravitational forces on the remaining matter, together with the compressional forces of the supernova explosion, are great enough to collapse nuclei, forcing protons and electrons together into neutrons, forming the core of a **neu-**

tron star. A neutron star is the very small (10 to 20 km diameter), superdense (10^{11} kg/cm^3 or greater) remains of a supernova with a center core of pure neutrons.

Because it is a superconcentrated form of matter, the neutron star also has an extremely powerful magnetic field, capable of becoming a pulsar. A **pulsar** is a very strongly magnetized neutron star that emits a uniform series of equally spaced electromagnetic pulses. Evidently, the magnetic field of a rotating neutron star makes it a powerful electric generator, capable of accelerating charged particles to very high energies. These accelerated charges are responsible for emitting a beam of electromagnetic radiation, which sweeps through space with amazing regularity. The pulsating radio signals from a pulsar were a big mystery when first discovered. For a time extraterrestrial life was considered as the source of the signals, so they were jokingly identified as LGM (for Little Green Men). Over 300 pulsars have been identified, and most emit radiation in the form of radio waves. Two, however, emit visible light, two emit beams of gamma radiation, and one emits X-ray pulses.

Another theoretical limit occurs if the remaining core has a mass of about 3 solar masses or more. At this limit the force of gravity overwhelms *all* nucleon forces, including the repulsive forces between like charged particles. If this theoretical limit is reached, nothing can stop the collapse, and the collapsed star will become so dense that even light cannot escape. The star is now a **black hole** in space. Since nothing can stop the collapsing star, theoretically a black hole would continue to collapse to a pinpoint and then to a zero radius called a *singularity*. This event seems contrary to anything that can be directly observed in the physical universe, but it does agree with the general theory of relativity and concepts about the curvature of space

Connections . . .

Navigation and Migration

Since animals move from place to place to meet their needs it is useful to be able to return to a nest, water hole, den, or favorite feeding spot. This requires some sort of memory of their surroundings (a mental map) and a way of determining direction. Often it is valuable to have information about distance as well. Direction can be determined by such things as magnetic fields, identifying landmarks, scent trails, or reference to the sun or stars. If the sun or stars are used for navigation, then some sort of time sense is also needed since these bodies move in the sky.

Like honeybees, some daytime-migrating birds use the sun to guide them. We need two instruments to navigate by the sun—an accurate clock and a sextant for measuring the angle between the sun and the horizon. Can a bird perform such measurements without instruments when we, with our much bigger brains, need these instruments to help us? It is unquestionably

true! For nighttime migration, some birds use the stars to help them find their way. In one interesting experiment, warblers, which migrate at night, were placed in a planetarium. The pattern of stars as they appear at any season could be projected onto a large domed ceiling. During autumn, when these birds would normally migrate southward, the stars of the autumn sky were shown on the ceiling. The birds responded with much fluttering activity at the south side of the cage, as if they were trying to migrate southward. Then the experimenters tried projecting the stars of the spring sky, even though it was autumn. Now the birds tended to try to fly northward, although there was less unity in their efforts to head north; the birds seemed somewhat confused. Nevertheless, the experiment showed that the birds recognized star patterns and were influenced by them.

produced by such massively dense objects. Black holes are theoretical and none has been seen, of course, because a black hole theoretically pulls in radiation of all wavelengths and emits nothing. Evidence for the existence of a black hole is sought by studying X rays that would be given off by matter as it is accelerated into a black hole.

Another form of evidence for the existence of a black hole has now been provided by the Hubble Space Telescope. Hubble pictured a disk of gas only about 60 light-years out from the center of a galaxy (M87), moving at more than 1.6 million km/h (about 1 million mi/h). The only known possible explanation for such a massive disk of gas moving with this velocity at the distance observed would require the presence of a 1–2 billion solar-mass black hole. This gas disk could only be resolved by the Hubble Space Telescope, so this telescope has provided the first observational evidence of a black hole.

GALAXIES

Stars are associated with other stars on many different levels, from double stars that orbit a common center of mass, to groups of tens or hundreds of stars that have gravitational links and a common origin, to the billions and billions of stars that form the basic unit of the universe, a **galaxy.** The sun is but one of an estimated one hundred billion stars that are held together by gravitational attraction in the Milky Way galaxy. The numbers of stars and vastness of the Milky Way galaxy alone seem almost beyond comprehension, but there is more to come. The Milky Way is but one of *billions* of galaxies that are associated with other galaxies in clusters, and these clusters are associated with one another in superclusters. Through a large telescope you can see more galaxies than individual stars in any direction, each galaxy with its own structure of billions of stars. Yet, there are similarities that point to a common origin. Some of the similarities and associations of stars will be introduced in this section along with the Milky Way galaxy, the vast, flat, spiraling arms of stars, gas, and dust where the sun is located (figure 12.11).

The Milky Way Galaxy

Away from city lights, you can clearly see the faint, luminous band of the Milky Way galaxy on a moonless night. Through a telescope or a good pair of binoculars, you can see that the luminous band is made up of countless numbers of stars. You may also be able to see the faint glow of nebulae, concentrations of gas and dust. There are dark regions in the Milky Way that also give an impression of something blocking starlight, such as dust. You can also see small groups of stars called **galactic clusters.** Galactic clusters are gravitationally bound subgroups of as many as one thousand stars that move together within the Milky Way. Other clusters are more symmetrical and tightly packed, containing as many as a million stars, and are known as **globular clusters.**

Viewed from a distance in space, the Milky Way would appear to be a huge, flattened cloud of spiral arms radiating out from the center. There are three distinct parts: (1) the spherical concentration of stars at the center of the disk called the *galactic nucleus;* (2) the rotating *galactic disk,* which contains most of the bright, blue stars along with much dust and gas; and (3) a spherical *galactic halo,* which contains some 150 globular clusters located outside the galactic disk (figure 12.12). The sun is located in one of the arms of the galactic disk, some 25,000 to 30,000 light-years from the center. The galactic disk rotates, and the sun completes one full rotation every 200 million years.

The diameter of the *galactic disk* is about 100,000 light-years. Yet, in spite of the one hundred billion stars in the Milky Way, it is mostly full of emptiness. By way of analogy, imagine reducing the size of the Milky Way disk until stars like the sun were reduced to the size of tennis balls. The distance between two of these tennis-ball-sized stars would now compare to the distance across the state of Texas. The space between the stars is not actually empty since it contains a thin concentration of gas, dust, and molecules of chemical compounds. The gas particles outnumber the dust particles about 10^{12} to 1. The gas is mostly hydrogen, and the dust is mostly solid iron, carbon, and silicon compounds. Over forty different chemical molecules have been discovered in the space between the stars, including many organic molecules. Some nebulae consist of clouds of molecules with a maximum density of about 10^6 molecules/cm^3. The gas, dust, and chemical compounds make up part of the mass of the galactic disk, and the stars make up the remainder. The gas plays an important role in the formation of new stars, and the dust and chemical compounds play an important role in the formation of planets.

Other Galaxies

Outside the Milky Way is a vast expanse of emptiness, lacking even the few molecules of gas and dust spread thinly through the galactic nucleus. There is only the light from faraway galaxies and the time that it takes for this light to travel across the vast vacuum of intergalactic space. How far away is the nearest galaxy? Recall that the Milky Way is so large that it takes light 100,000 years to travel the length of its diameter.

The nearest galactic neighbor to the Milky Way is a dwarf spherical galaxy only 80,000 light-years from the solar system. The nearby galaxy is called a dwarf because it has a diameter of only about 1,000 light-years. It is apparently in the process of being pulled apart by the gravitational pull of the Milky Way, which now is known to have eleven satellite galaxies.

The nearest galactic neighbor similar to the Milky Way is Andromeda, about 2 million light-years away. Andromeda is similar to the Milky Way in size and shape, with about one hundred billion stars, gas, and dust turning in a giant spiral pinwheel (figure 12.13). Other galaxies have other shapes and other characteristics. The American astronomer Edwin Hubble developed a classification scheme for the structure of galaxies based on his 1926 study of some six hundred different galaxies. The basic galactic structures were identified as elliptical, spiral, barred, and irregular.

Figure 12.11

A wide-angle view toward the center of the Milky Way galaxy. Parts of the white, milky band are obscured from sight by gas and dust clouds in the galaxy.

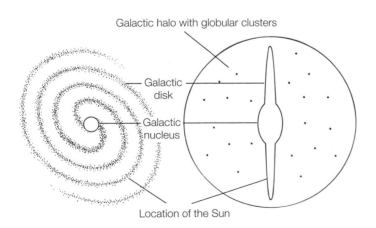

Galactic halo with globular clusters

Galactic disk

Galactic nucleus

Location of the Sun

Figure 12.12

The structure of the Milky Way galaxy.

Extraterrestrial is a descriptive term, meaning a thing or event outside the earth or its atmosphere. The term is also used to describe a being, a life-form that originated away from the earth. This reading is concerned with the search for extraterrestrials, intelligent life that might exist beyond the earth and outside the solar system.

Why do people believe that extraterrestrials might exist? The affirmative answer comes from a mixture of current theories about the origin and development of stars, statistical odds, and faith. Considering the statistical odds, note that our sun is one of the some 300 billion stars that make up our Milky Way galaxy. The Milky Way galaxy is one of some 10 billion galaxies in the observable universe. Assuming an average of about 300 billion stars per galaxy, this means there are some 300 billion times 10 billion stars, or 3×10^{21} stars in the observable universe. There is nothing special or unusual about our sun, and astronomers believe it to be quite ordinary among all the other stars (all 3,000,000,000,000,000,000,000 or so).

So the sun is an ordinary star, but what about our planet? Not too long ago most people, including astronomers, thought our solar system with its life-supporting planet (Earth) to be unique. Evidence collected over the past decade or so, however, has strongly suggested that this is not so. Planets are now believed to be formed as a natural part of the star-forming process. Evidence of planets around other stars has also been detected by astronomers. One of the stars with planets is "only" 53 light-years from Earth.

Even with a very low probability of planetary systems forming with the development of stars, a population of 3×10^{21} stars means there are plenty of planetary systems in existence, some with the conditions necessary to support life (Note:

Box Figure 12.1
The 300 m (984 ft) diameter radio telescope at Arecibo, Puerto Rico, is the largest fixed-dish radio telescope in the world.

If 1 percent have planetary systems, this means 3×10^{19} stars have planets). Thus, it is a statistical observation that suitable planets for life are very likely to exist. In addition, radio astronomers have found that many organic molecules exist, even in the space between the stars. Based on statistics alone, there should be life on other planets, life that may have achieved intelligence and developed into a technological civilization.

If extraterrestrial life exists, why have we not detected them or why have they not contacted us? The answer to this question is found in the unbelievable distances involved in interstellar space. For example, a logical way to search for extraterrestrial intelligence is to send radio signals to space and analyze those coming from space. Modern radio telescopes can send powerful radio beams and present-day computers and

data processing techniques can now search through incoming radio signals for the patterns of artificially generated radio signals (box figure 12.1). Radio signals, however, travel through space at the speed of light. The diameter of our Milky Way galaxy is about 100,000 light-years, which means 100,000 years would be required for a radio transmission to travel across the galaxy. If we were to transmit a super, super strong radio beam from the earth, it would travel at the speed of light and cross the distance of our galaxy in 100,000 years. If some extraterrestrials on the other side of the Milky Way galaxy did detect the message and send a reply, it could not arrive at Earth until 200,000 years after the message was sent. Now consider the fact that of all the 10 billion other galaxies in the observable universe, our nearest galactic neighbor similar to the Milky Way is Andromeda. Andromeda is 2 million light-years from the Milky Way galaxy.

In addition to problems with distance/time, there are questions about in which part of the sky you should send and look for radio messages, questions about which radio frequency to use, and problems with the power of present-day radio transmitters and detectors. Realistically, the hope for any exchange of radio-transmitted messages would be restricted to within several hundred light-years of Earth.

Considering all the limitations, what sort of signals should we expect to receive from extraterrestrials? Probably a series of pulses that somehow indicate counting, such as 1, 2, 3, 4, and so on, repeated at regular intervals. This is the most abstract, while at the same time the simplest concept that an intelligent being anywhere would have. It could provide the foundation for communications between the stars.

The Life of a Galaxy

Hubble's classification of galaxies into distinctly different categories of shape was an exciting accomplishment because it suggested that some relationship or hidden underlying order might exist in the shapes. Finding underlying order is important because it leads to the discovery of the physical laws that govern the universe. Soon after Hubble published his classification results in 1926, two models of galactic evolution were proposed. One model, which was suggested by Hubble, had extremely slowly spinning spherical galaxies forming first, which gradually flattened out as their rate of spin increased while they condensed. This is a model of spherical galaxies flattening out to increasingly elliptical shapes, eventually spinning

off spirals until they finally broke up into irregular shapes over a long period of time.

Among many uses, the Hubble Space Telescope is used to study young galaxies and galaxies on collision courses. Based on these studies, astronomers today recognize that the different shapes of galaxies do not represent an evolutionary sequence. Their shapes are understood to be a result of the various conditions under which the galaxies were formed.

The current model of how galaxies form is based on the **big bang theory** of the creation of the universe. The big bang theory considers the universe to have had an explosive beginning. According to this theory all matter in the universe was located together in an arbitrarily dense state from which it began to expand, an expansion that continues today. Evidence that supports the big

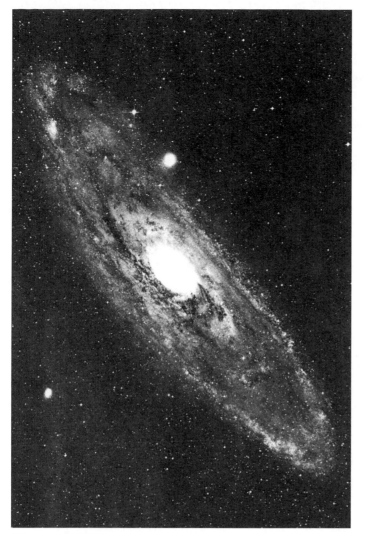

Figure 12.13

The Andromeda galaxy, which is believed to be similar in size, shape, and structure to the Milky Way galaxy.

bang theory comes from (1) present-day microwave radiation from outer space, (2) current data on the expansion of the universe, (3) the relative abundance of elements that were altered in the core of older stars is in agreement with predictions based on analysis of the big bang, and (4) the Cosmic Background Explorer (COBE) spacecraft which studied diffuse cosmic background radiation. Findings announced from WMAPS in 2003 confirmed several important predictions from the big bang theory.

The initial evidence for the big bang theory came from Edwin Hubble and his earlier work with galaxies. Hubble had determined the distances to some of the galaxies that had red-shifted spectra. From this expansive red shift it was known that these galaxies were moving away from the Milky Way. Hubble found a relationship between the distance to a galaxy and the velocity with which it was moving away. He found the velocity to be directly proportional to the distance; that is, the greater the distance to a galaxy the greater the velocity. This means that a galaxy twice as far from the Milky Way as a second galaxy

is moving away from the Milky Way at twice the speed as the second galaxy. Since this relationship was seen in all directions, it meant that the universe is expanding uniformly. The same effect would be viewed from any particular galaxy; that is, all the other galaxies are moving away with a velocity proportional to the distance to the other galaxies. This points to a common beginning, a time when all matter in the universe was together.

How old is the universe? Astronomical and physical "clocks" indicate that the universe was created in a "big bang" some 13 billion years ago (give or take a billion), expanding as an intense and brilliant explosion from a primeval fireball with a temperature of some 10^{12} kelvins. This estimate of the age is based on precise measurements of the rate at which galaxies are moving apart, an expansion that started with the big bang. Astronomers used data on the expansion to back-calculate the age, much like running a movie backward, to arrive at the age estimate. At first, this technique found the universe was about 18 to 20 billion years old. But then astronomers found that the universe is not expanding at a constant rate. Instead, the separation of galaxies is actually accelerating, pushed by a mysterious force known as **dark energy.** By adding in calculations for this poorly understood force, the estimate of 13 to 14 billion years was developed.

Other age-estimating techniques agree with this approximation. Studies of white dwarfs, for example, established their rate of cooling. By looking at the very faintest and oldest white dwarfs with the Hubble Space Telescope, astronomers were able to use the cooling rate to estimate the age of the universe. The result found the dimmest of the white dwarfs to be about 13 billion years old, plus or minus about half a billion years. An age of 13 billion years may not be the final answer for the universe's age, but when two independent measurement techniques agree closely the answer becomes more believable.

An often-used analogy for the movement of galaxies after the big bang is a loaf of rising raisin bread. Consider galaxies as raisins in a loaf of raisin bread dough as it rises (figure 12.14). As the dough expands in size it carries along the raisins, which are moved farther and farther apart by the expansion. If you were on a raisin you would see all the other raisins moving away from you, and the speed of this movement would be directly proportional to their distance away. It would not matter which raisin you were on since all the raisins would appear to be moving away from all the other raisins. In other words, the galaxies are not expanding into space that is already there. It is space itself that is expanding, and the galaxies move with the expanding space (figure 12.15).

The ultimate fate of the universe will strongly depend on the mass of the universe and if the expansion is slowing or continuing. All the evidence tells us it began by expanding with a big bang. Will it continue to expand, becoming increasingly cold and diffuse? Will it slow to a halt, then start to contract to a fiery finale of a big crunch (figure 12.16)? Researchers continue searching for answers with experimental data and theoretical models, looking for clues in matches between the data and models.

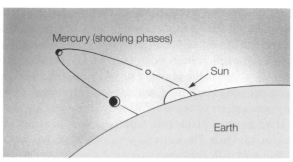

Figure 13.3

Mercury is close to the Sun and is visible only briefly before or after sunrise or sunset, showing phases. Mercury actually appears much smaller in an orbit that is not tilted as much as shown in this figure.

rotates once every fifty-nine days. This means that Mercury rotates on its axis three times every two orbits.

The long Mercury day with a nearby large, hot Sun means high temperatures on the surface facing the sun. High temperatures mean higher gas kinetic energies, and with a low gravity, gases easily escape from Mercury so it has only trace gases for an atmosphere. The lack of an atmosphere to even the heat gains from the long days and heat losses from the long nights results in some very large temperature differences. The temperature of the surface of Mercury ranges from above the melting point of lead on the sunny side to below the temperature of liquid oxygen on the dark side.

Mercury has been visited by only one spacecraft, *Mariner 10*, which flew by three times in 1973 and 1974. The photographs transmitted by *Mariner 10* revealed that the surface of Mercury is covered with craters, and it very much resembles the surface of Earth's Moon. There are large craters, small craters, superimposed craters, and craters with lighter colored rays coming from them just like the craters on the Moon. Also as on Earth's Moon, there are hills and smooth areas with light and dark colors that were covered by lava in the past, some time after most of the impact craters were formed (figure 13.4).

Mercury has no natural satellites or moons, it has a weak magnetic field, and it has an average density more similar to Venus or Earth than to the Moon. The presence of the magnetic field and the relatively high density for such a small body probably means that Mercury has a relatively large core of iron with at least part of the core molten. Because of its high density it is believed that Mercury lost much of its less dense, outer layer of rock materials sometime during its formation.

Figure 13.4

A photomosaic of Mercury made from pictures taken by the *Mariner 10* spacecraft. The surface of Mercury is heavily cratered, looking much like the surface of Earth's Moon. All the interior planets and the Moon were bombarded early in the life of the solar system.

Venus

Venus is the brilliant evening and morning "star" that appears near sunrise or sunset, sometimes shining so brightly that you can see it while it is still daylight. Venus orbits the Sun at an average distance of about 0.7 AU. Venus is sometimes to the left of the Sun, appearing as the evening star, and sometimes to the right of the Sun, appearing as the morning star. Venus also has phases just as the Moon does. When Venus is in the full phase it is small and farthest away from Earth. A crescent Venus appears much larger and thus the brightest because it is closest to Earth when it is in its crescent phase. You can see the phases of Venus with a good pair of binoculars.

Venus shines brightly because it is covered with clouds that reflect about 80 percent of the sunlight, making it the brightest object in the sky after the Sun and the Moon. These same clouds prevented any observations of the surface of Venus until the early 1960s, when radar astronomers were able to penetrate the clouds and measure the planet's spin rate. Venus was found to spin slowly, so slowly that each day on Venus is longer than a Venus year! Also a surprise, Venus was found to spin in the *opposite* direction to its direction of movement in its orbit. On Venus, you would observe the Sun to rise in the west and set in the east, if you could see it, that is, through all the clouds.

In addition to early studies by radio astronomers, Venus has been the target of many American and Soviet spacecraft probes (table 13.2).

Venus has long been called Earth's sister planet since its mass, size, and density are very similar. That is where the similarities with Earth end, however, as Venus has been found to have a very hostile environment. Spacecraft probes found a hot, dry surface under tremendous atmospheric pressure (figure 13.5). The atmosphere consists mostly of carbon dioxide, a few percent of nitrogen, and traces of water vapor and other gases. The atmospheric pressure at the surface of Venus is almost one hundred times the pressure at the surface of Earth, a pressure many times beyond what a human could tolerate. The average surface temperature is comparable to the surface temperature on Mercury, which is hot enough to melt lead. The hot temperature on Venus, which is nearly twice the distance from the Sun as Mercury, is a result of the greenhouse effect. Sunlight filters through the atmosphere of Venus, warming the surface. The surface reemits the energy in the form of infrared radiation, which is absorbed by the almost pure carbon dioxide atmosphere. Carbon dioxide molecules absorb the infrared radiation, increasing their kinetic energy and the temperature.

The surface of Venus is mostly a flat, rolling plain. There are several raised areas, or "continents," on about 5 percent of the surface. There is also a mountain larger than Mount Everest, a great valley deeper and wider than the Grand Canyon, and many large, old impact craters. In general, the surface of Venus appears to have evolved much as did the surface of Earth, but without the erosion caused by ice, rain, and running water.

Venus, like Mercury, has no satellites. Venus also does not have a magnetic field, as might be expected. Two condi-

tions seem to be necessary in order for a planet to generate a magnetic field: a molten center and a relatively rapid rate of rotation. Since Venus takes 243 days to complete one rotation, the slowest of all the planets, it does not have a magnetic field even if some of the interior of Venus is still liquid as in Earth.

Table 13.2	Spacecraft Missions to Venus		
Date	Name	Owner	Remark
Feb 12, 1961	*Venera 1*	U.S.S.R.	Flyby
Aug 27, 1962	*Mariner 2*	U.S.	Flyby
Apr 2, 1964	*Zond 1*	U.S.S.R.	Flyby
Nov 12, 1965	*Venera 2*	U.S.S.R.	Flyby
Nov 16, 1965	*Venera 3*	U.S.S.R.	Crashed on Venus
June 12, 1967	*Venera 4*	U.S.S.R.	Impacted Venus
June 14, 1967	*Mariner 5*	U.S.	Flyby
Jan 5, 1969	*Venera 5*	U.S.S.R.	Impacted Venus
Jan 10, 1969	*Venera 6*	U.S.S.R.	Impacted Venus
Aug 17, 1970	*Venera 7*	U.S.S.R.	Venus landing
Mar 27, 1972	*Venera 8*	U.S.S.R.	Venus landing
Nov 3, 1973	*Mariner 10*	U.S.	Venus, Mercury flyby photos
June 8, 1975	*Venera 9*	U.S.S.R.	Lander/orbiter
June 14, 1975	*Venera 10*	U.S.S.R.	Lander/orbiter
May 20, 1978	*Pioneer 12* (also called *Pioneer Venus 1* or *Pioneer Venus*)	U.S.	Orbital studies of Venus
Aug 8, 1978	*Pioneer 13* (also called *Pioneer Venus 2* or *Pioneer Venus*)	U.S.	Orbital studies of Venus
Sept 9, 1978	*Venera 11*	U.S.S.R.	Lander; sent photos
Sept 14, 1978	*Venera 12*	U.S.S.R.	Lander; sent photos
Oct 30, 1981	*Venera 13*	U.S.S.R.	Lander; sent photos
Nov 4, 1981	*Venera 14*	U.S.S.R.	Lander; sent photos
June 2, 1983	*Venera 15*	U.S.S.R.	Radar mapper
June 7, 1983	*Venera 16*	U.S.S.R.	Radar mapper
Dec 15, 1984	*Vega 1*	U.S.S.R.	Venus/Comet Halley probe
Dec 21, 1984	*Vega 2*	U.S.S.R.	Venus/Comet Halley probe
May 4, 1989	*Magellan*	U.S.	Orbital radar mapper
Oct 18, 1989	*Galileo*	U.S.	Flyby measurements and photos

Figure 13.5

This is an image of an 8 km (5 mile) high volcano on the surface of Venus. The image was created by a computer using *Magellan* radar data, simulating a viewpoint elevation of 1.7 km (1 mile) above the surface. The lava flows extend for hundreds of kilometers across the fractured plains shown in the foreground. The simulated colors are based on color images recorded by the Soviet *Venera 13* and *14* spacecraft.

Earth's Moon

Next to the Sun, the Moon is the largest, brightest object in the sky. The Moon is Earth's nearest neighbor and surface features can be observed with the naked eye. With the aid of a telescope or a good pair of binoculars, you can see light-colored mountainous regions called the *lunar highlands,* smooth dark areas called *maria,* and many sizes of craters, some with bright streaks extending from them (figure 13.6). The smooth dark areas are called maria after a Latin word meaning "sea." They acquired this name from early observers who thought the dark areas were oceans and the light areas were continents. Today, the maria are

---

CONCEPTS APPLIED

A Bigger Moon?

Why does the Moon appear so large when it is on the horizon? The truth is that the Moon is actually *smaller* when it is on the horizon than when it is overhead, because on the horizon it is farther away by a distance of one Earth radius. It is an optical illusion that causes the Moon to appear larger on the horizon. You can test this illusion by bending a paper clip so that it is "one moon" wide when held at arm's lenth. Repeat the paper clip measurement when the Moon is overhead.

---

understood to have formed from ancient floods of molten lava that poured across the surface and solidified to form the "seas" of today. There is no water and no atmosphere on the Moon.

Many facts known about the Moon were established during the *Apollo* missions, the first human exploration of a place away from Earth. A total of twelve *Apollo* astronauts walked on the Moon, taking thousands of photographs, conducting hundreds of experiments, and returning to Earth with over 380 kg (about 840 lb) of moon rocks (table 13.3). In addition, instruments were left on the Moon that continued to radio data back to Earth after the *Apollo* program ended in 1972. As a result of the *Apollo* missions, many questions were answered about the Moon, but unanswered questions still remain.

The *Apollo* astronauts found that the surface of the Moon is covered by a 3 m (about 10 ft) layer of fine gray dust that contains microscopic glass beads. The dust and beads were formed from millions of years of continuous bombardment of micrometeorites. These very small meteorites generally burn up in Earth's atmosphere. The Moon does not have an atmosphere, so meteorites have continually fragmented and pulverized the surface in a slow, steady rain. The glass beads are believed to have formed when larger meteorite impacts melted part of the surface, which was immediately forced into a fine spray that cooled rapidly while above the surface.

The rocks on the surface of the Moon were found to be mostly *basalts,* a type of rock formed on Earth from the cooling and solidification of molten lava. The dark-colored rocks from the maria are similar to Earth's basalts but contain greater amounts of titanium and iron oxides. The light-colored rocks from the highlands are mostly *breccias,* a kind of rock made up of rock fragments that have been compacted together. On the Moon, the compacting was done by meteorite impacts. The rocks from the highlands contain more aluminum and less iron than the maria basalts and thus have a lower density than the darker mare rocks.

All the moon rocks contained a substantial amount of radioactive elements, which made it possible to precisely measure their age. The light-colored rocks from the highlands were formed about 4 billion years ago. The dark-colored rocks from the maria were much younger, with ages ranging from 3.1 to 3.8 billion years. This indicates a period of repeated volcanic eruptions and lava flooding over a 700-million-year period that ended about 3 billion years ago.

The moon rocks brought back to Earth, the results of the lunar seismographs, and all the other data gathered through the *Apollo* missions have increased our knowledge about the Moon, leading to new understandings of how it formed. This model pictures the present Moon developing through four distinct stages.

The *origin stage* describes how the moon originally formed. The Moon is believed to have originated from the impact of Earth with a very large object, perhaps as large as Mars or larger. The Moon formed from ejected material produced by this collision. The impact theory is supported by the evidence found in the moon rocks and is widely accepted, but not all parts of this theory are fully developed.

Figure 13.6
You can easily see the light-colored lunar highlands, smooth and dark maria, and many craters on the surface of Earth's nearest neighbor in space.

Table 13.5 Completed Spacecraft Missions to Study Comets

Date	Name	Owner	Remark
Sep 11, 1985	*ISSE 3* (or *ICE*)	U.S.	Studies of electric and magnetic fields around Giacobini-Zinner comet from 7,860 km (4,880 mi)
Mar 6, 1986	*Vega 1*	U.S.S.R.	Photos and studies of nucleus of Halley's comet from 8,892 km (5,525 mi)
Mar 8, 1986	*Suisei*	Japan	Studied hydrogen halo of Halley's comet from 151,000 km (93,800 mi)
Mar 9, 1986	*Vega 2*	U.S.S.R.	Photos and studies of nucleus of Halley's comet from 8,034 km (4,992 mi)
Mar 11, 1986	*Sakigake*	Japan	Studied solar wind in front of Halley's comet 7.1 million km (4.4 million mi)
Mar 28, 1986	*Giotto*	ESA	Photos and studies of Halley's comet from 541 km (336 mi)
Mar 28, 1986	*ISSE 3* (or *ICE*)	U.S.	Studies of electric and magnetic fields around Halley's comet from 32 million km (20 million mi)

There is also a disk-shaped region of small icy bodies, which ranges from about 30 to 100 AU from the Sun, called the **Kuiper Belt.** The small icy bodies in the Kuiper Belt are understood to be the source of short-period comets, with orbital periods of less than 200 years. There are thousands of Kuiper Belt objects that are larger than 100 km in diameter, and six are known to be orbiting between Jupiter and Neptune. Called *Centaurs,* these objects are believed to have escaped the Kuiper Belt. Centaurs might be small, icy bodies similar to Pluto. Indeed, some speculate that Pluto is a large Kuiper Belt object.

The current theory of the origin of comets was developed by the Dutch astronomer Jan Oort in 1950. According to the theory, the huge cloud and belt of icy, dusty aggregates are leftovers from the formation of the solar system and have been slowly circling the solar system ever since it formed. Something, perhaps a gravitational nudge from a passing star, moves one of the icy bodies enough that it is pulled toward the Sun in what will become an extremely elongated elliptical orbit. The icy, dusty body forms the only substantial part of a comet, and the body is called the comet *nucleus.*

Observations by the *Vega* and *Giotto* spacecrafts found the nucleus of Halley's comet to be an elongated mass of about 8 by 11 km (about 5 by 7 mi) with an overall density less than one-fourth that of solid water ice. As the comet nucleus moves toward the Sun, it warms from the increasing intense solar radiation. Somewhere between Jupiter and Mars the ice and frozen gases begin to vaporize, releasing both grains of dust and evaporated

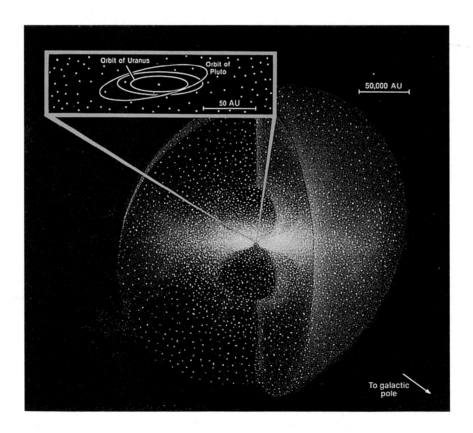

Figure 13.17

Although there may be as many as 1 trillion comets in the Oort cloud, the volume of space that the Oort cloud occupies is so immense that the comets are separated from one another by distances that are typically about 10 AU.

ices. These materials form a large hazy head around the comet called a *coma*. The coma grows larger with increased vaporization, perhaps several hundred or thousands of kilometers across. The coma reflects sunlight as well as producing its own light, making it visible from Earth. The coma generally appears when a comet is within about 3 AU of the Sun. It reaches its maximum diameter about 1.5 AU from the Sun. The nucleus and coma together are called the *head* of the comet. In addition, a large cloud of invisible hydrogen gas surrounds the head and this hydrogen *halo* may be hundreds of thousands of kilometers across.

As the comet nears the Sun, the solar wind and solar radiation ionize gases and push particles from the coma, pushing both into the familiar visible *tail* of the comet. Comets may have two types of tails, (1) ionized gases, and (2) dust. The dust is pushed from the coma by the pressure from sunlight. It is visible because of reflected sunlight. The ionized gases are pushed into the tail by magnetic fields carried by the solar wind. The ionized gases of the tail are fluorescent, emitting visible light because they are excited by ultraviolet radiation from the Sun. The tail generally points away from the Sun, so it follows the comet as it approaches the Sun, but leads the comet as it moves away from the Sun (figure 13.18).

Comets are not very massive or solid, and the porous, snow-like mass has a composition more similar to the giant planets than to the terrestrial planets in comparison. Each time a comet passes near the Sun, it loses some of its mass through evaporation of gases and loss of dust to the solar wind. After passing the Sun the surface forms a thin, fragile crust covered with carbon and other dust particles. Each pass by the Sun means a loss of matter, and the coma and tail are dimmer with each succeeding pass. About 20 percent of the approximately six hundred comets that are known have orbits that return them to the Sun within a two-hundred-year period, some of which return as often as every five or ten years. The other 80 percent have long elliptical orbits that return them at intervals exceeding two hundred years. The famous Halley's comet has a smaller elliptical orbit and returns about every seventy-six years. Halley's comet, like all other comets, may eventually break up into a trail of gas and dust particles that orbit the Sun.

Asteroids

Between the orbits of Mars and Jupiter is a belt, or circular region of thousands of small rocky bodies called **asteroids** (figure 13.19). This belt contains thousands of *asteroids* that range in size from 1 km or less up to the largest asteroid, named Ceres, which has a diameter of about 1,000 km (over 600 mi). The asteroids are thinly distributed in the belt, 1 million km or so apart (about 600,000 mi), but there is evidence of collisions occurring in the past. Most asteroids larger than 50 km (about 30 mi) have been studied by analyzing the sunlight reflected from their surfaces. These spectra provide information about the composition of the asteroids. Asteroids on the inside of the belt are made of stony materials, and those on the outside of the belt are dark with carbon minerals. Still other asteroids are metallic, containing iron and nickel. These spectral composition studies, analyses of the orbits of asteroids, and studies of meteorites that have fallen to

Figure 13.18

As a comet nears the Sun it grows brighter, with the tail always pointing away from the Sun.

Figure 13.19

Most of the asteroids in the asteroid belt are about halfway between the Sun and Jupiter.

Earth all indicate that the asteroids are not the remains of a planet or planets that were broken up. The asteroids are now believed to have formed some 4.6 billion years ago from the original solar nebula. During their formation, or shortly thereafter, their interiors were partly melted, perhaps from the heat of short-lived radioactive decay reactions. Their location close to Jupiter, with its gigantic gravitational field, prevented the slow gravitational clumping-together process that would have formed a planet.

Jupiter's gigantic gravitational field also captured some of the asteroids, pulling them into its orbit. Today there are two groups of asteroids, called the *Trojan asteroids,* which lead and follow Jupiter in its orbit. They lead and follow at a distance where the gravitational forces of Jupiter and the Sun balance to keep them in the orbit. A third group of asteroids, called the *Apollo asteroids,* has orbits that cross the orbit of Earth. It is possible that one of the Apollo asteroids could collide with Earth. One theory about what happened to the dinosaurs is based on evidence that such a collision indeed did occur some sixty-five million years ago. The chemical and physical properties of the two satellites of Mars, Phobos and Deimos, are more similar to the asteroids than to Mars. It is probable that the Martian satellites are captured asteroids.

Meteors and Meteorites

Comets leave trails of dust and rock particles after encountering the heat of the Sun, and collisions between asteroids in the past have ejected fragments of rock particles into space. In space, the remnants of comets and asteroids are called **meteoroids.** When a meteoroid encounters Earth moving through space, it accelerates toward the surface with a speed that depends on its direction of travel and the relative direction that Earth is moving. It soon begins to heat from air friction in the upper atmosphere, melting into a visible trail of light and smoke. The streak of light and smoke in the sky is called a **meteor.** The "falling star" or "shooting star" is a meteor. Most meteors burn up or evaporate completely within seconds after reaching an altitude of about 100 km (about 60 mi) because they are nothing more than a speck of dust. A **meteor shower** occurs when Earth passes through a stream of particles left by a comet in its orbit. Earth might meet the stream of particles concentrated in such an orbit on a regular basis as it travels around the Sun, resulting in predictable meteor showers (table 13.6). In the third week of October, for example, Earth crosses the orbital path of Halley's comet, resulting in a shower of some ten to fifteen meteors per hour. Meteor showers are named for the constellation in which they appear to originate. The October meteor shower resulting from an encounter with the orbit of Halley's comet, for example, is called the Orionid shower because it appears to come from the constellation Orion.

Did you know that atom-bomb-sized meteoroid explosions often occur high in Earth's atmosphere? Most smaller meteors melt into the familiar trail of light and smoke. Larger meteors may fragment upon entering the atmosphere, and the smaller fragments will melt into multiple light trails. Still larger meteors may actually explode at altitudes of about 32 km (about 20 mi) or so. Military satellites that watch Earth for signs of rockets blasting off or nuclear explosions record an average of eight meteor explosions a year. These are big explosions, with an energy equivalent estimated to be similar to a small nuclear bomb. Actual explosions, however, may be ten times larger than the estimation. Based on statistical data, scientists have estimated that every 10 million years Earth should be hit by a very, very large meteor. The catastrophic explosion and aftermath would devastate life over much of the planet, much like the theoretical dinosaur-killing impact of 65 million years ago.

If a meteoroid survives its fiery trip through the atmosphere to strike the surface of Earth, it is called a **meteorite.** Most meteors are from fragments of comets, but most meteorites generally come from particles that resulted from collisions between asteroids that occurred long ago. Meteorites are classified into three basic groups according to their composition: (1) *iron meteorites,* (2) *stony meteorites,* and (3) *stony-iron meteorites* (figure 13.20). The most common meteorites are stony, composed of the same minerals that make up rocks on Earth. The stony meteorites are further subdivided into two groups according to their structure, the *chondrites* and the *achondrites.* Chondrites have a structure of small spherical lumps of silicate minerals or glass, called *chondrules,* held together by a fine-grained cement. The achondrites do not have the chondrules, as their name implies, but have a homogeneous texture more like volcanic rocks such as basalt that cooled from molten rock.

The iron meteorites are about half as abundant as the stony meteorites. They consist of variable amounts of iron and nickel, with traces of other elements. In general, there is proportionally much more nickel than is found in the rocks of Earth. When cut, polished, and etched, beautiful crystal patterns are observed on the surface of the iron meteorite. The patterns mean that the iron was originally molten, then cooled very slowly over millions of years as the crystal patterns formed.

A meteorite is not, as is commonly believed, a ball of fire that burns up the landscape where it lands. The iron or rock has been in the deep freeze of space for some time, and it travels rapidly through the earth's atmosphere. The outer layers become hot enough to melt, but there is insufficient time for this heat to be conducted to the inside. Thus, a newly fallen iron meteorite will be hot since metals are good heat conductors, but it will not be hot enough to start a fire. A stone meteorite is a poor conductor of heat so it will be merely warm.

Table 13.6 Some Annual Meteor Showers		
Name	**Date of Maximum**	**Hour Rate**
Quadrantid	January 3	30
Aqaurid	May 4	5
Perseid	August 12	40
Orionid	October 22	15
Taurids	November 1, 16	5
Leonid	November 17	5
Geminid	December 12	55

A   B

Figure 13.20

(*A*) A stony meteorite. The smooth, black surface was melted by friction with the atmosphere. (*B*) An iron meteorite that has been cut, polished, and etched with acid. The pattern indicates that the original material cooled from a molten material over millions of years.

A Closer Look   Strange Meteorites

Meteorites are a valuable means of learning about the solar system since they provide direct evidence about objects away from Earth in space. Over two thousand meteorites have been collected, classified, and analyzed for this purpose. Most appear to have remained unchanged since they were formed in the asteroid belt between Jupiter and Mars, along with the rest of the solar system, some five billion years ago. Yet a few meteorites are very unusual or strange, not fitting in with the rest. The mystery of two types of these strange meteorites is the topic of this reading.

The first example of a meteorite mystery is the case of the carbonaceous chondrites, which are soft, black stony meteorites with a high carbon content. Fewer than twenty-five of these soft and sooty black fragments have been recovered since the first one was identified in France in 1806. The carbonaceous chondrite is similar to the ordinary chondrite except that it contains carbon compounds, including amino acids. Amino acids are organic compounds basic to life. The mystery of these meteorites is in the meaning of the presence of amino acids. Are they the remains of some former life? Or are they precursors to life,

organic compounds produced naturally when the solar system formed? If this is true, what is the relationship between the amino acids and life on Earth, and what are the implications for elsewhere? The answers to these questions will probably not be available until more of these strange meteorites are available for study.

The second example of a meteorite mystery is the case of the shergottites, which are meteorites of volcanic origin with strange properties. The shergottites are named after a town in India where the first one identified was observed to fall in 1865. A second shergottite fell in Africa in 1962, and a third was recently found on an ice sheet in Antarctica. The shergottites are known to be meteorites because the first two were actually observed to fall and all three have the typical glassy crust that forms from partial melting during the fall.

The strangeness of the shergottites comes from their age, chemical composition, and physical structure as compared to all the other meteorites. Radioactive dating shows that the shergottites are the youngest meteorites known, cooling from a molten state somewhere between 650 million and 1 billion years ago. All of

the other meteorites were originally formed some 4.6 billion years ago in the asteroid belt. In addition, the shergottites do not have the same composition as the asteroids according to spectral analysis. The conclusion so far is that they are not from the asteroid belt. Their chemical composition is more similar to that of a planet such as Earth or Mars than it is to an asteroid.

One interesting feature of the shergottites is the special kind of glass that comprises about half of their masses. This special glass forms only when certain minerals are subjected to tremendous pressures by impact shock. Evidently these meteorites were subjected to a tremendous shock sometime after cooling from the molten state.

The last piece of the puzzle to the mystery comes from the quantitative analysis of the soil on Mars that was performed by the *Viking* landers. The analysis showed that the chemical composition of the shergottites is more similar to Mars than it is to Earth. Large volcanoes were

—Continued top of next page

observed on Mars by the *Viking* spacecraft, and calculations indicate that Mars was volcanically active when the shergottites were formed.

Meteorites are pieces of the solar system that have fallen to Earth. Most come from as-teroids, but fifteen have been identified as orig-inating on the Moon and eighteen from Mars. One of the Martian meteorites, known as Allen Hills 84001, was found to contain possible remnants of life. Today, these particular indica-tions are not accepted as solid evidence. The Martian meteorites, and other meteorites as well, are being carefully studied for information about how they formed, their history, and evi-dence of life.

ORIGIN OF THE SOLAR SYSTEM

Any model of how the solar system originated presents a prob-lem in testing or verification. This problem is that the solar sys-tem originated a long time ago, some 5 billion years ago according to a number of different independent sources of evi-dence. Also, there are no other planetary systems that can be di-rectly observed for comparison either in existence or in the process of being formed. At the distance they occur, even the Hubble Space Telescope would not be able to directly observe planets around their suns. Astronomers have identified about one hundred extrasolar planets. They identify the presence of planets by measuring the very slight wobble of a central star and then use the magnitude of this motion to determine the pres-ence of orbiting planets, the size and shape of their orbits, and their mass. The technique works only for larger planets and can-not detect those much smaller than about half the mass of Sat-urn. The technique does not provide a visual image of the planets, but only measures the gravitational effect of the planets on the star.

The most widely accepted theory of the origin of the solar system is called the **protoplanet nebular model.** A *protoplanet* is the earliest stage in the formation of a planet. The model can be considered in stages, which are not really a part of the model but are simply a convenient way to organize the total picture (figure 13.21).

Stage A. The first important event in the formation of our so-lar system involves stars that disappeared billions of years ago, long before the Sun was born. Earth, the other planets, and all the members of the solar system are composed of elements that were manufactured by these former stars. In a sequence of nu-clear reactions hydrogen fusion in the core of large stars results in the formation of the elements up to iron. Elements heavier than iron are formed in rare supernova explosions of dying mas-sive stars. Thus *stage A* of the formation of the solar system con-sisted of the formation of elements heavier than hydrogen in many, many previously existing stars, including the supernovas of more massive stars. Many stars had to live out their life cycles to provide the raw materials of the solar system. The death of each star, including supernovas, added newly formed elements to the accumulating gases and dust in interstellar space. Over a long period of time these elements began to concentrate in one region of space as dust, gases, and chemical compounds, but hy-drogen was still the most abundant element in the nebula that was destined to become the solar system.

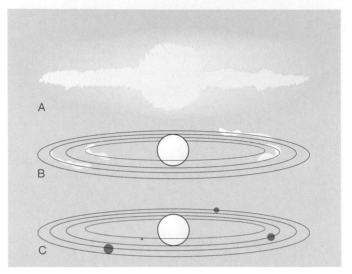

Figure 13.21

Formation of the solar system according to the protoplanet nebular model, not drawn to scale. (*A*) The process starts with a nebula of gas, dust, and chemical elements from previously existing stars. (*B*) The nebula is pulled together by gravity, collapsing into the protosun and protoplanets. (*C*) As the planets form, they revolve around the Sun in newly formed orbits.

Stage B. During *stage B*, the hydrogen gas, dust, elements, and chemical compounds from former stars began to form a large, slowly rotating nebula that was much, much larger than the present solar system. Under the influence of gravity, the large but diffuse, slowly rotating nebula began to contract, increasing its rate of spin. The largest mass pulled together in the center, contracting to the protostar, which eventually would become the Sun. The remaining gases, elements, and dusts formed an enormous, fat, bulging disk called an *accretion disk,* which would eventually form the planets and smaller bodies. The frag-ments of dust and other solid matter in the disk began to stick together in larger and larger accumulations from numerous col-lisions over the first million years or so. All of the present-day el-ements of the planets must have been present in the nebula along with the most abundant elements of hydrogen and he-lium. The elements and the familiar chemical compounds accu-mulated into basketball-sized or larger chunks of matter.

Did the planets have an icy slush beginning? Over a period of time, perhaps 100 million years or so, huge accumulations of frozen water, frozen ammonia, and frozen crystals of methane began to accumulate, together with silicon, aluminum, and iron oxide plus other metals in the form of rock and mineral grains. Such a slushy mixture would no doubt have been surrounded by an atmosphere of hydrogen, helium, and other vapors thinly interspersed with smaller rocky grains of dust. Local concentrations of certain minerals might have occurred throughout the whole accretion disk, with a greater concentration of iron, for example, in the disk where the protoplanet Mars was forming compared to where the protoplanet Earth was forming. Evidence for this part of the model is found in Mars today, with its greater abundance of iron. The abundant iron oxides are what color Mars the red planet.

All of the protoplanets might have started out somewhat similarly as huge accumulations of a slushy mixture with an atmosphere of hydrogen and helium gases. Gravitational attraction must have compressed the protoplanets as well as the protosun. During this period of contraction and heating, gravitational adjustments continue, and about a fifth of the disk nearest to the protosun must have been pulled into the central body of the protosun, leaving a larger accumulation of matter in the outer part of the accretion disk.

Stage C. During *stage C,* the warming protosun became established as a star, perhaps undergoing an initial flare-up that has been observed today in other newly forming stars. Such a flare-up might have been of such a magnitude that it blasted away the hydrogen and helium atmospheres of the interior planets (Mercury, Venus, Earth, and Mars) out past Mars, but it did not reach far enough out to disturb the hydrogen and helium atmospheres of the outer planets. The innermost of the outer planets, Jupiter and Saturn, might have acquired some of the matter blasted away from the inner planets, becoming the giants of the solar system by comparison. This is just speculation, however, and the two giants may have simply formed from greater concentrations of matter in that part of the accretion disk.

The evidence, such as separation of heavy and light mineral matter, shows that the protoplanets underwent heating early in their formation. Much of the heating may have been provided by gravitational contraction, the same process that gave the protosun sufficient heat to begin its internal nuclear fusion reactions. Heat was also provided from radioactive decay processes inside the protoplanets, and the initial greater heating from the Sun may have played a role in the protoplanet heating process. Larger bodies were able to retain this heat better than smaller ones, which radiated it to space more readily. Thus, the larger bodies underwent a more thorough heating and melting, perhaps becoming completely molten early in their history. In the larger bodies the heavier elements, such as iron, were pulled to the center of the now molten mass, leaving the lighter elements near the surface. The overall heating and cooling process took millions of years as the planets and smaller bodies were formed. Gases from the hot interiors formed secondary atmospheres of water vapor, carbon dioxide, and nitrogen on the larger interior planets.

Interestingly, the asteroid belt was discovered from a prediction made by the German astronomer Bode at the end of the eighteenth century. Bode had noticed a pattern of regularity in the spacing of the planets that were known at the time. He found

Table 13.7 Distances from the Sun to Planets Known in the 1790s

Planet	n	Distance Predicted by $(n + 4)/10$ (AU)	Actual Distance (AU)
Mercury	0	0.4	0.39
Venus	3	0.7	0.72
Earth	6	1.0	1.0
Mars	12	1.6	1.5
(Asteroid belt)	24	2.8	—
Jupiter	48	5.2	5.2
Saturn	96	10.0	9.5
Uranus	192	19.6	19.2

that by expressing the distances of the planets from the Sun in astronomical units, these distances could be approximated by the relationship $(n + 4)/10$, where n is a number in the sequence 0, 3, 6, 12, and so on where each number (except the first) is doubled in succession. When these calculations were done the distances turned out to be very close to the distances of all the planets known at that time, but the numbers also predicted a planet between Mars and Jupiter where there was none. Later a belt of asteroids was found where the Bode numbers predicted there should be a planet. This suggested to some people that a planet had existed between Mars and Jupiter in the past, and somehow this planet was broken into pieces, perhaps by a collision with another large body (table 13.7).

Such patterns of apparent regularity in the spacing of planetary orbits were of great interest because, if a true pattern existed, it could hold meaning about the mechanism that determined the location of planets at various distances from the Sun. Many attempts have been made to explain the mechanism of planetary spacing and why a belt of asteroids exists where the Bode numbers predict there should be a planet. The most successful explanations concern Jupiter and the influence of its gigantic gravitational field on the formation of clumps of matter at certain distances from the Sun. In other words, a planet does not exist today between Mars and Jupiter because there never was a planet there. The gravitational influence of Jupiter prevented the clumps of matter from joining together to form a planet, and a belt of asteroids formed instead.

SUMMARY

The planets can be classified into two major groups: (1) the *terrestrial planets* of Mercury, Venus, Mars, and Earth and (2) the *giant planets* of Jupiter, Saturn, Uranus, and Neptune.

Comets are porous aggregates of water ice, frozen methane, frozen ammonia, dry ice, and dust. The solar system is surrounded by the *Kuiper Belt* and the *Oort cloud* of these objects. Something nudges one of the icy bodies and it falls into a long elliptical orbit around the Sun. As it approaches the Sun, increased radiation evaporates ices and pushes ions and dust into a long visible tail. *Asteroids* are rocky or metallic bodies that are mostly located in a belt between Mars and Jupiter. The remnants of comets, fragments of asteroids, and dust are called *meteoroids*. A meteoroid that falls through Earth's atmosphere and melts to a visible trail of light and smoke is called a *meteor*. A meteoroid that survives the trip through the atmosphere to strike the surface of Earth is called a *meteorite*. Most meteors are fragments and pieces of dust from comets. Most meteorites are fragments that resulted from collisions between asteroids.

The *protoplanet nebular model* is the most widely accepted theory of the origin of the solar system, and this theory can be considered as a series of events, or stages. *Stage A* is the creation of all the elements heavier than hydrogen in previously existing stars. *Stage B* is the formation of a nebula from the raw materials created in stage A. The nebula contracts from gravitational attraction, forming the *protosun* in the center with a fat, bulging *accretion disk* around it. The Sun will form from the protosun, and the planets will form in the accretion disk. *Stage C* begins as the protosun becomes established as a star. The icy remains of the original nebula are the birthplace of *comets*. *Asteroids* are other remains that did undergo some melting.

KEY TERMS

asteroids (p. **269**)

astronomical unit (p. **254**)

comet (p. **267**)

giant planets (p. **255**)

Kuiper Belt (p. **268**)

meteor (p. **270**)

meteorite (p. **270**)

meteoroids (p. **270**)

meteor shower (p. **270**)

Oort cloud (p. **267**)

protoplanet nebular model (p. **272**)

terrestrial planets (p. **255**)

APPLYING THE CONCEPTS

1. Earth, other planets, and all the members of the solar system
   - (a) have always existed.
   - (b) formed thousands of years ago from elements that have always existed.
   - (c) formed millions of years ago, when the elements and each body were created at the same time.
   - (d) formed billions of years ago from elements that were created in many previously existing stars.

2. The belt of asteroids between Mars and Jupiter is probably
   - (a) the remains of a planet that exploded.
   - (b) clumps of matter that condensed from the accretion disk, but never got together as a planet.
   - (c) the remains of two planets that collided.
   - (d) the remains of a planet that collided with an asteroid or comet.

3. Which of the following planets would be mostly composed of hydrogen, helium, and methane and have a density of less than 2 g/cm³?
   - (a) Uranus
   - (b) Mercury
   - (c) Mars
   - (d) Venus

4. Which of the following planets probably still has its original atmosphere?
   - (a) Mercury
   - (b) Venus
   - (c) Mars
   - (d) Jupiter

5. Venus appears the brightest when it is in the
   - (a) full phase.
   - (b) half phase.
   - (c) quarter phase.
   - (d) crescent phase.

6. The largest planet is
   - (a) Saturn.
   - (b) Jupiter.
   - (c) Uranus.
   - (d) Neptune.

7. The small body with a composition and structure closest to the materials that condensed from the accretion disk is a (an)
   - (a) asteroid.
   - (b) meteorite.
   - (c) comet.
   - (d) None of the above are correct.

8. A small body from space that falls on the surface of the earth is a
   - (a) meteoroid.
   - (b) meteor.
   - (c) meteor shower.
   - (d) meteorite.

Answers

1. d **2.** b **3.** a **4.** d **5.** d **6.** b **7.** c **8.** d

QUESTIONS FOR THOUGHT

1. Describe the surface and atmospheric conditions on Mars.

2. What evidence exists that Mars at one time had abundant liquid water? If Mars did have liquid water at one time, what happened to it and why?

3. Describe the internal structure of Jupiter and Saturn.

4. Describe some of the unusual features found on the moons of Jupiter, Saturn, and Neptune.

5. What are the similarities and the differences between the Sun and Jupiter?

6. What evidence exists today that the number of rocks and rock particles floating around in the solar system was much greater in the past soon after the planets formed?

7. Explain why oxygen is a major component of Earth's atmosphere but not the atmospheres of Venus or Mars.

8. Using the properties of the planets other than Earth, discuss the possibilities of life on each.

9. What are "shooting stars"? Where do they come from? Where do they go?

10. What is an asteroid? What evidence indicates that asteroids are parts of a broken-up planet? What evidence indicates that asteroids are not parts of a broken-up planet?

11. Where do comets come from? Why are astronomers so interested in studying the physical and chemical structure of a comet?

12. What is a meteorite? What is the most likely source of meteorites?

The Earth

These rose-red rhodochrosite crystals are a naturally occurring form of manganese carbonate.
Rhodochrosite is but one of about 2,500 minerals that are known to exist, making up the solid materials of the earth's crust.

O U T L I N E

The separation of the earth sciences into independent branches—such as geology, oceanography, and meteorology—was traditionally done for convenience. This made it easier to study a large and complex Earth. In the past, scientists in each branch studied their field without considering the earth as an interacting whole. Today, most earth scientists consider changes in the earth as taking place in an overall dynamic system. The parts of the earth's interior, the rocks on the surface, the oceans, the atmosphere, and the environmental conditions are today understood to be parts of a complex, interacting system with a cyclic movement of materials from one part to another.

How can materials cycle through changes from the interior to the surface, and to the atmosphere and back? The answer to this question is found in the unique combination of fluids of the earth. No other known planet has Earth's combination of (1) an atmosphere consisting mostly of nitrogen and oxygen, (2) a surface that is mostly covered with liquid water, and (3) an interior that is partly fluid, partly semifluid, and partly solid. Earth's atmosphere is unique both in terms of its composition and in terms of interactions with the liquid water surface (figure 14.1). These interactions have cycled materials, such as carbon dioxide, from the atmosphere to the land and oceans of the earth. The internal flow of rock materials, on the other hand, produces the large-scale motion of the earth's continents and the associated phenomena of earthquakes and volcanoes. Volcanoes cycle carbon dioxide back into the atmosphere and the movement of land cycles rocks from the earth's interior to the surface and back to the interior again.

Altogether, the earth's atmosphere, liquid water, and motion of its landmasses make up a dynamic cycling system that is found only on the planet Earth.

Earth also seems to be unique because there is life on Earth, but apparently not on the other planets. The cycling of atmospheric gases and vapors, waters of the surface, and flowing interior rock materials sustain a wide diversity of life on Earth. There are millions of different species of plants, animals, and other kinds of organisms. Yet, there is no evidence of even one species of life existing outside the earth. The existence of life on Earth must be related to Earth's unique, dynamic system of interacting fluids. This chapter is concerned with earth materials, the internal structure of the earth, the movement of landmasses across the surface of the earth, and how earth materials are recycled.

EARTH MATERIALS

The earth, like all other solid matter in the universe, is made up of chemical elements. The different elements are not distributed equally throughout the mass of the earth, nor are they equally abundant. Chemical analysis of thousands of rocks from the earth's surface found that only eight elements make up about 98.6 percent of the crust. All the other elements make up the remaining 1.4 percent of the crust. Oxygen is the most abundant element, making up about 50 percent of the weight of the crust. Silicon makes up over 25 percent, so oxygen and silicon alone make up about 75 percent of the earth's solid surface. Figure 14.2 shows the eight most abundant elements that occur as elements or combine to form the chemical compounds of the earth's crust. The earth was apparently molten during an early stage in its development, as evidenced by distribution of the elements. During the molten stage the heavier abundant elements, such as iron and nickel, apparently sank to the deep interior of the earth, leaving a thin layer of lighter elements on the surface. This relatively thin layer is called the *crust*. The rocks and rock materials that you see on the surface and the materials sampled in the deepest mines and well holes are all materials of the earth's crust. The bulk of the earth's mass lies below the crust and has not been directly sampled.

Minerals

In everyday usage the word "mineral" can have several different meanings. It can mean something your body should have (vitamins and minerals), something a fertilizer furnishes for a plant (nitrogen, potassium, and phosphorus), or sand, rock, and coal taken from the earth for human use (mineral resources). In the earth sciences, a **mineral** is defined as a naturally occurring, inorganic solid element or compound with a crystalline structure

(figure 14.3). This definition means that the element or compound cannot be synthetic (must be naturally occurring), must not be directly produced by a living organism (must be inorganic), and must have atoms arranged in a regular, repeating pattern (a crystal structure). Note that the crystal structure of a mineral can be present on the microscopic scale and it is not necessarily obvious to the unaided eye. Even crystals that could be observed with the unaided eye are sometimes not noticed (figure 14.4).

The crystal structure of a mineral can be made up of atoms of one or more kinds of elements. Diamond, for example, is a mineral with only carbon atoms in a strong crystal structure. Quartz, on the other hand, is a mineral with atoms of silicon and oxygen in a different crystal structure (figure 14.5). No matter how many kinds of atoms are present, each mineral has its own defined chemical composition or range of chemical compositions. A range of chemical composition is possible because the composition of some minerals can vary with the substitution of chemically similar elements. For example, some atoms of magnesium might be substituted for some chemically similar atoms of calcium. Such substitutions might slightly alter some properties, but not enough to make a different mineral.

Crystals can be classified and identified on the basis of the symmetry of their surfaces. This symmetry is an outward expression of the internal symmetry in the arrangement of the atoms making up the crystal. Thus, on the basis of symmetry, crystalline substances are classified into six major systems, which, in turn, are subdivided into smaller groups.

Silicon and oxygen are the most abundant elements in the earth's crust and, as you would expect, the most common minerals contain these two elements. All minerals are classified on the basis of whether the mineral structure contains these two elements or not. The two main groups are thus called the *silicates*

Figure 14.1

No other planet in the solar system has the unique combination of fluids of Earth. Earth has a surface that is mostly covered with liquid water, water vapor in the atmosphere, and both frozen and liquid water on the land.

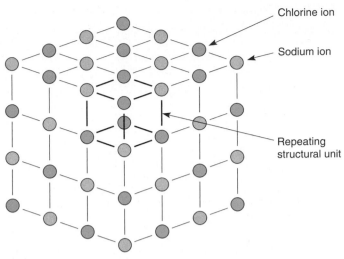

Figure 14.3

A crystal is composed of a structural unit that is repeated in three dimensions. This is the basic structural unit of a crystal of sodium chloride, the mineral halite.

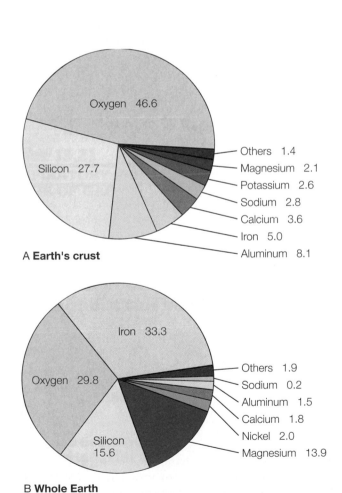

A Earth's crust

B Whole Earth

Figure 14.2

(A) The percentage by weight of the elements that make up the earth's crust.
(B) The percentage by weight of the elements that make up the whole earth.

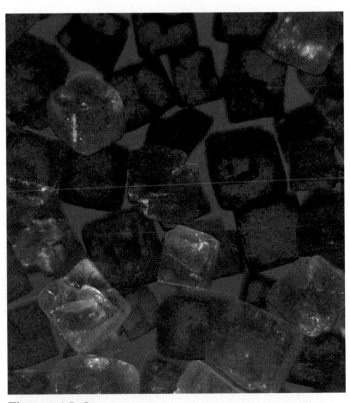

Figure 14.4

The structural unit for a crystal of table salt, sodium chloride, is cubic, as you can see in the individual grains.

and the *nonsilicates* (table 14.1). Note, however, that the silicates can contain some other elements in addition to silicon and oxygen. The silicate minerals are by far the most abundant, making up about 92 percent of the earth's crust. When an atom of silicon (Si^{+4}) combines with four oxygen atoms (O^{-2}), a tetrahedral structure of $(SiO_4)^{-4}$ forms (see figure 14.6). All **silicates**

Table 14.1 Classification Scheme of Some Common Minerals

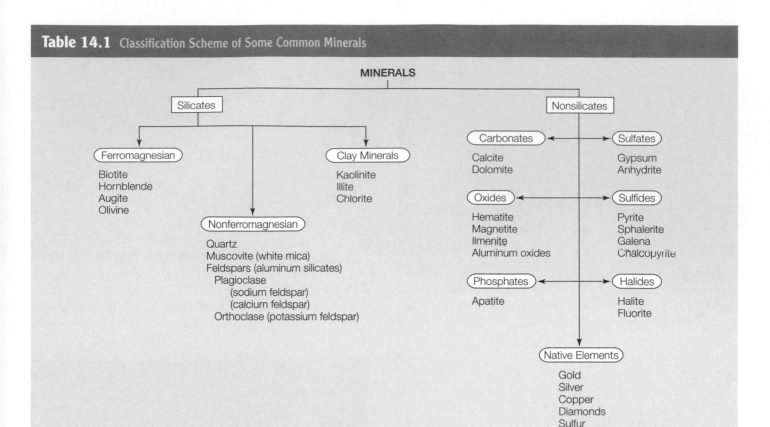

MINERALS
- Silicates
  - Ferromagnesian
    - Biotite
    - Hornblende
    - Augite
    - Olivine
  - Nonferromagnesian
    - Quartz
    - Muscovite (white mica)
    - Feldspars (aluminum silicates)
      - Plagioclase
        - (sodium feldspar)
        - (calcium feldspar)
      - Orthoclase (potassium feldspar)
  - Clay Minerals
    - Kaolinite
    - Illite
    - Chlorite
- Nonsilicates
  - Carbonates
    - Calcite
    - Dolomite
  - Sulfates
    - Gypsum
    - Anhydrite
  - Oxides
    - Hematite
    - Magnetite
    - Ilmenite
    - Aluminum oxides
  - Sulfides
    - Pyrite
    - Sphalerite
    - Galena
    - Chalcopyrite
  - Phosphates
    - Apatite
  - Halides
    - Halite
    - Fluorite
  - Native Elements
    - Gold
    - Silver
    - Copper
    - Diamonds
    - Sulfur

Figure 14.5

These quartz crystals are hexagonal prisms.

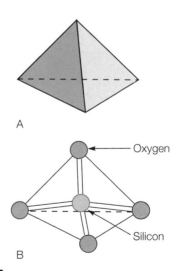

Figure 14.6

(*A*) The geometric shape of a tetrahedron with four equal sides. (*B*) A silicon and four oxygen atoms are arranged in the shape of a tetrahedron with the silicon in the center. This is the basic building block of all silicate minerals.

have a basic silicon-oxygen tetrahedral unit either isolated or joined together in the crystal structure. The structure has four unattached electrons on the oxygen atoms that can combine with metallic ions such as iron or magnesium. They can also combine with the silicon atoms of *other* tetrahedral units. Some silicate minerals are thus made up of single tetrahedral units combined with metallic ions. Other silicate minerals are combi-

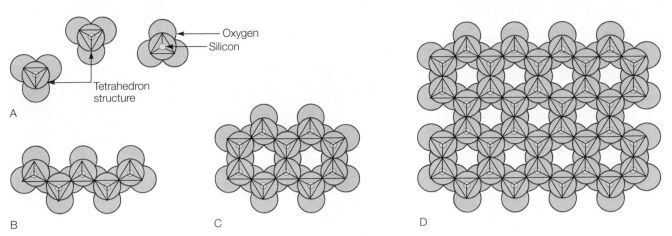

Figure 14.7

(*A*) Isolated silicon-oxygen tetrahedra do not share oxygens. This structure occurs in the mineral olivine. (*B*) Single chains of tetrahedra are formed by each silicon ion having two oxygens all to itself and sharing two with other silicons at the same time. This structure occurs in augite. (*C*) Double chains of tetrahedra are formed by silicon ions sharing either two or three oxygens. This structure occurs in hornblende. (*D*) The sheet structure in which each silicon shares three oxygens occurs in the micas, resulting in layers that pull off easily because of cleavage between the sheets.

nations of tetrahedral units combined in single chains, double chains, or sheets (figure 14.7).

The silicate minerals can be conveniently subdivided into two groups based on the presence of iron and magnesium. The basic tetrahedral structure joins with ions of iron, magnesium, calcium, and other elements in the *ferromagnesian silicates*. Examples of ferromagnesian silicates are *olivine, augite, hornblende*, and *biotite* (figure 14.8). They have a greater density and a darker color than the other silicates because of the presence of the metal ions. Augite, hornblende, and biotite are very dark in color, practically black, and olivine is light green.

The *nonferromagnesian silicates* do not contain iron or magnesium ions. These minerals have a light color and a low density compared to the ferromagnesians. This group includes the minerals *muscovite* (*white mica*), the *feldspars*, and *quartz* (figure 14.9).

Figure 14.8

Compare the dark colors of the ferromagnesian silicates augite (right), hornblende (left), and biotite to the light-colored nonferromagnesian silicates in figure 14.9.

Figure 14.9

Compare the light colors of the nonferromagnesian silicates mica (front center), white and pink orthoclase (top and center), and quartz (left center) to the dark-colored ferromagnesian silicates in figure 14.8.

Asbestos is a common name for any of several minerals that can be separated into fireproof fibers, fibers that will not melt or ignite. The fibers can be woven into a fireproof cloth, used directly as fireproof insulation material, or mixed with plaster or other materials. People now have a fear of all asbestos because it is presumed to be a health hazard (box figure 14.1). However, there are about six commercial varieties of asbestos. Five of these varieties are made from an amphibole mineral and are commercially called "brown" and "blue" asbestos. The other variety is made from *chrysotile*, a *serpentine* family of minerals and is commercially called "white" asbestos. White asbestos is the asbestos mined and most commonly used in North America. It is only the amphibole asbestos (brown and blue asbestos) that has been linked to cancer, even for a short exposure time. There is, however, no evidence that exposure to white asbestos results in an increased health hazard. It makes sense to ban the use of and remove all the existing amphibole asbestos from public buildings. It does not make sense to ban or remove the serpentine asbestos since it is not a proven health hazard.

Box Figure 14.1

Did you know there are different kinds of asbestos? Are all kinds of asbestos a health hazard?

The remaining 8 percent of minerals making up the earth's crust that do not have silicon-oxygen tetrahedrons in their crystal structure are called *nonsilicates*. There are eight subgroups of nonsilicates: (1) carbonates, (2) sulfates, (3) oxides, (4) sulfides, (5) halides, (6) phosphates, (7) hydroxides, and (8) native elements. Some common nonsilicates are identified in table 14.1. The carbonates are the most abundant of the nonsilicates, but others are important as fertilizers, sources of metals, and sources of industrial chemicals.

Rocks

Elements are *chemically* combined to make minerals. Minerals are *physically* combined to make rocks. A **rock** is defined as an aggregation of one or more minerals and perhaps other materials that have been brought together into a cohesive solid. These materials include volcanic glass, a silicate that is not considered a mineral because it lacks a crystalline structure. Thus, a rock can consist of one or more kinds of minerals that are somewhat "glued" together by other materials such as glass. Most rocks are composed of silicate minerals, as you might expect since most minerals are silicates.

There is a classification scheme that is based on the way the rocks were formed. There are three main groups: (1) *igneous rocks* formed as a hot, molten mass of rock materials cooled and solidified; (2) *sedimentary rocks* formed from particles or dissolved materials from previously existing rocks; and (3) *metamorphic rocks* formed from rocks that were subjected to high temperatures and pressures that deformed or recrystallized the rock without complete melting.

Igneous Rocks

The word *igneous* comes from the Latin *ignis*, which means "fire." This is an appropriate name for **igneous rocks** that are defined as rocks that formed from a hot, molten mass of melted rock materials. The first step in forming igneous rocks is the creation of some very high temperature, hot enough to melt rocks. A mass of melted rock materials is called a *magma*. A magma may cool and crystallize to solid igneous rock either below or on the surface of the earth. The earth has had a history of molten materials, and all rocks of the earth were at one time igneous rocks. Today, about two-thirds of the outer layer, or crust, is made up of igneous rocks. This is not apparent in many locations because the surface is covered by other kinds of rocks and rock materials (sand, soil, etc.). When they are found, igneous

Figure 14.10
Granite is a coarse-grained igneous rock composed mostly of light-colored, light-density, nonferromagnesian minerals. The earth's continental areas are dominated by granite and by rocks with the same mineral composition of granite.

CONCEPTS APPLIED
Collect Your Own

Make a collection of rocks and minerals that can be found in your location, showing the name and location found for each. What will determine the total number of rocks and minerals possible to collect in your particular location?

rocks are usually granite or basalt. *Granite* is a light colored igneous rock that is primarily made up of three silicate minerals: quartz, mica, and feldspar. You can see the grains of these three minerals in a freshly broken surface of most samples of granite (figure 14.10). *Basalt* is a dark, more dense igneous rock with mineral grains too tiny to see. Basalt is the most common volcanic rock found on the surface of the earth.

Sedimentary Rocks

Sedimentary rocks are rocks that formed from particles or dissolved materials from previously existing rocks. Chemical reactions with air and water tend to break down and dissolve the less chemically stable parts of existing rocks, freeing more stable particles and grains in the process. The remaining particles are usually transported by moving water and deposited as sediments. *Sediments* are accumulations of silt, clay, sand, or other materials that settle out of water. Weathered rock fragments and dissolved rock materials both contribute to sediment deposits (table 14.2).

Most sediments are deposited as many separate particles that accumulate as loose materials. Such accumulations of rock fragments, chemical deposits, or animal shells must become consolidated into a solid, coherent mass to become sedimentary rock. There are two main parts to this *lithification,* or

Table 14.2 Classification Scheme for Sedimentary Rocks

Sediment Type	Particle or Composition	Rock
Fragment	Larger than sand	Conglomerate or breccia
Fragment	Sand	Sandstone
Fragment	Silt and Clay	Siltstone, claystone, or shale
Chemical	Calcite	Limestone
Chemical	Dolomite	Dolomite
Chemical	Gypsum	Gypsum
Chemical	Halite (sodium chloride)	Salt

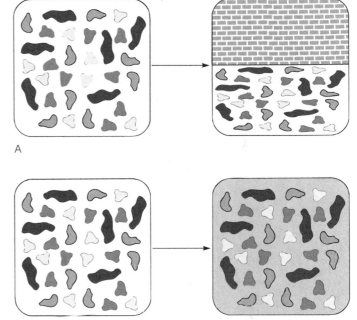

Figure 14.11
(*A*) In compaction, the sediment grains are packed more tightly together, often by overlying sediments, as represented by the bricks. (*B*) In cementation, dissolved minerals from fluids are precipitated in the space between the grains, cementing them together into a rigid, solid mass.

rock-forming process, (1) compaction and (2) cementation (figure 14.11).

The weight of an increasing depth of overlying sediments causes an increasing pressure on the sediments below. This pressure squeezes the deeper sediments together, gradually reducing the pore space between the individual grains. This *compaction* of the grains reduces the thickness of a sediment deposit, squeezing out water as the grains are packed more tightly together. Compaction alone is usually not enough to make loose sediment into solid rock. Cementation is needed to hold the compacted grains together.

In *cementation* the spaces between the sediment particles are filled with a chemical deposit. The chemical deposit binds the particles together into the rigid, cohesive mass of a sedimentary rock.

Table 14.3	Simplified Classification Scheme for Sediment Fragments and Rocks	
Sediment Name	**Size Range**	**Rock**
Boulder	Over 256 mm (10 in)	
Gravel	2 to 256 mm (0.08–10 in)	Conglomerate or breccia*
Sand	1/16 to 2 mm (0.025–0.08 in)	Sandstone
Silt (or dust)	1/256 to 1/16 mm (0.00015–0.025 in)	Siltstone**
Clay (or dust)	Less than 1/256 mm (less than 0.00015 in)	Claystone**

*Conglomerate has a rounded fragment; breccia has an angular fragment.
**Both also known as mudstone; called shale if it splits along parallel planes.

Compaction and cementation may occur at the same time, but the cementing agent must have been introduced before compaction restricts the movement of the fluid through the open spaces. Many soluble minerals can serve as cementing agents, and calcite (calcium carbonate) and silica (silicon dioxide) are common.

Sediments accumulate from rocks that are in various stages of being broken down, so there is a wide range of sizes of particles in sediments. The size and shape of the particles are used as criteria to name the sedimentary rocks (table 14.3).

C O N C E P T S   A P P L I E D

Sand Sort

Collect dry sand from several different locations. Use a magnifying glass to determine the minerals found in each sample.

Metamorphic Rocks

The third group of rocks is called metamorphic. **Metamorphic rocks** are previously existing rocks that have been changed by heat, pressure, or hot solutions into a distinctly different rock. The heat, pressure, or hot solutions that produced the changes are associated with geologic events of (1) movement of the crust, which will be discussed in chapter 15, and with (2) heating and hot solutions from magma. Pressures from movement of the crust can change the rock texture by flattening, deforming, or realigning mineral grains. Temperatures from magma must be just right to produce a metamorphic rock. They must be high enough to disrupt the crystal structures to cause them to recrystallize, but not high enough to melt the rocks and form igneous rocks (figure 14.12).

The Rock Cycle

Earth is a dynamic planet with a constantly changing surface and interior. As you will see in the next chapters, internal changes alter the earth's surface by moving the continents and, for example, building mountains that are eventually worn away

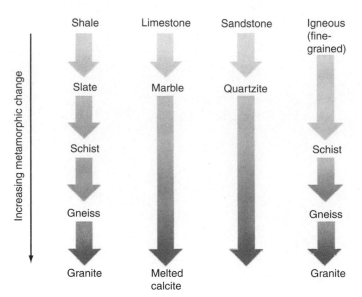

Figure 14.12
Increasing metamorphic change occurs with increasing temperatures and pressures. If the melting point is reached, the change is no longer metamorphic, and igneous rocks are formed.

by weathering and erosion. Seas advance and retreat over the continents as materials are cycled from the atmosphere to the land and from the surface to the interior of the earth and then back again. Rocks are transformed from one type to another through this continual change. There is not a single rock on the earth's surface today that has remained unchanged through the earth's long history. The concept of continually changing rocks through time is called the *rock cycle.* The rock cycle concept views an igneous, a sedimentary, or a metamorphic rock as the present, but temporary stage in the ongoing transformation of rocks to new types. Any particular rock sample today has gone through countless transformations in the 4.6-billion-year history of the earth and will continue to do so in the future.

THE EARTH'S INTERIOR

Earth is not a solid, stable sphere that is "solid as a rock." All rocks and rock materials can be made to flow, behaving as warm wax or putty that can be molded. Pressure and temperature just a few kilometers below the surface can be high enough so that rock flows very slowly. Rock flowage takes place as hot, buoyant material deep within the earth moves slowly upward toward the cooler surface. Elsewhere, cold, denser material moves downward.

The earth's internal heat and rock movement are related to some things that happen on the surface, such as moving continents, earthquakes, and volcanoes. Before considering how this happens, we will look at the earth's interior and some internal structures. Information about the earth's interior is deducted mostly from a study of earthquake waves as they move through different parts of the earth. Earthquake waves furnish a sort of "X-ray" machine to view the earth's interior.

The sudden movement of very large sections of rock can produce vibrations that move out as waves throughout the entire earth. These vibrations are called *seismic waves.* Strong seismic waves are what people feel as a shaking, quaking, or vibrating during an earthquake. Seismic waves are generated when a huge mass of rock breaks and slides into a different position.

Seismic waves radiate outward from an earthquake, spreading in all directions through the solid earth's interior like sound waves from an explosion. There are basically three kinds of waves:

1. A longitudinal (compressional) wave called a *P-wave.* P-waves are the fastest and move through surface rocks and solid and liquid materials below the surface.
2. A transverse (shear) wave called an *S-wave.* S-waves are second fastest after the P-wave. S-waves do not travel through liquids since liquids do not have the cohesion necessary to transmit a shear, or side-to-side, motion.
3. An up-and-down (crest and trough) wave that travels across the surface called a *surface wave* that is much like a wave on water that moves across the solid surface of the earth. Surface waves are the slowest and occur where S- or P-waves reach the surface.

Using data from seismic waves that move through the earth, scientists were able to plot three main parts of the interior (figure 14.13). The *crust* is the outer layer of rock that forms a thin shell around the earth. Below the crust is the *mantle,* a much thicker shell than the crust. The mantle separates the crust from the center part, which is called the *core.* The following section starts on the earth's surface, at the crust, and then digs deeper and deeper into the earth's interior.

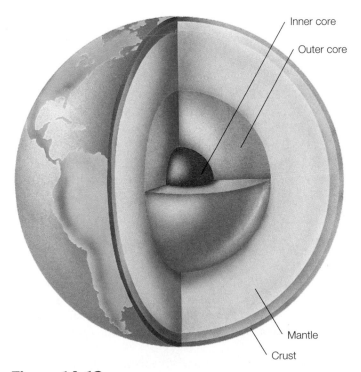

Figure 14.13
The structure of the earth's interior.

Table 14.4	Comparison of Oceanic Crust and Continental Crust	
	Oceanic Crust	**Continental Crust**
Age	Less than 200 million years	Up to 3.8 billion years old
Thickness	5 to 8 km (3 to 5 mi)	10 to 70 km (6 to 45 mi)
Density	3.0 g/cm^3	2.7 g/cm^3
Composition	Basalt	Granite, schist, gneiss

The Crust

The earth's **crust** is a thin layer of rock that covers the entire earth, existing below the oceans as well as making up the continents. According to seismic waves, there are differences in the crust making up the continents and the crust beneath the oceans (table 14.4). These differences are (1) the oceanic crust is much thinner than the continental crust and (2) seismic waves move through the oceanic crust faster than they do through the continental crust. The two types of crust vary because they are made up of different kinds of rock.

The boundary between the crust and the mantle is marked by a sharp increase in the velocity of seismic waves as they pass from the crust to the mantle. Today, this boundary is called the *Mohorovicic discontinuity,* or the "Moho" for short. The boundary is a zone where seismic P-waves increase in velocity because of changes in the composition of the materials. The increase occurs because the composition on both sides of the boundary is different. The mantle is richer in ferromagnesian minerals and poorer in silicon than the crust.

Studies of the Moho show that the crust varies in thickness around the earth's surface. It is thicker under the continents and the crust under the oceans is much thinner.

The age of rock samples from the continent has been compared with the age samples of rocks taken from the seafloor by oceanographic ships. This sampling has found the continental crust to be much older, with parts up to 3.8 billion years old. By comparison, the oldest oceanic crust is less than 200 million years old.

Comparative sampling also found that continental crust is a less dense, granite-type rock with a density of about 2.7 g/cm^3. Oceanic crust, on the other hand, is made up of basaltic rock with a density of about 3.0 g/cm^3. The less dense crust behaves as if it were floating on the mantle, much as less dense ice floats on water. There are explainable exceptions, but in general the thicker, less dense continental crust "floats" in the mantle above sea level and the thin, dense oceanic crust "floats" in the mantle far below sea level (figure 14.14).

The Mantle

The middle part of the earth's interior is called the **mantle.** The mantle is a thick shell between the core and the crust. This shell takes up about 80 percent of the total volume of the earth and accounts for about two-thirds of the earth's total mass. Information

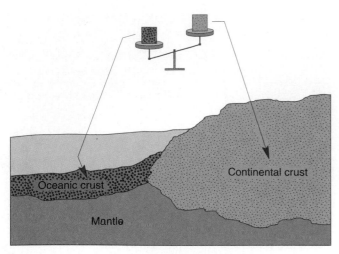

Figure 14.14

Continental crust is less dense, granite-type rock, while the oceanic crust is more dense, basaltic rock. Both types of crust behave as if they were floating on the mantle, which is more dense than either type of crust.

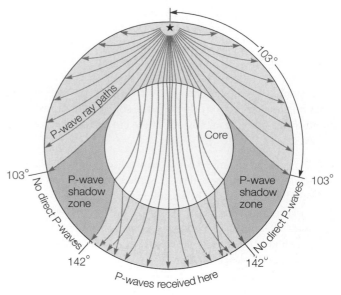

Figure 14.15

The P-wave shadow zone, caused by refraction of P-waves within the earth's core.

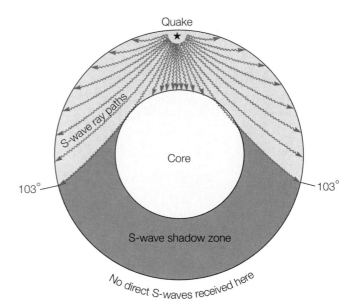

Figure 14.16

The S-wave shadow zone. Since S-waves cannot pass through a liquid, at least part of the core is either a liquid or has some of the same physical properties as a liquid.

about the composition and nature of the mantle comes from (1) studies of seismological data, (2) studies of the nature of meteorites, and (3) studies of materials from the mantle that have been ejected to the earth's surface by volcanoes. The evidence from these separate sources all indicates that the mantle is composed of silicates, predominantly the ferromagnesian silicate *olivine.* Meteorites, as mentioned earlier, are basically either iron meteorites or stony meteorites. Most of the stony meteorites are silicates with a composition that would produce the chemical composition of olivine if they were melted and the heavier elements were separated by gravity. This chemical composition also agrees closely with the composition of basalt, the most common volcanic rock found on the surface of the earth.

The Core

Information about the nature of the **core,** the center part of the earth, comes from several sources of information. Seismological data provide the primary evidence for the structure of the core of the earth. Seismic P-waves spread through the earth from a large earthquake. Figure 14.15 shows how the P-waves spread out, soon arriving at seismic measuring stations all around the world. However, there are places between 103° and 142° of arc from the earthquake that do not receive P-waves. This region is called the *P-wave shadow zone.* The P-wave shadow zone is explained by P-waves being refracted by the core, leaving a shadow. The paths of P-waves can be accurately calculated, so the size and shape of the earth's core can also be accurately calculated.

Seismic S-waves leave a different pattern at seismic receiving stations around the earth. S- (sideways or transverse) waves can travel only through solid materials. An *S-wave shadow zone* also exists and is larger than the P-wave shadow zone (figure 14.16). S-waves are not recorded in the entire region more than 103° away from the epicenter. The S-wave shadow zone seems to indicate that S-waves do not travel through the core at all. If this

is true, it implies that the core of the earth is a liquid, or at least acts like a liquid.

Analysis of P-wave data suggests that the core has two parts, a *liquid outer core* and a *solid inner core.* Both the P-wave and S-wave data support this conclusion. Overall, the core makes up about 15 percent of the earth's total volume and about one-third of its mass.

Evidence from the nature of meteorites indicates that the earth's core is mostly iron. The earth has a strong magnetic field that has its sources in the turbulent flow of the liquid part

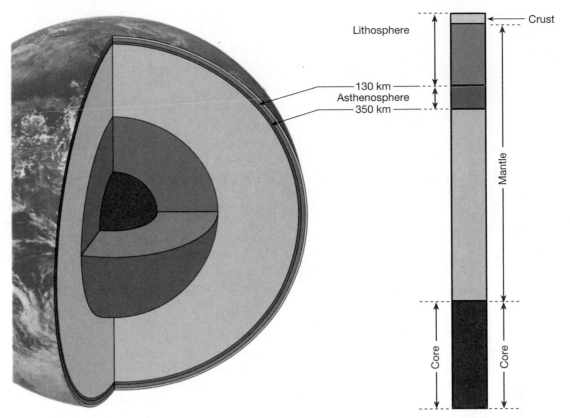

Figure 14.17

The earth's interior, showing the weak, plastic layer called the *asthenosphere*. The rigid, solid layer above the asthenosphere is called the *lithosphere*. The lithosphere is broken into plates that move on the upper mantle like giant ice sheets floating on water. This arrangement is the foundation for plate tectonics, which explains many changes that occur on the earth's surface such as earthquakes, volcanoes, and mountain building.

of the earth's core. To produce such a field, the material of the core would have to be an electrical conductor, that is, a metal such as iron. There are two general kinds of meteorites that fall to the earth, (1) stony meteorites that are made of silicate minerals and (2) iron meteorites that are made of iron or of a nickel-iron alloy. Since the earth has a silicate-rich crust and mantle, by analogy the earth's core must consist of iron or a nickel and iron alloy.

A More Detailed Structure

There is strong evidence that the earth has a layered structure with a core, mantle, and crust. This description of the structure is important for historical reasons and for understanding how the earth evolved over time. There is another, more detailed structure that can be described. This structure is far more important in understanding the history and present appearance of the earth's surface, including the phenomena of earthquakes and volcanoes.

The important part of this different detailed description of the earth's interior was first identified from seismic data. There is a thin zone in the mantle, where seismic waves undergo a sharp decrease in velocity (figure 14.17). This low-velocity zone is evidently a hot, elastic semiliquid layer that extends around the entire earth. It is called the **asthenosphere** after the Greek word for "weak shell." The asthenosphere is weak because it is

plastic, mobile, and yields to stresses. In some regions the asthenosphere is completely liquid, containing pockets of magma.

The rocks above and below the asthenosphere are rigid, solid, and brittle. The solid layer above the asthenosphere is called the **lithosphere** after the Greek word for "stone shell." The lithosphere is also known as the "strong layer" in contrast to the "weak layer" of the asthenosphere. The lithosphere includes the entire crust, the Moho, and the upper part of the mantle. As you will see in the next section, the asthenosphere is one important source of magma that reaches the earth's surface. It is also a necessary part of the mechanism involved in the movement of the crust. The lithosphere is made up of comparatively rigid plates that are moving, floating in the upper mantle like giant ice sheets floating in the ocean.

PLATE TECTONICS

If you observe the shape of the continents on a world map or a globe, you will notice that some of the shapes look as if they would fit together like the pieces of a puzzle. The most obvious is the eastern edge of North and South America, which seem to fit the western edge of Europe and Africa in a slight S-shaped curve. Such patterns between continental shapes seem to suggest that the continents were at one time together, breaking apart and moving to their present positions some time in the past (figure 14.18).

A

B

Figure 14.18

(*A*) Normal position of the continents on a world map. (*B*) A sketch of South America and Africa, showing how they once might have been joined together and subsequently separated by a continental drift.

In the early 1900s a German scientist named Alfred Wegener became enamored with the idea that the continents had shifted positions, and published papers on the subject for nearly two decades. Wegener supposed that at one time there was a single large landmass that he called "Pangaea," which is from the Greek word meaning "all lands." He pointed out that similar fossils found in landmasses on both sides of the Atlantic Ocean today must be from animals and plants that lived in Pangaea, which later broke up and split into smaller continents. Wegener's concept came to be known as *continental drift,* the idea that individual continents could shift positions on the earth's surface. Some people found the idea of continental drift plausible, but most had difficulty imagining how a huge and massive continent could "drift" around on a solid earth. Since Wegener had provided no explanations of why or how continents might do this, most scientists found the concept unacceptable. The concept of continental drift was dismissed as an interesting, but odd idea. Then new evidence indicated that the continents had indeed moved. The first of this evidence came from the bottom of the ocean and led to a new, broader theory about movement of the earth's crust.

Evidence from Earth's Magnetic Field

The earth's magnetic field is probably created by electric currents within the slowly circulating liquid part of the iron core. However, there is nothing static about the earth's magnetic poles. Geophysical studies have found that the magnetic poles are moving slowly around the geographic poles. Studies have also found that the earth's magnetic field occasionally undergoes magnetic reversal. Magnetic reversal is the flipping of polarity of the earth's magnetic field. During a magnetic reversal, the north magnetic pole and the south magnetic pole exchange positions. The present magnetic field orientation has persisted for the past 700,000 years, and according to the evidence, is now preparing for another reversal. The evidence, such as the magnetized iron particles found in certain Roman ceramic artifacts, shows that the magnetic field was 40 percent stronger 2,000 years ago than it is today. If the present rate of weakening were to continue, the earth's magnetic field would be near zero by the end of the next 2,000 years—if it weakens that far before reversing orientation and then increases to its usual value.

Many igneous rocks contain a record of the strength and direction of the earth's magnetic field at the time the rocks formed. Iron minerals, such as magnetite (Fe_3O_4), crystallize as particles in a cooling magma and become magnetized and oriented to the earth's magnetic field, like tiny compass needles, at the time they were formed. Such rocks thus provide evidence of the direction and distance to the earth's ancient magnetic poles. The study of ancient magnetism, called *paleomagnetics,* provides the information that the earth's magnetic field has undergone twenty-two magnetic reversals during the past 4.5 million years (figure 14.19).

The record shows the time between pole flips is not consistent, sometimes reversing in as little as 10,000 years and sometimes taking as long as 25 million years. Once a reversal starts, however, it takes about 5,000 years to complete the process.

Evidence from the Ocean

The first important studies concerning the movement of continents came from studies of the ocean basin, the bottom of the ocean floor. The basins are covered by 4 to 6 km (about 3 to 4 mi) of water and were not easily observed during Wegener's time. It was not until the development and refinement of sonar and other new technologies that scientists began to learn about the nature of the ocean basin. They found that it was not the flat, featureless plain that many had imagined. There are valleys, hills, mountains, and mountain ranges. Long, high, and continuous chains of mountains that seem to run clear around the earth were discovered, and these chains are called **oceanic ridges.** The *Mid-Atlantic Ridge* is one such oceanic ridge, which is located in the center of the Atlantic Ocean basin. The Mid-Atlantic Ridge divides the Atlantic Ocean into two nearly equal

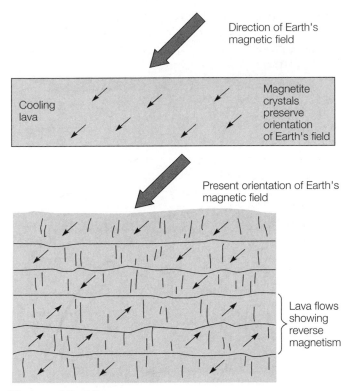

Figure 14.19
Magnetite mineral grains align with the earth's magnetic field and are frozen into position as the magma solidifies. This magnetic record shows the earth's magnetic field has reversed itself in the past.

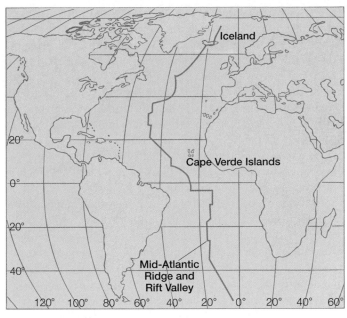

Figure 14.20
The Mid-Atlantic Ridge divides the Atlantic Ocean into two nearly equal parts. Where the ridge reaches above sea level, it makes oceanic islands, such as Iceland.

parts. Where it is high enough to reach sea level, it makes oceanic islands such as Iceland (figure 14.20). The basins also contain **oceanic trenches.** These trenches are long, narrow, and deep troughs with steep sides. Oceanic trenches always run parallel to the edge of continents.

Studies of the Mid-Atlantic Ridge found at least three related groups of data and observations: (1) submarine earthquakes were discovered and measured, but the earthquakes were all observed to occur mostly in a narrow band under the crest of the Mid-Atlantic Ridge; (2) a long, continuous, and deep valley was observed to run along the crest of the Mid-Atlantic Ridge for its length. This continuous, crest-running valley is called a **rift;** and (3) a large amount of heat was found to be escaping from the rift. One explanation of the related groups of findings is that the rift might be a crack in the earth's crust, a fracture through which lava flowed to build up the ridge. The evidence of excessive heat flow, earthquakes along the crest of the ridge, and the very presence of the ridge all led to a **seafloor spreading** hypothesis. This hypothesis explained that hot, molten rock moved up from the interior of the earth to emerge along the rift, flowing out in both directions to create new rocks along the ridge. The creation of new rock like this would tend to spread the seafloor in both directions, thus the name. The test of this hypothesis would come from further studies on the ages and magnetic properties of the seafloor along the ridge (figure 14.21).

Evidence of the age of sections of the seafloor was obtained by drilling into the ocean floor from a research ship. From these

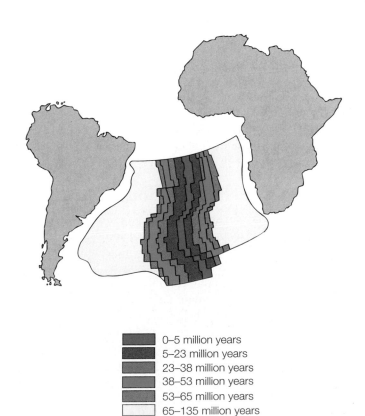

■	0–5 million years
■	5–23 million years
■	23–38 million years
■	38–53 million years
■	53–65 million years
□	65–135 million years

Figure 14.21
The pattern of seafloor ages on both sides of the Mid-Atlantic Ridge reflects seafloor spreading activity. Younger rocks are found closer to the ridge.

drillings, scientists were able to obtain samples of fossils and sediments at progressive distances outward from the Mid-Atlantic Ridge. They found thin layers of sediments near the ridge that became progressively thicker toward the continents. This is a pattern you would expect if the seafloor were spreading, because older layers would have more time to accumulate greater depths of sediments. The fossils and sediments in the bottom of the layer were also progressively older at increasing distances from the ridge. The oldest, which were about 150 million years old, were near the continents. This would seem to indicate that the Atlantic Ocean did not exist until 150 million years ago. At that time a fissure formed between Africa and South America and new materials have been continuously flowing, adding new crust to the edges of the fissure.

More convincing evidence for the support of seafloor spreading came from the paleomagnetic discovery of patterns of magnetic strips in the rocks of the ocean floor. The earth's magnetic field has been reversed many times in the last 150 million years. The periods of time between each reversal were not equal, ranging from thousands to millions of years. Since iron minerals in molten basalt formed, became magnetized, then frozen in the orientation they had when the rock cooled, they made a record of reversals in the earth's ancient magnetic field (figure 14.22). Analysis of the magnetic pattern in the rocks along the Mid-Atlantic Ridge found identical patterns of magnetic bands on both sides of the ridge. This is just what you would expect if molten rock flowed out of the rift, cooled to solid basalt, then moved away from the rift on both sides. The pattern of magnetic bands also matched patterns of reversals measured elsewhere, providing a means of determining the age of the basalt. This showed that the oceanic crust is like a giant conveyer belt that is moving away from the Mid-Atlantic Ridge in both directions. It is moving at an average 5 cm (about 2 inches) a year, which is about how fast your fingernails grow. This means that in 50 years the seafloor will have moved 5 cm/yr × 50 yr, or 2.5 meters (about 8 feet). This slow rate is why most people do not recognize that the seafloor—and the continents—move.

Lithosphere Plates and Boundaries

The strong evidence for seafloor spreading soon led to the development of a new theory called **plate tectonics.** According to plate tectonics, the lithosphere is broken into a number of fairly rigid plates that move on the asthenosphere. Some plates, as shown in figure 14.23, contain continents and part of an ocean basin, while other plates contain only ocean basins. The plates move, and the movement is helping to explain why mountains form where they do, the occurrence of earthquakes and volcanoes, and in general, the entire changing surface of the earth.

Earthquakes, volcanoes, and most rapid changes in the earth's crust occur at the edge of a plate, which is called a *plate boundary.* There are three general kinds of plate boundaries that describe how one plate moves relative to another:

1. **Divergent boundaries** occur between two plates moving away from each other. Magma forms as the plates separate, decreasing pressure on the mantle below. This molten material from the

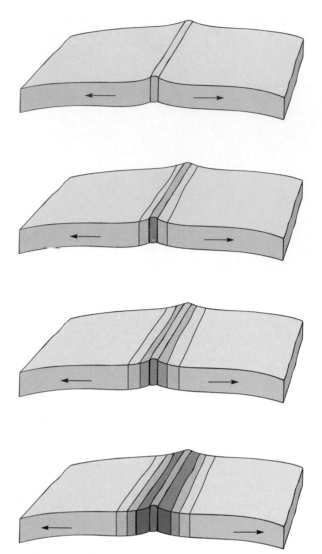

Figure 14.22

Formation of magnetic strips on the seafloor. As each new section of seafloor forms at the ridge, iron minerals become magnetized in a direction that depends on the orientation of the earth's field at that time. This makes a permanent record of reversals of the earth's magnetic field.

asthenosphere rises, cools, and adds new crust to the edges of the separating plates. The new crust tends to move horizontally from both sides of the divergent boundary, usually known as an oceanic ridge. A divergent boundary is thus a **new crust zone.** Most new crust zones are presently on the seafloor, producing seafloor spreading (figure 14.24).

The Mid-Atlantic Ridge is a divergent boundary between the South American and African Plates, extending north between the North American and Eurasian Plates (see figure 14.23). This ridge is one segment of the global mid-ocean ridge system that encircles the earth. The results of divergent plate movement can be seen in Iceland, where the Mid-Atlantic Ridge runs as it separates the North American and Eurasian Plates. In the northeastern part of Iceland ground cracks are widening, often accompanied by volcanic activity. The movement was measured extensively between 1975 and 1984, when displacements caused a total separation of about 7 meters (about 23 feet).

Figure 14.23

The major plates of the lithosphere that move on the asthenosphere.

Source: After W. Hamilton, U.S. Geological Survey.

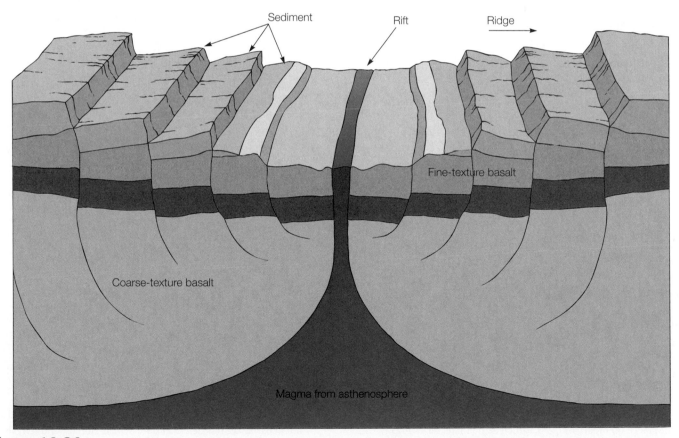

Figure 14.24

A divergent boundary is a new crust zone where molten magma from the asthenosphere rises, cools, and adds new crust to the edges of the separating plates. Magma that cools at deeper depths forms a coarse-grained basalt, while surface lava cools to a fine-grained basalt. Note that sediment deposits are deeper farther from the spreading rift.

The measured rate of spreading along the Mid-Atlantic Ridge ranges from 1 to 6 centimeters per year. This may seem slow, but the process has been going on for millions of years and has caused a tiny inlet of water between the continents of Europe, Africa, and the Americas to grow into the vast Atlantic Ocean that exists today.

Another major ocean may be in the making in East Africa, where a divergent boundary has already moved Saudi Arabia away from the African continent, forming the Red Sea. If this spreading between the African Plate and the Arabian Plate continues, the Indian Ocean will flood the area and the easternmost corner of Africa will become a large island.

2. **Convergent boundaries** occur between two plates moving toward each other. The creation of new crust at a divergent boundary means that old crust must be destroyed somewhere else at the same rate, or else the earth would have a continuously expanding diameter. Old crust is destroyed by returning to the asthenosphere at convergent boundaries. The collision produces an elongated belt of down-bending called a **subduction zone.** The lithosphere of one plate, which contains the crust, is subducted beneath the second plate and partially melts, then becoming part of the mantle. The more dense components of this may become igneous materials that remain in the mantle. Some of it may eventually migrate to a spreading ridge to make new crust again. The less dense components may return to the surface as a silicon, potassium, and sodium-rich lava, forming volcanoes on the upper plate, or it may cool below the surface to form a body of granite. Thus, the oceanic lithosphere is being recycled through this process, which explains why ancient seafloor rocks do not exist. Convergent boundaries produce related characteristic geologic features depending on the nature of the materials in the plates, and there are three general possibilities: (1) converging continental and oceanic plates, (2) converging oceanic plates, and (3) converging continental plates.

As an example of *ocean-continent plate convergence,* consider the plate containing the South American continent (the South American Plate) and its convergent boundary with an oceanic plate (the Nazca Plate) along its western edge. Continent-oceanic plate convergence produces a characteristic set of geologic features as the oceanic plate of denser basaltic material is subducted beneath the less dense granite-type continental plate (figure 14.25). The subduction zone is marked by an oceanic trench (the Peru-Chile Trench), deep-seated earthquakes, and volcanic mountains on the continent (the Andes Mountains). The trench is formed from the down-bending associated with subduction and the volcanic mountains from subducted and melted crust that rise up through the overlying plate to the surface. The earthquakes are associated with the movement of the subducted crust under the overlying crust.

Ocean-ocean plate convergence produces another set of characteristics and related geologic features (figure 14.26). The northern boundary of the oceanic Pacific Plate, for example, converges with the oceanic part of the North American Plate near the Bering Sea. The Pacific Plate is subducted, forming the Aleutian oceanic trench with a zone of earthquakes that are shallow near the trench and progressively more deep-seated

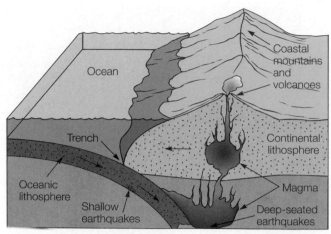

Figure 14.25
Ocean-continent plate convergence. This type of plate boundary accounts for shallow and deep-seated earthquakes, an oceanic trench, volcanic activity, and mountains along the coast.

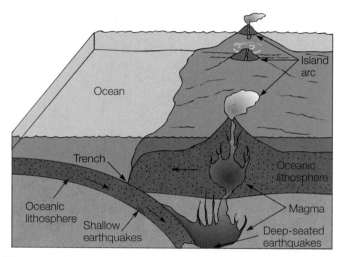

Figure 14.26
Ocean-ocean plate convergence. This type of plate convergence accounts for shallow and deep-focused earthquakes, an oceanic trench, and a volcanic arc above the subducted plate.

toward the continent. The deeper earthquakes are associated with the movement of more deeply subducted crust into the mantle. The Aleutian Islands are typical **island arcs,** curving chains of volcanic islands that occur over the belt of deep-seated earthquakes. These islands form where the melted subducted material rises up through the overriding plate above sea level. The Japanese, Marianas, and Indonesians are similar groups of arc islands associated with converging oceanic-oceanic plate boundaries.

During *continent-continent plate convergence,* subduction does not occur as the less dense, granite-type materials tend to resist subduction (figure 14.27). Instead, the colliding plates pile up into a deformed and thicker crust of the lighter material. Such a collision produced the thick, elevated crust known as the Tibetan Plateau and the Himalayan Mountains.

The CAT scan is a common diagnostic imaging procedure that combines the use of X rays with computer technology. CAT stands for "Computed Axial Tomography" and the word "tomography" means "drawing slices." The CAT scan shows organs of interest by making X-ray images from many different angles as the source of the X rays moves around the patient. CAT scan images are assembled by a computer into a three-dimensional picture that can show organs, bones, and tissues in great detail.

The CAT scan is applied to the earth's interior in a *seismic tomography* procedure. It works somewhat like the medical CAT scan, but seismic tomography uses seismic waves instead of X rays. The velocities of S and P seismic waves vary with depth, changing with density, temperature, pressure, and the composition of the earth's interior. Interior differences in temperature, pressure, and composition are what cause the motion of tectonic plates on the earth's surface. Thus a picture of the earth's interior can be made by mapping variations in seismic wave speeds.

Suppose an earthquake occurs and a number of seismic stations record when the S and P waves arrive. From these records you could compare the seismic velocity data and identify if there were low seismic velocities between the source and some of the receivers. The late arrival of seismic waves could mean some structural difference in the structure of the earth between the source and the observing station. When an earthquake occurs in a new location more data will be collected at different observing stations, which provides additional information about the shape and structure of whatever is slowing the seismic waves. Now, imagine repeating these measurements for many new earthquakes until you have enough data to paint a picture of what is beneath the surface.

Huge amounts of earthquake data—perhaps 10 million data points in 5 million groups—are needed to construct a picture of the earth's interior. A really fast computer may take days to process this much data and construct cross sections through some interesting place, such as a subduction zone where an oceanic plate dives into the mantle.

Seismic tomography has also identified massive plumes of molten rock rising toward the surface from deep within the earth. These *superplumes* originate from the base of the mantle, rising to the lithosphere. The hot material was observed to spread out horizontally under the lithosphere toward midocean ridges. This may contribute to tectonic plate movement. Regions above the superplumes tend to bulge upward and other indications of superplumes, such as variations in gravity, have been measured. Scientists continue to work with new seismic tomography data with more precise images and higher resolutions to better describe the interior structure of the earth.

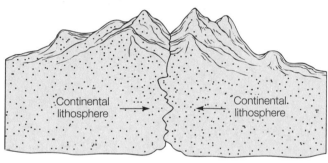

Figure 14.27

Continent-continent plate convergence. Rocks are deformed, and some lithosphere thickening occurs, but neither plate is subducted to any great extent.

CONCEPTS APPLIED

New Earthquakes and Volcanoes

Locate and label the major plates of the lithosphere on an outline map of the world according to the most recent findings in plate tectonics. Show all types of boundaries and associated areas of volcanoes and earthquakes.

---

3. **Transform boundaries** occur between two plates sliding by each other. Crust is neither created nor destroyed at transform boundaries as one plate slides horizontally past another along a long, vertical fault. The movement is neither smooth nor equal along the length of the fault, however, as short segments move independently with sudden jerks that are separated by periods without motion. The Pacific Plate, for example, is moving slowly to the northwest, sliding past the North American Plate. The San Andreas Fault is one boundary along the California coastline. Vibrations from plate movements along this boundary are the famous California earthquakes.

Present-Day Understandings

The theory of plate tectonics, developed during the late 1960s and early 1970s, is new compared to most major scientific theories. Measurements are still being made, evidence is being gathered and evaluated, and the exact number of plates and their boundaries are yet to be determined with certainty. The major question that remains to be answered is what drives the plates, moving them apart, together, and past each other? The most widely favored explanation is that slowly turning *convective cells* in the plastic asthenosphere drive the plates (figure 14.28). According to this hypothesis, hot fluid materials rise at the diverging boundaries. Some of the material escapes to form new crust, but most of it spreads out beneath the lithosphere. As it moves beneath the lithosphere, it drags the overlying plate with it. Eventually, it cools and sinks back inward under a subduction zone.

There is uncertainty about the existence of convective cells in the asthenosphere and their possible role because of a lack of clear evidence. Seismic data is not refined enough to show convective cell movement beneath the lithosphere. In addition, deep-seated earthquakes occur to depths of about 700 km, which means that descending materials—parts of a subducted plate—must extend to that depth. This could mean that a convective cell might operate all the way down to the core-mantle boundary some 2,900 km below the surface. This presents

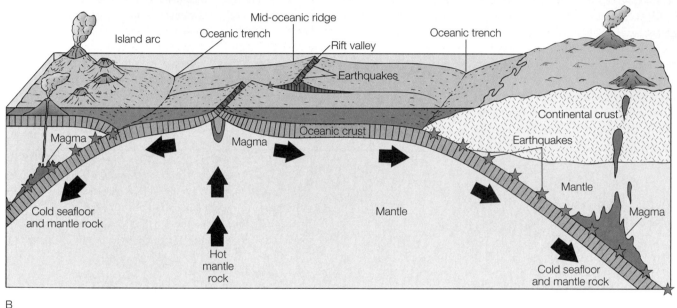

Figure 14.28

Not to scale. One idea about convection in the mantle has a convection cell circulating from the core to the lithosphere, dragging the overlying lithosphere laterally away from the oceanic ridge.

another kind of problem because little is known about the lower mantle and how it interacts with the upper mantle. Theorizing without information is called speculation, and that is the best that can be done with existing data. The full answer may include the role of heat and the role of gravity. However, understanding the mechanism that drives the plates will probably have to await a better understanding of how the lower mantle behaves.

What is generally accepted about the plate tectonic theory is the understanding that the solid materials of the earth are engaged in a continual cycle of change. Oceanic crust is subducted, melted, then partly returned to the crust as volcanic igneous rocks in island arcs and along continental plate boundaries. Other parts of the subducted crust become mixed with the upper mantle, returning as new crust at diverging boundaries. The materials of the crust and the mantle are thus cycled back and forth in a mixing that may include the deep mantle and the core as well. There is more to this story of a dynamic earth that undergoes a constant change. The story continues in the next chapters with different cycles to consider.

SUMMARY

The elements silicon and oxygen make up 75 percent of all the elements in the outer layer, or *crust*, of the earth. The elements combine to make crystalline chemical compounds called minerals. A *mineral* is defined as a naturally occurring, inorganic solid element or compound with a crystalline structure.

About 92 percent of the minerals of the earth's crust are composed of silicon and oxygen, the *silicate minerals*. The basic unit of the silicates is a *tetrahedral structure* that combines with positive metallic ions or with other tetrahedral units to form chains, sheets, or an interlocking framework. The *ferromagnesian silicates* are tetrahedral structures combined with ions of iron, magnesium, calcium, and other elements. The ferromagnesian silicates are darker in color and more dense than other silicates. The *nonferromagnesian silicates* do not have irons or magnesium ions and they are lighter in color and less dense than the ferromagnesians. The *nonsilicate minerals* do not contain silicon and are carbonates, sulfates, oxides, halides, sulfides, and native elements.

According to the theory of plate tectonics, the earth's outer shell is made up of moving plates. The plates making up the continents are about 100 km thick and are gradually drifting at a rate of about 0.5 to 10 cm (0.2 to 4 in) per year. This reading is about one way that scientists know that the earth's plates are moving and how this movement is measured.

The very first human lunar landing mission took place in July, 1969. Astronaut Neil Armstrong stepped onto the lunar surface, stating, "That's one small step for a man, one giant leap for mankind." In addition to fulfilling a dream, the *Apollo* project carried out a program of scientific experiments. The *Apollo 11* astronauts placed a number of experiments on the lunar surface in the Sea of Tranquility. Among the experiments was the first Laser Ranging Retro-reflector Experiment, which was designed to reflect pulses of laser light from the earth. Later, three more reflectors were placed on the moon, including two by other *Apollo* astronauts and one by an unmanned Soviet *Lunakhod 2* lander.

The McDonald Observatory in Texas, the Lure Observatory on the island of Maui, Hawaii, and a third observatory in southern France have regularly sent laser beams through optical telescopes to the reflectors. The return signals, which are too weak to be seen with the unaided eye, are detected and measured by sensitive detection equipment at the observatories. The accuracy of these measurements, according to NASA reports, is equivalent to determining the distance between a point on the east and a point on the west coast of the United States to an accuracy of 0.5 millimeter, about the size of the period at the end of this sentence.

Reflected laser light experiments have found that the moon is pulling away from the earth at about 4 cm/yr (about 1.6 in/yr), that the shape of the earth is slowly changing, undergoing isostatic adjustment from the compression by the glaciers during the last ice age, and that the observatory in Hawaii is slowly moving away from the one in Texas. This provides a direct measurement of the relative drift of two of the earth's tectonic plates. Thus, one way that changes on the surface of the earth are measured is through lunar ranging experiments. Results from lunar ranging, together with laser ranging to the artificial satellites in the earth's orbit, have revealed the small but constant drift rate of the plates making up the earth's dynamic surface.

A *rock* is defined as an aggregation of one or more minerals that have been brought together into a cohesive solid. *Igneous rocks* formed as hot, molten *magma* cooled and crystallized to firm, hard rocks. *Sedimentary rocks* are formed from *sediments,* accumulations of weathered rock materials that settle out of the atmosphere or out of water. Sediments become sedimentary rocks through a rock-forming process that involves both the *compaction* and *cementation* of the sediments. *Metamorphic rocks* are previously existing rocks that have been changed by heat, pressure, or hot solution into a different kind of rock without melting. The *rock cycle* is a concept that an igneous, a sedimentary, or a metamorphic rock is a temporary stage in the ongoing transformation of rocks to new types.

The earth has a layered interior that formed as the earth's materials underwent *differentiation,* the separation of materials while in the molten state. The center part, or *core,* is predominantly iron with a solid inner part and a liquid outer part. The core makes up about 15 percent of the earth's total volume and about a third of its total mass. The *mantle* is the middle part of the earth's interior that accounts for about two-thirds of the earth's total mass and about 80 percent of its total volume. The *Mohorovicic discontinuity* separates the outer layer, or *crust,* of the earth from the mantle. The crust of the *continents* is composed mostly of less dense granite-type rock. The crust of the *ocean basins* is composed mostly of the more dense basaltic rocks.

Another way to consider the earth's interior structure is to consider the weak layer in the upper mantle, the *asthenosphere* that extends around the entire earth. The rigid, solid, and brittle layer above the asthenosphere is called the *lithosphere.* The lithosphere includes the entire crust, the Moho, and the upper part of the mantle.

Evidence from the ocean floor revived interest in the idea that continents could move. The evidence for *seafloor spreading* came from related observations concerning oceanic ridge systems, sediment and fossil dating of materials outward from the ridge, and from magnetic patterns of seafloor rocks. Confirmation of seafloor spreading led to the *plate tectonic theory.* According to plate tectonics, new basaltic crust is added at *diverging boundaries* of plates, and old crust is *subducted* at *converging boundaries.* Mountain building, volcanoes, and earthquakes are seen as *related geologic features* that are caused by plate movements. The force behind the movement of plates is uncertain, but it may involve *convection* in the deep mantle.

KEY TERMS

asthenosphere (p. **285**)	new crust zone (p. **288**)
convergent boundaries (p. **290**)	oceanic ridges (p. **286**)
core (p. **284**)	oceanic trenches (p. **287**)
crust (p. **283**)	plate tectonics (p. **288**)
divergent boundaries (p. **288**)	rift (p. **287**)
igneous rocks (p. **280**)	rock (p. **280**)
island arcs (p. **290**)	seafloor spreading (p. **287**)
lithosphere (p. **285**)	sedimentary rocks (p. **281**)
mantle (p. **283**)	silicates (p. **277**)
metamorphic rocks (p. **282**)	subduction zone (p. **290**)
mineral (p. **276**)	transform boundaries (p. **291**)

APPLYING THE CONCEPTS

1. Sedimentary rocks are formed by the processes of compaction and
   (a) pressurization.
   (b) melting.
   (c) cementation.
   (d) heating, but not melting.

2. Which type of rock probably existed first, starting the rock cycle?
   (a) metamorphic
   (b) igneous
   (c) sedimentary
   (d) All of the above are correct.

3. From seismological data, the earth's shadow zone indicates that part of the earth's interior must be
   (a) liquid.
   (b) solid throughout.
   (c) plastic.
   (d) hollow.

4. The Mohorovicic discontinuity is a change in seismic wave velocity that is believed to take place because of
   (a) structural changes in minerals of the same composition.
   (b) changes in the composition on both sides of the boundary.
   (c) a shift in the density of minerals of the same composition.
   (d) changes in the temperature with depth.

5. The oldest rocks are found in (the)
   (a) continental crust.
   (b) oceanic crust.
   (c) neither, since both are the same age.

6. The least dense rocks are found in (the)
   (a) continental crust.
   (b) oceanic crust.
   (c) neither, since both are the same density.

7. The idea of seafloor spreading along the Mid-Atlantic Ridge was supported by evidence from
   (a) changes in magnetic patterns and ages of rocks moving away from the ridge.
   (b) faulting and volcanoes on the continents.
   (c) the observation that there was no relationship between one continent and another.
   (d) all of the above.

8. According to the plate tectonics theory, seafloor spreading takes place at a
   (a) convergent boundary.
   (b) subduction zone.
   (c) divergent boundary.
   (d) transform boundary.

9. The presence of an oceanic trench, a chain of volcanic mountains along the continental edge, and deep-seated earthquakes is characteristic of a (an)
   (a) ocean-ocean plate convergence.
   (b) ocean-continent plate convergence.
   (c) continent-continent plate convergence.
   (d) None of the above are correct.

10. The presence of an oceanic trench with shallow earthquakes and island arcs with deep-seated earthquakes is characteristic of a (an)
    (a) ocean-ocean plate convergence.
    (b) ocean-continent plate convergence.
    (c) continent-continent plate convergence.
    (d) None of the above are correct.

11. The ongoing occurrence of earthquakes without seafloor spreading, oceanic trenches, or volcanoes is most characteristic of a
    (a) convergent boundary between plates.
    (b) subduction zone.
    (c) divergent boundary between plates.
    (d) transform boundary between plates.

12. The evidence that the earth's core is part liquid or acts like a liquid comes from (the)
    (a) P-wave shadow zone.
    (b) S-wave shadow zone.
    (c) meteorites.
    (d) all of the above.

Answers

1. c 2. b 3. a 4. b 5. a 6. a 7. a 8. c 9. b 10. a 11. d 12. b

QUESTIONS FOR THOUGHT

1. Briefly describe the rock-forming process that changes sediments into solid rock.

2. What are metamorphic rocks? What limits the maximum temperatures possible in metamorphism? Explain.

3. What evidence provides information about the nature of the earth's core?

4. What is the asthenosphere? Why is it important in modern understandings of the earth?

5. Describe the origin of the magnetic strip patterns found in the rocks along an oceanic ridge.

6. Explain why ancient rocks are not found on the seafloor.

7. Describe the three major types of plate boundaries and what happens at each.

8. Briefly describe the theory of plate tectonics and how it accounts for the existence of certain geologic features.

9. What is an oceanic trench? What is its relationship to major plate boundaries? Explain this relationship.

10. Describe the probable source of all the earthquakes that occur in southern California.

11. The northwestern coast of the United States has a string of volcanoes running along the coast. According to plate tectonics, what does this mean about this part of the North American Plate? What geologic feature would you expect to find on the seafloor off the northwestern coast? Explain.

12. Explain how the crust of the earth is involved in a dynamic, ongoing recycling process.

CHAPTER FIFTEEN

The Earth's Surface

Folding, faulting, and lava flows, such as the one you see here, tend to build up, or elevate, the earth's surface.

OUTLINE

The central idea of plate tectonics, which was discussed in chapter 14, is that the earth's surface is made up of rigid plates that are moving slowly across the surface. Since the plates and the continents riding on them are in constant motion, any given map of the world is only a snapshot that shows the relative positions of the continents at a given time. The continents occupied different positions in the distant past. They will occupy different positions in the distant future. The surface of the earth, which seems so solid and stationary, is in fact mobile.

Plate tectonics has changed the accepted way of thinking about the solid, stationary nature of the earth's surface and ideas about the permanence of the surface as well. The surface of the earth is no longer viewed as having a permanent nature, but is understood to be involved in an ongoing cycle of destruction and renewal. Old crust is destroyed as it is plowed back into the mantle through subduction, becoming mixed with the mantle. New crust is created as molten materials move from the mantle through seafloor spreading and volcanoes. Over time, much of the crust must cycle into and out of the mantle.

The movement of plates, the crust-mantle cycle, and the rock cycle all combine to produce a constantly changing surface. There are basically two types of surface changes: (1) changes that originate within the earth, resulting in a building up of the surface (figure 15.1) and (2) changes that occur when rocks are exposed to the atmosphere and water, resulting in a sculpturing and tearing down of the surface. This chapter is about changes in the land. The concepts of this chapter will provide you with something far more interesting about the earth's surface than the scenic aspect. The existence of different features (such as mountains, folded hills, islands) and the occurrence of certain events (such as earthquakes, volcanoes, faulting) are all related. The related features and events also have a story to tell about the earth's past, a story about the here and now, and yet another story about the future.

INTERPRETING EARTH'S SURFACE

Because many geologic changes take place slowly it is difficult for a person to see significant change occur to mountains, canyons, and shorelines in the brief span of a lifetime. Given a mental framework based on a lack of appreciation of change over geologic time, how do you suppose people interpreted the existence of features such as mountains and canyons? Some believed, as they had observed in their lifetimes, that the mountains and canyons had "always" been there. Statements such as "unchanging as the hills" or "old as the hills" illustrate this lack of appreciation of change over geologic time. Others did not believe the features had always been there, but believed they were formed by a sudden, single catastrophic event (figure 15.2). A catastrophe created a feature of the earth's surface all at once, with little or no change occurring since that time. The Grand Canyon, for example, was not interpreted as the result of incomprehensibly slow river erosion, but as the result of a giant crack or rip that appeared in the surface. The canyon that you see today was interpreted as forming when the earth split open and the Colorado River fell into the split. This interpretation was used to explain the formation of major geologic features based on the little change observed during a person's lifetime.

About two hundred years ago the idea of unchanging, catastrophically formed landscapes was challenged by James Hutton, a Scottish physician. Hutton, who is known today as the founder of modern geology, traveled widely throughout the British Isles. Hutton was a keen observer of rocks, rock structures, and other features of the landscape. He noted that sandstone, for example, was made up of rock fragments that appeared to be (1) similar to the sand being carried by rivers and (2) similar to the sand making up the beaches next to the sea. He also noted fossil shells of sea animals in sandstone on the land, while the living relatives of these animals were found in the shallow waters of the sea. This and other evidence led Hutton to realize that rocks were being ground into fragments, then carried by rivers to the sea. He surmised that these particles would be reformed into rocks later, then lifted and shaped into the hills and mountains of the land.

He saw all this as quiet, orderly change that required only *time* and the ongoing work of the water and some forces to make the sediments back into rocks. With Hutton's logical conclusion came the understanding that the earth's history could be interpreted by tracing it backwards, from the present to the past. This tracing required a frame of reference of slow, uniform change, not the catastrophic frame of reference of previous thinkers. The frame of reference of uniform changes is today called the **principle of uniformity** (also called *uniformitarianism*). The principle of uniformity is often represented by a statement that "the present is the key to the past." This statement means that the same geologic processes you see changing rocks today are the very same processes that changed them in the ancient past, although not necessarily at the same rate. The principle of uniformity does not *exclude* the happening of sudden or catastrophic events on the surface of the earth. A violent volcanic explosion, for example, is a catastrophic event that most certainly modifies the surface of the earth. What the principle of uniformity does state is that the physical and chemical laws we observe today operated exactly the same in the past. The rates of operation may or may not have been the same in the past, but the events you see occurring today are the same events that occurred in the past. Given enough time, you can explain the formation of the structures of the earth's surface with known events and concepts.

The principle of uniformity has been used by geologists since the time of Hutton. The concept of how the constant changes occur has evolved with the development of plate tectonics, but the basic frame of reference is the same. You will see how the principle of uniformity is applied by first considering what can happen to rocks and rock layers that are deeply buried.

PROCESSES THAT BUILD UP THE SURFACE

All the possible movements of the earth's plates, including drift toward or away from other plates, and any process that deforms the earth's surface are included in the term *diastrophism.*

Figure 15.1
An aerial view from the south of the eruption of Mount St. Helens volcano on May 18, 1980.

Figure 15.2
Would you believe that this rock island has "always" existed where it is? Would you believe it was formed by a sudden, single event? What evidence would it take to convince you that the rock island formed ever so slowly, starting as a part of southern California and moving very slowly, at a rate of centimeters per year, to its present location near the coast of Alaska?

Diastrophism is the process of deformation that changes the earth's surface. It produces many of the basic structures you see on the earth's surface, such as plateaus, mountains, and folds in the crust. The movement of magma is called *vulcanism* or *volcanism*. Diastrophism, volcanism, and earthquakes are closely related, and their occurrence can usually be explained by events involving plate tectonics. The results of diastrophism are discussed in the following section, which is followed by a discussion of earthquakes, volcanoes, and mountain chains.

Stress and Strain

Any solid material responds to a force in a way that depends on the extent of coverage (force per unit area, or pressure), the nature of the material, and other variables such as the temperature. Consider, for example, what happens if you place the point of a ball-point pen on the side of an aluminum pop (soda) can and apply an increasing pressure. With increasing pressure you can observe at least four different and separate responses:

1. At first, the metal successfully resists a slight pressure and *nothing happens.*
2. At a somewhat greater pressure you will be able to deform, or bend the metal into a concave surface. The metal will return to its original shape, however, when the pressure is removed. This is called an *elastic deformation* since the metal was able to spring back into its original shape.
3. At a still greater pressure the metal is deformed to a concave surface, but this time the metal does not return to its original shape. This means the *elastic limit* of the metal has been exceeded, and it has now undergone a *plastic deformation.* Plastic deformation permanently alters the shape of a material.
4. Finally, at some great pressure the metal will rupture, resulting in a *break* in the material.

Many materials, including rocks, respond to increasing pressures in this way, showing (1) no change, (2) an elastic change with recovery, (3) a plastic change with no recovery, and (4) finally breaking from the pressure.

A **stress** is a force that tends to compress, pull apart, or deform a rock. Rocks in the earth's solid outer crust are subjected to forces as the earth's plates move into, away from, or alongside each other. Thus, there are three types of forces that cause rock stress:

1. *Compressive stress* is caused by two plates moving together or by one plate pushing against another plate that is not moving.
2. *Tensional stress* is the opposite of compressional stress. It occurs when one part of a plate moves away, for example, and another part does not move.
3. *Shear stress* is produced when two plates slide past one another or by one plate sliding past another plate that is not moving.

Just like the metal in the soda can, a rock is able to withstand stress up to a limit. Then it might undergo elastic deformation, plastic deformation, or breaking with progressively greater pressures. The adjustment to stress is called **strain.** A rock unit might respond to stress by changes in volume, changes in shape, or by breaking. Thus, there are three types of strain: elastic, plastic, and fracture.

1. In *elastic strain,* rock units recover their original shape after the stress is released.
2. In *plastic strain,* rock units are molded or bent under stress and do not return to their original shape after the stress is released.
3. In *fracture strain,* rock units crack or break as the name suggests.

The relationship between stress and strain, that is, exactly how the rock responds, depends on at least four variables. They are (1) the nature of the rock, (2) the temperature of the rock, (3) how slowly or quickly the stress is applied over time, and (4) the confining pressure on the rock. The temperature and confining pressure are generally a function of how deeply the rock is buried. In general, rocks are better able to withstand compressional rather than pulling-apart stresses. Cold rocks are more likely to break than warm rocks, which tend to undergo plastic deformation. In addition, a stress that is applied quickly tends to break the rock, where stress applied more slowly over time, perhaps thousands of years, tends to result in plastic strain.

In general, rocks at great depths are under great pressure at higher temperatures. These rocks tend to undergo plastic deformation, then plastic flow, so rocks at great depths are bent and deformed extensively. Rocks at less depth can also bend, but they have a lower elastic limit and break more readily (figure 15.3). Rock deformation often results in recognizable surface features called folds and faults, the topics of the next sections.

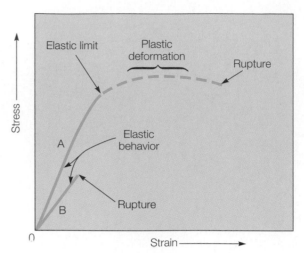

Figure 15.3

Stress and deformation relationships for deeply buried, warm rocks under high pressure (*A*) and cooler rocks near the surface (*B*). Breaking occurs when stress exceeds rupture strength.

Folding

Sediments that form most sedimentary rocks are deposited in nearly flat, horizontal layers at the bottom of a body of water. Conditions on the land change over time, and different mixtures of sediments are deposited in distinct layers of varying thickness. Thus, most sedimentary rocks occur naturally as structures of horizontal layers, or beds (figure 15.4).

A

B

Figure 15.4

(*A*) Rock bedding on a grand scale in the Grand Canyon. (*B*) A closer example of rock bedding can be seen in this roadcut.

Figure 15.5
These folded rock layers are in the Calico Hills, California. Can you figure out what might have happened here to fold flat rock layers like this?

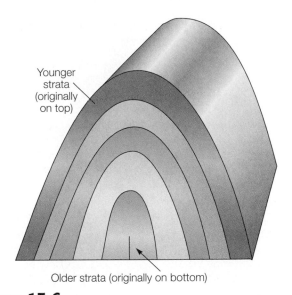

Younger strata (originally on top)

Older strata (originally on bottom)

Figure 15.6
An anticline, or arching fold, in layered sediments. Note that the oldest strata are at the center.

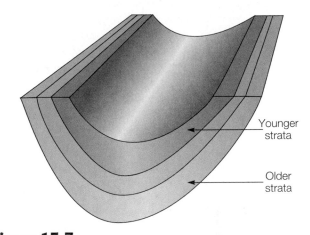

Younger strata

Older strata

Figure 15.7
A syncline, showing the reverse age pattern.

A sedimentary rock layer that is not horizontal may have been subjected to some kind of compressive stress. The source of such a stress could be from colliding plates, from the intrusion of magma from below, or from a plate moving over a rising superplume of magma. Seismic tomography has identified massive plumes of molten rock rising toward the surface from deep within the earth. These superplumes originate from the base of the mantle, rising to the lithosphere. The hot material was observed to spread out horizontally, causing bulges as it exerts forces under the lithosphere.

Stress on buried layers of horizontal rocks can result in plastic strain, resulting in a wrinkling of the layers into *folds*. **Folds** are bends in layered bedrock (figure 15.5). They are analogous to layers of rugs or blankets that were stacked horizontally, then pushed into a series of arches and troughs. Folds in layered bedrock of all shapes and sizes can occur from plastic strain, depending generally on the regional or local nature of the stress and other factors. Of course, when the folding occurred the rock layers were in a plastic condition, and were probably under considerable confining pressure from deep burial. However, you see the results of the folding when the rocks are under very different conditions at the surface.

The most common regional structures from deep plastic deformation are arch-shaped and trough-shaped folds. In general, an arch-shaped fold is called an **anticline** (figure 15.6). The corresponding trough-shaped fold is called a **syncline** (figure 15.7). Anticlines and synclines sometimes alternate across the land like waves on water. You can imagine that a great compressional stress must have been involved over a wide region to wrinkle the land like this.

Anticlines, synclines, and other types of folds are not always visible as such on the earth's surface. The ridges of anticlines are constantly being weathered into sediments. The sediments, in turn, tend to collect in the troughs of synclines, filling them in. The Appalachian Mountains have ridges of rocks that are more resistant to weathering, forming hills and mountains. The San Joaquin Valley, on the other hand, is a very large syncline in California.

Note that any kind of rock can be folded. Sedimentary rocks are usually the best example of folding, however, since the fold structures of rock layers are easy to see and describe. Folding is much harder to see in igneous or metamorphic rocks that are blends of minerals without a layered structure.

Faulting

Rock layers do not always respond to stress by folding. Rocks near the surface are cooler and under less pressure, so they tend to be more brittle. A sudden stress on these rocks may reach the rupture point, resulting in a cracking and breaking of the rock structure. When there is relative movement between the rocks

A

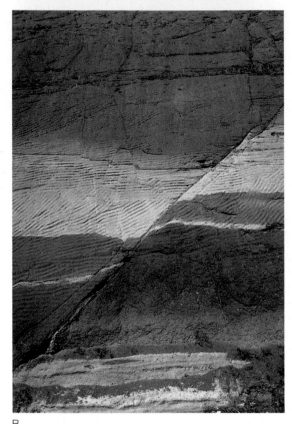

B

Figure 15.8

(*A*) This sketch shows the relationship between the hanging wall, the footwall, and a fault. (*B*) A photo of a fault near Kingman, Arizona, showing how the hanging wall has moved relative to the footwall.

One classification scheme for faults is based on an orientation referent borrowed from mining (many ore veins are associated with fault planes). Imagine a mine with a fault plane running across a horizontal shaft. Unless the plane is perfectly vertical, a miner would stand on the mass of rock below the fault plane and look up at the mass of rock above. Therefore, the mass of rock below is called the *footwall* and the mass of rock above is called the *hanging wall* (figure 15.8). How the footwall and hanging wall have moved relative to one another describes three basic classes of faults, the (1) normal, (2) reverse, and (3) thrust faults. A **normal fault** is one in which the hanging wall has moved downward relative to the footwall. This seems "normal" in the sense that you would expect an upper block to slide *down* a lower block along a slope (figure 15.9A). Sometimes a huge block of rock bounded by normal faults will drop down, creating a *graben* (figure 15.9B). The opposite of a graben is a *horst*, which is a block bounded by normal faults that is uplifted (figure 15.9C). A very large block lifted sufficiently becomes a fault-block mountain. Many parts of the western United States are characterized by numerous fault-block mountains separated by adjoining valleys.

In a **reverse fault,** the hanging wall block has moved upward relative to the footwall block. As illustrated in figure 15.10A, a reverse fault is probably the result of horizontal compressive stress.

A reverse fault with a low-angle fault plane is also called a **thrust fault** (figure 15.10B). In some thrust faults the hanging wall block has completely overridden the lower footwall for 10 to 20 km (6 to 12 mi). This is sometimes referred to as an "overthrust."

A

B

C

Figure 15.9

How tensional stress could produce (*A*) a normal fault, (*B*) a graben, and (*C*) a horst.

on either side of a fracture, the crack is called a **fault.** When faulting occurs, the rocks on one side move relative to the rocks on the other side along the surface of the fault, which is called the *fault plane.* Faults are generally described in terms of (1) the steepness of the fault plane, that is, the angle between the plane and imaginary horizontal plane and (2) the direction of relative movement. There are basically three ways that rocks on one side of a fault can move relative to the rocks on the other side: (1) up and down (called "dip"), (2) horizontally, or sideways (called "strike"), and (3) with elements of both directions of movement (called "oblique").

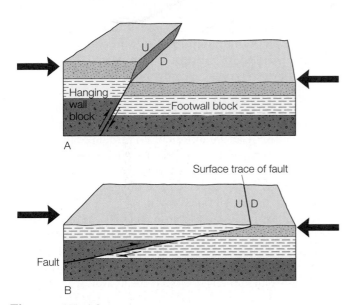

Figure 15.10

How compressive stress could produce (*A*) a reverse fault, and (*B*) a thrust fault.

As shown in figures 15.9 and 15.10, the relative movement of blocks of rocks along a fault plane provides information about the stresses that produced the movement. Reverse and thrust faulting result from compressional stress in the direction of the movement. Normal faulting, on the other hand, results from a pulling-apart stress that might be associated with diverging plates. It might also be associated with the stretching and bulging up of the crust over a hot spot.

EARTHQUAKES

What is an earthquake? An **earthquake** is a quaking, shaking, vibrating, or upheaval of the ground. Earthquakes are the result of the sudden release of energy that comes from *stress* on rock beneath the earth's surface. In the previous section you learned that rock units can bend and become deformed in response to stress, but there are limits as to how much stress rock can take before it fractures. When it does fracture, the sudden movement of blocks of rock produces vibrations that move out as waves throughout the earth. These vibrations are called **seismic waves.** It is strong seismic waves that people feel as a shaking, quaking, or vibrating during an earthquake.

Seismic waves are generated when a huge mass of rock breaks and slides into a different position. As you learned in the previous section, the plane between two rock masses that have moved into new relative positions is called a *fault*. Major earthquakes occur along existing fault planes or when a new fault is formed by the fracturing of rock. In either case, most earthquakes occur along a fault plane when there is displacement of one side relative to the other.

Most earthquakes occur along a fault plane and they occur near the earth's surface. You might expect this to happen since the rocks near the surface are brittle, and those deeper are more plastic from increased temperature and pressure. Shallow earthquakes are typical of those that occur at the boundary of the North American Plate, which is moving against the Pacific Plate. In California, the boundary between these two plates is known as the *San Andreas fault* (figure 15.11). The San Andreas fault runs north-south through California, with the Pacific Plate moving on one side and the North American Plate moving on the other. The two plates are tightly pressed against each other, and friction between the rocks along the fault prevents them from moving easily. Stress continues to build along the entire fault as one plate attempts to move along the other. Some elastic deformation does occur from the stress, but eventually the rupture strength of the rock (or the friction) is overcome. The

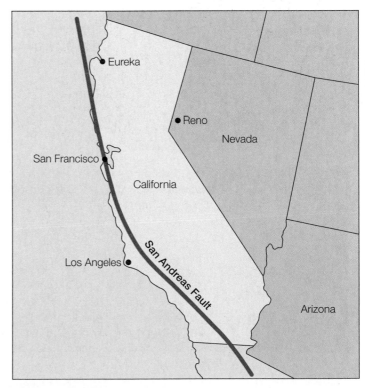

Figure 15.11

The San Andreas fault runs north-south for about 1,300 km (800 mi) through California.

stressed rock, now released of the strain, snaps suddenly into new positions in the phenomenon known as **elastic rebound** (figure 15.12). The rocks are displaced to new positions on either side of the fault, and the vibrations from the sudden movement are felt as an earthquake. The elastic rebound and movement tend to occur along short segments of the fault at different times rather than along long lengths. Thus, the resulting earthquake tends to be a localized phenomenon rather than a regional one.

Most earthquakes occur near plate boundaries. They do happen elsewhere, but are rare. The actual place where seismic waves originate beneath the surface is called the *focus* of the earthquake. The focus is considered to be the center of the earthquake and the place of initial rock movement on a fault. The point on the earth's surface directly above the focus is called the earthquake *epicenter* (figure 15.13).

Seismic waves radiate outward from an earthquake focus, spreading in all directions through the solid earth's interior like the sound waves from an explosion. As introduced in the previous chapter, there are three kinds of waves: *P-waves, S-waves,* and *surface waves.* S- and P-waves provide information about the location and magnitude of an earthquake as well as information about the earth's interior.

Seismic S- and P-waves leave the focus of an earthquake at essentially the same time. As they travel away from the focus, they gradually separate because the P-waves travel faster than the S-waves. To locate an epicenter, at least three recording stations measure the time lag between the arrival of the P-waves and the slower S-waves. The difference in the speed between the two waves is a constant. Therefore, the farther they travel, the greater the time lag between the arrival of the faster P-waves and the slower S-waves (figure 15.14A). By measuring the time lag and knowing the speed of the two waves, it is possible to calculate the distance to their source. However, the calculated distance provides no information about the direction or location of the source of the waves. The location is found by first using the calculated distance as the radius of a circle drawn on a map. The place where the circles from the three recording stations intersect is the location of the source of the waves (figure 15.14B).

Earthquakes occur in a wide range of intensities, from the many that are barely detectable to the few that cause widespread destruction. Destruction is caused by the seismic waves, which cause the land and buildings to vibrate or shake. Vibrations during small quakes can crack windows and walls, but vibrations in strong quakes can topple bridges and buildings. Injuries and death are usually the result of falling debris, crumbling buildings, or falling bridges. Fire from broken gas pipes was a problem in the 1906 and 1989 earthquakes in San Francisco and the 1994 earthquake in Los Angeles. Broken water mains made it difficult to fight the 1906 San Francisco fires, but in 1989 fire-

Figure 15.12

The elastic rebound theory of the cause of earthquakes. (*A*) Rock with stress acting on it. (*B*) Stress has caused strain in the rock. Strain builds up over a long period of time. (*C*) Rock breaks suddenly, releasing energy, with rock movement along a fault. Horizontal motion is shown; rocks can also move vertically. (*D*) Horizontal offset of rows in a lettuce field, 1979, El Centro, California.

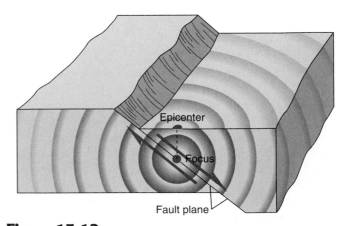

Figure 15.13

Simplified diagram of a fault, illustrating component parts and associated earthquake terminology.

A

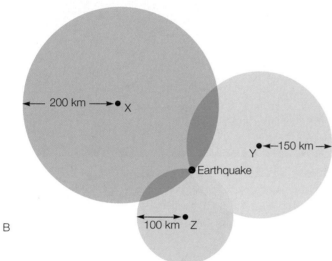

B

Figure 15.14

Use of seismic waves in locating earthquakes. (*A*) Difference in times of first arrivals of P-waves and S-waves is a function of the distance from the focus. (*B*) Triangulation using data from several seismograph stations allows location of the earthquake.

boats and fire hoses using water pumped from the bay were able to extinguish the fires. Table 15.1 lists some recommendations of things to do during and after an earthquake.

Other effects of earthquakes include landslides, displacement of the land surface, and tsunamis. Vertical, horizontal, or both vertical and horizontal displacement of the land can occur during a quake. People sometimes confuse cause and effect when they see a land displacement saying things like, "Look what the earthquake did!" The fact is that the movement of the land probably produced the seismic waves (the earthquake). The seismic waves did not produce the land displacement. Displacements from a single earthquake can be up to 10 to 15 m (about 30 to 50 ft), but such displacements rarely happen.

Tsunami is a Japanese term used to describe the very large ocean waves that can be generated by an earthquake, landslide, or volcanic explosion. Such large waves were formerly called "tidal waves." Since the large, fast waves were not associated with

tides or tidal forces in any way, the term *tsunami* or *seismic sea wave* is preferred.

A tsunami, like other ocean waves, is produced by a disturbance. Most common ocean waves are produced by winds, travel at speeds of 90 km/h (55 mi/h), and produce a wave height of 0.6 to 3 m (2 to 10 ft) when they break on the shore. A tsunami, on the other hand, is produced by some strong disturbance in the seafloor, travels at speeds of 725 km/h (450 mi/h), and can produce a wave height of over 8 m (about 25 ft) when it breaks on the shore. A tsunami may have a very long wavelength of up to 200 km (about 120 mi) compared to the wavelength of ordinary wind-generated waves of 400 m (1,300 ft). Because of its great wavelength, a tsunami does not just break on the shore, then withdraw. Depending on the seafloor topography, the water from a tsunami may continue to rise for 5 to 10 minutes, flooding the coastal region before the wave withdraws. A gently sloping seafloor and a funnel-shaped bay can force tsunamis to great heights as they break on the shore. The size of a particular tsunami depends on the nature of how the seafloor was disturbed. Generally, an earthquake that causes the seafloor suddenly to rise or fall favors the generation of large tsunami.

The size of an earthquake (if it was a "big one" or a "little one") can be measured in terms of vibrations, in terms of displacement, or in terms of the amount of energy released at the site of the earthquake. The energy of the vibrations or motion of the land associated with an earthquake is called its *magnitude*. Earthquake magnitude is often reported by the media using the *Richter scale* (table 15.2). The Richter scale was developed by

Table 15.2 Effects of Earthquakes of Various Magnitudes

Richter Magnitudes	Description
0–2	Smallest detectable earthquake.
2–3	Detected and measured but not generally felt.
3–4	Felt as small earthquake but no damage occurs.
4–5	Minor earthquake with local damage.
5–6	Moderate earthquake with structural damage.
6–7	Strong earthquake with destruction.
7–8	Major earthquake with extensive damage and destruction.
8–9	Great earthquake with total destruction.

Figure 15.15

The folded structure of the Appalachian Mountains, revealed by weathering and erosion, is obvious in this *Skylab* photograph of the Virginia-Tennessee-Kentucky boundary area. The clouds are over the Blue Ridge Mountains.

Charles Richter, who was a seismologist at the California Institute of Technology in the early 1930s. The scale was based on the widest swing in the back-and-forth line traces of seismograph recording. This scale assigns a number that increases with the magnitude of an earthquake. The numbers have meaning about (1) the severity of the ground-shaking vibrations and (2) the energy released by the earthquake. Each higher number indicates about ten times more ground movement and about thirty times more energy released than the preceding number. An earthquake measuring below 3 on the scale is usually not felt by people near the epicenter. The largest earthquake measured so far had a magnitude near 9, but there is actually no upper limit to the scale. Today, professional seismologists rate the size of earthquakes in different ways, depending on what they are comparing and why, but each way results in logarithmic scales similar to the Richter scale.

ORIGIN OF MOUNTAINS

Folding and faulting have created most of the interesting features of the earth's surface, and the most prominent of these features are **mountains.** Mountains are elevated parts of the earth's crust that rise abruptly above the surrounding surface. Most mountains do not occur in isolation, but occur in chains or belts. These long, thin belts are generally found along the edges of continents rather than in the continental interior. There are a number of complex processes involved in the origin of mountains and mountain chains, and no two mountains are exactly alike. For convenience, however, mountains can be classified according to three basic origins: (1) folding, (2) faulting, and (3) volcanic activity.

The major mountain ranges of the earth—the Appalachian, Rocky, and Himalayan Mountains, for example—have a great height that involves complex folding on a very large scale. The crust was thickened in these places as compressional forces produced tight, almost vertical folds. Thus, folding is a major feature of the major mountain ranges, but faulting and igneous intrusions are invariably also present. Differential weathering of different rock types produced the parallel features of the Ap-

palachian Mountains that are so prominent in satellite photographs (figure 15.15). The folded sedimentary rocks of the Rockies are evident in the almost upright beds along the flanks of the front range.

A broad arching fold, which is called a dome, produced the Black Hills of South Dakota. The sedimentary rocks from the top of the dome have been weathered away, leaving a somewhat circular area of more resistant granite hills surrounded by upward-tilting sedimentary beds (figure 15.16). The Adirondack Mountains of New York are another example of this type of mountain formed from folding, called domed mountains.

Compression and relaxation of compressional forces on a regional scale can produce large-scale faults, shifting large crustal blocks up or down relative to one another. Huge blocks of rocks can be thrust to mountainous heights, creating a series of fault block mountains. Fault block mountains rise sharply from the surrounding land along the steeply inclined fault plane. The mountains are not in the shape of blocks, however, as weathering has carved them into their familiar mountain-like shapes (figure 15.17). The Teton Mountains of Wyoming and the Sierra Nevadas of California are classic examples of fault block mountains that rise abruptly from the surrounding land. The various mountain ranges of Nevada, Arizona, Utah, and southeastern California have large numbers of fault block mountains that generally trend north and south.

Oldest formation exposed on the surface

Youngest formation

A

B

Figure 15.16

(*A*) A sketch of an eroded structural dome where all the rock layers dip away from the center. (*B*) A photo of a dome named Little Sundance Mountain (Wyoming), showing the more resistant sedimentary layers that dip away from the center.

Original sharply bounded fault blocks softened by erosion and sedimentation

Sediments

Figure 15.17

Fault block mountains are weathered and eroded as they are elevated, resulting in a rounded shape and sedimentation rather than sharply edged fault blocks.

Lava and other materials from volcanic vents can pile up to mountainous heights on the surface. These accumulations can form local volcano-formed mountains near mountains produced by folding or faulting. Such mixed-origin mountains are common in northern Arizona, New Mexico, and in western Texas. The Cascade Mountains of Washington and Oregon are a series of towering volcanic peaks, most of which are not active today. As a source of mountains, volcanic activity has an overall limited impact on the continents. The major mountains built by volcanic activity are the mid-oceanic ridges formed at diverging plate boundaries.

Deep within the earth previously solid rock melts at high temperatures to become *magma,* a pocket of molten rock. Magma is not just melted rock alone, however, as the melt contains variable mixtures of minerals (resulting in different types of lava flows). It also includes gases such as water vapor, sulfur dioxide, hydrogen sulfide, carbon dioxide, and hydrochloric acid. You can often smell some of these gases around volcanic vents and hot springs. Hydrogen sulfide smells like rotten eggs or sewer gas. The sulfur smells like a wooden match that has just been struck.

The gases dissolved in magma play a major role in forcing magma out of the ground. As magma nears the surface it comes under less pressure, and this releases some of the dissolved gases from the magma. The gases help push the magma out of the ground. This process is similar to releasing the pressure on a can of warm soda, which releases dissolved carbon dioxide.

Magma works its way upward from its source below to the earth's surface, here to erupt into a lava flow or a volcano. A **volcano** is a hill or mountain formed by the extrusion of lava or rock fragments from magma below. Some lavas have a lower viscosity than others, are more fluid, and flow out over the land rather than forming a volcano. Such *lava flows* can accumulate into a plateau of basalt, the rock that the lava formed as it cooled and solidified. The Columbia Plateau of the states of Washington, Idaho, and Oregon is made up of layer after layer of basalt that accumulated from lava flows. Individual flows of lava formed basalt layers up to 100 m (about 330 ft) thick, covering an area of hundreds of square kilometers. In places, the Columbia Plateau is up to 3 km (about 2 mi) thick from the accumulation of many individual lava flows.

The hill or mountain of a volcano is formed by ejected material that is deposited in a conical shape. The materials are deposited around a central *vent,* an opening through which an eruption takes place. The *crater* of a volcano is a basin-like depression over a vent at the summit of the cone. Figure 15.18 is an air view of Mount St. Helens, looking down into the crater at a volcanic dome that formed as magma periodically welled upward into the floor of the crater. This photo was taken several years after Mount St. Helens erupted in May 1980. Since that time the volcano has been quiet. It is not known when Mount St. Helens will erupt again. Geologists can tell when an eruption is near, since they are preceded by thousands of small earthquakes. Mount Garibaldi, Mount Rainier, Mount Hood, Mount Baker, and about ten other volcanic cones of the Cascade Range in Washington and Oregon could erupt next year, in the next decade, or in the next century. There are strong reasons that

A volcanic eruption changes the local landscape; that much is obvious. What is not so obvious are the worldwide changes that can happen just because of the eruption of a single volcano. Perhaps the most discussed change brought about by a volcano occurred back in 1815–16 after the eruption of Tambora in Indonesia. The Tambora eruption was massive, blasting huge amounts of volcanic dust, ash, and gas high into the atmosphere. Most of the ash and dust fell back to the earth around the volcano, but some dust particles and sulfur dioxide gas were pushed high into the stratosphere. It is known today that the sulfur dioxide from such explosive volcanic eruptions reacts with water vapor in the stratosphere, forming tiny droplets of diluted sulfuric acid. In the stratosphere there is no convection, so the droplets of acid and dust from the volcano eventually form a layer around the entire globe. This forms a haze that remains in the stratosphere for years, reflecting and scattering sunlight.

What were the effects of a volcanic haze around the entire world? There were fantastic, brightly colored sunsets from the added haze in the stratosphere. On the other hand, it was also cooler than usual, presumably because of the reflected sunlight that did not reach the earth's surface. It snowed in New England in June of 1816, and the cold continued into July. Crops failed, and 1816 became known as the "year without summer."

More information is available about the worldwide effects of present-day volcanic eruptions because there are now instruments to make more observations in more places. However, it is still necessary to do a great deal of estimating because of the relative inaccessibility of the worldwide stratosphere. It was estimated, for example, that the 1982 eruption of El Chichon in Mexico created enough haze in the stratosphere to reflect 5 percent of the solar radiation away from the earth. Researchers also estimated that

the effects of the El Chichon eruption cooled the global temperatures by a few tenths of a degree for two or three years. The cooling did take place, but the actual El Chichon contribution to the cooling is not clear because of other interactions. The earth may have been undergoing global warming from the greenhouse effect, for example, so the El Chichon cooling effect could have actually been much greater. Other complicating factors such as the effects of El Niño make changes difficult to predict.

In June 1991, the Philippine volcano Mount Pinatubo erupted, blasting twice as much gas and dust into the stratosphere as El Chichon had about a decade earlier. The haze from such eruptions has the potential to cool the climate about 0.5°C (1°F). The overall result, however, will always depend on a possible greenhouse effect, a possible El Niño effect, and other complications.

Figure 15.18

This is the top of Mount St. Helens several years after the 1980 explosive eruption.

make it very likely that any one of the volcanic cones of the Cascade Range will indeed erupt again.

A volcano forms when magma breaks out at the earth's surface. Only a small fraction of all the magma generated actually reaches the earth's surface. Most of it remains below the ground, cooling and solidifying to form igneous rocks that were described in chapter 14. A large amount of magma that has crys-

tallized below the surface is known as a *batholith.* A small protrusion from a batholith is called a *stock.* By definition a stock has less than 100 km² (40 mi²) of exposed surface area, and a batholith is larger. Both batholiths and stocks become exposed at the surface through erosion of the overlying rocks and rock materials, but not much is known about their shape below. The sides seem to angle away with depth, suggesting that they become larger with depth. The intrusion of a batholith sometimes tilts rock layers upward, forming a *hogback,* a ridge with equal slopes on its sides (figure 15.19). Other forms of intruded rock were formed as moving magma took the paths of least resistance, flowing into joints, faults, and planes between sedimentary bodies of rock. An intrusion that has flowed into a joint or fault that cuts across rock bodies is called a *dike.* A dike is usually tabular in shape, sometimes appearing as a great wall when exposed at the surface. One dike can occur by itself, but frequently dikes occur in great numbers, sometimes radiating out from a batholith like spokes around a wheel. If the intrusion flowed into the plane of contact between sedimentary rock layers, it is called a *sill.* A *laccolith* is similar to a sill, but has an arched top where the intrusion has raised the overlying rock into a blister-like uplift (figure 15.19).

Overall, the origin of mountain systems and belts of mountains such as the Cascades involves a complex mixture of volcanic activity as well as folding and faulting. An individual mountain, such as Mount St. Helens, can be identified as having a volcanic origin. The overall picture is best seen, however, from generalizations about how the mountains have grown along the edge of plates that are converging. Such converging boundaries are the places of folding, faulting, and associated earthquakes. They are also the places of volcanic activities, events that build

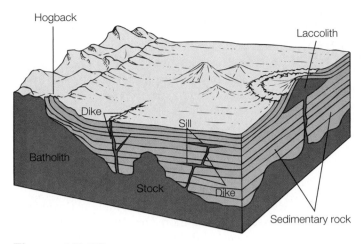

Figure 15.19
Here are the basic intrusive igneous bodies that form from volcanic activity.

and thicken the earth's crust. Thus, plate tectonics explains that mountains are built as the crust thickens at a convergent boundary between two plates. These mountains are slowly weathered and worn down as the next belt of mountains begins to build at the new continental edge.

PROCESSES THAT TEAR DOWN THE SURFACE

Sculpturing of the earth's surface takes place through agents and processes acting so gradually that humans are usually not aware that it is happening. Sure, some events such as a landslide or the movement of a big part of a beach by a storm are noticed. But the continual, slow, downhill drift of all the soil on a slope or the constant shift of grains of sand along a beach are outside the awareness of most people. People do notice the muddy water moving rapidly downstream in the swollen river after a storm, but few are conscious of the slow, steady dissolution of limestone by acid rain percolating through it. Yet, it is the processes of slow moving, shifting grains and bits of rocks, and slow dissolving that will wear down the mountains, removing all the features of the landscape that you can see.

A mountain of solid granite on the surface of the earth might appear to be a very solid, substantial structure, but it is always undergoing slow and gradual changes. Granite on the earth's surface is exposed to and constantly altered by air, water, and other agents of change. It is altered both in appearance and in composition, slowly crumbling, and then dissolving in water. Smaller rocks and rock fragments are moved downhill by gravity or streams, exposing more granite that was previously deeply buried. The process continues, and ultimately—over much time—a mountain of solid granite is reduced to a mass of loose rock fragments and dissolved materials. The photograph in figure 15.20 is a snapshot of a mountain-sized rock mass in a stage somewhere between its formation and its eventual destruction to rock fragments. Can you imagine the length of time that such a process requires?

Figure 15.20
The piles of rocks and rock fragments around a mass of solid rock is evidence that the solid rock is slowly crumbling away. This solid rock that is crumbling to rock fragments is in the Grand Canyon, Arizona.

Weathering

The slow changes that result in the breaking up, crumbling, and destruction of any kind of solid rock are called **weathering.** The term implies changes in rocks from the action of the weather, but it actually includes chemical, physical, and biological processes. These weathering processes are important and necessary in (1) the rock cycle, (2) the formation of soils, and (3) the movement of rock materials over the earth's surface. Weathering is important in the rock cycle because it produces sediment, the raw materials for new rocks. It is important in the formation of soils because soil is an accumulation of rock fragments and organic matter. Because weathering reduces the sizes of rock particles, it is important in preparing the particles to be transported by wind or moving water.

Weathering breaks down rocks physically and chemically, and this breaking down can occur while the rocks are stationary or while they are moving. The process of physically removing weathered materials is called **erosion.** *Weathering* prepares the way for erosion by breaking solid rock into fragments. The fragments are then *eroded,* physically picked up by an agent such as a stream or a glacier. After they are eroded, the materials are then removed by *transportation.* Transportation is the movement of eroded materials by agents such as rivers, glaciers, wind, or waves. The weathering process continues during transportation. A rock being tumbled downstream, for example, is physically worn down as it bounces from rock to rock. It may be chemically altered as well as it is bounced along by the moving water. Overall, the combined action of weathering and erosion wears away and lowers the elevated parts of the earth and sculpts their surfaces.

There are two basic kinds of weathering that act to break down rocks: *mechanical weathering* and *chemical weathering. Mechanical weathering* is the physical breaking up of rocks without any changes in their chemical composition. Mechanical weathering results in the breaking up of rocks into smaller and smaller pieces, so it is also called *disintegration.* If you smash a sample of granite into smaller and smaller pieces you are mechanically weathering the granite. *Chemical weathering* is the alteration of minerals by chemical reactions with water, gases of the atmosphere, or solutions. Chemical weathering results in the dissolving or breaking down of the minerals in rocks, so it is also called *decomposition.* If you dissolve a sample of limestone in a container of acid, you are chemically weathering the limestone.

Examples of mechanical weathering in nature include the disintegration of rocks caused by (1) *wedging effects,* and (2) the *effects of reduced pressure.* Wedging effects are often caused by the repeated freezing and thawing of water in the pores and small cracks of otherwise solid rock. If you have ever seen what happens when water in a container freezes, you know that freezing water expands and exerts a pressure on the sides of its container. As water in a pore or a crack of a rock freezes, it also expands, exerting a pressure on the walls of the pore or crack, making it slightly larger. The ice melts and the enlarged pore or crack again becomes filled with water for another cycle of freezing and thawing. As the process is repeated many times, small pores and cracks become larger and larger, eventually forcing pieces of rock to break off. This process is called **frost wedging** (figure 15.21A). It is an important cause of mechanical weathering in mountains and other locations where repeated cycles of water freezing and thawing occur. The roots of trees and shrubs can also mechanically wedge rocks apart as they grow into cracks. You may have noticed the work of roots when trees or shrubs have grown next to a sidewalk for some period of time (figure 15.22).

The other example of mechanical weathering is believed to be caused by the reduction of pressure on rocks. As more and more weathered materials are removed from the surface, the downward pressure from the weight of the material on the rock below becomes less and less. The rock below begins to expand upward, fracturing into concentric sheets from the effect of reduced pressure. These curved, sheet-like plates fall away later in the mechanical weathering process called *exfoliation* (figure 15.21B). Exfoliation is the term given to the process of spalling off of layers of rock, somewhat analogous to peeling layers from an onion. Granite commonly weathers by exfoliation, producing characteristic dome-shaped hills and rounded boulders. Stone Mountain, Georgia, is a well-known example of an exfoliation-shaped dome. The onion-like structure of exfoliated granite is a common sight in the Sierras, Adirondacks, and any mountain range where older exposed granite is at the surface (figure 15.23).

Examples of chemical weathering include (1) oxidation, (2) carbonation, and (3) hydration. *Oxidation* is a chemical reaction between oxygen and the minerals making up rocks. The ferromagnesian minerals contain iron, magnesium, and other metal ions in a silicate structure. Iron can react with oxygen to produce several different iron oxides, each with its own characteristic color. The most common iron oxide (hematite) has a deep red color. Other oxides of iron are brownish to yellow-brownish. It is the presence of such iron oxides that color many sedimentary rocks and soils. The red soils of Oklahoma, Georgia, and many other places are colored by the presence of iron oxides produced by chemical weathering.

Carbonation is a chemical reaction between carbonic acid and the minerals making up rocks. Rainwater is naturally somewhat acidic because it dissolves carbon dioxide from the air. This forms a weak acid known as carbonic acid (H_2CO_3), the same acid that is found in your carbonated soda pop. Carbonic acid rain falls on the land, seeping into cracks and crevices where it reacts with minerals. Limestone, for example, is easily weathered to a soluble form by carbonic acid. The limestone caves of Missouri, Kentucky, New Mexico, and elsewhere were produced by the chemical weathering of limestone by carbonation (figure 15.24). Minerals containing calcium, magnesium, sodium, potassium, and iron are chemically weathered by carbonation to produce salts that are soluble in water.

Hydration is a reaction between water and the minerals of rocks. The process of hydration includes (1) the dissolving of a mineral and (2) the combining of water directly with a mineral. Some minerals, for example halite (which is sodium chloride), dissolve in water to form a solution. The carbonates formed from carbonation are mostly soluble, so they are easily leached from a rock by dissolving. Water also combines directly with some minerals to form new, different minerals. The feldspars, for example, undergo hydration and carbonation to produce (1) water-soluble potassium carbonate, (2) a chemical product that combines with water to produce a clay mineral, and (3) silica. The silica, which is silicon dioxide (SiO_2), may appear as a suspension of finely divided particles or in solution.

Mechanical and chemical weathering are interrelated, working together in breaking up and decomposing solid rocks of the earth's surface. In general, mechanical weathering results in cracks in solid rocks and broken-off coarse fragments. Chemical weathering results in finely pulverized materials and ions in

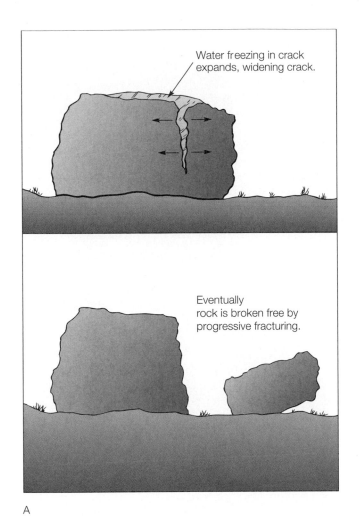

Water freezing in crack expands, widening crack.

Eventually rock is broken free by progressive fracturing.

A

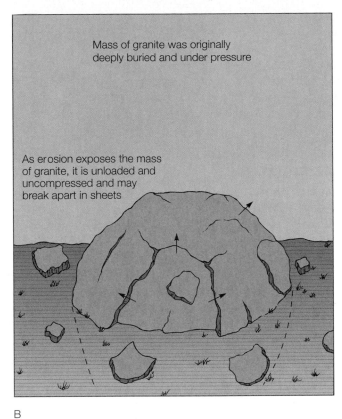

Mass of granite was originally deeply buried and under pressure

As erosion exposes the mass of granite, it is unloaded and uncompressed and may break apart in sheets

B

Figure 15.21

(*A*) Frost wedging and (*B*) exfoliation are two examples of mechanical weathering, or disintegration, of solid rock.

Figure 15.22

Growing trees can break, separate, and move solid rock. (*A*) Note how this tree has raised the sidewalk. (*B*) This tree is surviving by growing roots into tiny joints and cracks, which become larger as the tree grows.

A

B

Figure 15.23
Spheroidal weathering of granite. The edges and corners of an angular rock are attacked by weathering from more than one side and retreat faster than flat rock faces. The result is rounded granite boulders, which often shed partially weathered minerals in onion-like layers.

solution, the ultimate decomposition of a solid rock. Consider, for example, a mountain of solid granite, the most common rock found on continents. In general, granite is made up of 65 percent feldspars, 25 percent quartz, and about 10 percent ferromagnesian minerals. Mechanical weathering begins the destruction process as exfoliation and frost wedging create cracks in the solid mass of granite. Rainwater, with dissolved oxygen and carbon dioxide, flows and seeps into the cracks and reacts with ferromagnesian minerals to form soluble carbonates and metal oxides. Feldspars undergo carbonation and hydration, forming clay minerals and soluble salts, which are washed away. Quartz is less susceptible to chemical weathering and remains mostly unchanged to form sand grains. The end products of the complete weathering of granite are quartz sand, clay minerals, metal oxides, and soluble salts.

Erosion

Weathering has prepared the way for erosion to pick up and for some agent of transportation to move or carry away the fragments, clays, and solutions that have been produced from solid rock. The weathered materials can be moved to a lower elevation by the direct result of gravity acting alone. They can also be moved to a lower elevation by gravity acting through some intermediate agent, such as running water, wind, or glaciers. The erosion of weathered materials as a result of gravity alone will be considered first.

Mass Movement

Gravity constantly acts on every mass of materials on the surface of the earth, pulling parts of elevated regions toward lower levels. Rocks in the elevated regions are able to temporarily resist this constant pull through their cohesiveness with a main rock mass or by the friction between the rock and the surface of

A

B

Figure 15.24
Limestone caves develop when slightly acidic groundwater dissolves limestone along joints and bedding planes, carrying away rock components in solution. (*A*) Joints and bedding planes in a limestone bluff. (*B*) This stream has carried away less-resistant rock components, forming a cave under the ledge.

Acid rain is a generic term used to describe any acidic substances, wet or dry, that fall from the atmosphere. Wet acidic deposition could be in the form of rain, but snow, sleet, and fog could also be involved. Dry acidic deposition could include gases, dust, or any solid particles that settle out of the atmosphere to produce an acid condition.

The acid or alkaline strength of a solution is measured on the pH scale, which is a powers of ten scale with values from 0 to 14. A pH of 0 is the strongest acid reading on the scale, 7 is neutral (neither acidic nor alkaline), and a pH of 14 is the strongest alkaline reading. Remember that the pH is not a simple linear scale, and each lower number between 7 and 0 is *ten times* more acidic than the next higher number. For example, a pH of 3 is ten times more acidic than a pH of 4; a pH of 2 is ten times more acidic than a pH of 3. A pH of 2 is ten times ten, or one-hundred times more acidic than a pH of 4. Values above 7 are alkaline, again increasing by powers of ten for each number on the pH scale.

Pure, unpolluted rain is naturally acidic. Carbon dioxide in the atmosphere is absorbed by rainfall, forming carbonic acid (H_2CO_3). Carbonic acid lowers the pH of pure rainfall to a range of 5.6 to 6.2. Decaying vegetation in local areas can provide more CO_2, making the pH even lower. A pH range of 4.5 to 5.0, for example, has been measured in remote areas of the Amazon jungle. Human-produced exhaust emissions of sulfur and nitrogen oxides can lower the pH of rainfall even more, to a 4.0 to 5.5 range. This is the pH range of acid rain (box table 15.1).

Box Table 15.1

The Approximate pH of Some Common Acidic Substances

Substance	pH (or pH Range)
Hydrochloric acid (4%)	0
Gastric (stomach) solution	1.6–1.8
Lemon juice	2.2–2.4
Vinegar	2.4–3.4
Carbonated soft drinks	2.0–4.0
Grapefruit	3.0–3.2
Oranges	3.2–3.6
Acid rain	**4.0–5.5**
Tomatoes	4.2–4.4
Potatoes	5.7–5.8
Natural rainwater	**5.6–6.2**
Milk	6.3–6.7
Pure water	7.0

The sulfur and nitrogen oxides that produce acid rain come from exhaust emissions of industries and electric utilities that burn fossil fuels and from the exhaust of cars, trucks, and buses. The oxides are the raw materials of acid rain and are not themselves acidic. They react with other atmospheric chemicals to form sulfates and nitrates, which combine with water vapor to form sulfuric acid and nitric acid. These are the chemicals of concern in acid rain.

Acid rain falls on the land, bodies of water, forests, crops, buildings, and people, so the concerns about acid rain center on its environmental impact on lakes, forests, crops, materials, and human health. It is believed, for example, that acid rain accelerates chemical weathering and also leaches essential nutrients from the soil. It can also cause more rapid chemical weathering of rocks and buildings and acidify lakes and streams, sometimes to the detriment of wildlife.

The type of rocks making up the local landscape can either moderate or aggravate the problems of acid rain. Limestone, and the soils of arid climates, tend to neutralize acid, while waters in granitic rocks and soils already tend to be somewhat acidic.

Natural phenomena, such as volcanoes, and human-produced emissions of sulfur and nitrogen oxides from burning fossil fuels are the cause of acid rain. The heavily industrialized northeastern part of the United States, from the Midwest through New England, releases sulfur and nitrogen emissions that result in a precipitation pH of 4.0 to 4.5. Unfortunately, the area downwind of major sulfur sources has a granitic bedrock, which means that the effects of acid rain will not be moderated as they would in the west or midwest. This region, the northeast section of North America, is the geographic center of the acid rain problem. The solution to the problem is being sought by (1) using fuels other than fossil fuels when possible and (2) reducing the thousands of tons of sulfur and nitrogen oxides that are dumped into the atmosphere per day when fossil fuels are used.

---

the slope. Whenever anything happens to reduce the cohesiveness or to reduce the friction, gravity pulls the freed material to a lower elevation. Thus, gravity acts directly on individual rock fragments and on large amounts of surface materials as a mass, pulling all to a lower elevation. Erosion caused by gravity acting directly is called **mass movement** (also called mass wasting). Mass movement can be so slow that it is practically imperceptible. *Creep*, the slow downhill movement of soil down a steep slope, for example, is detectable only from the peculiar curved growth patterns of trees growing in the slowly moving soil (figure 15.25). At the other extreme, mass movement can be as sudden and swift as a rock bounding and clattering down a slope below a cliff. A *landslide* is a generic term used to describe any slow to rapid movement of any type or mass of materials, from the short slump of a hillside to the slide of a whole mountainside. Either slow or sudden, mass move-

ment is a small victory for gravity in the ongoing process of leveling the landmass of the earth.

Running Water

Running water is the most important of all the erosional agents of gravity that remove rock materials to lower levels. Erosion by

> ### C O N C E P T S   A P P L I E D
> #### City Weathering
>
> Prepare a display of photographs that show weathering and erosion processes at work in a *city* environment. Provide location, agent of weathering or erosion, and a description of what could be done to slow or stop each process illustrated.

Figure 15.25
The slow creep of soil is evidenced by the strange growth pattern of these trees.

Figure 15.26
Moving streams of water carry away dissolved materials and sediments as they slowly erode the land.

running water begins with rainfall. Each raindrop impacting the soil moves small rock fragments about, but also begins to dissolve some of the soluble products of weathering. If the rainfall is heavy enough, a shallow layer, or sheet of water forms on the surface, transporting small fragments and dissolved materials across the surface. This *sheet erosion* picks up fragments and dissolved material, then transports them to small streams at lower levels (figure 15.26). The small streams move to larger channels and the running water transports materials three different ways, (1) as dissolved rock materials carried in solution, (2) as clay minerals and small grains carried in suspension, and (3) as sand and larger rock fragments that are rolled, bounced, and slid along the bottom of the stream bed. Just how much material is eroded and transported by the stream depends on the volume of water, its velocity, and the load that it is already carrying.

Streams and major rivers are at work, for the most part, 24 hours a day every day of the year moving rock fragments and dissolved materials from elevated landmasses to the oceans. Any time you see mud, clay, and sand being transported by a river, you know that the river is at work moving mountains, bit by bit, to the ocean. It has been estimated that rivers remove enough dissolved materials and sediments to lower the whole surface of the United States flat in a little over 20 million years, a very short time compared to the 4.6-billion-year age of the earth.

In addition to transporting materials that were weathered and eroded by other agents of erosion, streams do their own erosive work. Streams can dissolve soluble materials directly from rocks and sediments. They also quarry and pluck fragments and pieces of rocks from beds of solid rock by hydraulic action. Most of the erosion accomplished directly by streams, however, is done by the more massive fragments that are rolled, bounced, and slid along the stream bed and against each other. This results in a grinding action on the fragments and a wearing away of the stream bed.

As a stream cuts downward into its bed, other agents of erosion such as mass movement begin to widen the channel as materials on the stream bank tumble into the moving water. The load that the stream carries is increased by this activity, which slows the stream. As the stream slows, it begins to develop bends, or *meanders* along the channel. Meanders have a dramatic effect on stream erosion because the water moves faster around an outside bank than it does around the inside bank downstream. This difference in stream velocity means that the stream has a greater erosion ability on the outside, downstream side and less on the sheltered area inside of curves. The stream begins to widen the floor of the valley through which it runs by eroding on the outside of the meander, then depositing the eroded material on the inside of another bend downstream. The stream thus begins to erode laterally, slowly working its way across the land. Sometimes two bends in the stream meet, forming a cutoff meander called an *oxbow lake*.

A stream, along with mass movement, develops a valley on a widening floodplain. A **floodplain** is the wide, level floor of a valley built by a stream (figure 15.27). It is called a floodplain because this is where the stream floods when it spills out of its channel. The development of a stream channel into a widening floodplain seems to follow a general, idealized aging pattern (figure 15.28). When a stream is on a recently uplifted landmass, it has a steep gradient, a vigorous, energetic ability to

Figure 15.27

A river usually stays in its channel, but during a flood it spills over and onto the adjacent flat land called the *floodplain.*

Figure 15.28

Three stages in the aging and development of a stream valley, (*A*) youth, (*B*) maturity, and (*C*) old age.

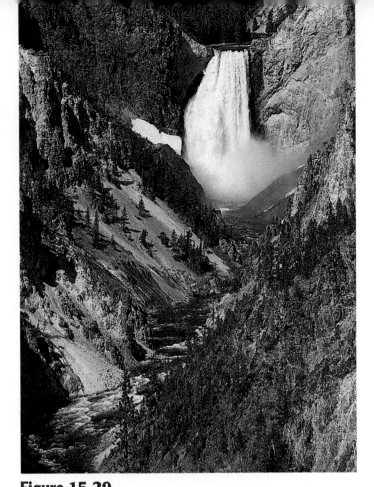

Figure 15.29

The waterfall and rapids on the Yellowstone River in Wyoming indicate that the river is actively down cutting. Note the V-shaped cross-profile and lack of floodplain, characteristics of a young stream valley.

erode the land, and characteristic features known as the stage of youth. *Youth* is characterized by a steep gradient, a V-shaped valley without a floodplain, and the presence of features that interrupt its smooth flow such as boulders in the stream bed, rapids, and waterfalls (figure 15.29). Stream erosion during youth is predominantly downward. The stream eventually erodes its way into *maturity* by eroding away the boulders, rapids, and waterfalls, and in general smoothing and lowering the stream gradient. During maturity, meanders form over a wide floodplain that now occupies the valley floor. The higher elevations are now more sloping hills at the edge of the wide floodplain rather than steep-sided walls close to the river channel. *Old age* is marked by a very low gradient in extremely broad, gently sloping valleys. The stream now flows slowly in broad meanders over the wide floodplain. Floods are more common in old age since the stream is carrying a full load of sediments and flows sluggishly.

Many assumptions are made in any generalized scheme of the erosional aging of a stream. Streams and rivers are dynamic systems that respond to local conditions, so it is possible to find an "old age feature" such as meanders in an otherwise youthful valley. This is not unlike finding a gray hair on an 18-year-old youth, and in this case the presence of the gray hair does not mean old

A

Figure 15.30

(A) Delta of Nooksack River, Washington. Note the sediment-laden water and how the land is being built outward by river sedimentation. (B) Cross section showing how a small delta might form. Large deltas are more complicated than this.

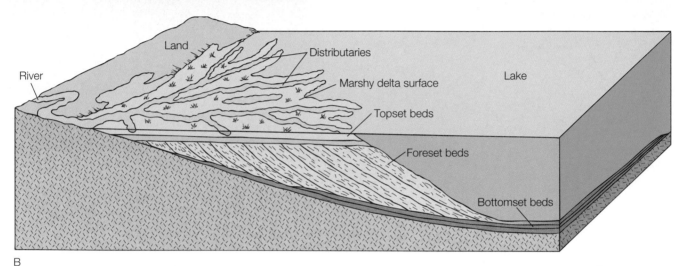

B

age. In general, old age characteristics are observed near the *mouth* of a stream where it flows into an ocean, lake, or another stream. Youthful characteristics are observed at the *source,* where the water collects to first form the stream channel. As the stream slowly lowers the land, the old age characteristics will move slowly but surely upstream from the mouth toward the source.

When the stream flows into the ocean or a lake, it loses all of its sediment-carrying ability. It drops the sediments, forming a deposit at the mouth called a **delta** (figure 15.30). Large rivers such as the Mississippi River have large and extensive deltas that actually extend the landmass more and more over time. In a way, you could think of the Mississippi River delta as being formed from pieces and parts of the Rocky Mountains, the Ozark Mountains, and other elevated landmasses that the Mississippi has carried there over time.

Glaciers

Glaciers presently cover only about 10 percent of the earth's continental land area, and since much of this is at higher latitudes, it might seem that glaciers would not have much of an overall effect in eroding the land. However, ice has sculptured much of the present landscape, and features attributed to glacial episodes are found over about three-quarters of the continental surface. Only a few tens of thousands of years ago, sheets of ice covered major portions of North America, Europe, and Asia. Today, the most extensive glaciers in the United States are those of Alaska, which cover about 3 percent of the state's land area. Less extensive glacier ice is found in the

Figure 15.31
Valley glacier on Mount Logan, Yukon Territory.

mountainous regions of Washington, Montana, California, Colorado, and Wyoming.

A **glacier** is a mass of ice on land that moves under its own weight. Glacier ice forms gradually from snow, but the quantity of snow needed to form a glacier does not fall in a single winter. Glaciers form in cold climates where some snow and ice persists throughout the year. The amount of winter snowfall must exceed the summer melting to accumulate a sufficient mass of snow to form a glacier. As the snow accumulates, it is gradually transformed into ice. The weight of the overlying snow packs it down, driving out much of the air, and causing it to recrystallize into a coarser, denser mass of interlocking ice crystals that appears to have a blue to deep blue color. Complete conversion of snow into glacial ice may take from 5 to 3,500 years, depending on such factors as climate and rate of snow accumulation at the top of the pile. Eventually the mass of ice will become large enough that it begins to flow, spreading out from the accumulated mass. Glaciers that form at high elevations in mountainous regions, which are called *alpine glaciers,* tend to flow downhill through a valley so they are also called *valley glaciers* (figure 15.31). Glaciers that cover a large area of a continent are called *continental glaciers.* Continental glaciers can cover whole continents and reach a thickness of 1 km or more. Today, the remaining continental glaciers are found on Greenland and the Antarctic.

Glaciers move slowly and unpredictably, spreading like a huge blob of putty under the influence of gravity. As an alpine glacier moves downhill through a V-shaped valley, the sides and bottom of the valley are eroded wider and deeper. When the glacier later melts, the V-shaped valley has been transformed into a U-shaped valley that has been straightened and deepened by the glacial erosion. The glacier does its erosional work using three different techniques, (1) by bulldozing, (2) by plucking, and (3) by abrasion. *Bulldozing,* as the term implies, is the pushing along of rocks, soil, and sediments by the leading edge of an advancing glacier. Deposits of bulldozed rocks and other materials that remain after the ice melts are called *moraines. Plucking* occurs as water seeps into cracked rocks and freezes, becoming a part of the solid glacial mass. As the glacier moves on, it pulls the fractured rock apart and plucks away chunks of it. The process is accelerated by the frost-wedging action of the freezing water. Plucking at the uppermost level of an alpine glacier, combined with weathering of the surrounding rocks, produces a rounded or bowl-like depression known as a *cirque* (figure 15.32). *Abrasion* occurs as the rock fragments frozen into the moving glacial ice scratch, polish, and grind against surrounding rocks at the base and along the valley walls. The result of this abrasion is the pulverizing of rock into ever finer fragments, eventually producing a powdery, silt-sized sediment called *rock flour.* Suspended rock flour in meltwater from a glacier gives the water a distinctive gray to bluegray color.

Glaciation is continuously at work eroding the landscape in Alaska and many mountainous regions today. The glaciation that formed the landscape features in the Rockies, the Sierras, and across the northeastern United States took place thousands of years ago.

Wind

Like running water and moving glaciers, wind also acts as an agent shaping the surface of the land. It can erode, transport, and deposit materials. However, wind is considerably less efficient than ice or water in modifying the surface. Air is much less dense and does not have the eroding or carrying power of water or ice. In addition, a stream generally flows most of the time but the wind blows only occasionally in most locations. Thus, on a worldwide average, winds move only a few percent as much material as do streams. Wind also lacks the ability to attack rocks chemically as water does through carbonation and other processes, and the wind cannot carry dissolved sediments in solution. Even in many deserts, more sediment is moved during the brief periods of intense surface runoff following the

A

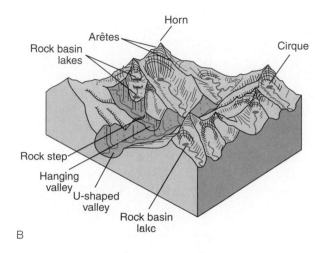

B

Figure 15.32

(*A*) A stream-carved mountainside before glaciation. (*B*) The same area after glaciation, with some of the main features of mountain glaciation labeled.

occasional rainstorms than is moved by wind during the prolonged dry periods.

Flowing air and moving water do have much in common as agents of erosion since both are fluids. Both can move larger particles by rolling them along the surface and can move finer particles by carrying them in suspension. Both can move larger and more massive particles with increased velocities. Water is denser and more viscous than air, so it is more efficient at transporting quantities of material than is the wind, but the processes are quite similar.

Two major processes of wind erosion are called (1) abrasion and (2) deflation. *Wind abrasion* is a natural sandblasting process that occurs when the particles carried along by the wind break off small particles and polish what they strike. Generally, the harder mineral grains such as quartz sand accomplish this best near the ground where the wind is bouncing them along. Wind abrasion can strip paint from a car exposed to the moving particles of a dust storm, eroding the paint along with rocks on the surface. Rocks and boulders exposed to repeated wind storms where the wind blows consistently from one or a few directions may be planed off from the repeated action of this natural sandblasting. Rocks sculptured by wind abrasion are called *ventifacts,* after the Latin meaning "wind-made" (figure 15.33).

Deflation, named after the Latin meaning "to blow away," is the widespread picking up of loose materials from the surface. Deflation is naturally most active where winds are unobstructed and the materials are exposed and not protected by vegetation. These conditions are often found on deserts, beaches, and on unplanted farmland between crops. During the 1930s several years of drought killed the native vegetation in the Plains states during a period of increased farming activity. Unusually strong winds eroded the unprotected surface, removing and transporting hundreds of millions of tons of soil. This period of prolonged drought, dust storms, and general economic disaster for farmers in the area is known as the Dust Bowl episode.

The most common wind-blown deposits are (1) dunes and (2) loess. A **dune** is a low mound or ridge of sand or other sediments. Dunes form when sediment-bearing wind encounters an

If wind is predominantly from one direction, rocks will be planed off or flattened on the upwind side.

With a persistent shift in wind direction, additional facets are cut in the rock.

Figure 15.33

Ventifact formation by abrasion from one or several directions.

obstacle that reduces the wind velocity. With a slower velocity the wind cannot carry as large a load, so sediments are deposited on the surface. This creates a windbreak, which results in a growing obstacle, a dune. Once formed, a dune tends to migrate, particularly if the winds blow predominantly from one direction. Dunes are commonly found in semiarid areas or near beaches.

Another common wind deposit is called *loess* (pronounced "luss"). Loess is a very fine dust, or silt, that has been deposited over a large area. One such area is located in the central part of the United States, particularly to the east sides of the major rivers of the Mississippi basin. Apparently this deposit originated from the rock flour produced during the last great ice age. The rock flour was probably deposited along the major river valleys and later moved eastward by the prevailing westerly winds. Since rock flour is produced by the mechanical grinding action of glaciers, it has not been chemically broken down. Thus, the loess deposit contains many minerals that were not leached out of the deposit as typically occurs with chemical weathering. It also has an open, porous structure since it does not have as much of the chemically-produced clay minerals. The good moisture-holding capacity from this open structure, together with the presence of minerals that serve as plant nutrients, make farming soils formed from loess deposits particularly productive.

SUMMARY

The *principle of uniformity* is the frame of reference that the same geologic processes you see changing rocks today are the same processes that changed them in the past.

Diastrophism is the process of deformation that changes the earth's surface, and the movement of magma is called *vulcanism. Diastrophism, vulcanism,* and *earthquakes* are closely related, and their occurrence can be explained most of the time by events involving *plate tectonics.*

Stress is a force that tends to compress, pull apart, or deform a rock, and the adjustment to stress is called *strain.* Rocks respond to stress by (1) withstanding the stress without change, (2) undergoing *elastic strain,* (3) undergoing *plastic strain,* or (4) by *breaking* in *fracture strain.* Exactly how a particular rock responds to stress depends on (1) the nature of the rock, (2) the temperature, and (3) how quickly the stress is applied over time.

Deeply buried rocks are at a higher temperature and tend to undergo *plastic deformation,* resulting in a wrinkling of the layers into *folds.* The most common are an arch-shaped fold called an *anticline* and a trough-shaped fold called a *syncline.* Anticlines and synclines are most easily observed in sedimentary rocks because they have bedding planes, or layers.

Rocks near the surface tend to break from a sudden stress. A break with movement on one side of the break relative to the other side is called a *fault.* The vibrations that move out as waves from the movement of rocks is called an *earthquake.* The actual place where an earthquake originates is called its *focus.* The place on the surface directly above a focus is called an *epicenter.* There are three kinds of waves that travel from the focus, *S-, P-,* and *surface waves.* The magnitude of earthquake waves is measured on the *Richter scale.*

Folding and faulting produce prominent features on the surface called *mountains. Mountains* can be classified as having an origin of *folding, faulting,* or *volcanic.* In general, mountains that occur in long narrow belts called *ranges* have an origin that can be explained by *plate tectonics.*

Weathering is the breaking up, crumbling, and destruction of any kind of solid rock. The process of physically picking up weathered rock materials is called *erosion.* After the eroded materials are picked up, they are removed by *transportation* agents. The combined action of weathering, erosion, and transportation wears away and lowers the surface of the earth.

The physical breaking up of rocks is called *mechanical weathering.* Mechanical weathering occurs by *wedging effects* and the *effects of reduced pressure. Frost wedging* is a wedging effect that occurs from repeated cycles of water freezing and thawing. The process of spalling off of curved layers of rock from reduced pressure is called *exfoliation.*

The breakdown of minerals by chemical reactions is called *chemical weathering.* Examples include *oxidation,* a reaction between oxygen and the minerals making up rocks; *carbonation,* a reaction between carbonic acid (carbon dioxide dissolved in water) and minerals making up rocks; and *hydration,* the dissolving or combining of a mineral with water. When the end products of complete weathering of rocks are removed directly by gravity, the erosion is called *mass movement.* Erosion and transportation also occur through the agents of *running water, glaciers,* or *wind.* Each create their own characteristic features of erosion and deposition.

KEY TERMS

anticline (p. **299**)
delta (p. **314**)
dune (p. **316**)
earthquake (p. **301**)
elastic rebound (p. **302**)
erosion (p. **308**)
fault (p. **300**)
floodplain (p. **312**)
folds (p. **299**)
frost wedging (p. **308**)
glacier (p. **315**)
mass movement (p. **311**)

mountains (p. **304**)
normal fault (p. **300**)
principle of uniformity (p. **296**)
reverse fault (p. **300**)
seismic waves (p. **301**)
strain (p. **297**)
stress (p. **297**)
syncline (p. **299**)
thrust fault (p. **300**)
tsunami (p. **303**)
volcano (p. **305**)
weathering (p. **307**)

APPLYING THE CONCEPTS

1. The basic difference in the frame of reference called the principle of uniformity and the catastrophic frame of reference used by previous thinkers is
   (a) the energy for catastrophic changes is much less.
   (b) the principle of uniformity requires more time.
   (c) catastrophic changes have a greater probability of occurring.
   (d) none of the above.

2. The difference between elastic deformation and plastic deformation of rocks is that plastic deformation (is)
   (a) permanently alters the shape of a rock layer.
   (b) always occurs just before a rock layer breaks.
   (c) returns to its original shape after the pressure is removed.
   (d) all of the above.

3. Whether a rock layer subjected to stress undergoes elastic deformation, plastic deformation, or rupture depends on
   (a) the temperature of the rock.
   (b) the confining pressure on the rock.
   (c) how quickly or slowly the stress is applied over time.
   (d) all of the above.

4. When subjected to stress, rocks buried at great depths are under great pressure at high temperatures, so they tend to undergo
   (a) no change because of the pressure.
   (b) elastic deformation because of the high temperature.
   (c) plastic deformation.
   (d) breaking or rupture.

5. Earthquakes that occur at the boundary between two tectonic plates moving against each other occur along
   (a) the entire length of the boundary at once.
   (b) short segments of the boundary at different times.
   (c) the entire length of the boundary at different times.
   (d) none of the above.

6. Each higher number of the Richter scale
   (a) increases with the magnitude of an earthquake.
   (b) means ten times more ground movement.
   (c) indicates about thirty times more energy released.
   (d) means all of the above.
7. Other than igneous activity, all mountain ranges have an origin resulting from
   (a) folding.
   (b) faulting.
   (c) stresses.
   (d) sedimentation.
8. The preferred name for the very large ocean waves that are generated by an earthquake, landslide, or volcanic explosion is
   (a) tidal wave.
   (b) tsunami.
   (c) tidal bore.
   (d) Richter wave.
9. Freezing water exerts pressure on the wall of a crack in a rock mass, making the crack larger. This is an example of
   (a) mechanical weathering.
   (b) chemical weathering.
   (c) exfoliation.
   (d) hydration.
10. Which of the following would have the greatest overall effect in lowering the elevation of a continent such as North America?
   (a) continental glaciers
   (b) alpine glaciers
   (c) wind
   (d) running water
11. Broad meanders on a very wide, gently sloping floodplain with oxbow lakes are characteristics you would expect to find in a river valley during what stage?
   (a) newborn
   (b) youth
   (c) maturity
   (d) old age
12. A glacier forms when
   (a) the temperature does not rise above freezing.
   (b) snow accumulates to form ice, which begins to flow.
   (c) a summer climate does not occur.
   (d) a solid mass of snow moves downhill under the influence of gravity.

Answers
1. b 2. a 3. d 4. c 5. b 6. d 7. c 8. b 9. a 10. d 11. d 12. b

QUESTIONS FOR THOUGHT

1. What is the principle of uniformity? What are the underlying assumptions of this principle?
2. Describe the responses of rock layers to increasing compressional stress when it increases slowly on deeply buried, warm layers, when it increases slowly on cold rock layers, and when it is applied quickly to rock layers of any temperature.
3. Describe the conditions that would lead to faulting as opposed to folding of rock layers.
4. Where would the theory of plate tectonics predict that earthquakes would occur?
5. Identify some areas of probable active volcanic activity today in the United States. Explain your reasoning for selecting these areas.
6. Granite is the most common rock found on continents. What are the end products after granite has been completely weathered? What happens to these weathering products?
7. What other erosion processes are important as a stream of running water carves a valley in the mountains? Explain.
8. Describe the characteristic features associated with stream erosion as the stream valley passes through the stages of youth, maturity, and old age.
9. Could a stream erode the land lower than sea level? Explain.
10. Explain why glacial erosion produces a U-shaped valley but stream erosion produces a V-shaped valley.
11. Name and describe as many ways as you can think of that mechanical weathering occurs in nature. Do not restrict your thinking to those discussed in this chapter.
12. Compare the features caused by stream erosion, wind erosion, and glacial erosion.

CHAPTER SIXTEEN

Earth's Weather

This cloud forms a thin covering over the mountaintop. Likewise, the earth's atmosphere forms a thin shell around the earth, with 99 percent of the mass within 32 km, or about 20 mi of the surface.

OUTLINE

The earth's atmosphere has a unique composition because of the cyclic flow of materials. Some of these cycles involve the movement of materials in and out of the earth's atmosphere. Carbon dioxide, for example, is a very minor part of the earth's atmosphere. It has been maintained as a minor component in a mostly balanced state for about the past 570 million years, cycling into and out of the atmosphere.

Water is also involved in a global cyclic flow between the atmosphere and the surface. Water on the surface is mostly in the ocean, with lesser amounts in lakes, streams, and underground. Not much water is found in the atmosphere at any one time on a worldwide basis, but billions of tons are constantly evaporating into the atmosphere each year and returning as precipitation in an ongoing cycle.

The cycling of carbon dioxide and water to and from the atmosphere takes place in a dynamic system that is energized by the sun. Radiant energy from the sun heats some parts of the earth more than others. Winds redistribute this energy with temperature changes, rain, snow, and other changes that are generally referred to as the *weather*.

Understanding and predicting the weather is the subject of *meteorology*. Meteorology is the science of the atmosphere and weather phenomena, from understanding everyday rain and snow to predicting not-so-common storms and tornadoes (figure 16.1). Understanding weather phenomena depends on a knowledge of the atmosphere and the role of radiant energy on a spinning Earth that is revolving around the Sun. This chapter is concerned with understanding the atmosphere of the earth, its cycles, and the influence of radiant energy on the atmosphere.

THE ATMOSPHERE

The atmosphere is a relatively thin shell of gases that surrounds the solid earth. If you could see the molecules making up the atmosphere, you would see countless numbers of rapidly moving particles, all undergoing a terrific jostling from the billions of collisions occurring every second. Since this jostling mass of tiny particles is pulled toward the earth by gravity, more are found near the surface than higher up. Thus, the atmosphere thins rapidly with increasing distance above the surface, gradually merging with the very diffuse medium of outer space.

To understand how rapidly the atmosphere thins with altitude, imagine a very tall stack of open boxes. At any given instant each consecutively higher box would contain fewer of the jostling molecules than the box below it. Molecules in the lowest box on the surface, at sea level, might be able to move a distance of only 1×10^{-8} m (about 3×10^{-6} in) before colliding with another molecule. A box moved to an altitude of 80 km (about 50 mi) above sea level would have molecules that could move perhaps 10^{-2} m (about 1/2 in) before colliding with another molecule. At 160 km (about 100 mi) the distance traveled would be about 2 m (about 7 ft). As you can see, the distance between molecules increases rapidly with increasing altitude. Since air density is defined by the number of molecules in a unit volume, the density of the atmosphere decreases rapidly with increasing altitude (figure 16.2).

It is often difficult to imagine a distance above the surface because there is nothing visible in the atmosphere for comparison. Recall our stack of boxes from the previous example. Imagine that this stack of boxes is so tall that it reaches from the surface past the top of the atmosphere. Now imagine that this tremendously tall stack of boxes is tipped over and carefully laid out horizontally on the surface of the earth. How far would you have to move along these boxes to reach the box that was in outer space, outside of the atmosphere? From the bottom box, you would cover a distance of only 5.6 km (about 3.5 mi) to reach the box that was above 50 percent of the mass of the earth's atmosphere. At 12 km (about 7 mi), you would reach the box that was above 75 percent of the earth's atmosphere. At 16 km (about 10 mi), you would reach the box that was above

Figure 16.1

The probability of a storm can be predicted, but nothing can be done to stop or slow a storm. Understanding the atmosphere may help in predicting weather changes, but it is doubtful that weather will ever be controlled on a large scale.

about 90 percent of the atmosphere. And, after only 32 km (about 20 mi), you would reach the box that was above 99 percent of the earth's atmosphere. The significance of these distances might be better appreciated if you can imagine the distances to some familiar locations; for example, from your campus to a store 16 km (about 10 mi) away would place you above 90 percent of the atmosphere if you were to travel this same distance straight up.

Since the average radius of the solid earth is about 6,373 km (3,960 mi), you can see that the atmosphere is a very thin shell with 99 percent of the mass within 32 km (about 20 mi) by comparison. The atmosphere is much closer to the earth than most people realize (figure 16.3).

Composition of the Atmosphere

A sample of pure, dry air is colorless, odorless, and composed mostly of the molecules of just three gases, nitrogen (N_2), oxy-

Figure 16.2

At greater altitudes, the same volume contains fewer molecules of the gases that make up the air. This means that the density of air decreases with increasing altitude.

Figure 16.3

The earth's atmosphere thins rapidly with increasing altitude and is much closer to the earth than most people realize.

gen (O_2), and argon (Ar). Nitrogen is the most abundant (about 78 percent of the total volume), followed by oxygen (about 21 percent), then argon (about 1 percent). The molecules of these three gases are well mixed, and this composition is nearly constant everywhere near the earth's surface (figure 16.4).

Nitrogen does not readily enter into chemical reactions with rocks, so it has accumulated in the atmosphere. Some nitrogen is involved in a nitrogen cycle, however, as it is removed from the atmosphere by certain bacteria in the soil and by lightning. Both form nitrogen compounds that are essential to the growth of plants. These nitrogen compounds are absorbed by plants and consequently utilized throughout the food chain. Eventually the nitrogen is returned to the atmosphere through the decay of plant and animal matter. Overall, these processes of nitrogen removal and release must be in balance since the amount of nitrogen in the atmosphere is essentially constant over time.

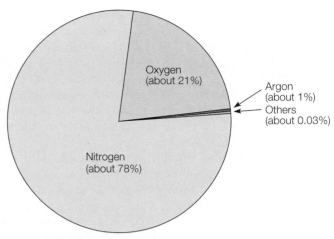

Figure 16.4

The earth's atmosphere has a unique composition of gases when compared to that of the other planets in the solar system.

Oxygen gas also cycles into and out of the atmosphere in balanced processes of removal and release. Oxygen is removed (1) by living organisms as food is oxidized to carbon dioxide and water and (2) by chemical weathering of rocks as metals and other elements combine with oxygen to form oxides. Oxygen is released by green plants as a result of photosynthesis and the amount released balances the amount removed by organisms and weathering. So oxygen, as well as nitrogen, is maintained in a state of constant composition through balanced chemical reactions.

The third major component of the atmosphere, argon, is inert and does not enter into any chemical reactions or cycles. It is produced as a product of radioactive decay and once released, remains in the atmosphere as an inactive filler.

In addition to the relatively fixed amounts of nitrogen, oxygen, and argon, the atmosphere contains variable amounts of water vapor. Water vapor is the invisible, molecular form of water in the gaseous state, which should not be confused with fog or clouds. Fog and clouds are tiny droplets of liquid water, not water in the single molecular form of water vapor. The amount of water vapor in the atmosphere can vary from a small fraction of a percent composition by volume in cold, dry air to about 4 percent in warm, humid air. This small, variable percentage of water vapor is essential in maintaining life on the earth. It enters the atmosphere by evaporation, mostly from the ocean, and leaves the atmosphere as rain or snow.

Apart from the variable amounts of water vapor, the relatively fixed amounts of nitrogen, oxygen, and argon make up about 99.97 percent of the volume of a sample of dry air. The remaining gases are mostly carbon dioxide (CO_2) and traces of inert gases. Carbon dioxide makes up only 0.03 percent of the atmosphere, but it is important to life on the earth and it is constantly cycled through the atmosphere by the process of photosynthesis that removes CO_2, and respiration that releases CO_2.

In addition to gases and water vapor, the atmosphere contains particles of dust, smoke, salt crystals, and tiny solid or liquid particles called *aerosols*. These particles become suspended and are dispersed among the molecules of the atmospheric gases. Aerosols are produced by combustion, often resulting in air pollution. Aerosols are also produced by volcanoes and forest fires. Volcanoes, smoke from combustion, and the force of the wind lifting soil and mineral particles into the air all contribute to dust particles larger than aerosols in the atmosphere. These larger particles, which range in size up to 500 micrometers, are not suspended as the aerosols are, and they soon settle out of the atmosphere as dust and soot.

Tiny particles of salt crystals that are suspended in the atmosphere come from the mist created by ocean waves and the surf. This mist forms an atmospheric aerosol of seawater that evaporates, leaving the solid salt crystals suspended in the air. The aerosol of salt crystals and dust becomes well mixed in the lower atmosphere around the globe, playing a large and important role in the formation of clouds.

Atmospheric Pressure

At the earth's surface (sea level), the atmosphere exerts a force of about 10.0 newtons on each square centimeter (14.7 lb/in^2). As you go to higher altitudes above sea level the pressure rapidly decreases with increasing altitude. At an altitude of about 5.6 km (about 3.5 mi) the air pressure is about half of what it is at sea level, about 5.0 newtons/cm^2 (7.4 lb/in^2). At 12 km (about 7 mi) the air pressure is about 2.5 newtons/cm^2 (3.7 lb/in^2). Compare this decreasing air pressure at greater elevations to figure 16.3. Again, you can see that most of the atmosphere is very close to the earth, and it thins rapidly with increasing altitude. Even a short elevator ride takes you high enough that the atmospheric pressure on your eardrum is reduced. You equalize the pressure by opening your mouth, allowing the air under greater pressure inside the eardrum to move through the eustachian tube. This makes a "pop" sound that most people associate with changes in air pressure.

Atmospheric pressure is measured by an instrument called a **barometer.** The mercury barometer was invented in 1643 by an Italian named Torricelli. He closed one end of a glass tube and then filled it with mercury. The tube was then placed, open end down, in a bowl of mercury while holding the mercury in the tube with a finger. When Torricelli removed his finger with the open end below the surface in the bowl, a small amount of mercury moved into the bowl leaving a vacuum at the top end of the tube. The mercury remaining in the tube was supported by the atmospheric pressure on the surface of the mercury in the bowl. The pressure exerted by the weight of the mercury in the tube thus balanced the pressure exerted by the atmosphere. At sea level, Torricelli found that atmospheric pressure balanced a column of mercury about 76.00 cm (29.92 in) tall (figure 16.5).

As the atmospheric pressure increases and decreases, the height of the supported mercury column moves up and down. Atmospheric pressure can be expressed in terms of the height of such a column of mercury. Public weather reports give the pressure by referring to such a mercury column, for example, "The pressure is 30 inches (about 76 cm) and rising." If the atmospheric pressure at sea level is measured many times over long periods of time, an average value of 76.00 cm (29.92 in) of mer-

Figure 16.5

The mercury barometer measures the atmospheric pressure from the balance between the pressure exerted by the weight of the mercury in a tube and the pressure exerted by the atmosphere. As the atmospheric pressure increases and decreases, the mercury rises and falls. This sketch shows the average height of the column at sea level.

cury is obtained. This average measurement is called the **standard atmospheric pressure** and is sometimes referred to as the *normal pressure.* It is also called *one atmosphere of pressure.*

Warming the Atmosphere

Radiation from the sun must pass through the atmosphere before reaching the earth's surface. The atmosphere filters, absorbs, and reflects incoming solar radiation as shown in figure 16.6. On the average, the earth as a whole reflects about 30 percent of the total radiation back into space, with two-thirds of the reflection occurring from clouds. The amount reflected at any one time depends on the extent of cloud cover, the amount of dust in the atmosphere, and the extent of snow and vegetation on the surface. Substantial changes in any of these influencing variables could increase or decrease the reflectivity, leading to increased heating or cooling of the atmosphere.

As figure 16.6 shows, only about one-half of the incoming solar radiation reaches the earth's surface. The reflection and selective filtering by the atmosphere allow a global average of about 240 watts per square meter to reach the surface. Wide variations from the average occur with latitude as well as with the season.

The incoming solar radiation that does reach the earth's surface is absorbed. Rocks, soil, water, and the ground become warmer as a result. These materials emit the absorbed solar energy as infrared radiation, wavelengths longer than the visible part of the electromagnetic spectrum. This longer-wavelength infrared radiation has a frequency that matches some of the natural frequencies of vibration of carbon dioxide and water molecules. This match means that carbon dioxide and water molecules readily absorb infrared radiation that is emitted from the surface of the earth. The absorbed infrared energy shows up as an increased kinetic energy of the molecules, which is indi-

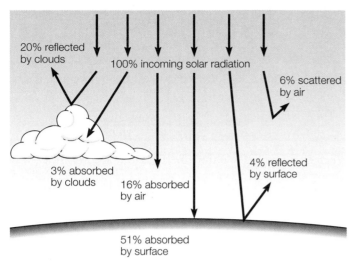

Figure 16.6
On the average, the earth's surface absorbs only 51 percent of the incoming solar radiation after it is filtered, absorbed, and reflected.

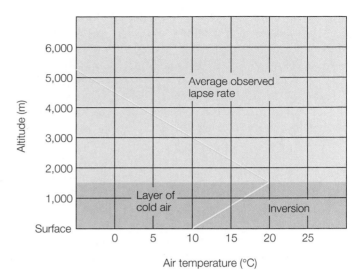

Figure 16.7
On the average, the temperature decreases about 6.5°C/km which is known as the *observed lapse rate.* An inversion is a layer of air in which the temperature increases with height.

cated by an increase in temperature. Carbon dioxide and water vapor molecules in the atmosphere now emit infrared radiation of their own, this time in all directions. Some of this reemitted radiation is again absorbed by other molecules in the atmosphere, some is emitted to space, and significantly, some is absorbed by the surface to start the process all over again. The net result is that less of the energy from the sun escapes immediately to space after being absorbed and emitted as infrared. It is retained through the process of being redirected to the surface, increasing the surface temperature more than it would have otherwise been. The more carbon dioxide that is present in the atmosphere, the more energy that will be bounced around and redirected back toward the surface, increasing the temperature near the surface. The process of heating the atmosphere in the lower parts by the absorption of solar radiation and reemission of infrared radiation is called the **greenhouse effect.** It is called the greenhouse effect because greenhouse glass allows the short wavelengths of solar radiation to pass into the greenhouse but does not allow all of the longer infrared radiation to leave. This analogy is misleading, however, because carbon dioxide and water vapor molecules do not "trap" infrared radiation, but they are involved in a dynamic absorption and downward reemission process that increases the surface temperature. The more carbon dioxide molecules that are involved in this dynamic process, the more infrared radiation will be directed back to the earth and the more the temperature will increase. More layers of glass on a greenhouse will not increase the temperature significantly. The significant heating factor in a real greenhouse is the blockage of convection by the glass, a process that does not occur from the presence of carbon dioxide and water vapor in the atmosphere.

Structure of the Atmosphere

Convection currents and the repeating absorption and reemission processes of the greenhouse effect tend to heat the atmosphere from the ground up. In addition, the higher-altitude gases

lose radiation to space more readily than the lower-altitude gases. Thus, the lowest part of the atmosphere is warmer, and the temperature decreases with increasing altitude. On the average, the temperature decreases about 6.5°C for each kilometer of altitude (3.5°F/1,000 ft). This change of temperature with altitude is called the *observed lapse rate.* The observed lapse rate applies only to air that is not rising or sinking, and the actual change with altitude can be very different from this average value. For example, a stagnant mass of very cold air may settle over an area, producing colder temperatures near the surface than in the air layers above. Such a layer where the temperature increases with height is called an **inversion** (figure 16.7). Inversions often occur on calm winter days after the arrival of a cold front. They also occur on calm, clear, and cool nights ("C" nights) when the surface rapidly loses radiant energy to space. In either case, the situation results in a "cap" of cooler, more dense air overlying the warmer air beneath. This often leads to an increase of air pollution because the inversion prevents dispersion of the pollutants.

Temperature decreases with height at the observed lapse rate until an average altitude of about 11 km (about 6.7 mi), where it then begins to remain more or less constant with increasing altitude. The layer of the atmosphere from the surface up to where the temperature stops decreasing with height is called the *troposphere.* Almost all weather occurs in the troposphere, which is derived from Greek words meaning "turning layer."

Above the troposphere is the second layer of the atmosphere called the *stratosphere.* This layer is derived from the Greek terms for "stratified layer." It is stratified, or layered, because the temperature increases with height. Cooler air below means that consecutive layers of air are denser on the bottom, which leads to a stable situation rather than the turning turbulence of the troposphere below. The stratosphere contains little moisture or dust and lacks convective turbulence, making it a desirable altitude for aircraft to fly. Temperature in the lower

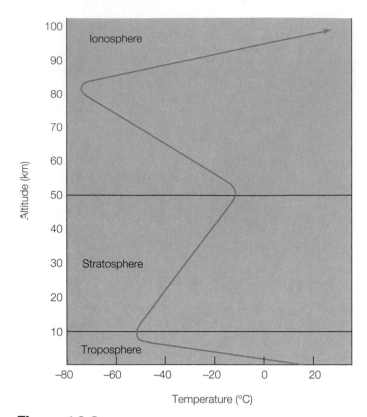

Figure 16.8

The structure of the atmosphere based on temperature differences.

stratosphere increases gradually with increasing altitude to a height of about 48 km (about 30 mi) (figure 16.8).

The temperature increases in the stratosphere as a result of interactions between high-energy ultraviolet radiation and ozone. Ozone is triatomic oxygen (O_3) that is concentrated mainly in the upper portions of the stratosphere. Diatomic molecules of oxygen (O_2) are concentrated in the troposphere and monatomic molecules of oxygen (O) are found in the outer edges of the atmosphere. Although the amount of ozone present in the stratosphere is not great, its presence is vital to life on the earth's surface. Without the **ozone shield** ultraviolet radiation would reach the surface of the earth at a level sufficient to damage both skin and eyes even more readily than it now does. Ultraviolet light causes mutations and in high enough doses destroys DNA, so the development of life on the earth was restricted to oceans until there had been sufficient buildup of oxygen to allow for the formation of ozone. Only then could terrestrial life develop. The incidence of skin cancer increases dramatically without the protection offered by the ozone.

Here is how the ozone shield works. The ozone concentration is not static because there is an ongoing process of ozone formation and destruction. For ozone to form, diatomic oxygen must first be broken down into the monatomic form. Short ultraviolet (UV) radiation is absorbed by ordinary oxygen, which breaks it down into the single-atom (O) form. This reaction is significant because of (1) the high-energy ultraviolet radiation that is removed from the sunlight and (2) the monatomic oxy-

gen that is formed, which will combine to make ozone that will absorb even more ultraviolet radiation. This initial reaction is

$$O_2 + UV \rightarrow O + O$$

When the O molecule collides with an O_2 molecule and any third, neutral molecule (NM), the following reaction takes place:

$$O_2 + O + NM \rightarrow O_3 + NM$$

When O_3 is exposed to ultraviolet, the ozone absorbs the UV radiation and breaks down to two forms of oxygen in the following reaction:

$$O_3 + UV \rightarrow O_2 + O$$

The monatomic molecule that is produced combines with an ozone molecule to produce two diatomic molecules,

$$O + O_3 \rightarrow 2\,O_2$$

and the process starts all over again. There is much concern about Freon from older refrigeration units and other similar chemicals that make their way to the stratosphere. These chemicals are broken down by UV radiation, releasing chlorine, which reacts with the oxygen. This forms a compound that is stable, so it effectively removes the oxygen from the ongoing ozone-forming and UV-radiation-absorbing process. Regional zones of decreased ozone in the stratosphere are referred to as a "hole in the ozone layer" since extreme UV radiation, can now reach the surface beneath. Concerns about the impact of chlorine-containing refrigerants like Freon on the ozone layer led to an international agreement known as the *Montreal Protocol,* which resulted in a phaseout of the use of Freon and similar compounds in most of the world. Subsequently the size of the "holes in the ozone layer" appear to be getting smaller.

The *exosphere* (Greek for "outer layer") is the outermost layer where the molecules merge with the diffuse vacuum of space. Molecules of this layer that have sufficient kinetic energy are able to escape and move off into space. This layer is sometimes called the *ionosphere* because of the free electrons and ions at this altitude. The electrons and ions here are responsible for reflecting radio waves around the earth and for the northern lights.

THE WINDS

The troposphere is heated from the bottom up as the surface of the earth absorbs sunlight. Uneven heating of the earth's surface sets the stage for *convection.* As a local region of air becomes heated the air expands, reducing its density. This less dense air is pushed upward by nearby cooler, more dense air. This results in three general motions of air: (1) the upward movement of air over a region of greater heating, (2) the sinking of air over a cooler region, and (3) a horizontal air movement between the cooler and warmer regions. In general, a horizontal movement of air is called **wind,** and the direction of a wind is defined as the direction from which it blows.

Air in the troposphere rises, moves as the wind, and sinks. All three of these movements are related, and all occur at the same time over regions of the landscape. During a day with gentle breezes on the surface, the individual, fluffy clouds you see are forming over areas where the air is moving upward. The clear air between the clouds is over areas where the air is moving downward. On a smaller scale, air can be observed moving from a field of cool grass toward an adjacent asphalt parking lot on a calm, sunlit day. Soap bubbles or smoke will often reveal the gentle air movement of this localized convection.

Local Wind Patterns

Considering average conditions, there are two factors that are important for a generalized model to help you understand local wind patterns. These factors are (1) the relationship between air temperature and air density and (2) the relationship between air pressure and the movement of air. The relationship between air temperature and air density is that cool air has a greater density than warm air because a mass of warming air expands and a mass of cooling air contracts. Warm, less dense air is buoyed upward by cooler, more dense air, which results in the upward, downward, and horizontal movement of air called a convection cell.

The upward and downward movement of air leads to the second part of the generalized model, that (1) the upward movement produces a "lifting" effect on the surface that results in an area of lower atmospheric pressure and (2) the downward movement produces a "piling up" effect on the surface that results in an area of higher atmospheric pressure. On the surface, air is seen to move from the "piled up" area of higher pressure horizontally to the "lifted" area of lower pressure (figure 16.9). In other words, air generally moves from an area of higher pressure to an area of lower pressure. The movement of air and the pressure differences occur together, and neither is the cause of the other. This is an important relationship in a working model of air movement that can be observed and measured on a very small scale, such as between an asphalt parking lot and a grass

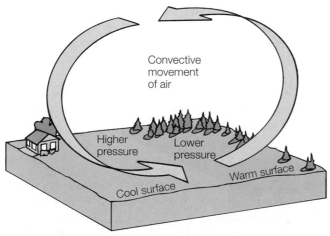

Higher pressure

Lower pressure

Convective movement of air

Warm surface

Cool surface

Figure 16.9

A model of the relationships between differential heating, the movement of air, and pressure difference in a convective cell. Cool air pushes the less dense, warm air upward, reducing the surface pressure. As the uplifted air cools and becomes more dense, it sinks, increasing the surface pressure.

field. It can also be observed and measured for local, regional wind patterns and for worldwide wind systems.

Adjacent areas of the surface can have different temperatures because of different heating or cooling rates. The difference is very pronounced between adjacent areas of land and water. Under identical conditions of incoming solar radiation, the temperature changes experienced by the water will be much less than the changes experienced by the adjacent land. There are three principal reasons for this difference: (1) The specific heat of water is about twice the specific heat of soil. This means that it takes more energy to increase the temperature of water than it does for soil. Equal masses of soil and water exposed to sunlight will result in the soil heating about 1°C while the water heats 1/2°C from absorbing the same amount of solar radiation. (2) Water is a transparent fluid that is easily mixed, so the incoming solar radiation warms a body of water throughout, spreading out the heating effect. Incoming solar radiation on land, on the other hand, warms a relatively thin layer on the top, concentrating the heating effect. (3) The water is cooled by evaporation, which helps keep a body of water at a lower temperature than an adjacent landmass under identical conditions of incoming solar radiation.

A local wind pattern may result from the resulting temperature differences between a body of water and adjacent landmasses. If you have ever spent some time along a coast, you may have observed that a cool, refreshing gentle breeze blows from the water toward the land during the summer. During the day the temperature of the land increases more rapidly than the water temperature. The air over the land is therefore heated more, expands, and becomes less dense. Cool, dense air from over the water moves inland under the air over the land, buoying it up. The air moving from the sea to the land is called a *sea breeze*. The sea breeze along a coast may extend inland several miles during the hottest part of the day in the summer. The same pattern is sometimes observed around the Great Lakes during the summer, but this breeze usually does not reach more than several city blocks inland. During the night the land surface cools more rapidly than the water, and a breeze blows from the land to the sea (figure 16.10).

Another pattern of local winds develops in mountainous regions. If you have ever visited a mountain in the summer, you may have noticed that there is usually a breeze or wind blowing up the mountain slope during the afternoon. This wind pattern develops because the air over the mountain slope is heated more than the air in a valley. As shown in figure 16.11, the air over the slope becomes warmer because it receives more direct sunlight than the valley floor. Sometimes this air movement is so gentle that it would be unknown except for the evidence of clouds that form over the peaks during the day and evaporate at night. During the night the air on the slope cools as the land loses radiant energy to space. As the air cools, it becomes denser and flows downslope, forming a reverse wind pattern to the one observed during the day.

During cooler seasons cold, dense air may collect in valleys or over plateaus, forming a layer or "puddle" of cold air. Such an accumulation of cold air often results in some very cold nighttime temperatures for cities located in valleys, temperatures that

The term *wind chill* is attributed to the Antarctic explorer Paul A. Siple. During the 1940s Siple and Charles F. Passel conducted experiments on how long it took a can of water to freeze at various temperatures and wind speeds at a height of 33 feet above the ground (typical height of an anemometer). They found that the time depended on the air temperature and the wind speed. From these data an equation was developed to calculate the **wind chill factor** for humans.

In 2001 the National Weather Service changed to a new method of computing wind chill temperatures. The new wind chill formula was developed by a year-long cooperative effort between the U.S. and Canadian governments and university scientists. The new standard is based on wind speeds at an average height of five feet

above the ground, which is closer to the height of a human face rather than the height of an anemometer and takes advantage of advances in science, technology, and computers. The new chart also highlights the danger of frostbite.

Here is the reason why the wind chill factor is an important consideration. The human body constantly produces heat to maintain a core temperature, and some of this heat is radiated to the surroundings. When the wind is not blowing (and you are not moving), your body heat is also able to warm some of the air next to your body. This warm blanket of air provides some insulation, protecting your skin from the colder air farther away. If the wind blows, however, it moves this air away from your body and you feel cooler. How much cooler depends on how fast the air is mov-

ing and on the outside temperature—which is what the wind chill factor tells you. Thus, wind chill is an attempt to measure the combined effect of low temperature and wind on humans (box figure 16.1). It is just one of the many factors that can affect winter comfort. Others include the type of clothes, level of physical exertion, amount of sunshine, humidity, age, and body type.

There is a wind chill calculator at the National Weather Service website, *www.nws.noaa.gov/om/windchill/*. All you need to do is enter the air temperature (in degrees Fahrenheit) and the wind speed (in miles per hour) and the calculator will give you the wind chill using both the old and the new formula.

Figure 16.10

The land warms and cools more rapidly than an adjacent large body of water. During the day, the land is warmer, and air over the land expands and is buoyed up by cooler, more dense air from over the water. During the night, the land cools more rapidly than the water, and the direction of the breeze is reversed.

Figure 16.11

Incoming solar radiation falls more directly on the side of a mountain, which results in differential heating. The same amount of sunlight falls on the areas shown in this illustration, with the valley floor receiving a more spread-out distribution of energy per unit area. The overall result is an upslope mountain breeze during the day. During the night, dense cool air flows downslope for a reverse wind pattern.

are much colder than anywhere in the surrounding region. Some weather disturbance, such as an approaching front, can disturb such an accumulation of cold air and cause it to pour out of its resting place and through canyons or lower valleys. Air

moving from a higher altitude like this becomes compressed as it moves to lower elevations under increasing atmospheric pressure. Compression of air increases the temperature by increasing the kinetic energy of the molecules. This creates a wind called a *Chinook*, which is common to mountainous and adjacent regions. A Chinook is a wind of compressed air with sharp temperature increases that can melt away any existing snow cover in a single day. The *Santa Ana* is a well-known compressional wind that occurs in southern California.

Wind (mph)	Temperature (°F)																	
	40	35	30	25	20	15	10	5	0	−5	−10	−15	−20	−25	−30	−35	−40	−45
5	36	31	25	19	13	7	1	−5	−11	−16	−22	−28	−34	−40	−46	−52	−57	−63
10	34	27	21	15	9	3	−4	−10	−16	−22	−28	−35	−41	−47	−53	−59	−66	−72
15	32	25	19	13	6	0	−7	−13	−19	−26	−32	−39	−45	−51	−58	−64	−71	−77
20	30	24	17	11	4	−2	−9	−15	−22	−29	−35	−42	−48	−55	−61	−68	−74	−81
25	29	23	16	9	3	−4	−11	−17	−24	−31	−37	−44	−51	−58	−64	−71	−78	−84
30	28	22	15	8	1	−5	−12	−19	−26	−33	−39	−46	−53	−60	−67	−73	−80	−87
35	28	21	14	7	0	−7	−14	−21	−27	−34	−41	−48	−55	−62	−69	−76	−82	−89
40	27	20	13	6	−1	−8	−15	−22	−29	−36	−43	−50	−57	−64	−71	−78	−84	−91
45	26	19	12	5	−2	−9	−16	−23	−30	−37	−44	−51	−58	−65	−72	−79	−86	−93
50	26	19	12	4	−3	−10	−17	−24	−31	−38	−45	−52	−60	−67	−74	−81	−88	−95
55	25	18	11	4	−3	−11	−18	−25	−32	−39	−46	−54	−61	−68	−75	−82	−89	−97
60	25	17	10	3	−4	−11	−19	−26	−33	−40	−48	−55	−62	−69	−76	−84	−91	−98

Frostbite Times   ☐ 30 minutes   ■ 10 minutes   ☐ 5 minutes

Box Figure 16.1
Wind chill chart.

Global Wind Patterns

Local wind patterns tend to mask the existence of the overall global wind pattern that is also present. The global wind pattern is not apparent if the winds are observed and measured for a particular day, week, or month. It does become apparent when the records for a long period of time are analyzed. These records show that the earth has a large-scale pattern of atmospheric circulation that varies with latitude. There are belts in which the winds average an overall circulation in one direction, belts of higher atmospheric pressure averages, and belts of lower atmospheric pressure averages. This has led to a generalized pattern of atmospheric circulation and a global atmospheric model. This model, as you will see, today provides the basis for the daily weather forecast for local and regional areas.

As with local wind patterns, it is temperature imbalances that drive the global circulation of the atmosphere. The earth receives more direct solar radiation in the equatorial region than it does at higher latitudes (figure 16.12). As a result, the temperatures of the lower troposphere are generally higher in the equatorial region, decreasing with latitude toward both poles. The lower troposphere from 10°N to 10°S of the equator is heated, expands, and becomes less dense. Hot air rises in this belt around the equator, known as the *intertropical convergence zone*. The rising air cools because it expands as it rises, resulting in heavy average precipitation. The tropical rain forests of the earth occur in this zone of high temperatures and heavy rainfall. As the now dry, rising air reaches the upper parts of the troposphere it begins to spread toward the north and toward the

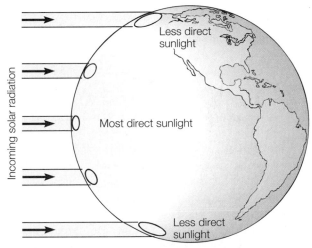

Figure 16.12
On a global, yearly basis, the equatorial region of the earth receives more direct incoming solar radiation than the higher latitudes. As a result, average temperatures are higher in the equatorial region and decrease with latitude toward both poles. This sets the stage for worldwide patterns of prevailing winds, high and low areas of atmospheric pressure, and climatic patterns.

south, sinking back toward the earth's surface (figure 16.13). The descending air reaches the surface to form a high-pressure belt that is centered about 30°N and 30°S of the equator. Air moving on the surface away from this high-pressure belt produces the prevailing northeast trade winds and the prevailing

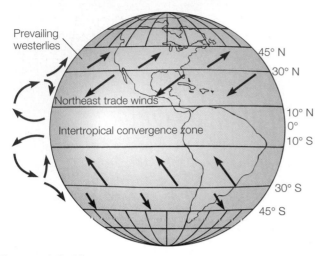

Figure 16.13

Part of the generalized global circulation pattern of the earth's atmosphere. The scale of upward movement of air above the intertropical convergence zone is exaggerated for clarity. The troposphere over the equator is thicker than elsewhere, reaching a height of about 20 km (about 12 mi).

westerly winds of the Northern Hemisphere. The great deserts of the earth are also located in this high-pressure belt of descending dry air.

Poleward of the belt of high pressure the atmospheric circulation is controlled by a powerful belt of wind near the top of the troposphere called a **jet stream.** Jet streams are sinuous, meandering loops of winds that tend to extend all the way around the earth, moving generally from the west in both hemispheres at speeds of 160 km/h (about 100 mi/h) or more. A jet stream may occur as a single belt, or loop, of wind but sometimes it divides into two or more parts. The jet stream develops north and south loops of waves much like the waves you might make on a very long rope. These waves vary in size, sometimes beginning as a small ripple but then growing slowly as the wave moves eastward. Waves that form on the jet stream bulge toward the poles (called a crest) or toward the equator (called a trough). Warm air masses move toward the poles ahead of a trough and cool air masses move toward the equator behind a trough as it moves eastward. The development of a wave in the jet stream is understood to be one of the factors that influences the movement of warm and cool air masses, a movement that results in weather changes on the surface.

The intertropical convergence zone, the 30° belt of high pressure, and the northward and southward migration of a meandering jet stream all shift toward or away from the equator during the different seasons of the year. The troughs of the jet stream influence the movement of alternating cool and warm air masses over the belt of the prevailing westerlies, resulting in frequent shifts of fair weather to stormy weather, then back again. The average shift during the year is about 6° of latitude, which is sufficient to control the overall climate in some locations. The influence of this shift of the global circulation of the

A Closer Look Wind Power

Four nonrenewable sources of energy—coal, oil, natural gas, and nuclear—provide more than 97 percent of the world's energy supply. The remainder is supplied by renewable energy sources, including solar energy, tidal power, and geothermal. The term *solar energy* describes a number of technologies that directly or indirectly utilize sunlight as an alternative energy source. In part, these include solar cells, power towers, biomass, and wind energy.

The earth's surface is warmed by the radiant energy from the sun, and some parts are warmed more than other parts. This differential heating sets up convection currents in the atmosphere, resulting in horizontal movements of air called wind. Millions of windmills were installed in the United States between the late 1800s and the late 1940s in rural areas. These windmills used wind energy to pump water, grind grain, or generate electricity. Some are still in use today, but most were no longer needed after inexpensive electric power became generally available in rural areas of the country.

In the 1970s wind energy made a comeback as a clean renewable energy alternative to fossil fuels. The windmills of the past were replaced by wind turbines of today. A wind turbine is usually mounted on a tower, with blades that are rotated by the wind. This rotary motion drives a generator that produces electricity. An area should have yearly wind speeds of about 12 miles per hour to provide enough wind energy for a turbine, and a greater yearly average means more energy is available. Farms, homes, and businesses in these areas can use smaller turbines, which are generally 50 kilowatts or less. Large turbines of 500 kilowatts or more are used in "wind farms," which are large clusters of interconnected wind turbines connected to a utility power grid.

Many areas of the United States have a high potential for wind-power use. The states of North Dakota, South Dakota, and Texas have enough wind resources to provide electricity for the entire nation. Today only California has extensively developed wind farms, with more than 13,000 wind turbines on three wind farms in the Altamont Pass region (east of San Francisco), Techachapi (southeast of Bakersfield), and San Gorgonio (near Palm Springs). With a total of 1,700 megawatts of installed capacity in California, wind energy generates enough electricity to more than meet the needs of a city the size of San Francisco.

The wind farms in California have a rated capacity that is comparable to two large coal-fired power plants, but without the pollution and limits of this nonrenewable energy source. Wind energy makes economic as well as environmental sense and new wind farms are being developed in the states of Minnesota, Oregon, and Wyoming. Other states with a strong wind-power potential include Kansas, Montana, Nebraska, Oklahoma, Iowa, Colorado, Michigan, and New York. All of these states, in fact, have a greater wind-energy potential than the state of California.

earth's atmosphere will be considered as a climatic influence after considering the roles of water and air masses in frequent weather changes.

WATER AND THE ATMOSPHERE

Water exists on the earth in all three states: (1) as a liquid when the temperature is generally above the freezing point of 0°C (32°F), (2) as a solid in the form of ice, snow, or hail when the temperature is generally below the freezing point, and (3) as the invisible, molecular form of water in the gaseous state, which is called *water vapor.*

Over 98 percent of all the water on the earth exists in the liquid state, mostly in the ocean, and only a small, variable amount of water vapor is in the atmosphere at any given time. Since so much water seems to fall as rain or snow at times, it may be a surprise that the overall atmosphere really does not contain very much water vapor. If the average amount of water vapor in the earth's atmosphere were condensed to liquid form, the vapor *and* all the droplets present in clouds would form a uniform layer around the earth only 3 cm (about 1 in) thick. Nonetheless, it is this small amount of water vapor that is eventually responsible for (1) contributing to the greenhouse effect, which helps make the earth a warmer planet, (2) serving as one of the principal agents in the weathering and erosion of the land, which creates soils and sculptures the landscape, and (3) maintaining life, for all organisms (bacteria, protozoa, algae, fungi, plants, and animals) cannot survive without water. It is the ongoing cycling of water vapor into and out of the atmosphere that makes all this possible. Understanding this cycling process and the energy exchanges involved is also closely related to understanding the earth's weather patterns.

Evaporation and Condensation

Water tends to undergo a liquid-to-gas or a gas-to-liquid phase change at any temperature. The phase change can occur in either direction at any temperature.

In *evaporation,* more molecules are leaving the liquid state than are returning. In *condensation,* more molecules are returning to the liquid state than are leaving. This is a dynamic, ongoing process with molecules leaving and returning continuously (figure 16.14). If the air were perfectly dry and still, more molecules would leave (evaporate) the liquid state than would return (condense). Eventually, however, an equilibrium would be reached with as many molecules returning to the liquid state per unit of time as are leaving. An equilibrium condition between evaporation and condensation occurs in **saturated air.** Saturated air occurs when the processes of evaporation and condensation are in balance.

Air will remain saturated as long as (1) the temperature remains constant and (2) the processes of evaporation and condensation remain balanced. Temperature influences the equilibrium condition of saturated air because increases or decreases in the temperature mean increases or decreases in the kinetic energy of water vapor molecules. Water vapor molecules usually undergo condensation when attractive forces between the mol-

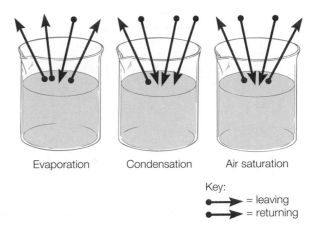

Evaporation Condensation Air saturation

Key:
●——➤ = leaving
●——➤ = returning

Figure 16.14
Evaporation and condensation are occurring all the time. If the number of molecules leaving the liquid state exceeds the number returning, the water is evaporating. If the number of molecules returning to the liquid state exceeds the number leaving, the water vapor is condensing. If both rates are equal, the air is saturated; that is, the relative humidity is 100 percent.

ecules can pull them together into the liquid state. Lower temperature means lower kinetic energies, and slow-moving water vapor molecules spend more time close to one another and close to the surface of liquid water. Spending more time close together means an increased likelihood of attractive forces pulling the molecules together. On the other hand, higher temperature means higher kinetic energies, and molecules with higher kinetic energy are less likely to be pulled together. As the temperature increases, there is therefore less tendency for water molecules to return to the liquid state. If the temperature is increased in an equilibrium condition, more water vapor must be added to the air to maintain the saturated condition. Warm air can therefore hold more water vapor than cooler air. In fact, warm air on a typical summer day can hold five times as much water vapor as cold air on a cold winter day.

Humidity

The amount of water vapor in the air is referred to generally as **humidity.** Damp, moist air is more likely to have condensation than evaporation, and this air is said to have a *high humidity.* Dry air is more likely to have evaporation than condensation, on the other hand, and this air is said to have a *low humidity.* A measurement of the amount of water vapor in the air at a particular time is called the **absolute humidity** (figure 16.15). At room temperature, for example, humid air might contain 15 grams of water vapor in each cubic meter of air. At the same temperature, air of low humidity might have an absolute humidity of only 2 grams per cubic meter. The absolute humidity can range from near zero up to a maximum that is determined by the temperature at the time. Since the temperature of the water vapor present in the air is the same as the temperature of the air, the maximum absolute humidity is usually said to be determined by the air temperature. What this really means is that the maximum absolute humidity is determined by the temperature of the water vapor, that is, the average kinetic energy of the water vapor.

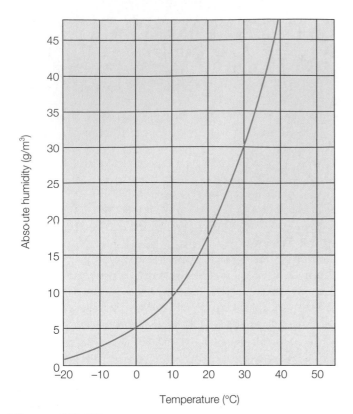

Figure 16.15
The maximum amount of water vapor that can be in the air at different temperatures. The amount of water vapor in the air at a particular temperature is called the *absolute humidity.*

The relationship between the *actual* absolute humidity at a particular temperature and the *maximum* absolute humidity that can occur at that temperature is called the **relative humidity.** Relative humidity is a ratio between (1) the amount of water vapor in the air and (2) the amount of water vapor needed to saturate the air at that temperature. The relationship is

$$\text{relative humidity} = \frac{\begin{array}{c}\text{acual absolute humidity}\\\text{at present temperature}\end{array}}{\begin{array}{c}\text{maximum absolute humidity}\\\text{at present temperature}\end{array}} \times 100\%$$

For example, suppose a measurement of the water vapor in the air at 10°C (50°F) finds an absolute humidity of 5.0 g/m³. According to figure 16.15, the maximum amount of water vapor that can be in the air when the temperature is 10°C is about 10 g/m³. The relative humidity is then

$$\frac{5.0 \text{ g/m}^3}{10 \text{ g/m}^3} \times 100\% = 50\%$$

If the absolute humidity had been 10 g/m³, then the air would have all the water vapor it could hold, and the relative humidity would be 100 percent. A humidity of 100 percent means that the air is saturated at the present temperature.

The important thing to understand about relative humidity is that the capacity of air to hold water vapor changes with the temperature. Cold air cannot hold as much water vapor, and

warming the air will increase its capacity. With the capacity *increased* the relative humidity *decreases* because you can now add more water vapor to the air than you could before. Just warming the air, for example, can reduce the relative humidity from 50 percent to 3 percent. Lower relative humidity results because warming the air increases the capacity of air to hold water vapor. This explains the need to humidify a home in the winter. Evaporation occurs very rapidly when the humidity is low. Evaporation is a cooling process since the molecules with higher kinetic energy are the ones to escape, lowering the average kinetic energy as they evaporate. Dry air will therefore cause you to feel cool even though the air temperature is fairly high. Adding moisture to the air will enable you to feel warmer at lower air temperatures, and thus lower your fuel bill.

The relationship between the capacity of air to hold water vapor and temperature also explains why the relative humidity increases in the evening after the sun goes down. A cooler temperature means less capacity of air to hold water vapor. With the same amount of vapor in the air, a reduced capacity means a higher relative humidity.

The Condensation Process

Condensation, depends on two factors: (1) the relative humidity and (2) the temperature of the air. During condensation, molecules of water vapor join together to produce liquid water on the surface as dew or in the air as the droplets of water making up fog or clouds. Water molecules may also join together to produce solid water in the form of frost or snow. Before condensation can occur, however, the air must be saturated, which means that the relative humidity must be 100 percent with the air containing all the vapor it can hold at the present temperature. A parcel of air can become saturated as a result of (1) water vapor being added to the air from evaporation, (2) cooling, which reduces the capacity of the air to hold water vapor and therefore increases the relative humidity, or (3) a combination of additional water vapor with cooling.

The process of condensation of water vapor explains a number of common observations. You are able to "see your breath" on a cold day, for example, because the high moisture content of your exhaled breath is condensed into tiny water droplets by cold air. The small fog of water droplets evaporates as it spreads into the surrounding air with a lower moisture content. The white trail behind a high-flying jet aircraft is also a result of condensation of water vapor. Water is one of the products of combustion, and the white trail is condensed water vapor, a trail of tiny droplets of water in the cold upper atmosphere. The trail of water droplets is called a *contrail* after "condensation trail." Back on the surface, a cold glass of beverage seems to "sweat" as water vapor molecules near the outside of the glass are cooled, moving more slowly. Slowly moving water vapor molecules spend more time closer together, and the molecular forces between the molecules pull them together, forming a thin layer of liquid water on the outside of the cold glass. This same condensation process sometimes results in a small stream of water from the cold air conditioning coils of an automobile or home mechanical air conditioner.

As air is cooled, its capacity to hold water vapor is reduced to lower and lower levels. Even without water vapor being added to the air, a temperature will eventually be reached at which saturation, 100 percent humidity, occurs. Further cooling below this temperature will result in condensation. The temperature at which condensation begins is called the **dew point temperature.** If the dew point is above 0°C (32°F) the water vapor will condense on surfaces as a liquid called **dew.** If the temperature is at or below 0°C the vapor will condense on surfaces as a solid called **frost.** Note that dew and frost form on the tops, sides, and bottoms of objects. Dew and frost condense directly on objects and do not "fall out" of the air. Note also that the temperature that determines if dew or frost forms is the temperature of the object where they condense. This temperature near the open surface can be very different from the reported air temperature, which is measured more at eye level in a sheltered instrument enclosure.

Observations of where and when dew and frost form can lead to some interesting things to think about. Dew and frost, for example, seem to form on "C" nights, nights that can be described by the three "C" words of clear, calm, and cool. Dew and frost also seem to form more (1) in open areas rather than under trees or other shelters, (2) on objects such as grass rather than on the flat, bare ground, and (3) in low-lying areas before they form on slopes or the sides of hills. What is the meaning of these observations?

Dew and frost are related to clear nights and open areas because these are the conditions best suited for the loss of infrared radiation. Air near the surface becomes cooler as infrared radiation is emitted from the grass, buildings, streets, and everything else that absorbed the shorter-wavelength radiation of incoming solar radiation during the day. Clouds serve as a blanket, keeping the radiation from escaping to space so readily. So a clear night is more conducive to the loss of infrared radiation and therefore to cooling. On a smaller scale, a tree serves the same purpose, holding in radiation and therefore retarding the cooling effect. Thus, an open area on a clear, calm night would have cooler air near the surface than would be the case on a cloudy night or under the shelter of a tree.

The observation that dew and frost form on objects such as grass before forming on flat, bare ground is also related to loss of infrared radiation. Grass has a greater exposed surface area than the flat, bare ground. A greater surface area means a greater area from which infrared radiation can escape, so grass blades cool more rapidly than the flat ground. Other variables, such as specific heat, may be involved, but overall frost and dew are more likely to form on grass and low-lying shrubs before they form on the flat, bare ground.

Dew and frost form in low-lying areas before forming on slopes and the sides of hills because of the density differences of cool and warm air. Cool air is more dense than warm air and is moved downhill by gravity, pooling in low-lying areas. You may have noticed the different temperatures of low-lying areas if you have ever driven across hills and valleys on a clear, calm, and cool evening. Citrus and other orchards are often located on slopes of hills rather than on valley floors because of the gravity drainage of cold air.

It is air near the surface that is cooled first by the loss of radiation from the surface. Calm nights favor dew or frost formation because the wind mixes the air near the surface that is being cooled with warmer air above the surface. If you have ever driven near a citrus orchard, you may have noticed the huge, airplanelike fans situated throughout the orchard on poles. These fans are used on "C" nights when frost is likely to form to mix the warmer, upper layers of air with the cooling air in the orchard (figure 16.16).

Condensation occurs on the surface as frost or dew when the dew point is reached. When does condensation occur in the air? Water vapor molecules in the air are constantly colliding and banging into each other, but they do not just join together to form water droplets, even if the air is saturated. The water molecules need something to condense upon. Condensation of water vapor into fog or cloud droplets takes place on tiny particles present in the air. The particles are called **condensation nuclei.** There are hundreds of tiny dust, smoke, soot, and salt crystals suspended in each cubic centimeter of the air that serve as condensation nuclei. Tiny salt crystals, however, are particularly effective condensation nuclei because salt crystals attract water molecules. You may have noticed that salt in a salt shaker becomes moist on a humid day because of the way it attracts water molecules. Tiny salt crystals suspended in the air act the same

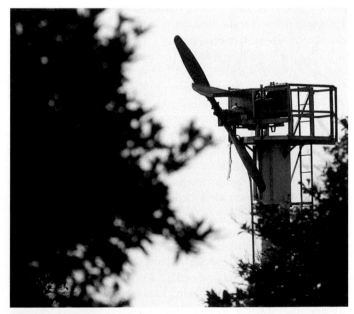

Figure 16.16
Fans like this one are used to mix the warmer, upper layers of air with the cooling air in the orchard on nights when frost is likely to form.

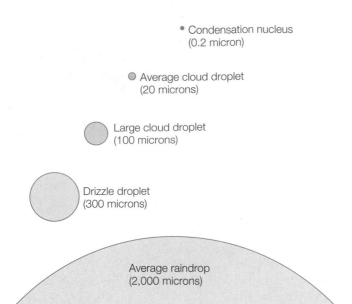

Figure 16.17

This figure compares the size of the condensation nuclei to the size of typical condensation droplets. Note that 1 micron is 1/1,000 mm.

way, serving as nuclei that attract water vapor into tiny droplets of liquid water.

After water vapor molecules begin to condense on a condensation nucleus, other water molecules will join the liquid water already formed, and the tiny droplet begins to increase in volume. The water droplets that make up a cloud are about 1,500 times larger than a condensation nuclei, and these droplets can condense out of the air in a matter of minutes. As the volume increases, however, the process slows, and hours and days are required to form the even larger droplets and drops. For comparison to the sizes shown in figure 16.17, consider that the average human hair is about 100 microns in diameter. This is about the same diameter as the large cloud droplet of water. Large raindrops have been observed falling from clouds that formed only a few hours previously, so it must be some process or processes other than the direct condensation of raindrops that form precipitation. These processes are discussed in the section on precipitation.

Fog and Clouds

Fog and clouds are both accumulations of tiny droplets of water that have been condensed from the air. These water droplets are very small and a very slight upward movement of the air will keep them from falling. If they do fall they usually evaporate. Fog is sometimes described as a cloud that forms at or near the surface. A fog, as a cloud, forms because air containing water vapor and condensation nuclei has been cooled to the dew point. Some types of fog form under the same "C" night conditions favorable for dew or frost to form, that is, on clear, cool, and calm nights when the relative humidity is high. Sometimes this type of fog forms only in valleys and low-lying areas where cool air accumulates (figure 16.18). This type of fog is typical of inland fogs, those that form away from bodies of water. Other types of

A

B

Figure 16.18

(*A*) An early morning aerial view of fog that developed close to the ground in cool, moist air on a clear, calm night. (*B*) Air moves from a warm current, then over a cool current forming fog. The fog often moves inland at night.

fog may form somewhere else, such as in the humid air over an ocean, and then move inland. Many fogs that occur along coastal regions were formed over the ocean and then carried inland by breezes. A third type of fog looks much like a steamy mist rising from melting snow on a street, over a body of water into cold air, or a steamy mist over streets after a summer rain shower. These are examples of a temporary fog that forms as a lot of water vapor is added to cool air. This is a cool fog, like other fogs, and is not hot as the steamlike appearance may lead you to believe.

Sometimes a news report states something about the sun "burning off" a fog. A fog does not burn, of course, because it is made up of droplets of water. What the reporter really means is that the sun's radiation will increase the temperature, which increases the air capacity to hold water vapor. With an increased capacity to hold water the relative humidity drops, and the fog simply evaporates back to the state of invisible water vapor molecules.

Clouds, like fogs, are made up of tiny droplets of water that have been condensed from the air. Luke Howard, an English weather observer, made one of the first cloud classification schemes in 1803. He used the Latin terms *cirrus* (curly), *cumulus* (piled up), and *stratus* (spread out) to identify the basic shapes of clouds (figure 16.19). The clouds usually do not occur just in these basic cloud shapes but in combinations of the different shapes. Later, Howard's system was modified by expanding the different shapes of clouds into ten classes by using the basic cloud shapes and altitude as criteria. Clouds give practical hints about the approaching weather. The relationship between the different cloud shapes and atmospheric conditions and what clouds can mean about the coming weather are discussed in the next section.

Clouds form when a mass of air above the surface is cooled to its dew point temperature. In general, the mass of air is cooled because something has given it an upward push, moving it to higher levels in the atmosphere. There are three major causes of upward air movement: (1) *convection* resulting from differential heating, (2) mountain ranges that serve as *barriers* to moving air masses, and (3) the meeting of *moving air masses* with different densities, for example, a cold, dense mass of air meeting a warm, less dense mass of air.

The three major causes of uplifted air sometimes result in clouds, but just as often they do not. Whether clouds form or not depends on the condition of the atmosphere at the time. As a parcel of warm air is moved upward it tends to stay together, mixing very little with the surrounding air. As it is forced upward it becomes cooler because it is expanding. Similarly, the temperature of a gas increases when it is compressed. So rising air is cooled and descending air is warmed.

What happens to a parcel of air that is pushed upward depends on the difference in density between the parcel and the surrounding air. Air temperature will tell you about air density since the density of air is determined by its temperature. Instruments attached to a weather balloon can measure the change of temperature with altitude. By comparing this change with the rate of cooling by expansion, the state of atmospheric stability can be determined. There are many different states of atmospheric stability, and the following is a simplified description of just a few of the possible states, first considering dry air only.

The atmosphere is in a state of *stability* when a lifted parcel of air is cooler than the surrounding air. Being cooler, the parcel of air will be more dense than the surroundings. If it is moved up to a higher level and released in a stable atmosphere, it will move back to its former level. A lifted parcel of air always returns to its original level when the atmosphere is stable. Any clouds that do develop in a stable atmosphere are usually arranged in the horizontal layers of stratus-type clouds.

The atmosphere is in a state of *instability* when a lifted parcel of air is warmer than the surrounding air. Being warmer, the parcel of air will be less dense than the surroundings. If it is

moved up to a higher level, it will continue moving after the uplifting force is removed. Cumulus clouds usually develop in an unstable atmosphere, and the rising parcels of air, called thermals, can result in a very bumpy airplane ride.

So far only dry air has been considered. As air moves upward and cools from expansion, sooner or later the dew point is reached and the air becomes saturated. As some of the water vapor in the rising parcel condenses to droplets, the latent heat of vaporization is released. The rising parcel of air now cools at a slower rate because of the release of this latent heat of vaporization. The release of latent heat warms the air in the parcel and decreases the density even more, accelerating the ascent. This leads to further condensation and the formation of towering cumulus clouds, often leading to rain.

Precipitation

Water that returns to the surface of the earth, in either the liquid or solid form, is called **precipitation** (figure 16.20). Note that dew and frost are not classified as precipitation because they form directly on the surface and do not fall through the air. Precipitation seems to form in clouds by one of two processes: (1) the *coalescence* of cloud droplets or (2) the *growth of ice crystals*. It would appear difficult for cloud droplets to merge, or coalesce, with one another since any air movement would seem to move them all at the same time, not bring them together. Condensation nuclei come in different sizes, however, and cloud droplets of many different sizes form on these different-sized nuclei. Larger cloud droplets are slowed less by air friction as they drift downward, and they collide and merge with smaller droplets as they fall. They may merge, or coalesce, with a million other droplets before they fall from the cloud as raindrops. This *coalescence process* of forming precipitation is thought to take place in warm cumulus clouds that form near the ocean in the tropics. These clouds contain giant salt condensation nuclei and have been observed to produce rain within about twenty minutes after forming.

Clouds at middle latitudes, away from the ocean, also produce precipitation, so there must be a second way that precipitation forms. The *ice-crystal process* of forming precipitation is important in clouds that extend high enough in the atmosphere to be above the freezing point of water. Water molecules are more strongly bonded to each other in an ice crystal than in liquid water. Thus, an ice crystal can capture water molecules and grow to a larger size while neighboring water droplets are evaporating. As they grow larger and begin to drift toward the surface, they may coalesce with other ice crystals or droplets of water, soon falling from the cloud. During the summer, they fall through warmer air below and reach the ground as raindrops. During the winter, they fall through cooler air below and reach the ground as snow.

Tiny water droplets do not freeze as readily as a larger mass of liquid water, and many droplets do not freeze until the temperature is below about −40°C (−40°F). Water that is still in the liquid state when the temperature is below the freezing temperature is said to be *supercooled*. Supercooled clouds of water droplets are common between the temperatures of −40°C and

Figure 16.19

(*A*) Cumulus clouds. (*B*) Stratus and stratocumulus. Note the small stratocumulus clouds forming from increased convection over each of the three small islands. (*C*) An aerial view between the patchy cumulus clouds below and the cirrus and cirrostratus above (the patches on the ground are clear-cut forests). (*D*) Altocumulus. (*E*) A rain shower at the base of a cumulonimbus. (*F*) Stratocumulus.

Figure 16.20
Precipitation is water in the liquid or solid form that returns to the surface of the earth. The precipitation you see here is liquid, and each raindrop is made from billions of the tiny droplets that make up the clouds. The tiny droplets of clouds become precipitation by merging to form larger droplets or by the growth of ice crystals that melt while falling.

0°C (−40°F and 32°F), a range of temperatures that is often found in the upper atmosphere. The liquid droplets at these temperatures need solid particles called **ice-forming nuclei** to freeze upon. Generally, dust from the ground serves as ice-forming nuclei that start the ice-crystal process of forming precipitation. Artificial rainmaking has been successful by (1) dropping crushed dry ice, which is cooler than 40°C, on top of a supercooled cloud and (2) by introducing "seeds" of ice-forming nuclei in supercooled clouds. Tiny crystals from the burning of silver iodide are effective ice-forming nuclei, producing ice crystals at temperatures as high as −4.0°C (about 25°F). Attempts at ground-based cloud seeding with silver iodide in the mountains of the western United States have suggested up to 15 percent more snowfall, but it is difficult to know how much snowfall would have resulted without the seeding.

In general, the basic form of a cloud has meaning about the general type of precipitation that can occur as well as the coming weather. Cumulus clouds usually produce showers or thunderstorms that last only brief periods of time. Longer periods of drizzle, rain, or snow usually occur from stratus clouds. Cirrus clouds do not produce precipitation of any kind, but they may have meaning about the coming weather, which is discussed in the following section.

WEATHER PRODUCERS

The idealized model of the general atmospheric circulation starts with the poleward movement of warm air from the tropics. The region between 10°N and 10°S of the equator receives more direct radiation, on the average, than other regions of the earth's surface. The air over this region is heated more, expands, and becomes less dense as a consequence of the heating. This less dense air is buoyed up by convection to heights up to 20 km (about 12 mi). As it rises it is cooled by radiation to less than −73°C (about −110°F). This accumulating mass of cool, dry air spreads north and south toward both poles (see figure 16.13), then sinks back toward the surface at about 30°N and 30°S. The descending air is warmed by compression and is warm and dry by the time it reaches the surface. Part of the sinking air then moves back toward the equator across the surface, completing a large convective cell. This giant cell has a low-pressure belt over the equator and high-pressure belts over the subtropics near latitudes of 30°N and 30°S. The other part of the sinking air moves poleward across the surface, producing belts of westerly winds in both hemispheres to latitudes of about 60°.

The overall pattern of pressure belts and belts of prevailing winds is seen to shift north and south with the seasons, resulting in a seasonal shift in the types of weather experienced at a location. This shift of weather is related to three related weather producers: (1) the movement of large bodies of air, called *air masses*, that have acquired the temperature and moisture conditions where they have been located, (2) the leading *fronts* of air masses when they move, and (3) the local *high- and low-pressure* patterns that are associated with air masses and fronts. These are the features shown on almost all daily weather maps, and they are the topics of this section.

Air Masses

An **air mass** is defined as a large, more or less uniform body of air with nearly the same temperature and moisture conditions. An air mass forms when a large body of air, perhaps covering millions of square kilometers, remains over a large area of land or water for an extended period of time. While it is stationary it acquires the temperature and moisture characteristics of the land or water through the heat transfer processes of conduction, convection, and radiation and through the moisture transfer processes of evaporation and condensation. For example, a large body of air that remains over the cold, dry, snow-covered surface of Siberia for some time will become cold and dry. A large body of air that remains over a warm tropical ocean, on the other hand, will become warm and moist. Knowledge about the condition of air masses is important because they tend to retain the acquired temperature and moisture characteristics when they finally break away, sometimes moving long distances. An air mass that formed over Siberia can bring cold, dry air to your location, while an air mass that formed over a tropical ocean will bring warm, moist air.

Air masses are classified according to the temperature and moisture conditions where they originate. There are two temperature extreme possibilities, a *polar air mass* from a cold region and a *tropical air mass* from a warm region. There are also two moisture extreme possibilities, a moist *maritime air mass* from over the ocean and a generally dry *continental air mass* from over the land. Thus, there are four main types of air masses that can influence the weather where you live: (1) continental polar,

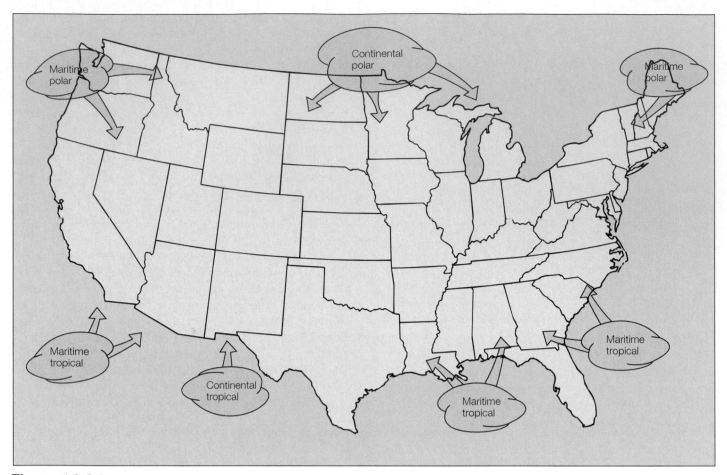

Figure 16.21

The general movement of the four main types of air masses that influence the weather over the contiguous United States. The tropical air masses visit most often in the summer, and the polar air masses visit most often during the winter. During other times, the polar and tropical air masses battle back and forth over the land.

(2) maritime polar, (3) continental tropical, and (4) maritime tropical. Figure 16.21 shows the general direction in which these air masses usually move over the mainland United States.

Once an air mass leaves its source region it can move at speeds of up to 800 km (about 500 mi) per day while mostly retaining the temperature and moisture characteristics of the source region (figure 16.22). If it slows and stagnates over a new location, however, the air may again begin to acquire a new temperature and moisture equilibrium with the surface. When a location is under the influence of an air mass, the location is having a period of *air mass weather*. This means that the weather conditions will generally remain the same from day to day with slow, gradual changes. Air mass weather will remain the same until a new air mass moves in or until the air mass acquires the conditions of the new location. This process may take days or several weeks, and the weather conditions during this time depend on the conditions of the air mass and conditions at the new location. For example, a polar continental air mass arriving over a cool, dry land area may produce a temperature inversion with the air colder near the surface than higher up. When the temperature increases with height the air is stable and cloudless, and cold weather continues with slow, gradual warming. The tem-

perature inversion may also result in hazy periods of air pollution in some locations. A polar continental air mass arriving over a generally warmer land area, on the other hand, results in a condition of instability. In this situation each day will start clear and cold, but differential heating during the day develops cumulus clouds in the unstable air. After sunset the clouds evaporate, and a clear night results because the thermals during the day carried away the dust and air pollution. Thus, a dry, cold air mass can bring different weather conditions, each depending on the properties of the air mass and the land it moves over.

Weather Fronts

The boundary between air masses of different temperatures is called a **front.** A front is actually a thin transition zone between two air masses that ranges from about 5 to 30 km thick (about 3 to 20 mi), and the air masses do not mix other than in this narrow zone. The density differences between the two air masses prevent any general mixing since the warm, less-dense air mass is forced upward by the cooler, more dense air moving under it. You may have noticed on a daily weather map that fronts are usually represented with a line bulging outward in the direction

Figure 16.22

This satellite photograph shows the result of a polar air mass moving southeast over the southern United States. Clouds form over the warmer waters of the Gulf of Mexico and the Atlantic Ocean, showing the state of atmospheric instability from the temperature differences.

of cold air mass movement (figure 16.23). A cold air mass is much like a huge, flattened bubble of air that moves across the land (figure 16.24). The line on a weather map represents the place where the leading edge of this huge, flattened bubble of air touches the surface of the earth.

A **cold front** is formed when a cold air mass moves into warmer air, displacing it in the process. A cold front is generally steep, and when it runs into the warmer air it forces it to rise quickly. If the warm air is moist, it is quickly cooled to the dew point temperature, resulting in large, towering cumulus clouds and thunderclouds along the front (figure 16.25). You may have

observed that thunderstorms created by an advancing cold front often form in a line along the front. These thunderstorms can be intense but are usually over quickly, soon followed by a rapid drop in temperature from the cold air mass moving past your location. The passage of the cold front is also marked by a rapid shift in the wind direction and a rapid increase in the barometric pressure. Before the cold front arrives, the wind is generally moving toward the front as warm, less dense air is forced upward by the cold, more dense air. The lowest barometric pressure reading is associated with the lifting of the warm air at the front. After the front passes your location, you are in the cooler,

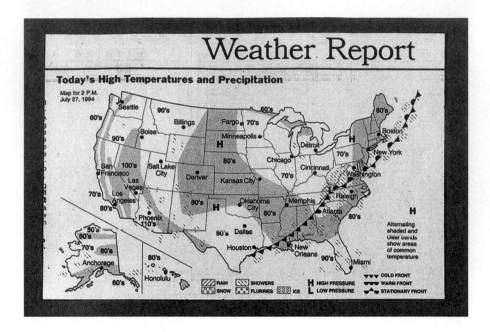

Figure 16.23
This weather map of the United States shows a cold front running from Houston, Texas to near Raleigh, North Carolina, where it becomes a stationary front that runs in a northeasterly direction. Note the areas of showers and the temperature predictions.

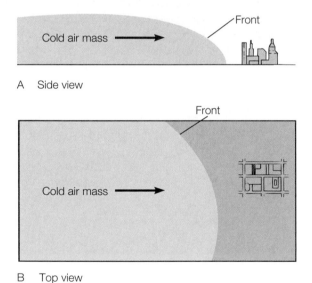

Figure 16.24
(*A*) A cold air mass is similar to a huge, flattened bubble of cold air that moves across the land. The front is the boundary between two air masses, a narrow transition zone of mixing. (*B*) A front is represented by a line on a weather map, which shows the location of the front at ground level.

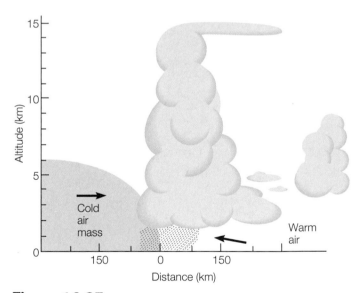

Figure 16.25
An idealized cold front, showing the types of clouds that might occur when an unstable cold air mass moves through unstable warm air. Stable air would result in more stratus clouds rather than cumulus clouds.

more dense air that is settling outward, so the barometric pressure increases and the wind shifts with the movement of the cold air mass.

A **warm front** forms when a warm air mass advances over a mass of cooler air. Since the advancing warm air is less dense than the cooler air it is displacing, it generally overrides the cooler air, forming a long, gently sloping front. Because of this, the overriding warm air may form clouds far in advance of the ground-level base of the front (figure 16.26). This may produce

high cirrus clouds a day or more in advance of the front, which are followed by thicker and lower stratus clouds as the front advances. Usually these clouds result in a broad band of drizzle, fog, and the continuous light rain associated with stratus clouds. This light rain (and snow in the winter) may last for days as the warm front passes.

Sometimes the forces influencing the movement of a cold or warm air mass lessen or become balanced, and the front stops advancing. When this happens a stream of cold air moves along the north side of the front, and a stream of warm air moves

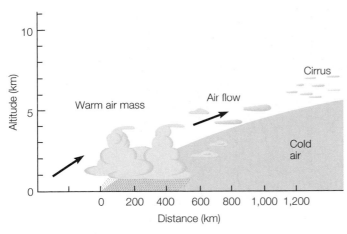

Figure 16.26

An idealized warm front, showing a warm air mass overriding and pushing cold air in front of it. Notice that the overriding warm air produces a predictable sequence of clouds far in advance of the moving front.

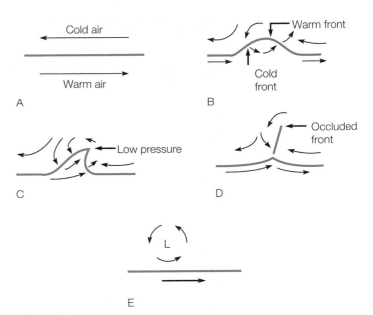

Figure 16.27

The development of a low-pressure center, or cyclonic storm, along a stationary front as seen from above. (*A*) A stationary front with cold air on the north side and warm air on the south side. (*B*) A wave develops, producing a warm front moving northward on the right side and a cold front moving southward on the left side. (*C*) The cold front lifts the warm front off the surface at the apex, forming a low-pressure center. (*D*) When the warm front is completely lifted off the surface, an occluded front is formed. (*E*) The cyclonic storm is now a fully developed low-pressure center.

along the south side in an opposite direction. This is called a **stationary front** because the edge of the front is not advancing. A stationary front may sound as if it is a mild frontal weather maker because it is not moving. Actually, a stationary front represents an unstable situation that can result in a major atmospheric storm. This type of storm is discussed in the following section.

Waves and Cyclones

A slowly advancing cold front and a stationary front often develop a bulge, or *wave*, in the boundary between cool and warm air moving in opposite directions (figure 16.27B). The wave grows as the moving air is deflected, forming a warm front moving northward on the right side and a cold front moving southward on the left side. Cold air is more dense than warm air, and the cold air moves faster than the slowly moving warm front. As the faster moving cold air catches up with the slower moving warm air, the cold air underrides the warm air, lifting it upward. This lifting action produces a low-pressure area at the point where the two fronts come together (figure 16.27C). The lifted air expands, cools, and reaches the dew point. Clouds form and precipitation begins from the lifting and cooling action. Within days after the wave first appears the cold front completely overtakes the warm front, forming an occlusion (figure 16.27D). An **occluded front** is one that has been lifted completely off the ground into the atmosphere. The disturbance is now a *cyclonic storm* with a fully developed low-pressure center. Since its formation, this low-pressure cyclonic storm has been moving, taking its associated stormy weather with it in a generally easterly direction. Such cyclonic storms usually follow principal tracks along a front. Since they are observed generally to follow these same tracks, it is possible to predict where the storm might move next.

A *cyclone* is defined as a low-pressure center where the winds move into the low-pressure center and are forced upward. As air moves in toward the center, the Coriolis effect and friction with the ground cause the moving air to change direction. In the Northern Hemisphere this produces a counterclockwise circulation pattern of winds around the low-pressure center (figure 16.28). The upward movement associated with the low-pressure center of a cyclone cools the air, resulting in clouds, precipitation, and stormy conditions.

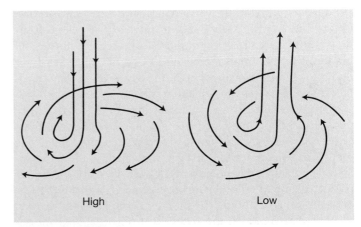

Figure 16.28

Air sinks over a high-pressure center and spreads over the surface in a clockwise pattern in the Northern Hemisphere. Air over the surface moves toward a low-pressure center in a counterclockwise pattern in the Northern Hemisphere.

Air is sinking in the center of a region of high pressure, producing winds that move outward. In the Northern Hemisphere, the Coriolis effect and frictional forces deflect this wind to the right, producing a clockwise circulation (figure 16.28). A high-pressure center is called an *anticyclone*, or simply a *high*. Since air in a high-pressure zone sinks, it is warmed, and the relative humidity is lowered. Thus, clear, fair weather is usually associated with a high. By observing the barometric pressure, you can watch for decreasing pressure, which can mean the coming of a cyclone and its associated stormy weather. You can also watch for increasing pressure, which means a high and its associated fair weather are coming. Consulting a daily weather map makes such projections a much easier job, however.

Major Storms

A wide range of weather changes can take place as a front passes, because there is a wide range of possible temperature, moisture, stability, and other conditions between the new air mass and the air mass that it is displacing. The changes that accompany some fronts may be so mild that they go unnoticed. Others are noticed only as a day with breezes or gusty winds. Still other fronts are accompanied by a rapid and violent weather change called a **storm.** A snowstorm, for example, is a rapid weather change that may happen as a cyclonic storm moves over a location. The most rapid and violent changes occur with three kinds of major storms: (1) thunderstorms, (2) tornadoes, and (3) hurricanes.

Thunderstorms

A **thunderstorm** is a brief but intense storm with rain, lightning and thunder, gusty and often strong winds, and sometimes hail. Thunderstorms usually develop in warm, very moist, and unstable air. These conditions set the stage for a thunderstorm to develop when something lifts a parcel of air, starting it moving upward. This is usually accomplished by the same three general causes that produce cumulus clouds: (1) differential heating, (2) mountain barriers, or (3) along an occluded or cold front. Thunderstorms that occur from differential heating usually occur during warm, humid afternoons after the sun has had time to establish convective thermals. In the Northern Hemisphere most of these convective thunderstorms occur during the month of July. Frontal thunderstorms, on the other hand, can occur any month and any time of the day or night that a front moves through warm, moist, and unstable air.

Frontal thunderstorms generally move with the front that produced them. Thunderstorms that developed in mountains or over flat lands from differential heating can move miles after they form, sometimes appearing to wander aimlessly across the land. These storms are not just one big rain cloud but are sometimes made up of cells that are born, grow to maturity, then die out in less than an hour. The thunderstorm, however, may last longer than an hour because new cells are formed as old ones die out. Each cell is about 2 to 8 km (about 1 to 5 mi) in diameter and goes through three main stages in its life: (1) cumulus, (2) mature, and (3) final (figure 16.29).

Damage from a thunderstorm is usually caused by the associated lightning, strong winds, or hail. As illustrated in figure

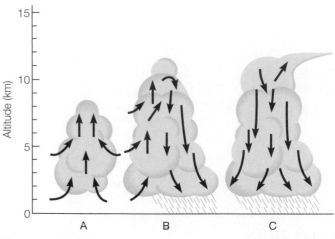

Figure 16.29

Three stages in the life of a thunderstorm cell. (*A*) The cumulus stage begins as warm, moist air is lifted in an unstable atmosphere. All the air movement is upward in this stage. (*B*) The mature stage begins when precipitation reaches the ground. This stage has updrafts and downdrafts side by side, which create violent turbulence. (*C*) The final stage begins when all the updrafts have been cut off, and only downdrafts exist. This cuts off the supply of moisture, and the rain decreases as the thunderstorm dissipates. The anvil-shaped top is a characteristic sign of this stage.

16.29, the first stage of a thunderstorm begins as convection, mountains, or a dense air mass slightly lifts a mass of warm, moist air in an unstable atmosphere. The lifted air mass expands and cools to the dew point temperature, and a cumulus cloud forms. The latent heat of vaporization released by the condensation process accelerates the upward air motion, called an *updraft,* and the cumulus cloud continues to grow to towering heights. Soon the upward-moving, saturated air reaches the freezing level and ice crystals and snowflakes begin to form. When they become too large to be supported by the updraft, they begin to fall toward the surface, melting into raindrops in the warmer air they fall through. When they reach the surface, this marks the beginning of the mature stage. As the raindrops fall through the air, friction between the falling drops and the cool air produces a downdraft in the region of the precipitation. The cool air accelerates toward the surface at speeds up to 90 km/h (about 55 mi/h), spreading out on the ground when it reaches the surface. In regions where dust is raised by the winds, this spreading mass of cold air from the thunderstorm has the appearance of a small cold front with a steep, bulging leading edge. This miniature cold front may play a role in lifting other masses of warm, moist air in front of the thunderstorm, leading to the development of new cells. This stage in the life of a thunderstorm has the most intense rainfall, winds, and possibly hail. As the downdraft spreads throughout the cloud, the supply of new moisture from the updrafts is cut off and the thunderstorm enters the final, dissipating stage. The entire life cycle, from cumulus cloud to the final stage, lasts for about an hour as the thunderstorm moves across the surface. During the mature stage of powerful updrafts, the top of the thunderstorm may reach all the way to the top of the troposphere, forming a cirrus

Figure 16.30
Different parts of a thunderstorm cloud develop centers of electric charge. Lightning is a giant electric spark that discharges the accumulated charges.

Figure 16.31
These hailstones fell from a thunderstorm in Iowa, damaging automobiles, structures, and crops.
© Telegraph Herald/Photo by Patti Carr

cloud that is spread into an anvil shape by the strong winds at this high altitude.

The updrafts, downdrafts, and falling precipitation separate tremendous amounts of electric charges that accumulate in different parts of the thundercloud. Large drops of water tend to carry negative charges, and cloud droplets tend to lose them. The upper part of the thunderstorm develops an accumulation of positive charges as cloud droplets are uplifted, and the middle portion develops an accumulation of negative charges from larger drops that fall. The voltage of these charge centers builds to the point that the electrical insulating ability of the air between them is overcome and a giant electrical discharge called *lightning* occurs (figure 16.30). Lightning discharges occur from the cloud to the ground, from the ground to a cloud, from one part of the cloud to another part, or between two different clouds. The discharge takes place in a fraction of a second and may actually consist of a number of strokes rather than one big discharge. The discharge produces an extremely high temperature around the channel, which may be only 6 cm or so wide (about 2 in). The air it travels through is heated quickly, expanding into a sudden pressure wave that you hear as *thunder*. A nearby lightning strike produces a single, loud crack. Farther away strikes sound more like a rumbling boom as the sound from the separate strokes become separated over distance. Echoing of the thunder produced at farther distances also adds to the rumbling sounds. Lightning can present a risk for people in the open, near bodies of water, or under a single, isolated tree during a thunderstorm. The safest place to be during a thunderstorm is inside a car or a building with a metal frame.

Updrafts are also responsible for **hail,** a frozen form of precipitation that can be very destructive to crops, automobiles, and other property. Hailstones can be irregular, somewhat spherical, or flattened forms of ice that range from the size of a BB to the size of a softball (figure 16.31). Most hailstones, however, are less than 2 cm (about 1 in) in diameter. The larger hailstones have alternating layers of clear and opaque, cloudy ice.

These layers are believed to form as the hailstone goes through cycles of falling then being returned to the upper parts of the thundercloud by updrafts. The clear layers are believed to form as the hailstone moves through heavy layers of supercooled water droplets, which accumulate quickly on the hailstone but freeze slowly because of the release of the latent heat of fusion. The cloudy layers are believed to form as the hailstone accumulates snow crystals or moves through a part of the cloud with less supercooled water droplets. In either case, rapid freezing traps air bubbles, which result in the opaque, cloudy layer. Thunderstorms with hail are most common during the month of May in the states of Colorado, Kansas, and Nebraska.

Tornadoes

A **tornado** is the smallest, most violent weather disturbance that occurs on the earth (figure 16.32). Tornadoes occur with intense thunderstorms, and resemble a long, narrow funnel or ropelike structure that drops down from a thundercloud that may or may not touch the ground. This ropelike structure is a rapidly whirling column of air, usually 100 to 400 m (about 330 to 1,300 ft) in diameter. An average tornado will travel 6 to 8 km (about 4 to 5 mi) on the ground, lasting only a few minutes. The bottom of the column moves across the ground, sometimes skipping into the air, then back down again at speeds that average about 50 km/h (about 30 mi/h) across the surface. The speed of the whirling air in the column has been estimated to be up to about 480 km/h (about 300 mi/h), but most tornadoes have winds of less than 180 km/h (112 mi/h). The destruction is produced by the powerful winds, the sudden drop in atmospheric pressure that occurs at the center of the funnel, and the debris that is flung through the air like projectiles. A passing tornado sounds like very loud, continuous rumbling thunder with cracking and hissing noises that are punctuated by the crashing of debris projectiles.

Figure 16.32

A tornado might be small, but it is the most violent storm that occurs on the earth. This tornado, moving across an open road, eventually struck Dallas, Texas.

On the average, several hundred tornadoes are reported in the United States every year. These occur mostly during spring and early summer afternoons over the Great Plains states. The states of Texas, Oklahoma, Kansas, and Iowa have such a high occurrence of tornadoes that the region is called "tornado alley." During the spring and early summer, this region has maritime tropical air from the Gulf of Mexico at the surface. Above this warm, moist layer is a layer of dry, unstable air that has just crossed the Rocky Mountains, moved along rapidly by the jet stream. The stage is now set for some event, such as a cold air mass moving in from the north, to shove the warm, moist air upward, and the result will be violent thunderstorms with tornadoes.

Hurricanes

What is the difference between a tropical depression, tropical storm, and a hurricane? In general, they are all storms with strong upward atmospheric motion and a cyclonic surface wind circulation (figure 16.33). They are born over tropical or sub-

Connections . . .

Tornado Damage

Tornadoes are rated on wind speed and damage. Here is the scale with approximate wind speeds:

0: Light damage, winds under 120 km/h (75 mi/h). Damage to chimneys, tree limbs broken, small trees pushed over, signs damaged.

1: Moderate damage, winds 121 to 180 km/h (76 to 112 mi/h). Roofing materials removed, mobile homes and moving autos pushed around or overturned.

2: Considerable damage, winds 181 to 250 km/h (113 to 155 mi/h). Roofs torn off homes, mobile homes demolished, boxcars overturned, large trees snapped or uprooted, light objects become missiles.

3: Severe damage, winds 251 to 330 km/h (156 to 205 mi/h). Roofs and some walls torn off homes, whole trains overturned, most trees uprooted, cars lifted off ground and thrown.

4: Devastating damage, winds 331 to 418 km/h (206 to 260 mi/h). Homes leveled, cars thrown, large missiles generated.

5: Incredible damage, winds 419 to 512 km/h (261 to 318 mi/h). Homes demolished and swept away, automobile-sized missiles fly through the air more than 100 yards, trees debarked.

tropical waters and are not associated with a weather front. The varieties of storm intensities are classified according to the *speed* of the maximum sustained surface winds.

A *tropical depression* is an area of low pressure around which the winds are generally moving 55 km/h (about 35 mi/h) or less. The tropical depression might dissolve into nothing or it might develop into a more intense disturbance. A *tropical storm* is a more intense low pressure area with winds between 56 and 120 km/h (about 35 to 75 mi/h). A **hurricane** is a very intense low-pressure area with winds greater than 120 km/h (about 75 mi/h). A strong storm of this type is called a hurricane if it occurs over the Atlantic Ocean or the Pacific Ocean east of the international date line. It is called a **typhoon** if it occurs over the North Pacific Ocean west of the international date line.

A tropical cyclone is similar to the wave cyclone of the mid-latitudes because both have low-pressure centers with a counterclockwise circulation in the Northern Hemisphere. They are different because a wave cyclone is usually about 2,500 km (about 1,500 mi) wide, has moderate winds, and receives its energy from the temperature differences between two air masses. A tropical cyclone, on the other hand, is often less than 200 km (about 125 mi) wide, has very strong winds, and receives its energy from the latent heat of vaporization released during condensation.

A fully developed hurricane has heavy bands of clouds, showers, and thunderstorms that rapidly rotate around a relatively clear, calm eye (figure 16.34). As a hurricane approaches a location the air seems unusually calm as a few clouds appear,

Figure 16.33

This is a satellite photo of hurricane John, showing the eye and counterclockwise motion.

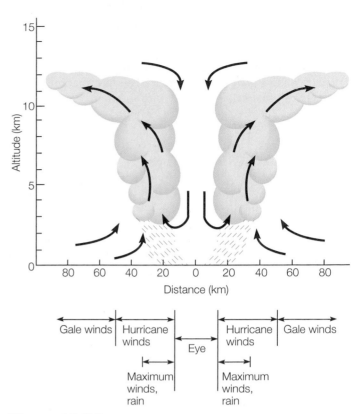

Figure 16.34

Cross section of a hurricane.

then thicken as the wind begins to gust. Over the next six hours or so the overall wind speed increases as strong gusts and intense rain showers occur. Thunderstorms, perhaps with tornadoes, and the strongest winds occur just before the winds suddenly die down and the sky clears with the arrival of the eye of the hurricane. The eye is an average of 10 to 15 km (about 6 to 9 mi) across, and it takes about an hour or so to cross a location. When the eye passes, the intense rain showers, thunderstorms, and hurricane-speed winds begin again, this time blowing from the opposite direction. The whole sequence of events may be over in a day or two, but hurricanes are unpredictable and sometimes stall in one location for days. In general, they move at a rate of 15 to 50 km/h (about 10 to 30 mi/h).

Most of the damage from hurricanes results from strong winds, flooding, and the occasional tornado. Flooding occurs from the intense, heavy rainfall but also from the increased sea level that results from the strong, constant winds blowing seawater toward the shore. The sea level can be raised some 5 m (about 16 ft) above normal, with storm waves up to 15 m (about

50 ft) high on top of this elevated sea level. Overall, large inland areas can be flooded with extensive property damage. A single hurricane moving into a populated coastal region has caused billions of dollars of damage and the loss of hundreds of lives in the past. Today, the National Weather Service tracks hurricanes by weather satellites. Warnings of hurricanes, tornadoes, and severe thunderstorms are broadcast locally over special weather alert stations located across the country.

Connections . . .

Hurricane Damage

Hurricanes are classified according to category and damage to be expected. Here is the classification scheme:

Category	Damage	Winds	
1	minimal	120–153 km/h	(75–95 mi/h)
2	moderate	154–177 km/h	(96–110 mi/h)
3	extensive	178–210 km/h	(111–130 mi/h)
4	extreme	211–250 km/h	(131–155 mi/h)
5	catastrophic	>250 km/h	(>155 mi/h)

WEATHER FORECASTING

Today, weather predictions are based on information about the characteristics, location, and rate of movement of air masses and associated fronts and pressure systems. This information is summarized as average values, then fed into a computer model of the atmosphere. The model is a scaled-down replica of the real atmosphere, and changes in one part of the model result in changes in another part of the model just as they do in the real atmosphere. Underlying the computer model are the basic scientific laws concerning solar radiation, heat, motion, and the gas laws. All these laws are written as a series of mathematical equations, which are applied to thousands of data points in a three-dimensional grid that represents the atmosphere. The computer is given instructions about the starting conditions at each data point, that is, the average values of temperature, atmospheric pressure, humidity, wind speed, and so forth. The computer is then instructed to calculate the changes that will take place at each data point, according to the scientific laws, within a very short period of time. This requires billions of mathematical calculations when the program is run on a worldwide basis. The new calculated values are then used to start the process all over again, and it is repeated some 150 times to obtain a one-day forecast (figure 16.35).

A problem with the computer model of the atmosphere is that small-scale events are inadequately treated, and this introduces errors that grow when predictions are attempted for farther and farther into the future. Small eddies of air, for example, or gusts of wind in a region have an impact on larger-scale atmospheric motions such as those larger than a cumulus cloud. But all of the small eddies and gusts cannot be observed without filling the atmosphere with measuring instruments. This lack of ability to observe small events that can change the large-scale events introduces uncertainties in the data, which, over time, will increasingly affect the validity of a forecast.

To find information about the accuracy of a forecast the computer model can be run several different times, with each run having slightly different initial conditions. If the results of all the runs are close to each other the forecasters can feel confident that the atmosphere is in a predictable condition, and this means the forecast is probably accurate. In addition, multiple computer runs can provide forecasts in the form of probabilities. For example, if 8 out of 10 forecasts indicate rain, the "averaged" forecast might call for an 80 percent chance of rain.

The use of new computer technology has improved the accuracy of next-day forecasts tremendously, and the forecasts up to three days are fairly accurate, too. For forecasts of more than five days, however, the number of calculations and the effect of uncertainties increase greatly. It has been estimated that the reductions of observational errors could increase the range of accurate forecasting up to two weeks. The ultimate range of accurate forecasting will require a better understanding—and thus an improved model—of patterns of changes that occur in the ocean as well as in the atmosphere. All of this increased understanding and reduction of errors leads to an estimated ultimate future forecast of three weeks, beyond which any pinpoint forecast would be only slightly better than a wild guess. In the meantime, regional and local daily weather forecasts are fairly accurate, and computer models of the atmosphere now provide the basis for extending the forecasts for up to about a week.

Figure 16.35

Supercomputers make routine weather forecasts possible by solving mathematical equations that describe changes in a mathematical model of the atmosphere. This "fish-eye" view was necessary to show all of this Cray supercomputer.

Connections . . .

Cities Change the Weather

Using satellite measurements, NASA researchers have confirmed that the presence of a city does influence the weather. First, because of all the concrete, asphalt, and buildings, cities are 0.6°C to 5.6°C (about 1°F to 10°F) warmer than the surrounding country. The added heat of this "urban heat island" tends to make the air less stable. Less stable air brings about the second influence of more rain falling downwind of a city. Evidently the less stable air and rougher surfaces cause air moving toward a city to rise, and the turbulence mixes in city pollutants that add nuclei for water vapor to condense upon. The result is an average 28 percent increase in the amount of rainfall 30 to 60 km (about 19 to 37 mi) downwind of cities. The NASA research team satellite measurements were verified by data from a large array of ground-based thermometers and rain gauges. The verification of these satellite-based findings are important for urban planning, water resource management, and decisions about where to farm the land. It may also show that local surface environments are more important in computer weather forecast models than had previously been believed.

CLIMATE

Changes in the atmospheric condition over a brief period of time, such as a day or a week, are referred to as changes in the *weather*. These changes are part of a composite, larger pattern called **climate.** Climate is the general pattern of the weather that occurs for a region over a number of years. Among other things, the climate determines what types of vegetation grow in a particular region, resulting in characteristic groups of plants associated with the region (figure 16.36). For example, orange, grapefruit, and palm trees grow in a region that has a climate with warm monthly temperatures throughout the year. On the other hand, blueberries, aspen, and birch trees grow in a region that has cool temperature patterns throughout the year. Climate determines what types of plants and animals live in a location, the types of houses that people build, and the lifestyles of people. Climate also influences the processes that shape the landscape, the type of soils that form, the suitability of the region for different types of agriculture, and how productive the agriculture will be in a region. This section is about climate, what determines the climate of a region, and how climate patterns are classified.

Major Climate Groups

The earth's atmosphere is heated directly by incoming solar radiation and by absorption of infrared radiation from the surface. The amount of heating at any particular latitude on the surface depends primarily on two factors: (1) the *intensity* of the incoming solar radiation, which is determined by the angle at which the radiation strikes the surface, and (2) the amount of *time* that the radiation is received at the surface, that is, the number of daylight hours compared to the number of hours of night.

The earth is so far from the sun that all rays of incoming solar radiation reaching the earth are essentially parallel. The earth, however, has a mostly constant orientation of its axis with respect to the stars as it moves around the sun in its orbit. Since the inclined axis points toward the sun part of the year and away from the sun the other part, radiation reaches different latitudes at different angles during different parts of the year. The orientation of the earth's axis to the sun during different parts of the year also results in days and nights of nearly equal length in the equatorial region but increasing differences at increasing latitudes to the poles. During the polar winter months, the night is twenty-four hours long, which means no solar radiation is received at all. The equatorial region receives more solar radiation during a year, and the amount received decreases toward the poles as a result of (1) yearly changes in intensity and (2) yearly changes in the number of daylight hours.

In order to generalize about the amount of radiation received at different latitudes, some means of organizing, or grouping, the latitudes is needed (figure 16.37). For this purpose the latitudes are organized into three groups:

1. The *low latitudes,* those that some time of the year receive *vertical* solar radiation at noon.
2. The *high latitudes,* those that some time of the year receive *no* solar radiation at noon.
3. The *middle latitudes,* which are between the low and high latitudes.

This definition of low, middle, and high latitudes means that the low latitudes are between the tropics of Cancer and Capricorn (between 23.5°N and 23.5°S latitudes) and that the high latitudes are above the Arctic and Antarctic circles (above 66.5°N and above 66.5°S latitudes).

In general:

1. The low latitudes receive a high amount of incoming solar radiation that varies little during a year. Temperatures are high throughout the year, varying little from month to month.

Figure 16.36

The climate determines what types of plants and animals live in a location, the types of houses that people build, and the lifestyles of people. This orange tree, for example, requires a climate that is relatively frost-free, yet it requires some cool winter nights to produce a sweet fruit.

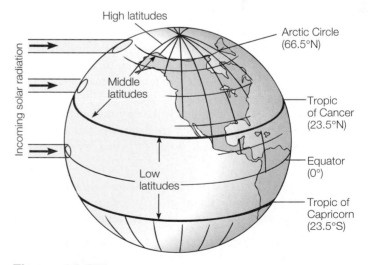

Figure 16.37

Latitude groups based on incoming solar radiation. The low latitudes receive vertical solar radiation at noon some time of the year, the high latitudes receive no solar radiation at noon during some time of the year, and the middle latitudes are in between.

Figure 16.38

The principal climate zones are defined in terms of yearly temperature averages, which are determined by the amount of solar radiation received at the different latitude groups.

2. The middle latitudes receive a higher amount of incoming radiation during one part of the year and a lower amount during the other part. Overall temperatures are cooler than in the low latitudes and have a wide seasonal variation.
3. The high latitudes receive a maximum amount of radiation during one part of the year and none during another part. Overall temperatures are low, with the highest range of annual temperatures.

The low, middle, and high latitudes provide a basic framework for describing the earth's climates. These climates are associated with the low, middle, and high latitudes illustrated in figure 16.37, but they are defined in terms of yearly temperature averages. Temperature and moisture are the two most important climate factors, and temperature will be considered first.

The principal climate zones are defined in terms of yearly temperature averages, which occur in broad regions (figure 16.38). They are

1. The *tropical climate zone* of the low latitudes (figure 16.39).
2. The *polar climate zone* of the high latitudes (figure 16.40).
3. The *temperate climate zone* of the middle latitudes (figure 16.41).

The tropical climate zone is near the equator and receives the greatest amount of sunlight throughout the year. Overall, the tropical climate zone is hot. Average monthly temperatures stay above 18°C (64°F), even during the coldest month of the year.

The other extreme is found in the polar climate zone, where the sun never sets during some summer days and never rises during some winter days. Overall, the polar climate zone is cold.

Average monthly temperatures stay below 10°C (50°F), even during the warmest month of the year.

The temperate climate zone is between the polar and tropical zones, with average temperatures that are neither very cold nor very hot. Average monthly temperatures stay between 10°C and 18°C (50°F and 64°F) throughout the year.

General patterns of precipitation and winds are also associated with the low, middle, and high latitudes. An idealized model of the global atmospheric circulation and pressure patterns was described earlier. Recall that this model described a huge convective movement of air in the low latitudes, with air being forced upward over the equatorial region. This air expands, cools to the dew point, and produces abundant rainfall throughout most of the year. On the other hand, air is slowly sinking over 30°N and 30°S of the equator, becoming warm and dry as it is compressed. Most of the great deserts of the world are near 30°N or 30°S latitude for this reason. There is another wet zone near 60° latitudes and another dry zone near the poles. These wet and dry zones are shifted north and south during the year with the changing seasons. This results in different precipitation patterns in each season. Figure 16.42 shows where the wet and dry zones are in winter and in summer seasons.

Regional Climatic Influence

Latitude determines the basic tropical, temperate, and polar climatic zones, and the wet and dry zones move back and forth over the latitudes with the seasons. If these were the only factors

Figure 16.39

A wide variety of plant life can grow in a tropical climate, as you can see here.

Figure 16.40

Polar climates occur at high elevations as well as high latitudes. This mountain location has a highland polar climate and tundra vegetation, but little else.

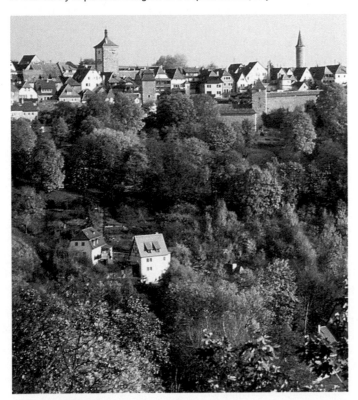

Figure 16.41

This temperate-climate deciduous forest responds to seasonable changes in autumn with a show of color.

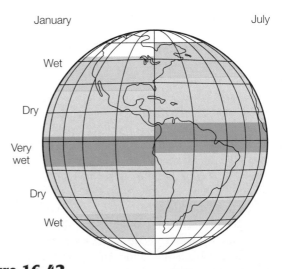

Figure 16.42

The idealized general rainfall patterns over the earth change with seasonal shifts in the wind and pressure areas of the planet's general atmospheric circulation patterns.

influencing the climate, you would expect to find the same climatic conditions at all locations with the same latitude. This is not what is found, however, because there are four major factors that affect a regional climate. These are (1) altitude, (2) mountains, (3) large bodies of water, and (4) ocean currents. The following describes how these four factors modify the climate of a region.

The first of the four regional climate factors is *altitude.* The atmosphere is warmed mostly by the greenhouse effect from the surface upward, and air at higher altitudes increasingly radiates more and more of its energy to space. Average air temperatures therefore decrease with altitude, and locations with higher altitudes will have lower average temperatures. This is why the tops of mountains are often covered with snow when none is found at lower elevations. St. Louis, Missouri and Denver, Colorado are located almost at the same latitude (within 1° of 39°N), so you might expect the two cities to have about the same average temperature. Denver, however, has an altitude of 1,609 m (5,280 ft), and the altitude of St. Louis is 141 m (465 ft). The yearly average temperature for Denver is about 10°C (about 50°F) and for St. Louis it is about 14°C (about 57°F). In general, higher altitude means lower average temperature.

The second of the regional climate factors is the influence of *mountains.* In addition to the temperature change caused by the altitude of the mountain, mountains also affect the conditions of a passing air mass. The western United States has mountainous regions along the coast. When a moist air mass from the Pacific meets these mountains it is forced upward and cools. Water vapor in the moist air mass condenses, clouds form, and the air mass loses much of its moisture as precipitation falls on the western side of the mountains. Air moving down the eastern slope is compressed and becomes warm and dry. As a result the western slopes of these mountains are moist and have forests of spruce, redwood, and fir trees. The eastern slopes are dry and have grassland or desert vegetation.

The third of the regional climate factors is the presence of a large body of *water.* Water, as discussed previously, has a higher specific heat than land material, it is transparent, and it loses energy through evaporation. All of these affect the temperature of a landmass located near a large body of water, making the temperatures more even from day to night and from summer to winter. San Diego, California and Dallas, Texas, for example, are at about the same latitude (both almost 33°N), but San Diego is at a seacoast and Dallas is inland. Because of its nearness to water, San Diego has an average summer temperature about 7°C (about 13°F) cooler and an average winter temperature about 5°C (about 9°F) warmer than the average temperatures in Dallas. Nearness to a large body of water keeps the temperature more even at San Diego.

The fourth of the regional climate factors is *ocean currents.* In addition to the evenness brought about by being near the ocean, currents in the ocean can bring water that has a temperature different from the land. For example, currents can move warm water northward or they can move cool water southward (figure 16.43). This can influence the temperatures of air masses that move from the water to the land and, thus,

Figure 16.43
Ocean currents can move large quantities of warm or cool water to influence the air temperatures of nearby landmasses.

the temperatures of the land. For example, the North Pacific current brings warm waters to the western coast of North America, which results in warmer temperatures for cities near the coast.

Describing the earth's climates presents a problem because there are no sharp boundaries that exist naturally between two adjacent regions with different climates. Even if two adjacent climates are very different, one still blends gradually into the other. For example, if you are driving from one climate zone to another you might drive for miles before becoming aware that the vegetation is now different than it was an hour ago. Since the vegetation is very different from what it was before, you know that you have driven from one regional climate zone to another. Chapter 22, Ecology and Environment, discusses the typical plants and animals associated with the various climate types.

Actually, no two places on the earth have exactly the same climate. Some plants will grow on the north or south side of a building, for example, but not on the other side. The two sides of the building could be considered as small, local climate zones within a larger, major climate zone.

SUMMARY

The earth's *atmosphere* thins rapidly with increasing altitude. Pure, dry air is mostly *nitrogen, oxygen,* and *argon,* with traces of *carbon dioxide* and other gases. Atmospheric air also contains a variable amount of *water vapor.* Water vapor cycles into and out of the atmosphere through evaporation and precipitation.

Atmospheric pressure is measured with a *mercury barometer.* At sea level, the atmospheric pressure will support a column of mercury

The term "El Niño" was originally used to describe an occurrence of warm, above-normal ocean temperatures off the South American coast. Fishermen along this coast learned long ago to expect associated changes in fishing patterns about every 3 to 7 years, which usually lasted for about 18 months. They called this event El Niño, which is Spanish for "the boy child" or "Christ child," because it typically began near Christmas. The El Niño event occurs when the trade winds along the equatorial Pacific become reduced or calm, allowing sea surface temperatures to increase much above normal. The warm water drives the fish to deeper waters or farther away from usual fishing locations.

Today "El Niño" is understood to be much more involved than just a warm ocean in the Pacific. It is more than a local event, and the bigger picture is sometimes called the "El Niño–Southern Oscillation," or ENSO. In addition to the warmer tropical Pacific water of "El Niño," the boy, the term "La Niña," the girl, has been used to refer to the times when the water of the tropical Pacific is *colder* than normal. The "Southern Oscillation" part of the name comes from observations that atmospheric pressure around Australia seems to be inversely linked to the atmospheric pressure in Tahiti. They seem to be linked because when the pressure is low in Australia, it is high in Tahiti. Conversely, when the atmospheric pressure is high in Australia, it is low in Tahiti. The strength of this Southern Oscillation is measured by the Southern Oscillation Index (SOI), which is defined as the pressure at Darwin, Australia, subtracted from that at Tahiti. Negative values of SOI are usually associated with El Niño events, so the Southern Oscillation and the El Niño are obviously linked. How ENSO can impact the weather in other parts of the world has only recently become better understood.

The atmosphere is a system that responds to incoming solar radiation, the spinning earth, and other factors, such as the amount of water vapor present. The ocean and atmospheric systems undergo changes by interacting with each other, most visibly in the tropical cyclone. The ocean system supplies water vapor, latent heat, and condensation nuclei, which are the essential elements of a tropical cyclone as well as everyday weather changes and climate. The atmosphere, on the other hand, drives the ocean with prevailing winds, moving warm or cool water to locations where they affect the climate on the land. There is a complex, interdependent relationship between the ocean and the atmosphere, and it is probable that even small changes in one system can lead to bigger changes in the other.

Normally, during non–El Niño times, the Pacific Ocean along the equator has established systems of prevailing wind belts, pressure systems, and ocean currents. In July, these systems push the surface seawater offshore from South America, westward along the equator and toward Indonesia. During El Niño times the trade winds weaken and the warm water moves back eastward, across the Pacific to South America, where it then spreads north and south along the coast. Why the trade winds weaken and become calm is unknown, the subject of ongoing research.

Warmer waters along the coast of South America bring warmer, more humid air and the increased possibility of thunderstorms. Thus the possibility of towering thunderstorms, tropical storms, or hurricanes increases along the Pacific coast of South America as the warmer waters move north and south. This creates the possibility of weather changes not only along the western coast, but elsewhere, too. The towering thunderstorms reach high into the atmosphere, adding tropical moisture and creating changes in prevailing wind belts. These wind belts carry or steer weather systems across the middle latitudes of North America, so typical storm paths are shifted. This shifting can result in

- increased precipitation in California during the fall to spring season;
- a wet winter with more and stronger storms along regions of the southern United States and Mexico;
- a warmer and drier than average winter across the northern regions of Canada and the United States;
- a variable effect on central regions of the United States, ranging from reduced snowfall to no effect at all; and
- other changes in the worldwide global complex of ocean and weather events, such as droughts in normally wet climates and heavy rainfall in normally dry climates.

One major problem in these predictions is a lack of understanding of what causes many of the links and a lack of consistency in the links themselves. For example, southern California did not always have an unusually wet season every time an El Niño occurred and in fact experienced a drought during one event.

Scientists have continued to study the El Niño since the mid-1980s, searching for patterns that will reveal consistent cause-and-effect links. Part of the problem may be that other factors, such as a volcanic eruption, may influence part of the linkage but not another part. Another part of the problem may be the influence of unknown factors, such as the circulation of water deep beneath the ocean surface, the track taken by tropical cyclones, or the energy released by tropical cyclones one year compared to the next.

The results so far have indicated that atmosphere-ocean interactions are much more complex than early theoretical models had predicted. Sometimes a new model will predict some weather changes that occur with El Niño, but no model is yet consistently correct in predicting the conditions that lead to the event and the weather patterns that result. All this may someday lead to a better understanding of how the ocean and the atmosphere interact on this dynamic planet.

Recent years in which El Niño events have occurred are 1951, 1953, 1957–1958, 1965, 1969, 1972–1973, 1976, 1982–1983, 1986–1987, 1991–1992, 1994, 1997–1998, and 2002–2003.

---

about 76.00 cm (about 29.92 in) tall. This is the average pressure at sea level, and it is called the *standard atmospheric pressure, normal pressure,* or *one atmosphere of pressure.*

Materials on the earth's surface absorb sunlight, emitting more and more *infrared radiation* as they are warmed. *Carbon dioxide* and *molecules* of water vapor in the atmosphere absorb infrared radiation, which then reemit the energy many times before it reaches outer space again. The overall effect warms the lower atmosphere from the bottom up in a process called the *greenhouse effect.*

The layer of the atmosphere from the surface up to where the temperature stops decreasing with height is called the *troposphere.* The *stratosphere* is the layer above the troposphere. Temperatures in the stratosphere increase because of the interaction between ozone (O_3) and ultraviolet radiation from the sun.

The surface of the earth is not heated uniformly by sunlight. This results in a *differential heating,* which sets the stage for *convection.* The horizontal movement of air on the surface from convection is called *wind.* A generalized model for understanding why the wind blows involves (1) the relationship between *air temperature and air density* and (2) the relationship between *air pressure and the movement of air.* This model explains local wind patterns and wind patterns observed on a global scale.

The amount of water vapor in the air at a particular time is called the *absolute humidity.* The *relative humidity* is a ratio between the amount of water vapor that is in the air and the amount needed to saturate the air at the present temperature.

When the air is saturated, condensation can take place. The temperature at which this occurs is called the *dew point temperature.* If the dew point temperature is above freezing, *dew* will form. If the temperature is below freezing, *frost* will form. Both dew and frost form directly on objects and do not fall from the air.

Water vapor condenses in the air on *condensation nuclei.* If this happens near the ground, the accumulation of tiny water droplets is called a *fog. Clouds* are accumulations of tiny water droplets in the air above the ground. In general, there are three basic shapes of clouds, *cirrus, cumulus,* and *stratus.* These basic cloud shapes have meaning about the atmospheric conditions and about the coming weather conditions.

Water that returns to the earth in liquid or solid form falls from the clouds as *precipitation.* Precipitation forms in clouds through two processes: (1) the *coalescence* of cloud droplets or (2) the *growth of ice crystals* at the expense of water droplets.

Weather changes are associated with the movement of large bodies of air called *air masses,* the leading *fronts* of air masses when they move, and local *high- and low-pressure* patterns that accompany air masses or fronts. Examples of air masses include (1) *continental polar,* (2) *maritime polar,* (3) *continental tropical,* and (4) *maritime tropical.*

When a location is under the influence of an air mass the location is having *air mass weather* with slow, gradual changes. More rapid changes take place when the *front,* a thin transition zone between two air masses, passes a location.

A *stationary front* often develops a bulge, or *wave,* that forms into a moving cold front and a moving warm front. The faster moving cold front overtakes the warm front, lifting it into the air to form an *occluded front.* The lifting process forms a low-pressure center called a *cyclone.* Cyclones are associated with heavy clouds, precipitation, and stormy conditions because of the lifting action.

A *thunderstorm* is a brief, intense storm with rain, lightning and thunder, gusty and strong winds, and sometimes hail. A *tornado* is the smallest, most violent weather disturbance that occurs on the earth. A *hurricane* is a *tropical cyclone,* a large, violent circular storm that is born over warm tropical waters near the equator.

The general pattern of the weather that occurs for a region over a number of years is called *climate.* The three principal climate zones are (1) the *tropical climate zone,* (2) the *polar climate zone,* and (3) the *temperate climate zone.* The climate in these zones is influenced by four factors that determine the local climate: (1) *altitude,* (2) *mountains,* (3) *large bodies of water,* and (4) *ocean currents.* The climate for a given location is described by first considering the principal climate zone, then looking at subdivisions within each that result from local influences.

KEY TERMS

absolute humidity (p. **329**)
air mass (p. **335**)
barometer (p. **322**)
climate (p. **345**)
cold front (p. **337**)
condensation nuclei (p. **331**)
dew (p. **331**)
dew point temperature (p. **331**)
front (p. **336**)
frost (p. **331**)
greenhouse effect (p. **323**)
hail (p. **341**)
humidity (p. **329**)
hurricane (p. **342**)
ice-forming nuclei (p. **335**)
inversion (p. **323**)

jet stream (p. **328**)
occluded front (p. **339**)
ozone shield (p. **324**)
precipitation (p. **333**)
relative humidity (p. **330**)
saturated air (p. **329**)
standard atmospheric pressure (p. **322**)
stationary front (p. **339**)
storm (p. **340**)
thunderstorm (p. **340**)
tornado (p. **341**)
typhoon (p. **342**)
warm front (p. **338**)
wind (p. **324**)
wind chill factor (p. **326**)

APPLYING THE CONCEPTS

1. Without adding or removing any water vapor, a sample of air experiencing an increase in temperature will have
   (a) a higher relative humidity.
   (b) a lower relative humidity.
   (c) the same relative humidity.
   (d) a changed absolute humidity.

2. Cooling a sample of air results in a (an)
   (a) increased capacity to hold water vapor.
   (b) decreased capacity to hold water vapor.
   (c) unchanged capacity to hold water vapor.

3. On a clear, calm, and cool night, dew or frost is most likely to form
   (a) under trees or other shelters.
   (b) on bare ground on the side of a hill.
   (c) under a tree on the side of a hill.
   (d) on grass in an open, low-lying area.

4. Longer periods of drizzle, rain, or snow usually occur from which basic form of a cloud?
   (a) stratus
   (b) cumulus
   (c) cirrus
   (d) None of the above are correct.

5. Brief periods of showers are usually associated with which type of cloud?
   (a) stratus
   (b) cumulus
   (c) cirrus
   (d) None of the above are correct.

6. The type of air mass weather that results from the arrival of polar continental air is
   (a) frequent snowstorms with rapid changes.
   (b) clear and cold with gradual changes.
   (c) unpredictable, but with frequent and rapid changes.
   (d) much the same from day to day, the conditions depending on the air mass and the local conditions.

7. The appearance of high cirrus clouds followed by thicker, lower stratus clouds, and then continuous light rain over several days probably means which of the following air masses has moved to your area?
   (a) continental polar
   (b) maritime tropical
   (c) continental tropical
   (d) maritime polar

8. A fully developed cyclonic storm is most likely to form
   (a) on a stationary front.
   (b) in a high-pressure center.
   (c) from differential heating.
   (d) over a cool ocean.

9. The basic difference between a tropical storm and a hurricane is
   (a) size.
   (b) location.
   (c) wind speed.
   (d) amount of precipitation.

10. Most of the great deserts of the world are located
    (a) near the equator.
    (b) 30° north or south latitude.
    (c) 60° north or south latitude.
    (d) anywhere, as there is no pattern to their location.

11. The average temperature of a location is made more even by the influence of
    (a) a large body of water.
    (b) elevation.
    (c) nearby mountains.
    (d) dry air.

12. The climate of a specific location is determined by
    (a) its latitude.
    (b) how much sunlight it receives.
    (c) its altitude and nearby mountains and bodies of water.
    (d) all of the above.

Answers

1. b 2. b 3. d 4. a 5. b 6. d 7. b 8. a 9. c 10. b 11. a 12. d

QUESTIONS FOR THOUGHT

1. Explain why frost is more likely to form on a clear, calm, and cool night than on nights with other conditions.
2. What is a cloud? Describe how a cloud forms.
3. Describe two ways that precipitation may form from the water droplets of a cloud.
4. What is an air mass?
5. What kinds of clouds and weather changes are usually associated with the passing of (a) a warm front? (b) a cold front?
6. Describe the wind direction, pressure, and weather conditions that are usually associated with (a) low-pressure centers and (b) high-pressure centers.
7. In which of the four basic types of air masses would you expect to find afternoon thunderstorms? Explain.
8. Describe the three main stages in the life of a thunderstorm cell, identifying the events that mark the beginning and end of each stage.
9. What is a tornado? When and where do tornadoes usually form?
10. What is a hurricane? Describe how the weather conditions change as a hurricane approaches, passes directly over, then moves away from a location.
11. Describe the average conditions found in the three principal climate zones.
12. Identify the four major factors that influence the climate of a region and explain how each does its influencing.

Earth's Waters

The earth's vast oceans cover more than 70 percent of the surface of the earth.
Freshwater is generally abundant on the land because the supply is replenished from ocean waters through the hydrologic cycle.

OUTLINE

Throughout history humans have diverted rivers and reshaped the land to ensure a supply of freshwater. There is evidence, for example, that ancient civilizations along the Nile River diverted water for storage and irrigation some five thousand years ago. The ancient Greeks and Romans built systems of aqueducts to divert streams to their cities some two thousand years ago. Some of these aqueducts are still standing today. More recent water diversion activities were responsible for the name of Phoenix, Arizona. Phoenix was named after a mythical bird that arose from its ashes after being consumed by fire. The city was given this name because it is built on a system of canals that were first designed and constructed by ancient Native Americans, then abandoned hundreds of years before settlers reconstructed the ancient canal system (figure 17.1). Water is and always has been an essential resource.

Where water is in short supply, humans have historically turned to extensive diversion and supply projects to meet their needs.

Precipitation is the basic source of the water supply found today in streams, lakes, and beneath the earth's surface. Much of the precipitation that falls on the land, however, evaporates back into the atmosphere before it has a chance to become a part of this supply. The water that does not evaporate mostly moves directly to rivers and streams, flowing back to the ocean, but some soaks into the land. The evaporation of water, condensation of water vapor, and the precipitation-making processes were introduced in chapter 16 as important weather elements. They are also part of the generalized *hydrologic cycle* of evaporation from the ocean, transport through the atmosphere by moving air masses, precipitation on the land, and movement of water back to the

ocean. Only part of this cycle was considered previously, however, and this was the part from evaporation through precipitation. This chapter is concerned with the other parts of the hydrologic cycle, that is, what happens to the water that falls on the land and makes it back to the ocean. It begins with a discussion of how water is distributed on the earth and a more detailed look at the hydrologic cycle. Then the travels of water across and into the land will be considered as streams, wells, springs, and other sources of usable water are discussed as limited resources. The tracing of the hydrologic cycle will be completed as the water finally makes it back to the ocean. This last part of the cycle will consider the nature of the ocean floor, the properties of seawater, and how waves and currents are generated. The water is now ready to evaporate, starting another one of earth's never-ending cycles.

WATER ON THE EARTH

Some water is tied up in chemical bonds deep in the earth's interior, but free water is the most abundant chemical compound near the surface. Water is five or six times more abundant than the most abundant mineral in the outer 6 km (about 4 mi) of the earth, so it should be no surprise that water covers about 70 percent of the surface. On the average, about 98 percent of this water exists in the liquid state in depressions on the surface and in sediments. Of the remainder, about 2 percent exists in the solid state as snow and ice on the surface in colder locations. Only a fraction of a percent exists as a variable amount of water vapor in the atmosphere at a given time. Water is continually moving back and forth between these "reservoirs," but the percentage found in each is assumed to be essentially constant.

As shown in figure 17.2, over 97 percent of the earth's water is stored in the earth's oceans. This water contains a relatively high level of dissolved salts, which will be discussed in a later section. These dissolved salts make ocean water unfit for human consumption and unfit for most agricultural purposes. All other water, which is fit for human consumption and agriculture, is called **freshwater.** About two-thirds of the earth's freshwater supply is locked up in the ice caps of Greenland and the Antarctic and in glaciers. This leaves less than 1 percent of all the water found on the earth as available freshwater. There is a generally abundant supply, however, because the freshwater supply is continually replenished by the hydrologic cycle.

Evaporation of water from the ocean is an important process of the hydrologic cycle because (1) water vapor leaves the dissolved salts behind, forming precipitation that is freshwater, and (2) the gaseous water vapor is easily transported in the atmosphere from one part of the earth to another. Over a year this natural desalination process produces and transports enough freshwater to cover the entire earth with a layer about

Figure 17.1

This is one of the water canals of the present-day system in Phoenix, Arizona. These canals were reconstructed from a system that was built by American Indians, then abandoned. Phoenix is named after a mythical bird that was consumed by fire and then arose from its ashes.

85 cm (about 33 in) deep. Precipitation is not evenly distributed like this, of course, and some places receive much more, while other places receive almost none. Considering global averages, more water is evaporated from the ocean than returns directly to it by precipitation. On the other hand, more water is precipitated over the land than evaporates from the land surface back to the atmosphere. The net amount evaporated and precipitated over the land and over the ocean is balanced by the return of water to the ocean by streams and rivers. This cycle of evaporation, precipitation, and return of water to the oceans is known as the

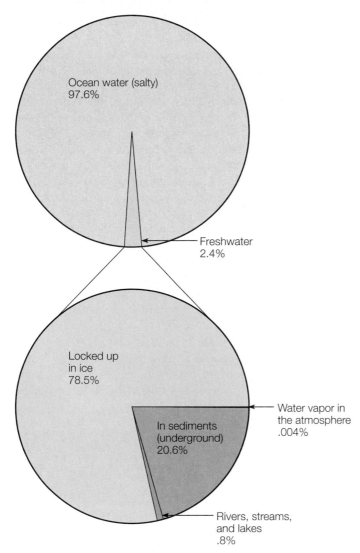

Figure 17.2
Estimates of the distribution of all the water found on the earth's surface.

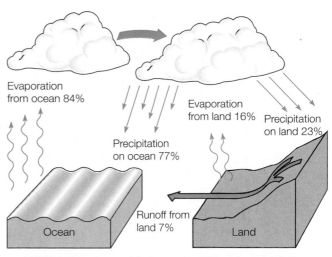

100% is based on a global average of 85 cm/yr precipitation.

Figure 17.3
On the average, more water is evaporated from the ocean than is returned by precipitation. More water is precipitated over the land than evaporates. The difference is returned to the ocean by rivers and streams.

hydrologic cycle (figure 17.3). This freshwater returning on and under the land is the source of freshwater. What happens to the freshwater during its return to the ocean is discussed in the following section.

Freshwater

The basic source of freshwater is precipitation, but not all precipitation ends up as part of the freshwater supply. Liquid water is always evaporating, even as it falls. In arid climates, rain sometimes evaporates completely before reaching the surface, even from a fully developed thunderstorm. Evaporation continues from the water that does reach the surface. Puddles and standing water on the hard surface of city parking lots and streets, for example, gradually evaporate back to the atmosphere after a rain and the surface is soon dry. There are many factors that determine how much of a particular rainfall evaporates, but in general more than two-thirds of the rain eventually returns to the atmosphere. The remaining amount either (1) flows downhill

across the surface of the land toward a lower place or (2) soaks into the ground. Water moving across the surface is called **runoff.** Runoff begins as rain accumulates in thin sheets of water that move across the surface of the land. These sheets collect into a small body of running water called a *stream*. A stream is defined as any body of water that is moving across the land, from one so small that you could step across it to the widest river. Water that soaks into the ground moves downward to a saturated zone and is now called *groundwater*. Groundwater moves through sediments and rocks beneath the surface, slowly moving downhill. Streams carry the runoff of a recent rainfall or melting snow, but otherwise most of the flow comes from groundwater that seeps into the stream channel. This explains how a permanent stream is able to continue flowing when it is not being fed by runoff or melting snow (figure 17.4). Where or when the source of groundwater is in low supply a stream may flow only part of the time, and it is designated as an *intermittent stream*.

Figure 17.4
Some of the precipitation soaks into the ground to become groundwater. Groundwater slowly moves underground, and some of it emerges in streambeds keeping the streams running during dry spells.

The amount of a rainfall that becomes runoff or groundwater depends on a number of factors, including (1) the type of soil on the surface, (2) how dry the soil is, (3) the amount and type of vegetation, (4) the steepness of the slope, and (5) if the rainfall is a long, gentle one or a cloudburst. Different combinations of these factors can result in from 5 percent to almost 100 percent of a rainfall event running off, with the rest evaporating or soaking into the ground. On the average, however, about 70 percent of all precipitation evaporates back into the atmosphere, about 30 percent becomes runoff, and less than 1 percent soaks into the ground.

Surface Water

If you could follow the smallest of streams downhill, you would find that it eventually merges with other streams until they form a major river. The land area drained by a stream is known as the stream's drainage basin, or **watershed.** Each stream has its own watershed, but the watershed of a large river includes all the watersheds of the smaller streams that feed into the larger river. Figure 17.5 shows the watersheds of the Columbia River, the Colorado River, and the Mississippi River. Note that the water from the Columbia River and the Colorado River watersheds empties into the Pacific Ocean. The Mississippi River

watershed drains into the Gulf of Mexico, which is part of the Atlantic Ocean.

Two adjacent watersheds are separated by a line called a *divide.* Rain that falls on one side of a divide flows into one watershed and rain that falls on the other side flows into the other watershed. A *continental divide* separates river systems that drain into opposite sides of a continent. The North American continental divide trends northwestward through the Rocky Mountains. Imagine standing over this line with a glass of water in each hand, then pouring the water to the ground. The water from one glass will eventually end up in the Atlantic Ocean, and the water from the other glass will end up in the Pacific Ocean. Sometimes the Appalachian Mountains are considered to be an eastern continental divide, but water from both sides of this divide ends up on the same side of the continent, in the Atlantic Ocean.

Water moving downhill is sometimes stopped by a depression in a watershed, a depression where water temporarily collects as a standing body of freshwater. A smaller body of standing water is usually called a *pond,* and one of much larger size is called a *lake.* A pond or lake can occur naturally in a depression, or it can be created by building a dam on a stream. A natural pond, a natural lake, or a pond or lake created by building a dam is called a *reservoir* if it is used for (1) water storage,

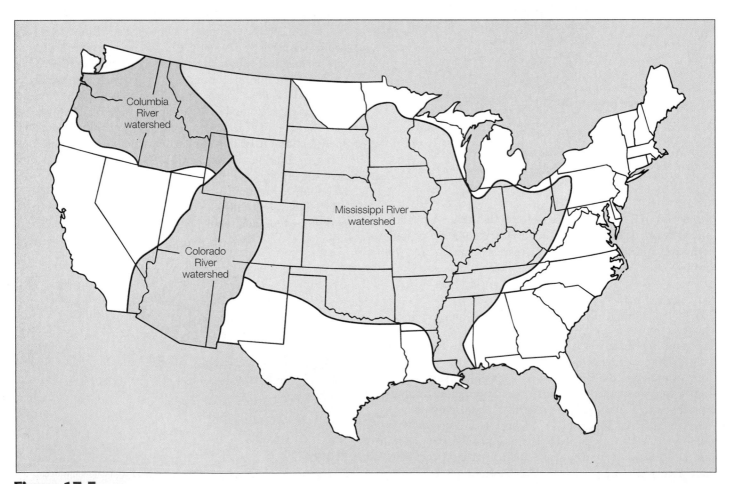

Figure 17.5

The approximate watersheds of the Columbia River, the Colorado River, and the Mississippi River.

(2) flood control, or (3) generating electricity. A reservoir can be used for one or two of these purposes but not generally for all three. A reservoir built for water storage, for example, is kept as full as possible to store water. This use is incompatible with use for flood control, which would require a low water level in the reservoir in order to catch runoff, preventing waters from flooding the land. In addition, extensive use of reservoir water to generate electricity requires the release of water, which could be incompatible with water storage. The water of streams, ponds, lakes, and reservoirs is collectively called *surface water,* and all serve as sources of freshwater. The management of surface water, as you can see, can present some complicated problems.

Groundwater

Precipitation soaks into the ground, or *percolates* slowly downward until it reaches an area, or zone, where the open spaces between rock and soil particles are completely filled with water. Water from such a saturated zone is called **groundwater.** There is a tremendous amount of water stored as groundwater, which makes up a supply about twenty-five times larger than all the surface water on the earth. Groundwater is an important source of freshwater for human consumption and for agriculture. Groundwater is often found within 100 m (about 330 ft) of the surface, even in arid regions where little surface water is found. Groundwater is the source of water for wells in addition to being the source that keeps streams flowing during dry periods.

Water is able to percolate down to a zone of saturation because sediments contain open spaces between the particles called *pore spaces.* The more pore space a sediment has, the more water it will hold. The total amount of pore spaces in a given sample of sediment is a measure of its *porosity.* Sand and gravel sediments, for example, have grains that have large pore spaces between them, so these sediments have a high porosity. In order for water to move through a sediment, however, the pore spaces must be connected. The ability of a given sample of sediment to transmit water is a measure of its *permeability.* Sand and gravel have a high permeability because the grains do not fit tightly together, allowing water to move from one pore space to the next. Sand and gravel sediments thus have a high porosity as well as a high permeability. Clay sediments, on the other hand, have small, flattened particles that fit tightly together. Clay thus has a low porosity, and when saturated or compressed, clay becomes *impermeable,* meaning water cannot move through it at all (figure 17.6).

The amount of groundwater available in a given location depends on a number of factors, such as the present and past climate, the slope of the land, and the porosity and permeability of the sediments beneath the surface. Generally sand and gravel sediments, along with solid sandstone, have the best porosity and permeability for transmitting groundwater. Other solid rocks, such as granite, can also transmit groundwater if they are sufficiently fractured by joints and cracks. In any case, groundwater will percolate downward until it reaches an area where pressure and other conditions have eliminated all pores, cracks, and joints. Above this impermeable layer it collects in all avail-

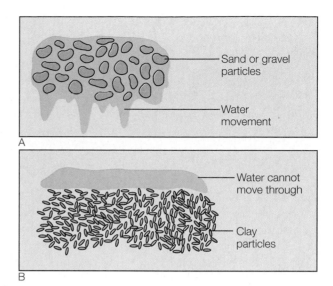

Figure 17.6

(*A*) Sand and gravel have large, irregular particles with large pore spaces, so they have a high porosity. Water can move from one pore space to the next, so they also have a high permeability. (*B*) Clay has small, flat particles, so it has a low porosity and is practically impermeable because water cannot move from one pore to the next.

able spaces to form a *zone of saturation.* Water from the zone of saturation is considered to be groundwater. Water from the zone above is not considered to be groundwater. The surface of the boundary between the zone of saturation and the zone above is called the **water table.** The surface of a water table is not necessarily horizontal, but it tends to follow the topography of the surface in a humid climate. A hole that is dug or drilled through the earth to the water table is called a well. The part of the well that is below the water table will fill with groundwater, and the surface of the water in the well is generally at the same level as the water table.

Precipitation falls on the land and percolates down to the zone of saturation, then begins to move laterally, or sideways, to lower and lower elevations until it finds its way back to the surface. This surface outflowing could take place at a stream, pond, lake, swamp, or spring (figure 17.7). Groundwater flows gradually and very slowly through the tiny pore spaces, moving at a rate that ranges from kilometers (miles) per day to meters (feet) per year. Surface streams, on the other hand, move much faster at rates up to about 30 km per hour (about 20 mi/h).

An **aquifer** is a layer of sand, gravel, sandstone, or other highly permeable material beneath the surface that is capable of producing water. In some places an aquifer carries water from a higher elevation, resulting in a pressure on water trapped by impermeable layers at lower elevations. Groundwater that is under such a confining pressure is in an *artesian* aquifer. "Artesian" refers to the pressure, and groundwater from an artesian well rises above the top of the aquifer but not necessarily to the surface. Some artesian wells are under sufficient pressure to produce a fountainlike flow or spring (figure 17.8). Some people call groundwater from any deep well "artesian water," which is technically incorrect.

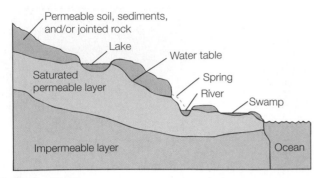

Figure 17.7
Groundwater from below the water table seeps into lakes, streams, and swamps and returns to the surface naturally at a spring. Groundwater eventually returns to the ocean, but the trip may take hundreds of years.

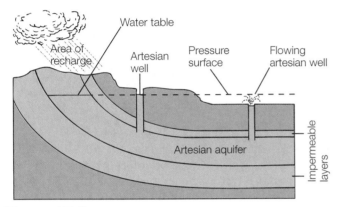

Figure 17.8
An artesian aquifer has groundwater that is under pressure because the groundwater is confined between two impermiable layers and has a recharge area at a higher elevation. The pressure will cause the water to rise in a well drilled into the aquifer, becoming a flowing well if the pressure is sufficiently high.

> **CONCEPTS APPLIED**
> **Table Top**
>
> After doing some research, make a drawing to show the depth of the water table in your area, and its relationship to local streams, lakes, swamps, and water wells.

Freshwater as a Resource

Water is an essential resource, not only because it is required for life processes but also because of its role in a modern industrialized society. Water is used in the home for drinking, cooking, and cleaning, as a carrier to remove wastes, and for maintaining lawns and gardens. These domestic uses lead to an equivalent consumption of about 570 liters per person each day (about 150 gal/person/day), but this is only about 10 percent of the total consumed. Average daily use of water in the United States amounts to some 5,700 liters per person each day (about 1,500 gal/person/day), or about enough water to fill a small swimming pool once a

week. The bulk of the water is used by agriculture (about 40 percent), for the production of electricity (about 40 percent), and for industrial purposes (about 10 percent). These overall percentages of use vary from one region of the country to another.

Most of the water supply is obtained from the surface water resources of streams, lakes, and reservoirs, and 37 percent of the municipal water supply comes from groundwater. If you then add farms, villages, and many suburban areas the percentage of groundwater used by humans is well above 40 percent. Surface water contains more sediments, bacteria, and possible pollutants than groundwater. This means that surface water requires filtering to remove suspended particles, treatment to kill bacteria, and sometimes processing to remove dissolved chemicals. Groundwater is naturally filtered as it moves through the pore spaces of an aquifer, so it is usually relatively free of suspended particles and bacteria. Thus the processing or treatment of groundwater is usually not necessary (figure 17.9). But groundwater, on the other hand, will cost more to use as a resource because it must be pumped to the surface. The energy required for this pumping can be very expensive. In addition, groundwater generally contains more dissolved minerals (hard water), which may require additional processing or chemical treatment to remove the troublesome minerals.

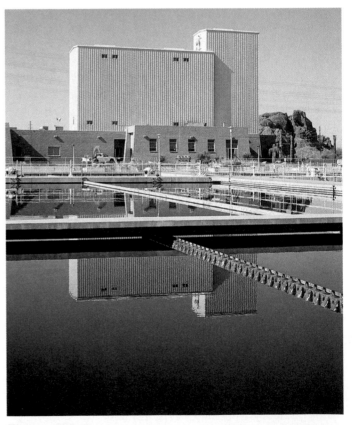

Figure 17.9
The filtering beds of a city water treatment facility. Surface water contains more sediments, bacteria, and other suspended materials because it is on the surface and is exposed to the atmosphere. This means that surface water must be filtered and treated when used as a domestic resource. Such processing is not required when groundwater is used as the resource.

The use of surface water as a source of freshwater means that the supply depends on precipitation. When a drought occurs, low river and lake levels may require curtailing water consumption. In some parts of the western United States, such as the Colorado River watershed, *all* of the surface water is already being used, with certain percentages allotted for domestic, industrial, and irrigation uses. Groundwater is also used in this watershed, and in some locations it is being pumped from the ground faster than it is being replenished by precipitation (figure 17.10). As the population grows and new industries develop, more and more demands are placed on the surface water supply, which has already been committed to other uses, and on the diminishing supply of groundwater. This raises some very controversial issues about how freshwater should be divided among agriculture, industries, and city domestic use. Agricultural interests claim they should have the water because they produce the food and fibers that people must have. Industrial interests claim they should have the water because they create the jobs and the products that people must have. Cities, on the other hand, claim that domestic consumption is the most important because people cannot survive without water. Yet, others claim that no group has a right to use water when it is needed to maintain habitats. Who should have the first priority for water use in such cases?

Some have suggested that people should not try to live and grow food in areas that have a short water supply, that plenty of freshwater is available elsewhere. Others have suggested that humans have historically moved rivers and reshaped the land to obtain water, so perhaps one answer to the problem is to find new sources of freshwater. Possible sources include the recycling of waste water and turning to the largest supply of water in the world, the ocean. About 90 percent of the water used by industries is presently dumped as a waste product. In some areas treated city waste water is already being recycled for use in power plants and for watering parks. A practically limitless supply of freshwater could be available by desalting ocean water, something which occurs naturally in the hydrologic cycle. The treatment of seawater to obtain a new supply of freshwater is presently too expensive because of the cost of energy to accom-

Figure 17.10
This is groundwater pumped from the ground for irrigation. In some areas, groundwater is being removed from the ground faster than it is being replaced by precipitation, resulting in a water table that is falling. It is thus possible that the groundwater resource will soon become depleted in some areas.

plish the task. New technologies, perhaps ones that use solar energy, may make this more practical in the future. In the meantime, the best sources of extending the supply of freshwater appear to be the control of pollution, the recycling of waste water, and conservation of the existing supply.

CONCEPTS APPLIED
Who Uses How Much?

Find out how much water is used in the industrial processes in your location. Compare this to the amount of water used in the home for drinking, cooking, cleaning, and so on.

A Closer Look Water Quality and Wastewater Treatment

What do you think of when you see a stream? Do you think about the water and how deep it is? Do you think of something to do with the water such as swimming, fishing, or having a good time? As you might imagine, not all people look at a stream and think about the same thing. A city engineer, for example, might wonder if the stream has enough water to serve as a source to supplement the city water supply. A rancher or farmer might wonder how the stream could be easily diverted to serve as a source of water for irrigation. An electric utility planner, on the other hand,

might wonder if the stream could serve as a source of power.

Water in a stream is a resource that can be used many different ways, but using it requires knowing about the water quality as well as quantity. We need to know if the quality of the water is good enough for the intended use—and different uses have different requirements. Water fit for use in an electric power plant, for example, might not be suitable for use as a city water supply. Indeed, water fit for use in a power plant might not be suitable for irrigation. Water quality is deter-

mined by the kinds and amounts of substances dissolved and suspended in the water and the consequences to users. If a source of water can be used for drinking water or not, for example, is regulated by stringent rules and guidelines about what cannot be in the water. These rules do not call for pure water, but rather are designed to protect human health.

The water of even the healthiest stream is not absolutely pure. All water contains many naturally

—Continued top of next page

Chapter Seventeen Earth's Waters   **359**

Continued—

occurring substances such as ions of bicarbonates, calcium, and magnesium. A *pollutant* is not naturally occurring, usually a waste material that contaminates air, soil, or water. There are basically two types of water pollutants: degradable and persistent. Examples of degradable pollutants include domestic sewage, fertilizers, and some industrial wastes. As the term implies, degradable pollutants can be broken down into simple, nonpolluting substances such as carbon dioxide and nitrogen. The chemical reactions or natural bacterial processes that break down the pollutants can lead to low oxygen levels if the pollution load is high, but this is usually reversible.

Examples of persistent pollutants include some pesticides (such as DDT), petroleum and petroleum products, plastic materials, leached chemicals from landfill sites, oil-based paints, heavy metals and metal compounds such as lead, mercury, and cadmium, and certain radioactive materials. The damage they cause is either irreversible or reparable only over long periods of time.

One of the most common forms of degradable pollution control in the United States is wastewater treatment. The United States has a vast system of sewer pipes, pumping stations, and treatment plants, and about 74 percent of all Americans are served by such wastewater systems. Sewer pipelines collect the wastewater from homes, businesses, and many industries, and deliver it to treatment plants. Most of these plants were designed to make wastewater fit for discharge into streams or other receiving waters.

Years ago raw sewage was dumped directly into waterways. The natural process of purification began as the larger volume of clean water in the stream diluted the wastes. How does water clean itself? Water is purified in large part by the normal actions of organisms. Energy from the sun drives the process of photosynthesis in aquatic plants, which produces oxygen as a by-product. Bacteria use oxygen to break down organic material, including sewage that may have been dumped into the water. This decomposition produces the carbon dioxide, nutrients, and other substances needed by plants living in the water and the cycle continues. Thus the water can purify itself biologically but only to a degree since the water can absorb only so much. There is a point where the natural cleaning processes can no longer cope. The greater populations of today and greater volume of domestic and industrial waste-

Primary Treatment

Screens   Grit chamber   Sedimentation tank

Box Figure 17.1

water require that communities provide treatment to give nature a hand.

The basic function of a waste treatment plant is to speed up the natural process of purifying the water. There are two basic stages in the treatment, called the *primary stage* and the *secondary stage*. The primary stage physically removes solids from the wastewater. The secondary stage uses biological processes to further purify wastewater. Sometimes, these stages are combined into one operation.

As raw sewage enters a treatment plant it first flows through a screen to remove large floating objects such as rags and sticks that might cause clogs. After this initial screening, it passes into a grit chamber where cinders, sand, and small stones settle to the bottom (box figure 17.1). A grit chamber is particularly important in communities with combined sewer systems where sand or gravel may wash into the system along with rain, mud, and other stuff, all with the storm water.

After screening and grit removal the sewage is basically a mixture of organic and inorganic matter along with other suspended solids. The solids are minute particles that can be removed in a sedimentation tank. The speed of the flow through the larger sedimentation tank is slower and suspended solids gradually sink to the bottom of the tank. They form a mass of solids called *raw primary sludge*, which is usually removed from the tank by pumping. The sludge may be further treated for use as a fertilizer or disposed of through incineration if necessary.

Once the effluent has passed through the primary treatment process, which is primarily physical filtering and settling, it enters a secondary stage that involves biological activities of mi-

croorganisms. The secondary stage of treatment removes about 85 percent of the organic matter in sewage by making use of the bacteria that are naturally a part of the sewage. There are two principal techniques used to provide secondary treatment; (1) trickling filters or (2) activated sludge. A trickling filter is simply a bed of stones from 1 to 2 meters (3 to 6 feet) deep through which the effluent from the sedimentation tank flows. Interlocking pieces of corrugated plastic or other synthetic media have also been used in trickling beds, but the important part is that it provides a place for bacteria to live and grow. Bacteria grow on the stones or synthetic media and consume most of the organic matter flowing by in the effluent. The now cleaner water trickles out through pipes to another sedimentation tank to remove excess bacteria. Disinfection of the effluent with chlorine is generally used to complete this secondary stage of basic treatment.

The trend today is toward the use of an activated sludge process instead of trickling filters. The activated sludge process speeds up the work of the bacteria by bringing air and sludge heavily laden with bacteria into close contact with the effluent (box figure 17.2). After the effluent leaves the sedimentation tank in the primary stage, it is pumped into an aeration tank, where it is mixed with air and sludge loaded with bacteria and allowed to remain for several hours. During this time, the bacteria break down the organic matter into harmless by-products.

The sludge, now activated with additional millions of bacteria, can be used again by returning it to the aeration tank for mixing with new effluent and ample amounts of air. As with trickling, the final step is generally the addition of chlorine

Secondary Treatment
Activated sludge process

Aeration tank

Air

Activated sludge

Box Figure 17.2

to the effluent which kills more than 99 percent of the harmful bacteria. Some municipalities are now manufacturing chlorine solution on site to avoid the necessity of transporting and storing large amounts of chlorine, sometimes in a gaseous form. Alternatives to chlorine disinfection, such as ultraviolet light or ozone, are also being used in situations where chlorine in sewage effluents can be harmful to fish and other aquatic life.

New pollution problems have placed additional burdens on wastewater treatment systems. Today's pollutants may be more difficult to remove from water. Increased demands on the water supply only aggravate the problem. These challenges are being met through better and more complete methods of removing pollutants at treatment plants, or through prevention of pollution at the source. Pretreatment of industrial waste, for example, removes many troublesome pollutants at the beginning, rather than at the end, of the pipeline.

The increasing need to reuse water calls for better and better wastewater treatment. Every use of water—whether at home, in the factory, or on a farm—results in some change in its quality. New methods for removing pollutants are being developed to return water of more usable quality to receiving lakes and streams. Advanced waste treatment techniques in use or under development range from biological treatment capable of removing nitrogen and phosphorus to physical-chemical separation techniques such as filtration, carbon adsorption, distillation, and reverse osmosis. These activities typically follow secondary treatment and are known as tertiary treatment.

These wastewater treatment processes, alone or in combination, can achieve almost any degree of pollution control desired. As waste effluents are purified to higher degrees by such treatment, the effluent water can be used for industrial, agricultural, or recreational purposes, or even drinking water supplies.

When pollution makes water unsuitable for drinking, recreation, agriculture, and industry, it eventually also diminishes the aesthetic quality of lakes and rivers. Even more seriously, when polluted water is able to destroy aquatic life it is able to menace human health. Nobody escapes the effects of water pollution. What can you do to improve water quality? Each individual effort can make a difference to water quality and the environment as a whole. You can start by not misusing your sewage system, especially by dumping toxic household products. Also consider:

- tossing items such as dental floss, disposable diapers, and plastic items into the wastebasket, not the toilet;
- using completely the contents of oven, toilet bowl, and sink drain cleaners, bleaches, paints, rust removers, and solvents;
- saving food scraps and composting them, not dumping them down the drain; and
- choosing water-based latex paint instead of oil-based paint, and using it up instead of storing or dumping it.

There is more you can do, and this begins by reading up on environmental issues for your local area.

Drawings, and some text, from *How Wastewater Treatment Works . . . The Basics*, The Environmental Protection Agency, Office of Water, www.epa.gov/owm/featinfo.htm

SEAWATER

More than 70 percent of the surface of the earth is covered by seawater, with an average depth of 3,800 m (about 12,500 ft). The land areas cover 30 percent, less than a third of the surface, with an average elevation of only about 830 m (about 2,700 ft). With this comparison, you can see that humans live on and fulfill most of their needs by drawing from a small part of the total earth. As populations continue to grow and as resources of the land continue to diminish, the ocean will be looked at more as a resource rather than a convenient place for dumping wastes. The ocean already provides some food and is a source of some minerals, but it can possibly provide freshwater, new sources of food, new sources of important minerals, and new energy sources in the future. There are vast deposits of phosphorus and man-

ganese minerals on the ocean bottom, for example, that can provide valuable resources. Phosphorus is an element that can be used to manufacture an important fertilizer needed in agriculture, and the land supplies are becoming depleted. Manganese nodules, which occur in great abundance on the ocean bottom, can be a source of manganese, iron, copper, cobalt, and nickel. Seawater contains enough deuterium to make it a feasible source of energy. One gallon of seawater contains about a spoonful of deuterium, with the energy equivalent of three hundred gallons of gasoline. It has been estimated there is sufficient deuterium in the oceans to supply power at one hundred times the present consumption for the next ten billion years. The development of controlled nuclear fusion is needed, however, to utilize this potential energy source. The sea may provide new sources of food through *aquaculture*, the farming of the sea the way that the land is presently farmed. Some aquaculture projects have already started with the farming of oysters, clams, lobsters, shrimp, and certain fishes, but these projects have barely begun to utilize the full resources that are possible.

Part of the problem of utilizing the ocean is that the ocean has remained mostly unexplored and a mystery until recent times. Only now are scientists beginning to understand the complex patterns of the circulation of ocean waters, the nature of the chemical processes at work in the ocean, and the interactions of the ocean and the atmosphere and to chart the topography of the ocean floor.

Oceans and Seas

The vast body of salt water that covers more than 70 percent of the earth's surface is usually called the *ocean* or the *sea*. In general, the **ocean** is a single, continuous body of salt water on the surface of the earth. Although there is really only one big ocean on the earth, specific regions have been given names for convenience in describing locations. For this purpose, three principal regions are recognized, the (1) Atlantic Ocean, (2) Indian Ocean, and (3) Pacific Ocean as shown in figure 17.11. Specific regions (Atlantic, Indian, and Pacific) are often subdivided further into North Atlantic Ocean, South Atlantic Ocean, and so on.

A *sea* is usually a smaller part of the ocean, a region with characteristics that distinguish it from the larger ocean of which it is a part. Often the term "sea" is used in the name of certain inland bodies of salty water.

The Pacific Ocean is the largest of the three principal ocean regions. It has the largest surface area, covering 180 million km^2 (about 70 million mi^2), and has the greatest average depth of 3.9 km (about 2.4 mi). The Pacific is circled by active converging plate boundaries, so it is sometimes described as being circled by a "rim of fire." It is called this because of the volcanoes associated with the converging plates. The "rim" also has the other associated features of converging plate boundaries such as oceanic trenches, island arcs, and earthquakes. The Atlantic Ocean is second in size, with a surface area of 107 million km^2 (about 41 million mi^2), and the shallowest average depth of only 3.3 km (about 2.1 mi). The Atlantic Ocean is bounded by nearly parallel continental margins with a diverging plate boundary between. It lacks the trench and island arc features of the Pacific, but it does have islands, such as Iceland, that are a part of the Mid-Atlantic Ridge at the plate boundary. The shallow seas of the Atlantic, such as the Mediterranean, Caribbean, and Gulf of Mexico, contribute to the shallow average depth of the Atlantic. The Indian Ocean has the smallest surface area, with 74 million km^2 (about 29 million mi^2), and an average depth of 3.8 km (about 2.4 mi).

As mentioned earlier, a "sea" is usually a part of an ocean that is identified because some characteristic sets it apart. For example, the Mediterranean, Gulf of Mexico, and Caribbean seas are bounded by land, and they are located in a warm, dry climate. Evaporation of seawater is greater than usual at these locations, which results in the seawater being saltier. Being bounded by land and having saltier seawater characterizes these locations as being different from the rest

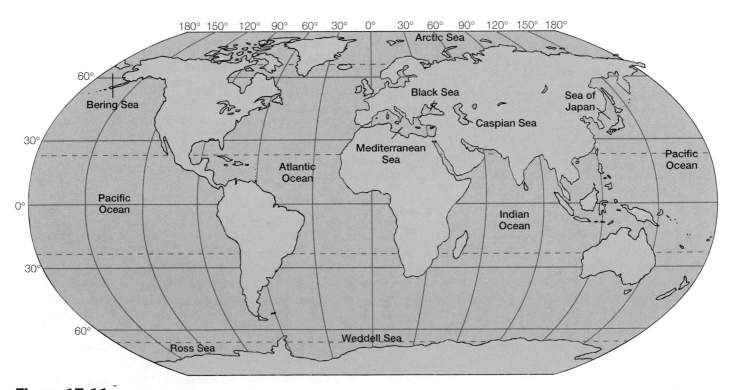

Figure 17.11

Distribution of the oceans and major seas on the earth's surface. There is really only one ocean; for example, where is the boundary between the Pacific, Atlantic, and Indian oceans in the Southern Hemisphere?

of the Atlantic. The Sargasso Sea, on the other hand, is a part of the Atlantic that is not bounded by land and has a normal concentration of sea salts. This sea is characterized by having an abundance of floating brown seaweeds that accumulate in this region because of the global wind and ocean current patterns. The Arctic Sea, which is also sometimes called the Arctic Ocean, is a part of the North Atlantic Ocean that is less salty. Thus, the terms "ocean" and "sea" are really arbitrary terms that are used to describe different parts of the earth's one continuous ocean.

The Nature of Seawater

According to one theory, the ocean is an ancient feature of the earth's surface, forming at least three billion years ago as the earth cooled from its early molten state. The seawater, and much of the dissolved materials, are believed to have formed from the degassing of water vapor and other gases from molten rock materials. The degassed water vapor soon condensed, and over a period of time it began collecting as a liquid in the depression of the early ocean basin. Ever since, seawater has continuously cycled through the hydrologic cycle, returning water to the ocean through the world's rivers. For millions of years, these rivers have carried large amounts of suspended and dissolved materials to the ocean. These dissolved materials, including salts, stay behind in the seawater as the water again evaporates, condenses, falls on the land, and then brings more dissolved materials much like a continuous conveyor belt.

You might wonder why the ocean basin has not become filled in by the continuous supply of sediments and dissolved materials that would accumulate over millions of years. The basin has not filled in because (1) accumulated sediments have been recycled to the earth's interior through plate tectonics and (2) dissolved materials are removed by natural processes just as fast as they are supplied by the rivers. Some of the dissolved materials, such as calcium and silicon, are removed by organisms to make solid shells, bones, and other hard parts. Other dissolved materials, such as iron, magnesium, and phosphorous, form solid deposits directly and also make sediments that settle to the ocean floor. Hard parts of organisms and solid deposits are cycled to the earth's interior along with suspended sediments that have settled out of the seawater. Studies of fossils and rocks indicate that the composition of seawater has changed little over the past 600 million years.

The dissolved materials of seawater are present in the form of ions because of the strong dissolving ability of water molecules. Almost all of the chemical elements are present, but only six ions make up more than 99 percent of any given sample of seawater. As shown in table 17.1, chlorine and sodium are the most abundant ions. These are the elements of sodium chloride, or common table salt. As a sample of seawater evaporates, the positive metal ions join with the different negative ions to form a complex mixture of ionic compounds known as *sea salt*. Sea salt is mostly sodium chloride, but it also contains salts of the four metal ions (sodium, magnesium, calcium, and potassium) combined with the different negative ions of chlorine, sulfate, bicarbonate, and so on.

Table 17.1 Major Dissolved Materials in Seawater

Ion	Percent (by weight)
Chloride (Cl^-)	55.05
Sodium (Na^+)	30.61
Sulfate (SO_4^{-2})	7.68
Magnesium (Mg^{+2})	3.69
Calcium (Ca^{+2})	1.16
Potassium (K^+)	1.10
Bicarbonate (HCO_3^-)	0.41
Bromine (Br^-)	0.19
Total	99.89

Figure 17.12
Salinity is defined as the mass of salts dissolved in 1.0 kg of seawater. Thus, if a sample of seawater has a salinity of 35 ‰, a 1,000 g sample would evaporate 965 g of water and leave 35 g of sea salts behind.

The amount of dissolved salts in seawater is measured as **salinity.** Salinity is defined as the mass of salts dissolved in 1.0 kg, or 1,000 g of seawater. Since the salt content is reported in parts per thousand, the symbol ‰ is used (% means parts per hundred). Thus 35 ‰ means that 1,000 g of seawater contains 35 g of dissolved salts (and 965 g of water). This is the same concentration as a 3.5 percent salt solution (figure 17.12). Oceanographers use the salinity measure because the mass of a sample of seawater does not change with changes in the water temperature. Other measures of concentration are based on the volume of a sample, and the volume of a liquid does vary as it expands and contracts with changes in the temperature. Thus, by using the salinity measure, any corrections due to temperature differences are eliminated.

The average salinity of seawater is about 35 ‰, but the concentration varies from a low of about 32 ‰ in some locations up to a high of about 36 ‰ in other locations. The salinity of seawater in a given location is affected by factors that tend to increase or decrease the concentration. The concentration is increased by two factors, evaporation and the formation of sea ice. Evaporation increases the concentration because it is water vapor only that evaporates, leaving the dissolved salts behind in a greater concentration. Ice that forms from freezing seawater increases the concentration because when ice forms the salts are excluded from the crystal structure. Thus, sea ice is freshwater, and the removal of this water leaves the dissolved salts behind in a greater concentration. The salinity of seawater is decreased by three factors: heavy precipitation, the melting of ice, and the addition of freshwater by a large river. All three of these factors tend to dilute seawater with freshwater, which lowers the concentration of salts.

Note that increases or decreases in the salinity of seawater are brought about by the addition or removal of freshwater. This changes only the amount of water present in the solution. The *kinds* or *proportions* of the ions present (table 17.1) in seawater does not change with increased or decreased amounts of freshwater. The same proportion, meaning the same chemical composition, is found in seawater of any salinity of any sample taken from any location anywhere in the world, from any depth of the ocean, or taken any time of the year. Seawater has a remarkably uniform composition that varies only in concentration. This means that the ocean is well mixed and thoroughly stirred around the entire earth. How seawater becomes so well mixed and stirred on a worldwide basis is discussed in the next section.

If you have ever allowed a glass of tap water to stand for a period of time, you may have noticed tiny bubbles collecting as the water warms. These bubbles are atmospheric gases, such as nitrogen and oxygen, that were dissolved in the water (figure 17.13). Seawater contains dissolved gases in addition to the dissolved salts. Near the surface, seawater contains mostly nitrogen and oxygen, in similar proportions to the mixture that is found in the atmosphere. There is more carbon dioxide than you would expect, however, as seawater contains a large amount of this gas. More carbon dioxide can dissolve in seawater because it reacts with water to form carbonic acid, H_2CO_3, the same acid that is found in a bubbly cola. In seawater, carbonic acid breaks down into bicarbonate and carbonate ions, which tend to remain in solution. Water temperature and the salinity have an influence on how much gas can be dissolved in seawater, and

Figure 17.13
Air will dissolve in water, and cooler water will dissolve more air than warmer water. The bubbles you see here are bubbles of carbon dioxide that came out of solution as the soda became warmer.

increasing either, or both, will reduce the amount of gases that can be dissolved. Cold, lower salinity seawater in colder regions will dissolve more gases than the warm, higher salinity seawater in tropical locations. Abundant seaweeds (large algae) and phytoplankton (microscopic algae) in the upper, sunlit water tends to reduce the concentration of carbon dioxide and increase the concentration of dissolved oxygen through the process of photosynthesis. With increasing depth, less light penetrates the water, and below about 80 m (about 260 ft) there is insufficient light for photosynthesis. Thus, more algae and phytoplankton and more dissolved oxygen are found above this depth. Below this depth there is more dissolved carbon dioxide, and less dissolved oxygen. The oxygen-poor, deep ocean water does eventually circulate back to the surface, but the complete process may take several thousand years.

Movement of Seawater

Consider the enormity of the earth's ocean, which has a surface area of some 361 million km^2 (about 139 million mi^2) and a volume of 1,370 million km^3 (about 328 million mi^3) of seawater. There must be a terrific amount of stirring in such an enormous amount of seawater to produce the well-mixed, uniform chemical composition that is found in seawater throughout the world. The amount of mixing required is more easily imagined if you consider the long history of the ocean, the very long period of time over which the mixing has occurred. Based on investigations of the movement of seawater, it has been estimated that there is a complete mixing of all the earth's seawater about every 2,000 years or so. With an assumed age of 3 billion years, this means that the earth's seawater has been mixed 3,000,000,000 ÷ 2,000, or 1.5 million times. With this much mixing, you would be surprised if seawater were *not* identical all around the earth.

How does seawater move to accomplish such a complete mixing? Seawater is in a constant state of motion, both on the surface and far below the surface. The surface has two types of motion: (1) *waves,* which have been produced by some disturbance, such as the wind, and (2) *currents,* which move water from one place to another. Waves travel across the surface as a series of wrinkles. Waves crash on the shore as booming breakers and make the surf. This produces local currents as water moves along the shore and back out to sea. There are also permanent, worldwide surface currents that move ten thousand times more water across the ocean than all the water moving in all the large rivers on the land. Beneath the surface there are currents that move water up in some places and move it down in other places. Finally, there are enormous deep ocean currents that move tremendous volumes of seawater. The overall movement of many of the currents on the surface and their relationship to the deep ocean currents are not yet fully mapped or understood. The surface waves are better understood. The general trend and cause of permanent, worldwide currents in the ocean can also be explained. The following is a brief description and explanation of waves, currents, and the deep ocean movements of seawater.

Pollution is usually understood to mean something that is not naturally occurring and contaminates air, soil, or water to interfere with human health, well being, or quality of the environment. An important factor in understanding pollution is the size of the human population and the amount of material that might become a pollutant. When the human population was small and produced few biological wastes, there was not a pollution problem. The decomposers broke down the material into simpler nonpolluting substances such as water and carbon dioxide and no harm was done. For example, suppose one person empties the tea leaves remaining from a cup of tea into a nearby river once a week. In this case decomposer organisms in the water would break down the tea leaves almost as fast as they are added to the river. But imagine 100,000 people doing this every day. In this case the tea leaves are released faster than they decompose and the leaves become pollutants.

The part of the wide lower course of a river where the freshwater of the river mixes with the saline water from the oceans is called a coastal *estuary.* Estuary waters include bays and tidal rivers that serve as nursery areas for many fish and shellfish populations, including shrimp, oysters, crabs, and scallops. Unfortunately, the rivers carry pollution from their watersheds and adjacent wetlands to the estuary, where it impacts the fish and shellfish industry, swimming, and recreation.

In 1996 the U.S. Environmental Protection Agency asked the coastal states to rate the general water quality in their estuaries. The states reported that pollutants impact aquatic life in 31 percent of the area surveyed, violate shellfish harvesting criteria in 27 percent of the area surveyed, and violate swimming-use criteria in 16 percent of the area surveyed.

The most common pollutants affecting the surveyed estuaries were excessive *nutrients,* which were found in 22 percent of all the estuaries surveyed. Excessive nutrients stimulate population explosions of algae. Fast-growing masses of algae block light from the habitat below, stressing the aquatic life. The algae die and eventually decompose and this depletes the available oxygen supply, leading to further fish and shellfish kills.

The second most common pollutant was the presence of *bacteria,* which pollute 16 percent of all the estuary waters surveyed. Because *Escherichia coli* is a bacterium commonly found in the intestines of humans and other warm-blooded animals, the presence of *E. coli* is evidence that sewage is polluting the water. Bacteria interfere with recreation activities of people and can contaminate fish and shellfish.

The states also reported that *toxic organic chemicals* pollute 15 percent of the surveyed waters, *oxygen-depleting chemicals* pollute 12 percent, and *petroleum products* pollute another 8 percent. These pollutants impact the fish and shellfish industry, swimming, and recreation activities that require contact with the water.

The leading sources of the pollutants were identified as industrial discharges, urban runoff, municipal wastewater, agriculture runoff, and wastes from landfills. Of these, agricultural runoff is the primary contributor to water pollution, followed by urban runoff. All the other sources together make up a distant third source of pollution.

Waves

Any slight disturbance will create ripples that move across a water surface. For example, if you gently blow on the surface of water in a glass you will see a regular succession of small ripples moving across the surface. These ripples, which look like small, moving wrinkles, are produced by the friction of the air moving across the water surface. The surface of the ocean is much larger, but a gentle wind produces patches of ripples in a similar way. These patches appear, then disappear as the wind begins to blow over calm water. If the wind continues to blow, larger and longer-lasting ripples are made, and the moving air can now push directly on the side of the ripples. A ripple may eventually grow into an **ocean wave,** a moving disturbance that travels across the surface of the ocean. In its simplest form, each wave has a ridge, or mound, of water called a *crest,* which is followed by a depression called a *trough.* Ocean waves are basically repeating series of these crests and troughs that move across the surface like wrinkles (figure 17.14).

The simplest form of an ocean wave can be described by measurements of three distinct characteristics: (1) the *wave height*, which is the vertical distance between the top of a crest and the bottom of the next trough, (2) the *wavelength*, which is the horizontal distance between two successive crests (or other successive parts of the wave), and (3) the *wave period*, which is the time required for two successive crests (or other successive parts) of the wave to pass a given point (figure 17.15).

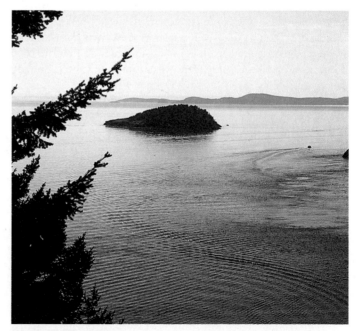

Figure 17.14

The surface of the ocean is rarely, if ever, still. Any disturbance can produce a wave, but most waves on the open ocean are formed by a local wind.

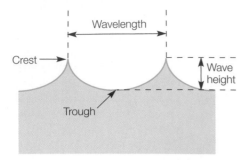

Figure 17.15

The simplest form of ocean waves, showing some basic characteristics. Most waves do not look like this representation because most are complicated mixtures of superimposed waves with a wide range of sizes and speeds.

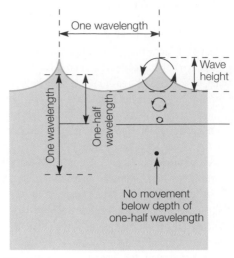

Figure 17.16

Water particles are moved in a circular motion by a wave passing in the open ocean. On the surface, a water particle traces out a circle with a diameter that is equal to the wave height. The diameters of the circles traced out by water particles decrease with depth to a depth that is equal to one-half the wavelength of the ocean wave.

The characteristics of an ocean wave formed by the wind depend on three factors: (1) the wind speed, (2) the length of time that the wind blows, and (3) the *fetch,* which is the distance the wind blows across the open ocean. As you can imagine, larger waves are produced by strong winds that blow for a longer time over a long fetch. In general, longer-blowing, stronger winds produce waves with greater wave heights, longer wavelengths, and longer periods, but a given wind produces waves with a wide range of sizes and speeds. In addition, the wind does not blow in just one direction, and shifting winds produce a chaotic pattern of waves of many different heights and wavelengths. Thus, the surface of the ocean in the area of a storm or strong wind has a complicated mixture of many sizes and speeds of superimposed waves. The smaller waves soon die out from friction within the water, and the larger ones grow as the wind pushes against their crests. Ocean waves range in height from a few centimeters up to more than 30 m (about 100 ft), but giant waves more than 15 m (about 50 ft) are extremely rare.

The larger waves of the chaotic, superimposed mixture of waves in a storm area last longer than the winds that formed them, and they may travel for hundreds or thousands of kilometers from their place of origin. The longer wavelength waves travel faster and last longer than the shorter wavelength waves, so the longer wavelength waves tend to outrun the shorter wavelength waves as they die out from energy losses to water friction. Thus, the irregular, superimposed waves created in the area of a storm become transformed as they travel away from the area. They become regular groups of long-wavelength waves with low wave height that are called **swell.** The regular waves of swell that you might observe near a shore may have been produced by a storm that occurred days before, thousands of kilometers across the ocean.

The regular crests and troughs of swell carry energy across the ocean, but they do not transport water across the open ocean. If you have ever been in a boat that is floating in swell, you know that you move in a regular pattern of up and forward on each crest, then backward and down on the following trough. The boat does not move along with the waves unless it is moved along by a wind or by some current. Likewise, a particle of water on the surface moves upward and forward with each wave

crest, then backward and down on the following trough, tracing out a nearly circular path through this motion. The particle returns to its initial position, without any forward movement while tracing out the small circle. Note that the diameter of the circular path is equal to the wave height (figure 17.16). Water particles farther below the surface also trace out circular paths as a wave passes. The diameters of these circular paths below the surface are progressively smaller with increasing depth. Below a depth equal to about half the wavelength there is no circular movement of the particles. Thus, you can tell how deeply the passage of a wave disturbs the water below if you measure the wavelength.

As swell moves from the deep ocean to the shore the waves pass over shallower and shallower water depths. When a depth is reached that is equal to about half the wavelength, the circular motion of the water particles begins to reach the ocean bottom. The water particles now move across the ocean bottom, and the friction between the two results in the waves moving slower as the wave height increases. These important modifications result in a change in the direction of travel and in an increasingly unstable situation as the wave height increases.

Most waves move toward the shore at some angle. As the wave crest nearest the shore starts to slow, the part still over deep water continues on at the same velocity. The slowing at the shoreward side *refracts,* or bends, the wave so it is more parallel to the shore. Thus, waves always appear to approach the shore head-on, arriving at the same time on all parts of the shore.

After the waves reach water that is less than one-half the wavelength, friction between the bottom and the circular motion of the water particles progressively slow the bottom part of the wave. The wave front becomes steeper and steeper as the top overruns the bottom part of the wave. When the wave front becomes too steep the top part breaks forward and the wave is now

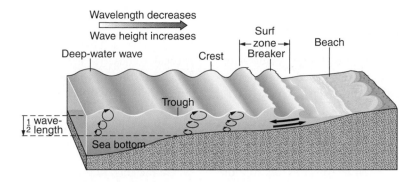

Figure 17.17

As a pattern of swell approaches a gently sloping beach, friction between the circular motion of the water particles and the bottom slows the wave, and the wave front becomes steeper and steeper. When the depth is about one and one-third times the wave height, the wave breaks forward, moving water toward the beach.

Figure 17.18

The white foam is in the surf zone, which is where the waves grow taller and taller, then break forward into a froth of turbulence. Do you see any evidence of rip currents in this picture (see figure 17.19)?

called a *breaker* (figure 17.17). In general, this occurs where the water depth is about one and one-third times the wave height. The zone where the breakers occur is called **surf** (figure 17.18).

Waves break in the foamy surf, sometimes forming smaller waves that then proceed to break in progressively shallower water. The surf may have several sets of breakers before the water is finally thrown on the shore as a surging sheet of seawater. The turbulence of the breakers in the surf zone and the final surge expend all the energy that the waves may have brought from thousands of kilometers away. Some of the energy does work in eroding the shoreline, breaking up rock masses into the sands that are carried by local currents back to the ocean. The rest of the energy goes into the kinetic energy of water molecules, which appears as a temperature increase.

Swell does not transport water with the waves over a distance, but small volumes of water are moved as a growing wave is pushed to greater heights by the wind over the open ocean. A strong wind can topple such a wave on the open ocean, producing a foam-topped wave known as a *whitecap*. In general, whitecaps form when the wind is blowing at 30 km/hr (about 20 mi/hr) or more.

Waves do transport water where breakers occur in the surf zone. When a wave breaks, it tosses water toward the shore, where the water begins to accumulate. This buildup of water tends to move away in currents, or streams, as the water returns to a lower level. Some of the water might return directly to the sea by moving beneath the breakers. This direct return of water forms a weak current known as *undertow*. Other parts of the accumulated water might be pushed along by the waves, producing a *longshore current* that moves parallel to the shore in the surf zone. This current moves parallel to the shore until it finds a lower place or a channel that is deeper than the adjacent bottom. Where the current finds such a channel, it produces a *rip current*, a strong stream of water that bursts out against the waves and returns water through the surf to the sea (figure 17.19). The rip current usually extends beyond the surf zone, and then diminishes. A rip current, or where rip currents are occurring, can usually be located by looking for the combination of (1) a lack of surf, (2) darker looking water, which means a deeper channel, and (3) a turbid, or muddy, streak of water that extends seaward from the channel indicated by the darker water that lacks surf.

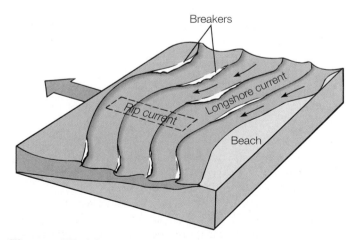

Figure 17.19

Breakers result in a buildup of water along the beach that moves as a longshore current. Where it finds a shore bottom that allows it to return to the sea, it surges out in a strong flow called a *rip current*.

In addition to waves created by winds, important waves are created by earthquakes and by tides. Movement of the earth's crust or underwater landslides can produce giant destructive waves called **tsunamis.** Tsunamis are sometimes incorrectly called "tidal waves" (they are in no way associated with tides or tide-making processes). Tsunamis can be the largest of all ocean waves, but their tremendous, destructive energy is often not apparent until they reach the shore. A tsunami traveling across the deep ocean can have wavelengths up to 200 km (about 120 mi) and wave heights up to 0.5 m (about 1.5 ft) while traveling in an impulse-produced series of five to ten or so waves. Because of the small wave height, a tsunami often moves unnoticed across the deep ocean. But the tsunami forms taller waves as it slows in the shallow water near the shore, waves that can reach over 8 m (about 25 ft) tall. Waves of this size can result in tremendous property damage and loss of life along low coastal areas, especially if the waves make giant breakers. Many low coastal areas of Japan, Alaska, Hawaii, Chile, and the Philippines have experienced the destructive effects of tsunamis.

Tides are basically produced by the gravitational pull of the moon and sun (figure 17.20). The basic pattern of two high tides and two low tides every twenty-four hours and fifty minutes can produce strong *tidal currents* in narrow bays. These are reversing currents, moving up the bay with the rising tide and out to sea with the falling tide. If the bay is long and narrow, such as some in Alaska and Nova Scotia, a *tidal bore* may be produced. A tidal bore is a wave that moves rapidly up the bay as the tide rises. Sometimes the tidal bore is easily observed, and when the conditions are right, the bore is large enough to carry a person on a surfboard for a great distance.

Ocean Currents

Waves generated by the winds, earthquakes, and tidal forces keep the surface of the ocean in a state of constant motion. Local, temporary currents associated with this motion, such as rip currents or tidal currents, move seawater over a short distance. Seawater also moves in continuous **ocean currents,** streams of water that stay in about the same path as they move through other seawater over large distances. Ocean currents can be difficult to observe directly since they are surrounded by water that looks just like the water in the current. Wind is likewise difficult to observe directly since the moving air looks just like the rest of the atmosphere. Unlike the wind, an ocean current moves *continuously* in about the same path, often carrying water with different chemical and physical properties than the water it is moving through. Thus, an ocean current can be identified and tracked by measuring the physical and chemical characteristics of the current and the surrounding water. This shows where the current is coming from and where in the world it is going. In general, ocean currents are produced by (1) density differences in seawater and (2) winds that blow persistently in the same direction.

Density Currents. The density of seawater is influenced by three factors: (1) the water temperature, (2) salinity, and (3) suspended sediments. Cold water is generally more dense than warm water, thus sinking and displacing warmer water. Seawater of a high salinity has a higher relative density than less salty water, so it sinks and displaces water of less salinity. Likewise, seawater with a larger amount of suspended sediments has a higher relative density than clear water, so it sinks and displaces clear water. The following describes how these three ways of changing the density of seawater result in the ocean current known as a *density current,* which is an ocean current that flows because of density differences.

The earth receives more incoming solar radiation in the tropics than it does at the poles, which establishes a temperature difference between the tropical and polar oceans. The surface water in the polar ocean is often at or below the freezing point of freshwater, while the surface water in the tropical ocean averages about 26°C (about 79°F). Seawater freezes at a temperature below that of freshwater because the salt content lowers the freezing point. Seawater does not have a set freezing point, however, because as it freezes the salinity is increased as salt is excluded from the ice structure. Increased salinity lowers the freezing point more, so the more ice that freezes from seawater, the lower the freezing point for the remaining seawater. Cold seawater near the poles is therefore the densest, sinking and creeping slowly as a current across the ocean floor toward the equator. Where and how such a cold, dense bottom current moves is influenced by the shape of the ocean floor, the rotation of the earth, and other factors. The size and the distance that cold bottom currents move can be a surprise. Cold, dense water from the Arctic, for example, moves in a 200 m (about 660 ft) diameter current on the ocean bottom between Greenland and Iceland. This current carries an estimated 5 million m^3 per second (about 177 million ft^3/s) of seawater to the 3.5 km (about 2.1 mi) deep water of the North Atlantic Ocean. This is a flow rate about 250 times larger than that of the Mississippi River. At about 30°N, the cold Arctic waters meet even denser water that has moved in currents all the way from the Antarctic to the deepest part of the North Atlantic Basin (figure 17.21).

A second type of density current results because of differences in salinity. The water in the Mediterranean, for example,

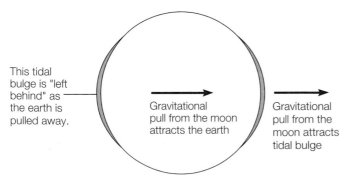

Figure 17.20

Gravitational attraction pulls on the earth's waters on the side of the earth facing the moon, producing a tidal bulge. A second tidal bulge on the side of the earth opposite the moon is produced when the earth, which is closer to the moon, is pulled away from the waters.

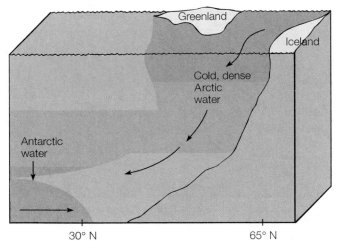

Figure 17.21

A cold density current carries about 250 times more water than the Mississippi River from the Arctic and between Greenland and Iceland to the deep Atlantic Ocean. At about 30°N latitude, it meets water that has moved by cold density currents all the way from the Antarctic.

has a high salinity because it is mostly surrounded by land in a warm, dry climate. The Mediterranean seawater, with its higher salinity, is more dense than the seawater in the open Atlantic Ocean. This density difference results in two separate currents that flow in opposite directions between the Mediterranean and the Atlantic. The greater density seawater flows from the bottom of the Mediterranean into the Atlantic, while the less dense

Atlantic water flows into the Mediterranean near the surface. The dense Mediterranean seawater sinks to a depth of about 1,000 m (about 3,300 ft) in the Atlantic, where it spreads over a large part of the North Atlantic Ocean. This increases the salinity of this part of the ocean, making it one of the more saline areas in the world.

The third type of density current occurs when underwater sediments on a slope slide toward the ocean bottom, producing a current of muddy or turbid water called a *turbidity current.* Turbidity currents are believed to be a major mechanism that moves sediments from the continents to the ocean basin. They may also be responsible for some undersea features, such as submarine canyons. Turbidity currents are believed to occur only occasionally, however, and none has ever been directly observed or studied. There is thus no data or direct evidence of how they form or what effects they have on the ocean floor.

Surface Currents. There are broad and deep-running ocean currents that slowly move tremendous volumes of water relatively near the surface. As shown in figure 17.22, each current is actually part of a worldwide system, or circuit, of currents. This system of ocean currents is very similar to the worldwide system of prevailing winds. This similarity exists because it is the friction of the prevailing winds on the seawater surface that drives the ocean currents. The currents are modified by other factors, such as the rotation of the earth and the shape of the ocean basins, but they are basically maintained by the wind systems.

Each ocean has a great system of moving water called a **gyre** that is centered in the mid-latitudes. The gyres rotate to the right

Figure 17.22

The earth's system of ocean currents.

KEY TERMS

abyssal plain (p. **370**)

aquifer (p. **357**)

continental shelf (p. **370**)

continental slope (p. **370**)

freshwater (p. **354**)

groundwater (p. **357**)

gyre (p. **369**)

ocean (p. **362**)

ocean basin (p. **370**)

ocean currents (p. **368**)

ocean wave (p. **365**)

runoff (p. **355**)

salinity (p. **363**)

surf (p. **367**)

swell (p. **366**)

tsunamis (p. **368**)

watershed (p. **356**)

water table (p. **357**)

APPLYING THE CONCEPTS

1. The most abundant chemical compound at the surface of the earth is
   (a) silicon dioxide.
   (b) nitrogen gas.
   (c) water.
   (d) minerals of iron, magnesium, and silicon.

2. Of the total supply, the amount of water that is available for human consumption and agriculture is
   (a) 97 percent.
   (b) about two-thirds.
   (c) about 3 percent.
   (d) less than 1 percent.

3. In a region of abundant rainfall a layer of extensively cracked, but otherwise solid, granite could serve as a limited source of groundwater because it has
   (a) limited permeability and no porosity.
   (b) average porosity and average permeability.
   (c) no permeability and no porosity.
   (d) limited porosity and no permeability.

4. How many different oceans are actually on the earth's surface?
   (a) 14
   (b) 7
   (c) 3
   (d) 1

5. The largest of the three principal ocean regions of the earth is the
   (a) Atlantic Ocean.
   (b) Pacific Ocean.
   (c) Indian Ocean.
   (d) South American Ocean.

6. The Gulf of Mexico is a shallow sea of the
   (a) Atlantic Ocean.
   (b) Pacific Ocean.
   (c) Indian Ocean.
   (d) South American Ocean.

7. Measurement of the salts dissolved in seawater taken from various locations throughout the world show that seawater has a
   (a) uniform chemical composition and a variable concentration.
   (b) variable chemical composition and a variable concentration.
   (c) uniform chemical composition and a uniform concentration.
   (d) variable chemical composition and a uniform concentration.

8. The salinity of seawater is *increased* locally by
   (a) the addition of water from a large river.
   (b) heavy precipitation.
   (c) the formation of sea ice.
   (d) none of the above.

9. Considering only the available light and the dissolving ability of gases in seawater, more abundant life should be found in a
   (a) cool, relatively shallow ocean.
   (b) warm, very deep ocean.
   (c) warm, relatively shallow ocean.
   (d) cool, very deep ocean.

10. The regular, low profile waves called swell are produced from
    (a) constant, prevailing winds.
    (b) small, irregular waves becoming superimposed.
    (c) longer wavelengths outrunning and outlasting shorter wavelengths.
    (d) all wavelengths becoming transformed by gravity as they travel any great distance.

11. If the wavelength of swell is 10.0 m, then you know that the fish below the surface feel the waves to a depth of
    (a) 5.0 m.
    (b) 10.0 m.
    (c) 20.0 m.
    (d) however deep it is to the bottom.

Answers

1. c 2. d 3. a 4. d 5. b 6. a 7. a 8. c 9. a 10. c 11. a

QUESTIONS FOR THOUGHT

1. Describe in general all the things that happen to the water that falls on the land.

2. Explain how a stream can continue to flow even during a dry spell.

3. What is the water table? What is the relationship between the depth to the water table and the depth that a well must be drilled? Explain.

4. Compare the advantages and disadvantages of using (a) surface water and (b) groundwater as a source of freshwater.

5. Prepare arguments for (a) agriculture, (b) industries, and (c) cities each having first priority in the use of a limited water supply. Identify one of these arguments as being the "best case" for first priority, then justify your choice.

6. Discuss some possible ways of extending the supply of freshwater.

7. The world's rivers and streams carry millions of tons of dissolved materials to the ocean each year. Explain why this does not increase the salinity of the ocean.

8. What is swell and how does it form?

9. Why do waves always seem to approach the shore head-on?

10. What factors determine the size of an ocean wave made by the wind?

11. Describe how a breaker forms from swell. What is surf?

12. Describe what you would look for to avoid where rip current occurs at a beach.

CHAPTER EIGHTEEN

Organic and Biochemistry

This is a computer-generated model of a benzene molecule, showing six carbon atoms (gold) and six hydrogen atoms (white). Benzene is a hydrocarbon, an organic compound made up of the elements carbon and hydrogen.

OUTLINE

The impact of ancient Aristotelian ideas on the development of understandings of motion, elements, and matter was discussed in earlier chapters. Historians also trace the "vitalist theory" back to Aristotle. According to Aristotle's idea, all living organisms are composed of the four elements (earth, air, fire, and water) and have in addition an *actuating force,* the life or soul that makes the organism different from nonliving things made of the same four elements. Plants, as well as animals, were considered to have this actuating, or vital, force in the Aristotelian scheme of things.

There were strong proponents of the vitalist theory as recently as the early 1800s. Their basic argument was that organic matter, the materials and chemical compounds recognized as being associated with life, could not be produced in the laboratory. Organic matter could only be produced in a living organism, they argued, because the organism had a vital force that is not present in laboratory chemicals.

Then, in 1828, a German chemist named Fredrich Wohler decomposed a chemical that was *not organic* to produce urea (N_2H_4CO), a known *organic* compound that occurs in urine. Wohler's production of an organic compound was soon followed by the production of other organic substances by other chemists. The vitalist theory gradually disappeared with each new reaction, and a new field of study, organic chemistry, emerged.

This chapter is an introductory survey of the field of organic chemistry, which is concerned with compounds and reactions of compounds that contain carbon. You will find this an interesting, informative introduction, particularly if you have ever wondered about synthetic materials, natural foods and food products, or any of the thousands of carbon-based chemicals you use every day. The survey begins with the simplest of organic compounds, those consisting of only carbon and hydrogen atoms, compounds known as hydrocarbons.

Hydrocarbons are the compounds of crude oil, which is the source of hundreds of petroleum products (figure 18.1).

Most common organic compounds can be considered derivatives of the hydrocarbons, such as alcohols, ethers, fatty acids, and esters. Some of these are the organic compounds that give flavors to foods, and others are used to make hundreds of commercial products, from face cream to oleo. The main groups, or classes, of derivatives will be briefly introduced, along with some interesting examples of each group. A very strong link exists between organic chemistry and the chemistry of living things, which is called biochemistry, or biological chemistry. Some of the important organic compounds of life, including proteins, carbohydrates, and fats, are discussed next. The chapter concludes with an introduction to synthetic polymers, what they are, and how they are related to the fossil fuel supply.

ORGANIC COMPOUNDS

Today, **organic chemistry** is defined as the study of compounds in which carbon is the principal element, whether the compound was formed by living things or not. The study of compounds that do not contain carbon as a central element is called **inorganic chemistry.** An *organic compound* is thus a compound in which carbon is the principal element, and an *inorganic compound* is any other compound.

Organic compounds, by definition, must contain carbon while all the inorganic compounds can contain all the other elements. Yet, the majority of known compounds are organic. Several million organic compounds are known and thousands of new ones are discovered every year. You use organic compounds every day, including gasoline, plastics, grain alcohol, foods, flavorings, and many others.

It is the unique properties of carbon that allow it to form so many different compounds. The carbon atom has a valence of four, and can combine with one, two, three, or four *other carbon atoms* in addition to a wide range of other kinds of atoms (figure 18.2). The number of possible molecular combinations is almost limitless, which explains why there are so many organic compounds. Fortunately, there are patterns of groups of carbon atoms and groups of other atoms that lead to similar chemical characteristics, making the study of organic chemistry less difficult. The key to success in studying organic chemistry is to recognize patterns and to understand the code and meaning of organic chemical names.

Figure 18.1

Refinery and tank storage facilities, like this one in Texas, are needed to change the hydrocarbons of crude oil into many different petroleum products. The classes and properties of hydrocarbons form one topic of study in organic chemistry.

Figure 18.3

A molecular formula (*A*) describes the numbers of different kinds of atoms in a molecule, and a structural formula (*B*) represents a two-dimensional model of how the atoms are bonded to each other. Each dash represents a bonding pair of electrons.

Figure 18.2

(*A*) The carbon atom forms bonds in a tetrahedral structure with a bond angle of 109.5°. (*B*) Carbon-to-carbon bond angles are 109.5°, so a chain of carbon atoms makes a zigzag pattern. (*C*) The unbranched chain of carbon atoms is usually simplified in a way that looks like a straight chain, but it is actually a zigzag, as shown in (*B*).

HYDROCARBONS

A **hydrocarbon** is an organic compound consisting of only two elements. As the name implies, these elements are hydrogen and carbon. The simplest hydrocarbon has one carbon atom and four hydrogen atoms (figure 18.3), but since carbon atoms can combine with one another, there are thousands of possible structures and arrangements. The carbon-to-carbon bonds are covalent and can be single, double, or triple (figure 18.4). Recall that the dash in a structural formula means one shared electron pair has a covalent bond. To satisfy the octet rule, this means that each satisfied carbon atom must have a total of four dashes around it each attached to another atom, no more and no less. Note that when the carbon atom has double or triple bonds, fewer hydrogen atoms can be attached as the octet rule is satisfied. There are four groups of hydrocarbons that are classified according to how the carbon atoms are put together, the (1) *alkanes*, (2) *alkenes*, (3) *alkynes*, and (4) *aromatic hydrocarbons*.

The **alkanes** are *hydrocarbons with single covalent bonds* between the carbon atoms. Alkanes that are large enough to form chains of carbon atoms occur with a straight structure, a branched structure, or a ring structure as shown in figure 18.5. (In 3-D the "straight" structure is actually a zigzag as shown in figure 18.2.) You are familiar with many alkanes, for they make up the bulk of petroleum and petroleum products.

The alkanes are also called the *paraffin series*. The alkanes are not as chemically reactive as the other hydrocarbons, and the term *paraffin* means "little affinity." They are called a series because *each higher molecular weight alkane has an additional CH_2*. The simplest alkane is methane, CH_4, and the next highest mo-

Figure 18.4

Carbon-to-carbon bonds can be single (*A*), double (*B*), or triple (*C*). Note that in each example, each carbon atom has four dashes, which represent four bonding pairs of electrons, satisfying the octet rule.

lecular weight alkane is ethane, C_2H_6. As you can see, C_2H_6 is CH_4 with an additional CH_2.

Note the names of the alkanes listed in table 18.1. After pentane the names have a consistent prefix and suffix pattern. The prefix and suffix pattern is a code that provides a clue about the compound. The Greek prefix tells you the *number of carbon atoms* in the molecule, for example, "oct-" means eight, so *octane* has eight carbon atoms. The suffix "-ane" tells you this hydrocarbon is a member of the alk*ane* series, so it has single bonds only. With the general alkane formula of C_nH_{2n+2}, you can now write the formula when you hear the name. Octane has eight carbon atoms with single bonds and $n = 8$. Two times 8 plus 2 ($2n + 2$) is 18, so the formula for octane is C_8H_{18}. Most organic chemical names provide clues like this.

A Straight chain for C_5H_{12}

B Branched chain for C_5H_{12}

C Ring chain for C_5H_{10}

Figure 18.5

Carbon-to-carbon chains can be (A) straight, (B) branched, or (C) in a closed ring. (Some carbon bonds are drawn longer, but are actually the same length.)

The alkanes in table 18.1 all have straight chains. A straight, continuous chain is identified with the term *normal,* which is abbreviated *n.* Figure 18.6A shows *n*-butane with a straight chain and a molecular formula of C_4H_{10}. Figure 18.6B shows a different branched structural formula that has the same C_4H_{10} molecular formula. Compounds with the same molecular formulas with different structures are called **isomers.** Since the straight-chained isomer is called *n*-butane, the branched isomer is called *isobutane.* The isomers of a particular alkane, such as butane, have different physical and chemical properties because they have different structures. Isobutane, for example, has a boiling point of $-10°C$. The boiling point of *n*-butane, on the other hand, is $-0.5°C$. In the next section you will learn that the various isomers of the octane hydrocarbon perform differently in automobile engines, requiring the "reforming" of *n*-octane to *iso*-octane before it can be used.

Methane, ethane, and propane can have only one structure each, and butane has two isomers. The number of possible isomers for a particular molecular formula increases rapidly as the number of carbon atoms increase. After butane, hexane has five isomers, octane eighteen isomers, and decane seventy-five isomers. Because they have different structures, each isomer has different physical properties. A different naming system is needed because there are just too many isomers to keep track of. The system of naming the branched-chain alkanes is described by rules agreed upon by a professional organization, the International Union of Pure and Applied Chemistry, or IUPAC.

Alkenes and Alkynes

The **alkenes** are *hydrocarbons with a double covalent carbon-to-carbon bond.* To denote the presence of a double bond the "-ane"

Table 18.1 Examples Of Straight-Chain Alkanes

Name	Molecular Formula	Structural Formula	Name	Molecular Formula	Structural Formula
Methane	CH_4		Butane	C_4H_{10}	
Ethane	C_2H_6		Heptane	C_7H_{16}	
Propane	C_3H_8		Octane	C_8H_{18}	

Figure 18.6
(*A*) A straight-chain alkane. (*B*) A branched-chain alkane isomer.

suffix of the alkanes is changed to "-ene" as in alk*ene* (table 18.2). Figure 18.4 shows the structural formula for (A) ethane, C_2H_6, and (B) ethene, C_2H_4. Alkenes have room for two fewer hydrogen atoms because of the double bond, so the general alkene formula is C_nH_{2n}. Note the simplest alkene is called ethene, but is commonly known as ethylene.

Ethylene is an important raw material in the chemical industry. Obtained from the processing of petroleum, about half of the commercial ethylene is used to produce the familiar polyethylene plastic. It is also produced by plants to ripen fruit, which explains why unripe fruit enclosed in a sealed plastic bag with ripe fruit will ripen more quickly. The ethylene produced by the ripe fruit acts on the unripe fruit. Commercial fruit packers sometimes use small quantities of ethylene gas to quickly ripen fruit that was picked while green.

Perhaps you have heard the terms "saturated" and "unsaturated" in advertisements for cooking oil and margarine. An organic molecule, such as a hydrocarbon, that does not contain the maximum number of hydrogen atoms is an **unsaturated** hydrocarbon. For example, ethylene can add more hydrogen atoms by reacting with hydrogen gas to form ethane:

The ethane molecule has all the hydrogen atoms possible, so ethane is a **saturated** hydrocarbon. Unsaturated molecules are less stable, which means that they are more chemically reactive than saturated molecules.

An **alkyne** is a *hydrocarbon with a carbon-to-carbon triple bond* and the general formula of C_nH_{2n-2}. The alkynes are highly reactive, and the simplest one, ethyne, has a common name of acetylene. Acetylene is commonly burned with oxygen gas in a welding torch because the flame reaches a temperature of about 3,000°C. Acetylene is also an important raw material in the production of plastics.

Cycloalkanes and Aromatic Hydrocarbons

The hydrocarbons discussed up until now have been straight or branched open-ended chains of carbon atoms. Carbon atoms can also bond to each other to form a ring, or *cyclic,* structure. Figure 18.7 shows the structural formulas for some of these cyclic structures. Note that the cycloalkanes have the same molecular formulas as the alkenes, and thus they are isomers of the alkenes. They are, of course, very different compounds, with different physical and chemical properties. This shows the importance of structural, rather than simply molecular, formulas in referring to organic compounds.

The six-carbon ring structure shown in figure 18.8A has three double bonds that do not behave like the double bonds in the alkenes. In this six-carbon ring the double bonds are not localized in one place but are spread over the whole molecule. Instead of alternating single and double bonds, all the bonds are something in between. This gives the C_6H_6 molecule increased stability. As a result, the molecule does not behave like other unsaturated compounds, that is, it does not readily react in order to add hydrogen to the ring. The C_6H_6 molecule is the organic compound named *benzene.* Organic compounds that are based on the benzene ring structure are called *aromatic hydrocarbons.* To denote the six-carbon ring with delocalized electrons, benzene is represented by the symbol shown in figure 18.8B.

The circle in the six-sided benzene symbol represents the delocalized electrons. Figure 18.8B illustrates how this benzene

Group	General Molecular Formula	Example Compound	Molecular Structure
Alkanes	C_nH_{2n+2}	Ethane	
Alkenes	C_nH_{2n}	Ethene	
Alkynes	C_nH_{2n-2}	Ethyne	$H-C\equiv C-H$

Table 18.2 The General Molecular Formulas and Molecular Structures of the Alkanes, Alkenes, and Alkynes

Figure 18.7

(*A*) The "straight" chain has carbon atoms that are able to rotate freely around their single bonds, sometimes linking up in a closed ring. (*B*) Ring compounds of the first four cycloalkanes.

Figure 18.8

(*A*) The bonds in C_6H_6 are something between single and double, which gives it different chemical properties than double-bonded hydrocarbons. (*B*) The six-sided symbol with a circle represents the benzene ring. Organic compounds based on the benzene ring are called *aromatic hydrocarbons* because of their aromatic character.

ring symbol is used to show the structural formula of some aromatic hydrocarbons. You may have noticed some of the names on labels of paints, paint thinners, and lacquers. Toluene and the xylenes are commonly used in these products as solvents. A benzene ring attached to another molecule or functional group is given the name *phenyl*.

PETROLEUM

Petroleum is a mixture of alkanes, cycloalkanes, and some aromatic hydrocarbons. The origin of petroleum is uncertain, but it is believed to have formed from the slow decomposition of buried marine life, primarily plankton and algae in the absence of oxygen (i.e., anaerobic). Time, temperature, pressure, and perhaps bacteria are considered important in the formation of petroleum. As the petroleum formed, it was forced through porous rock until it reached a rock type or rock structure that stopped it. Here, it accumulated to saturate the porous rock, forming an accumulation called an **oil field.** The composition of petroleum varies from one oil field to the next. The oil from a

given field might be dark or light in color, and it might have an asphalt base or a paraffin base. Some oil fields contain oil with a high quantity of sulfur, referred to as "sour crude." Because of such variations, some fields have oil with more desirable qualities than oil from other fields.

Early settlers found oil seeps in the eastern United States and collected the oil for medicinal purposes. One enterprising oil peddler tried to improve the taste by running the petroleum through a whiskey still. He obtained a clear liquid by distilling the petroleum and, by accident, found that the liquid made an excellent lamp oil. This was fortunate timing, for the lamp oil used at that time was whale oil, and whale oil production was declining. This clear liquid obtained by distilling petroleum is today known as *kerosene.*

Wells were drilled, and crude oil refineries were built to produce the newly discovered lamp oil. Gasoline was a by-product of the distillation process and was used primarily as a spot remover. With Henry Ford's automobile production and Edison's

Figure 18.9

Crude oil from the ground is separated into usable groups of hydrocarbons at this Louisiana refinery. Each petroleum product has a boiling point range, or "cut," of distilled vapors that collect in condensing towers.

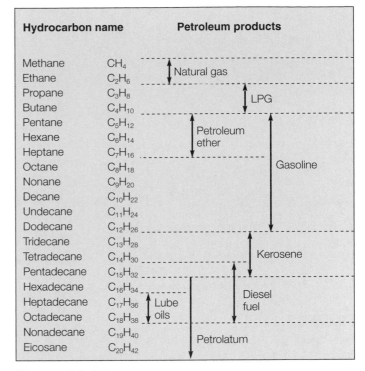

Figure 18.10

Petroleum products and the ranges of hydrocarbons in each product.

Table 18.3 Petroleum Products

Name	Boiling Range (°C)	Carbon Atoms per Molecule
Natural gas	Less than 0	C_1 to C_4
Petroleum ether	35–100	C_5 to C_7
Gasoline	35–215	C_5 to C_{12}
Kerosene	35–300	C_{12} to C_{15}
Diesel fuel	300–400	C_{15} to C_{18}
Motor oil, grease	350–400	C_{16} to C_{18}
Paraffin	Solid, melts at about 55	C_{20}
Asphalt	Boiler residue	C_{40} or more

electric light invention, the demand for gasoline increased, and the demand for kerosene decreased. The refineries were converted to produce gasoline, and the petroleum industry grew to become one of the world's largest industries.

Crude oil is petroleum that is pumped from the ground, a complex and variable mixture of hydrocarbons with an upper limit of about fifty carbon atoms. This thick, smelly black mixture is not usable until it is refined, that is, separated into usable groups of hydrocarbons called petroleum products. Petroleum products are separated by distillation. The larger the molecule, the higher the boiling point. Thus, each product has a boiling point range, or "cut" of the distilled vapors (figure 18.9). Thus, each product, such as gasoline or heating oil, is made up of hydrocarbons within a range of carbon atoms per molecule (figure 18.10). The products, their boiling ranges, and ranges of carbon atoms per molecule are listed in table 18.3.

The hydrocarbons that have one to four carbon atoms (CH_4 to C_4H_{10}) are gases at room temperature. They can be pumped from certain wells as a gas, but they also occur dissolved in crude oil. *Natural gas* is a mixture of hydrocarbon gases, but it is about 95 percent methane (CH_4). Propane (C_3H_8) and butane (C_4H_{10}) are liquified by compression and cooling and are sold as liquified petroleum gas, or *LPG*. LPG is used where natural gas is not available for cooking or heating and is widely used as a fuel in barbecue grills and camp stoves.

Gasoline is a mixture of hydrocarbons that may have five to twelve carbon atoms per molecule. Gasoline distilled from crude oil consists mostly of straight-chain molecules not suitable for use as an automotive fuel. Straight-chain molecules burn too rapidly in an automobile engine, producing more of an explosion than a smooth burn. You hear these explosions as a knocking or pinging in the engine, and they indicate poor efficiency and could damage the engine. On the other hand, branched-chain molecules burn comparatively slower, without the pinging

or knocking explosions. The burning rate of gasoline is described by the *octane number* scale. The scale is based on pure *n*-heptane, straight-chain molecules that are assigned an octane number of 0, and a multiple branched isomer of octane, trimethylpentane, which is assigned an octane number of 100 (figure 18.11). Most unleaded gasolines have an octane rating of 87, which could be obtained with a mixture that is 87 percent trimethylpentane and 13 percent heptane. Gasoline, however, is a much more complex mixture.

A heptane, C_7H_{16}

B trimethylpentane (or iso-octane), C_8H_{18}

Figure 18.11

The octane rating scale is a description of how rapidly gasoline burns. It is based on (A) heptane, with an assigned octane number of 0, and (B) trimethylpentane, with an assigned number of 100.

Kerosene is a mixture of hydrocarbons that have from twelve to fifteen carbon atoms. The petroleum product called kerosene is also known by other names, depending on its use. Some of these names are lamp oil (with coloring and odorants added), jet fuel (with a flash flame retardant added), heating oil, #1 fuel oil, and in some parts of the country, "coal oil."

Diesel fuel is a mixture of a group of hydrocarbons that have from fifteen to eighteen carbon atoms per molecule. Diesel fuel also goes by other names, again depending on its use, i.e., distillate fuel oil, or # 2 fuel oil.

Motor oil and *lubricating oils* have sixteen to eighteen carbon atoms per molecule. Lubricating grease is heavy oil that is thickened with soap. *Petroleum jelly,* also called petrolatum (or Vaseline), is a mixture of hydrocarbons with sixteen to thirty-two carbon atoms per molecule. *Mineral oil* is a light lubricating oil that has been decolorized and purified.

Depending on the source of the crude oil, varying amounts of *paraffin* wax (C_{20} or greater) or *asphalt* (C_{36} or more) may be present. Paraffin is used for candles, waxed paper, and home canning. Asphalt is mixed with gravel and used to surface roads.

HYDROCARBON DERIVATIVES

The hydrocarbons account for only about 5 percent of the known organic compounds, but the other 95 percent can be considered as hydrocarbon derivatives. **Hydrocarbon derivatives** are formed when *one or more hydrogen atoms on a hydrocarbon have been replaced by some element or group of elements*

other than hydrogen. For example, the halogens (F_2, Cl_2, Br_2) react with an alkane in sunlight or when heated, replacing a hydrogen:

In this particular *substitution reaction* a hydrogen atom on methane is replaced by a chlorine atom to form methyl chloride. Replacement of any number of hydrogen atoms is possible, and a few *organic halides* are illustrated in figure 18.12.

If a hydrocarbon molecule is unsaturated (has a multiple bond), a hydrocarbon derivative can be formed by an *addition reaction:*

The bromine atoms add to the double bond on propene, forming dibromopropane.

Alkene molecules can also add to each other in an addition reaction to form a very long chain consisting of hundreds of molecules. A long chain of repeating units is called a **polymer** (*poly* = many; *mer* = segment), individual pieces are called **monomers** (*mono* = single; *mer* = segment or piece), and the reaction is called *addition polymerization*. Ethylene, for example,

Chloroform ($CHCl_3$)

Carbon tetrachloride (CCl_4)

Dichlorodifluoromethane (a Freon, CCl_2F_2)

Vinyl chloride (C_2H_3Cl)

Figure 18.12

Common examples of organic halides.

is heated under pressure with a catalyst to form *polyethylene.* Heating breaks the double bond,

which provides sites for single covalent bonds to join the ethylene units together,

which continues the addition polymerization until the chain is hundreds of units long.

The atom or group of atoms in an organic molecule that is the site of a chemical reaction is identified as a **functional group.** *It is the functional group that is responsible for the chemical properties of an organic compound.* Functional groups usually have (1) multiple bonds or (2) lone pairs of electrons that cause them to be sites of reactions. Table 18.4 lists some of the common hydrocarbon functional groups. Look over this list, comparing the structure of the functional group with the group name. Some of the more interesting examples from a few of these groups will be considered next. Note that the R and R′ (pronounced, "R prime") stand for one or more of the hydrocarbon groups from table 18.5. For example, in the reaction between methane and chlorine, the product is methyl chloride. In this case the R in RCl stands for methyl, but it could represent any hydrocarbon group.

Alcohols

An *alcohol* is an organic compound formed by replacing one or more hydrogens on an alkane with a hydroxyl functional group (−OH). The hydroxyl group should not be confused with the hydroxide ion, OH^-. The hydroxyl group is attached to an organic compound and does not form ions in solution as the hydroxide ion does. It remains attached to a hydrocarbon group (R), giving the compound its set of properties that are associated with alcohols.

The name of the hydrocarbon group (table 18.5) determines the name of the alcohol. If the hydrocarbon group in ROH is methyl, for example, the alcohol is called *methyl alcohol.* Using the IUPAC naming rules, the name of an alcohol has the suffix "-ol" (figure 18.13).

All alcohols have the hydroxyl functional group, and all are chemically similar. Alcohols are toxic to humans, for example,

Table 18.4 Selected Organic Functional Groups

Name of Functional Group	General Formula	General Structure
Organic Halide	RCl	R—C̈l:
Alcohol	ROH	R—Ö—H
Ether	ROR′	R—Ö—R′
Aldehyde	RCHO	R—C—H ‖ :O:
Ketone	RCOR′	R—C—R′ ‖ :O:
Organic Acid	RCOOH	R—C—Ö—H ‖ :O:
Ester	RCOOR′	R—C—Ö—R′ ‖ :O:
Amine	RNH₂	R—N̈—H | H

Table 18.5 Alkane Hydrocarbons and Corresponding Hydrocarbon Groups

Alkane Name	Molecular Formula	Hydrocarbon Group	Molecular Formula
Methane	CH_4	Methyl	$-CH_3$
Ethane	C_2H_6	Ethyl	$-C_2H_5$
Propane	C_3H_8	Propyl	$-C_3H_7$
Butane	C_4H_{10}	Butyl	$-C_4H_9$
Pentane	C_5H_{12}	Amyl	$-C_5H_{11}$
Hexane	C_6H_{14}	Hexyl	$-C_6H_{13}$
Heptane	C_7H_{16}	Heptyl	$-C_7H_{15}$
Octane	C_8H_{18}	Octyl	$-C_8H_{17}$

Note: $-CH_3$ means
$$* \;-\; \overset{\textstyle H}{\underset{\textstyle H}{C}} \;-\; H$$
where * denotes unattached. The attachment takes place on a base chain or functional group.

Methanol

H
|
H — C — OH
|
H

(methyl alcohol)

Ethanol

H   H
| |
H — C — C — OH
| |
H   H

(ethyl alcohol)

propanol

H   H   H
| | |
H — C — C — C — OH
| | |
H   H   H

(propyl alcohol)

propanol

H   OH   H
| | |
H — C — C — C — H
| | |
H   H   H

(isopropyl alcohol)

Figure 18.13
Four different alcohols.

except that ethanol can be consumed in limited quantities. Consumption of other alcohols such as propanol (isopropyl alcohol, a component of rubbing alcohol) can result in serious gastric illness. Consumption of methanol can result in blindness and death. Ethanol, C_2H_5OH, is produced by the action of yeast or by a chemical reaction of ethylene derived from petroleum refining. Yeast acts on sugars to produce ethanol and CO_2. When beer, wine, and other such beverages are the desired products, the CO_2 escapes during fermentation, and the alcohol remains in solution. In baking, the same reaction utilizes the CO_2 to make the dough rise, and the alcohol is evaporated during baking. Most alcoholic beverages are produced by the yeast fermentation reaction:

sugar → carbon dioxide and ethyl alcohol,

but some are made from ethanol derived from petroleum refining.

Alcohols with six or fewer carbon atoms per molecule are soluble in both alkanes and water. A solution of ethanol and gasoline is called *gasohol* (figure 18.14). Alcoholic beverages are a solution of ethanol and water. The *proof* of such a beverage is double the ethanol concentration by volume. Therefore, a solution of 40 percent ethanol by volume in water is 80 proof,

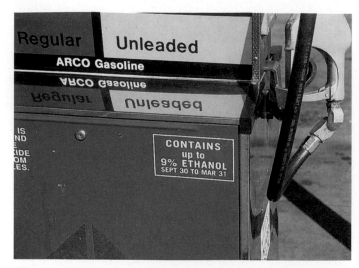

Figure 18.14
Gasoline is a mixture of hydrocarbons (C_8H_{18} for example) that contain no atoms of oxygen. Gasohol contains ethyl alcohol, C_2H_5OH, which does contain oxygen. The addition of alcohol to gasoline, therefore, adds oxygen to the fuel. Since carbon monoxide forms when there is an insufficient supply of oxygen, the addition of alcohol to gasoline helps cut down on carbon monoxide emissions. An atmospheric inversion, with increased air pollution, is likely during the dates shown on the pump, so that is when the ethanol is added.

and wine that is 12 percent alcohol by volume is 24 proof. Distillation alone will produce a 190 proof concentration, but other techniques are necessary to obtain 200 proof absolute alcohol. *Denatured alcohol* is ethanol with acetone, formaldehyde, and other chemicals in solution that are difficult to separate by distillation. Since these denaturants make consumption impossible, denatured alcohol is sold without the consumption tax.

Methanol, ethanol, and isopropyl alcohol all have one hydroxyl group per molecule. An alcohol with two hydroxyl groups per molecule is called a *glycol*. Ethylene glycol is perhaps the best-known glycol since it is used as an antifreeze. An alcohol with three hydroxyl groups per molecule is called *glycerol* (or glycerin). Glycerol is a building block of fat molecules and a by-product in the making of soap. It is added to toothpastes, lotions, and some candies to retain moisture and softness. Ethanol, ethylene glycol, and glycerol are compared in figure 18.15.

Glycerol reacts with nitric acid in the presence of sulfuric acid to produce glyceryl trinitrate, commonly known as *nitroglycerine*. Nitroglycerine is a clear oil that is violently explosive, and when warmed, it is extremely unstable. In 1867, Alfred Nobel discovered that a mixture of nitroglycerine and siliceous earth was more stable than pure nitroglycerine but was nonetheless explosive. Siliceous earth is also known as diatomaceous earth. It is composed of the skeletons of algae known as diatoms and provides an enormous surface area upon which the explosive reaction can take place. The cell walls of diatoms are composed of silicon dioxide (SiO_2). The mixture is packed in a tube and is called *dynamite*. The name dynamite comes from the Greek word meaning 'power.'

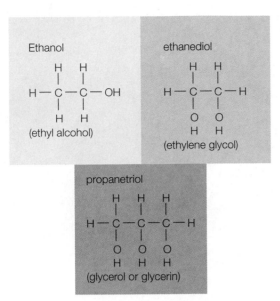

Figure 18.15
Common examples of alcohols with one, two, and three hydroxyl groups per molecule.

Figure 18.16
The carbonyl group (*A*) is present in both aldehydes and ketones, as shown in (*B*). (*C*) The simplest example of each with the more technical name above the common name below each formula.

Ethers, Aldehydes, and Ketones

An *ether* has a general formula of ROR′, and the best-known ether is diethylether. In a molecule of diethylether, both the R and the R′ are ethyl groups. Diethylether is a volatile, highly flammable liquid that was used as an anesthetic in the past. Today, it is used as an industrial and laboratory solvent.

Aldehydes and *ketones* both have a functional group of a carbon atom doubly bonded to an oxygen atom called a *carbonyl group.* The *aldehyde* has a hydrocarbon group, R (or a hydrogen in one case), and a hydrogen attached to the carbonyl group. A *ketone* has a carbonyl group with two hydrocarbon groups attached (figure 18.16).

The simplest aldehyde is *formaldehyde.* Formaldehyde is soluble in water, and a 40 percent concentration called *formalin* has been used as an embalming agent and to preserve biological specimens. Formaldehyde is also a raw material used to make plastics such as Bakelite. All the aldehydes have odors, and the odors of some aromatic hydrocarbons include the odors of almonds, cinnamon, and vanilla. The simplest ketone is *acetone.* Acetone has a fragrant odor and is used as a solvent in paint removers and nail polish removers.

Organic Acids and Esters

Mineral acids, such as hydrochloric acid and sulfuric acid, are made of inorganic materials. Acids that were derived from organisms are called **organic acids.** Because many of these organic acids can be formed from fats, they are sometimes called *fatty acids.* Chemically, they are known as the *carboxylic acids* because they contain the carboxyl functional group, −COOH, and have a general formula of RCOOH.

The simplest carboxylic acid has been known since the Middle Ages, when it was isolated by the distillation of ants. The Latin word *formica* means "ant," so this acid was given the name *formic acid* (figure 18.17A). Formic acid is

$$H - C - OH$$
$$\parallel$$
$$O$$

It is formic acid, along with other irritating materials, that causes the sting of bees, ants, and certain plants, for example the Stinging Nettle (*Urtica dioica*) (figure 18.17B).

Acetic acid, the acid of vinegar, has been known since antiquity. Acetic acid forms from the oxidation of ethanol. An oxidized bottle of wine contains acetic acid in place of the alcohol, which gives the wine a vinegar taste. Before wine is served in a restaurant, the person ordering is customarily handed the bottle cork and a glass with a small amount of wine. You first break the cork in half to make sure it is dry, which tells you that the wine has been sealed from oxygen. The small sip is to taste for vinegar before the wine is served. If the wine has been oxidized, the reaction is

Organic acids are common in many foods. The juice of citrus fruit, for example, contains citric acid, which relieves a

A

B

Figure 18.17

These red ants (*A*) are classified by biologists as: Class: Insecta, Order: Hymenoptera, Family: Formicidae, Genus: *Solenopis*. Species = *Solenopsis invicta*. Like other ants, they make the simplest of the organic acids, formic acid. The sting of bees, ants, and some plants contains formic acid, along with some other irritating materials. Formic acid is HCOOH. The plant known as the Stinging Nettle (*B*) (*Urtica dioica*) has small, hollow hairs that contain several irritating substances, including formic acid. These hairs have the ability to scratch the skin and mucous membranes, resulting in almost immediate burning, itching, and irritation.

CONCEPTS APPLIED
Household Organic Compounds

Pick a household product that has ingredients sounding like they could be organic compounds. On a separate sheet of paper write the brand name of the product and the type of product (Example: Oil of Olay; skin moisturizer), then list the ingredients one under the other (column 1). In a second column beside each name put the type of compound if you can figure it out from its name or find it in any reference (Example: cetyl palmitate—an ester of cetyl alcohol and palmitic acid.) In a third column put the structural formula if you can figure it out or find it in any reference such as a CRC handbook or the Merck Index. Finally, in a fourth column put whatever you know or can find out about the function of that substance in the product.

---

thirsty feeling by stimulating the flow of saliva. Lactic acid is found in sour milk, buttermilk, sauerkraut, and pickles. Lactic acid also forms in your muscles as a product of anaerobic carbohydrate metabolism, causing a feeling of fatigue. Citric and lactic acids are small molecules compared to some of the carboxylic acids that are formed from fats. Palmitic acid, for example, is $C_{16}H_{32}O_2$ and comes from palm oil. The structure of palmitic acid is a chain of fourteen CH_2 groups with CH_3- at one end and $-COOH$ at the other. Again, it is the functional carboxyl group, $-COOH$, that gives the molecule its acid properties. Organic acids are also raw materials used in the making of polymers of fabric, film, and paint.

Esters are common in both plants and animals, giving fruits and flowers their characteristic odor and taste. Esters are also used in perfumes and artificial flavorings. A few of the flavors that particular esters are responsible for are listed in table 18.6.

Mother Vinegar

Vinegar is an important acidifier and preservative used in the food industry and the kitchen. It is regularly used for preserving and acidifying vegetables, sauces, and salads. It is also used for making pickles and disinfecting. There is evidence that vinegar was made in ancient Egypt, Greece, and Rome. This organic acid is the product of microbial fermentation. Several members of the bacterial genus *Acetobacter* (the '*mother of vinegar*') oxidize ethyl alcohol to acetic acid (4 to 12 percent) under aerobic conditions. This fermentation is called *acetification*. Grape, apple, and date wine wastes are typically used in the commercial preparation of vinegar. White distilled vinegar is produced in a fermenter by spraying alcohol over the surfaces of an *Acetobacter*-inoculatated material that provides a large surface area for acetification. Beech wood chips, corncobs, or charcoal can be in the fermenter, called a vinegar *generator*. As the solution trickles over the surfaces, the *Acetobacter* converts the alcohol to acetic acid, and vinegar is drained from the tank. Using this controlled acetification method produces a high-quality consistent product. Today many industrial producers of vinegar use another type of fermenter that is fully automated and needs no

Box Figure 18.1

packing material such as beech wood shavings. Unlike the old vinegar generator, this large-scale operation can handle a variety of raw alcoholic materials (box figure 18.1).

Table 18.6 Flavors and Esters

Ester Name	Formula	Flavor
Amyl Acetate	$CH_3-\overset{\overset{\textstyle O}{\|}}{C}-O-C_5H_{11}$	Banana
Octyl Acetate	$CH_3-\overset{\overset{\textstyle O}{\|}}{C}-O-C_8H_{17}$	Orange
Ethyl Butyrate	$C_3H_7-\overset{\overset{\textstyle O}{\|}}{C}-O-C_2H_5$	Pineapple
Amyl Butyrate	$C_3H_7-\overset{\overset{\textstyle O}{\|}}{C}-O-C_5H_{11}$	Apricot
Ethyl Formate	$H-\overset{\overset{\textstyle O}{\|}}{C}-O-C_2H_5$	Rum

These liquid esters can be obtained from natural sources or they can be chemically synthesized. Whatever the source, amyl acetate, for example, is the chemical responsible for what you identify as the flavor of banana. Natural flavors, however, are complex mixtures of these esters along with other organic compounds. Lower molecular weight esters are fragrant-smelling liquids, but higher molecular weight esters are odorless oils and fats.

ORGANIC COMPOUNDS OF LIFE

Aristotle and the later proponents of the vitalist theory were *partly* correct in their concept that living organisms are different from inorganic substances made of the same elements. Living organisms, for example, have the ability to (1) exchange matter and energy with their surroundings and (2) transform matter and energy into different forms as they (3) respond to changes in their surroundings. In addition, living organisms can use the transformed matter and energy to (4) grow and (5) reproduce. Living organisms are able to do these things through a great variety of organic reactions that are catalyzed by enzymes, however, and not through some mysterious "vital force." These enzyme-regulated organic reactions take place because living organisms are highly organized and have an incredible number of relationships between many different chemical processes.

Organic molecules in living things have molecular weights of thousands or millions of atomic mass units (u's) and are therefore referred to as **macromolecules.** Many types of macromolecules are comprised of similar repeated monomers. Monomers are usually combined to form a polymer by a chemical reaction (*dehydration synthesis*) that results in a water molecule being removed from between two monomers. The reverse of this reaction (*hydrolysis*) is the process of splitting a larger polymer into its parts by the addition of water. Digestion of food molecules in the stomach and small intestine is an important

Glucose (an aldehyde sugar)

Fructose (a ketone sugar)

Figure 18.18

Glucose (blood sugar) is an aldehyde, and fructose (fruit sugar) is a ketone. Both have a molecular formula of $C_6H_{12}O_6$.

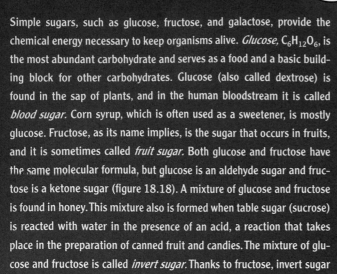
Table 18.7 The Relative Sweetness of Various Sugars and Sugar Substitutes	
Type of Sugar or Artificial Sweetener	**Relative Sweetness**
Lactose (milk sugar)	0.16
Galactose	0.30
Maltose (malt sugar)	0.33
Glucose	0.75
Sucrose (table sugar)	**1.00**
Fructose (fruit sugar)	1.75
Cyclamate	30.00
Aspartame	150.00
Saccharin	350.00

example of this reaction. There are three main types of polymeric macromolecules: (1) carbohydrates, (2) proteins, and (3) nucleic acids. In addition, there is a fourth category, the lipids, that are not true polymers but are macromolecules.

Carbohydrates

One class of organic molecules, **carbohydrates,** is composed of carbon, hydrogen, and oxygen atoms linked together to form monomers called *simple sugars* or *monosaccharides* (*mono* = single; *saccharine* = sweet, sugar) (table 18.7). The name *carbohydrate* literally means, "watered carbon," and the empirical formula (CH_2O) for most carbohydrates indicates one carbon (C) atom for each water (H_2O).

The formula for a simple sugar is easy to recognize because there are equal numbers of carbons and oxygens and twice as many hydrogens—for example, $C_3H_6O_3$ or $C_5H_{10}O_5$. We usually describe simple sugars by the number of carbons in the molecule. The ending -*ose* indicates that you are dealing with a carbohydrate. A tri*ose* has three carbons, a pent*ose* has five, and a hex*ose* has six. If you remember that the number of carbons equals the number of oxygen atoms and that the number of hydrogens is double that number, these names tell you the formula for the simple sugar. Carbohydrates play a number of roles in living things. They serve as an immediate source of energy (sugars), provide shape to certain cells (cellulose in plant cell walls), are components of many antibiotics and coenzymes, and are an essential part of genes (DNA).

Simple sugars can be combined with each other to form **complex carbohydrates** (figure 18.19). When two simple sugars bond to each other, a *disaccharide* (*di-* = two) is formed; when three bond together, a *trisaccharide* (*tri-* = three) is formed. Generally we call a complex carbohydrate that is larger than this a *polysaccharide* (many sugar units). In all cases, the complex carbohydrates are formed by the removal of water from between the sugars. For example, when glucose and fructose are joined together they form a *disaccharide (sucrose)* with the loss of a water molecule,

$$C_6H_{12}O_6 \; + \; C_6H_{12}O_6 \; \rightarrow \; C_{12}H_{22}O_{11} \; + \; H_2O$$

glucose   fructose   sucrose   water

Sucrose, ordinary table sugar, is the most common disaccharide and occurs in high concentrations in sugar cane and sugar beets. It is extracted by crushing the plant materials, then dissolving the sucrose from the materials with water. The water is evaporated and the crystallized sugar is decolorized with charcoal to produce white sugar. Other common disaccharides include *lactose* (milk sugar) and *maltose* (malt sugar). All three disaccharides have similar properties, but maltose tastes only about one-third as sweet as sucrose. Lactose tastes only about one-sixth as sweet as sucrose. No matter which disaccharide sugar is consumed (sucrose, lactose, or maltose), it is converted into glucose and transported by the bloodstream for use by the body.

Charcoal Briquettes from Wood

Charcoal, tar, acetic acid, acetone, turpentine oil, wood pitch, and methanol are all the products of a process called the *pyrolysis* of wood. Charcoal has been one of the most widely used of these products throughout history. During the Bronze Age shallow pits of charcoal were used to smelt tin. Besides its primary use as a heating agent, charcoal can also be used as a purifying agent. *Activated* charcoal is processed in such a manner that it develops an enormous surface area in comparison to briquettes. Because of this, it can be used for filtering chemicals and odors from water, aquariums, sugars, and the air.

Pyrolysis (destruction by fire) is the anaerobic degradation of wood by heat. This reaction results in the loss of hydrogen and oxygen from the carbon atoms. The pyrolysis of wood uses a combination of pressure and temperature to produce charcoal. Wood charcoal has a heating value of over 13,300 BTU/lb, comparable to the highest grade of hard coal. Charcoal is made from small, chipped pieces of wood or large logs. It takes about 5 kg of wood to produce 1 kg of charcoal.

To make briquettes, the charcoal is crushed into fine particles and passed through a variety of screens to ensure their small size. Starch and water are added as a binder. The charcoal comprises 75 percent of the briquette mixture. As a heat source, briquettes have several advantages over other materials:

1. a low sulfur content,
2. relatively few and unreactive inorganic impurities,
3. a large surface area, and
4. little smoke discharge.

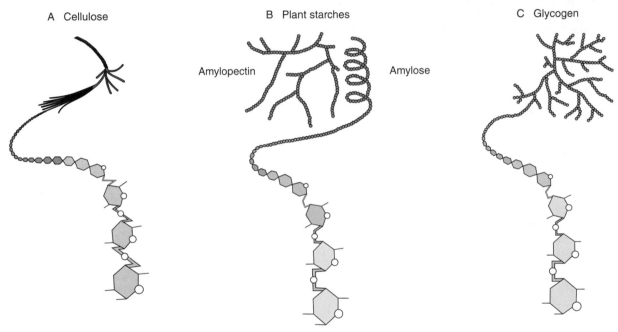

A   Cellulose   B   Plant starches   C   Glycogen

Amylopectin   Amylose

Figure 18.19

Simple sugars are attached to each other by the removal of water from between them. Three common complex carbohydrates are (*A*) cellulose (wood fibers), (*B*) plant starch (amylopectin and amylose), and (*C*) glycogen (sometimes called animal starch). Glycogen is found in muscle cells. Notice how each is similar in that they are all polymers of simple sugars, but differ from one another in how they are joined together. While many organisms are capable of digesting (hydrolyzing) the bonds that are found in glycogen and plant starch molecules, few are able to break those that link together the monosaccharides of cellulose.

Some common examples of polysaccharides are **cellulose, starch** and **glycogen.** Cellulose is an important polysaccharide used in constructing plant cell walls. Humans cannot digest this complex carbohydrate, so we are not able to use it as an energy source. On the other hand, animals known as ruminants (e.g., cows and sheep) and termites have microorganisms within their digestive tracts that do digest cellulose, making it an energy source for them. Plant cell walls add bulk or fiber to our diet, but no calories. Fiber is an important addition to the diet because it helps control weight, reduces the risks of colon cancer, and con-trols constipation and diarrhea. Starch is also a plant product digestible by most other organisms. Once broken down, the monosaccharide can be used as an energy source or building materials. A close but structurally different polysaccharide is glycogen. This macromolecule is found in the muscle cells of many animals as a storage form of polysaccharide.

Many types of sugars can be used by cells as components in other, more complex molecules. Sugar molecules are a part of molecules such as DNA (deoxyribonucleic acid), RNA (ribonucleic acid), or ATP (adenosine triphosphate). The ATP molecule is important in energy transfer. It has a simple sugar (ribose) as part of its structural makeup. RNA is important in the synthesis of cell proteins and also contains the monosaccharide ribose. The building blocks of the genetic material (DNA) also have a sugar component, deoxyribose.

Proteins

Proteins play many important roles. Enzymes are catalysts that speed the rate of chemical reactions. Proteins like hemoglobin serve as carriers of other molecules such as oxygen. Others like collagen provide shape and support, and several kinds of protein in muscle cells are responsible for movement. Proteins also act as chemical messengers (certain hormones) and help defend the body against dangerous microbes and chemicals (antibodies). Chemically, proteins are polymers made up of monomers known as *amino acids.* An **amino acid** is a short carbon skeleton that contains an amino group (a nitrogen and two hydrogens) on one end of the skeleton and a carboxylic acid group at the other end (figure 18.20). In addition, the carbon skeleton may have one of several different *side chains* on it. These vary in their composition and are generally noted as the amino acid's *R-group.* About twenty common amino acids are important to cells and each differs from one another in the nature of its R-group.

Any amino acid can form a bond with any other amino acid. They fit together in a specific way, with the amino group of one bonding to the acid group of the next. You can imagine that by using twenty different amino acids as building blocks, you can construct millions of different combinations. There are over 3 million possible combinations for a molecule five amino acids long. Each of these combinations is termed a **polypeptide chain.** A specific polypeptide is composed of a specific sequence of amino acids bonded end to end.

There are four levels or degrees of protein structure. A listing of the amino acids in their proper order within a particular polypeptide constitutes its *primary* structure. The specific sequence of amino acids in a polypeptide is controlled by the genetic information of an organism. Genes are specific messages that tell the cell to link particular amino acids in a specific order, that is, they determine a polypeptide's primary structure.

The string of amino acids in a polypeptide is likely to twist into particular shapes (a coil or a pleated sheet), whereas other portions remain straight. These twisted forms are referred to as the *secondary structure* of polypeptides. For example, at this secondary level some proteins (e.g., hair) take the form of a helix: a shape like that of a coiled spring. Other polypeptides form hydrogen bonds that cause them to make several flat folds that resemble a pleated skirt. This is called a pleated sheet. The way a particular protein folds is important to its function. In bovine spongiform encephalitis (mad cow disease) and Creutzfeldt-Jakob diseases, protein structures are not formed correctly, resulting in characteristic nervous system symptoms.

It is also possible for a single polypeptide to contain one or more coils and pleated sheets along its length. As a result, these different portions of the molecule can interact to form an even more complex globular structure. The complex three-dimensional structure formed in this manner is the polypeptide's *tertiary* (third-degree) *structure.* A good example of a protein with tertiary structure is the myoglobin molecule. This is an oxygen-holding protein found in muscle cells.

Frequently, several different polypeptides, each with its own tertiary structure, twist around each other and chemically combine. The larger, globular structure formed by these interacting polypeptides is referred to as the protein's *quaternary* (fourth-degree) *structure* (figure 18.21). Quaternary structure is displayed by the protein molecules called immunoglobulins or *antibodies,* which are involved in fighting diseases such as mumps and chicken pox (A Closer Look: Some Interesting Amino Acid Information). The protein portion of the hemoglobin molecule (globin is globular in shape) also demonstrates quaternary structure.

Energy in the form of heat or light may break the bonds within protein molecules. When this occurs, the chemical and physical properties of the protein are changed and the protein is said to be **denatured.** (Keep in mind, a protein is a molecule, not a living thing, and therefore cannot be "killed.") A common example of this occurs when the gelatinous, clear portion of an egg is cooked and the protein changes to a white solid. Some medications are proteins and must be protected from denaturation so as not to lose their effectiveness. Insulin is an example. For protection, such medications may be stored in brown-colored bottles or kept under refrigeration.

The thousands of kinds of proteins can be placed into three categories. Some proteins are important for maintaining the shape of cells and organisms—they are usually referred to as **structural proteins.** The proteins that make up the cell membrane, muscle cells, tendons, and blood cells are examples of structural proteins. The protein collagen is found throughout the human body and gives tissues shape, support, and strength. The second category of proteins, **regulator proteins,** helps determine what activities will occur in the organism. These regulator proteins include enzymes and some hormones. These molecules help control the chemical activities of cells and organisms. Some

Figure 18.20

An amino acid is composed of a short carbon skeleton with three functional groups attached: an amino group, a carboxylic acid group (acid group), and an additional variable group (R-group). It is the variable group that determines which specific amino acid is constructed.

A Primary structure

α helix
(e.g., hair)

β sheet
(e.g., silk)

B Secondary structure

C Tertiary structure
(e.g., myoglobin)

D Quaternary structure
(e.g., antibody)

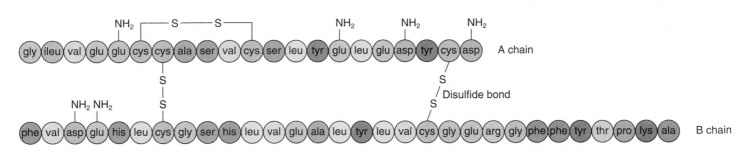

E Structure of insulin

Figure 18.21

(*A*) The primary structure of a molecule is simply a list of its component amino acids in the order in which they occur. (*B*) This figure illustrates the secondary structure of protein molecules or how one part of the molecule initially attached to another part of the same molecule. (*C*) If already folded parts of a single molecule attach at other places, the molecule is said to display tertiary (third-degree) structure. (*D*) Quaternary (fourth-degree) structure is displayed by molecules that are the result of two separate molecules (each with its own tertiary structure) combining into one large macromolecule. (*E*) The protein insulin is composed of two polypeptide chains bonded together at specific points by reactions between the side chains of particular sulfur-containing amino acids. The side chains of one interact with the side chains of the other and form a particular three-dimensional shape. The bonds that form between the polypeptide chains are called disulfide bonds.

Connections . . .

Protein Structure and Sickle-Cell Anemia

Any changes in the arrangement of amino acids within a protein can have far-reaching effects on its function. For example, normal hemoglobin found in red blood cells consists of two kinds of polypeptide chains called the alpha and beta chains. The beta chain is 146 amino acids long. If just one of these amino acids is replaced by a different one, the hemoglobin molecule may not function properly. A classic example of this results in a condition known as *sickle-cell anemia*. In this case, the sixth amino acid in the beta chain, which is normally glutamic acid, is replaced by valine. This minor change causes the hemoglobin to fold differently, and the red blood cells that contain this altered hemoglobin assume a sickle shape when the body is deprived of an adequate supply of oxygen.

examples of enzymes are the digestive enzymes in the stomach. Two hormones that are regulator proteins are insulin and oxytocin. Insulin is produced by the pancreas and controls the amount of glucose in the blood. If insulin production is too low, or if the molecule is improperly constructed, glucose molecules are not removed from the bloodstream at a fast enough rate. The excess sugar is then eliminated in the urine. Other symptoms of excess sugar in the blood include excessive thirst and even loss of consciousness. The disease caused by improperly functioning insulin is known as *diabetes*. Oxytocin, a second protein hormone, stimulates the contraction of the uterus during childbirth. It is also an example of an organic molecule that has been produced artificially (e.g., pitocin) and is used by physicians to induce labor. The third category of proteins is **carrier proteins.** Proteins in this category pick up and deliver molecules at one place and transport them to another. For example, proteins regularly attach to cholesterol entering the system from the diet-forming molecules called lipoproteins, which are transported through the circulatory system. The cholesterol is released at a distance from the digestive tract and the proteins return to pick up more dietary cholesterol entering the system.

A Closer Look Some Interesting Amino Acid Information

Nine Essential Amino Acids (those not able to be manufactured by the body and required in the diet)

Threonine
Lysine
Valine
Leucine
Methionine

Tryptophan
Phenylalanine
Histidine
Isoleucine

The Structure and Function of Four of the Essential Amino Acids

Lysine

Found in such foods as yogurt, fish, chicken, brewer's yeast, cheese, wheat germ, pork, and other meats; improves calcium uptake; in concentrations higher than arginine helps to control cold sores (herpes virus infection).

Tryptophan

Found in turkey, dairy products, eggs, fish, and nuts; required for the manufacture of hormones such as serotonin, prolactin, and growth hormone; has been shown to be of value in controlling depression, PMS (premenstrual syndrome), insomnia, migraine headaches, and immune function disorders.

Asparagine

Found in asparagus—some individuals excrete this amino acid in aromatically noticeable amount after eating asparagus.

Glutamic Acid

Found in animal and vegetable proteins; used in monosodium glutamate (MSG), a flavor-enhancing salt; required for the synthesis of folic acid found to accumulate in and damage brain cells following stroke; the only amino acid metabolized in the brain.

Nucleic Acids

Nucleic acids are complex polymeric molecules that store and transfer information within a cell. There are two types of nucleic acids, DNA and RNA. DNA serves as genetic material determining which proteins will be manufactured, while RNA plays a vital role in the process of protein manufacture. All nucleic acids are constructed of fundamental monomers known as **nucleotides.** Each nucleotide is composed of three parts: (1) a 5-carbon simple sugar molecule that may be ribose or deoxyribose, (2) a phosphate group, and (3) a nitrogenous base. There are five types of nitrogenous bases. Two of the bases are the larger, double-ring molecules, *Adenine* and *Guanine.* The smaller bases are the single-ring bases, *Thymine, Cytosine,* and *Uracil.* Nucleotides (monomers) are linked together in long sequences (polymers) so that the sugar and phosphate sequence forms a "backbone" and the nitrogenous bases stick out to the side. DNA has deoxyribose sugar and the bases A, T, G, and C, while RNA has ribose sugar and the bases A, U, G, and C (figure 18.22).

DNA (*deoxyribo nucleic acid*) is composed of two strands to form a ladder-like structure thousands of nucleotide bases long. The two strands are attached between their protruding bases according to *the base pair rule,* that is, *Adenine* protruding from one strand always pairs with *Thymine* protruding from the other (in the case of RNA, *Adenine* always pairs with *Uracil*). *Guanine* always pairs with *Cytosine.*

A T (or A U) and G C

One strand of DNA is called the *coding strand* because it has a meaningful genetic message written using the nitrogenous bases as letters (e.g., the base sequence CATTAGACT) (figure 18.23). This is the basis of the genetic code for all organisms. If these bases are read in groups of three, they make sense to us (i.e., "cat," "tag," and "act"). The opposite strand is called *non-coding* since it makes no "sense" but protects the coding strand from chemical and physical damage. Both strands are twisted into a helix—that is, a molecule turned around a tubular space. Strands of helical DNA may contain tens or thousands of base pairs (AT and GC combinations) that an organism reads as a sequence of chapters in a book. Each gene is a chapter. Just as chapters in a book are identified by beginning and ending statements, different genes along a DNA strand have beginning and ending signals. They tell when to start and when to stop reading a particular gene. Human body cells contain 46 strands (books) of helical DNA, each containing thousands of genes (chapters). These strands are called **chromosomes** when they become super coiled in preparation for cellular reproduction. Before cell reproduction, the DNA makes copies of the coding and non-coding strands ensuring that the

A DNA single strand   **B** RNA

Figure 18.22

(*A*) A single strand of DNA is a polymer composed of nucleotides. Each nucleotide consists of deoxyribose sugar, phosphate, and one of four nitrogenous bases: A,T, G, or C. Notice the backbone of sugar and phosphate. (*B*) RNA is also a polymer but each nucleotide is composed of ribose sugar, phosphate, and one of four nitrogenous bases: A, U, G, or C.

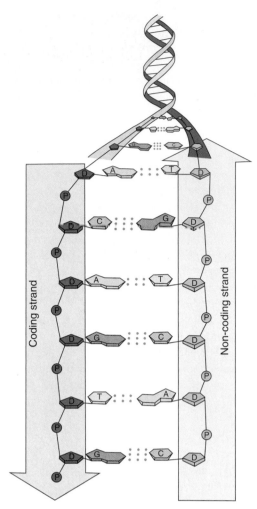

Figure 18.23

The genetic material is really double-stranded DNA molecules comprised of sequences of nucleotides that spell out an organism's genetic code. The coding strand of the double molecule is the side that can be translated by the cell into meaningful information. The genetic code has the information for telling the cell what proteins to make, which in turn become the major structural and functional components of the cell. The non-coding strand is unable to code for such proteins.

offspring or *daughter cells* will each receive a full complement of the genes required for their survival. Each chromosome is comprised of a sequence of genes. A **gene** is a segment of DNA that is able to (1) replicate by directing the manufacture of copies of itself; (2) mutate, or chemically change, and transmit these changes to future generations; (3) store information that determines the characteristics of cells and organisms; and (4) use this information to direct the synthesis of structural and regulatory proteins.

RNA (*ribonucleic acid*) is found in three forms. **Messenger RNA (mRNA)** is a single-strand copy of a portion of the coding strand of DNA for a specific gene. When mRNA is formed on the surface of the DNA, the base pair rule (A pairs with U and G pairs with C) applies. After mRNA is formed and peeled off, it moves to a cellular structure called the *ribosome* where the genetic message can be translated into a protein molecule. Ribosomes con-

tain another type of RNA, **ribosomal RNA (rRNA).** rRNA is also an RNA copy of DNA, but after being formed it becomes twisted and covered in protein to form a ribosome. The third form of RNA, **transfer RNA (tRNA),** is also made from copies of different segments of DNA, but when peeled off the surface, each takes the form of a cloverleaf. tRNA molecules are responsible for transferring or carrying specific amino acids to the ribosome where all three forms of RNA come together and cooperate in the manufacture of protein molecules (figure 18.24).

Lipids

We generally call molecules in this group the **fats.** However, there are three different types of **lipids:** *true fats* (pork chop fat or olive oil), *phospholipids* (the primary component of cell membranes), and *steroids* (most hormones). In general, lipids are large, nonpolar, organic molecules that do not easily dissolve in polar solvents such as water. They are soluble in nonpolar substances such as ether or acetone. Just like carbohydrates, the lipids are composed of carbon, hydrogen, and oxygen. Lipids generally have very small amounts of oxygen in comparison to the amounts of carbon and hydrogen. Simple lipids cannot be digested into smaller, similar subunits. Complex lipids such as true fats and phospholipids can be digested into smaller, similar units.

True (Neutral) Fats

True (neutral) fats are important, complex organic molecules that are used to provide, among other things, energy. The building blocks of a fat are a glycerol molecule and fatty acids. Recall that **glycerol** is a carbon skeleton that has three alcohol groups attached to it. Its chemical formula is $C_3H_5(OH)_3$. At room temperature, glycerol looks like clear, lightweight oil. It is used under the name glycerin as an additive to many cosmetics to make them smooth and easy to spread.

$$
\begin{array}{ccccccc}
 & \text{OH} & & \text{OH} & & \text{OH} & \\
 & | & & | & & | & \\
\text{H}-&\text{C}&-&\text{C}&-&\text{C}&-\text{H} \\
 & | & & | & & | & \\
 & \text{H} & & \text{H} & & \text{H} & \\
\end{array}
$$

Glycerol

The **fatty acids** found in fats are a long-chain carbon skeleton that has a carboxylic acid functional group. If the carbon skeleton has as much hydrogen bonded to it as possible,

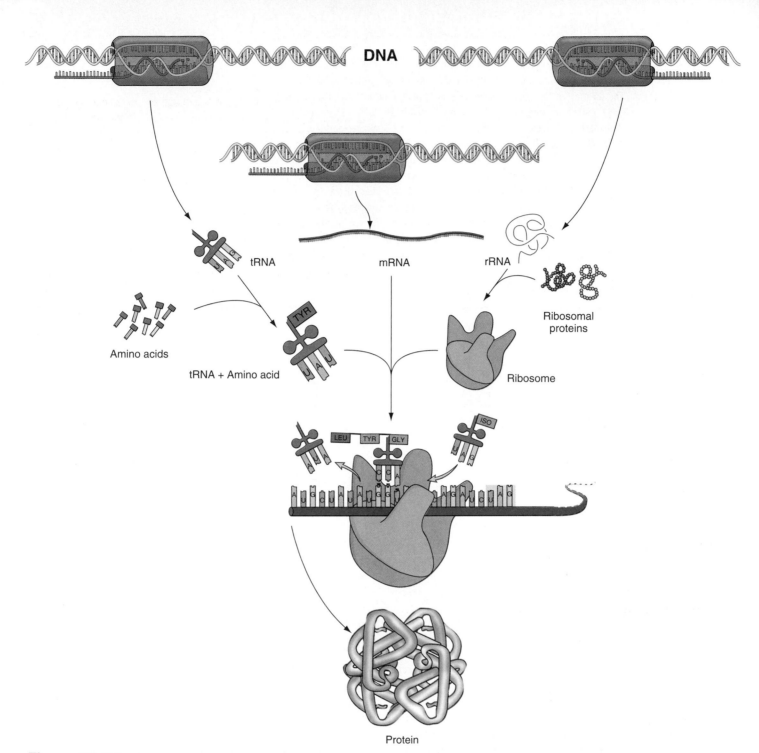

DNA

tRNA

mRNA

rRNA

Amino acids

Ribosomal proteins

tRNA + Amino acid

Ribosome

Protein

Figure 18.24

All forms of RNA (messenger, transfer, and ribosomal) are copies of different sequences of coding strand DNA. However, each plays a different role in the manufacture of proteins. When the protein synthesis process is complete, the RNA can be reused to make more of the same protein coded for by the mRNA. This is similar to replaying a cassette in a tape machine. The tape is like the mRNA and the tape machine is like the ribosome. Eventually the tape and machine will wear out and must be replced. In a cell, this involves the synthesis of new RNA molecules from food.

we call it *saturated*. The saturated fatty acid in figure 18.25A is stearic acid, a component of solid meat fats such as mutton tallow. Notice that at every point in this structure the carbon has as much hydrogen as it can hold. Saturated fats are generally found in animal tissues and tend to be solids at room temper-

atures. Some examples of saturated fats are butter, whale blubber, suet, lard, and fats associated with such meats as steak or pork chops.

If the carbons in a fatty acid are double-bonded to each other at one or more points, the fatty acid is said to be

The occurrence of a double bond in a fatty acid is indicated by the Greek letter ω (omega) followed by a number indicating the location of the first double bond in the molecule. Oleic acid, one of the fatty acids found in olive oil, is comprised of 18 carbons with a single (I) double bond between carbons 9 and 10. Therefore, it is chemically designated C18:1ω9 and is a monounsaturated fatty acid. This fatty acid is commonly referred to as an omega-9 fatty acid. The unsaturated fatty acid in figure 18.25B is linoleic acid, a component of sunflower and safflower oils. Notice that there are two double bonds between the carbons and fewer hydrogens than in the saturated fatty acid. Linoleic acid is chemically a polyunsaturated fatty acid with two double (2) bonds and is designated C18:2ω6, an omega-6 fatty acid. This indicates that the first double bond of this 18-carbon molecule is between carbons 6 and 7. Since the human body cannot make this fatty acid, it is called an *essential fatty acid* and must be taken in as a part of the diet. The other essential fatty acid, linolenic acid, is C18:3ω3 and has three (3) double bonds. This fatty acid is commonly referred to as an omega-3 fatty acid. One key function of these essential fatty acids is the synthesis of prostaglandin hormones that are necessary in controlling cell growth and specialization.

Sources of Omega-3 Fatty Acids	Sources of Omega-6 Fatty Acids
Certain fish oil (salmon, sardines, herring)	Corn oil
	Peanut oil
	Cottonseed oil
Flaxseed oil	Soybean oil
Soybeans	
Soybean oil	Sesame oil
	Safflower oil
Walnuts	Sunflower oil
Walnut oil	

A   Stearic acid

B   Linoleic acid

Figure 18.25

Structure of saturated and unsaturated fatty acids. (*A*) Stearic acid is an example of a saturated fatty acid. (*B*) Linoleic acid is an example of an unsaturated fatty acid.

unsaturated. Unsaturated fats are frequently plant fats or oils and they are usually liquids at room temperature. Peanut, corn, and olive oil are examples of unsaturated fats that are mixtures of different triglycerides. When glycerol and three fatty acids are combined, a fat is formed. This reaction is almost exactly the same as the reaction that causes simple sugars to bond together.

Fats are important molecules for storing energy. There is more than twice as much energy in a gram of fat as in a gram of sugar, 9 Calories versus 4 Calories. This is important to an organism because fats can be stored in a relatively small space and still yield a high amount of energy. Fats in animals also provide protection from heat loss. The thick layer of blubber in whales, walruses, and seals prevents the loss of internal body heat to the cold, watery environment in which they live. This same layer of fat, together with the fat deposits around some internal organs—such as the kidneys and heart—serve as a cushion that protects these organs from physical damage. If a fat is formed from a glycerol molecule and three attached fatty acids, it is called a **triglyceride;** if two, a *diglyceride;* and if one,

a *monoglyceride* (figure 18.26). Triglycerides account for about 95 percent of the fat stored in human tissue.

Phospholipids

Phospholipids are a class of complex water-insoluble organic molecules that resemble fats but contain a phosphate group (PO_4) in their structure (figure 18.27). One of the reasons phospholipids are important is that they are a major component of membranes in cells. Without these lipids in our membranes, the cell contents would not be separated from the exterior environment. Some of the phospholipids are better known as the *lecithins.* Lecithins are found in cell membranes and also help in the emulsification of fats. They help separate large portions of fat into smaller units. This allows the fat to mix with other materials. Lecithins are added to many types of food for this purpose (chocolate bars, for example). Some people take lecithin as nutritional supplements because they believe it leads to healthier hair and better reasoning ability; but once inside your intestines, lecithins are destroyed by enzymes, just like any other phospholipid.

A Closer Look Fat and Your Diet

When triglycerides are eaten in fat-containing foods, digestive enzymes hydrolyze them into glycerol and fatty acids. These molecules are absorbed by the intestinal tract and coated with protein to form lipoprotein, as shown in the accompanying diagram.

The five types of lipoproteins found in the body are: (1) chylomicrons, (2) very-low-density lipoproteins (VLDL), (3) low-density lipoproteins (LDL), (4) high-density lipoproteins (HDL), and (5) lipoprotein a—Lp(a). Chylomicrons are very large particles formed in the intestine and are between 80 percent and 95 percent triglycerides. As the chylomicrons circulate through the body, cells remove the triglycerides in order to make hormones, store energy, and build new cell parts. When most of the triglycerides have been removed, the remaining portions of the chylomicrons are harmlessly destroyed.

The VLDLs and LDLs are formed in the liver. VLDLs contain all types of lipid, protein, and 10 percent to 15 percent cholesterol, while the LDLs are about 50 percent cholesterol. As with the chylomicrons, the body uses these molecules for fats they contain. However, in some people, high levels of LDLs and Lp(a) in the blood are associated with the diseases atherosclerosis, stroke, and heart attack. While in the blood, LDLs may stick to the insides of the vessels, forming deposits that restrict blood flow and contribute to high blood pressure, strokes, and heart attacks. Even though they are 30 percent cholesterol, a high level of HDLs (made in the intestine) in comparison to LDLs and Lp(a) is associated with a lower risk of atherosclerosis. One way to reduce the risk of this disease is to lower your intake of LDLs and Lp(a). Reducing your consumption of saturated fats can do this since the presence of saturated fats disrupts the removal of LDLs from the bloodstream. An easy way to remember the association between LDLs and HDLs is "L = lethal" and "H = Healthy" or "Low = Bad" and "High = Good." The federal government's new cholesterol guidelines recommend that all adults get a full lipoprotein profile (total cholesterol, HDL, and LDL and triglycerides) once every five years. They also recommend a sliding scale for desirable LDL levels. The higher your heart attack risk, the lower your LDL should be.

Box Figure 18.2
Diagram of a Lipoprotein

Reprinted with permission, Best Foods, a division of CPC International, Inc.

Normal HDL Values	Normal LDL Values	Normal VLDL Values
Men: 40–70 mg/dL	Men: 91–100 mg/dL	Men: 0–40 mg/dL
Women: 40–85 mg/dL	Women: 69–100 mg/dL	Women: 0–40 mg/dL
Children: 30–65 mg/dL		
Minimum desirable: 40 mg/dL	High risk: no higher than 100 mg/dL Moderate risk: no higher than 130 mg/dL Low risk: at or below 160 mg/dL	Desirable: below 40 mg/dL

For Total Cholesterol Levels:

Desirable: below 200 mg/dL	Borderline: 200–240	Undesirable: above 240

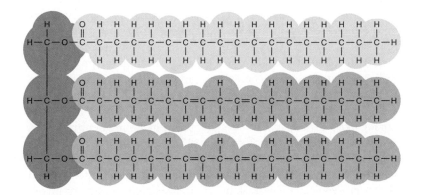

Figure 18.26
The arrangement of the three fatty acids attached to a glycerol molecule is typical of the formation of a fat. The structural formula of the fat appears to be very cluttered until you dissect the fatty acids from the glycerol; then it becomes much more manageable. This example of a triglyceride contains a glycerol molecule, two unsaturated fatty acids (linoleic acid), and a third saturated fatty acid (stearic acid).

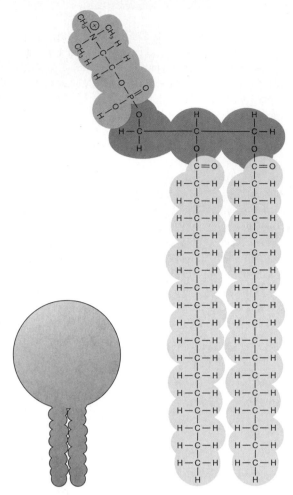

Figure 18.27

This molecule is similar to a fat but has a phosphate group in its structure. The phosphate group and two fatty acids are bonded to the glycerol by a dehydration synthesis reaction. Molecules like these are also known as lecithin. The diagram of phospholipid molecules is shown as a balloon with two strings. The balloon portion is the glycerol and phosphate group, while the strings are the fatty acid segments of the molecule.

Steroids

Steroids, another group of lipid molecules, are characterized by an arrangement of interlocking rings of carbon. They often serve as hormones that aid in regulating body processes. We have already mentioned one familiar steroid molecule: cholesterol. Although serum cholesterol (the kind found in your blood associated with lipoproteins) has been implicated in many cases of atherosclerosis, this steroid is made by your body for use as a component of cell membranes. It is also used by your body to make bile acids. These products of your liver are channeled into your intestine to emulsify fats. Cholesterol is also necessary for the manufacture of vitamin D. Cholesterol molecules in the skin react with ultraviolet light to produce vitamin D, which assists in the proper development of bones and teeth. Figure 18.28 illustrates some of the steroid compounds that are typically man-

ufactured by organisms. A large number of steroid molecules are hormones. Some of them regulate reproductive processes such as egg and sperm production; others regulate such things as salt concentration in the blood.

SYNTHETIC POLYMERS

Natural polymers are huge, chainlike molecules made of hundreds or thousands of smaller, repeating molecular units called monomers. Cellulose and wool are examples of natural polymers made of glucose monomers. *Synthetic polymers* are manufactured from a wide variety of substances, and you are familiar with these polymers as synthetic fibers such as nylon and the inexpensive light plastic used for wrappings and containers (figure 18.29).

The first synthetic polymer was a modification of the naturally existing cellulose polymer. Cellulose was chemically modified in 1862 to produce celluloid, the first *plastic.* The term "plastic" means that celluloid could be molded to any desired shape. Celluloid was produced by first reacting cotton with a mixture of nitric and sulfuric acids. The cellulose in the cell walls of the cotton plant reacts with the acids to produce the ester cellulose nitrate. This ester is an explosive compound known as "guncotton," or smokeless gunpowder. When made with ethanol and camphor, the product is less explosive and can be formed and molded into useful articles. This first plastic, celluloid, was used to make dentures, combs, eyeglass frames, and photographic film. Before the discovery of celluloid, many of these articles, including dentures, were made from wood. Today, only Ping-Pong balls are made from cellulose nitrate.

Cotton reacted with acetic acid and sulfuric acid produces a cellulose acetate ester. This polymer, through a series of chemical reactions, produces rayon filaments when forced through small holes. The filaments are twisted together to form viscose rayon thread. When forced through a thin slit, a sheet is formed rather than filaments, and the transparent sheet is called *cello-*

Figure 18.28

(A) Cholesterol is produced by the human body and is found in your cells' membranes. (B) Vitamin D_2 is important to the normal growth of teeth and bones. (C) Cholic acid is a bile salt produced by the liver and used to break down fats. (D) Cortisol controls the metabolism of food and helps control inflammation. Excess amount of cortisol can also be responsible for a person's feelings of sadness, stress, and anxiety. (E) Testosterone increases during puberty, causing the male sex organs to mature. (F) Estradiol, a form of estrogen, is a female sex hormone necessary for many processes in the body. (G) Progesterone is a female sex hormone produced by the ovaries and placenta. (H) Prostaglandin E_1 (PGE_1) is a chemical messenger in the body and is associated with the immune system.

phane. Both rayon and cellophane, as celluloid, are manufactured by modifying the natural polymer of cellulose.

The first truly synthetic polymer was produced in the early 1900s by reacting two chemicals with relatively small molecules rather than modification of a natural polymer. Phenol, an aromatic hydrocarbon, was reacted with formaldehyde, the simplest aldehyde, to produce the polymer named *Bakelite.* Bakelite is a *thermosetting* material that forms cross-links between the polymer chains. Once the links are formed during production, the plastic becomes permanently hardened and cannot be softened or made to flow. Some plastics are *thermoplastic* polymers and soften during heating and harden during cooling because they do not have cross-links.

Polyethylene is a familiar thermoplastic polymer used for vegetable bags, dry cleaning bags, grocery bags, and plastic squeeze bottles. Polyethylene is a polymer produced by a polymerization reaction of ethylene, which is derived from petroleum. Polyethylene was invented just before World War II and was used as an electrical insulating material during the war. Today, there are many variations of polyethylene that are produced by different reaction conditions or by substitution of one or more hydrogen atoms in the ethylene molecule. When soft polyethylene near the melting point is rolled in alternating perpendicular directions or expanded and compressed as it is cooled, the polyethylene molecules become ordered in a way that improves the rigidity and tensile strength. This change in the microstructure produces *high-density polyethylene* with a superior rigidity and tensile strength compared to *low-density polyethylene.* High-density polyethylene is used in milk jugs, as liners in screw-on jar tops and bottle caps, and as a material for toys.

Name	Chemical unit	Uses	Name	Chemical unit	Uses
Polyethylene	(—CH₂—CH₂—)ₙ structure	Squeeze bottles, containers, laundry and trash bags, packaging	Polyvinyl acetate	structure	Mixed with vinyl chloride to make vinylite; used as an adhesive and resin in paint
Polypropylene	structure with CH₃	Indoor-outdoor carpet, pipe valves, bottles	Styrene-butadiene rubber	structure	Automobile tires
Polyvinyl chloride (PVC)	structure with Cl	Plumbing pipes, synthetic leather, plastic tablecloths, phonograph records, vinyl tile	Polychloroprene (Neoprene)	structure with Cl	Shoe soles, heels
Polyvinylidene chloride (Saran)	structure with Cl, Cl	Flexible food wrap	Polymethyl methacrylate (Plexiglas, Lucite)	structure	Moldings, transparent surfaces on furniture, lenses, jewelry, transparent plastic "glass"
Polystyrene (Styrofoam)	structure with benzene ring	Coolers, cups, insulating foam, shock-resistant packing material, simulated wood furniture	Polycarbonate (Lexan)	structure	Tough, molded articles such as motorcycle helmets
Polytetrafluoroethylene (Teflon)	structure with F	Gears, bearings, coating for nonstick surface of cooking utensils	Polyacrylonitrile (Orlon, Acrilan, Creslan)	structure with CN	Textile fibers

Figure 18.29

Synthetic polymers, the polymer unit, and some uses of each polymer.

The properties of polyethylene are changed by replacing one of the hydrogen atoms in a molecule of ethylene. If the hydrogen is replaced by a chlorine atom the compound is called vinyl chloride, and the polymer formed from vinyl chloride is

polyvinyl chloride (PVC). Polyvinyl chloride is used to make plastic water pipes, synthetic leather, and other vinyl products. It differs from the waxy plastic of polyethylene because of the chlorine atom that replaces hydrogen on each monomer.

Replacement of a hydrogen atom with a benzene ring makes a monomer called *styrene*. Styrene is

and polymerization of styrene produces *polystyrene*. Polystyrene is puffed full of air bubbles to produce the familiar Styrofoam coolers, cups, and insulating materials.

If all hydrogens of an ethylene molecule are replaced with atoms of fluorine, the product is polytetrafluoroethylene, a tough plastic that resists high temperatures and acts more like a metal than a plastic. Since it has a low friction it is used for bearings, gears, and as a nonstick coating on frying pans. You probably know of this plastic by its trade name of *Teflon*.

There are many different polymers in addition to PVC, Styrofoam, and Teflon, and the monomers of some of these are shown in figure 18.29. There are also polymers of isoprene, or synthetic rubber, in wide use. Fibers and fabrics may be polyamides (such as nylon), polyesters (such as Dacron), or polyacrylonitriles (Orlon, Acrilan, Creslan), which have a CN in place of a hydrogen atom on an ethylene molecule and are called acrylic materials. All of these synthetic polymers have added much to practically every part of your life. It would be impossible to list all of their uses here; however, they present problems since (1) they are manufactured from raw materials obtained from coal and a dwindling petroleum supply and (2) they do not readily decompose when dumped into rivers, oceans, or other parts of the environment. However, research in the polymer sciences is beginning to reflect new understandings learned from research on biological tissues. This could lead to whole new molecular designs for synthetic polymers that will be more compatible with the ecosystems.

SUMMARY

Organic chemistry is the study of compounds that have carbon as the principal element. Such compounds are called *organic compounds,* and all the rest are *inorganic compounds.* There are millions of organic compounds because a carbon atom can link with other carbon atoms as well as atoms of other elements.

A *hydrocarbon* is an organic compound consisting of hydrogen and carbon atoms. The simplest hydrocarbon is one carbon atom and four hydrogen atoms, or CH_4. All hydrocarbons larger than CH_4 have one or more carbon atoms bonded to another carbon atom. The bond can be single, double, or triple, and may be found as chains or rings of carbon. The *alkanes* are hydrocarbons with single carbon-to-carbon bonds, the *alkenes* have a double carbon-to-carbon bond, and the *alkynes* have a triple carbon-to-carbon bond. The alkanes, alkenes, and alkynes can have straight- or branched-chain molecules. When the number of carbon atoms is greater than three, there are different arrangements that can occur for a particular number of carbon atoms. The different arrangements with the same molecular formula are called isomers. *Isomers* have different physical properties, so each isomer is given its own name.

The alkanes have all the hydrogen atoms possible, so they are *saturated* hydrocarbons. The alkenes and the alkynes can add more hydrogens to the molecule, so they are *unsaturated* hydrocarbons. Unsaturated hydrocarbons are more chemically reactive than saturated molecules.

Hydrocarbons that occur in a ring or cycle structure are cyclohydrocarbons. A six-carbon cyclohydrocarbon with three double bonds has different properties than the other cyclohydrocarbons because the double bonds are not localized. This six-carbon molecule is *benzene,* the basic unit of the *aromatic hydrocarbons.*

Petroleum is a mixture of alkanes, cycloalkanes, and a few aromatic hydrocarbons that formed from the slow decomposition of buried marine plankton and algae. Petroleum from the ground, or *crude oil,* is distilled into petroleum products of *natural gas, LPG, petroleum ether,* *gasoline, kerosene, diesel fuel,* and *motor oils.* Each group contains a range of hydrocarbons and is processed according to use.

In addition to oxidation, hydrocarbons react by *substitution, addition,* and *polymerization* reactions. Reactions take place at sites of multiple bonds or lone pairs of electrons on the *functional groups.* The functional group determines the chemical properties of organic compounds. Functional group results in the *hydrocarbon derivatives* of *alcohols, ethers, aldehydes, ketones, organic acids, esters,* and *amines.*

Living organisms have an incredible number of highly organized chemical reactions that are catalyzed by *enzymes,* using food and energy to grow and reproduce. These biochemical processes involve building large *macromolecules* such as proteins, carbohydrates, and lipids.

Carbohydrates are a class of organic molecules composed of CHO. The *monosaccharides* are simple sugars such as *glucose* and *fructose.* Glucose is *blood sugar,* a source of energy. The disaccharides are *sucrose* (table sugar), *lactose* (milk sugar), and *maltose* (malt sugar). The polysaccharides are polymers or glucose in straight or branched chains used as a near-term source of stored energy. Plants store the energy in the form of *starch,* and animals store it in the form of *glycogen. Cellulose* is a polymer similar to starch that humans cannot digest.

Proteins are macromolecular polymers of *amino acids* held together by *peptide bonds.* There are twenty amino acids that are used in various polymer combinations to build structural, carrier, and functional proteins.

Nucleic acids are polymers composed of nucleotide units. Two forms are recognized, DNA and RNA. In most organisms DNA serves as the genetic material, RNA plays crucial roles in the synthesis of proteins.

Lipids are a fourth important class of biochemicals. *True* or *neutral fats* and oils belong in the category known as *lipids* and are esters formed from three fatty acids and glycerol into a *triglyceride.* True fats are usually solid triglycerides associated with animals, and *oils* are liquid triglycerides associated with plant life, but both represent a high-energy storage material. The other two subgroups are the *phospholipids* used in cell membranes and *steroids,* which primarily serve as hormones.

Polymers occur naturally in plants and animals, and *synthetic polymers* are made today from variations of the ethylene-derived monomers. Among the more widely used synthetic polymers derived from ethylene are polyethylene, polyvinyl chloride, polystyrene, and Teflon. Problems with the synthetic polymers include that (1) they are manufactured from fossil fuels that are also used as the primary energy supply, and (2) they do not readily decompose and tend to accumulate in the environment.

KEY TERMS

alkanes (p. **375**)	functional group (p. **381**)
alkenes (p. **376**)	glycerol (p. **392**)
alkyne (p. **377**)	glycogen (p. **387**)
amino acid (p. **388**)	gene (p. **392**)
carbohydrates (p. **386**)	hydrocarbon (p. **375**)
carrier proteins (p. **389**)	hydrocarbon derivatives (p. **380**)
cellulose (p. **387**)	inorganic chemistry (p. **374**)
chromosome (p. **391**)	isomers (p. **376**)
complex carbohydrate (p. **386**)	lipid (p. **392**)
denatured (p. **388**)	macromolecule (p. **385**)
deoxyribonucleic acid (DNA) (p. **391**)	messenger RNA (mRNA) (p. **392**)
fats (p. **392**)	monomer (p. **380**)
fatty acids (p. **392**)	nucleic acid (p. **391**)

APPLYING THE CONCEPTS

1. An organic compound is a one that
   (a) contains carbon and was formed only by a living organism.
   (b) is a natural compound that has not been synthesized.
   (c) contains carbon, no matter if it was formed by a living thing or not.
   (d) was formed by a plant.

2. There are millions of organic compounds but only thousands of inorganic compounds because
   (a) organic compounds were formed by living things.
   (b) there is more carbon on the earth's surface than any other element.
   (c) atoms of elements other than carbon never combine with themselves.
   (d) carbon atoms can combine with up to four other atoms, including other carbon atoms.

3. An alkane with 4 carbon atoms would have how many hydrogen atoms in each molecule?
   (a) 4
   (b) 8
   (c) 10
   (d) 16

4. Isomers are compounds with the same
   (a) molecular formula with different structures.
   (b) molecular formula with different atomic masses.
   (c) atoms, but different molecular formulas.
   (d) structures, but different formulas.

5. The hydrocarbons with a double covalent carbon-carbon bond are called
   (a) alkanes.
   (b) alkenes.
   (c) alkynes.
   (d) none of the above.

6. Petroleum is believed to have formed mostly from the anaerobic decomposition of buried
   (a) dinosaurs.
   (b) fish.
   (c) pine trees.
   (d) plankton and algae.

7. Ethylene molecules can add to each other in a reaction to form a long chain called a
   (a) monomer.
   (b) dimer.
   (c) trimer.
   (d) polymer.

8. Chemical reactions usually take place on an organic compound at the site of (a)
   (a) double bond.
   (b) lone pair of electrons.
   (c) functional group.
   (d) any of the above.

9. The R in ROH represents
   (a) a functional group.
   (b) a hydrocarbon group with a name ending in "-yl."
   (c) an atom of an inorganic element.
   (d) a polyatomic ion that does not contain carbon.

10. A protein is a polymer formed from the linking of many
    (a) glucose units.
    (b) DNA molecules.
    (c) amino acid molecules.
    (d) monosaccharides.

11. Which of the following is *not* converted to blood sugar by the human body?
    (a) lactose
    (b) dextrose
    (c) cellulose
    (d) glycogen

12. Many synthetic polymers become a problem in the environment because they
    (a) decompose to nutrients, which accelerates plant growth.
    (b) do not readily decompose and tend to accumulate.
    (c) do not contain vitamins as natural materials do.
    (d) become a source of food for fish, but ruin the flavor of fish meat.

Answers

1. c **2.** d **3.** c **4.** a **5.** b **6.** d **7.** d **8.** d **9.** b **10.** c **11.** c **12.** b

QUESTIONS FOR THOUGHT

1. What is an organic compound?
2. There are millions of organic compounds but only thousands of inorganic compounds. Explain why this is the case.
3. Is it possible to have an isomer of ethane? Explain.
4. Suggest a reason that ethylene is an important raw material used in the production of plastics but ethane is not.
5. Give an example of each of the following classes of organic molecules: carbohydrates, proteins, nucleic acids, lipids.
6. What are (a) natural gas, (b) LPG, and (c) petroleum ether?
7. What does the octane number of gasoline describe? On what is the number based?
8. What is a functional group? What is it about the nature of a functional group that makes it the site of chemical reactions?
9. Draw a structural formula for alcohol. Describe how alcohols are named.
10. A soft drink is advertised to "contain no sugar." The label lists ingredients of carbonated water, dextrose, corn syrup, fructose, and flavorings. Evaluate the advertising and the list of ingredients.
11. What are the three subgroups of lipids? What are the differences between saturated and unsaturated fats?
12. What is a polymer? Give an example of a naturally occurring plant polymer. Give an example of a synthetic polymer.

CHAPTER NINETEEN

The Nature of Living Things

The cell is the simplest structure capable of existing as an individual living unit. Within this unit, many chemical reactions are required for maintaining life. These reactions do not occur at random, but are associated with specific parts of the many kinds of cells.

OUTLINE

What does it mean to be alive? You would think that a science textbook could answer this question easily. However, this is more than just a theoretical question because in recent years it has been necessary to construct legal definitions of what life is and especially of when it begins and ends. The legal definition of death is important because it may determine whether a person will receive life insurance benefits or if body parts may be used in transplants. In the case of heart transplants, the person donating the heart may be legally "dead," but the heart certainly isn't since it can be removed while it still has "life." In other words, there are different kinds of death. There is the death of the whole living unit and the death of each cell within the living unit. A person actually "dies" before every cell has died. Death, then, is the absence of life, but that still doesn't tell us what life is. At this point, we won't try to define life but we will describe some of the basic characteristics of living things.

CHARACTERISTICS OF LIFE

The science of **biology** is, broadly speaking, the study of living things. It draws on chemistry and physics for its foundation and applies these basic physical laws to living things. The ability to manipulate energy and matter is unique to living things. Understanding just how this is accomplished will help you understand how living things differ from nonliving objects. Living things show five characteristics that the nonliving do not display: (1) metabolic processes, (2) generative processes, (3) responsive processes, (4) control processes, and (5) a unique structural organization. This section gives a brief introduction to the basic characteristics of living things that will be expanded upon later.

1. **Metabolic processes** are all the chemical reactions and associated energy changes taking place within an organism. Energy is necessary for movement, growth, and many other activities. The energy that organisms use comes from energy stored in the chemical bonds of complex molecules. This energy becomes available through a controlled sequence of chemical reactions.

   There are three essential aspects to metabolic processes: (1) *nutrient uptake,* (2) *nutrient processing,* and (3) *waste elimination.* All living things expend energy to take in nutrients (raw materials) from their environment. Many animals take in these materials by eating or swallowing other organisms. Microorganisms and plants absorb raw materials into their cells to maintain their lives. Once inside, raw materials are used in a series of chemical reactions to manufacture new parts, make repairs, reproduce, and provide energy for essential activities. However, not all the raw materials entering a living thing are valuable to it. There may be portions that are useless or even harmful. Organisms eliminate these portions as waste. Metabolic processes also produce unusable heat energy that may also be considered a waste product.

2. The second group of characteristics of life, **generative processes,** are reactions that result in an increase in the size of an individual organism—*growth*—or an increase in the number of individuals in a population of organisms—*reproduction.* During growth, living things add to their structure, repair parts, and store nutrients for later use. Growth and reproduction are directly related to metabolism, since neither can occur without the acquisition and processing of nutrients.

   Reproduction of individuals must occur because each organism eventually dies. Thus reproduction is the only way that living things can perpetuate themselves. There are a number of different ways that various kinds of organisms reproduce and guarantee their continued existence. Some reproductive processes known as sexual reproduction involve two organisms contributing to the creation of a unique, new organism. Asexual reproduction occurs when organisms make identical copies of themselves.

3. Organisms also respond to changes within their bodies and in their surroundings in a meaningful way. These **responsive processes** have been organized into three categories: *irritability, individual adaptation,* and *population adaptation* or *evolution.* The term irritability is not used in the everyday context of an individual being angry. Here, irritability means an individual's rapid response to a stimulus, such as your response to a loud noise, beautiful sunset, or noxious odor. This type of response occurs only in the individual receiving the stimulus and is rapid because the mechanism that allows the response to occur (i.e., muscles, bones, and nerves) is already in place. Individual adaptation is also an individual response but is slower since it requires a growth or some other fundamental change in an organism. For example, a weasel's fur color will change from its brown summer coat to its white winter coat when genes responsible for the production of brown pigment are "turned off." Or the response of our body to disease organisms requires a change in the way cells work that eventually gets control of the organism causing the disease. Population adaptation involves changes in the kinds of characteristics displayed by individuals within the population. It is also known as *evolution,* which is a change in the genetic makeup of a *population* of organisms. This process occurs over long periods of time and enables a species (specific kind of organism) to adapt and better survive long-term changes in its environment over many generations. For example, the structures that give birds the ability to fly long distances allow them to respond to a world in which the winter season presents severe conditions that would threaten survival. Similarly, the ability of humans to think and use tools allows them to survive and be successful in a great variety of environmental conditions.

4. The **control processes** of *coordination* and *regulation* constitute the fourth characteristic of life. Control processes are mechanisms that ensure that an organism will carry out all metabolic activities in the proper sequence (coordination) and at the proper rate (regulation). All the chemical reactions of an organism are coordinated and linked together in specific

Figure 19.1

Each individual organism, whether it is simple or complex, is able to independently carry on metabolic, generative, responsive, and control processes. Some organisms, like yeast or the protozoan *Euplotes,* consist of single cells while others, like humans and orchids, consist of many cells organized into complex structures.

pathways. The orchestration of all the reactions ensures that there will be specific stepwise handling of the nutrients needed to maintain life. The molecules responsible for coordinating these reactions are known as *enzymes.* **Enzymes** are molecules produced by organisms, molecules that are able to increase and control the rate at which life's chemical reactions occur. Enzymes also regulate the amount of nutrients processed into other forms.

Many of the internal activities of organisms are interrelated and coordinated to maintain a constant internal environment. The process of maintaining a constant internal environment is called **homeostasis.** For example, when we begin to exercise we use up oxygen more rapidly so the amount of oxygen in the blood falls. In order to maintain a "constant internal environment" the body must obtain more oxygen. This involves more rapid contractions of the muscles that cause breathing and a more rapid and forceful pumping of the heart to get blood to the lungs. These activities must occur together at the right time and at the correct rate, and when they do, the level of oxygen in the blood will remain normal while supporting the additional muscular activity.

5. In addition to these four basic processes that are typical of living things, living things also share some basic **structural similarities.** All living things are organized into complex structural units called *cells.* The cell units have an outer limiting membrane and internal structural units that have specific functions. Some living

things, like you, consist of trillions of cells with specialized abilities that interact to provide the independently functioning unit called an *organism* (figure 19.1). Typically, in such large, multicellular organisms as humans, cells cooperate with one another in units called *tissues* (e.g., muscle, nervous). Groups of tissues are organized into larger units known as *organs* (e.g., heart), and in turn, into *organ systems* (e.g., circulatory system). Other organisms, such as bacteria or yeast, carry out all four of the life processes within a single cell. Nonliving materials, such as rocks, water, or gases, do not share a structurally complex common subunit.

THE CELL THEORY

The concept of a *cell* is one of the most important ideas in biology because it applies to all living things. It did not emerge all at once, but has been developed and modified over many years. It is still being modified today.

Several individuals made key contributions to the cell concept. Anton van Leeuwenhoek (1632–1723) was one of the first to make use of a *microscope* to examine biological specimens. When van Leeuwenhoek discovered that he could see things moving in pond water using his microscope, his curiosity stimulated him to look at a variety of other things. He studied blood, semen, feces, pepper, and tartar, for example. He was the first to

To transfer the right amount of chemical-bond energy from energy-releasing to energy-requiring reactions, cells use the molecule ATP. Adenosine triphosphate (ATP) is a handy source of the right amount of usable chemical-bond energy. Each ATP molecule used in the cell is like a recharge-able AAA battery used to power small toys and electronic equipment. Each contains just the right amount of energy to power the job. When the power has been drained, it can be recharged numerous times before it must be recycled. Recharging the AAA battery requires getting a small amount of energy from a source of high energy such as a hydroelectric power plant (box figure 19.1). Energy from the electric plant is too powerful to directly run a small flashlight or portable recording device. If you plug your recorder directly into the power plant, the recorder would be destroyed. However, the recharged AAA battery delivers just the right amount of energy at the right time and place. ATP

Box Figure 19.1
When rechargeable batteries in a flashlight have been drained of their power, they can be recharged by placing them in a specially designed battery charger. This enables the right amount of power from a power plant to be packed into the batteries for reuse. Cells operate in much the same manner. When the cell's "batteries," ATPs, become drained as a result of powering a job like muscle contraction, these discharged "batteries," ADPs, can be recharged back to full ATP power.

see individual cells and recognize them as living units, but he did not call them cells. The name he gave to these "little animals" that he saw moving around in the pond water was *animalcules.*

The first person to use the term "cell" was Robert Hooke (1635–1703) of England, who was also interested in how things looked when magnified. He chose to study thin slices of cork from the bark of a cork oak tree. He saw a mass of cubicles fitting neatly together, which reminded him of the barren rooms in a monastery. Hence, he called them *cells.* As it is currently used, the term **cell** refers to the basic structural unit that makes up all living things. When Hooke looked at cork, the tiny boxes he saw were, in fact, only the cell walls that surrounded the living portions of plant cells. We now know that the **cell wall** is composed of the complex carbohydrate cellulose, which provides strength and protection to the living contents of the cell. The cell wall appears to be a rigid, solid layer of material, but in reality it is composed of many interwoven strands of cellulose molecules. Its structure allows certain very large molecules to pass through it readily, but it acts as a screen to other molecules.

Hooke's use of the term *cell* in 1666 in his publication *Micrographia* was only the beginning, for nearly two hundred years passed before it was generally recognized that all living things are made of cells and that these cells can reproduce themselves. In 1838, Mathias Jakob Schleiden stated that all plants are made up of smaller cellular units. In 1839, Theodor Schwann published the idea that all animals are composed of cells.

Soon after the term *cell* caught on, it was recognized that the cell's vitally important portion is inside the cell wall. This living material was termed **protoplasm,** which means *first-formed substance.* The term *protoplasm* allowed scientists to distinguish between the living portion of the cell and the nonliving cell wall. Very soon microscopists were able to distinguish two different regions of protoplasm. One type of protoplasm was more viscous and darker than the other. This region, called the **nucleus** or core, is a central body within a more fluid material surrounding it. **Cytoplasm** is the name given to the colloidal fluid portion of the protoplasm (figure 19.2). While the term *proto-*

functions in much the same manner. After the chemical-bond energy has been drained by breaking one of its bonds:

ATP → ADP + P + energy

→ Used to power chemical reactions

→ Lost as heat to the environment

the discharged molecule (ADP) is recharged by "plugging it in" to a high-powered energy source. This source may be (1) sunlight (photosynthesis) or (2) chemical-bond energy (released from cellular respiration):

Sunlight (photosynthesis)

Energy + ADP + P → ATP

Chemical-bond energy (cellular respiration)

An ATP molecule is formed from adenine (nitrogenous base), ribose (sugar), and phosphates (box figure 19.2). These three are chemically bonded to form AMP, adenosine monophosphate (one phosphate). When a second phosphate group is added to the AMP, a molecule of ADP (diphosphate) is formed. The ADP, with the addition of more energy, is able to bond to a third phosphate group and form ATP. The covalent bond that attaches the second phosphate to the AMP molecule is easily broken to release energy for energy-requiring cell processes. Because the energy in this bond is so easy for a cell to use, it is called a *high-energy phosphate bond*. ATP has two high-energy phosphate bonds represented by curved solid lines.

Both ADP and ATP, because they contain high-energy bonds, are very unstable molecules and readily lose their phosphates. When this occurs, the energy held in the high-energy bonds of the phosphate can be transferred to another molecule or released to the environment.

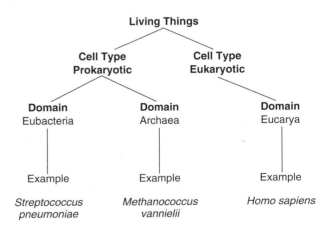

Monophosphate

Diphosphate

Triphosphate

High-energy bonds

Adenine base   Ribose sugar   Phosphate Phosphate Phosphate

Box Figure 19.2

A macromolecule of ATP consists of a molecule of adenine, a molecule of ribose, and three phosphate groups. The two end phosphate groups are bonded together by high-energy bonds. When these bonds are broken, they release an unusually great amount of energy; therefore, they are known as *high-energy bonds.* These bonds are represented by curved, solid lines. The ATP molecule is considered an energy carrier.

plasm is seldom used today, the term *cytoplasm* is still very common in the vocabulary of cell biologists.

The development of better microscopes and better staining techniques revealed that protoplasm contains many tiny structures called **organelles.** It has been determined that certain functions are performed in certain organelles. The essential job an organelle does is related to its structure. Each organelle is dynamic in its operation, changing shape and size as it works. Organelles move throughout the cell, and some even self-duplicate.

All living things are cells or composed of cells. To date, most biologists recognize two major cell types, *prokaryotes* and *eukaryotes.* Whether they are **prokaryotic cells** or **eukaryotic cells,** all have certain things in common: (1) cell membranes, (2) cytoplasm, (3) genetic material, (4) energy currency, (5) enzymes and coenzymes. These are all necessary in order to carry out life's functions. Should any of these not function properly, a cell would die.

The differences among cell types are found in the details of their structure. While prokaryotic cells lack most of the complex internal organelles typical of eukaryotes, they are cells and can carry out life's functions. To better understand and focus on the nature and differences among cell types, biologists have further classified organisms into large categories called **domains.** The following diagram illustrates this level of organization:

Living Things

Cell Type Prokaryotic

Cell Type Eukaryotic

Domain Eubacteria

Domain Archaea

Domain Eucarya

Example

Example

Example

Streptococcus pneumoniae

Methanococcus vannielii

Homo sapiens

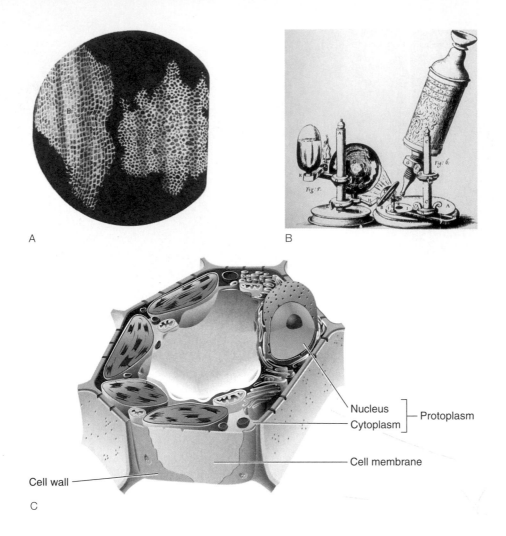

A

B

Nucleus ⎤
Cytoplasm ⎦ Protoplasm

Cell membrane

Cell wall

C

Figure 19.2

The cell concept has changed considerably over the last three hundred years. Robert Hooke's idea of a cell (*A*) was based on his observation of slices of cork (cell walls of the bark of the cork oak tree). Hooke invented the compound microscope and illumination system shown above (*B*), one of the best such microscopes of his time. One of the first subcellular differentiations was to divide the protoplasm into cytoplasm and nucleus as shown in this plant cell (*C*). We now know that cells are much more complex than this; they are composed of many kinds of subcellular structures, some components numbering in the thousands.

Most single-celled organisms that we commonly refer to as *bacteria* are prokaryotic cells and are classified in the domain Eubacteria. Other less well known prokaryotes display significantly different traits that have caused biologists to create a second category of prokaryotes, the domain Archaea or the Archaebacteria. All other living things are based on the eukaryotic cell plan. Members of the kingdoms Protista (algae and protozoa), Fungi (molds, mushrooms), Plantae (plants), and Animalia (animals) are all comprised of eukaryotic cells (figure 19.3).

Notice that viruses are not included in this classification system. That is because they are not cellular in nature. Viruses are not composed of the basic cellular structural components. They are composed of a core of nucleic acid (DNA or RNA, never both) and a surrounding coat or capsid composed of protein. For this reason, the viruses are called acellular or noncellular.

CELL MEMBRANES

One feature common to all cells and many of the organelles they contain is a thin layer of material called *membrane.* Membrane can be folded and twisted into many different structures, shapes, and forms. The particular arrangement of membrane of an organelle is related to the functions that it is capable of performing. This is similar to the way a piece of fabric can be fashioned into a pair of pants, a shirt, sheets, pillowcases, or a rag doll. All cellular membranes have a fundamental molecular structure that allows them to be fashioned into a variety of different organelles.

Cell membranes are thin sheets composed primarily of phospholipids and proteins. The current hypothesis of how membranes are constructed is known as the *fluid-mosaic model,* which proposes that the various molecules of the mem-

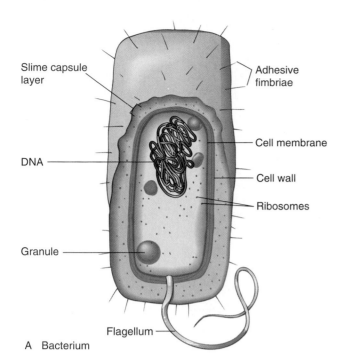

Slime capsule layer

DNA

Granule

Flagellum

A Bacterium

Adhesive fimbriae

Cell membrane

Cell wall

Ribosomes

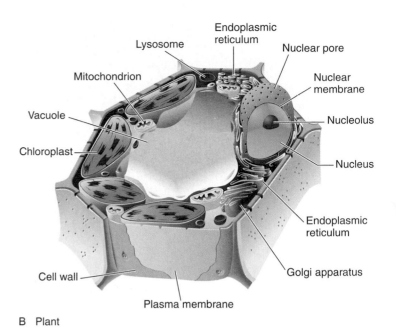

Lysosome

Endoplasmic reticulum

Nuclear pore

Mitochondrion

Nuclear membrane

Vacuole

Nucleolus

Chloroplast

Nucleus

Endoplasmic reticulum

Golgi apparatus

Cell wall

Plasma membrane

B Plant

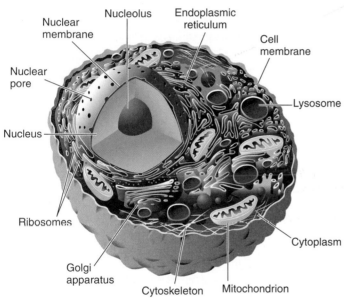

Nuclear membrane

Nucleolus

Endoplasmic reticulum

Cell membrane

Nuclear pore

Lysosome

Nucleus

Ribosomes

Golgi apparatus

Cytoskeleton

Mitochondrion

Cytoplasm

C Animal

Figure 19.3

There are two major types of cells, the prokaryotes and the eukaryotes. Prokaryotic cells are represented by the (*A*) bacteria, and eukaryotic cells by (*B*) plant and (*C*) animal cells.

brane are able to flow and move about (figure 19.4). The membrane maintains its form because of the physical interaction of its molecules with its surroundings. The phospholipid molecules of the membrane are polar molecules since they have one end (the glycerol portion) that is soluble in water and is therefore called *hydrophilic* (water loving). The other end that is not water soluble, called *hydrophobic* (water hating), is comprised of fatty acid. When phospholipid molecules are placed in water, they form a double-layered sheet, with the water-soluble (hydrophilic) portions of the molecules facing away from each other. This is commonly referred to as a *phospholipid bilayer*.

The protein component of cellular membranes can be found on either surface of the membrane or in the membrane, among the phospholipid molecules. Many of the membrane proteins are capable of moving from one side to the other, while others aid in the movement of molecules across the membrane by forming channels through which substances may travel or by acting as transport molecules. Some protein molecules found on the outside surfaces of cellular membranes have carbohydrates or fats attached to them. These combination molecules are important in determining the "sidedness" (inside-outside) of the membrane and also help organisms recognize differences between types of cells. Your body can recognize disease-causing

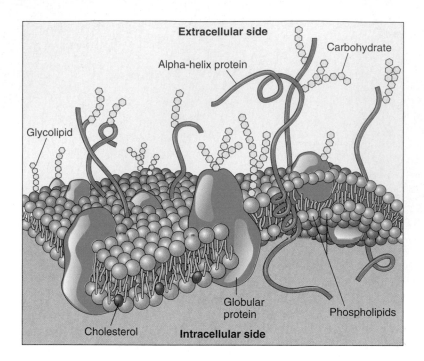

Extracellular side

Carbohydrate

Alpha-helix protein

Glycolipid

Cholesterol

Globular protein

Phospholipids

Intracellular side

Figure 19.4

Notice in this section of a generalized human cell that there is no surrounding cell wall as pictured in Hooke's cell, figure 19.2A. Membranes in all cells are composed of protein and phospholipids. Two layers of phospholipid are oriented so that the hydrophobic fatty ends extend toward each other and the hydrophilic glycerol portions are on the outside. The phosphate-containing chain of the phospholipid is coiled near the glycerol portion. Buried within the phospholipid layer and/or floating on it are the globular proteins. Some of these proteins accumulate materials from outside the cell; others act as sites of chemical activity. Carbohydrates are often attached to one surface of the membrane.

Connections . . .

Casual Contact, Catching AIDS/HIV, and Cell Receptor Sites

Cell membranes have molecules on their surface known as *clusters of differentiation* or *CD markers*. These are designated by numbered groups, that is, CD1, CD2, CD3, CD4, etc. Different cell types have different markers on their surfaces. Those with CD4 markers include macrophages, nerve cells, and certain cells of the immune system known as T-helper (T_H) and T-delayed-type (T_D) hypersensitivity cells. CD4 cells are typically found deep inside the body (e.g., in blood) and not on the skin. The CD4 marker acts as the attachment site for human immunodeficiency virus (HIV) which is responsible for *A*cquired *I*mmuno*D*eficiency *S*yndrome (AIDS). In order to cause AIDS, the virus must enter the CD4 cell and take command of its metabolism. HIV cannot attach to cells that do not have CD4 cell surface receptors. Since CD4 cells do not normally occur on the surface of the skin, casual contact (e.g., shaking hands, touching) is not likely to be the way this virus is transmitted from one person to another.

organisms because their surface proteins are different from those of its own cellular membranes. Some of these molecules also serve as attachment sites for specific chemicals, bacteria, protozoa, white blood cells, and viruses. Many dangerous agents cannot stick to the surface of cells and therefore cannot cause harm. For this reason cell biologists explore the exact structure and function of these molecules. They are also attempting to identify molecules that can interfere with the binding of such agents as viruses and bacteria in the hope of controlling infections.

Other types of molecules found in cell membranes are cholesterol and carbohydrates. Cholesterol appears to play a role in stabilizing the membrane and keeping it flexible. Carbohydrates are usually found on the outside of the membrane, where they are bound to proteins or lipids. These carbohydrates appear to play a role in cell-to-cell interactions and are involved in binding with regulatory molecules.

GETTING THROUGH MEMBRANES

If a cell is to stay alive it must meet the characteristics of life outlined earlier. This includes taking in nutrients and eliminating wastes and other by-products of metabolism. Several mechanisms allow cells to carry out the processes characteristic of life. They include diffusion, osmosis, facilitated diffusion, active transport, and phagocytosis.

Connections . . .

The Other Outer Layer—The Cell Wall

The *cell walls* of microorganisms, plants, and fungi appear to be rigid, solid layers of material but are really loosely woven layers. And like water pouring through a leaky straw basket, many types of molecules easily pass through a cell wall. The cell wall lends strength and protection to the contents of the cell but hampers flexibility and movement. There are three kinds of materials typically used for cell walls. All the higher plants and most algae have *cellulose* as their wall material. When found in large amounts, it is known as *wood*. Another cell wall material, *chitin*, is found in the fungi.

Chitin also constitutes the exoskeleton material in insects, and the outer shell of a beetle or a shrimp is chitin. But in those animals, the chitin surrounds masses of tissue instead of individual cells. The fungi are not crunchy or brittle like shrimp skeletons because the chitin is very thin. Many bacteria have their walls composed of *peptidoglycan*. It is composed of both amino acids and carbohydrates. The lengths of peptidoglycan chains and how they are interlinked may determine the shape of a bacterium.

Diffusion

There is a natural tendency in gases and liquids for molecules of different types to completely mix with each other. This is because they are constantly moving about with various levels of kinetic energy. Consider two types of molecules. As the molecules of one type move about, they tend to scatter from the place where they are most concentrated. The other type of molecule also tends to disperse in the same way. The result of this random motion is that the two types of molecules eventually become mixed throughout.

Remember that the motion of the molecules is completely random. If you follow the paths of molecules from a sugar cube placed in a glass of water, you will find that some of the sugar molecules move away from the cube while others move in the opposite direction. However, more sugar molecules would move away from the original cube because there were more there to start with.

We generally are not interested in the individual movement but rather in the overall movement, which is called the *net movement*. The direction of greatest movement (net movement) is determined by the relative concentration of the molecules. **Diffusion** is defined as the net movement of a kind of molecule from a place where that molecule is in higher concentration to a place where that molecule is scarcer. When a kind of molecule is completely dispersed, and movement is equal in all directions, we say that the system has reached a state of *dynamic equilibrium*. There is no longer a net movement because movement in one direction equals movement in the other. It is dynamic, however, because the system still has energy, and the molecules are still moving.

Because the cell membrane is composed of phospholipid and protein molecules that are in constant motion, temporary openings are formed that allow small molecules to cross from one side of the membrane to the other. Molecules close to the membrane are in constant motion as well. They are able to move into and out of a cell by passing through these openings in the membrane.

The rate of diffusion is related to the kinetic energy and size of the molecules. Since diffusion occurs only when molecules are unevenly distributed, the relative concentration of the molecules is important in determining how fast diffusion occurs.

The difference in concentration of the molecules is known as a *concentration gradient* or *diffusion gradient*. When the molecules are equally distributed, no such gradient exists.

Diffusion can take place only as long as there are no barriers to the free movement of molecules. In the case of a cell, the membrane permits some molecules to pass through, while others are not allowed to pass or are allowed to pass more slowly. This permeability is based on size, ionic charge, and solubility of the molecules involved. The membrane does not, however, distinguish direction of movement of molecules; therefore, the membrane does not influence the direction of diffusion. The direction of diffusion is determined by the relative concentration of specific molecules on the two sides of the membrane, and the energy that causes diffusion to occur is supplied by the kinetic energy of the molecules themselves (figure 19.5).

Diffusion is an important means by which materials are exchanged between a cell and its environment. Since the movement of the molecules is random, the cell has little control over the process; thus, diffusion is considered a passive process, that

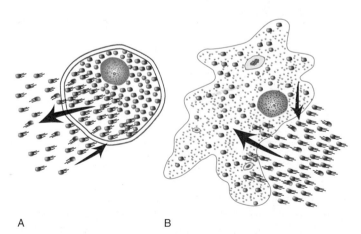

A   B

Figure 19.5

As a result of molecular motion, molecules move from areas where they are concentrated to areas where they are less concentrated. This figure shows (*A*) molecules leaving an animal cell by diffusion and (*B*) molecules entering a cell by diffusion. The direction is controlled by concentration (always high-to-low concentration), and the energy necessary is supplied by the kinetic energy of the molecules themselves.

Connections . . .

Kidney Machines in Action

When the kidneys do not function properly, the toxic waste urea builds up in the blood because the ailing kidneys cannot filter it out. This eventually leads to a condition known as *uremia*, "urine in the blood." Unless the urea is removed from the body, the concentration of urea increases and will eventually lead to death. Kidney or dialysis machines use tubular cellulose membranes to separate the urea from the blood.

Cellulose is a polymer of glucose molecules and contains openings through which urea molecules pass. In such a machine, cellulose membrane tubes are immersed in a large volume of "cleansing" or dialysis solution. The blood is pumped from the vein of a person with uremia through this tubing, and then returned to the patient. While in the machine, the membrane allows most of the urea in the blood to pass through the tubing walls but not the larger blood proteins and blood cells. Therefore, the patient does not loose vital proteins or blood cells. This system works because the dialysis solution in which the tubing is immersed has a lower amount of urea than does the blood. Because of this concentration gradient, urea in the blood passes from the high concentration in the blood through the cellulose membrane tubing pores into the surrounding dialysis solution. This results in a decrease in urea concentration in the blood. As the concentration of the urea increases in the dialysis solution, the solution is flushed out. This enables the dialysis solution to maintain a constant, low concentration of urea and maintain the movement of urea molecules out of the person's body. To maintain the blood's concentration of other vital ions, the solution is made with the same concentration of those ions as found in the blood. This means that these other ions are in dynamic equilibrium, and so their concentration does not change during kidney dialysis (box figure 19.3).

Box Figure 19.3

Blood inlet from patient

Dialysis solution outlet

Tubular cellulose membranes

Dialysis solution inlet

Blood outlet return to patient

---

is, chemical bond energy does not have to be expended. For example, animals are constantly using oxygen in various chemical reactions. Consequently, the oxygen concentration in cells always remains low. The cells, then, contain a lower concentration of oxygen than the oxygen level outside of the cells. This creates a diffusion gradient, and the oxygen molecules diffuse from the outside of the cell to the inside of the cell.

Osmosis

Another characteristic of all membranes is that they are selectively permeable. **Selectively permeable** means that a membrane will allow certain molecules to pass across it and will prevent others from doing so. Molecules that are able to dissolve in phospholipids, such as vitamins A and D, can pass through the membrane rather easily; however, many molecules cannot pass through at all. In certain cases, the membrane differentiates on the basis of molecular size; that is, the membrane allows small molecules, such as water, to pass through and prevents the passage of larger molecules. The membrane may also regulate

the passage of ions. If a particular portion of the membrane has a large number of positive ions on its surface, positively charged ions in the environment will be repelled and prevented from crossing the membrane.

Water molecules easily diffuse through cell membranes. The net movement (diffusion) of water molecules through a selectively permeable membrane is known as **osmosis.** In any osmotic situation, there must be a selectively permeable membrane separating two solutions. For example, a solution of 90 percent water and 10 percent sugar separated by a selectively permeable membrane from a different sugar solution, such as one of 80 percent water and 20 percent sugar, demonstrates osmosis. The membrane allows water molecules to pass freely but prevents the larger sugar molecules from crossing. There is a higher concentration of water molecules in one solution compared to the concentration of water molecules in the other, so more of the water molecules move from the solution with 90 percent water to the other solution with 80 percent water. Be sure that you recognize that osmosis is really diffusion in which the diffusing substance is water, and that the regions of different concentrations are separated by a membrane that is more permeable to water.

A proper amount of water is required if a cell is to function efficiently. Too much water in a cell may dilute the cell contents and interfere with the chemical reactions necessary to keep the cell alive. Too little water in the cell may result in a buildup of poisonous waste products. As with the diffusion of other molecules, osmosis is a passive process because the cell has no control over the diffusion of water molecules. This means that the cell can remain in balance with an environment only if that environment does not cause the cell to lose or gain too much water.

Connections . . .

Cell Trivia

Here are some tidbits of information about cells:

- The smallest cells are the bacteria, some of which are only 0.1 to 0.2 micrometers in diameter. *E. coli,* a very common bacterium in human intestines, is only 0.5 micrometer long.
- The largest cells are the egg cells of birds and mammals. The yolk of a chicken egg is a single cell, which can be 50 millimeters in diameter. The nerve cell from the human spine to the big toe can be 1 meter or more in length.
- The animal cells that live the longest are the nerve or muscle cells of long-lived reptiles and mammals. The sea tortoise, which lives two hundred years, has the same nerve cells at birth that it does at death.
- Cells of certain bacteria can divide every 15 minutes under optimum conditions. This means that if you start with only one bacterial cell and it divides as often as possible, in only twelve hours there could be as many as 281 trillion individual cells.

Plant cells also experience osmosis. If the water concentration outside the plant cell is higher than the water concentration inside, more water molecules enter the cell than leave. This creates internal pressure within the cell. But plant cells do not burst because they are surrounded by a rigid cell wall. Lettuce cells that are crisp are ones that have gained water so that there is high internal pressure. Wilted lettuce has lost some of its water to its surroundings so that it has only slight internal cellular water pressure. Osmosis occurs when you put salad dressing on a salad. Because the dressing has a very low water concentration, water from the lettuce diffuses from the cells into the surroundings. Salad that has been "dressed" too long becomes limp and unappetizing.

So far, we have considered only situations in which cells have no control over the movement of molecules. Cells cannot rely solely on diffusion and osmosis, however, because many of the molecules they require either cannot pass through the cell membranes or occur in relatively low concentrations in the cells' surroundings.

Controlled Methods of Transporting Molecules

Some molecules move across the membrane by combining with specific carrier proteins. When the rate of diffusion of a substance is increased in the presence of a carrier, we call this **facilitated diffusion.** Since this is diffusion, the net direction of movement is in accordance with the concentration gradient. Therefore, this is considered a passive transport method, although it can occur only in living organisms with the necessary carrier proteins. One example of facilitated diffusion is the movement of glucose molecules across the membranes of certain cells. In order for the glucose molecules to pass into these cells, specific proteins are required to carry them across the membrane. The action of the carrier does not require an input of energy other than the kinetic energy of the molecules (figure 19.6).

When molecules are moved across the membrane from an area of *low* concentration to an area of *high* concentration, the cell must expend energy. The process of using a carrier protein to move molecules up a concentration gradient is called **active transport** (figure 19.7). Active transport is very specific: Only certain molecules or ions can be moved in this way, and specific proteins in the membrane must carry them. The action of the carrier requires an input of energy other than the kinetic energy of the molecules; therefore, this process is termed *active* transport. For example, some ions, such as sodium and potassium, are actively pumped across cell membranes. Sodium ions are pumped out of cells up a concentration gradient. Potassium ions are pumped into cells up a concentration gradient.

In addition to active transport, materials can be transported into a cell by *endocytosis* and out by *exocytosis*. **Phagocytosis** is another name for one kind of endocytosis that is the process cells use to wrap membrane around a particle (usually food) and engulf it (figure 19.8). This is the process leukocytes (white blood cells) use to surround invading bacteria, viruses, and

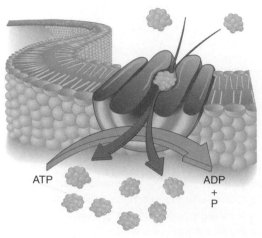

Figure 19.7

One possible method whereby active transport could cause materials to accumulate in a cell is illustrated here. Notice that the concentration gradient is such that if simple diffusion were operating, the molecules would leave the cell. The action of the carrier protein requires an active input of energy (the compound ATP) other than the kinetic energy of the molecules; therefore, this process is termed *active* transport.

Figure 19.6

This method of transporting materials across membranes is a diffusion process (i.e., a movement of molecules from a high to a low concentration). However, the process is helped (facilitated) by a particular membrane protein. No chemical-bond energy in the form of ATP is required for this process. The molecules being moved through the membrane attach to a specific transport carrier protein in the membrane. This causes a change in its shape that propels the molecule or ion through to the other side.

other foreign materials. Because of this, these kinds of cells are called *phagocytes*. When phagocytosis occurs, the material to be engulfed touches the surface of the phagocyte and causes a portion of the outer cell membrane to be indented. The indented cell membrane is pinched off inside the cell to form a sac containing the engulfed material. This sac, composed of a single membrane, is called a **vacuole**. Once inside the cell, the membrane of the vacuole is broken down, releasing its contents inside the cell, or it may combine with another vacuole containing destructive enzymes.

ORGANELLES COMPOSED OF MEMBRANES

Now that you have some background concerning the structure and the function of membranes, let's turn our attention to the way cells use membranes to build the structural components of their protoplasm. The outer boundary of the cell is termed the cell membrane or **plasma membrane.** It is associated with a great variety of metabolic activities including taking up and releasing molecules, sensing stimuli in the environment, recognizing other cell types, and attaching to other cells and nonliving objects. In addition to the cell membrane, many other organelles are composed of membranes. Each of these membranous organelles has a unique shape or structure that is associated with particular functions.

One of the most common organelles found in cells is the *endoplasmic reticulum*. The **endoplasmic reticulum,** or *ER,* is a set of folded membranes and tubes throughout the cell. This system of membranes provides a large surface upon which chemical activities take place (figure 19.9). Since the ER has an enormous surface area, many chemical reactions can be carried out in an extremely small space.

The **Golgi apparatus** is another organelle composed of membrane. Even though this organelle is also composed of membrane, the way in which it is structured enables it to perform jobs different from those performed by the ER. The typical Golgi is composed of from five to twenty flattened, smooth, membranous sacs, which resemble a stack of pancakes. The Golgi apparatus is the site of the synthesis and packaging of certain molecules produced in the cell. It is also the place where particular chemicals are concentrated prior to their release from the cell or distribution within the cell.

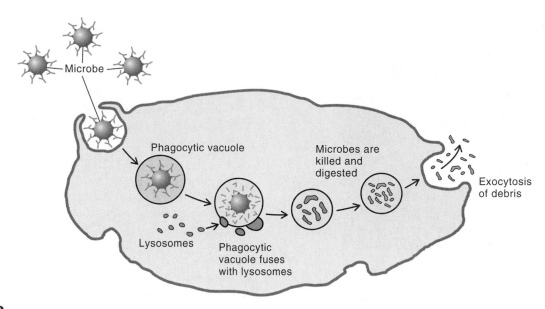

Figure 19.8

Phagocytosis. The sequence illustrates a cell engulfing a microbe and surrounding it with a membrane. Once encased in a portion of the cell membrane (now called a phagocytic vacuole) a lysosome adds its digestive enzymes to it, which speeds the breakdown of the dangerous microbes. Finally, the hydrolyzed (digested) material moves from the vacuole to the inner surface of the cell membrane where the contents are discharged by exocytosis.

Figure 19.9

Membranous cytoplasmic organelles. Certain structures in the cytoplasm are constructed of membranes. Membranes are composed of protein and phospholipids. The four structures here—lysosomes, endoplasmic reticulum, vacuoles, and the Golgi body—are constructed of simple membranes.

Figure 19.10

This person displays the trait known as syndactylism (*syn* = connected; *dactyl* = finger or toe). In most people, lysosomal enzymes break down the cells between the toes, allowing the toes to separate. In this genetic abnormality, these enzymes fail to do their job.

An important group of molecules that is necessary to the cell includes the hydrolytic enzymes. This group of enzymes is capable of destroying carbohydrates, nucleic acids, proteins, and lipids. Since cells contain large amounts of these molecules, these enzymes must be controlled in order to prevent the destruction of the cell. The Golgi apparatus is the site where these enzymes are converted from their inactive to their active forms and packaged in membranous sacs. These vesicles are pinched off from the outside surfaces of the Golgi sacs and given the spe-

cial name **lysosomes,** or "bursting body." Cells use the lysosomes in four major ways:

1. Decompose dead and dying cells.
2. Selectively destroy cells to fashion a developing organism from its immature to mature state (figure 19.10).
3. Digest macromolecules that have been ingested into the cell.
4. Kill dangerous microorganisms that have been taken into the cell by phagocytosis.

A nucleus is a place in a cell—not a solid mass. Just as a room is a place created by walls, a floor, and a ceiling, the nucleus is a place in the cell created by the **nuclear membrane.** This membrane separates the *nucleoplasm,* liquid material in the nucleus, from the cytoplasm. Because they are separated, the cytoplasm and nucleoplasm can maintain different chemical compositions. If the membrane were not formed around the genetic material, the organelle we call the nucleus would not exist. The nuclear membrane is formed from many flattened sacs fashioned into a hollow sphere around the genetic material, DNA (**d**eoxyribo**n**ucleic **a**cid). It also has large openings called nuclear pores that allow thousands of relatively large molecules to pass into and out of the nucleus each minute. These pores are held open by donut-shaped molecules that resemble the "eyes" in shoes through which the shoelace is strung.

All of the membranous organelles just described can be converted from one form to another (figure 19.11). For example, phagocytosis results in the formation of vacuolar membrane from cell membrane that fuses with lysosomal membrane, which in turn came from Golgi membrane. Two other organelles composed of membranes are chemically different and are incapable of interconversion. Both types of organelles are as-

Figure 19.11

Eukaryotic cells contain a variety of organelles composed of phospholipids and proteins. Each has a unique shape and function. Many of these organelles are interconverted from one to another as they perform their essential functions. Cell membranes can become vacuolar membrane or endoplasmic reticulum, which can become vesicular membrane, which in turn can become Golgi or nuclear membrane. However, mitochondria cannot exchange membrane parts with other membranous organelles.

Mitochondrion

Inner membrane

Cristae

Outer membrane

Stroma

Thylakoid
membrane

Granum

A

B

Figure 19.12

(*A*) The chloroplast, the container of the pigment chlorophyll, is the site of photosynthesis. The chlorophyll, located in the grana, captures light energy that is used to construct organic molecules in the stroma. (*B*) The mitochondria with their inner folds, called cristae, are the site of aerobic cellular respiration, where food energy is converted to usable cellular energy. Both organelles are composed of phospholipid and protein membranes.

sociated with energy conversion reactions in the cell. These organelles are the *mitochondrion* and the *chloroplast* (figure 19.12).

The **mitochondrion** is an organelle resembling a small bag with a larger bag inside that is folded back on itself. These inner folded surfaces are known as the *cristae.* Located on the surface of the cristae are particular proteins and enzymes involved in *aerobic cellular respiration.* **Aerobic cellular respiration** is the series of reactions involved in the release of usable energy from food molecules, which requires the participation of oxygen molecules. Enzymes that speed the breakdown of simple nutrients are arranged in a sequence on the mitochondrial membrane. The average human cell contains upwards of ten thousand mitochondria. Cells that are involved in activities that require large amounts of energy, such as muscle cells, contain

many more mitochondria. When properly stained, they can be seen with a compound light microscope. When cells are functioning aerobically, the mitochondria swell with activity. But when this activity diminishes, they shrink and appear as threadlike structures.

A second energy-converting organelle is the **chloroplast.** This membranous, saclike organelle is found only in plants and algae—organisms that carry out photosynthesis—and contains the green pigment *chlorophyll.* Some cells contain only one large chloroplast, while others contain hundreds of smaller chloroplasts. In this organelle light energy is converted to chemical-bond energy in a process known as *photosynthesis.* Chemical-bond energy is found in food molecules. A study of the ultrastructure—that is, the structures seen with an electron

Photosynthesis

Ultimately the energy to power most organisms comes from the sun. Chlorophyll-containing plants, algae, and certain bacteria have the ability to capture and transform light energy into chemical bond energy through the process of photosynthesis. In eukaryotic cells, chlorophyll is found on membranes within chloroplasts. In prokaryotic cells photosynthetic pigments are on membranes in the cytoplasm. Light energy is converted to chemical-bond energy in the form of the molecule called ATP. Organisms then use ATP to produce organic molecules such as glucose. These are the organic molecules that organisms use as an energy source through the process of cellular respiration.

The following equation summarizes the chemical reactions green plants and many other photosynthetic organisms use to make ATP and organic molecules:

$$\text{light energy} + CO_2 + H_2O \rightarrow$$
$$C_6H_{12}O_6 + H_2O + O_2$$

There are three stages in the photosynthetic pathway:

1. **Light-capturing stage.** In eukaryotic cells photosynthetic pigments such as chlorophyll are clustered together on chloroplast membranes. When enough of the right kind of light is available, the pigment electrons absorb extra energy and become "excited." With this added energy they are capable of entering into the chemical reactions responsible for the production of ATP.

$$\text{light energy} + \text{photosynthetic pigments} \rightarrow$$
$$\text{higher-energy-containing electrons}$$
$$\text{called "excited electrons"} +$$
$$\text{electron-deficient pigments}$$

2. **Light-dependent reaction stage.** Since this stage depends on the presence of light, it is also called light dependent or the *light reaction.* During this stage "excited" electrons from the light-capturing stage use their additional energy in a

chemical reaction that combines smaller components to form a molecule of ATP:

$$\text{phosphate} + \text{ADP (adenosine diphosphate)} \rightarrow$$
$$\text{ATP (adenosine triphosphate)}$$

In addition, water is broken down to hydrogen and oxygen. The oxygen is released to the environment as O_2 and the hydrogens are transferred to electron-carrier coenzymes, $NADP^+$ (nicotinamide adenine dinucleotide phosphate). ($NADP^+$ is similar to NAD used in aerobic cellular respiration.) Light-dependent reactions also take place on the thylakoid membranes.

$$H_2O + NADP^+ + \text{"excited" electrons} \rightarrow$$
$$NADPH + O_2 + ATP$$

3. **Light-independent reaction stage.** This stage is also known as the *dark reaction* since light is not needed for the reactions to take place. During these reactions, ATP and NADPH from the light-dependent

| Light-capturing Stage | + | Light-dependent Stage | + | Light-independent Stage $\longrightarrow C_6H_{12}O_6 + 6\,H_2O + 6\,O_2$ |

Light energy + photosynthetic pigments → 'Excited' electrons

$H_2O + NADP^+ + $ 'excited' electrons → $NADPH + O_2 + ATP$

$ATP + NADPH + CO_2 + 5$ carbon starter $\longrightarrow C_6H_{12}O_6 + 6\,H_2O + 6\,O_2$

$$\text{Light energy} + H_2O + CO_2 \longrightarrow C_6H_{12}O_6 + 6\,H_2O + 6\,O_2$$

Box Figure 19.4

---

microscope—of a chloroplast shows that the entire organelle is enclosed by a membrane, while other membranes are folded and interwoven throughout. As shown in figure 19.12A, in some areas concentrations of these membranes are stacked up or folded back on themselves. Chlorophyll molecules are attached to these membranes. These areas of concentrated chlorophyll are called the *grana* of the chloroplast. The space between the grana, which has no chlorophyll, is known as the *stroma*.

Mitochondria and chloroplasts are different from other kinds of membranous structures in several ways. First, their membranes are chemically different from those of other membranous organelles. Second, they are composed of double layers of membrane—an inner and an outer membrane. Third, both of these structures have ribosomes and DNA that are similar to those of bacteria. Finally, these two structures have a certain de-

gree of independence from the rest of the cell—they have a limited ability to reproduce themselves but must rely on DNA from the cell's nucleus for assistance.

All of the organelles just described are composed of membranes. Many of these membranes are modified for particular functions. Each membrane is composed of the double phospholipid layer with protein molecules associated with it.

NONMEMBRANOUS ORGANELLES

Suspended in the cytoplasm and associated with the membranous organelles are various kinds of structures that are not composed of phospholipids and proteins arranged in sheets.

In the cytoplasm are many very small structures called **ribosomes** that are composed of ribonucleic acid (RNA) and reaction stage are used to attach CO_2 to 5 carbon starter molecules (already present in the cell) to manufacture new larger organic molecules, for example, glucose ($C_6H_{12}O_6$). These reactions take place in the light or dark as long as substrates are available from the light-dependent stage (box figure 19.4).

$$ATP + NADPH + CO_2 + 5\ carbon\ starter \rightarrow larger\ organic\ molecules\ (e.g., glucose)$$

Aerobic Cellular Respiration

In eukaryotic cells the process of releasing energy from food molecules begins in the cytoplasm and is completed in the mitochondrion:

$$C_6H_{12}O_6 + O_2 + H_2O \rightarrow CO_2 + H_2O + energy\ (ATP + heat)$$

The major parts of the cellular respiration process are:

1. *Glycolysis* (*glyco* = carbohydrate; *lys* = splitting; *sis* = the process of) breaks the 6-carbon sugar (glucose) into two smaller 3-carbon molecules named pyruvic acid; ATP is produced. Hydrogens and their electrons are carried by the coenzyme nicotinamide adenine dinucleotide (NAD) and are sent to the electron-transport system (ETS) for processing.

$$C_6H_{12}O_6\ (glucose) \rightarrow CH_3COCOOH\ (pyruvic\ acid) + NADH + energy\ (ATP + heat)$$

2. The *Krebs cycle* removes the remaining hydrogen, electrons, and carbon from pyruvic acid. ATP is produced for cell use. The hydrogens and their electrons are carried by NAD and another coenzyme, flavin adenine dinucleotide (FAD), and are sent to the ETS for processing. The remaining carbon and oxygen is released as carbon dioxide (CO_2).

$$CH_3COCOOH\ (pyruvic\ acid) \rightarrow CO_2 + NADH + FADH_2 + energy\ (ATP + heat)$$

3. The *electron-transport system (ETS)* converts the kinetic energy of hydrogen electrons received from glycolysis and the Krebs cycle to the high-energy phosphate bonds of the molecule ATP, as the hydrogen ions and electrons are ultimately bonded with oxygen to form water (box figure 19.5).

$$NADH + FADH_2 + O_2 \rightarrow H_2O + energy\ (ATP + heat) + NAD + FAD$$

Box Figure 19.5

protein. Ribosomes function in the manufacture of protein. Many ribosomes are also found floating freely in the cytoplasm (figure 19.13), wherever proteins are being assembled. Cells that are actively producing protein (e.g., liver cells) have great numbers of free and attached ribosomes.

Among the many types of nonmembranous organelles found there are elongated protein structures that are known as *microtubules, microfilaments,* and *intermediate filaments.* Their various functions are as complex as those provided by the structural framework and cables of a high-rise office building, geodesic dome, or skeletal and muscular systems of a large animal. All three types of organelles interconnect and some are attached to the inside of the cell membrane, forming what is known as the *cytoskeleton* of the cell (figure 19.14). These cellular components provide the cell with shape, support, and ability to move about the environment.

An arrangement of two sets of microtubules at right angles to each other makes up the structures known as the *centrioles,* and they operate by organizing microtubules into a complex of strings called *spindle fibers.* The *spindle* is the structure upon which chromosomes are attached in order that they may be properly separated during cell division. Each set is composed of nine groups of short microtubules arranged in a cylinder (figure 19.15). One curious fact about centrioles is that they are present in most animal cells but not in many types of plant cells.

Many cells have microscopic, hairlike structures projecting from their surfaces; these are *cilia* or *flagella* (figure 19.16). In general, we call them flagella if they are long and few in number, and cilia if they are short and more numerous. The cell has the ability to control the action of these microtubular structures, enabling them to be moved in a variety of different ways. Their

Ribosome

Figure 19.13

Each ribosome is constructed of two subunits of protein and ribonucleic acid. These globular organelles are associated with the construction of protein molecules from individual amino acids. They are sometimes located individually in the cytoplasm where protein is being assembled, or they may be attached to endoplasmic reticulum (ER). They are so obvious on the ER when using electron micrograph techniques that, when they are present, we label this *rough ER.*

coordinated actions either propel the cell through the environment or the environment past the cell surface. The protozoan *Paramecium* is covered with thousands of cilia that actively beat a rhythmic motion to move the cell through the water. The cilia on the cells that line your trachea move mucus-containing particles from deep within your lungs.

NUCLEAR COMPONENTS

When nuclear structures were first identified, it was noted that certain dyes stained some parts more than others. The parts that stained more heavily were called **chromatin,** which means colored material. Chromatin is composed of long molecules of deoxyribonucleic acid (DNA) in association with proteins. Chromatin is loosely organized DNA/protein strands in the nucleus. When the chromatin is tightly coiled into shorter, denser structures, we call them **chromosomes.** Chromatin and chromosomes are really the same molecules but differ in structural arrangement. In addition to chromosomes, the nucleus may also contain one, two, or several *nucleoli.* A **nucleolus** is the site of ribosome manufacture. Nucleoli are composed of specific granules and fibers in association with the cell's DNA used in the manufacture of ribosomes. These regions, together with the completed or partially completed ribosomes, are called nucleoli.

The final component of the nucleus is its liquid matrix called the *nucleoplasm.* It is a colloidal mixture composed of water and the molecules used in the construction of ribosomes, nucleic acids, and other nuclear material (figure 19.17).

A

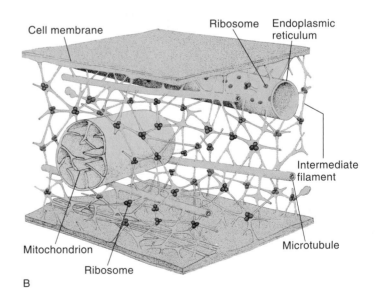

Cell membrane

Ribosome Endoplasmic reticulum

Intermediate filament

Mitochondrion

Ribosome

Microtubule

B

Figure 19.14

A complex array of microfilaments, microtubules, and intermediate filaments provides an internal framework structure for the cell. The cellular skeleton is not a rigid, fixed-in-place structure, but is dynamic and changes as the microfilament and microtubule component parts assemble and disassemble. (*A*) Elements of the cytoskeleton have been labeled with a fluorescent dye to make them visible. The microtubules have fluorescent red dye and actin filaments are green. Part (*B*) shows how the various parts of the cytoskeleton are interconnected.

Source for (*B*) from "The Ground Substance of the Living Cell" by Keith R. Porter and Jonathan B. Tucker, in *Scientific American,* March 1981. Copyright © 1981 by Scientific American, Inc. All Rights Reserved.

Figure 19.15

These two sets of short microtubules that constitute a centriole are located just outside the nuclear membrane in many types of cells. The micrograph shows an end view of one of these sets. Magnification is about 160,000 times.

MAJOR CELL TYPES

Not all of the cellular organelles we have just described are located in every cell. Some cells typically have combinations of organelles that differ from others. For example, some cells have a nuclear membrane, mitochondria, chloroplasts, ER, and Golgi; others have mitochondria, centrioles, Golgi, ER, and nuclear membrane. Other cells are even more simple and lack the complex membranous organelles described in this chapter. Because of this fact, biologists have been able to classify cells into two major types: prokaryotic and eukaryotic (figure 19.18).

The Prokaryotic Cell Structure

Prokaryotic cells, the *bacteria* and *archaea,* do not have a typical nucleus bound by a nuclear membrane, nor do they contain mitochondria, chloroplasts, Golgi, or extensive networks of ER. However, prokaryotic cells contain DNA and enzymes and are able to reproduce and engage in metabolism. They perform all of the basic functions of living things with fewer and more simple organelles. Members of the Archaea are of little concern to the medical profession because none have been identified as disease causing. They are typically found growing in extreme environments where the pH, salt concentration, or temperatures make it impossible for most other organisms to survive. The other prokaryotic cells are called bacteria and about 5 percent cause diseases such as tuberculosis, strep throat, gonorrhea, and acne.

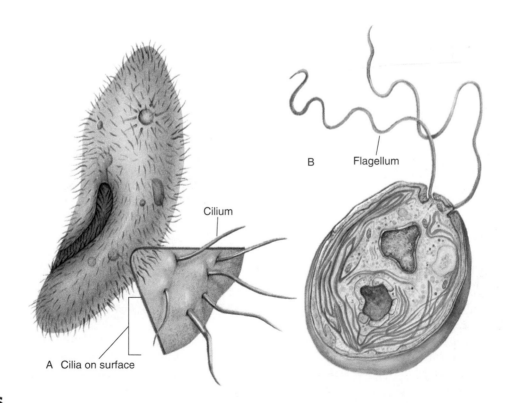

Figure 19.16

(A) Cilia and (B) flagella function like oars or propellers that move the cell through its environment or move the environment past the cell. Flagella are usually less numerous and longer than cilia.

Nuclear membrane

Nuclear pore

Nucleolus

Chromosomal material

Figure 19.17

One of the two major regions of protoplasm, the nucleus has its own complex structure. It is bounded by two layers of membrane that separate it from the cytoplasm. Inside the nucleus are the nucleoli, chromosomes or chromatin material composed of DNA and protein, and the liquid matrix (nucleoplasm). Magnification is about 20,000 times.

A Closer Look   Antibiotics and Cell Structural Differences

One significant difference between prokaryotic and eukaryotic cells is in the chemical makeup of their ribosomes. The ribosomes of prokaryotic cells contain different proteins from those found in eukaryotic cells. Prokaryotic ribosomes are also smaller. This discovery was important to medicine because many cellular forms of life that cause common diseases are bacterial. As soon as differences in the ribosomes were noted, researchers began to look for ways in which to interfere with the prokaryotic ribosome's function but *not* interfere with the ribosomes of eukaryotic cells. Antibiotics, such as streptomycin, are the result of this research. This drug combines with prokaryotic ribosomes and causes the death of the prokaryote by preventing the production of proteins essential to its survival. Because eukaryotic ribosomes differ from prokaryotic ribosomes, streptomycin does not interfere with the normal function of ribosomes in human cells.

Prokaryotic Cells
Characterized by few membranous organelles; DNA not separated from the cytoplasm by a membrane

Eukaryotic Cells
Cells larger than prokaryotic cells; DNA found within nucleus with a membrane separating it from the cytoplasm; many complex organelles composed of many structures including phospholipid bilayer membranes

	Prokaryotic Cells		Eukaryotic Cells			
	Domain Eubacteria	**Domain Archaea**	**Domain Eucarya**			
			Kingdom Protista	Kingdom Fungi	Kingdom Plantae	Kingdom Animalia
	Kingdoms not specified	Kingdoms Euryarchaeota, Korarchaeota, Krenarchaeota				
	Unicellular microbes; typically associated with bacterial "diseases," but 90%–95% are ecologically important and not pathogens	Unicellular microbes; typically associated with extreme environments including low pH, high salinity, and extreme temperatures	Unicellular microbes; some in colonies; both photosynthetic and heterotropic nutrition; a few are parasites	Multicellular organisms or loose colonial arrangements of cells; organism is a row or filament of cells; decay fungi and parasites	Multicellular organisms; cells supported by a rigid cell wall of cellulose; some cells have chloroplasts; complex arrangement into tissues	Multicellular organisms with division of labor into complex tissues; no cell wall present; acquire food from the environment; some are parasites
	Examples: Gram-positive bacteria such as *Steptococcus pneumonia* and Gram-negative bacteria such as *E. coli*	Examples: *Methanococcus*, halophiles, and *Thermococcus*	Examples: protozoans such as *Amoeba* and *Paramecium* and algae such as *Chlamydomonas* and *Euglena*	Examples: yeast such as baker's yeast, molds such as *Penicillium*; morels, mushrooms, and rusts	Examples: moss, ferns, cone-bearing trees, and flowering plants	Examples: worms, insects, starfish, frogs, reptiles, birds, and mammals

Figure 19.18

The two types of cells (prokaryotic and eukaryotic) are described in relationship to the major patterns found in all living things. Note the similarities and the subtle differences among them.

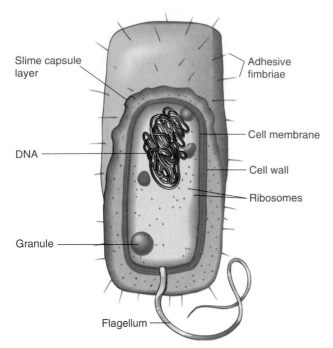

Slime capsule layer

Adhesive fimbriae

DNA

Cell membrane

Cell wall

Ribosomes

Granule

Flagellum

Figure 19.19

All bacteria are prokaryotic cells. While smaller and less complex than eukaryotic cells, each is capable of surviving on its own. Most bacteria are involved with decay and decomposition, but a small percentage are pathogens responsible for disease such as strep throat, TB, syphilis, and gas gangrene. The cell illustrated here is a bacillus because it has a rod shape.

Other bacteria are responsible for the decay and decomposition of dead organisms. Although some bacteria have a type of green photosynthetic pigment and carry on photosynthesis, they do so without chloroplasts and use different chemical reactions.

Most prokaryotic cells are surrounded by a *capsule* or slime layer that can be composed of a variety of compounds (figure 19.19). In certain bacteria this layer is responsible for their ability to stick to surfaces (including host cells) and to resist phagocytosis. Many bacteria also have fimbriae, hairlike protein structures, which help the cell stick to objects. Those with flagella are capable of propelling themselves through the environment. Below the capsule is the rigid cell wall composed of a unique protein/carbohydrate complex called peptidoglycan. This complex provides the cell with the strength to resist osmotic pressure changes and gives the cell shape. Just beneath the wall is the cell membrane. Thinner and with a slightly different chemical composition from eukaryotes, it carries out the same functions as the cell membranes in eukaryotes. Most bacteria are either rods (bacilli), spherical (cocci), or curved (spirilla). The genetic material within the cytoplasm is DNA in the form of a loop.

The Eukaryotic Cell Structure

Eukaryotic cells contain a true nucleus and most of the membranous organelles described earlier. Eukaryotic organisms can be further divided into several categories based on the specific

Table 19.1 Comparison of General Plant and Animal Cell Structure

Plant Cells	Animal Cells
CELL WALL	
Cell membrane	Cell membrane
Cytoplasm	Cytoplasm
Nucleus	Nucleus
Mitochondria	Mitochondria
CHLOROPLASTS	**CENTRIOLE**
Golgi apparatus	Golgi apparatus
Endoplasmic reticulum	Endoplasmic reticulum
Lysosomes	Lysosomes
Vacuoles/vesicles	Vacuoles/vesicles
Ribosomes	Ribosomes
Nucleolus	Nucleolus
Inclusions	Inclusions
Cytoskeleton	Cytoskeleton

combination of organelles they contain. The cells of plants, fungi, protozoa and algae, and animals are all eukaryotic. The most obvious characteristic that sets the plants and algae apart from other organisms is their green color, which indicates that the cells contain chlorophyll. Chlorophyll is necessary for the process of photosynthesis—the conversion of light energy into chemical-bond energy in food molecules. These cells, then, are different from the other cells in that they contain chloroplasts in their cytoplasm. Another distinguishing characteristic of plants and algae is the presence of cellulose in their cell walls (Table 19.1).

The group of organisms that has a cell wall but lacks chlorophyll in chloroplasts is collectively known as *fungi*. They were previously thought to be either plants that had lost their ability to make their own food or animals that had developed cell walls. Organisms that belong in this category of eukaryotic cells include yeasts, molds, mushrooms, and the fungi that cause such human diseases as athlete's foot, "jungle rot," and ringworm. Now we have come to recognize this group as different enough from plants and animals to place them in a separate kingdom.

Eukaryotic organisms that lack cell walls and cannot photosynthesize are placed in separate groups. Organisms that consist of only one cell are called protozoans—examples are *Amoeba* and *Paramecium*. They have all the cellular organelles described in this chapter except the chloroplast; therefore, protozoans must consume food as do the fungi and the multicellular animals.

Although the differences in these groups of organisms may seem to set them worlds apart, their similarity in cellular structure is one of the central themes unifying the field of biology. You can obtain a better understanding of how cells operate in general by studying specific examples. Because the organelles have the same general structure and function regardless of the kind of cell in which they are found, you can learn more about how mitochondria function in plants by studying how mitochondria function in animals. There is a commonality among all living things with regard to their cellular structure and function.

THE IMPORTANCE OF CELL DIVISION

The process of cell division replaces dead cells with new ones, repairs damaged tissues, and allows living organisms to grow. For example, you began as a single cell that resulted from the union of a sperm and an egg. One of the *first* activities of this single cell was to divide. As this process continued, the number of cells in your body increased, so that as an adult your body consists of several trillion cells.

The *second* function of cell division is to maintain the body. Certain cells in your body, such as red blood cells and cells of the gut lining and skin, wear out. As they do, they must be replaced with new cells. Altogether, you lose about 50 million cells per second; this means that millions of cells are dividing in your body at any given time.

A *third* purpose of cell division is repair. When a bone is broken, the break heals because cells divide, increasing the number of cells available to knit together the broken pieces. If some skin cells are destroyed by a cut or abrasion, cell division produces new cells to repair the damage.

During cell division, two events occur. The replicated genetic information of a cell is equally distributed to two daughter nuclei in a process called **mitosis.** As the nucleus goes through its division, the cytoplasm also divides into two new cells. This division of the cell's cytoplasm is called **cytokinesis**—cell splitting. Each new cell gets one of the two daughter nuclei so that both have a complete set of genetic information.

THE CELL CYCLE

All eukaryotic cells go through the same basic life cycle, but they vary in the amount of time they spend in the different stages. A generalized picture of a cell's life cycle may help you understand it better (figure 19.20). Once begun, cell division is a continuous process without a beginning or an end. It is a cycle in which cells continue to grow and divide. There are five stages to the life cycle of a eukaryotic cell: (1) G_1, gap (growth)—phase one; (2) S, synthesis; (3) G_2, gap (growth)—phase two; (4) cell division (mitosis and cytokinesis); and (5) G_0, gap (growth)—mitotic dormancy or differentiation.

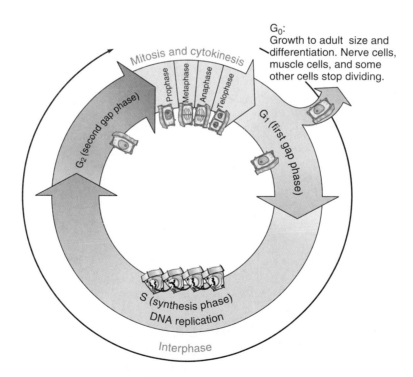

Figure 19.20

During the cell cycle, tRNA, mRNA, ribosomes, and enzymes are produced in the G_1 stage. DNA replication occurs in the S stage. Proteins required for the spindles are synthesized in the G_2 stage. The nucleus is replicated in mitosis and two cells are formed by cytokinesis. Once some organs, such as the brain, have completely developed, certain types of cells, such as nerve cells, enter the G_0 stage. The time periods indicated are relative and vary depending on the type of cell and the age of the organism.

During the G_0 phase, cells are not considered to be in the cycle of division but become differentiated or specialized in their function. It is at this time that they "mature" to play the role specified by their genetic makeup. Whereas some cells entering the G_0 phase remain there more-or-less permanently (e.g., nerve cells), others have the ability to move back into the cell cycle of mitosis—G_1, S, and G_2—with ease (e.g., skin cells).

The first three phases of the cell cycle—G_1, S, and G_2—occur during a period of time known as interphase. **Interphase** is the stage between cell divisions. During the G_1 stage, the cell grows in volume as it produces tRNA, mRNA, ribosomes, enzymes, and other cell components. During the S stage, DNA replication occurs in preparation for the distribution of genes to daughter cells. During the G_2 stage that follows, final preparations are made for mitosis with the synthesis of spindle-fiber proteins.

During interphase, the cell is not dividing but is engaged in metabolic activities such as muscle-cell contractions, photosynthesis, or glandular-cell secretion. During interphase, the nuclear membrane is intact and the individual chromosomes are not visible. The individual loosely coiled *chromatin* strands are too thin and tangled to be seen. Remember that chromosomes are highly coiled strands of DNA and contain coded information that tells the cell such things as (1) how to produce enzymes required for the digestion of nutrients, (2) how to manufacture enzymes that will metabolize the nutrients and eliminate harmful wastes, (3) how to repair and assemble cell parts, (4) how to reproduce healthy offspring, (5) when and how to react to favorable and unfavorable changes in the environment, and (6) how to coordinate and regulate all of life's essential functions. The double helix of DNA and the nucleosomes are arranged as a **chromatid,** and there are two attached chromatids for each replicated chromosome. These chromatids (chromosomes) are what will be distributed during mitosis.

THE STAGES OF MITOSIS

All stages in the life cycle of a cell are continuous; there is no precise point when the G_1 stage ends and the S stage begins, or when the interphase period ends and mitosis begins. Likewise, in the individual stages of mitosis there is a gradual transition from one stage to the next. However, for purposes of study and communication, scientists have divided the process of mitosis into four stages based on recognizable events. These four phases are prophase, metaphase, anaphase, and telophase. A summary of the events that occur during the cell cycle are presented in figure 19.21.

Interphase		As the cell moves from G_0 into mitosis, the chromosomes replicate during the S phase of interphase.
Prophase		1. The replicated chromatin strands begin to coil into recognizable chromosomes. 2. The nuclear membrane fragments. 3. Centrioles move to form the cell's poles; spindle fibers form. As prophase proceeds and as the chromosomes become more visible, we recognize that each chromosome is made of two parallel, threadlike parts lying side by side. Each parallel thread is called a chromatid.
Metaphase		1. Chromosomes move to the equator of the cell. 2. They attach to the spindle fibers at a point on the chromosomes known as the centromeres. 3. In late metaphase, each chromosome splits as the centromeres replicate and the cell enters the next phase, anaphase. At this stage in mitosis, each chromosome still consists of two chromatids attached at a centromere. In a human cell, there are 46 chromosomes, or 92 chromatids, aligned at the cell's equatorial plane during metaphase.
Anaphase		1. Centromeres complete DNA replication. 2. The chromatids separate toward the poles. As this separation of chromatids occurs, the chromatids are called daughter chromosomes. Daughter chromosomes contain identical genetic information.
Telophase		1. Two daughter cells are formed from the divided cells. 2. The nuclear membranes and nucleoli re-form. 3. Spindle fibers fragment. 4. The chromosomes unwind and change from chromosomes to chromatin. These cells can grow, replicate their DNA, and enter another round of mitosis and cytokinesis to continue the cell cycle.

Figure 19.21
The Stages of Mitosis.

When cells lose their ability to control mitotic divisions, a group of cells forms a *tumor*. A **tumor** is a mass of undifferentiated cells not normally found in a certain portion of the body. A *benign tumor* is a cell mass that does not fragment and spread beyond its original area of growth. A benign tumor can become harmful by growing large enough to interfere with normal body functions. Some tumors are malignant. *Malignant tumors* are nonencapsulated growths of tumor cells that are harmful; they may spread or invade other parts of the body. Cells of these tumors move from the original site (**metastasize**) and establish new colonies in other regions of the body. Cells break off from the original tumor and enter the bloodstream. When they get stuck to the inside of a capillary, these cells move through the wall of the blood vessel and invade the tissue, where they begin to reproduce by mitosis. This tumor causes new blood vessels to grow into this new site, which will carry nutrients to this growing mass. These vessels can also bring even more spreading cells to the new tumor site. **Cancer** is the term used to refer to any abnormal growth of cells that has a malignant potential. Agents responsible for causing cancer are called **carcinogens**.

CONCEPTS APPLIED
What Is Cancer?

Cancer is a group of many diseases characterized by the uncontrolled growth and spread of abnormal cells. If the spread is not controlled or checked, it results in death. However, many cancers can be cured if detected and treated promptly.

Cancer's Seven Warning Signals

1. A persistent change in bowel or bladder habits
2. A sore that does not heal
3. Unusual bleeding or discharge
4. A lump or thickening in the breast or elsewhere
5. Persistent indigestion or difficulty in swallowing
6. An obvious change in a wart or mole
7. A nagging cough or hoarseness

What percent of Americans will eventually have cancer? Cancer will strike three out of four families. If you notice one of the preceding symptoms, see your doctor.

To become better informed about cancer, contact your local Cancer Society office or write the American Cancer Society, Inc., 90 Park Ave., New York, NY 10016. Request a copy of the latest publication, *Cancer Facts and Figures*. Go to the Internet and search some of the boldface terms used in this chapter to learn more about this topic. Be certain that the sources you access are "good science."

Connections . . .

Keep Your Lawn Crabgrass Free with a Mitosis Inhibitor

The crabgrass control agent dithiopyr appears to bind to a cellular protein of the spindle. It is thought that dithiopyr acts on the chemical Microtubule Associated Protein (MAP), altering spindle stability. When this occurs, spindle formation does not occur and the cell dies in metaphase.

---

large organic molecules necessary to maintain the forms of life we know.

In the light-capturing reaction stage of photosynthesis, plants use chlorophyll to trap the energy of sunlight. During the light-dependent stage they manufacture a source of chemical energy, ATP, and a source of hydrogen, NADPH. Atmospheric oxygen is released in this stage. In the light-independent reaction stage of photosynthesis, the ATP energy is used in a series of reactions to join the hydrogen from the NADPH to a molecule of carbon dioxide and form a simple carbohydrate. In subsequent reactions, plants use this molecule as a source of energy and raw materials to make complex carbohydrates, fats, and other organic molecules.

The *cell* is the common unit of life. We study individual cells and their structures to understand how they function as *individual living organisms* and as parts of *many-celled beings*. Knowing how *prokaryotic* and *eukaryotic cell* types resemble or differ from each other helps physicians control some organisms dangerous to humans.

Cell division is necessary for *growth, repair,* and *reproduction*. Cells go through a *cell cycle* that includes *cell division* (*mitosis* and *cytokinesis*) and *interphase*. *Interphase* is the period of growth and preparation for division. Mitosis is divided into four stages: *prophase, metaphase, anaphase,* and *telophase*. During mitosis, *two daughter nuclei* are formed from *one parent nucleus*. These nuclei have *identical sets of chromosomes and genes* that are *exact copies* of those of the parent. Although the process of *mitosis* has been presented as a series of phases, you should realize that it is a *continuous, flowing process* from *prophase* through *telophase*. Following mitosis, *cytokinesis* divides the *cytoplasm*, and the cell returns to *interphase*.

SUMMARY

Living things show the *characteristics* of (1) *metabolic processes*, (2) *generative processes*, (3) *responsive processes*, (4) *control processes*, and (5) *a unique structural organization*. The concept of the *cell* has developed over a number of years. Initially, only two regions, the *cytoplasm* and the *nucleus*, could be identified. At present, numerous *organelles* are recognized as essential components of both *prokaryotic* and *eukaryotic* cell types.

In the process of respiration, organisms convert foods into energy (ATP) and waste materials (carbon dioxide and water). Organisms that have oxygen (O_2) available can employ the Krebs cycle and electron-transport system (ETS). Plants use the waste materials of respiration. Therefore, there is a constant cycling of materials between plants and animals. Sunlight supplies the essential initial energy for making the

KEY TERMS

active transport (p. 411)
adenosine triphosphate (ATP) (p. 405)
aerobic cellular respiration (p. 415)
biology (p. 402)
cancer (p. 425)
carcinogens (p. 425)
cell (p. 404)
cell membrane (p. 406)
cell wall (p. 404)
chloroplast (p. 415)
chromatid (p 424)
chromatin (p. 418)
chromosomes (p. 418)
control processes (p. 402)
cytokinesis (p. 423)
cytoplasm (p. 404)
diffusion (p. 409)
domain (p. 405)
endoplasmic reticulum (p. 412)
enzymes (p. 403)
eukaryotic cells (p. 405)
facilitated diffusion (p. 411)
generative processes (p. 402)
Golgi apparatus (p. 412)
homeostasis (p. 403)
interphase (p. 424)
lysosomes (p. 414)
metabolic processes (p. 402)
metastasize (p. 425)
mitochondrion (p. 415)
mitosis (p. 423)
nuclear membrane (p. 414)
nucleolus (p. 418)
nucleus (p. 404)
organelles (p. 405)
osmosis (p. 410)
phagocytosis (p. 411)
plasma membrane (p. 412)
prokaryotic cells (p. 405)
protoplasm (p. 404)
responsive processes (p. 402)
ribosomes (p. 416)
selectively permeable (p. 410)
structural similarities (p. 403)
tumor (p. 425)
vacuole (p. 412)

APPLYING THE CONCEPTS

1. Which one of the following represents a generative process?
   (a) enzymes
   (b) individual adaptation
   (c) nutrient uptake
   (d) cell division

2. When dealing with responsive processes:
   (a) metabolic processes decrease.
   (b) populations evolve through time.
   (c) organisms grow.
   (d) individuals coordinate activities.

3. Metabolic processes include
   (a) nutrient processing.
   (b) waste elimination.
   (c) nutrient uptake.
   (d) all of these.

4. The centromeres split during
   (a) anaphase.
   (b) prophase.
   (c) metaphase.
   (d) interphase.

5. The normal state of chromosomes in prophase of mitosis is as
   (a) daughter chromosomes.
   (b) chromosomes composed of two chromatids.
   (c) chromatids composed of two chromosomes.
   (d) chromosomes consisting of single chromatids.

6. The presence of cell walls in plants is associated with _____ in telophase.
   (a) cleavage furrows
   (b) spindle fiber formation
   (c) differentiation
   (d) cytokinesis

7. During which stage of the cell cycle does DNA replication occur?
   (a) the S stage of interphase
   (b) anaphase of mitosis
   (c) G_2 stage of metaphase
   (d) prophase

8. Nerve cells do not normally undergo mitosis. This means that
   (a) the brain is unimportant.
   (b) your brain cannot grow.
   (c) cytokinesis will be common in nerve tissue.
   (d) transcription of DNA will not occur.

9. The correct order for the stages of mitosis is
   (a) prophase, anaphase, metaphase, telophase.
   (b) metaphase, prophase, anaphase, telophase.
   (c) prophase, metaphase, anaphase, telophase.
   (d) prophase, metaphase, telophase, anaphase.

10. Compared to the mother cell, daughter cells at the end of mitosis have _____ number of chromosomes.
    (a) half the
    (b) the same
    (c) twice the
    (d) none of the above

11. During anaphase
    (a) chromosomes become visible.
    (b) daughter chromosomes migrate to the poles.
    (c) chromosomes line up at the equatorial plane.
    (d) cytokinesis is completed.

12. Cell division is needed for
    (a) growth.
    (b) replacement of worn-out cells.
    (c) healing of damaged tissue.
    (d) all of these.

Answers
1. d **2.** b **3.** d **4.** a **5.** b **6.** d **7.** a **8.** b **9.** c **10.** b **11.** b **12.** d

QUESTIONS FOR THOUGHT

1. Make a list of the membranous organelles of a eukaryotic cell and describe the function of each.
2. How do diffusion, facilitated diffusion, osmosis, and active transport differ?
3. What are the differences between the cell wall and the cell membrane?
4. Diagram a cell and show where proteins, nucleic acids, carbohydrates, and lipids are located.
5. Make a list of the nonmembranous organelles of the cell and describe their functions.
6. Why does putting salt on meat preserve it from spoilage by bacteria?
7. In what ways do mitochondria and chloroplasts resemble one another?
8. Name the four stages of mitosis and describe what occurs in each stage.
9. What is meant by cell cycle?
10. During which stage of a cell's cycle does DNA replication occur?
11. At what phase of mitosis does the DNA become most visible?
12. What are the differences between plant and animal mitosis?

The Origin and Evolution of Life

Early in the earth's history, inorganic materials were converted to organic materials. These organic materials have become combined into living units called cells. Cells are life. While we know a great deal about cells and the organisms they make up, scientists still ask these two questions: What is the nature of life? And how did life originate?

O U T L I N E

The complexity and diversity of living things has challenged thinkers for thousands of years. Has the earth always been filled with living things? Were there living things that have ceased to exist? Did living thing originate on the earth? Do new living things originate today? As our understanding of the laws of nature and the nature of living things have developed, the way we approach these questions has changed. However, the basic questions remain the same: How did life originate? and How and why does it change?

In earlier times, no one ever doubted that life originated from nonliving things. The Greeks, Romans, Chinese, and many other ancient peoples believed that maggots arose from decaying meat; mice developed from wheat stored in dark, damp places; lice formed from sweat; and frogs originated from damp mud. This theory was widely believed until the seventeenth century (figure 20.1); however, there were some who doubted it. These people subscribed to an opposing concept that life originates only from preexisting life.

One of the earliest challenges to the idea that life can be generated from nonliving matter came in 1668. Francesco Redi, an Italian physician, set up a controlled experiment (figure 20.2). He used two sets of jars that were identical except for one aspect. Both sets of jars contained decaying meat, and both were exposed to the atmosphere; however, one set of jars was covered by gauze, and the other was uncovered. Redi observed that flies settled on the meat in the open jar, but the gauze blocked their access to the covered jars. When maggots appeared on the meat in the uncovered jars but not on the meat in the covered ones, Redi concluded that the maggots arose from the eggs of the flies and were not produced from the decaying meat.

CURRENT THINKING ABOUT THE ORIGIN OF LIFE

The concept that life could be generated from nonliving matter is known as **spontaneous generation** and the concept that life can come only from other living things is known as **biogenesis.** Today, when we look at the question of how life on earth originated, we still have basically the same two theories. One holds that life arrived on the earth from some extraterrestrial source and the other maintains that life was created on the earth from nonliving material. It is important to recognize that we will probably never know for sure how life on the earth came to be, but it is interesting to speculate and examine the evidence related to this fundamental question.

Extraterrestrial Origin for Life on Earth

The concept that life arrived from extraterrestrial sources received some support in 1996 when a meteorite from Antarctica was analyzed. It has been known for many years that meteorites often contain organic molecules and this suggested that life might have existed elsewhere in the solar system. The chemical makeup of the Antarctic meteorite suggests that it was a portion of the planet Mars which was ejected from Mars as a result of a collision between the planet and an asteroid. Analysis of the meteorite shows the presence of complex organic molecules and small structures that resemble those found on the earth that are thought to be the result of the activity of ancient microorganisms. While scientists no longer think these structures are from microorganisms, many still think it is quite possible Mars may have had living things in the past. Since Mars currently has some water as ice and shows features that resemble dried-up river systems, Mars may have had much more water in the past. For these reasons many feel it is reasonable to consider that life of a nature similar to that presently found on the earth could have existed on Mars. While previous exploration of Mars' surface has failed to find evidence for life, many researchers are not satisfied with

Figure 20.1

Many works of art explore the idea that living things could originate from very different types of organisms or even from nonliving matter. M. C. Escher's work entitled "The Reptiles, 1943" shows the life cycle of a little alligator. Amid all kinds of objects, a drawing book lies open at a drawing of a mosaic of reptilian figures in three contrasting shades. Evidently, one of them is tired of lying flat and rigid among its fellows, so it puts one plastic-looking leg over the edge of the book, wrenches itself free, and launches out into "real" life. It climbs up the back of the zoology book and works its way laboriously up the slippery slope of the set square to the highest point of its existence. Then after a quick snort, tired but fulfilled, it goes downhill again, via an ashtray, to the level surface, to that flat drawing paper, and meekly rejoins its erstwhile friends, taking up once more its function as one element of surface division.

Figure 20.2

The two sets of jars here are identical in every way except one—the gauze covering. The set on the right is called the control group; the set on the left is the experimental group. Any differences seen between the control and the experimental groups are the result of a single variable. In this manner, Redi concluded that the presence of maggots in meat was due to flies laying their eggs on the meat and not spontaneous generation.

the results. Thus, future exploration of the surface of Mars will continue to look for evidence of current or past life.

Earth Origin for Life on Earth

The alternative view, that life originated on the earth, has also received support. There has been much research and speculation about what conditions would have been necessary for life to originate on the earth and the kinds of problems that early living things would have had to solve. Let us look at several lines of evidence.

1. Earth is the only planet in our solar system that has a temperature range that allows for water to exist as a liquid on its surface and water is the most common compound in most kinds of living things.
2. Analysis of the atmospheres of other planets shows that they all lack oxygen. The oxygen in Earth's atmosphere is the result of current biological activity. Therefore, before life on Earth the atmosphere probably lacked oxygen. This could have led to a buildup of organic molecules in the oceans.
3. Experiments demonstrate that organic molecules can be generated in an atmosphere that lacks oxygen.
4. Since it is assumed that all of the planets have been cooling off as they age, it is very likely that Earth was much hotter in the past. The large portions of Earth's surface that are of volcanic origin strongly support this idea. There is also evidence that various large bodies collided with Earth early in its history and that they could have led to increased temperatures at least in the site of the collision.
5. Recognition that prokaryotic cells are relatively simple and that certain prokaryotic organisms live in extreme environments of high temperature, high salinity, low pH, or the absence of oxygen suggests that they may have been adapted to life in a world that is

very different from today's Earth. These kinds of organisms are found today in unusual locations such as hot springs and around thermal vents in the ocean floor and may be descendants of the first organisms formed on the primitive Earth.

If we put all of these ideas together we can create the following scenario. A hot planet Earth with an atmosphere that lacked oxygen could have allowed organic molecules to form from inorganic molecules. These organic molecules could have collected in the oceans and provided a dilute organic "soup." These organic molecules could have served as "building blocks" for simple organisms as well as a source of energy. The earliest forms of life would have been adapted to conditions that are very different from those that exist on the earth today. In order to "invent" life there are several problems that would have to have been solved. By focusing on alternative ways of solving these problems, scientists have developed several different ideas about what the first living thing might have been like, regardless of whether life originated on the earth or arrived from an extraterrestrial source.

CONCEPTS APPLIED

Spontaneous Generation??

At one time people believed that life could be generated spontaneously from nonliving matter. To get an idea of why they thought this was possible you can do the following: Take a clear bottle or jar and place a small amount of soil in it. Fill it with water. Place the jar in a warm well-lighted place. A sunny window is ideal. Observe the changes in it over the next two to three weeks. How many different kinds of organisms can you identify? Examine the water under a microscope if you have access to one.

---

had one other important outcome: It opened the door for the evolution of aerobic organisms.

It appears that an oxidizing atmosphere began to develop about 2 billion years ago. Although various chemical reactions released small amounts of molecular oxygen into the atmosphere, it was photosynthesis that generated most of the oxygen. The oxygen molecules also reacted with one another to form ozone (O_3). Ozone collected in the upper atmosphere and acted as a screen to prevent most of the ultraviolet light from reaching the earth's surface. The reduction of ultraviolet light diminished the spontaneous formation of complex organic molecules. It also reduced the number of mutations in cells. In an oxidizing atmosphere, it was no longer possible for organic molecules to accumulate over millions of years to be later incorporated into living material.

The appearance of oxygen in the atmosphere also allowed for the evolution of aerobic respiration. Since the first heterotrophs were, of necessity, anaerobic organisms, they did not derive large amounts of energy from the organic materials available as food. With the evolution of aerobic heterotrophs, there could be a much more efficient conversion of food into usable energy. Aerobic organisms would have a significant advantage over anaerobic organisms: They could use the newly generated oxygen for aerobic respiration and, therefore, generate many more energy-rich ATP molecules from the food molecules they consumed.

The Establishment of Three Major Domains of Life

In 1977 Carl Woese published the idea that the "bacteria" (organisms that lack a nucleus), which had been considered a group of similar organisms, were really made up of two very different kinds of organisms, the Eubacteria and Archaea. Furthermore the Archaea shared some characteristics with eukaryotic organisms. Subsequent investigations have supported these ideas and led to an entirely different way of looking at the classification and evolution of living things. Although biologists have traditionally divided organisms into kingdoms based on their structure and function, it was very difficult to do this with microscopic organisms. With the newly developed ability to decode the sequence of nucleic acids, it became possible to look at the genetic nature of organisms without being confused by their external structures. Woese studied the sequences of ribosomal RNA and compared similarities and differences. As a result of his studies and those of many others, a new concept of the relationships between various kinds of organisms has emerged.

The three main kinds of living things, Eubacteria, Archaea, and Eucarya, have been labeled "domains." Table 20.1 gives the general characteristics of each of these groups. Within each domain there are several kingdoms. Previously, all members of the Eubacteria and Archaea were lumped into a single kingdom known as the Prokaryotae. However, as more extensive studies of these two groups take place, each will be further subdivided into several kingdoms. At this point there is no consensus on what these kingdom designations should be. Although there is greater diversity within each of these two domains than within the Eucarya, the Eucarya are better understood and are divided into four kingdoms: Animalia, Plantae, Fungi (mushrooms, molds, yeasts), and Protista (algae and protozoa).

This new picture of living things requires us to reorganize our thinking. It appears that the oldest organisms may have been prokaryotic organisms that were able to live in hot situations and that they gave rise to the Archaea, many of which still require extreme environments. Perhaps most interesting is the

Table 20.1 Major Domains of Life

Eubacteria	Archaea	Eucarya
No nuclear membrane.	No nuclear membrane.	Nuclear membrane present.
Chlorophyll-based photosynthesis is a bacterial invention but most can function anaerobically.	None have been identified as pathogenic.	Many kinds of membranous organelles present in cells.
Oxygen-generating photosynthesis was an invention of the cyanobacteria.	Probably have a common ancestor with Eucarya.	Chloroplasts are probably derived from cyanobacteria.
Large number of heterotrophs oxidize organic molecules for energy.	Many obtain energy from inorganic reactions to make organic matter.	Mitochondria probably derived from certain aerobic bacteria.
Some use energy from inorganic chemical reactions to produce organic molecules.	Typically found in extreme environments.	Probably have a common ancestor with Archaea.
Some live at high temperatures and may be ancestral to Archaea.	Few heterotrophs.	Common evolutionary theme is the development of complex cells through symbiosis with other organisms.
Much metabolic diversity among closely related organisms.	Example: *Pyrolobus fumarii,* deep-sea hydrothermal vents, hot springs, volcanic areas, growth to 113°C	Example: *Homo sapiens,* all cells of the human body
Example: *Streptococcus pneumoniae,* one cause of pneumonia		

idea that the Archaea and Eucarya share many characteristics suggesting that they are more closely related to each other than either is to the Eubacteria. If we incorporate this information along with information about the nature of genetic material, we can develop a diagram that helps us see how these various kinds of organisms are related to one another.

It appears that each domain developed specific abilities. The Archaea are primarily organisms that use inorganic chemical reactions to generate the energy they need to make organic matter. Often these reactions result in the production of methane. Others use sulfur and produce H_2S. Most of these organisms are found in extreme environments such as hot springs or in extremely salty or acidic environments.

The Eubacteria developed many different metabolic abilities. Today many are able to use organic molecules as a source of energy, some are able to carry on photosynthesis, and still others are able to get energy from inorganic chemical reactions similar to Archaea.

The Eucarya (animals, plants, fungi, protozoa, and algae) are familiar to most people and appear to have exploited the metabolic abilities of other organisms by incorporating other organisms into their own structure. Chloroplasts and mitochondria are both bacteria-like structures found inside eukaryotic cells.

THE ORIGIN OF EUKARYOTIC CELLS

The earliest fossils appear to be similar in structure to that of present-day Eubacteria and Archaea. Therefore, it is likely that the early heterotrophs and autotrophs were probably simple one-celled prokaryotic organisms that lacked nuclear membranes and other membranous organelles, such as mitochondria, an endoplasmic reticulum, chloroplasts, and a Golgi apparatus. Organisms with this kind of simple structure are called **prokaryotes.** Present-day bacteria and blue-green bacteria and archaea are prokaryotes. All other forms of life are made up of cells that have a nucleus and many kinds of membrane structures and are called **eukaryotes.** All members of the domain Eucarya are eukaryotes.

Biologists generally believe that the eukaryotes evolved from the prokaryotes. The **endosymbiotic theory** attempts to explain this evolution. This theory suggests that eukaryotic cells came into being as a result of the combining of several different types of primitive prokaryotic cells. It is thought that some organelles found in eukaryotic cells may have originated as free-living prokaryotes. Since mitochondria and chloroplasts contain bacteria-like DNA and ribosomes, control their own reproduction, and synthesize their own enzymes, it has been suggested that they originated as free-living prokaryotic bacteria. These

bacterial cells could have established a symbiotic relationship with another primitive nuclear-membrane-containing cell type (figure 20.3). When this theory was first suggested it met with a great deal of criticism. However, continuing research has uncovered several other instances of the probable joining of two different prokaryotic cells to form one.

If these cells adapted to one another and were able to survive and reproduce better as a team, it is possible that this relationship may have evolved into present-day eukaryotic cells. If this relationship had included only a nuclear-membrane-containing cell and aerobic bacteria, the newly evolved cell would have been similar to present-day heterotrophic protozoa, fungi, and animal cells. If this relationship had included both aerobic bacteria and photosynthetic bacteria, the newly formed cell would have been similar to present-day autotrophic algae and plant cells. In addition it is likely that endosymbiosis occurred among eukaryotic organisms as well. Several kinds of eukaryotic red and brown algae contain chloroplast-like structures that appear to have originated as free-living eukaryotic cells.

Let us return for a moment to the question that perplexed early scientists and caused the controversies surrounding the opposing theories of spontaneous generation and biogenesis. From our modern perspective we can see that all life we experience comes into being as a result of reproduction. Life is generated from other living things, the process of biogenesis. However, reproduction does not answer the question: Where did life come from in the first place? We can speculate, test hypotheses, and discuss various possibilities, but we will probably never know for sure. Life either always was or it started at some point in the past. If it started, then spontaneous generation of some type had to occur at least once, but it is not happening today.

Figure 20.4 summarizes several of the major events in the formation of the planet Earth and the development of life on it.

THE ROLE OF NATURAL SELECTION IN EVOLUTION

In many cultural contexts, the word *evolution* means progressive change. We talk about the evolution of economies, fashion, or musical tastes. From a biological perspective, the word has a more specific meaning. **Evolution** is the continuous genetic adaptation of a population of organisms to its environment over time. Evolution involves changes in the genes that are present in a population. By definition individual organisms are not able to evolve—only populations can. The organism's surroundings determine which characteristics favor survival and reproduction (i.e., which characteristics best fit the organism to its environment). The mechanism by which evolution occurs involves the selective passage of genes from one generation to the next through sexual reproduction. The various processes that encourage the passage of beneficial genes to future generations and discourage the passage of harmful or less valuable genes are collectively known as **natural selection.**

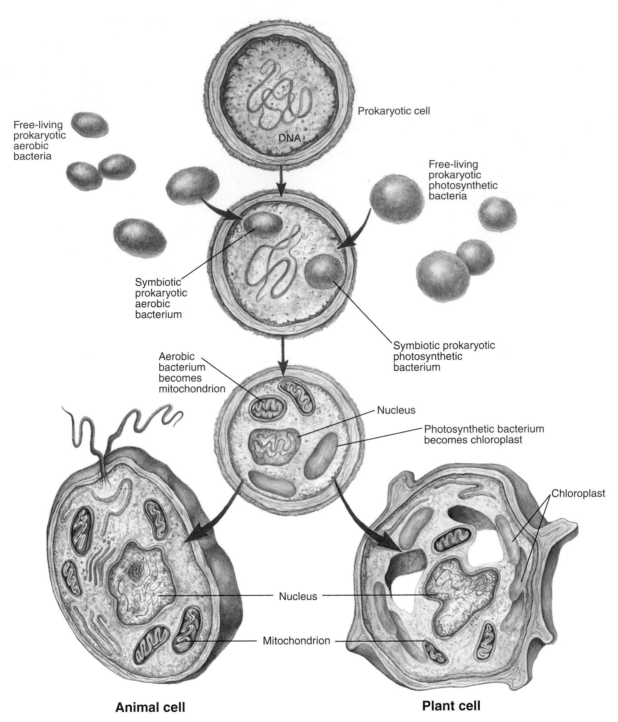

Free-living prokaryotic aerobic bacteria

Prokaryotic cell

DNA

Free-living prokaryotic photosynthetic bacteria

Symbiotic prokaryotic aerobic bacterium

Symbiotic prokaryotic photosynthetic bacterium

Aerobic bacterium becomes mitochondrion

Nucleus

Photosynthetic bacterium becomes chloroplast

Chloroplast

Nucleus

Mitochondrion

Animal cell

Plant cell

Figure 20.3

The endosymbiotic theory proposes that some free-living prokaryotic bacteria developed symbiotic relationships with a host cell. When some aerobic bacteria developed into mitochondria and photosynthetic bacteria developed into chloroplasts, a eukaryotic cell evolved. These cells evolved into eukaryotic plant and animal cells.

The idea that some individuals whose gene combinations favor life in their surroundings will be most likely to survive, reproduce, and pass their genes on to the next generation is known as the **theory of natural selection.** The *theory of evolution,* however, states that populations of organisms become genetically adapted to their surroundings over time. Natural selection is the process that brings about evolution by "selecting" which genes will be passed to the next generation.

A theory is a well-established generalization supported by many different kinds of evidence. The theory of natural selection was first proposed by Charles Darwin and Alfred Wallace and was clearly set forth in 1859 by Darwin in his book *On the*

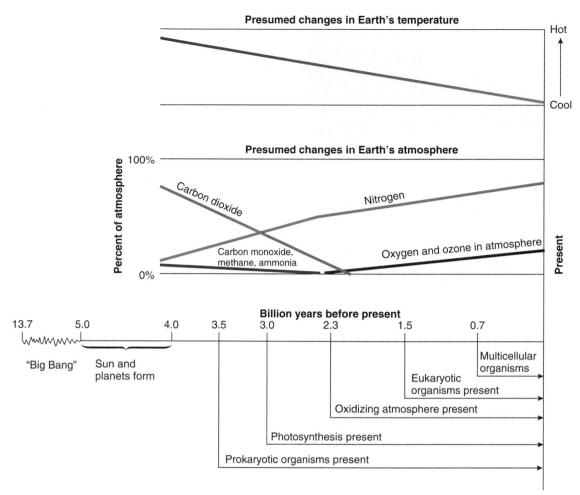

Presumed changes in Earth's temperature

Presumed changes in Earth's atmosphere

Billion years before present

Figure 20.4

Current thinking suggests that several factors shaped the nature of Earth and its atmosphere. It is probable that as Earth cooled, water remained on the surface as a liquid. Additional water probably arrived in the form of comets that collided with Earth. Conceivably, the initial atmosphere was primarily carbon dioxide (CO_2) with smaller amounts of nitrogen (N_2), ammonia (NH_3), methane (CH_4), and carbon monoxide (CO). Carbon dioxide decreased as a result of chemical reactions at the surface of Earth. Living organisms have had a significant effect on the atmosphere because once photosynthesis began carbon dioxide was further reduced and oxygen (O_2) began to increase. Carbon dioxide was further reduced by the accumulation of carbonates in the shells of many kinds of marine protozoa and animals. The presence of oxygen allowed for the increase in ozone (O_3), which reduces the amount of ultraviolet light penetrating the atmosphere to reach the surface of Earth. Because ultraviolet light causes mutations, a protective layer of ozone probably has reduced the frequency of mutations from that which was present on primitive Earth.

Origin of Species by Means of Natural Selection, or the Preservation of Favored Races in the Struggle for Life. Since the time it was first proposed, the theory of natural selection has been subjected to countless tests and remains the core concept for explaining how evolution occurs.

There are two common misinterpretations associated with the process of natural selection. The first involves the phrase "survival of the fittest." Individual survival is certainly important because those that do not survive will not reproduce. But the more important factor is the number of descendants an organism leaves. An organism that has survived for hundreds of years but has not reproduced has not contributed any of its genes to the next generation and so has been selected against. The key, therefore, is not survival alone but survival and reproduction of the more fit organisms.

Second, the phrase "struggle for life" does not necessarily refer to open conflict and fighting. It is usually much more subtle than that. When a resource such as nesting material, water, sunlight, or food is in short supply, some individuals survive and reproduce more effectively than others. For example, many kinds of birds require holes in trees as nesting places (see figure 20.5). If these are in short supply, some birds will be fortunate and find a top-quality nesting site, others will occupy less suitable holes, and some may not find any. There may or may not be fighting for possession of a site. If a site is already occupied, a bird may not necessarily try to dislodge its occupant but may just continue to search for suitable but less valuable sites. Those that successfully occupy good nesting sites will be much more successful in raising young than will those that must occupy poor sites or those that do not find any.

passed to the next generation and natural selection cannot occur. Genetic variety is generated in two ways. First of all, mutations may alter existing genes, resulting in the introduction of entirely new genetic information into a species' gene pool. **Spontaneous mutations** are changes in DNA that cannot be tied to a particular causative agent. It is suspected that cosmic radiation or naturally occurring mutagenic chemicals might be the cause of many of these mutations. It is known that subjecting organisms to high levels of radiation or to certain chemicals increases the rate at which mutations occur. It is for this reason that people who work with radioactive materials or other mutagenic agents take special safety precautions (figure 20.6).

Naturally occurring mutation rates are low (perhaps one chance in one hundred thousand that a gene will be altered), and mutations usually result in a gene that is harmful. However, in populations of millions of individuals, each of whom has thousands of genes, over thousands of generations it is quite possible that a new beneficial piece of genetic information could come about as a result of mutation. When we look at the various genes that exist in humans or in any other organism, we should remem-

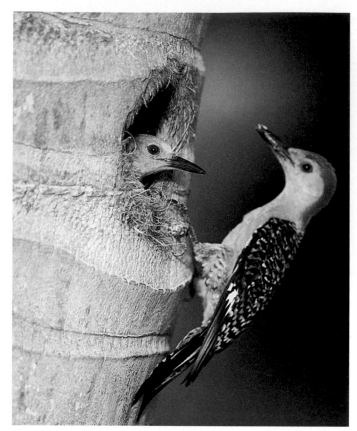

Figure 20.5
Many kinds of birds, like this red-bellied woodpecker, nest in holes in trees. If old and dead trees are not available they may not be able to breed. Many people build birdhouses that provide artificial "tree holes" to encourage such birds to nest near their homes.

Similarly, on a forest floor where there is little sunlight, some small plants may grow fast and obtain light while shading out plants that grow more slowly. The struggle for life in this instance involves a subtle difference in the rate at which the plants grow. But the plants are indeed engaged in a struggle, and a superior growth rate is the weapon for survival.

WHAT INFLUENCES NATURAL SELECTION?

Now that we have a basic understanding of how natural selection works, we can look in more detail at factors that influence it. Genetic variety within a species, genetic recombination as a result of sexual reproduction, the degree to which genes are expressed, and the ability of most species to reproduce excess offspring all exert an influence on the process of natural selection.

Genetic Variety Resulting from Mutation

In order for natural selection to occur, there must be genetic differences among the many individuals of an interbreeding population of organisms. If all individuals are identical genetically, it does not matter which ones reproduce—the same genes will be

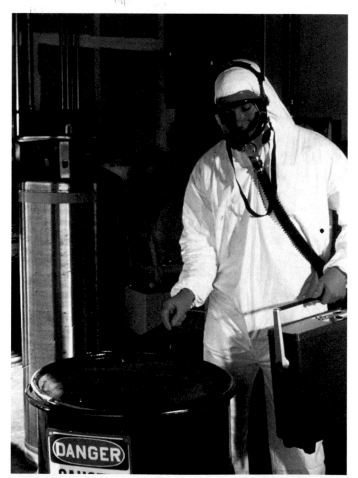

Figure 20.6
Because radiation and certain chemicals increase the likelihood of mutations, people who work in hazardous environments receive special training and use protective measures to reduce their exposure to mutagenic agents. It is particularly important to protect the ovaries and testes, since mutations that alter the DNA of eggs and sperm can be passed on to their children.

ber that every gene originated as a modification of a previously existing gene. For example, the gene for blue eyes may be a mutated brown-eye gene, or blond hair may have originated as a mutated brown-hair gene. Thus, mutations have been very important for introducing new genetic material into species over time.

In order for mutations to be important in the evolution of organisms, they must be in cells that will become sex cells. Mutations to the cells of the skin or liver will affect only those specific cells and will not be passed to the next generation.

Genetic Variety Resulting from Sexual Reproduction

A second very important process involved in generating genetic variety is sexual reproduction. While sexual reproduction does not generate new genetic information the way mutation does, it allows for the recombination of genes into mixtures that did not occur previously. Each individual entering a population by sexual reproduction carries a unique combination of genes; half are donated by the mother and half are donated by the father. During the formation of sex cells the manipulation of chromosomes results in new combinations of genes. This means that there are millions of possible combinations of genes in the sex cells of any individual. When fertilization occurs, one of the millions of possible sperm unites with one of the millions of possible eggs, resulting in a genetically unique individual. The gene mixing that occurs during sexual reproduction is known as **genetic recombination.** The new individual has a complete set of genes that is different from that of any other organism that ever existed.

Acquired Characteristics Do Not Influence Natural Selection

Many organisms survive because they have characteristics that are not genetically determined. These **acquired characteristics** are gained during the life of the organism; they are not genetically determined and, therefore, cannot be passed on to future generations through sexual reproduction. Consider an excellent tennis player's skill. Although this person may have inherited characteristics that are beneficial to a tennis player, the ability to play a good game of tennis is acquired through practice, not through genes. An excellent tennis player's offspring will not automatically be excellent tennis players. They may inherit some of the genetically determined physical characteristics necessary to become excellent tennis players, but the skills are still acquired through practice.

We often desire a specific set of characteristics in our domesticated animals. For example, the breed of dog known as boxers are "supposed" to have short tails. However, the genes for short tails are rare in this breed. Consequently, the tails of these dogs are amputated—a procedure called docking. Similarly, the tails of lambs are also usually amputated. These acquired characteristics are not passed on to the next generation. Removing the tails of these animals does not remove the genes for tail production from their genetic information and each generation of puppies is born with tails.

PROCESSES THAT DRIVE NATURAL SELECTION

Several mechanisms allow for selection of certain individuals for successful reproduction. The specific environmental factors that favor certain characteristics are called **selecting agents.** If predators must pursue swift prey organisms, then the faster predators will be selected for, and the selecting agent is the swiftness of available prey. If predators must find prey that are slow but hard to see, then the selecting agent is the camouflage coloration of the prey, and keen eyesight is selected for. If plants are eaten by insects, then the production of toxic materials in the leaves is selected for. All selecting agents influence the likelihood that certain characteristics will be passed to subsequent generations.

Differential Survival

As stated previously, the phrase "survival of the fittest" is often associated with the theory of natural selection. Although this is recognized as an oversimplification of the concept, survival is an important factor in influencing the flow of genes to subsequent generations. If a population consists of a large number of genetically and structurally different individuals it is likely that some of them will possess characteristics that make their survival difficult. Therefore, they are likely to die early in life and not have an opportunity to pass their genes on to the next generation.

The English peppered moth provides a classic example. Two color types are found in the species: One form is light-colored and one is dark-colored. These moths rest on the bark of trees during the day, where they may be spotted and eaten by birds. The birds are the selecting agents. About 150 years ago, the light-colored moths were most common. However, with the advance of the Industrial Revolution in England, which involved an increase in the use of coal, air pollution increased. The fly ash in the air settled on the trees, changing the bark to a darker color. Because the light moths were more easily seen against a dark background, the birds ate them (figure 20.7). The darker ones were less conspicuous; therefore, they were less frequently eaten and more likely to reproduce successfully. The light-colored moth, which was originally the more common type, became much less common. This change in the frequency of light- and dark-colored forms occurred within the short span of 50 years. Scientists who have studied this situation have estimated that the dark-colored moths had a 20 percent better chance of reproducing than did the light-colored moths. This study is continuing today. As England has reduced its air pollution and tree bark has become lighter in color, the light-colored form of the moth increased in frequency again.

As another example of how differential survival can lead to changed gene frequencies, consider what has happened to many insect populations as we have subjected them to a variety of insecticides. Since there is genetic variety within all species of insects, an insecticide that is used for the first time on a particular species kills all those that are genetically susceptible. However, the insecticide may not kill individuals with slightly different genetic compositions.

Figure 20.7

This photo of the two variations of the peppered moth shows that the light-colored moth is much more conspicuous against the dark tree trunk. (The two dark moths are indicated by arrows.) The trees are dark because of an accumulation of pollutants from the burning of coal. The more conspicuous light-colored moths are more likely to be eaten by bird predators, and the genes for light color should become more rare in the population.

Suppose that in a population of a particular species of insect, 5 percent of the individuals have genes that make them resistant to a specific insecticide. The first application of the insecticide could, therefore, kill 95 percent of the population. However, tolerant individuals would then constitute the majority of the breeding population that survived. This would mean that many insects in the second generation would be tolerant. The second use of the insecticide on this population would not be as effective as the first. With continued use of the same insecticide, each generation would become more tolerant.

Many species of insects produce a new generation each month. In organisms with a short generation time, 99 percent of the population could become resistant to the insecticide in just five years. As a result, the insecticide would no longer be useful in controlling the species. As a new factor (the insecticide) was introduced into the environment of the insect, natural selection resulted in a population that was tolerant of the insecticide. Figure 20.8 indicates that more than five hundred species of insects have populations that are resistant to many kinds of insecticides.

Differential Reproductive Rates

Survival alone does not always ensure reproductive success. For a variety of reasons, some organisms may be better able to utilize available resources to produce offspring. If one individual leaves 100 offspring and another leaves only 2, the first organism has passed more copies of its genetic information to the next generation than has the second. If we assume that all 102 individual offspring have similar survival rates, the first organism has been selected for and its genes have become more common in the subsequent population.

Scientists have conducted studies of the frequencies of genes for the height of clover plants (figure 20.9). Two identical fields of clover were planted and cows were allowed to graze in one of them. Cows acted as a selecting agent by eating the taller

Pest species resistant to insecticides

Figure 20.8

The continued use of insecticides has constantly selected for genes that give resistance to a particular insecticide. As a result, many species of insects and other arthropods are now resistant to many kinds of insecticides, and the number continues to increase.

Source: Data from Georghiou, University of California at Riverside.

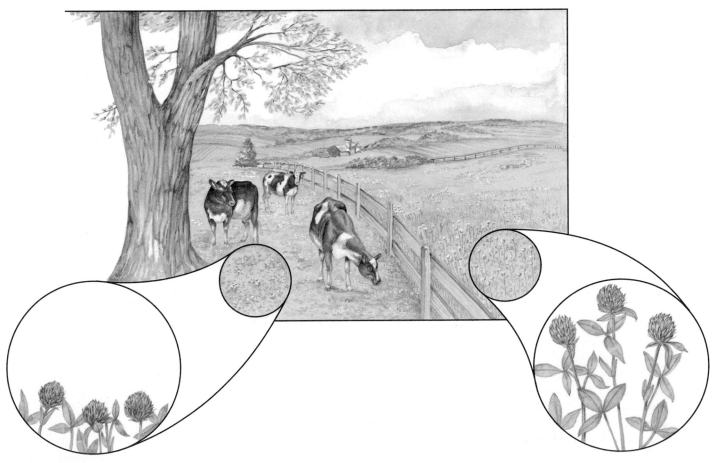

Figure 20.9

The clover field to the left of the fence is undergoing natural selection: the grazing cattle are eating the tall plants and causing them to reproduce less than do the short plants. The other field is not subjected to this selection pressure, so its clover population has more genes for tallness.

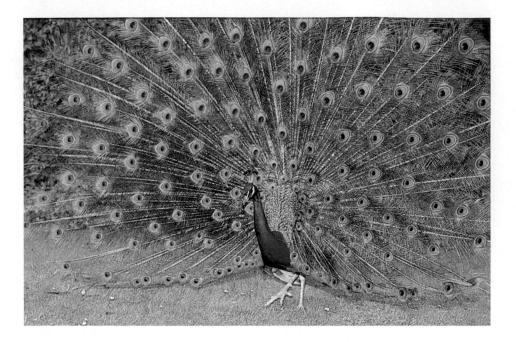

Figure 20.10

In many animal species the males display very conspicuous characteristics that are attractive to females. Because the females choose the males they will mate with, those males with the most attractive characteristics will have more offspring and, in future generations, there will be a tendency to enhance the characteristic. With peacocks, those individuals with large colorful displays are more likely to mate.

plants first. These tall plants rarely got a chance to reproduce. Only the shorter plants flowered and produced seeds. After some time, seeds were collected from both the grazed and ungrazed fields and grown in a greenhouse under identical conditions. The average height of the plants from the ungrazed field was compared to that of the plants from the grazed field. The seeds from the ungrazed field produced some tall, some short, but mostly medium-sized plants. However, the seeds from the grazed field produced many more shorter plants than medium or tall ones. The cows had selectively eaten the plants that had the genes for tallness. Since the flowers are at the tip of the plant, tall plants were less likely to successfully reproduce, even though they might have been able to survive grazing by cows.

Differential Mate Selection

Within animal populations, some individuals may be chosen as mates more frequently than others. Obviously, those that are frequently chosen have an opportunity to pass on more copies of their genes than those that are rarely chosen. Characteristics of the more frequently chosen individuals may involve general characteristics, such as body size or aggressiveness, or specific conspicuous characteristics attractive to the opposite sex.

For example, male red-winged blackbirds establish territories in cattail marshes where females build their nests. A male will chase out all other males but not females. Some males have large territories, some have small territories, and some are unable to establish territories. Although it is possible for any male to mate, it has been demonstrated that those who have no territory are least likely to mate. Those who defend large territories may have two or more females nesting in their territories and are

very likely to mate with those females. It is unclear exactly why females choose one male's territory over another, but the fact is that some males are chosen as mates and others are not.

In other cases, it appears that the females select males that display conspicuous characteristics. Male peacocks have very conspicuous tail feathers. Those with spectacular tails are more likely to mate and have offspring (figure 20.10). Darwin was puzzled by such cases as the peacock, in which the large and conspicuous tail should have been a disadvantage to the bird. Long tails require energy to produce, make it more difficult to fly, and make it more likely that predators will capture the individual. The current theory that seeks to explain this paradox involves female choice. If the females have an innate (genetic) tendency to choose the most elaborately decorated males, genes that favor such plumage will be regularly passed to the next generation. Such special cases in which females choose males with specific characteristics have been called sexual selection.

A SUMMARY OF THE CAUSES OF EVOLUTIONARY CHANGE

Earlier in this chapter, evolution was described as the change in gene frequency over time. We can now see that several different mechanisms operate to bring about this change. Mutations can either change one genetic message into another or introduce an entirely new piece of genetic information into the population. Immigration can introduce new genetic information if the entering organisms have unique genes. Emigration and death remove genes from the gene pool. Natural selection systematically filters some genes from the population, allowing other genes to

Humans have designed several kinds of plants and animals for their own purposes through the process of domestication. Most modern cereal grains are special plants that rely on human activity for their survival; most would not live without fertilizer, cultivation, and other helps. These grains are the descendants of wild plants. Initially the process of domestication of plants by humans probably was not a conscious effort to develop crops, but the unconscious selection of individual plants that had particularly useful characteristics from among all the plants of that species. For example, the seeds may have been larger than those of other plants of the same species or they may have tasted better or they may have been easier to chew. The unconscious selection of these seeds could have led to the dispersal of these particular seeds along traditional travel routes and eventually trading with other groups of people. Wheat, barley, rice, and corn are all plants that have been domesticated and currently supply a major part of the food for the world's population. However, modern genetic techniques have resulted in domesticated plants that only superficially resemble their wild ancestors.

Most domesticated animals are mammals or birds that are herbivores that grow rapidly, have a decent disposition, and will reproduce in captivity. Examples are cattle, sheep, pigs, goats, horses, camels, chickens, turkeys, ducks, and geese. (The carnivorous dog is an obvious exception to the general herbivore rule.) Obviously animals that do not reproduce in captivity would not be good candidates for domestication and those that had behaviors that made them dangerous to be around would not be chosen to be domesticated. Again, the choices by humans would have been "unconscious." They would not have planned to domesticate an animal for a particular purpose (milk, eggs, power, meat) and modern science has greatly modified domesticated animals to something that is quite different from their wild ancestors. In fact, many of the wild ancestors of modern domesticated plants and animals are extinct. Thousands of generations of selection have, in effect, caused the development of species that are known only by their domesticated remnants.

In recent years the genetic manipulation of organisms has expanded into a new arena. With the development of biotechnology techniques, genes can be moved from one species to another. The transfer of genes can be between plants and animals, bacteria and plants, or bacteria and animals. Genes from certain kinds of bacteria produce a compound that kills insects. This gene has been inserted into several kinds of plants so that the plants produce the natural insecticide and are protected from some of their insect pests. In addition animal genes have been inserted into plants and human genes have been inserted into bacteria. It appears that gene transfer between species occurs naturally as well. This knowledge requires us to rethink our concept of what a species is and how evolution occurs.

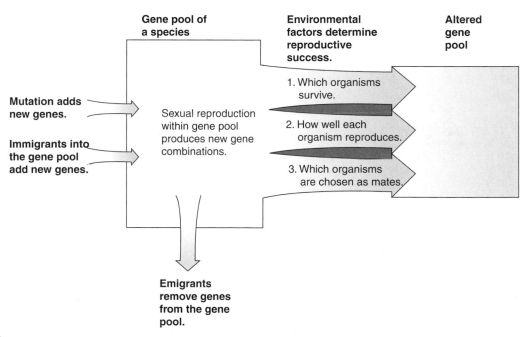

Figure 20.11

Several different processes cause gene frequencies to change. Genes enter populations through immigration and mutation. Genes leave populations through death and emigration. Natural selection operates within populations through death and rates of reproduction.

remain and become more common. The primary mechanisms involved in natural selection are differences in death rates, reproductive rates, and the rate at which individuals are selected as mates (figure 20.11). In addition, gene frequencies are more easily changed in small populations since events such as death, immigration, emigration, and mutation can have a greater impact on a small population than on a large population.

ACCUMULATING EVIDENCE OF EVOLUTION

The theory of evolution has become the major unifying theory of the biological sciences. Medicine recognizes the dangers of mutations, the similarity in function of the same organ in related species, and the way in which the environment can interfere with the preprogrammed process of embryological development. Agricultural science recognizes the importance of selecting specific genes for passage into new varieties of crop plants and animals. The concepts of mutation, selection, and evolution are so fundamental to understanding what happens in biology that we often forget to take note of the many kinds of observations that support the theory of evolution. The following list describes some of the more important pieces of evidence that support the idea that evolution has been and continues to be a major force in shaping the nature of living things.

1. Species and populations are not genetically fixed. Change occurs in individuals and populations.
   a) Mutations cause slight changes in the genetic makeup of an individual organism.
   b) Different populations of the same species show adaptations suitable for their local conditions.
   c) Changes in the characteristics displayed by species can be linked to environmental changes.
   d) Selective breeding of domesticated plants and animals indicates that the shape, color, behavior, metabolism, and many other characteristics of organisms can be selected for.
   e) Extinction of poorly adapted species is common.

2. New organisms are formed by the modification of ancestral species, not by major changes. All evidence suggests that, once embarked on a particular evolutionary road, the system is not abandoned, only modified. The following list supports the concept that evolution proceeds by modification of previously existing structures and processes rather than by catastrophic change.
   a) All species use the same DNA code.
   b) All species use the same left-handed, amino acid building blocks in their proteins.

c) It is difficult to eliminate a structure when it is part of a developmental process controlled by genes. Vestigial structures like the appendix and tailbone in humans are evidence of genetic material from previous stages in evolution.
   d) Embryological development of related animals is similar regardless of the peculiarities of adult anatomy. The embryos of all vertebrates have an early stage that contains gill slits.
   e) Species of organisms that are known to be closely related show greater similarity in their DNA than those that are distantly related.

3. The fossil record supports the concept of evolution.
   a) The nature of the earth has changed significantly over time.
   b) The fossil record shows vast changes in the kinds of organisms present on the earth. New species appear and most go extinct. This is evidence that living things change in response to changes in their environment.
   c) The fossils found in old rocks do not reappear in younger rocks. Once an organism goes extinct it does not reappear, but new organisms arise that are modifications of previous organisms.

4. New techniques and discoveries invariably support the theory of evolution.
   a) The recognition that the earth was formed billions of years ago supports the slow development of new kinds of organisms.
   b) The recognition that the continents of the earth have separated and drifted apart helps explain why organisms on Australia are so different from those found elsewhere.
   c) The discovery of DNA and how it works helps explain mutation and allows us to demonstrate the genetic similarity of closely related species.

A Closer Look   Common Misconceptions About the Theory of Evolution

1. *Evolution happened only in the past and is not occurring today.* In fact, we see lots of evidence that genetic changes are occurring in the populations of current species (antibiotic resistance, pesticide resistance, and moth color).

2. *Evolution has a specific goal.* Evolution does not move toward a specific goal. Natural selection selects those organisms that best fit the current environment. As the environment changes so do the characteristics that have value. Random events such as changes in sea level, major changes in climate such as ice ages, or collisions with asteroids have had major influences on subsequent natural selection and evolution. Evolution

results in organisms that "fit" the current environment.

3. *Changes in the environment cause mutations that are needed by an organism to survive under the new environmental conditions.* Mutations are random events and are not necessarily adaptive. However, when the environment changes, mutations that were originally detrimental may have greater value. The gene did not change but the environmental conditions did. In some cases the mutation rate may increase or there may be more frequent exchanges of genes between individuals when the environment changes, but the mutations are still random. They are not directed to a particular goal.

4. *Individual organisms evolve.* Individuals are stuck with the genes they inherited from their parents. Although individuals may adapt by changing their behavior or physiology, they cannot evolve; only populations can change gene frequencies.

5. *Today's species frequently can be shown to be derived from other present-day species. For example, it is commonly stated that apes gave rise to humans.* There are few examples in which it can be demonstrated that one current species gave rise to another. Apes did not become humans but apes and humans had a common ancestor several million years ago.

SPECIES: A WORKING DEFINITION

The smallest irreversible step in the evolutionary process is the invention of a new species from previous species. Before we consider how new species are produced, let's establish how one species is distinguished from another. A **species** is commonly defined as a population of organisms whose members have the potential to interbreed naturally to produce fertile offspring but do not interbreed with other groups. This is a working definition since it applies in most cases but must be interpreted to encompass some exceptions. There are two key ideas within this definition. First, a species is a population of organisms. An individual—you, for example—is not a species. You can only be a member of a group that is recognized as a species. The human species, *Homo sapiens,* consists of more than 6 billion individuals, whereas the endangered California condor species, *Gymnogyps californianus,* consists of about 100 individuals.

Second, the definition involves the ability of individuals within the group to produce fertile offspring. Obviously, we cannot check every individual to see if it is capable of mating with any other individual that is similar to it, so we must make some judgment calls. Do most individuals within the group potentially have the capability of interbreeding to produce fertile offspring? In the case of humans we know that some individuals are sterile and cannot reproduce, but we don't exclude them from the human species because of this. If they were not sterile, they would have the potential to interbreed. We recognize that, although humans normally choose mating partners from their subpopulation, humans from all parts of the world are potentially capable of inter-

breeding. We know this to be true because of the large number of instances of reproduction involving people of different ethnic and racial backgrounds. The same is true for many other species that have local subpopulations but have a wide geographic distribution.

Another way to look at this question is to think about **gene flow.** Gene flow is the movement of genes from one generation to the next or from one region to another. Two or more populations that demonstrate gene flow between them constitute a single species. Conversely, two or more populations that do not demonstrate gene flow between them are generally considered to be different species. Some examples will clarify this working definition.

The mating of a male donkey and a female horse produces young that grow to be adult mules, incapable of reproduction (figure 20.12). Since mules are nearly always sterile, there can be no gene flow between horses and donkeys and they are considered to be separate species. Similarly, lions and tigers can be mated in zoos to produce offspring. However, this does not happen in nature and so gene flow does not occur naturally; thus they are considered to be two separate species.

Still another way to try to determine if two organisms belong to different species is to determine their genetic similarity. The recent advances in molecular genetics allow scientists to examine the structure of the genes present in individuals from a variety of different populations. Those that have a great deal of similarity are assumed to have resulted from populations that have exchanged genes through sexual reproduction in the recent past. If there are significant differences in the genes present in individuals from two populations they have not exchanged genes recently and are more likely to be members of separate species. Interpretation of the results obtained by examining genetic differences still requires the judgment of experts. It will not unequivocally settle every dispute related to the identification of species, but it is another tool that helps to clarify troublesome situations. This concept that species can be distinguished from one another by their inability to interbreed is often called the **biological species concept.**

Often, organisms that are known to belong to distinct species differ in one or more ways that allow us to recognize them as separate species. Therefore, it is common to differentiate species on the basis of key structural characteristics. This method of using structural characteristics to identify species is called the **morphological species concept.** Structural differences are useful but not foolproof ways to distinguish species. However, we must rely on such indirect ways to identify species because we cannot possibly test every individual by breeding it with another to see if they will have fertile offspring. Furthermore, many kinds of organisms reproduce primarily by asexual means. Because organisms that reproduce exclusively by asexual methods do not exchange genes with any other individuals, they do not fit our *biological species* definition very well. In addition the study of fossil species must rely on structural characteristics to make species distictions since it is impossible to breed extinct organisms.

A

B

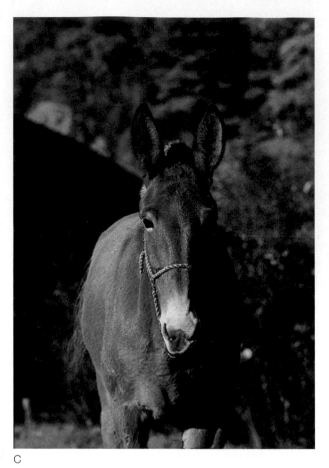

C

Figure 20.12

Even though they don't do so in nature, (*A*) horses and (*B*) donkeys can be mated. The offspring is called a (*C*) mule and is sterile. Because the mule is sterile, the horse and the donkey are considered to be of different species.

A Closer Look Is the Red Wolf a Species?

The red wolf (*Canis rufus*) is listed as an endangered species, so the U.S. Fish and Wildlife Service has instituted a captive breeding program to preserve the animal and reintroduce it to a suitable habitat in the southeastern United States, where it was common into the 1800s. Biologists have long known that red wolves will hybridize with both the coyote, *Canis latrans,* and the gray wolf, *Canis lupus,* and many suspect that the red wolf is really a hybrid between the gray wolf and the coyote. Gray wolf–coyote hybrids are common in nature where one or the other species is rare. Some have argued that the red wolf does not meet the definition of a species and should not be protected under the Endangered Species Act.

Museums have helped shed light on this situation by providing skulls of all three kinds of an-imals preserved in the early 1900s. It is known that during the early 1900s as the number of red wolves in the southeastern United States declined, they readily interbred with coyotes, which were very common. The gray wolf had been exterminated by the early 1900s. Some scientists believe that the skulls of the few remaining "red wolves" might not represent the true red wolf but a "red wolf" with many coyote characteristics. Studies of the structure of the skulls of red wolves, coyotes, and gray wolves show that the red wolves were recognizably different and inter-mediate in structure between coyotes and gray wolves. This supports the hypothesis that the red wolf is a distinct species.

DNA studies were performed using DNA from preserved red wolf pelts. The red wolf DNA was compared to coyote and gray wolf DNA. These studies show that red wolves contain DNA sequences typical of both gray wolves and coyotes but do not appear to have dis-tinct base sequences found only in the red wolf. These studies support the hypothesis that the red wolf is not a species but a population that re-sulted from hybridization between gray wolves and coyotes.

There is still no consensus on the status of the red wolf. Independent researchers disagree with one another and with Fish and Wildlife Ser-vice scientists, who have been responsible for de-veloping and administering a captive breeding program and planning reintroductions of the red wolf.

HOW NEW SPECIES ORIGINATE

The geographic area over which a species can be found is known as its **range.** The range of the human species is the entire world, while that of a bird known as a snail kite is a small region of southern Florida. As a species expands its range or environmental conditions change in some parts of the range, portions of the population can become separated from the rest. Thus, many species consist of partially isolated populations that display characteristics that differ significantly from other local populations. Many of the differences observed may be directly related to adaptations to local environmental conditions. This means that new colonies or isolated populations may have infrequent gene exchange with their geographically distant relatives. These genetically distinct populations are known as subspecies.

A portion of a species can become totally isolated from the rest of the gene pool by some geographic change, such as the for-

mation of a mountain range, river valley, desert, or ocean. When this happens the portion of the species is said to be in **geographic isolation** from the rest of the species. If two populations of a species are geographically isolated they are also reproductively isolated, and gene exchange is not occurring between them. The geographic features that keep the different portions of the species from exchanging genes are called **geographic barriers.** The uplifting of mountains, the rerouting of rivers, and the formation of deserts all may separate one portion of a gene pool from another. For example, two kinds of squirrels are found on opposite sides of the Grand Canyon. Some people consider them to be separate species, while others consider them to be different isolated subpopulations of the same species (figure 20.13). Even small changes may cause geographic isolation in species that have little ability to move. A fallen tree, a plowed field, or even a new freeway may effectively isolate populations within such species. Snails in two valleys separated by a high ridge have been

Kaibab squirrel

Aberts squirrel

Figure 20.13

These two squirrels are found on opposite sides of the Grand Canyon. Some people consider them to be different species; others consider them to be distinct populations of the same species.

Connections . . .

Dogs and Wolves Are the Same Species

Scientists compared the mitochondrial DNA sequences from 67 breeds of dogs and 162 wolves and coyotes and jackals. Mitochondrial DNA passes only from the mother to the offspring. All the wolf and dog sequences were similar and both differed significantly from coyotes and jackals. Careful analysis of the differences in mitochondrial DNA suggests that there were two major domestication events from different parent wolf populations and that there have been incidences of interbreeding of wolves and dogs several times. Furthermore, the various "breeds" of dogs could not be distinguished by the sequence of DNA. Dogs that superficially resemble one another may have more differences in mitochondrial DNA than those that appear quite different. The breeds of dogs are really rather superficial differences in the basic animal. The unmistakable conclusion is that dogs are simply domesticated wolves.

found to be closely related but different species. The snails cannot get from one valley to the next because of the height and climatic differences presented by the ridge (figure 20.14).

The separation of a species into two or more isolated subpopulations is not enough to generate new species. Even after many generations of geographic isolation, these separate groups may still be able to exchange genes (mate and produce fertile offspring) if they overcome the geographic barrier, because they have not accumulated enough genetic differences to prevent reproductive success. Differences in environments and natural selection play very important roles in the process of forming new species. Following separation from the main portion of the gene pool by geographic isolation, the organisms within the small, local population are likely to experience different environmental conditions. If, for example, a mountain range has separated a species into two populations, one of them may receive more rain or more sunlight than the other (figure 20.15). These environmental differences act as natural selecting agents on the two gene pools and acting over a long period of time account for dif-

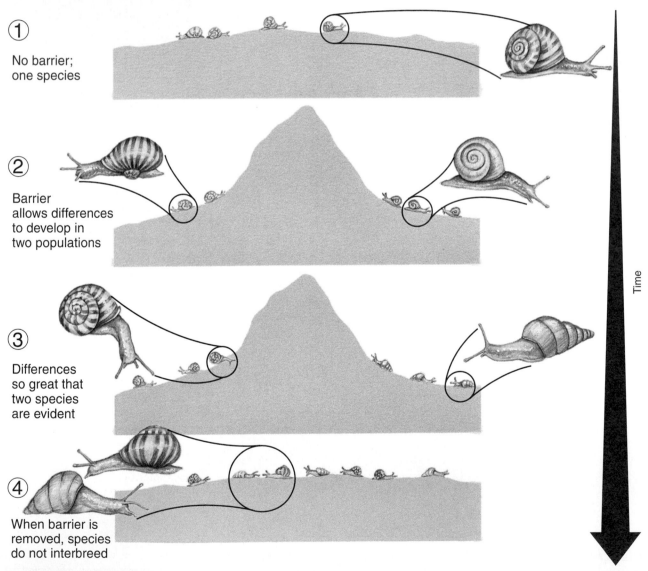

① No barrier; one species

② Barrier allows differences to develop in two populations

③ Differences so great that two species are evident

④ When barrier is removed, species do not interbreed

Time

Figure 20.14

If a single species of snail were to be divided into two different populations by the development of a ridge between them, the two populations could be subjected to different environmental conditions. This could result in a slow accumulation of changes that could ultimately result in two populations that would not be able to interbreed even if the ridge between them were to erode. They would be different species.

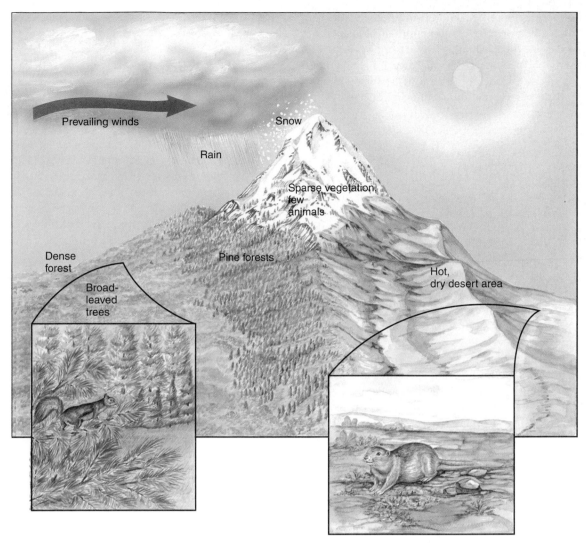

Figure 20.15

Most mountain ranges affect the local environment. Because of the prevailing winds, most rain falls on the windward side of the mountain. This supports abundant vegetation. The other side of the mountain receives much less rain and is drier. Often a desert may exist. Both plants and animals must be adapted to the kind of climate typical for their regions. Cactus and ground squirrels would be typical of the desert, and pine trees and tree squirrels would be typical of the windward side of the mountain.

ferent genetic combinations in the two places. Furthermore, different mutations may occur in the two isolated populations, and each may generate different random combinations of genes as a result of sexual reproduction. This would be particularly true if one of the populations was very small. As a result, the two populations may show differences in color, height, enzyme production, time of seed germination, or many other characteristics.

Over a long period of time, the genetic differences that accumulate may result in regional populations called **subspecies** that are significantly modified structurally, physiologically, or behaviorally. The differences among some subspecies may be so great that they reduce reproductive success when the subspecies mate. **Speciation** is the process of generating new species. This process has occurred only if gene flow between isolated populations does not occur even after barriers are removed. In other words, the process of speciation can begin

with the geographic isolation of a portion of the species, but new species are generated only if isolated populations become separate from one another genetically. Speciation by this method is really a three-step process. It begins with geographic isolation, is followed by the action of selective agents that choose specific genetic combinations as being valuable, and ends with the genetic differences becoming so great that reproduction between the two groups is impossible.

It is also possible to envision ways in which speciation could occur without geographic isolation being necessary. Any process that could result in the reproductive isolation of a portion of a species could lead to the possibility of speciation. For example, within populations, some individuals may breed or flower at a somewhat different time of the year. If the difference in reproductive time is genetically based, different breeding populations could be established, which could eventually lead to speciation. Among animals, variations in the genetically determined behaviors related to courtship and mating could effectively separate one species into two or more separate breeding populations. In plants, genetically determined incompatibility of the pollen of one population of flowering plants with the flowers of other populations of the same species could lead to separate species.

THE DEVELOPMENT OF EVOLUTIONARY THOUGHT

Today, most scientists consider speciation an important first step in the process of evolution. However, this was not always the case. For centuries people believed that the various species of plants and animals were fixed and unchanging—that is, they were thought to have remained unchanged from the time of their creation. This was a reasonable assumption because people knew nothing about DNA, how sperm and eggs were formed, or population genetics. Furthermore, the process of evolution is so slow that the results of evolution were usually not evident during a human lifetime. It is even difficult for modern scientists to recognize this slow change in many kinds of organisms. In the mid-1700s, Georges-Louis Buffon, a French naturalist, expressed some curiosity about the possibilities of change (evolution) in animals, but he did not suggest any mechanism that would result in evolution.

In 1809, Jean-Baptiste de Lamarck, a student of Buffon's, suggested a process by which evolution could occur. He proposed that acquired characteristics could be transmitted to offspring. For example, he postulated that giraffes originally had short necks. Since giraffes constantly stretched their necks to obtain food, their necks got slightly longer. This slightly longer neck acquired through stretching could be passed to the offspring, who were themselves stretching their necks, and over time, the necks of giraffes would get longer and longer. Although we now know Lamarck's theory was wrong (because acquired characteristics are not inherited), it stimulated further thought as to how evolution could occur. All during this period, from the mid-1700s to the mid-1800s, lively arguments continued about the possibility of evolutionary change. Some, like Lamarck and others, thought that change did take place, while many others said that it was not possible. It was the thinking of two English scientists that finally provided a mechanism that explained how evolution could occur.

In 1858, Charles Darwin and Alfred Wallace suggested the theory of natural selection as a mechanism for evolution. They based their theory on the following assumptions about the nature of living things:

1. All organisms produce more offspring than can survive.
2. No two organisms are exactly alike.
3. Among organisms, there is a constant struggle for survival.
4. Individuals that possess favorable characteristics for their environment have a higher rate of survival and produce more offspring.

5. Favorable characteristics become more common in the species, and unfavorable characteristics are lost.

Using these assumptions, the Darwin-Wallace theory of evolution by natural selection offers a different explanation for the development of long necks in giraffes (figure 20.16):

1. In each generation, more giraffes would be born than the food supply could support.
2. In each generation, some giraffes would inherit longer necks, and some would inherit shorter necks.
3. All giraffes would compete for the same food sources.
4. Giraffes with longer necks would obtain more food, have a higher survival rate, and produce more offspring.
5. As a result, succeeding generations would show an increase in the neck length of the giraffe species.

This logic seems simple and obvious today, but remember that at the time Darwin and Wallace proposed their theory, the processes of sperm and egg formation and fertilization were poorly understood, and the concept of the gene was only beginning to be discussed. Nearly 50 years after Darwin and Wallace suggested their theory, the rediscovery of the work of Gregor Mendel provided an explanation for how characteristics could be transmitted from one generation to the next. Not only did Mendel's idea of the gene provide a means of passing traits from one generation to the next, it also provided the first step in understanding mutations, gene flow, and the significance of reproductive isolation. All of these ideas are interwoven into the modern concept of evolution. If we look at the same five ideas from the thinking of Darwin and Wallace and update them with modern information, they might look something like this:

1. An organism's capacity to overproduce results in surplus organisms.
2. Because of mutation, new genes enter the gene pool. Because of sexual reproduction, involving sex-cell formation and fertilization, new combinations of genes are present in every generation. These processes are so powerful that each individual in a sexually reproducing population is genetically unique. The genes present are expressed in the structure and function of the organism.
3. Resources such as food, soil nutrients, water, mates, and nest materials are in short supply, so some individuals will do without. Other environmental factors such as disease organisms, predators, or helpful partnerships with other species also affect survival. All of these factors that affect survival are called selecting agents.
4. Selecting agents favor individuals with the best combination of genes. They will be more likely to survive and reproduce, passing more of their genes on to the next generation. An organism is selected against if it has fewer offspring than other individuals that have a more favorable combination of genes. It does not need to die to be selected against.
5. Therefore, genes or gene combinations that produce characteristics favorable to survival will become more common, and the species will become better adapted to its environment.

Figure 20.16

(*A*) Lamarck thought that acquired characteristics could be passed on to the next generation. Therefore, he postulated that as giraffes stretched their necks to get food, their necks got slightly longer. This characteristic was passed on to the next generation, which would have longer necks. (*B*) The Darwin-Wallace theory states that there is variation within the population and that those with longer necks would be more likely to survive and reproduce and pass on their genes for long necks to the next generation.

THE TENTATIVE NATURE OF THE EVOLUTIONARY HISTORY OF ORGANISMS

It is important to understand that thinking about the concept of evolution can take us in several different directions. First, it is clear that genetic changes do occur. Mutations introduce new genes into a species. This has been demonstrated repeatedly with chemicals and radiation. Our recognition of this danger is evident by the ways we protect ourselves against excessive exposure to mutagenic agents. We also recognize that species can change. We purposely manipulate the genetic constitution of our domesticated plants and animals and change their characteristics to suit our needs. We also recognize that different populations of the same species show genetic differences. Examination of fossils shows that species of organisms that once existed are no longer in existence. We even have historical examples of plants and animals that are now extinct. We can also demonstrate that new species come into existence, and this is easily done with plants. It is clear from this evidence that species are not fixed, unchanging entities.

However, when we try to piece together the evolutionary history of organisms over long periods of time, we must use much indirect evidence, and it becomes difficult to state definitively the specific sequence of steps that the evolution of a species followed. Although there is little question that evolution occurs, it is not possible to state unconditionally that evolution of a particular group of organisms has followed a specific path. There will always be new information that will require changes in thinking, and equally reputable scientists will disagree on the evolutionary processes or the sequence of events that led to a specific group of organisms. But there can be no question that evolution occurred in the past and continues to occur today.

SUMMARY

Current theories about the *origin of life* include the two competing ideas (1) that the primitive earth environment led to the *spontaneous organization of organic chemicals* into primitive cells and (2) that primitive forms of life *arrived on the earth from space*. Basic units of life were probably similar to present-day *prokaryotes*. These primitive cells could have *changed through time* as a result of *mutation* and *in response to a changing environment*.

All sexually reproducing organisms naturally exhibit genetic variety among the individuals in the population as a result of *mutations* and the *genetic recombination* resulting from sexual reproduction. The *genetic differences* are reflected in differences among individuals. These genetic differences are important for the survival of the species because *natural selection* must have *genetic variety* to select from. Natural selection by the *environment* results in better-suited individual organisms that have greater numbers of offspring than those that are less well-off genetically.

Organisms with wide *geographic distribution* often show different *gene frequencies* in different parts of their *range*. A *species* is a group of *organisms* that can interbreed to produce fertile offspring. This definition of a species is also known as the *biological species concept*. People

often distinguish species from one another based on structural differences. This is known as a *morphological species concept*. The process of *speciation* usually involves the *geographic separation* of the species into two or more isolated *populations*. While they are separated, natural selection operates to *adapt each population to its environment*.

At one time, people thought that all organisms had remained *unchanged* from the time of their creation. Lamarck suggested that *change did occur* and thought that *acquired characteristics could be passed from generation to generation*. Darwin and Wallace proposed the *theory of natural selection* as the mechanism that drives *evolution*.

KEY TERMS

acquired characteristics (p. 437)
autotroph (p. 430)
biogenesis (p. 428)
biological species concept
   (p. 443)
endosymbiotic theory (p. 433)
eukaryote (p. 433)
evolution (p. 433)
gene flow (p. 443)
genetic recombination (p. 437)
geographic barriers (p. 445)
geographic isolation (p. 445)
heterotroph (p. 430)
morphological species concept
   (p. 443)

natural selection (p. 433)
oxidizing atmosphere (p. 431)
prokaryote (p. 433)
range (p. 445)
selecting agents (p. 437)
speciation (p. 447)
species (p. 443)
spontaneous generation (p. 428)
spontaneous mutation (p. 436)
subspecies (p. 447)
theory of natural selection
   (p. 434)

APPLYING THE CONCEPTS

1. Which was *not* a major component of the early planet Earth's reducing atmosphere?
   (a) H_2
   (b) CO_2
   (c) NH_3
   (d) O_2

2. Prokaryotic cells hypothetically came into existence approximately _____ years ago?
   (a) 20 billion
   (b) 4–5 billion
   (c) 3.5 billion
   (d) 1.5 billion

3. Which one of the following researchers found evidence not to support spontaneous generation?
   (a) Redi
   (b) Darwin
   (c) Lamarck
   (d) Woese

4. The three main kinds of living things, Eubacteria, Archaea, and Eucarya, have been labeled as
   (a) species.
   (b) kingdoms.
   (c) domains.
   (d) families.

5. Hybrid animals like mules are not considered to be a species because they
   (a) do not reproduce through many generations.
   (b) are not common enough.
   (c) can be maintained only by humans.
   (d) look different from both parents.
6. Two closely related organisms are not considered to be separate species unless they
   (a) look different.
   (b) are reproductively isolated.
   (c) are able to interbreed.
   (d) are in different geographic parts of the world.
7. The Darwin-Wallace theory of natural selection differs from Lamarck's ideas in that
   (a) Lamarck understood the role of genes, and Darwin and Wallace did not.
   (b) Lamarck assumed that characteristics obtained during an organism's lifetime could be passed to the next generation; Darwin and Wallace did not.
   (c) Lamarck did not think that evolution took place; Darwin and Wallace did.
   (d) Lamarck developed the basic ideas of speciation, which Darwin and Wallace refined.
8. Oxygen present in our atmosphere is the result of
   (a) organisms carrying on respiration.
   (b) gases from volcanoes.
   (c) the breakdown of carbon dioxide into carbon and oxygen.
   (d) organisms carrying on photosynthesis.
9. Which of the following is *not* necessary for speciation?
   (a) genetic isolation from other species
   (b) genetic variety within a species
   (c) hundreds of millions of years
   (d) reproduction
10. It is assumed that the original earth had an atmosphere that was different from today. One way it is different today is that it has
    (a) less ozone.
    (b) less oxygen.
    (c) less nitrogen.
    (d) less carbon dioxide.
11. Two groups of organisms belong to different species if
    (a) gene flow between the two groups is not possible, even in the absence of physical barriers.
    (b) physical barriers separate the two groups, thereby preventing cross matings.
    (c) the two groups of organisms have a different physical appearance.
    (d) individuals from the two groups, when mated, produce fertile offspring.

12. An organism carries on photosynthesis, lacks a nucleus, and has DNA. It is a member of the Domain
    (a) Eucarya.
    (b) Archaea.
    (c) Eubacteria.
    (d) Protista.

Answers
1. d 2. c 3. a 4. c 5. a 6. b 7. b 8. d 9. c 10. d 11. a 12. c

QUESTIONS FOR THOUGHT

1. In what sequence did the following things happen: living cell, oxidizing atmosphere, eukaryotes developed, reducing atmosphere, first organic molecule?
2. What is meant by *spontaneous generation?* What is meant by *biogenesis?*
3. List two important effects caused by the increase of oxygen in the atmosphere.
4. What evidence supports the theory that eukaryotic cells arose from the development of a symbiotic relationship between primitive prokaryotic cells?
5. Why are acquired characteristics of little interest to evolutionary biologists?
6. What is natural selection? How does it work?
7. Give two examples of selecting agents and explain how they operate.
8. Why is geographic isolation important in the process of speciation?
9. Why aren't mules considered a species?
10. Why has Lamarck's theory been rejected?
11. List the series of events necessary for speciation to occur.
12. "Evolution is a fact." "Evolution is a theory." Explain how both of these statements can be true.

CHAPTER TWENTY-ONE

The History of Life on Earth

These are modern-day stromatolites from Hamlin Pool in Western Australia. The dome-shaped structures shown in the photograph are composed of layers of cyanobacteria and materials they secrete and grow up to 60 centimeters tall. Some of the oldest fossils are of ancient stromatolites that developed in shallow marine environments about 3.5 billion years ago. When samples from fossil stromatolites are cut into slices, microscopic images can be produced that show the fossil remains of some of the world's oldest cells.

O U T L I N E

At one time people assumed that the earth and its inhabitants were fixed and unchanging. However, as they began to understand the nature of events such as volcanic eruptions, earthquakes, and erosion it became obvious that the earth was changeable. Furthermore, the presence of the fossilized remains of extinct organisms showed that the nature of the kinds of things that inhabited the earth had changed as well. These ideas challenged core beliefs and began a long process of re-forming the way we look at the history of the earth and life on it. In particular, people needed to develop an understanding that the history of the earth and its life was very long and should be measured in billions of years rather than hundreds or thousands of years.

GEOLOGIC TIME

Reading history from the rocks of the earth's crust requires both a feel for the immensity of geologic time and an understanding of geologic processes. By "geologic time" we mean the age of the earth; the very long span of the earth's history. This span of time is difficult for most of us to comprehend. Human history is measured in units of hundreds and thousands of years, and even the events that can take place in a thousand years are hard to imagine. Geologic time, on the other hand, is measured in units of millions and billions of years.

The understanding of geologic processes has been made possible through the development of various means of measuring ages and time spans in geologic systems. An understanding of geologic time leads to an understanding of geologic processes, which then leads to an understanding of the environmental conditions that must have existed in the past. Thus, the mineral composition, texture, and sedimentary structure of rocks are clues about past events, events that make up the history of the earth.

Early Attempts at Earth Dating

Most human activities are organized by time intervals of minutes, hours, days, months, and years. These time intervals are based on the movements of the earth and the moon. Short time intervals are measured in minutes and hours, which are typically tracked by watches and clocks. Longer time intervals are measured in days, months, and years, which are typically tracked by calendars. How do you measure and track time intervals for something as old as the earth? First, you would need to know the age of the earth; then you would need some consistent, measurable events to divide the overall age into intervals. Questions about the age of the earth have puzzled people for thousands of years, dating back at least to the time of the ancient Greek philosophers. Many people have attempted to answer this question and understand geologic time, but with little success until the last few decades.

One early estimate of the age of the earth was attempted by Archbishop Ussher of Ireland in the seventeenth century. He painstakingly counted up the generations of people mentioned in biblical history, added some numerological considerations, and arrived at the conclusion that the earth was created at 9:00 A.M. on Tuesday, October 26, in the year 4004 B.C. On the authority of biblical scholars this date was generally accepted for the next century or so, even though some people thought that the geology of the earth seemed to require far longer to develop. The date of 4004 B.C. meant that the earth and all of the surface features on the earth had formed over a period of about 6,000 years. This required a model of great cataclysmic catastrophes to explain how all the earth's features could possibly have formed over a span of 6,000 years.

Near the end of the eighteenth century, James Hutton reasoned out the concept that the geologic processes we see today have operated since the earth began. This is known as the *principle of uniformity* and since many geologic processes such as erosion and sedimentation are slow, people began to assume a much older earth. The problem then became one of finding some uniform change or process that could serve as a geologic clock to measure the age of the earth. To serve as a geologic clock a process or change would need to meet three criteria: (1) the process must have been operating since the earth began, (2) the process must be uniform or at least subject to averaging, and (3) the process must be measurable.

During the nineteenth century many attempts were made to find earth processes that would meet the criteria to serve as a geologic clock. Among others, the processes explored were (1) the rate that salt is being added to the ocean, (2) the rate that sediments are being deposited, and (3) the rate that the earth is cooling. Comparing the load of salts being delivered to the ocean by all the rivers, and assuming the ocean was initially pure water, it was calculated that about 100 million years would be required for the present salinity to be reached. The calculations did not consider the amount of materials being removed from the ocean by organisms and by chemical sedimentation, however, so this technique was considered to be unacceptable. Even if the amount of materials removed were known, it would actually result in the age of the ocean, not the age of the earth.

A number of separate and independent attempts were made to measure the rate of sediment deposition, then compare that rate to the thickness of sedimentary rocks found on the earth. Dividing the total thickness by the rate of deposition resulted in estimates of an earth age that ranged from about 20 million to 1,500 million years. The wide differences occurred because there are gaps in many sedimentary rock sequences, periods when sedimentary rocks were being eroded away to be deposited again elsewhere as sediments. There were just too many unknowns for this technique to be considered as acceptable.

The idea of measuring the rate that the earth is cooling for use as a geologic clock assumed that the earth was initially a molten mass that has been cooling ever since. Calculations estimating the temperature that the earth must have been to be molten were compared to the earth's present rate of cooling.

This resulted in an estimated age of 20 to 40 million years. These calculations were made back in the nineteenth century before it was understood that natural radioactivity is adding heat to the earth's interior, so it has required much longer to cool down to its present temperature.

Modern Techniques for Determining the Age of the Earth

Soon after the beginning of the twentieth century the discovery of the radioactive decay process in the elements of minerals and rocks led to the development of a new, accurate geologic clock. This clock finds the *radiometric age* of rocks in years by measuring the radioactive decay of unstable elements within the crystals of certain minerals. Since radioactive decay occurs at a constant, known rate, the ratio of the remaining amount of a radioactive element to the amount of decay products present can be used to calculate the time that the radioactive element has been a part of that crystal (see chapter 11). Certain radioactive isotopes of potassium, uranium, and thorium are often included in the minerals of rocks, so they are often used as "radioactive clocks." By using radiometric aging techniques along with other information, we arrive at an age for the earth of about 4.5 billion years.

A recently developed geologic clock is based on the magnetic orientation of magnetic minerals. These minerals become aligned with the earth's magnetic field when the igneous rock crystallizes, making a record of the field at that time. The earth's magnetic field is global and has undergone a number of reversals in the past. A *geomagnetic time scale* has been established from the number and duration of magnetic field reversals occurring during the past 6 million years. Combined with radiometric age dating, the geomagnetic time scale is making possible a worldwide geologic clock that can be used to determine local chronologies.

Arranging Events in Order

The clues provided by thinking about geologic processes that must have occurred in the past are interpreted within a logical frame of reference that can be described by several basic principles. The following is a summary of these basic guiding principles that are used to read a story of geologic events from the rocks.

The cornerstone of the logic used to guide thinking about geologic time is the *principle of uniformity*. As described earlier, this principle is sometimes stated as "the present is the key to the past." This means that the geologic features that you see today have been formed in the past by the same processes of crustal movement, erosion, and deposition that are observed today. By studying the processes now shaping the earth, you can understand how it has evolved through time. This principle establishes the understanding that the surface of the earth has been continuously and gradually modified over the immense span of geologic time.

The *principle of original horizontality* is a basic principle that is applied to sedimentary rocks. It is based on the observation that, on a large scale, sediments are commonly deposited in flat-lying layers. Old rocks are continually being changed to new

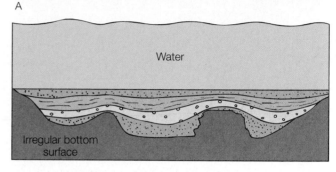

A

B

Figure 21.1

The principle of original horizontality. (*A*) Sediments tend to be deposited in horizontal layers. (*B*) Even where the sediments are draped over an irregular surface, they tend toward the horizontal.

ones in the continuous processes of crustal movement, erosion, and deposition. As sediments are deposited in a basin of deposition, such as a lake or ocean, they accumulate in essentially flat-lying, approximately horizontal layers (figure 21.1). Thus, any layer of sedimentary rocks that is not horizontal has been subjected to forces that have deformed the earth's surface.

The *principle of superposition* is another logical and obvious principle that is applied to sedimentary rocks. Layers of sediments are usually deposited in succession in horizontal layers, which later are compacted and cemented into layers of sedimentary rock. An undisturbed sequence of horizontal layers is thus arranged in chronological order with the oldest layers at the bottom. Each consecutive layer will be younger than the one below it (figure 21.2). This is true, of course, only if the layers have not been turned over by deforming forces.

The *principle of crosscutting relationships* is concerned with igneous and metamorphic rock, in addition to sedimentary rock layers. Any geologic feature that cuts across or is intruded into a rock mass must be younger than the rock mass. Thus, if a fault cuts across a layer of sedimentary rocks, the fault is the youngest feature. Faults, folds, and igneous intrusions are always younger than the rocks in which they occur. Often, there is a further clue to the correct sequence: The hotter igneous rock may have "baked," or metamorphosed, the surrounding rock immediately adjacent to it, so again the igneous rock must have come second (figure 21.3).

Shifting sites of erosion and deposition: The principle of uniformity states that the earth processes going on today have always been occurring. This does not mean, however, that they always occur in the same place. As erosion wears away the rock

Figure 21.2

The Grand Canyon, Arizona, provides a majestic cross section of horizontal sedimentary rocks. According to the principle of superposition, traveling deeper and deeper into the Grand Canyon means that you are moving into older and older rocks.

Rocks adjacent to intruding magma may also be metamorphosed by its heat.

Figure 21.3

A granite intrusion cutting across older rocks. The granite intrusion is younger than the surrounding sedimentary rock.

layers at a site, the sediments produced are deposited someplace else. Later, the sites of erosion and deposition may shift, and the sediments are deposited on top of the eroded area. When the new sediments later are formed into new sedimentary rocks, there will be a time lapse between the top of the eroded layer and the new layers. A time break in the rock record is called an **unconformity.** The unconformity is usually shown by a surface within a sedimentary sequence on which there was a lack of sediment deposition, or where active erosion may even have occurred for some period of time. When the rocks are later examined, that time span will not be represented in the record, and if the unconformity is erosional, some of the record once present will have been lost. An unconformity may occur within a sedimentary sequence of the same kind or between different kinds of rocks. The most obvious kind of unconformity to spot is an *angular unconformity*. An angular unconformity, as illustrated in figure 21.4, is one in which the bedding planes above and below the unconformity are not parallel. An angular un-

Deposition

Erosional surface

Rocks tilted, eroded

Water

Angular unconformity established

Subsequent deposition

Figure 21.4

Angular unconformity. Development of an angular unconformity usually involves some deformation and erosion before sedimentation is resumed.

conformity usually implies some kind of tilting or folding, followed by a significant period of erosion, which in turn was followed by a period of deposition (figure 21.5).

GEOLOGIC TIME AND THE FOSSIL RECORD

A **fossil** is *any* evidence of former life, so the term means more than fossilized remains such as those pictured in figure 21.6. Evidence can include actual or altered remains of plants and animals. It could also be just simple evidence of former life such as the imprint of a leaf, the footprint of a dinosaur, or droppings from bats in a cave.

What would you think if you were on the top of a mountain, broke open a rock, and discovered the fossil fish pictured in figure 21.6? How could you explain what you found? There are several ways that a fish could end up on a mountaintop as a fossil. For example, perhaps the ocean was once much deeper and covered the mountaintop. On the other hand, maybe the mountaintop was once below sea level and pushed its way up to its

Figure 21.5

A time break in the rock record in the Grand Canyon, Arizona. The horizontal sedimentary rock layers overlie almost vertically oriented metamorphic rocks. Metamorphic rocks form deep in the earth, so they must have been uplifted, and the overlying material eroded away before being buried again.

Figure 21.6

The fossil record of the hard parts is beautifully preserved, along with a carbon film, showing a detailed outline of the fish and some of its internal structure.

present high altitude. Another explanation might be that someone left the fossil on the mountain as a practical joke. What would you look for to help you figure out what actually happened? In every rock and fossil there are fascinating clues that help you read what happened in the past, including clues that tell you if an ocean had covered the area or if a mountain pushed its way up from lower levels. There are even clues that tell you if a rock has been brought in from another place. This section is about some of the clues found in fossils and rocks, and what the clues mean.

Early Ideas About Fossils

The story about finding a fossil fish on a mountain is not as far-fetched as it might seem, and in fact one of the first recorded evidences of understanding the meaning of fossils took place in a similar setting. The ancient Greek historian Herodotus was among the first to realize that fossil shells found in rocks far from any ocean were remnants of organisms left by a bygone sea. Other Greek philosophers were not convinced that this conclusion was as obvious as it might seem today. Aristotle, for example, could see no connection between the shells of organisms of his time and the fossils, which he believed to have formed inside the rocks. Note that he also believed that living organisms could arise by spontaneous generation from mud. A belief that the fossils must have "grown" in place in rocks would seem to be consistent with a belief in spontaneous generation.

It was a long time before it was generally recognized that fossils had anything to do with living things. Even when people started to recognize some similarities between living organisms and certain fossils, they did not make a connection. Fossils were considered to be the same as quartz crystals, or any other mineral crystals, meaning they were either formed with the earth, or grew there later (depending on the philosophical view of the interpreter). Fossils of marine organisms that were well preserved and very similar to living organisms were finally accepted as remains of once-living organisms—that were buried in Noah's flood. By the time of the Renaissance some people were starting to think of other fossils, too, as the remains of former life-forms. Leonardo da Vinci, like other Renaissance scholars, argued that fossils were the remains of organisms that had lived in the past.

By the early 1800s the true nature of fossils was becoming widely accepted. William "Strata" Smith, an English surveyor, discovered at this time that sedimentary rock strata could be identified by the fossils they contain. Smith grew up in a region of England where fossils were particularly plentiful. He became a collector, keeping careful notes on where he found each fossil and in which type of sedimentary rock layers. During his travels he discovered that the succession of rock layers on the south coast of England was the same as the succession of rock layers on the east coast. Through his keen observations, Smith found that each kind of sedimentary rock had a distinctive group of fossils that was unlike the group of fossils in other rock layers. Smith amazed his friends by telling them where and in what type of rock they had found their fossils.

Today, the science of discovering fossils, studying the fossil record, and deciphering the history of life from fossils is known as **paleontology.** The word "paleontology" was invented in 1838 by the British geologist Charles Lyell to describe this newly established branch of geology. It is derived from classic Greek roots and means "study of ancient life," and this means a study of fossils.

People sometimes blur the distinction between *paleontology* and *archaeology.* **Archaeology** is the study of past human life and culture from material evidence of artifacts, such as graves, buildings, tools, pottery, landfills, and so on ("artifact" literally means "something made"). The artifacts studied in archaeology can be of any age, from the garbage added to the city landfill yes-terday to the pot shards of an ancient tribe that disappeared hundreds of years ago. The word "fossil" originally meant "anything dug up," but today carries the meaning of any *evidence of ancient organisms in the history of life.* Artifacts are therefore not fossils, as you can see from the definitions.

Types of Fossils

Considering all the different things that can happen to the remains of an organism, and considering the conditions needed to form a fossil, it seems amazing that any fossils are formed and then found. For example, the animals you see killed beside the road or the dead trees that fall over in a forest rarely become fossils because scavengers eat the remains of dead animals, and decay organisms break down the organic remains of plants and animals. As a result, very little digestible organic matter escapes destruction, but indigestible skeletal material, such as shells, bones, and teeth, has a much better chance of not being destroyed.

Thus a fossil is not likely to form unless there is rapid burial of a recently deceased organism. The presence of hard parts, such as a shell or a skeleton, will also favor the formation of a fossil if there is rapid burial.

Several different kinds of fossils can be differentiated based on the kind of material that is preserved and the processes of preservation. Occasionally entire organisms are preserved and in rare cases even the unaltered remains of an organism's *soft parts* are found. In order for this to occur the organism must be quickly protected from scavengers and decomposers following death. There are several conditions that allow this to occur. The best examples of this uncommon method of fossilization include protection by freezing or entombing in tree resin. Mammoths, for example, have been found frozen and preserved by natural refrigeration in the ices of Alaska and Siberia. The bodies of ice age humans have also been discovered in glaciers, and early inhabitants of South America have been found in ice caves high in mountains. Insects and spiders, complete with delicate appendages, have been found preserved in amber, which is fossilized tree resin (figure 21.7). In each case—ice and resin—the soft parts were protected from scavengers, insects, and bacteria. In addition, if organisms are buried in acid bogs or quickly desiccated following death, they may be preserved in a relatively unaltered state.

Although fossils of soft parts are occasionally formed, the *remains of hard parts* such as shells, bones, and teeth of animals or the pollen and spores of plants are much more common. The fish fossil pictured in figure 21.6 is from Wyoming's Green River Formation. This freshwater fish died in its natural environment and was soon covered by fine-grained sediment. The sediment preserved the complete articulated skeleton, along with some carbon traces of soft tissue.

Because hard parts are more easily preserved, organisms that have such parts are more often found as fossils than those that lack hard parts. Worms of various kinds are hard to find as fossils for this reason. The hard parts of various organisms are composed of different kinds of materials. Calcium carbonate is present in the shells of most corals, mollusks, certain kinds of

Figure 21.7

This fly was stuck in and covered over with plant pitch. When the pitch fossilized to form amber, the fly was preserved as well. The entire fly is preserved in the amber.

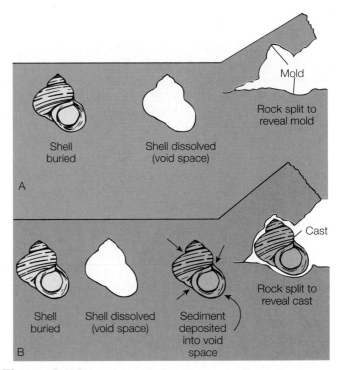

Figure 21.8

Origin of molds and casts. (*A*) Formation of a mold. (*B*) Formation of a cast.

protozoans, and the skeletal structures of some sponges. Some thick deposits of limestone are composed of the calcium carbonate shells of tiny marine protozoans known as foraminiferans. Calcium phosphate was found in the shells of some kinds of extinct brachiopods. Silica is found in the skeletal structures of certain protozoans and sponges. The silica shells of the marine microorganisms known as diatoms, have accumulated to such a degree that ancient deposits of the shells are mined as diatomaceous earth and used for filters and certain buffing compounds. Chitin is the tough organic material that makes up the exoskeletons of insects, crabs, and lobsters and resists decomposition. Organisms that have exoskeletons of chitin are common in the fossil record.

A carbon trace is an example of another kind of fossil, one that does not preserve the detail of internal structures but gives a good idea of the general shape of the organism. Plant fossils are often found as carbon traces, sometimes looking like a photograph of a leaf on a slab of shale or limestone. Other fossils that give a good idea of the general shape of an organism are molds and casts. Calcium carbonate shell material may be dissolved by groundwater in certain buried environments, leaving an empty *mold* in the rock. Sediment, or groundwater deposits may fill the mold and make a *cast* of the organism (figure 21.8).

Some fossils are formed by the modification of the original chemical nature of the organism. Figure 21.9 is a photograph of part of the Petrified Forest National Park in Arizona. Petrified trees are not trees that have been "turned to stone." They are stony *replicas* that were formed as the wood was altered by circulating groundwater carrying elements in solution. There are two processes involved in the making of petrified fossils, and they are not restricted to wood. The processes involve (1) *mineralization,* which is the filling of pore spaces within the structure of the buried organism with deposits of calcium carbonate, silica, or pyrite, and/or (2) *replacement,* which is the

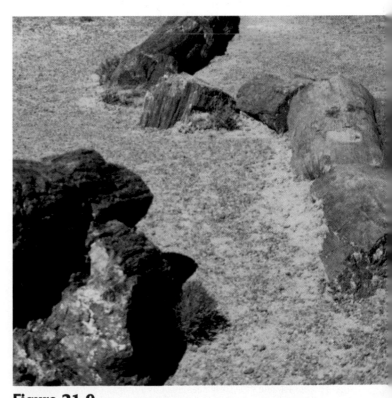

Figure 21.9

These logs of petrified wood are in the Petrified Forest National Park, Arizona.

dissolving of the original material and depositing of new material an ion at a time. Petrified wood is formed by both processes over a long period of time. As it decayed, the original wood was replaced by mineral matter an ion at a time. Over time the "mix" of minerals being deposited changed resulting in various colors. Since the mineralization or replacement processes take place an ion at a time, a great deal of detail can be preserved in such fossils. The size and shape of the cells and growth rings in petrified wood are preserved well enough that they can be compared to modern plants. The skeletons or shells of many extinct organisms are also typically preserved in this way. The "fossilized bones" of dinosaurs or the "fossilized shells" of many invertebrates are examples.

Finally, there are many kinds of fossils that are not the preserved remains of an organism but are preserved indications of the activities of organisms. Some of the most interesting of such fossils are the footprints of various kinds of animals. There are examples of footprints of dinosaurs and various kinds of extinct mammals including ancestors of humans. If an animal walked through mud or other soft substrates and the substrates were covered with silt or volcanic ash, the pattern of the footprints can be preserved (figure 21.10). The tunnels of burrowing animals, the indentations left by crawling worms, and the nests of dinosaurs have been preserved in a similar manner. Even the dung of dinosaurs and other animals has been preserved in a mineralized form.

As you can see, there are many different ways in which a fossil can be formed, but it must be *found and studied* if it is to reveal its part in the history of life. This means the rocks in which the fossil formed must somehow make it back to the surface of the earth. This usually involves movement and uplift of the rock, and weathering and erosion of the surrounding rock to release or reveal the fossil. Most fossils are found in recently eroded sedimentary rocks—sometimes atop mountains that were under the ocean a long, long time ago. The complete record of what has happened in the past is not found in the fossil alone but requires an understanding of the layers of rocks present, the relationship of the layers to each other, and their age.

Figure 21.10
This dinosaur footprint is in shale near Tuba City, Arizona. It tells you something about the relative age of the shale, since it must have been soft mud when the dinosaur stepped here.

Correlating Fossils with Geologic Time

The principle of superposition, the principle of crosscutting relationships, and the presence of an unconformity all have meaning about the order of geologic events that have occurred in the past. This order can be used to unravel a complex sequence of events such as the one shown in figure 21.11. The presence of fossils can help too. The *principle of faunal succession* recognizes that life-forms have changed through time. Old life-forms disappear from the fossil record and new ones appear, but the same form is never exactly duplicated independently at two different times in history. This principle implies that the same type of fossil organisms that lived only a brief geologic time should occur only in rocks that are the same age. According to the principle of faunal succession, then, once the basic sequence of fossil forms in the rock record is determined, rocks can be placed in their correct relative chronological position on the basis of the fossils contained in them. The principle also means that if the same

Figure 21.11
Deciphering a complex rock sequence. The limestones must be oldest (law of superposition), followed by the shales. The granite and basalt must both be younger than the limestone they crosscut (note the metamorphosed zone around the granite). It is not possible to tell whether the igneous rocks (granite and basalt) predate or postdate the shales or to determine whether the sedimentary rocks were tilted before or after the igneous rocks were emplaced. After the limestones and shales were tilted, they were eroded, and then the sandstones were deposited on top. Finally, the lava flow covered the entire sequence.

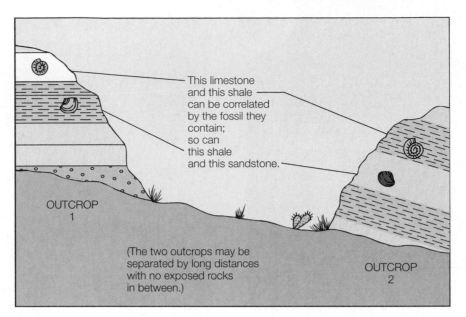

Figure 21.12
Similarity of fossils suggests similarity of ages, even in different rocks widely separated in space.

type of fossil organism is preserved in two different rocks, the rocks should be the same age. This is logical even if the two rocks have very different compositions and are from places far, far apart (figure 21.12).

Distinctive fossils of plant or animal species that were distributed widely over the earth but lived only a brief time are called **index fossils.** Index fossils, together with the other principles used in reading rocks, make it possible to compare the ages of rocks exposed in two different locations. This is called *age correlation* between rock units. Correlations of exposed rock units separated by a few kilometers are easier to do, but correlations have been done with exposed rock units that are separated by an ocean. Correlation allows the ordering of geologic events according to age. Since this process is only able to determine the

age of a rock unit or geologic event relative to some other unit or event, it is called *relative dating*. Dates with numerical ages are determined by means different from correlation.

The usefulness of correlation and relative dating through the concept of faunal succession is limited because the principles can be applied only to rocks in which fossils are well preserved, which are almost exclusively sedimentary rocks. Correlation also can be based on the occurrence of unusual rock types, distinctive rock sequences, or other geologic similarities. All this is useful in clarifying relative age relationships among rock units. It is not useful in answering questions about the age of rocks or the time required for certain events, such as the eruption of a volcano, to occur. Questions requiring numerical answers went unanswered until the twentieth century.

A Closer Look What Is Carbon-14 Dating?

Carbon is an element that occurs naturally as several isotopes. The most common form is carbon-12. A second, heavier radioactive isotope, carbon-14, is constantly being produced in the atmosphere by cosmic rays. Radioactive elements are unstable and break down into other forms of matter. Hence, radioactive carbon-14 naturally decays. The rate at which carbon-14 is formed and the rate at which it decays is about the same, therefore, the concentration of carbon-14 on the earth stays relatively constant. All living things contain large quantities of the element carbon. Plants take in carbon in the form of carbon dioxide from the atmosphere, and animals obtain carbon from the food they eat. While an organism is alive, the proportion of carbon-14 to carbon-12 within its body is equal to its surroundings. When an organism dies, the carbon-14 within its tissues disintegrates, but no new carbon-14 is added. Therefore, the age of plant and animal remains can be determined by the ratio of carbon-14 to carbon-12 in the tissues. The older the specimen, the less carbon-14 present. Radioactive decay rates are measured in half-life. One half-life is the amount of time it takes for one-half of a radioactive sample to decay.

The half-life of carbon-14 is 5,730 years. Therefore, a bone containing one-half the normal proportion of carbon-14 is 5,730 years old. If the bone contains one-quarter of the normal proportion of carbon-14, it is 2 × 5,730 = 11,460 years old, and if it contains one-eighth of the naturally occurring proportion of carbon-14, it is 3 × 5,730 = 17,190 years old. As the amount of carbon-14 in a sample becomes smaller it becomes more difficult to measure the amount remaining. Therefore, carbon-14 dating is generally only useful for dating things that are less than 50,000 years old.

Many books have illustrations of various kinds of extinct animals. Have you ever wondered how anyone can know what these animals looked like? Certainly the general size and shape can be determined by assembling fossils of individual skeletal parts, but most of the shape of an animal is determined by the size and distribution of its muscles. An understanding of basic principles of physics provides part of the answer. The skeleton of an animal is basically arranged like a large number of levers. For example, a leg and foot consists of several bones linked end to end. The size of a bone gives a clue to the amount of mass that particular bone had to support. Various bumps on the bones are places where muscles and tendons attach. By knowing where the muscles are attached and analyzing the kinds of levers they represent, the size of the muscle can be calculated. Thus the general shape of the animal can be estimated. However, this doesn't provide any information about the color or surface texture of the animal. That information must come from other sources.

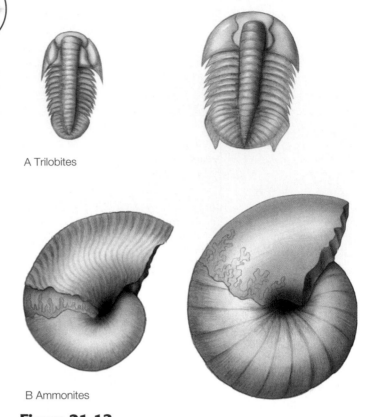

A Trilobites

B Ammonites

Figure 21.13

(*A*) Fossil trilobites from Cambrian rocks. (*B*) Fossil ammonites from Permian rocks.

The Geologic Time Scale

A yearly calendar helps you keep track of events over periods of time by dividing the year into months, weeks, and days. In a similar way, the **geologic time scale** helps you keep track of events that have occurred in the earth's geologic history. The first development of this scale came from the work of William "Strata" Smith, the English surveyor described in the section on fossils earlier in this chapter. Recall that Smith discovered that certain rock layers in England occurred in the same order, top to bottom, wherever they were located. He also found that he could correlate and identify each layer by the kinds of fossils in the rocks of the layers. In 1815, he published a geologic map of England, identifying the rock layers in a sequence from oldest to youngest. Smith's work was followed by extensive geological studies of the rock layers in other countries. Soon it was realized that similar, distinctive index fossils appeared in rocks of the same age when the principle of superposition was applied. For example, the layers at the bottom contained fossils of trilobites (figure 21.13A), but trilobites were not found in the upper levels. (Trilobites are extinct marine arthropods that may be closely related to living crustaceans, spiders, and horseshoe crabs.) On the other hand, fossil shells of ammonites (figure 21.13B) appeared in the middle levels, but not the lower nor upper levels of the rocks. The topmost layer was found to contain the fossils of animals identified as still living today. The early appearance and later disappearance of fossils in progressively younger rocks is explained by organic evolution and extinction, events that could be used to mark the time boundaries of the earth's geologic history.

The major blocks of time in the earth's geologic history are divided into several different categories. The largest are called **eons.** The *Phanerozoic eon* spans the period of time from about 540 million years ago when the various groups of animals we currently see on earth developed. The *Proterozoic eon* encompasses the period of time (2,500 to 540 million years ago) when most living things were simple, marine organisms. The first eukaryotic organisms developed during this time. The *Archean eon* (3,800 to 2,500 million years ago) constitutes a time when all organisms were prokaryotic and marine. The *Hadean eon* is the earlier period before life and before the earth solidified. Eons are divided into periods of time called **eras,** and each era is identified by the appearance and disappearance of particular fossils in the sedimentary rock record. The eras are: (1) *Cenozoic,* which refers to the time of recent life. Recent life means that the fossils for this time period are similar to the life found on earth today. (2) *Mesozoic,* which refers to the time of middle life. Middle life means that some of the fossils for this time period are similar to the life found on earth today, but many are different from anything living today. (3) *Paleozoic,* which refers to the time of ancient life. Ancient life means that the fossils for this time period are very different from anything living on the earth today. (4) Sometimes the Hadean, Proterozoic, and Archean eons are referred to as the *Precambrian,* which refers to the time before the time of ancient life. This means that the rocks for this time period contain very few fossils. The eras were divided into blocks of time called **periods,** and the periods were further subdivided into smaller blocks of time called **epochs** (table 21.1).

The geologic time scale developed in the 1800s was without dates because geologists did not yet have a way to measure the length of the eras, periods, or epochs. This was a *relative time scale,* based on the superposition of rock layers and fossil records of organic evolution. With the development of radiometric dating, it became possible to attach numbers to the time scale. When this was accomplished it became apparent that geologic history spanned very long periods of time. In addition, the part of the time scale with evidence of life makes up about 15 percent of all of the earth's history. This does not mean that life appeared suddenly in the Cambrian period. The fossil record is incomplete since it is the hard parts of animals or plants that form fossils, usually after rapid burial. Thus, the soft-bodied life-forms that existed during the Precambrian would make Precambrian fossils exceedingly rare. The Precambrian fossils that have been found are chiefly those of deposits from bacteria, algae, a few fungi, unusual soft-bodied animals, and the burrow holes of worms. Another problem in finding ancient fossils of soft-bodied life-forms is that heat and pressure have altered many of the ancient rocks over time, destroying any fossil evidence that may have been present.

In general, the earliest abundant fossils are found in rocks from the Cambrian period at the beginning of the Paleozoic era. They show an abundance of oceanic life that represents all the major groups of marine animals found today. There is no fossil evidence of life of any kind living on the land during the Cambrian. The dominant life-forms of the Cambrian ocean were echinoderms, mollusks, trilobites, and brachiopods. The trilobites, now extinct, made up more than half of the kinds of living things during the Cambrian.

During the Ordovician and Silurian periods, most living things were still marine organisms with various kinds of jawless fish becoming prominent. Also, by the end of the Silurian, some primitive plants were found on land. The Devonian period saw the further development of different kinds of fish including those that had jaws and many kinds of land plants and animals. Coral reefs were also common in the Devonian. The Carboniferous period was a time of vast swamps of ferns, horsetails, and other primitive nonseed plants that would form great coal deposits. Fossils of the first reptiles and the first winged insects are found in rocks from this age. The Paleozoic era closed with the extinction of about 90 percent of plant and animal life of that time.

The Mesozoic era was a time when the development of life on land flourished. The dinosaurs first appeared in the Triassic period, outnumbering all the other reptiles until the close of the Mesozoic. Fossils of the first mammals, and modern forms of gymnosperms (cone-bearing plants) developed in the Triassic. The first flowering plants, the first deciduous trees, and the first birds appeared in the Jurassic period. The Cretaceous is the final period of the Mesozoic era and is characterized by the dominance of dinosaurs and extensive evolution of flowering plants and the insects that pollinate them. Birds and mammals also increased in variety. Like the close of the Paleozoic, the Mesozoic era ended with a great dying of land and marine life that resulted in the extinction of many species, including the dinosaurs.

Throughout the Mesozoic era the reptiles dominated life on the earth. As the Cenozoic era opened, the dinosaurs were extinct and the mammals became the dominant life-form. The Cenozoic is thus called The Age of the Mammals.

You may wonder about the major extinctions of life on the earth during the close of the Paleozoic and Mesozoic eras. There are several theories about what could possibly cause such world-wide extinctions. The terminal Paleozoic extinctions are thought to be the result of plate tectonics: The joining of all the continents into one supercontinent, Pangaea, destroyed much of the suitable shallow-water, maritime, continental shelf area. The cause of the terminal Cretaceous extinctions is debated by geologists. Based on evidence of a thin clay layer marking the boundary between the Cretaceous and Tertiary periods, one theory proposes that a huge (16 km diameter and 10^{15} kg mass) meteorite struck the earth. The impact would have thrown a tremendous amount of dust into the atmosphere, obscuring the sun and significantly changing the climate and thus the conditions of life on the earth. The resulting colder climate may have led to the extinction of many plant and animal species, including the dinosaurs. This theory is based on the clay layer, which theoretically formed as the dust settled, and its location in the rock record at the time of the extinctions. The layer is enriched with a rare metal, iridium, which is not found on the earth in abundance, but occurs in certain meteorites in greater abundance. Perhaps the great extinctions of the past have occurred when heavy bombardments occur every 100 million years or so.

When interpreting the fossil record in any part of the world there are several things to keep in mind:

1. *We are dealing with long periods of time.* The history of life on earth goes back to 3.5 billion years ago, and the evolution of humans took place over a period of several million years. It is important to understand that many of the processes of sedimentation, continental drift, and climate change took place slowly over many millions of years.

2. *The earth has changed greatly over its history.* There have been repeated periods of warming and cooling that resulted in the formation of glaciers. Since glaciers tie up water, they had the effect of lowering sea level, which in turn exposed more land and changed the climate of any particular continent. In addition, the continents were not fixed in position. Changes in position affected the climate that the continent experienced. For example, at one time what is now North America was attached to Antarctica near the South Pole.

3. *There have been many periods in the history of the earth when most of the organisms went extinct.* Cooling climates, changes in sea level, and meteorite impacts are all suspected of causing mass extinctions. However, it should not be thought that these were sudden extinctions. Most took place over millions of years. Some of the extinctions affected as much as 90 percent of the things living at the time.

4. *New forms of life evolved that replaced those that went extinct.* The earliest organisms we see in the fossil record were prokaryotic marine organisms similar to present-day bacteria. The oldest fossils of these organisms date to about 3.5 billion years ago. The next major step was the development of eukaryotic marine organisms about 1.8 billion years ago. The first eukaryotic multicellular organisms were present by about 1 billion years ago. The development of multicellular organisms ultimately led to the colonization of land by plants and animals, with plants colonizing about 500 million years ago and animals at about 450 million years ago.

Table 21.1 Geologic Time and the History of Life (To trace the history of life start with the oldest time period, the Hadean.)

Eon	Era	Period	Epoch	Million Years Ago	Major Physical Events	Major Biological Events
	Cenozoic	Quaternary	Holocene (Recent)	0.01 to present	—Much of the earth has been modified by human activity	—Dominance of modern humans
			Pleistocene	1.8–0.01	—Extinction of many kinds of large mammals in late Pleistocene perhaps due to human activity —Most recent periods of glaciation	—Origin of modern humans —Many kinds of large mammals present in early Pleistocene
		Tertiary	Pliocene	5–1.8	—Cool, dry period —North and South America join —The Indian plate collides with Asia to form the Himalaya Mountains	—Origin of ancestors of humans —Grasslands and grazing mammals widespread —All modern groups of mammals present
			Miocene	23–5	—Warm, moist period —Current arrangement of continents present	—Grasslands and grazers common —Mammals widespread —Marine kelp forests common
			Oligocene	38–23	—Cool period —India, South America, and Australia are separate continents	—Tropics diminish —Woodlands and grasslands expand —Many new grazing mammals
			Eocene	54–38	—Australia and South America separate from Antarctica —North America separates from Europe	—All modern forms of flowering plants present —Forests common —All major groups of mammals present —Grasses found for the first time —First primates
Phanerozoic			Paleocene	65–54	—Warm period	—Many new kinds of mammals —Many new kinds of birds
		Cretaceous		144–65	—Major extinction at the end of Cretaceous affected both land and ocean organisms, probably because of meteorite impact —Warm, moist period —Continents continue to separate	—Dinosaurs dominant —Evolution of birds —Coevolution of flowering plants and their insect pollinators —The three modern mammal groups found
		Jurassic		208–144	—Pangea begins to split up —North and South America split —South America and Africa split —Climate cooler and wetter as Pangea splits	—Dinosaurs dominant —First birds —First flowering plants —Many small primitive mammals
		Triassic		245–208	—Supercontinent, Pangea, stable —Warm, dry climate	—Explosion in reptile diversity —Evolution of modern gymnosperms —Evolution of modern corals —First dinosaurs —First mammals —Mollusks are dominant aquatic invertebrates

464

Eon	Era	Period	Million Years Ago	Major Physical Events	Major Biological Events
		Permian	286–245	—Largest of all extinctions occurred at the end of the Permian (90% of all species) but no clear cause for this extinction —New giant continent of Pangea forms —Much of Pangea was probably desert —Modern levels of oxygen in atmosphere	—Gymnosperms dominant —Many new species of insects —Amphibians and reptiles dominant
		Carboniferous	360–286	—Major coal deposits formed —Two major continents moving toward each other to form Pangea	—Vast swamps of club mosses, horsetails, and ferns common near equator —First winged insects, some very large —Amphibians common —Armored fish extinct —Gymnosperms (cone-bearing plants) and reptiles present by end of period
		Devonian	408–360	—Glaciation probable cause of mass extinction at end of Devonian —Extinction primarily affected warm-water, marine life	—First seed plants (seed ferns) —Both jawless and jawed fish abundant —Coral reefs abundant —Land plants abundant (first forests) —Many new kinds of insects —Amphibians on land
		Silurian	436–408	—Melting of glaciers led to higher sea level	—Numerous coral reefs —Fungi present —First vascular plants —Widespread nonvascular land plants —First land animals—arthropods —First jawed fish —Jawless armored fish abundant
		Ordovician	505–436	—Glaciation caused sea level to drop, which is the probable cause of mass extinction at the end of Ordovician	—Primitive, nonvascular land plants present —Jawless, armored fish common
		Cambrian	540–505	—Rodinia breaks up into pieces —Glaciation probable cause of mass extinction at the end of Cambrian	—All major groups of animals present —Mollusks, brachiopods, echinoderms, trilobites, and nautiloids common
Proterozoic	This period of time is also known as the Precambrian		2,500–540	—Single large supercontinent (Rodinia) exists about 1.1 billion years ago —Oxidation of terrestrial iron deposits forms red beds that indicate presence of atmospheric oxygen (oxidized iron is red)	—Primitive multicellular, soft-bodied marine animals present about 600 million years ago —Algae present about 1 billion years ago —Fossil record poorly preserved —First eukaryotic cells present about 1.8 billion years ago
Archaean			3,800–2,500	—Formation of continents —Atmosphere lacked oxygen	—Fossil stromatolites (layers of bacteria) —Fossil bacteria 3.5 billion years old —Origin of life about 3.8 billion years ago
Hadean			4,500–3,800	—Origin of the earth —No rocks of this age	—Crust of the earth in the process of solidifying

5. *Although there were massive extinctions, there are many examples of the descendants of early life-forms present today.* Bacteria and many kinds of simple eukaryotic organisms are extremely common today as are various kinds of algae and primitive forms of plants. In the oceans many kinds of marine animals such as starfish, jelly fish, and clams are descendants of earlier forms.

6. *The kinds of organisms present have changed the nature of the earth.* The oxygen in the atmosphere is the result of the process of photosynthesis. Its presence has altered the amount of ultraviolet radiation reaching the earth. Plants tend to reduce the erosive effects of running water and humans have significantly changed the surface of the earth.

Table 21.1 lists the various geologic time periods and major events that occurred during those time periods.

PALEONTOLOGY, ARCHAEOLOGY, AND HUMAN EVOLUTION

There is intense curiosity about how our species (*Homo sapiens*) came to be and the evolution of the human species remains an interesting and controversial topic. We recognize that humans show genetic diversity, experience mutations, and are subject to the same evolutionary forces as other organisms. We also recognize that some individuals have genes that make them subject to early death or unable to reproduce. On the other hand, since all of our close evolutionary relatives are extinct, it is difficult for us to visualize our evolutionary development and we tend to think we are unique and not subject to the laws of nature.

We use several kinds of evidence to try to sort out our evolutionary history. Fossils of various kinds of human and prehuman ancestors have been found, but these are often fragmentary and hard to date. Stone tools of various kinds have also been found associated with human and prehuman sites. Finally, other aspects of the culture of our human ancestors have been found in burial sites, cave paintings, and the creation of ceremonial objects. Various methods have been used to age these findings. Some can be dated quite accurately, while others are more difficult to pinpoint.

When fossils are examined, anthropologists can identify differences in the structures of bones that are consistent with changes in species. Based on the amount of change they see and the ages of the fossils, these scientists make judgments about the species to which the fossil belongs. As new discoveries are made, opinions of experts will change and our evolutionary history may become clearer as old ideas are replaced. It is also clear from the fossil record that humans are relatively recent additions to the forms of life. Assembling all of these bits of information into a clear picture is not possible at this point, but several points are well accepted.

1. There is a great deal of fossil evidence that several species of hominids of the genera *Australopithecus* and *Paranthropus* were among the earliest hominid fossils. These organisms are often referred to collectively as australopiths.

2. Based on fossil evidence, it appears that the climate of Africa was becoming drier during the time that hominid evolution was occurring.

3. The earliest *Australopithecus* fossils are from about 4.2 million years ago. Earlier fossils such as *Ardipithecus* may be ancestral to *Australopithecus. Australopithecus* and *Paranthropus* were herbivores and walked upright. Their fossils and the fossils of earlier organisms like *Ardipithecus* are found only in Africa.

4. The australopiths were sexually dimorphic with the males much larger than the females and had relatively small brains (cranial capacity 530 cm^3 or less).

5. Several species of the genus *Homo* became prominent in Africa and appear to have made a change from a primarily herbivorous diet to a carnivorous or omnivorous diet.

6. All members of the genus *Homo* have relatively large brains (cranial capacity 650 cm^3 or more) and are associated with various degrees of stone tool construction and use. It is possible that some of the australopiths may have constructed stone tools.

7. Fossils of several later species of the genus *Homo* are found in Africa, Europe, and Asia, but not in Australia or the Americas. Only *Homo sapiens* is found in Australia and the Americas.

8. Since the fossils of *Homo* species found in Asia and Europe are generally younger than the early *Homo* species found in Africa, it is assumed that they moved to Europe and Asia from Africa.

9. Differences in size between males and females are less prominent in members of the genus *Homo,* so perhaps there were fewer sexual difference in activities than with the australopiths.

When we try to put all of these bits of information together we can construct the following scenario for the evolution of our species. Monkeys, apes, and other primates are adapted to living in forested areas where their grasping hands, opposable thumbs and big toes, and wide range of movement of the shoulders allow them to move freely in the trees. As the climate became dryer, grasslands replaced the forests and, as is always the case, some organisms became extinct while others adapted to the change.

The First Hominids—The Australopiths

Various species of *Australopithecus* and *Paranthropus* were present in Africa from over 4 million years ago until about 1 million years ago. It is important to recognize that there are few fossils of these early humanlike organisms and that often they are fragments of the whole organism. This has led to much speculation and argument among experts about the specific position each fossil has in the evolutionary history of humans. However, from examining the fossil bones of the leg, pelvis, and foot it is apparent that the australopiths were relatively short. Males were 1.5 m (5 ft) or less and females about 1.1 m (3.5 ft). They were stocky and walked upright like humans. Anthropologists recognize several different species of *Australopithecus* and *Paranthropus*.

An upright posture had several advantages in a world that was becoming drier. It allowed for more rapid movement over long distances, the ability to see longer distances, and reduced the amount of heat gained from the sun. In addition, upright posture freed the arms for other uses such as carrying and manipulating objects, and using tools. The various species of *Australopithecus* and *Paranthropus* shared these characteristics and, based on the structure of their skulls, jaws, and teeth, appeared to have been herbivores with relatively small brains.

An ongoing controversy surrounds the relationship between Neandertals and other forms of prehistoric humans. One position is that Neandertals were a small separate race or subspecies of human that could have interbred with other humans and may have disappeared because their subspecies was eliminated by interbreeding with more populous, successful groups. (Many small, remote tribes have been eliminated in the same way in recent history.) Others maintain that the Neandertals show such great difference from other early humans that they must be of different species and became extinct because they could not compete with more successful immigrants from Africa. (The names of these ancient people typically are derived from the place where the fossils were first discovered. For example, Neandertals were first found in the Neander Valley of Germany and Cro-Magnons were initially found in the Cro-Magnon caves in France.)

The use of molecular genetic technology has shed some light on the relationship of Neandertals to other kinds of humans. Examination of the mitochondrial DNA obtained from the bones of a Neandertal individual reveals that there are significant differences between Neandertals and other kinds of early humans. This greatly strengthens the argument that Neandertals were a separate species, *Homo neanderthalensis.*

Later Hominids—the Genus *Homo*

About 2.5 million years ago the first members of the genus *Homo* appeared on the scene. There is considerable disagreement about how many species there were, but *Homo habilis* is one of the earliest. *H. habilis* had a larger brain (650 cm^3) and smaller teeth than australopiths and made much more use of stone tools. Some people feel that it was a direct descendant of *Australopithecus africanus.* Many experts feel that *H. habilis* was a scavenger that made use of group activities, tools, and higher intelligence to commandeer the kills made by other carnivores. The higher-quality diet would have supported the metabolic needs of the larger brain.

About 1.8 million years ago *Homo ergaster* appeared on the scene. It was much larger, up to 1.6 m (6 ft) tall, than *H. habilis* who was about 1.3 m (4 ft) tall and also had a much larger brain (cranial capacity of 850 cm^3). A little later a similar species (*Homo erectus*) appears in the fossil record. Some people consider *H. ergaster* and *H. erectus* to be variations of the same species. The larger brain of *H. ergaster* and *H. erectus* appears to be associated with extensive use of stone tools. Hand axes were manufactured and used to cut the flesh of prey and crush the bones for marrow. These organisms appear to have been predators, whereas *H. habilis* was a scavenger. The use of meat as food allows animals to move about more freely, because appropriate food is available almost everywhere. By contrast, herbivores are often confined to places that have foods appropriate to their use: fruits for fruit eaters, grass for grazers, forests for browsers, etc. In fact, fossils of *H. erectus* have been found in the Middle East and Asia as well as Africa. Most experts feel that *H. erectus* originated in Africa and migrated through the Middle East to Asia.

At about 800,000 years ago another hominid classified as *Homo heidelbergensis* appears in the fossil record. Since fossils of this species are found in Africa, Europe, and Asia, it appears that they constitute a second wave of migration of early *Homo* from Africa to other parts of the world. Both *H. erectus* and *H. heidelbergensis* disappear from the fossil record as two new species (*Homo neanderthalensis* and *Homo sapiens*) become common.

The Neandertals were primarily found in Europe and adjoining parts of Asia and were not found in Africa. Therefore many people feel Neandertals are descendants of *H. heidelbergensis,* which was common in Europe.

The Origin of *Homo sapiens*

Homo sapiens is found throughout the world and is now the only hominid species remaining of a long line of ancestors. There are two different theories that seek to explain the origin of *Homo sapiens.* One theory, knows as the **out-of-Africa hypothesis,** states that modern humans, *H. sapiens,* originated in Africa as had several other hominid species and migrated from Africa to Asia and Europe and displaced species such as *H. erectus* and *H. heidelbergensis* that had migrated into these areas previously. The other theory, known as the **multiregional hypothesis,** states that *H. erectus* evolved into *H. sapiens.* During a period of about 1.7 million years, fossils of *H. erectus* showed a progressive increase in the size of the cranial capacity and reduction in the size of the jaw, so that it becomes difficult to distinguish *H. erectus* from *H. heidelbergensis* and *H. heidelbergensis* from *H. sapiens.* Proponents of this hypothesis believe that *H. heidelbergensis* is not a distinct species but an intermediate between the earlier *H. erectus* and *H. sapiens.* According to this theory, various subgroups of *H. erectus* existed throughout Africa, Asia, and Europe and interbreeding among the various groups gave rise to the various races of humans we see today.

Another continuing puzzle is the relationship of humans that clearly belong to the species *Homo sapiens* with a contemporary group known as Neandertals. Some people consider Neandertals to be a race of *H. sapiens* specially adapted to life in the harsh conditions found in postglacial Europe. Others consider them to be a separate species (*Homo neanderthalensis*). The Neandertals were muscular, had a larger brain capacity than modern humans, and had many elements of culture, including burials. The cause of their disappearance from the fossil record at about 25,000 years ago remains a mystery. Perhaps climate change to a warmer climate was responsible. Perhaps contact with *H. sapiens* resulted in their elimination either through hostile interactions or, if they were able to interbreed with *H. sapiens,* they could have been absorbed into the larger *H. sapiens* population.

Large numbers of fossils of prehistoric humans have been found in all parts of the world. Many of these show evidence of a collective group memory we call *culture.* Cave paintings, carvings in wood and bone, tools of various kinds, and burials are examples. These are also evidence of a capacity to think and invent, and "free time" to devote to things other than gathering

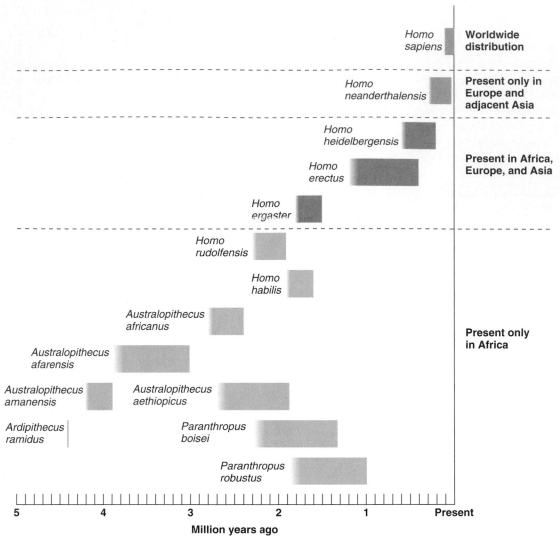

Figure 21.14

This diagram shows the various organisms thought to be relatives of humans. The bars represent approximate times the species are thought to have existed. Notice that (1) all species are extinct today except for modern humans; (2) several different species of organisms coexisted for extensive periods; (3) all the older species are found only in Africa; and (4) more recent species of *Homo* are found in Europe and Asia as well as Africa.

food and other necessities of life. We may never know how we came to be but we will always be curious and will continue to search and speculate about our beginnings. Figure 21.14 summarizes the current knowledge of the historical record of humans and their relatives.

THE CLASSIFICATION OF ORGANISMS

Every day you see a great variety of living things. Just think of how many different species of plants and animals you have observed. Recently, biologists at the Smithsonian Institution estimated that there are over 30 million species in the world; approximately 1.5 million of these have been named. What names do you assign to each? Is the name you use the same as that used in other sections of the country or regions of the world? In much of the United States and Canada, the fish pictured in figure 21.15A is known as a largemouth black bass, but in sections of the southern United States it is called a trout. This use of local names can lead to confusion. If a student in Mississippi writes to a friend in Wisconsin about catching a 6-pound trout, the person in Wisconsin thinks that the friend caught the kind of fish pictured in figure 21.15B. In the scientific community, accuracy is essential; local names cannot be used. When a biologist is writing about a species, all biologists in the world who read that article must know exactly what that species is.

Taxonomy is the science of naming organisms and grouping them into logical categories. Various approaches have been used to classify organisms. The Greek philosopher Aristotle (384–322 B.C.) had an interest in nature and was the first person to attempt a logical classification system. The root word for *taxonomy* is the Greek word *taxis*, which means *arrangement*. Aristotle used the size of plants to divide them into the categories of trees, shrubs, and herbs.

A *Micropterus salmoides* (Largemouth bass)

B *Salmo trutta* (Brown trout)

Figure 21.15

Using the scientific name *Micropterus salmoides* for largemouth black bass (*A*) and *Salmo trutta* for brown trout (*B*) correctly indicates which of these two species of fish a biologist is talking about. Both fish are called trout in some parts of the world.

The modern system of classification began in 1758 when Carolus Linnaeus (1707–1778), a Swedish doctor and botanist, published his tenth edition of *Systema Naturae* (figure 21.16). (Linnaeus's original name was Karl von Linne, which he "latinized" to Carolus Linnaeus.) In the previous editions, Linnaeus had used a polynomial (many-names) Latin system. However, in the tenth edition he introduced the **binomial** (two-name) **system of nomenclature.** This system used two Latin names, genus and specific epithet (epithet = descriptive word), for each species of organism.

Recall that a species is a population of organisms capable of interbreeding and producing fertile offspring. Individual organisms are members of a species. A **genus** (plural, *genera*) is a group of closely related species of organisms; the specific epithet is a word added to the genus name to identify which one of several species within the genus we are discussing. It is similar to the naming system we use with people. When you look in the phone book you look for the last name (surname), which gets you in the correct general category. Then you look for the first name (given name) to identify the individual you wish to call. The unique name given to a particular type of organism is called its species name or scientific name. In order to clearly identify the scientific name, binomial names are either *italicized* or <u>underlined</u>. The first letter of the genus name is capitalized. The specific epithet is always written in lowercase. *Micropterus salmoides* is the binomial name for the largemouth black bass.

When biologists adopted Linnaeus's binomial method, they eliminated the confusion that was the result of using common

Figure 21.16

Carolus Linnaeus (1707–1778), a Swedish doctor and botanist, originated the modern system of taxonomy.

local names. For example, with the binomial system the white water lily is known as *Nymphaea odorata*. Regardless of which of the 245 common names is used in a botanist's local area, when botanists read *Nymphaea odorata,* they know exactly which plant is being referred to. The binomial name cannot be changed unless there is compelling evidence to justify doing so. The rules that govern the worldwide classification and naming of species are expressed in the International Rules for Botanical Nomenclature, the International Rules for Zoological Nomenclature, and the International Bacteriological Code of Nomenclature.

In addition to assigning a specific name to each species, Linnaeus recognized a need for placing organisms into groups. This system divides all forms of life into **kingdoms,** the largest grouping used in the classification of organisms. Originally all organisms were classified into one of two kingdoms, Plantae and Animalia. Subsequently, as people began to recognize that there were major differences among the organisms, they were divided among five kingdoms: Plantae, Animalia, Fungi, Protista (protozoa and algae), and Prokaryotae (bacteria and archaea) (figure 21.17). Today we recognize that the bacteria and archaea are quite different from one another and this has caused a reorganization of the classification system into three domains: Eubacteria, Archaea, and Eucarya. Domains are groupings larger than kingdoms. The original kingdoms of Plantae, Animalia, Fungi, and Protista retained their original status within the domain Eucarya. However, there is no clear understanding at this time about how many kingdoms there are in the Eubacteria and Archaea, and the terms Eubacteria and Archaea are used as kingdom names as well as domain names. Each of the kingdoms is divided into smaller units and given specific names. The taxonomic subdivision under each kingdom is usually called a **phylum,** although microbiologists and botanists replace this term with the word *division.*

A *Streptococcus pyogenes*

Figure 21.17

(*A*) The two prokaryotic domains (Eubacteria and Archaea) are represented by the bacterium *Streptococcus pyogenes;* (*B*) Fungi, by the ascomycete *Morchella esculenta;* (*C*) Protista, by the one-celled *Amoeba proteus;* (*D*) Animalia, by the animal *Homo sapiens;* and (*E*) Plantae, by the tree *Acer saccharum.*

C *Amoeba proteus*

B *Morchella esculenta*

D *Homo sapiens*

E *Acer saccharum*

All kingdoms have more than one phylum. For example, the kingdom Plantae contains several phyla, including flowering plants, conifer trees, mosses, ferns, and several other groups. Organisms are placed in phyla based on careful investigation of the specific nature of their structure, metabolism, and biochemistry. An attempt is made to identify natural groups rather than artificial or haphazard arrangements. For example, while nearly all plants are green and carry on photosynthesis, only flowering plants have flowers and produce seeds; conifers lack flowers but have seeds in cones; ferns lack flowers, cones, and seeds; and mosses lack tissues for transporting water.

A **class** is a subdivision within a phylum. For example, within the phylum Chordata there are seven classes: mammals, birds, reptiles, amphibians, and three classes of fishes. An **order** is a category within a class. Carnivora is an order of meat-eating animals within the class Mammalia. There are several other or-

ders of mammals, including horses and their relatives, cattle and their relatives, rodents, rabbits, bats, seals, whales, and many others. A **family** subdivision of an order consists of a group of closely related genera, which in turn are composed of groups of closely related species. The cat family, Felidae, is a subgrouping of the order Carnivora and includes many species in several genera, including the Canada lynx and bobcat (genus *Lynx*), the cougar (genus *Puma*), the leopard, tiger, jaguar, and lion (genus *Panthera*), the house cat (genus *Felis*), and several other genera. Thus, in the present-day science of taxonomy, each organism that has been classified has its own unique binomial name. In turn, it is assigned to larger groupings that are thought to have a common evolutionary history. Table 21.2 uses the classification of humans to show how the various categories are used.

Phylogeny is the science that explores the evolutionary relationships among organisms and seeks to reconstruct evolu-

Table 21.2 Classification of Humans

Taxonomic Category	Human Classification	Representative Organisms in the Same Category
Domain	Eucarya	All eukaryotic organisms
Kingdom	Animalia	Heterotrophic organisms with specialized tissues that are usually mobile; insects, snails, starfish, worms, snakes, fish, dogs
Phylum	Chordata	Animals with stiffening rod in the back; reptiles, amphibians, birds, fish
Class	Mammalia	Animals with hair and mammary glands; dogs, whales, mice
Order	Primates	Animals with large brains and opposable thumbs; apes, squirrel monkeys, chimpanzees, baboons
Family	Hominidae	Lack tail and have upright posture; humans and extinct relatives (Neandertal)
Genus	*Homo*	Humans are the only surviving member of the genus; although other members of this genus existed in the past (*H. erectus*)
Species	*Homo sapiens*	Humans

tionary history. Taxonomists and phylogenists work together so that the products of their work are compatible. A taxonomic ranking should reflect the evolutionary relationships among the organisms being classified. Although taxonomy and phylogeny are sciences, there is no complete agreement as to how organisms are classified or how they are related. Just as there was dissension two hundred years ago when biologists disagreed on the theories of spontaneous generation and biogenesis, there are differences in opinion about the evolutionary relationships of organisms. People arrive at different conclusions because they use different kinds of evidence or interpret this evidence differently. Phylogenists use several lines of evidence to develop evolutionary histories: fossils, comparative anatomy, life-cycle information, and biochemical/molecular evidence.

Earlier in this chapter fossils were defined as physical evidence of previously existing life. From that earlier discussion you know that there are many different kinds of fossils and they are formed in different ways. Evidence obtained from the discovery and study of fossils allows biologists to place organisms in a time sequence. As geological time passes and new layers of sediment are laid down, the older organisms should be in deeper layers, providing the sequence of layers has not been disturbed. A comparison of the layers gives an indication of the relative age of the fossils found in the rocks. Therefore, fossils found in the same layer must have been alive during the same geological period.

It is also possible to compare subtle changes in particular kinds of fossils over time. For example, the size of the leaf of a specific fossil plant has been found to change extensively through long geological periods. A comparison of the extremes, the oldest with the newest, would lead to their classification into different categories. However, the fossil links between the extremes clearly show that the younger plant is a descendant of the older.

The comparative anatomy of fossils or currently living organisms can be very useful in developing a phylogeny. Since the structures of an organism are determined by its genes and developmental processes, those organisms having similar structures are thought to be related. Plants can be divided into several categories: All plants that have flowers are thought to be more closely related to one another than to plants like ferns, which do not have flowers. In the animal kingdom, all organisms that nurse their young from mammary glands are grouped together, and all animals in the bird category have feathers and beaks and lay eggs with shells. Reptiles also have shelled eggs but differ from birds in that reptiles lack feathers and have scales covering their bodies. The fact that these two groups share this fundamental eggshell characteristic implies that they are more closely related to each other than they are to other groups.

Another line of evidence useful to phylogenists and taxonomists comes from the field of developmental biology. Many organisms have complex life cycles that include many completely different stages. After fertilization, some organisms grow into free-living developmental stages that do not resemble the adults of their species. These are called *larvae* (singular, *larva*). Larval stages often provide clues to the relatedness of organisms. For example, barnacles live attached to rocks and other solid marine objects and look like small, hard cones. Their outward appearance does not suggest that they are related to shrimp; however, the larval stages of barnacles and shrimp are very similar. Detailed anatomical studies of barnacles confirm that they share many structures with shrimp; their outward appearance tends to be misleading (figure 21.18). This same kind of evidence is available in the plant kingdom. Many kinds of plants, such as peas, peanuts, and lima beans, produce large, two-parted seeds in pods (you can easily split the seeds into two parts). Even though peas grow as vines, lima beans grow as bushes, and peanuts have their seeds underground, all these plants are considered to be related.

Like all aspects of biology, the science of taxonomy is constantly changing as new techniques develop. Recent advances in DNA analysis are being used to determine genetic similarities among species. In the field of ornithology, which deals with the study of birds, there are those who believe that storks and flamingos are closely related; others believe that flamingos are more closely related to geese. An analysis of the DNA points to a higher degree of affinity between flamingos and storks than between flamingos and geese. This is interpreted to mean that

A Barnacle

B Shrimp

Figure 21.18

The adult barnacle (*A*) and shrimp (*B*) are very different from each other, but the early larval stages look very much alike.

the closest relationship is between flamingos and storks. Algae and plants have several different kinds of chlorophyll: chlorophyll *a, b, c, d,* and *e.* Most photosynthetic organisms contain a combination of two of these chlorophyll molecules. Members of the kingdom Plantae have chlorophyll *a* and *b.* The large seaweeds, like kelp, superficially resemble terrestrial plants like trees and shrubs. However, a comparison of the chlorophylls present shows that kelp has chlorophyll *a* and *d.* When another group of algae, called the *green algae,* are examined, they are found to have chlorophyll *a* and *b.* Along with other anatomical and developmental evidence, this biochemical information has helped to establish an evolutionary link between the green algae and plants. All of these kinds of evidence (fossils, comparative anatomy, developmental stages, and biochemical evidence) have been used to develop the various taxonomic categories, including kingdoms.

Given all these sources of evidence, biologists have developed a hypothetical picture of how all organisms are related (figure 21.19). At the base of this evolutionary scheme is the biochemical evolution of cells. These first cells are thought to be the origin of all organisms. While protocells no longer exist, their descendants have diversified over millions of years. Of these groups, the Eubacteria and Archaea have the simplest structure and are probably most similar to some of the first cellular organisms on earth.

Domains Eubacteria and Archaea

Members of the domains Archaea and Eubacteria are commonly known as bacteria. Some are disease-causing, such as *Streptococcus pneumoniae,* but most are not. In addition, many are able to photosynthesize while others live in extremely hostile environments. Members of these domains differ from one another in their cellular structures and chemistry. Although they are both prokaryotic, the members of the two domains are very different from one another and have been assigned specific positions on the evolutionary tree. Evidence gained from studying DNA and RNA nucleotide sequences and a comparison of the amino acid sequences of proteins indicates that the first ancesters probably gave rise to today's Eubacteria followed by the Archaea and finally the Eucarya.

Eubacteria

The "true bacteria" (*eu* = true) are small, prokaryotic, single-celled organisms. Their cell walls typically contain complex organic molecules, such as peptidoglycan, polymers of unique sugars, and unusual amino acids not found in other kinds of organisms. These characteristics are used to distinguish them from the Archaea and Eucarya.

There are thousands of species of Eubacteria that can be divided into groups based on their metabolic abilities. The cyanobacteria (blue-green bacteria) were probably the first organisms to carry on photosynthesis. They contain a blue-green pigment that allows them to capture sunlight and carry on a kind of photosynthesis. These bacteria are extremely common in marine and freshwater and contribute significantly to the production of oxygen in the atmosphere. Some bacteria have abilities similar to some of the Archaea in that they are able to use inorganic chemical reactions to provide the energy needed to build new organic molecules.

A large number of Eubacteria are heterotrophs that are important decomposers of dead organic matter. A few kinds are

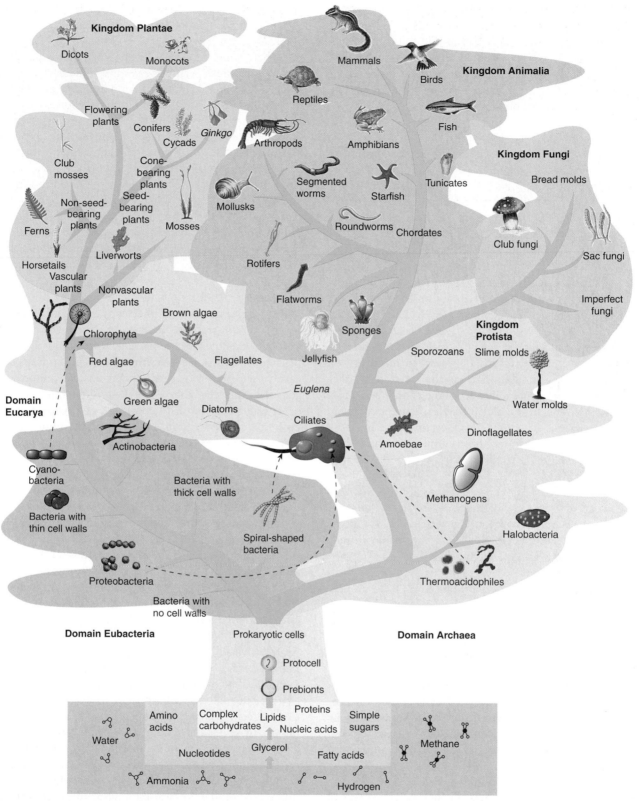

Kingdom Plantae

Dicots

Monocots

Flowering plants

Conifers

Cycads

Ginkgo

Club mosses

Cone-bearing plants

Seed-bearing plants

Non-seed-bearing plants

Ferns

Liverworts

Horsetails

Vascular plants

Nonvascular plants

Mosses

Brown algae

Chlorophyta

Red algae

Green algae

Diatoms

Flagellates

Actinobacteria

Domain Eucarya

Cyano-bacteria

Bacteria with thin cell walls

Bacteria with thick cell walls

Spiral-shaped bacteria

Proteobacteria

Bacteria with no cell walls

Domain Eubacteria

Prokaryotic cells

Mammals

Reptiles

Kingdom Animalia

Birds

Arthropods

Amphibians

Fish

Segmented worms

Starfish

Mollusks

Roundworms

Tunicates

Chordates

Rotifers

Flatworms

Sponges

Jellyfish

Euglena

Ciliates

Amoebae

Methanogens

Kingdom Fungi

Bread molds

Club fungi

Sac fungi

Imperfect fungi

Kingdom Protista

Sporozoans

Slime molds

Water molds

Dinoflagellates

Halobacteria

Thermoacidophiles

Domain Archaea

Protocell

Prebionts

Amino acids

Complex carbohydrates

Lipids

Proteins

Simple sugars

Water

Nucleic acids

Nucleotides

Glycerol

Fatty acids

Methane

Ammonia

Hydrogen

Figure 21.19

The theory of chemical evolution proposes that the molecules in the early atmosphere and early oceans accumulated to form prebionts—nonliving structures composed of carbohydrates, proteins, lipids, and nucleic acids. The prebionts are believed to have been the forerunners of the protocells—the first living cells. These protocells probably evolved into prokaryotic cells, on which the domains Eubacteria and Archaea are based. Some prokaryotic cells probably gave rise to eukaryotic cells. The organisms formed from these early eukaryotic cells were probably similar to members of the kingdom Protista. Members of this kingdom are thought to have given rise to the kingdoms Animalia, Plantae, and Fungi. Thus, all present-day organisms evolved from the protocells.

parasites that live on the tissues of other organisms and cause disease. In addition, heterotrophic bacteria can be differentiated based on their ability to use oxygen. Some are aerobic (use oxygen to break down organic food molecules), while others are anaerobic (do not use oxygen.) Since the early atmosphere is thought to have been a reducing atmosphere, the first Eubacteria were probably anaerobic organisms.

Archaea

The term Archaea comes from the Greek term *archaios* meaning "ancient." This group of prokaryotic cells is thought to have branched off from earlier prokaryotic organisms in the neighborhood of 1.3 to 2.6 million years ago. Since the Eubacteria are actually older as a group than the Archaea, the naming of this group may seem confusing. However, the Archaea have many chemical similarities to the Eucarya. These bacteria have been found in many shapes including rods, spheres, spirals, filaments, and flat plates. Because they are found in many kinds of extreme environments, they have become known as extremophiles. Based on this fact, the Archaea are divided into three groups. *Methanogens* are methane-producing bacteria that are anaerobic. They can be found in the intestinal tracts of humans, sewage, and swamps. *Halobacteria* (*halo* = salt) are found growing in very salty environments such as the Great Salt Lake (Utah), salt ponds, and brine solutions. Some contain a special kind of chlorophyll and are therefore capable of generating their ATP by photosynthesis. The *thermophilic* Archaea live in environments that normally have very high temperatures and high concentrations of sulfur (e.g., hot sulfur springs or around deep-sea hydrothermal vents). Over 500 species of thermophiles have been identified at the openings of hydrothermal vents in the ocean. One such thermophile, *Pyrolobus fumarii*, grows in a hot spring in Yellowstone National Park. Its maximum growth temperature is 113°C, its optimum is 106°C, and its minimum is 90°C.

Domain Eucarya

Some biologists hypothesize that eukaryotic cells evolved from prokaryotic cells by a process of endosymbiosis. This hypothesis proposes that structures like mitochondria, chloroplasts, and other membranous organelles originated from separate cells that were ingested by larger, more primitive cells. Once inside, these structures and their functions became integrated with the host cell and ultimately became essential to its survival. This new type of cell was the forerunner of present-day eukaryotic cells. Single-celled eukaryotic organisms are members of the kingdom Protista.

Kingdom Protista

The changes in cell structure that led to eukaryotic organisms most probably gave rise to single-celled organisms similar to those currently grouped in the kingdom Protista. Most members of this kingdom are one-celled organisms, although there are some colonial forms. Eukaryotic cells are usually much larger than the prokaryotes, typically having more than a thousand times the volume of prokaryotic cells. Their larger size was made possible by the presence of specialized membranous organelles, such as mitochondria, the endoplasmic reticulum, chloroplasts, and nuclei.

There is a great deal of diversity within the 60,000 known species of Protista. Many species live in freshwater; others are found in marine or terrestrial habitats. Some are parasitic and live on the tissues of other living things. Some are *commensal* organisms that live in another organisms without causing harm, and some are *mutualistic* organisms that live in partnership with another organism in a relationship in which both organisms benefit. All species can undergo mitosis, resulting in asexual reproduction. Some species can also produce sex cells and reproduce sexually. Many contain chlorophyll in chloroplasts and are autotrophic; others require organic molecules as a source of energy and are heterotrophic. Both autotrophs and heterotrophs have mitochondria and respire aerobically. Some specialized parasitic types lack mitochondria and other cellular structures.

Because members of this kingdom are so diverse with respect to form, metabolism, and reproductive methods, most biologists do not feel that the Protista form a valid phylogenetic unit. However, it is still a convenient taxonomic grouping. By placing these organisms together in this group it is possible to gain a useful perspective on how they relate to other kinds of organisms. After the origin of eukaryotic organisms, evolution proceeded along several different pathways. Three major lines of evolution can be seen today in the plantlike autotrophs (algae), animal-like heterotrophs (protozoa), and the funguslike heterotrophs (slime molds). *Amoeba* and *Paramecium* are commonly encountered examples of protozoa. Many seaweeds and pond scums are collections of large numbers of algal cells. Slime molds are less frequently seen because they live in and on the soil in moist habitats; they are most often encountered as slimy masses on decaying logs.

Through the process of evolution, the plantlike autotrophs probably gave rise to the kingdom Plantae, the animal-like heterotrophs probably gave rise to the kingdom Animalia, and the funguslike heterotrophs were probably the forerunners of the kingdom Fungi.

Kingdom Fungi

Fungus is the common name for members of the kingdom Fungi. The majority of fungi are not able to move about. They have a rigid, thin cell wall, which in most species is composed of chitin, a complex carbohydrate containing nitrogen. This is an important diagnostic feature, since plants have cell walls of cellulose. Members of the kingdom Fungi are nonphotosynthetic, eukaryotic organisms. The majority (mushrooms and molds) are multicellular, but a few, like yeasts, are single-celled. In the multicellular fungi the basic structural unit is a network of multicellular filaments. Because all of these organisms are heterotrophs, they must obtain nutrients from organic sources. Most secrete enzymes that digest large molecules into smaller units that are absorbed. They are very important as decomposers in all ecosystems. They feed on a variety of nutrients ranging from dead organisms to such products as shoes, foodstuffs, and clothing. Most synthetic organic molecules are not attacked as readily by fungi; this is one reason plastic bags, foam cups, and organic pesticides are slow to decompose.

Some fungi are parasitic and others are mutualistic. Many of the parasitic fungi are important plant pests. Some attack and kill plants (chestnut blight, Dutch elm disease); others injure the fruit, leaves, roots, or stems and reduce yields. The fungi that are human parasites are responsible for athlete's foot, vaginal yeast infections, valley fever, "ringworm," and other diseases. Mutualistic fungi are important in lichens and in combination with the roots of certain kinds of plants.

Kingdom Plantae

Another major group that has its roots in the kingdom Protista is the green, photosynthetic plants. The ancestors of plants were most likely specific kinds of algae commonly called *green algae*. Members of the kingdom Plantae are nonmotile, terrestrial, multicellular organisms that contain chlorophyll and produce their own organic compounds. All plant cells have a cellulose cell wall. Over 300,000 species of plants have been classified; about 85 percent are flowering plants, 14 percent are mosses and ferns, and the remaining 1 percent are cone-bearers and several other small groups within the kingdom.

A wide variety of plants exist on the earth today. Members of the kingdom Plantae range from simple mosses to vascular plants with stems, roots, leaves, and flowers. Most biologists feel that the evolution of this kingdom began about 400 million years ago. The nonvascular plants like the mosses evolved as one type of plant and the vascular plants like the ferns evolved as a second type (figure 21.20). Some of the vascular plants evolved into seed-producing plants, which today are the cone-bearing and flowering plants, while the ferns lack seeds. The development of vascular plants was a major step in the evolution of plants from an aquatic to a terrestrial environment.

Plants have a unique life cycle. There is a gametophyte stage that produces sex cells by mitosis. There is also a sporophyte stage that produces spores by meiosis (figure 21.21). In addition to sexual reproduction, plants are able to reproduce asexually.

Kingdom Animalia

Like fungi and plants, animals are thought to have evolved from the Protista. Over a million species of animals have been classified.

Figure 21.20

Two lines of plants are thought to have evolved from the plantlike Protista, the algae. The nonvascular mosses evolved as one type of plant. The second type, the vascular plants, evolved into the seed and nonseed plants.

Several organisms have been suggested as record holders for the title of oldest and largest organisms. Several of these are plants. The bristlecone pine (*Pinus longaeva*) in the White Mountains of California have been aged to over 5,000 years. Creosote bush (*Larrea divaricata*) forms clones that grow out from the center as the central portion of the plant dies. Several clones of creosote bush in the Mojave Desert have been estimated to have an age of 12,000 years. The title of the largest organism can be determined in several ways. The Giant Redwood (*Sequoiadendron giganteum*) is the tree with the single largest stem. The General Sherman tree probably weighs about 1,400 tons. However, a clone of trembling aspen (*Populus tremuloides*) consists of many individual stems that are probably all joined together by roots. One such clone in the Rocky Mountains covers about 0.4 square kilometer and probably weighs about 6,000 tons. However, it may be a clone of a fungus (*Armillaria*) that lives in the soil that holds the record for the largest organism. A clone in the state of Washington is estimated to cover about 3 square kilometers.

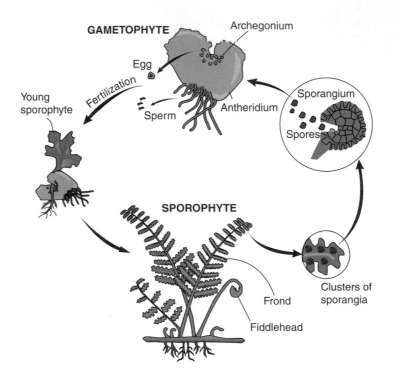

Figure 21.21

Plants have a multicelluar structure (sporophyte generation) that undergoes meiosis to form spores—haploid cells. These spores give rise to haploid organisms (gametophyte generation) that form the gametes—sperm cells and egg cells. The union of these gametes forms the diploid stage, which develops into the sporophyte. In this fern life cycle the large leafy part of the life cycle is the sporophyte generation and the gametophyte is a tiny structure. All plants show this alternation of sporophyte and gametophyte generations but the details are different for each major group of plants.

These range from microscopic types, like mites or aquatic larvae of marine animals, to huge animals like elephants or whales. Regardless of their type, all animals have some common traits. They all are composed of eukaryotic cells, and all species are heterotrophic and multicellular. Most animals are motile; however, some, like the sponges, barnacles, mussels, and corals are sessile (not able to move). All animals are capable of sexual reproduction, but many of the less complex animals are also able to reproduce asexually.

It is thought that animals originated from certain kinds of Protista that had flagella. This idea proposes that colonies of flagellated Protista gave rise to simple multicellular forms of animals like the ancestors of present-day sponges. These first animals lacked specialized tissues and organs. As cells became more specialized, organisms developed special organs and systems of organs and the variety of kinds of animals increased.

ACELLULAR INFECTIOUS PARTICLES

All of the groups discussed so far fall under the category of cellular forms of life. They all have at least the following features in common: They have (a) cell membranes, (b) nucleic acids as their genetic material, (c) cytoplasm, (d) enzymes and coenzymes, (e) ribosomes, and (f) use ATP as their source of chemical-bond energy. Since the three groups to follow lack this cellular organization, they are referred to as *acellular* (the prefix *a* = lacking) or are known as infectious particles. In order for these to make more of their own kind, they must make their way into true cells where they become parasites eventually causing harm or death to their host cells. Since evolutionary biologists can only speculate on the origin of acellular infectious particles, they are not classified using the same methods outlined previously. However, there are three generally recognized kinds of acellular infectious particles known as *viruses, viroids,* and *prions.*

Viruses

Although viruses lack all cellular structures, they still show some characteristics of life. In fact, some scientists consider them to be highly specialized parasites that have lost some of their complexity as they developed as parasites. Others consider them to be the simplest of living organisms. Some even consider them to be nonliving. For these reasons viruses are considered separately from the six kingdoms.

A **virus** consists of a nucleic acid core surrounded by a coat of protein (figure 21.22). Viruses are **obligate intracellular parasites,** which means they are infectious particles that can function only when inside a living cell. Due to their unusual characteristics, viruses are not a member of any kingdom. Biologists do not consider them to be living because they are not capable of living and reproducing by themselves and show the characteristics of life only when inside living cells.

Viruses are typically host-specific, which means that they usually attack only one kind of cell. The host is a specific kind of cell that provides what the virus needs to function. Viruses can infect only those cells that have the proper receptor sites to which the virus can attach. This site is usually a glycoprotein molecule on the surface of the cell membrane. For example, the virus responsible for measles attaches to the membranes of skin cells, hepatitis viruses attach to liver cells, and mumps viruses attach to cells in the salivary glands. Host cells for the HIV virus include some types of human brain cells and several types belonging to the immune system.

Once it has attached to the host cell, the virus either enters the cell or injects its nucleic acid into the cell. If it enters the cell, the virus loses its protein coat, releasing the nucleic acid. Once released into the cell, the nucleic acid of the virus may remain free in the cell or it may link with the host's genetic material. Some viruses contain as few as three genes, others contain as many as five hundred. A typical eukaryotic cell contains tens of thousands of genes. Most viruses need only a small number of genes since they rely on the host to perform most of the activities necessary for viral reproduction.

Some viruses have DNA as their genetic material but many have RNA. Viral genes are able to take command of the host's metabolic pathways and direct it to carry out the work of making new copies of the original virus. The virus makes use of the host's available enzymes and ATP for this purpose. When enough new viral nucleic acid and protein coat are produced, complete virus particles are assembled and released from the host (figure 21.23). The number of viruses released ranges from ten to thousands. The virus that causes polio releases about 10,000 new virus particles from each human host cell. Some viruses remain in cells and are only occasionally triggered to reproduce, causing symptoms of disease. Herpes viruses, which cause cold sores, genital herpes, and shingles, reside in nerve cells.

Almost all of the species in the six kingdoms serve as hosts to some form of virus (table 21.3).

Figure 21.22

Viruses consist of a core of nucleic acid, either DNA or RNA depending on the kind of virus. Some are surrounded by an envelope.

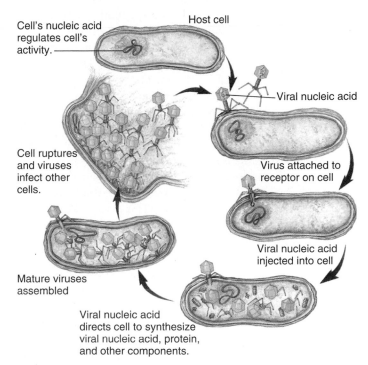

Figure 21.23

The viral nucleic acid takes control of the activities of the host cell. Because the virus has no functional organelles of its own, it can become metabolically active only while it is within a host cell.

Table 21.3 Viral Diseases	
Type of Virus	**Disease**
Papovaviruses	Warts in humans
Paramyxoviruses	Mumps and measles in humans; distemper in dogs
Adenoviruses	Respiratory infections in most mammals
Poxviruses	Smallpox
Wound-tumor viruses	Diseases in corn and rice
Potexviruses	Potato diseases
Bacteriophage	Infections in many types of bacteria

Viroids: Infectious RNA

Viroids are infectious particles that are composed solely of a small, single strand of RNA. To date no viroids have been found that parasitize animals. The hosts in which they have been found are cultivated crop plants such as potatoes, tomatoes, and cucumbers. Viroid infections result in stunted or distorted growth and may sometimes cause the plant to die. Pollen, seeds, or farm machinery can transmit viroids from one plant to another. Some scientists believe that viroids may be parts of normal RNA that have gone wrong.

Prions: Infectious Proteins

Several kinds of brain diseases appear to be caused by proteins that can be passed from one individual to another. These infectious agents are called **prions**. All the diseases of this type currently known cause changes in the brain that result in a spongy appearance called spongiform encephalopathies. The symptoms typically involve abnormal behavior and eventually death. In animals the most common examples are scrapie in sheep and goats and mad cow disease in cattle. Scrapie got its name because one of the symptoms of the disease is an itching of the skin associated with nerve damage that causes the animals to rub against objects and scrape off their hair.

The occurrence of mad cow disease (BSE—bovine spongiform encephalitis) in Great Britain was apparently caused by the spread of prions from sheep to cattle. This occurred because of the practice of processing unusable parts of sheep carcasses into a protein supplement that was fed to cattle. Other similar diseases are known from mink, cats, dogs, elk, and deer. A human form of this disease is called Creutzfeldt-Jakob disease (CJD). It now appears that the original form of BSE has changed to a variety that is able to infect humans. This new form is called vCJD, which makes scientists believe that BSE and CJD (Creutzfeldt-Jakob disease) are in fact the same prion.

Now it is well-established that these proteins can be spread from one animal to another. The multiplication of the prions appears to result from the disease-causing prion protein coming in contact with a normal body protein and converting it into the disease-causing form.

SUMMARY

Geologic time is measured through the radioactive decay process, determining the *radiometric age* of rocks in years. A *geomagnetic time scale* has been established from the number and duration of reversals in the magnetic field of the earth's past.

Correlation and the determination of the numerical ages of rocks and events have led to the development of a *geologic time scale*. The major blocks of time on this calendar are called eons. The Hadean eon, 3.8 to 4.5 billion years ago, is the period of time before the earth's surface solidified. The Archean eon, 2.5 to 3.8 billion years ago, is a time when the atmosphere lacked oxygen and only certain kinds of prokaryotic organisms existed. The Proterozoic eon, 2.5 billion to 540 million years ago, saw the development of an oxidizing atmosphere, the first eukaryotic cells, and the first multicellular organisms. The time encompassed

by the Hadean, Archaean, and Proterozoic is also referred to as the Precambrian. The Phanerozoic eon, 540 million years ago to the present, is a period of time that has seen the elaboration of the various kinds of living things we see today as well as many groups that have gone extinct. The Phanerozoic eon is divided into smaller units of time known as *eras*. The eras are the (1) *Cenozoic*, the time of recent life; (2) *Mesozoic*, the time of middle life; and (3) *Paleozoic*, the time of ancient life. The eras are divided into smaller blocks of time called *periods*, and the periods are further subdivided into *epochs*.

There are many different kinds of *fossils*—evidence of former living things. They are formed when organisms are covered over and prevented from being destroyed. Most are modified chemically but still give information about past living things. The fossil record is seen to change over geologic time with certain kinds of fossils being associated with certain periods of time. There have been several *great extinctions* of living things. Some of the extinctions eliminated more than 80 percent of the organisms alive at the time. These extinctions appear to be related to changes in climate that may have been initiated by changes in the location and arrangement of continents, volcanic activity, or meteorite impact.

The early evolution of humans has been difficult to piece together because of the fragmentary evidence. Beginning about 4.4 million years ago the earliest forms of *Australopithecus* and *Paranthropus* showed upright posture and other humanlike characteristics. The structure of the jaw and teeth indicates that the various kinds of australopiths were herbivores. *Homo habilis* had a larger brain and appears to have been a scavenger. Several other species of the genus *Homo* arose in Africa. These forms appear to have been carnivores. Some of these migrated to Europe and Asia. The origin of *Homo sapiens* is in dispute. It may have arisen in Africa and migrated throughout the world or evolved from earlier ancestors found throughout Africa, Asia, and Europe.

To facilitate accurate *communication*, biologists assign a *specific name* to each *species* that is cataloged. The various species are cataloged into larger groups on the basis of similar traits. *Taxonomy* is the science of classifying and naming organisms. *Phylogeny* is the science of trying to figure out the evolutionary history of a particular organism. The *taxonomic ranking of organisms* reflects their *evolutionary relationships*. *Fossil evidence, comparative anatomy, developmental stages*, and *biochemical evidence* are employed in the sciences of taxonomy and phylogeny.

The first organisms thought to have evolved were single-celled, prokaryotic organisms. These simple organisms gave rise to two different prokaryotic domains: the Archaea and the Eubacteria. These subsequently gave rise to the eukaryotic organisms in the domain Eucarya and established the kingdoms *Protista, Fungi, Plantae*, and *Animalia*.

KEY TERMS

order (p. **470**)
out-of-Africa hypothesis (p. **467**)
paleontology (p. **458**)
parasite (p. **474**)
period (p. **462**)
phylogeny (p. **470**)

phylum (p. **469**)
prions (p. **478**)
taxonomy (p. **468**)
unconformity (p. **456**)
viroids (p. **478**)
virus (p. **477**)

APPLYING THE CONCEPTS

1. Some of the oldest fossils are about how many years old?
   (a) 4.55 billion
   (b) 3.5 billion
   (c) 250 million
   (d) 10,000
2. A fossil of a jellyfish, if found, would most likely be
   (a) a preserved fossil.
   (b) one formed by mineralization.
   (c) a carbon film.
   (d) a cast or a mold.
3. In any sequence of sedimentary rock layers that has not been subjected to stresses, you would expect to find
   (a) essentially horizontal stratified layers.
   (b) the oldest layers at the bottom and the youngest at the top.
   (c) younger faults, folds, and intrusions in the rock layers.
   (d) all of the above.
4. You would expect to find the least number of fossils in rocks from which geologic time?
   (a) Cenozoic
   (b) Mesozoic
   (c) Paleozoic
   (d) Precambrian
5. The numerical dates associated with events on the geologic time scale were determined by
   (a) relative dating of the rate of sediment deposition.
   (b) radiometric dating using radioactive decay.
   (c) the temperature of the earth.
   (d) the rate that salt is being added to the ocean.
6. The three main kinds of living things, Eubacteria, Archaea, and Eucarya, have been labeled as
   (a) species.
   (b) kingdoms.
   (c) domains.
   (d) families.
7. Which is commonly used as sources of information when developing a phylogeny?
   (a) fossil evidence
   (b) biochemical information
   (c) embryological development
   (d) All of the above are correct.
8. Which of the following kingdoms contain members that are autotrophs?
   (a) Protista
   (b) Plantae
   (c) Eubacteria
   (d) All of the above are correct.

9. Alternation of generations is a characteristic of
   (a) Animalia.
   (b) Eubacteria.
   (c) Fungi.
   (d) Plantae.
10. Viruses
    (a) can be free-living or parasitic.
    (b) belong to the domain Archaea.
    (c) can function and reproduce only inside a host cell.
    (d) All of the above are correct.
11. In the evolution of humans, the fossil evidence suggests that
    (a) the earliest humanlike primates were present in Africa, Europe, and North America.
    (b) *Homo sapiens* has been around for about 2 million years.
    (c) there were several different species of human ancestors in Africa.
    (d) large numbers of fossils clearly show the way humans evolved.
12. *Homo sapiens* is the only member of its genus today,
    (a) however, at one time there were several members of the genus.
    (b) and the only other member of the genus is *Homo neanderthalensis*.
    (c) because *Homo neanderthalensis* evolved into *Homo sapiens*.
    (d) None of the above are true.

Answers
1. b 2. c 3. d 4. d 5. b 6. c 7. d 8. d 9. d 10. c 11. c 12. a

QUESTIONS FOR THOUGHT

1. Why does the rock record go back only 3.8 million years? If this missing record were available, what do you think it would show? Explain.
2. What major event marked the end of the Paleozoic and Mesozoic eras according to the fossil record? Describe one theory that proposes to account for this.
3. Describe how the principles of superposition, horizontality, and faunal succession are used in the relative dating of sedimentary rock layers.
4. Describe four different kinds of fossils and how they were formed.
5. Describe four things that fossils can tell you about the earth's history.
6. What are some of the major steps thought to have been involved in the evolution of humans?
7. How do the domains Eubacteria and Archaea differ?
8. What is the value of taxonomy?
9. How do viruses reproduce?
10. Why are Latin names used for genus and species?

Ecology and Environment

Even organisms of the same species affect one another in the course of their normal daily activities.

O U T L I N E

Today we recognize that environmental problems are a worldwide concern. Poor agricultural practices in Africa and China result in dust storms that affect the local people and are also carried to the rest of the world. Concern about global warming has led to the convening of international conferences and the drafting of treaties whose goal is to reduce the impact of energy use. Many are concerned about protecting endangered species and the forests, grasslands, and oceans that provide habitats for them. In many ways these problems are the result of an incredible increase in the size of the human population. As the human population has increased, humans have sought to use land for agriculture and other purposes. Converting land to agricultural use results in its loss for other purposes such as habitat and watershed protection. In order to understand the nature of environmental problems and steps we can take to solve them, we need to be familiar with some of the central ideas of the science of ecology.

A DEFINITION OF ENVIRONMENT

Ecology is the branch of biology that studies the relationships between organisms and their environments. This is a very simple definition for a very complex branch of science. Most ecologists define the word **environment** very broadly as anything that affects an organism during its lifetime. Environmental influences can be divided into two categories: **biotic factors** are other living things that affect an organism and **abiotic factors** are nonliving influences that affect an organism (Figure 22.1). If we consider a fish in a stream, we can identify many environmental factors that are important to its life. The temperature of the water is extremely important as an abiotic factor that may be influenced by the presence of trees (biotic factor) along the stream bank that shade the stream and prevent the sun from heating it. Obviously, the kind and number of food organisms in the stream are important biotic factors as well. The type of material that makes up the stream bottom and the amount of oxygen dissolved in the water are other important abiotic factors, both of which are related to how rapidly the water is flowing.

Characterizing the environment of an organism is a complex and challenging process; everything seems to be influenced or modified by other factors. A plant is influenced by many different factors during its lifetime: the types and amounts of minerals in the soil, the amount of sunlight hitting the plant, the animals that eat the plant, and the wind, water, and temperature. Each item on this list can be further subdivided into other areas of study. For instance, water is important in the life of plants, so rainfall is studied in plant ecology. But even the study of rainfall is not simple. The rain could come during one part of the year, or it could be evenly distributed throughout the year. The rainfall could be hard and driving, or it could come as gentle, misty showers of long duration. The water could soak into the soil for later use or it could run off into streams and be carried away.

The animals in an area are influenced as much by abiotic factors as are the plants. If nonliving factors do not favor the growth of plants, there will be little food and few hiding places for animal life. Two types of areas that support only small numbers of living animals are deserts and polar regions. Near the polar regions of the earth, the low temperature and short growing season inhibit growth; therefore, there are relatively few species of animals with relatively small numbers of individuals. Deserts receive little rainfall and therefore have poor plant growth and low concentrations of animals. On the other hand, tropical rain-

A   B

Figure 22.1

(*A*) The woodpecker feeding its young in the hole in this tree is influenced by several biotic factors. The tree itself is a biotic factor as is the disease that weakened it, causing conditions that allowed the woodpecker to make a hole in the rotting wood. (*B*) The irregular shape of the trees is the result of wind and snow, both abiotic factors. Snow driven by the prevailing winds tends to "sandblast" one side of the tree and prevent limb growth.

forests have high rates of plant growth and large numbers of animals of many kinds.

As you can see, abiotic and biotic factors are interrelated. Ecologists have developed several concepts that help us understand how biotic and abiotic factors interrelate in a complex system.

THE ORGANIZATION OF ECOLOGICAL SYSTEMS

Ecologists can study ecological relationships at several different levels. The smallest living unit is the individual **organism.** Groups of organisms of the same species are called **populations.**

Interacting groups of populations of different species are called **communities.** An **ecosystem** consists of all the interact-

Figure 22.2
The same organism can be looked at from several different perspectives. We can study it as an individual, as a member of a population, or as a participant in a community or ecosystem.

ing organisms in an area and their interactions with their abiotic surroundings. Figure 22.2 shows how these different levels of organization are related to one another.

All living things require a continuous supply of energy to maintain life. Therefore, many people like to organize living systems by the energy relationships that exist among the different kinds of organisms present. An ecosystem contains several different kinds of organisms. Those that trap sunlight for photosynthesis, resulting in the production of organic material from inorganic material, are called **producers.** Green plants and other photosynthetic organisms such as algae and cyanobacteria are, in effect, converting sunlight energy into the energy contained within the chemical bonds of organic compounds. There is a flow of energy from the sun into the living matter of plants.

The energy that plants trap can be transferred through a number of other organisms in the ecosystem. Since all of these organisms must obtain energy in the form of organic matter, they are called **consumers.** Consumers cannot capture energy from the sun as plants do. All animals are consumers. They either eat plants directly or eat other sources of organic matter derived from plants. Each time the energy enters a different organism, it is said to enter a different **trophic level,** which is a step, or stage, in the flow of energy through an ecosystem (figure

22.3). The plants (producers) receive their energy directly from the sun and are said to occupy the *first trophic level.*

Various kinds of consumers can be divided into several categories, depending on how they fit into the flow of energy through an ecosystem. Animals that feed directly on plants are called **herbivores,** or **primary consumers,** and occupy the *second trophic level.* Animals that eat other animals are called **carnivores,** or **secondary consumers,** and can be subdivided into different trophic levels depending on what animals they eat. Animals that feed on herbivores occupy the *third trophic level* and are known as **primary carnivores.** Animals that feed on the primary carnivores are known as **secondary carnivores** and occupy the *fourth trophic level.* For example, a human may eat a fish that ate a frog that ate a spider that ate an insect that consumed plants for food.

This sequence of organisms feeding on one another is known as a **food chain.** Figure 22.4 shows the six different trophic levels in this food chain. Obviously, there can be higher categories, and some organisms don't fit neatly into this theoretical scheme. Some animals are carnivores at some times and herbivores at others; they are called **omnivores.** They are classified into different trophic levels depending on what they happen to be eating at the moment. If an organism dies, the energy

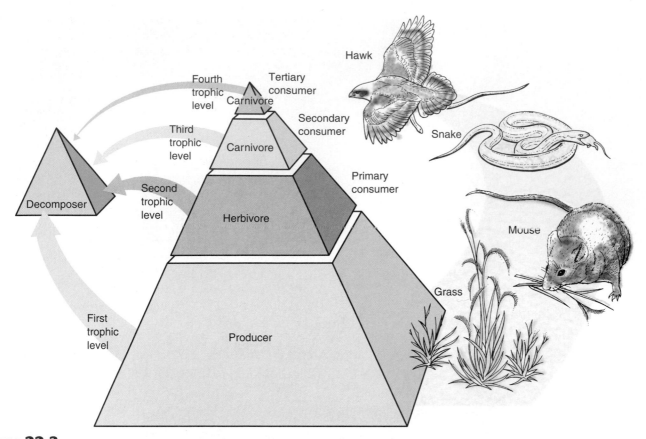

Figure 22.3

Organisms within ecosystems can be divided into several different trophic levels on the basis of how they obtain energy. Several different sets of terminology are used to identify these different roles. This illustration shows how the different sets of terminology are related to one another.

contained within the organic compounds of its body is finally released to the environment as heat by organisms that decompose the dead body into carbon dioxide, water, ammonia, and other simple inorganic molecules. Organisms of decay, called **decomposers,** are things such as bacteria, fungi, and other organisms that use dead organisms as sources of energy.

This group of organisms efficiently converts nonliving organic matter into simple inorganic molecules that can be used by producers in the process of trapping energy. Decomposers are thus very important components of ecosystems that cause materials to be recycled. As long as the sun supplies the energy, elements are cycled through ecosystems repeatedly. Table 22.1

Table 22.1 Roles in an Ecosystem

Classification	Description	Examples
Producers	Organisms that convert simple inorganic compounds into complex organic compounds by photosynthesis	Trees, flowers, grasses, ferns, mosses, algae, cyanobacteria
Consumers	Organisms that rely on other organisms as food; animals that eat plants or other animals	
Herbivore	Eats plants directly	Deer, goose, cricket, vegetarian human, many snails
Carnivore	Eats meat	Wolf, pike, dragonfly
Omnivore	Eats plants and meat	Rat, most humans
Scavenger	Eats food left by others	Coyote, skunk, vulture, crayfish
Parasite	Lives in or on another organism, using it for food	Tick, tapeworm, many insects
Decomposers	Organisms that return organic compounds to inorganic compounds; important components in recycling elements	Bacteria, fungi

Figure 22.4

As one organism feeds on another organism, there is a flow of energy from one trophic level to the next. This illustration shows six trophic levels.

summarizes the various categories of organisms within an ecosystem. Now that we have a better idea of how ecosystems are organized, we can look more closely at energy flow through ecosystems.

ENERGY FLOW IN ECOSYSTEMS

All living things need a constant source of energy. There are two fundamental physical laws of energy that are important when looking at ecological systems from an energy point of view. The law of conservation of energy states that energy is neither created nor destroyed. That means that we should be able to describe the amounts of energy in each tropic level and follow energy as it flows through successive trophic levels. The second law of thermodynamics states that heat flows from high-temperature objects to low-temperature objects, a natural process that results in less usable energy. This means that as energy passes from one trophic level to the next there will be a reduction in the amount of available energy in living things and an increase in the amount of heat in their surroundings.

The total energy of an ecosystem can be measured in several ways. The total producer trophic level can be harvested and burned. The number of kilocalories of heat energy produced by burning is equivalent to the energy content of the organic material of the plants. Another way of determining the energy present is to measure the rate of photosynthesis and respiration and calculate the amount of energy being trapped in the living material of the plants.

Since only the plants, algae, and cyanobacteria in the producer trophic level are capable of capturing energy from the sun, all other organisms are directly or indirectly dependent on the producer trophic level. The second trophic level consists of herbivores that eat the producers. This trophic level has significantly less energy in it for several reasons. *In general, there is about a 90 percent loss of available energy as we proceed from one trophic level to the next higher level.* Actual measurements will vary from one ecosystem to another, but 90 percent is a good rule of thumb. This loss in energy content at the second and subsequent trophic levels is predicted by the second law of thermodynamics. When the energy in producers is converted to the energy of herbivores much of the energy is lost as heat to the surroundings.

In addition to this loss of available energy there is an additional loss involved in the capture and processing of food material by herbivores. Although herbivores don't need to chase their food, they do need to travel to where food is available, then gather, chew, digest, and metabolize it. All these processes require energy.

Just as the herbivore trophic level experiences a 90 percent loss in energy content, the higher trophic levels of primary carnivores, secondary carnivores, and tertiary carnivores also experience a reduction in the energy available to them. Finally, decomposers act on dead organisms with this same kind of available energy loss. Figure 22.5 shows a diagram in which the energy content decreases by 90 percent as we pass from one trophic level to the next.

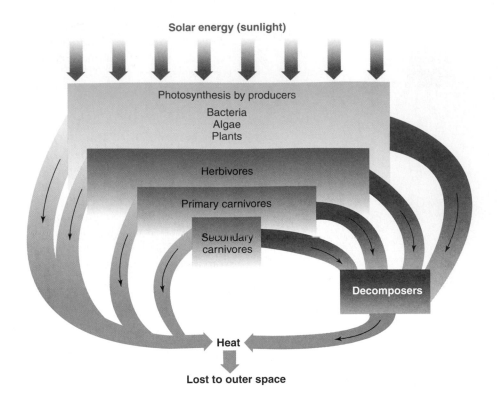

Figure 22.5

As energy flows from one trophic level to the next, approximately 90 percent of it is lost. This means that the amount of energy at the producer level must be ten times larger than the amount of energy at the herbivore level.

COMMUNITY INTERACTIONS

In the previous section we looked at ecological relationships from the point of view of ecosystems and the way energy flows through them. But we can also study relationships at the community level and focus on the kinds of interactions that take place among organisms.

As you know, one of the ways that organisms interact is by feeding on one another. A community includes many different food chains and many organisms may be involved in several of the food chains at the same time, so the food chains become interwoven into a *food web* (figure 22.6). In a community, the interacting food chains usually result in a relatively stable combination of populations. Although communities are relatively stable, we need to recognize that they are also dynamic collections of organisms: As one population increases, another decreases. This might occur over several years, or even in the period of one year. This happens because most ecosystems are not constant. There may be differences in rainfall throughout the year or changes in the amount of sunlight and in the average temperature. We should expect populations to fluctuate as abiotic factors change. A change in the size of one population will trigger changes in other populations as well.

Some communities, like tropical rainforests, have large numbers of different kinds of organisms present. Such communities have high **biodiversity.** Others, such as tundra communities, have low biodiversity. People also talk about a loss of biodiversity when a specific kind of organism is eliminated from a region. For example, when wolves were eliminated from much of North America there was a loss of biodiversity.

If numbers of a particular kind of organism in a community increase or decrease significantly, some adjustment usually occurs in the populations of other organisms within the community. For example, the populations of many kinds of small mammals fluctuate from year to year. This results in changes in the numbers of their predators or the predators must switch to other prey species and impact other parts of the community.

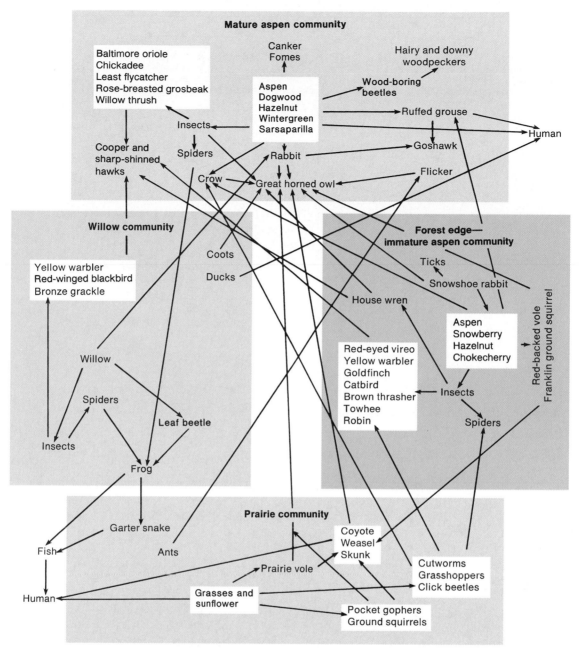

Figure 22.6

When many different food chains are interlocked with one another, a food web results. The arrows indicate the direction of energy flow. Notice that some organisms are a part of several food chains—the great horned owl in particular. Because of the interlocking nature of the food web, changing conditions may shift the way in which food flows through this system.

Not all organisms within a community have the same impact. Obviously plants are key organisms that supply the energy for all other organisms. However, even certain animal species can have a very high impact on the nature of a community. For example, sea otters feed on sea urchins that feed on a seaweed called kelp. When sea otter populations were reduced significantly, sea urchin populations increased, they ate much of the kelp, and other species of animals had fewer places to hide. Thus the sea otters played a key role in shaping the nature of the community. Such organisms are often referred to as **keystone species.**

TYPES OF TERRESTRIAL COMMUNITIES

Although each community of organisms is unique, similar communities can be combined into broad categories based on specific characteristics of the physical environment and the kinds of organisms that live in the area. The kind of terrestrial community that develops in an area is determined primarily by climatic factors of precipitation patterns and temperature ranges. These large regional communities are known as **biomes.** The map in figure 22.7 shows the distribution of the major terrestrial biomes of the world.

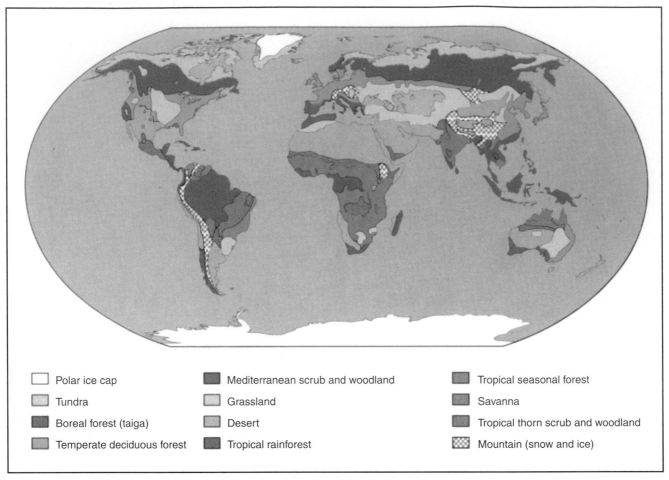

Figure 22.7

Major climatic differences determine the kind of vegetation that can live in a region of the world. Associated with specialized groups of plants are particular kinds of animals. These regional ecosystems are called biomes.

Legend:
- Polar ice cap
- Tundra
- Boreal forest (taiga)
- Temperate deciduous forest
- Mediterranean scrub and woodland
- Grassland
- Desert
- Tropical rainforest
- Tropical seasonal forest
- Savanna
- Tropical thorn scrub and woodland
- Mountain (snow and ice)

Temperate Deciduous Forest

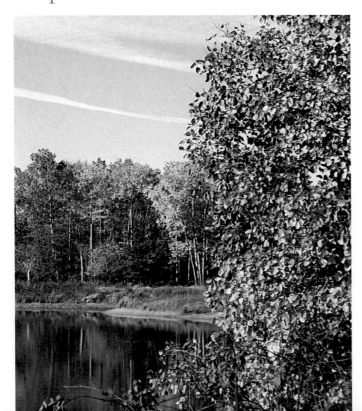

The *temperate deciduous forest* biome is found in parts of the world that have 100 cm (40 in) or more of rainfall and cold weather for a significant part of the year. The predominant plants are large trees that lose their leaves more or less completely during the fall of the year and are therefore called *deciduous*. Aspen, birch, cottonwood, oak, hickory, beech, and maple are typical trees found in this geographic region. Typical animals of this biome are skunks, porcupines, deer, frogs, opossums, owls, mosquitoes, and beetles.

In much of this region, the natural vegetation has been removed to allow for agriculture, so the original character of the biome is gone except where farming is not practical or the original forest has been preserved.

Grassland or Prairie

Grasslands are common in western North America and parts of Eurasia, Africa, Australia, and South America. The dominant vegetation in this region is made up of various species of grasses. The annual rainfall is 25 to 50 cm (10 to 20 in), an amount that is not adequate to support the growth of dense forests. Trees in grasslands grow primarily along streams where they can obtain sufficient water. Animals found in this area include the prairie dog, pronghorn antelope, prairie chicken, grasshopper, rattlesnake, and meadowlark. Most of the original grasslands, like the temperate deciduous forest, have been converted to agricultural uses.

Savanna

Savannas are found in tropical regions of the world that have pronounced rainy and dry seasons. During the dry season, fires are frequent. The fires kill tree seedlings and prevent the establishment of forests.

Savannas are typical biomes of central Africa and parts of South America and consist of grasses with scattered trees. Grazing animals such as antelope, zebra, and buffalo are common as are termites and many other insects, which serve as food for many birds and reptiles.

Desert

Very dry areas known as *deserts* are found throughout the world wherever rainfall is low and irregular. They receive less than 25 cm (10 in) of rain per year. Some deserts are extremely hot, while others can be quite cold during much of the year. The distinguishing characteristic of desert biomes is low rainfall, not high temperature. Deserts are characterized by scattered, thorny plants that lack leaves or have reduced leaves. Since leaves tend to lose water rapidly, the lack of leaves is an adaptation to dry conditions. Although a desert is a very harsh environment, many kinds of insects, reptiles, birds, and mammals can live in this biome. The animals usually avoid the hottest part of the day by staying in burrows or other shaded, cooler areas. Staying underground or in the shade also allows the animals to conserve water.

Boreal Forest

The *boreal forest* (also known as the *northern coniferous forest,* or *taiga*) is found through parts of southern Canada, extending southward along the mountains of the United States, and in much of northern Asia. The dominant vegetation consists of evergreen trees that are especially adapted to withstand long, cold winters with abundant snowfall. Spruces and firs are common tree species that are intermingled with many other kinds of vegetation. Small lakes and bogs are common. Characteristic animals in this biome include mice, bears, wolves, squirrels, moose, midges, and flies.

Temperate Rainforest

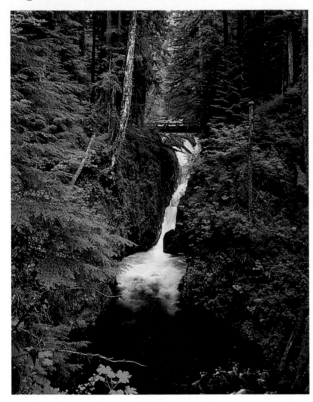

The coastal areas of northern California, Oregon, Washington, British Columbia, and southern Alaska contain an unusual set of environmental conditions that supports a *temperate rainforest.* The prevailing winds from the west bring moisture-laden air to the coast. As the air meets the coastal mountains and is forced to rise, it cools and the moisture falls as rain or snow. Most of these areas receive 200 cm (80 in) or more of precipitation per year. This abundance of water, along with fertile soil and mild temperatures, results in a lush growth of plants.

Sitka spruce, Douglas fir, and western hemlock are typical evergreen coniferous trees in the temperate rainforest. Undisturbed (old growth) forests of this region have trees as old as 800 years that are nearly 100 m (about 300 ft) tall. Deciduous trees of various kinds (red alder, big leaf maple, black cottonwood) also exist in open areas where they can get enough light. All trees are covered with mosses, ferns, and other plants that grow on the surface of the trees. The kinds of animals present are similar to those of a temperate deciduous forest.

Tundra

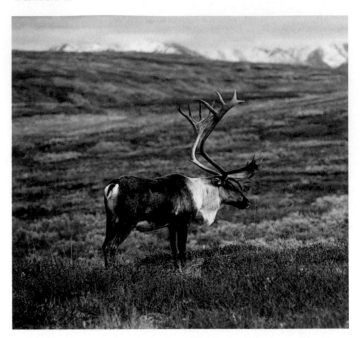

North of the coniferous forest biome is an area known as the *tundra.* It is characterized by extremely long, severe winters and short, cool summers. The deeper layers of the soil remain permanently frozen, forming a layer called the *permafrost.* Because the deeper layers of the soil are frozen, when the surface thaws the water forms puddles. Under these conditions of low temperature and short growing season, very few kinds of animals and plants can survive. No trees can live in this region. Typical plants and animals of the area are dwarf willow and some other shrubs, reindeer moss (actually a lichen), some flowering plants, caribou, wolves, musk oxen, fox, snowy owls, mice, and many kinds of insects. Many kinds of birds are summer residents only.

Tropical Rainforest

Tropical rainforests are found primarily near the equator in Central and South America, Africa, parts of southern Asia, and some Pacific Islands. The temperature is high, rain falls nearly every day, and there are thousands of species of plants in a small area. Balsa (a very light wood), teak (used in furniture), and ferns the size of trees are examples of plants from the tropical rainforest. Typically, every plant has other plants growing on it. Tree trunks are likely to be covered with orchids, many kinds of vines, and mosses. Tree frogs, bats, lizards, birds, monkeys, and an almost infinite variety of insects inhabit the rainforest. These forests are very dense and little sunlight reaches the forest floor. When the forest is opened up (by a hurricane or the death of a large tree) and sunlight reaches the forest floor, the opened area is rapidly overgrown with vegetation.

TYPES OF AQUATIC COMMUNITIES

Terrestrial biomes are determined by the amount and kind of precipitation and by temperature ranges. Other factors, such as soil type and wind, also play a part. Aquatic ecosystems also are shaped by key environmental factors. Several important factors are the ability of the sun's rays to penetrate the water, the depth of the water, the nature of the bottom substrate, the water temperature, and the amount of dissolved salts. Aquatic systems are often divided into **marine communities,** which have a high salt content, and **freshwater communities,** which have low salt concentrations.

Marine Communities

In the open ocean, many kinds of organisms float or swim actively and are called **pelagic.** Among the pelagic organisms there are two types: larger organisms that actively swim and smaller, weakly swimming organisms. The term **plankton** is used to describe aquatic organisms that are so small and weakly swimming that they are simply carried by currents. The planktonic organisms that carry on photosynthesis are called *phytoplankton.* In the open ocean, a majority of these organisms are microscopic floating algae and bacteria. The upper layer of the ocean, where the sun's rays penetrate, is known as the *euphotic zone.* It is in this euphotic zone where phytoplankton are most common. Small, weakly swimming animals of many kinds, known as *zooplankton,* feed on the phytoplankton. The zooplankton are in turn eaten by larger animals like fish and larger shrimp, which are eaten by larger fish like salmon, tuna, sharks, and mackerel (figure 22.8).

Those marine organisms that live on the bottom are called **benthic.** Many seaweeds, some fish, clams, oysters, various crustaceans, sponges, sea anemones, and many other kinds of organisms live on the bottom. In shallow water, sunlight can penetrate to the bottom, and a variety of attached seaweeds trap this energy as they carry on photosynthesis. Many of the benthic animals graze on the seaweeds and are in turn eaten by other larger animals. Some benthic animals, such as clams, filter water to obtain plankton for food.

Freshwater Communities

Freshwater ecosystems differ from marine ecosystems in several ways. In comparison to marine communities the amount of salt present is much less, the temperature of the water can change greatly, the water is in the process of moving to the ocean, oxygen can often be in short supply, and the organisms that inhabit freshwater systems are different. Freshwater ecosystems can be divided into two categories: those in which the water is relatively stationary, such as lakes, ponds, and reservoirs, and those in which the water is running downhill, such as streams and rivers.

Large lakes have many of the same characteristics as the ocean. If the lake is deep, there is a euphotic zone at the top, with many kinds of phytoplankton, and zooplankton that feed on the phytoplankton. Small fish feed on the zooplankton, which are in turn eaten by larger fish. The species of organisms found in freshwater lakes are different from those found in the ocean, but the roles played are similar, so the same terminology is used.

Along the shore and in the shallower parts of lakes, many kinds of flowering plants are rooted in the bottom. Some are completely submerged while others have leaves that float on the surface or protrude above the water. Cattails, bulrushes, arrowhead plants, and water lilies are examples. In addition, many kinds of freshwater algae are present

Associated with the plants and algae are a large number of different kinds of animals. Fish, crayfish, clams, and many kinds of aquatic insects are common inhabitants of this mixture of plants and algae. This region, with rooted vegetation, is known as the *littoral zone,* and the portion of the lake that does not have rooted vegetation is called the *limnetic zone* (figure 22.9).

Streams and rivers present a different set of conditions. Since the water is moving, planktonic organisms are less important than are attached organisms. Most algae grow attached to rocks and other objects on the bottom. Since the water is shallow, light can penetrate easily to the bottom (except for large or extremely muddy rivers). Even so, it is difficult for photosynthetic organisms to accumulate the nutrients necessary for growth, and most streams are not very productive. As a matter of fact, the major input of nutrients is from organic matter that falls into the stream from terrestrial sources. These are primarily the leaves from trees and other vegetation, as well as the bodies of living and dead insects.

Within the stream is a community of organisms specifically adapted to use the debris as a source of food. Bacteria and fungi colonize the organic matter, and many kinds of insects shred and eat this organic matter as well as the fungi and bacteria living on it. The feces (intestinal wastes) of these insects and the tiny particles produced during the eating process become food for other insects that build nets to capture the tiny bits of organic matter that drift their way. These herbivorous insects are in turn eaten by carnivorous insects and fish.

Estuaries

An **estuary** is a special category of aquatic ecosystem that consists of shallow, partially enclosed areas where freshwater enters the ocean. The saltiness of the water in the estuary changes with

High tide   Intertidal zone   **Low tide**

Sea Level

Euphotic Zone

Depth meters
0
200
400
600
800
1000
2000
4000

Benthic

Pelagic

Seaweed

Phytoplankton

Zooplankton

Figure 22.8

All of the photosynthetic activity of the ocean occurs in shallow water called the euphotic zone, either by attached algae near the shore or by minute phytoplankton in the upper levels of the open ocean. Consumers are either free-swimming pelagic organisms or benthic organisms that live on the bottom. Small animals that feed on phytoplankton are known as zooplankton.

tides and the flow of water from rivers. The organisms that live here are specially adapted to this set of physical conditions, and the number of species is less than in the ocean or in freshwater. Estuaries are especially productive ecosystems because of the large quantity of nutrients introduced into the basin from the rivers that run into them. This productivity is further enhanced by the fact that the shallow water allows light to penetrate to most of the water in the basin. Phytoplankton and attached algae and plants are able to use the sunlight and the nutrients for rapid growth. This photosynthetic activity supports many kinds of organisms in the estuary. Estuaries are especially important as nursery sites for fish and crustaceans like flounder and shrimp. The adults enter these productive, sheltered areas to reproduce and then return to the ocean. The young spend their early life in the estuary and eventually leave as they get larger and are more able to survive in the ocean.

COMMUNITY, HABITAT, AND NICHE

People approach the study of organism interactions in two major ways. Many people look at interrelationships from the broad ecosystem point of view, while others focus on individual organisms and the specific things that affect them in their daily lives. The first approach involves the study of all the organisms that interact with one another—the community—and usually looks at general relationships among them. For example, organisms are lumped into categories such as producers, consumers, and decomposers because they perform different functions in a community.

A second way of looking at interrelationships is to study in detail the ecological relationships of a certain species of organism. Each organism has particular requirements for life and lives where the environment provides what it needs. The environ-

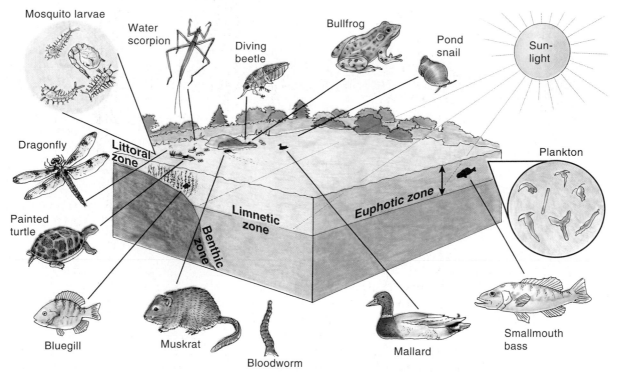

Figure 22.9

Lakes are similar in structure to oceans except that the species are different because most marine organisms cannot live in freshwater. Insects are common organisms in freshwater lakes, as are many kinds of fish, zooplankton, and phytoplankton.

mental requirements of a whale include large expanses of ocean, but with seasonally important feeding areas and protected locations used for giving birth. The kind of place, or part of an ecosystem, occupied by an organism is known as its **habitat.** Habitats are usually described in terms of conspicuous or particularly significant features in the area where the organism lives. For example, the habitat of a prairie dog is usually described as a grassland, while the habitat of a tuna is described as the open ocean. The key thing to keep in mind when you think of habitat is the *place* in which a particular kind of organism lives. When describing the habitats of organisms, we sometimes use the terminology of the major biomes of the world, such as desert, grassland, or savanna. It is also possible to describe the habitat of the bacterium *Escherichia coli* as the gut of humans and other mammals, or the habitat of a fungus as a rotting log.

Each species has particular requirements for life and places specific demands on the habitat in which it lives. The specific functional role of an organism is its **niche.** Its niche is the way it goes about living its life. Just as the word *place* is the key to understanding the concept of habitat, the word *function* is the key to understanding the concept of a niche. Understanding the niche of an organism involves a detailed understanding of the impacts an organism has on its biotic and abiotic surroundings as well as all of the factors that affect the organism. For example, the niche of an earthworm includes abiotic items such as soil particle size, soil texture, and the moisture, pH, and temperature of the soil. The earthworm's niche also includes biotic impacts such as serving as food for birds, moles, and shrews; as bait for

anglers; or as a consumer of dead plant organic matter (figure 22.10). In addition, an earthworm serves as a host for a variety of parasites, transports minerals and nutrients from deeper soil layers to the surface, incorporates organic matter into the soil, and creates burrows that allow air and water to penetrate the soil more easily. And this is only a limited sample of all of the aspects of its niche.

Some organisms have rather broad niches; others, with very specialized requirements and limited roles to play, have niches that are quite narrow. The opossum (figure 22.11A) is an animal with a very broad niche. It eats a wide variety of plant and animal foods, can adjust to a wide variety of climates, is used as food by many kinds of carnivores (including humans), and produces large numbers of offspring. By contrast, the koala of Australia (figure 22.11B) has a very narrow niche. It can live only in areas of Australia with specific species of *Eucalyptus* trees, because it eats the leaves of only a few kinds of these trees. Furthermore, it cannot tolerate low temperatures and does not produce large numbers of offspring. As you might guess, the opossum is expanding its range and the koala is endangered in much of its range.

It is often easy to overlook important roles played by some organisms. For example, when Europeans introduced cattle into Australia—a continent where there had previously been no large, hoofed mammals—they did not think about the impact of cow manure or the significance of a group of beetles called *dung beetles.* These beetles rapidly colonize fresh dung and cause it to be broken down. No such beetles existed in Australia;

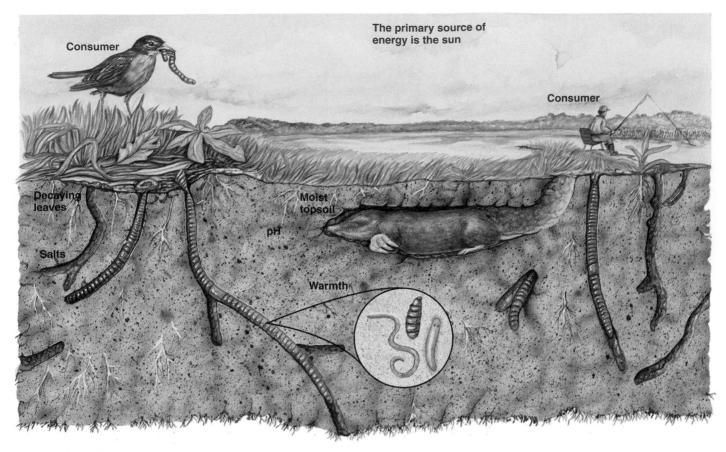

Figure 22.10

The niche of an earthworm involves a great many factors. It includes the fact that the earthworm is a consumer of dead organic matter, a source of food for other animals, or host to parasites, and bait for an angler. Furthermore, the earthworm loosens the soil by its burrowing and "plows" the soil when it deposits materials on the surface. In addition, the pH, texture, and moisture content of the soil have an impact on the earthworm. Keep in mind that this is but a small part of what the niche of the earthworm includes.

therefore, in areas where cattle were raised, a significant amount of land became covered with accumulated cow dung. This reduced the area where grass could grow and reduced productivity. The problem was eventually solved by the importation of several species of dung beetles from Africa, where large, hoofed mammals are common. The dung beetles made use of what the cattle did not digest, returning it to a form that plants could more easily recycle into plant biomass.

A Closer Look The Importance of Habitat Size

Many people interested in songbird populations have documented a significant decrease in the numbers of certain songbird species. Species particularly affected are those that migrate between North America and South America and require relatively large areas of undisturbed forest in both their northern and southern homes. Many of these species are being hurt by human activities that fragment large patches of forest into many smaller patches, creating more edges between different habitat types. Bird and other an-

imal species that thrive in edge habitats replace the songbirds, which require large patches of undisturbed forest. Cowbirds that normally live in open areas are a particular problem. Cowbirds do not build nests but lay their eggs in the nests of other birds after removing the eggs of the host species. When forests are broken into small patches, cowbirds reach a larger percentage of forest-nesting birds because the forest birds must nest closer to the edge.

The species most severely affected are those that have both their northern and southern habitats disturbed. A study of migrating sharp-shinned hawks indicated that their numbers have also been greatly reduced in recent years. It is thought that since sharp-shinned hawks use small songbirds as their primary source of food, the reduction in hawks is directly related to the reduction in migratory songbirds.

A Opossum

Figure 22.11

(*A*) The opossum has a very broad niche. It eats a variety of foods, is able to live in a variety of habitats, and has a large reproductive capacity. It is generally extending its range in the United States. (*B*) The koala has a narrow niche. It feeds on the leaves of only a few species of *Eucalyptus* trees, is restricted to relatively warm, forested areas of Australia, and is generally endangered in much of its habitat.

B Koala

KINDS OF ORGANISM INTERACTIONS

When organisms encounter one another, they can influence one another in numerous ways. Some interactions are harmful to one or both of the organisms. Others are beneficial. Ecologists have classified kinds of interactions between organisms into several broad categories.

Predation

Predation occurs when one animal captures, kills, and eats another animal. The organism that is killed is called the *prey* and the one that does the killing is called the *predator.* The predator obviously benefits from the relationship, while the prey organism is harmed. Most predators are relatively large compared to their prey and have specific adaptations that aid them in catching prey. Many spiders build webs that serve as nets to catch flying insects. The prey are quickly paralyzed by the spider's bite and wrapped in a tangle of silk threads. Other rapidly moving spiders, like wolf spiders and jumping spiders, have large eyes that help them find prey without using webs. Dragonflies patrol areas where they can capture flying insects. Hawks and owls have excellent eyesight that allows them to find their prey. Many predators, like leopards, lions, and cheetahs, use speed to run down their prey, while others such as frogs, toads, and many kinds of lizards blend in with their surroundings and strike quickly when a prey organism happens by (figure 22.12).

Parasitism

Another kind of interaction in which one organism is harmed and the other aided is the relationship of parasitism. In fact, there are more species of parasites in the world than there are nonparasites, making this a very common kind of relationship. **Parasitism** involves one organism living in or on another living organism from which it derives nourishment. The *parasite* derives the benefit and harms the *host,* the organism it lives in or on (figure 22.13). Many kinds of fungi live on trees and other kinds of plants, including those that are commercially valuable. Dutch elm disease is caused by a fungus that infects the living, sap-carrying parts of the tree. Mistletoe is a common plant that is a parasite on other plants. The mistletoe plant invades the tissues of the tree it is living on and derives nourishment from the tree.

Many kinds of worms, protozoa, bacteria, and viruses are important parasites. Parasites that live on the outside of their hosts are called *external parasites.* For example, fleas live on the outside of the bodies of mammals like rats, dogs, cats, and humans, where they suck blood and do harm to their host. At the same time, the host could also have a tapeworm in its intestine. Since the tapeworm lives inside the host, it is called an *internal parasite.* Bacterial and virus parasites are common causes of disease in humans and other organisms.

Commensalism

Predation and parasitism both involve one organism benefiting while the other is harmed. Another common relationship is one in which one organism benefits while the other is not affected. This is known as **commensalism.** For example, sharks often

A

B

Figure 22.12

(A) Many predators capture prey by making use of speed. The cheetah can reach estimated speeds of 100 kilometers per hour (about 70 mi/h) during sprints to capture its prey. (B) Other predators, like this veiled chameleon, blend in with their surroundings, lie in wait, and ambush their prey. Because strength is needed to kill the prey, the predator is generally larger than the prey. Obviously, predators benefit from the food they obtain to the detriment of the prey organism.

A Tapeworm

B Lamprey

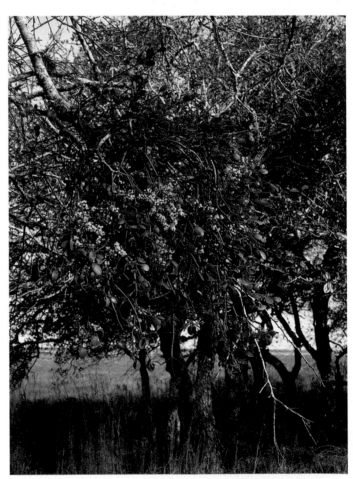

C Mistletoe

Figure 22.13

Parasites benefit from the relationship because they obtain nourishment from the host. Tapeworms (A) are internal parasites in the guts of their host, where they absorb food from the host's gut. The lamprey (B) is an external parasite that sucks body fluids from its host. Mistletoe (C) is a photosynthesizing plant that also absorbs nutrients from the tissues of its host tree. The host in any of these three situations may not be killed directly by the relationship, but it is often weakened, thus becoming more vulnerable to predators or diseases. There are more species of parasites in the world than species of organisms that are not parasites.

A Remora

B Epiphytes

Figure 22.14

In the relationship called commensalism, one organism benefits and the other is not affected. (*A*) The remora fish shown here hitchhike a ride on the shark. They eat scraps of food left over by the messy eating habits of the shark. The shark does not seem to be hindered in any way. (*B*) The epiphytic plants growing on this tree do not harm the tree but are aided by using the tree surface as a place to grow.

have another fish, the remora, attached to them. The remora has a sucker on the top side of its head that allows it to attach to the shark and get a free ride (figure 22.14A). While the remora benefits from the free ride and by eating leftovers from the shark's meals, the shark does not appear to be troubled by this uninvited guest, nor does it benefit from the presence of the remora.

Another example of commensalism is the relationship between trees and epiphytic plants. Epiphytes are plants that live on the surface of other plants but do not derive nourishment from them (figure 22.14B). Many kinds of plants (e.g., orchids, ferns, and mosses) use the surface of trees as a place to live. These kinds of organisms are particularly common in tropical rainforests. Many epiphytes derive benefit from the relationship because they are able to be located in the top of the tree, where they receive more sunlight and moisture. The trees derive no benefit from the relationship, nor are they harmed; they simply serve as a support surface for epiphytes.

Mutualism

So far in our examples, only one species has benefited from the association of two species. There are also many situations in which two species live in close association with one another, and both benefit. This is called **mutualism.** One interesting example of mutualism involves digestion in animals like rabbits, cows, and termites. These animals eat plant material that is high in cellulose, even though they do not produce the enzymes capable of breaking down cellulose molecules into simple sugars. They manage to get energy out of these cellulose molecules with the help of special microorganisms living in their digestive tracts. The microorganisms produce cellulose-digesting enzymes, called *cellulases,* which break down cellulose into smaller carbohydrate molecules that the host's digestive enzymes can break down into smaller glucose molecules. The microorganisms benefit because the host's gut provides them with a moist, warm,

Figure 22.15

Mutualism is an interaction between two organisms in which both benefit. The plant benefits because cross-fertilization (exchange of gametes from a different plant) is more probable; the butterfly benefits by acquiring nectar for food.

nourishing environment in which to live. The host benefits because the microorganisms provide them with a source of food.

Another kind of mutualistic relationship exists between flowering plants and bees. Undoubtedly you have observed bees and other insects visiting flowers to obtain nectar from the blossoms. Usually the flowers are constructed in such a manner that the bees pick up pollen (sperm-containing packages) on their hairy bodies and transfer it to the female part of the next flower they visit (figure 22.15). Because bees normally visit many individual flowers of the same species for several minutes and ignore other species of flowers, they can serve as pollen carriers between two flowers of the same species. Plants pollinated in this manner produce less pollen than do plants that rely on the wind to transfer pollen. This saves the plant energy because it doesn't need to produce huge quantities of pollen. It does, however, need to transfer some of its energy savings into the production of showy flowers and nectar to attract the bees. The bees benefit from both the nectar and pollen; they use both for food.

Competition

So far in our discussion of organism interactions we have left out the most common one. It is reasonable to envision every organism on the face of the earth being involved in *competitive* interactions. **Competition** is a kind of interaction between organisms in which both organisms are harmed to some extent. Competition occurs whenever two organisms both need a vital resource that is in short supply (figure 22.16). The vital resource could be food, shelter, nesting sites, water, mates, or space. It may involve a snarling tug-of-war between two dogs over a scrap of food, or it can be a silent struggle between plants for access to available light. If you have ever started tomato seeds (or other garden plants) in a garden and failed to eliminate the weeds, you have witnessed competition. If the weeds are not removed, they compete with the garden plants for available sunlight, water, and nutrients, resulting in poor growth of both the garden plants and the weeds.

It is important to recognize that although competition results in harm to both organisms there can still be winners and losers. The two organisms may not be harmed to the same extent, which results in one having greater access to the limited resource. Furthermore, even the loser can continue to survive if it migrates to an area where competition is less intense, or evolves to exploit a different niche. Thus, competition provides a major mechanism for natural selection. With the development of slight differences between niches, the intensity of competition is reduced. For example, many birds catch flying insects as food. However, they do not compete directly with each other because some feed at night, some feed high in the air, some feed only near the ground, and still others perch on branches and wait for insects to fly past. The insect-eating niche can be further subdivided by specializing on particular sizes or kinds of insects.

Many of the relationships just described involve the transfer of nutrients from one organism to another (predation, parasitism, mutualism). Another important way scientists look at ecosystems is to look at how materials are cycled from organism to organism.

THE CYCLING OF MATERIALS IN ECOSYSTEMS

Although some new atoms are being added to the earth from cosmic dust and meteorites, this amount is not significant in relation to the entire mass of the earth. Therefore, the earth can be considered to be a closed ecosystem as far as matter is concerned. Only sunlight energy comes to the earth in a continuous stream, and even this is ultimately returned to space as radiant energy. How-

Figure 22.16
Whenever a needed resource is in limited supply, organisms compete for it. This competition may be between members of the same species (*intraspecific*), illustrated by the vultures shown in the photograph, or may involve different species (*interspecific*).

ever, it is this flow of energy that drives all biological processes. Living systems have evolved ways of using this energy to continue life through growth and reproduction and the continual reuse of existing atoms. In this recycling process, inorganic molecules are combined to form the organic compounds of living things. If there were no way of recycling this organic matter back into its inorganic forms, organic material would build up as the bodies of dead organisms. This is thought to have occurred millions of years ago when the present deposits of coal, oil, and natural gas were formed. However, under most conditions decomposers are available to break down organic material to inorganic material that can then be reused by other organisms to rebuild organic material. One way to get an appreciation of how various kinds of organisms interact to cycle materials is to look at a specific kind of atom and follow its progress through an ecosystem.

The Carbon Cycle

All living things contain organic molecules that are composed of long chains of carbon atoms. Carbon and oxygen atoms combine to form the molecule carbon dioxide (CO_2), which is a gas found in small quantities in the atmosphere. During photosynthesis, carbon dioxide (CO_2) and water (H_2O) are used to form complex organic molecules like sugar ($C_6H_{12}O_6$). This process of reusing carbon atoms is often called the **carbon cycle.** At the same time, oxygen molecules (O_2) are released into the atmosphere.

The organic matter in the bodies of plants may be used by herbivores as food. When a herbivore eats a plant, it breaks down the complex organic molecules into more simple molecules, like simple sugars, amino acids, glycerol, and fatty acids. These can be used as building blocks in the construction of its own body.

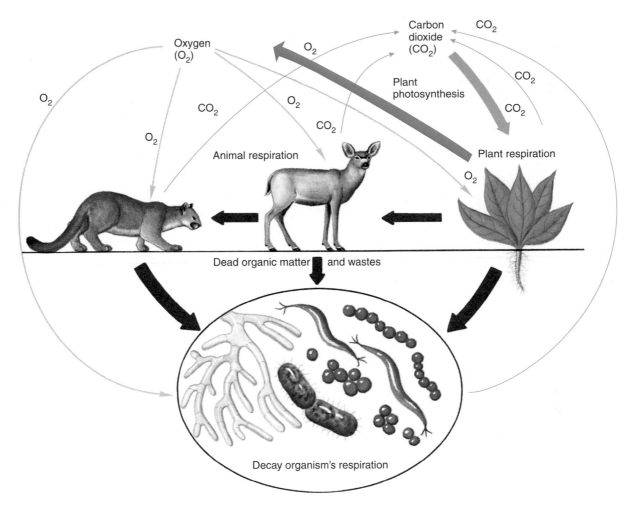

Figure 22.17

Carbon atoms are cycled through ecosystems. Carbon dioxide (green arrows) produced by respiration is the source of carbon that plants incorporate into organic molecules when they carry on photosynthesis. These carbon-containing organic molecules (black arrows) are passed to animals when they eat plants and other animals. Organic molecules in waste or dead organisms are consumed by decay organisms in the soil when they break down organic molecules into carbon dioxide and water. All organisms (plants, animals, and decomposers) return carbon atoms to the atmosphere as carbon dioxide when they carry on cellular respiration. Oxygen (blue arrows) is being cycled at the same time that carbon is. The oxygen is released into the atmosphere and into the water during photosynthesis and taken up during cellular respiration.

Thus, the atoms in the body of the herbivore can be traced back to the plants that were eaten. Similarly, when carnivores eat herbivores these same atoms are transferred to them. Finally, the waste products of plants and animals and the remains of dead organisms are used by decomposer organisms as sources of carbon and other atoms they need for survival. In addition, all the organisms in this cycle—plants, herbivores, carnivores, and decomposers—obtain energy (ATP) from the process of respiration, in which oxygen (O_2) is used to break down organic compounds into carbon dioxide (CO_2) and water (H_2O). Thus, the carbon atoms that started out as components of carbon dioxide (CO_2) molecules have passed through the bodies of living organisms as parts of organic molecules and returned to the atmosphere as carbon dioxide, ready to be cycled again. Similarly, the oxygen atoms (O) released as oxygen molecules (O_2) during photosynthesis have been used during the process of respiration (figure 22.17).

The Nitrogen Cycle

Another important element for living things is nitrogen (N). Nitrogen is essential in the formation of amino acids, which are needed to form proteins, and in the formation of nitrogenous bases, which are a part of ATP and the nucleic acids DNA and RNA. Nitrogen (N) is found as molecules of nitrogen gas (N_2) in the atmosphere. Although nitrogen gas (N_2) makes up approximately 80 percent of the earth's atmosphere, only a few kinds of bacteria are able to convert it into nitrogen compounds that other organisms can use. Therefore, in most terrestrial ecosystems, the amount of nitrogen available limits the amount of plant biomass that can be produced. (Most aquatic ecosystems are limited by the amount of phosphorus rather than the amount of nitrogen.) Plants utilize several different nitrogen-containing compounds to obtain the nitrogen atoms they need to make amino acids and other compounds (figure 22.18).

Humans have significantly altered the carbon cycle. As we burn fossil fuels, the amount of carbon dioxide in the atmosphere continually increases. Carbon dioxide allows light to enter the atmosphere but does not allow heat to exit. Since this is similar to what happens in a greenhouse, carbon dioxide and other gases that have similar effects are called greenhouse gases. Therefore, increased amounts of carbon dioxide in the atmosphere could lead to a warming of the earth. Many are concerned that increased carbon dioxide levels will lead to a warming of the planet that would cause major changes in weather and climate, leading to the flooding of coastal cities and major changes in agricultural production. The Intergovernmental Panel on Climate Change (IPCC) established by the United Nations concluded that there has been an increase in average temperature of the earth, and humans are the cause because of the burning of fossil fuels and the destruction of forests. There is no doubt that the amount of carbon dioxide in the atmosphere has been increasing. Despite this fact and the conclusions of the IPCC, there is still controversy about this topic and some doubt that warming is occurring. At a meeting in Kyoto, Japan, in 1998, many countries agreed to reduce the amount of carbon dioxide and other greenhouse gases they release into the atmosphere. However, emissions of carbon dioxide are directly related to economic activity and the energy usage that fuels it. Therefore, it remains to be seen if countries will meet their goals or will succumb to economic pressures to allow continued use of large amounts of fossil fuels. However, some countries have sought to control the change in the amount of carbon dioxide in the atmosphere by planting millions of trees or preventing the destruction of forests. The thought is that the trees carry on photosynthesis, grow, and store carbon in their bodies, leading to reduced carbon dioxide levels. At the same time, people in other parts of the world continue to destroy forests at a rapid rate. Tree planting does not offset deforestation. In addition, the trees that have been planted will ultimately die and decompose, releasing carbon dioxide back into the atmosphere, so it is not clear that this is an effective means of reducing atmospheric carbon dioxide over the long term.

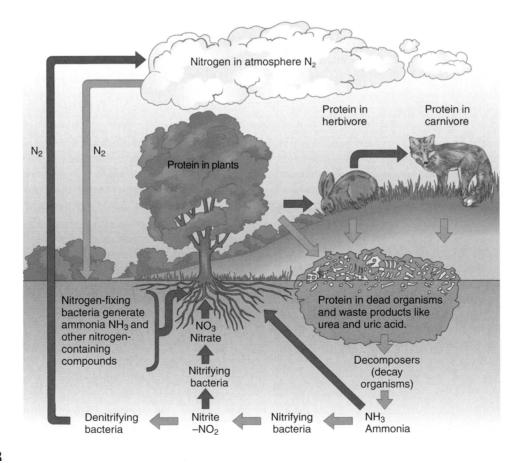

Figure 22.18

Nitrogen atoms are cycled through ecosystems. Atmospheric nitrogen is converted by nitrogen-fixing bacteria to nitrogen-containing compounds that plants can use to make proteins and other compounds. Proteins are passed to other organisms when one organism is eaten by another. Dead organisms and their waste products are acted upon by decay organisms to form ammonia, which may be reused by plants and converted to other nitrogen compounds by nitrifying bacteria. Denitrifying bacteria return nitrogen as a gas to the atmosphere.

Chemical Fertilizers and the Nitrogen Cycle

Modern agriculture depends on the use of fertilizers to replace the elements removed from the soil by the plants that are harvested. The three most important elements in fertilizers are nitrogen, phosphorus, and potassium. Although molecular nitrogen makes up 80 percent of the atmosphere, it is not available to plants. From a chemical point of view it is very difficult to get molecular nitrogen, which consists of two nitrogen atoms bonded to each other, to react with other atoms. Although some bacteria are able to attach hydrogen to nitrogen to form ammonia, a chemical method for producing ammonia was not determined until 1909 when Fritz Haber determined the conditions under which nitrogen and hydrogen would combine to form ammonia. The process of attaching hydrogen to ammonia requires a great deal of energy. This was an important discovery because it led to a practical way to produce a variety of nitrogen-containing compounds that can be used for fertilizer. Farmers apply ammonia directly into the soil to provide nitrogen for many kinds of plants. Ammonia can be reacted with other materials to produce a variety of compounds that are used as nitrogen fertilizers, including urea, ammonium nitrate, and ammonium sulfate.

Symbiotic nitrogen-fixing bacteria live in the roots of certain kinds of plants, where they convert nitrogen gas molecules into compounds that the plants can use to make amino acids and nucleic acids. The most common plants that enter into this mutualistic relationship with bacteria are the legumes, such as beans, clover, peas, alfalfa, and locust trees. Some other organisms, such as alder trees and even a kind of aquatic fern, can also participate in this relationship. There are also **free-living nitrogen-fixing bacteria** in the soil that provide nitrogen compounds that can be taken up through the roots, but the bacteria do not live in a close physical union with plants.

Another way plants get usable nitrogen compounds involves a series of different bacteria. *Decomposer bacteria* convert organic nitrogen-containing compounds into ammonia (NH_3). *Nitrifying bacteria* can convert ammonia (NH_3) into nitrite-containing (NO_2^-) compounds, which in turn can be converted into nitrate-containing (NO_3^-) compounds. Many kinds of plants can use either ammonia (NH_3) or nitrate (NO_3^-) from the soil as building blocks for amino acids and nucleic acids.

All animals obtain their nitrogen from the food they eat. The ingested proteins are broken down into their component amino acids during digestion. These amino acids can then be reassembled into new proteins characteristic of the animal. All dead organic matter and waste products of plants and animals are acted upon by decomposer organisms, and the nitrogen is released as ammonia (NH_3), which can be taken up by plants or acted upon by nitrifying bacteria to make nitrate (NO_3^-).

Finally, other kinds of bacteria called *denitrifying bacteria* are capable of converting nitrite (NO_2^-) to nitrogen gas (N_2), which is released into the atmosphere. Thus, in the **nitrogen cycle,** nitrogen from the atmosphere is passed through a series of organisms, many of which are bacteria, and ultimately returns to the atmosphere to be cycled again. However, there is also a secondary cycle in which nitrogen compounds are recycled without returning to the atmosphere.

POPULATION CHARACTERISTICS

A **population** is a group of organisms of the same species located in the same place at the same time. Examples are the number of dandelions in your front yard, the rat population in the sewers of your city, or the number of people in New York City. On a larger scale, all the people of the world constitute the human population. The terms *species* and *population* are interrelated because a species is a population—the largest possible population of a particular kind of organism. The term *population,* however, is often used to refer to portions of a species by specifying a space and time. For example, the size of the human population in a city changes from hour to hour during the day and varies according to where you set the boundaries of the city.

Since each local population is a small portion of a species, we should expect distinct populations of the same species to show differences. Some of these differences will be genetic. *Gene frequency* is a measure of how often a specific gene for a characteristic shows up in the individuals of a population. Two populations of the same species often have quite different gene frequencies. For example, many populations of mosquitoes have high frequencies of insecticide-resistant genes, whereas others do not. The frequency of the genes for tallness in humans is greater in certain African tribes than in any other human population. Figure 22.19 shows that the frequency of the genetic information for type B blood differs significantly from one human population to another.

Another feature of a population is its *age distribution,* which is the number of organisms of each age in the population. In addition, organisms are often grouped into the following categories:

- prereproductive juveniles—insect larvae, plant seedlings, or babies;
- reproductive adults—mature insects, plants producing seeds, or humans in early adulthood; or
- postreproductive adults no longer capable of reproduction—annual plants that have shed their seeds, salmon that have spawned, and many elderly humans.

A population is not necessarily divided into equal thirds (figure 22.20). In some situations, a population may be made up of a majority of one age group. If the majority of the population is prereproductive, then a "baby boom" should be anticipated in the future. If a majority of the population is reproductive, the population should be growing rapidly. If the majority of the population is postreproductive, a population decline should be anticipated. Many organisms that only live a short time and have

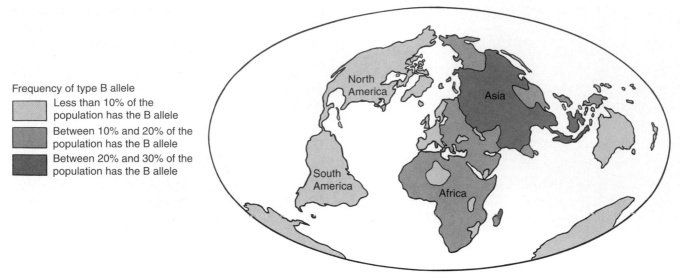

Figure 22.19

The allele for type B blood is not evenly distributed in the world. This map shows that the type B allele is most common in parts of Asia and has been dispersed to the Middle East and parts of Europe and Africa. There has been very little flow of the allele to the Americas.

Frequency of type B allele

Less than 10% of the population has the B allele

Between 10% and 20% of the population has the B allele

Between 20% and 30% of the population has the B allele

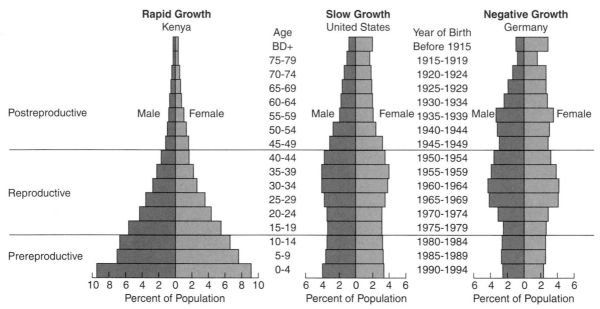

Figure 22.20

The relative number of individuals in each of the three categories (prereproductive, reproductive, and postreproductive) can give a good clue to the future of the population. Kenya has a large number of young individuals who will become reproducing adults. Therefore this population will grow rapidly and will double in about 27 years. The United States has a declining proportion of prereproductive individuals but a relatively large reproductive population. Therefore it will continue to grow for a time but will probably stabilize in the future. Germany's population has a large proportion of postreproductive individuals and a small proportion of prereproductive individuals. Its population is beginning to fall.

Source: U.S. Bureau of the Census and the United Nations, as reported in Joseph A. McFalls, Jr., "Population: A Lively Introduction," *Population Bulletin*, vol. 46, no. 2. Washington, D.C.: Population Reference Bureau, Inc., October 1991.

high reproductive rates can have age distributions that change significantly in a matter of weeks or months. For example, many birds have a flurry of reproductive activity during the summer months. Therefore, if you sample the population of a specific species of bird at different times during the summer you would have widely different proportions of reproductive and prereproductive individuals.

Populations can also differ in their sex ratios. The *sex ratio* is the number of males in a population compared to the number of females. In bird and mammal species where strong pair-bonding occurs, the sex ratio may be nearly one to one (1:1). Among mammals and birds that do not have strong pair-bonding, sex ratios may show a larger number of females than males. This is particularly true among species that are hunted, because more males

Figure 22.21

Some male animals defend a harem of females; therefore the sex ratio in these groups is several females per male.

than females are shot, resulting in a higher proportion of females surviving. Since one male can fertilize several females, the population can remain large even though the females outnumber the males. Many species of animals like bison, horses, and elk have

mating systems in which one male maintains a harem of females. The sex ratio in these small groups is quite different from a 1:1 ratio (figure 22.21). There are very few situations in which the number of males exceeds the number of females.

Regardless of the sex ratio of a population, most species can generate large numbers of offspring, producing a concentration of organisms in an area. *Population density* is the number of organisms of a species per unit area. Some populations are extremely concentrated into a limited space, while others are well dispersed. As the population density increases, competition among members of the population for the necessities of life increases. This increases the likelihood that some individuals will explore new habitats and migrate to new areas. Increases in the intensity of competition that cause changes in the environment and lead to dispersal are often referred to as *population pressure.* The dispersal of individuals to new areas can relieve the pressure on the home area and lead to the establishment of new populations. Among animals, it is often the juveniles who participate in this dispersal process.

In plant populations, dispersal is not very useful for relieving population density; instead, the death of weaker individuals usually results in reduced population density. In the lodgepole pine, seedlings become established in areas following fire and dense thickets of young trees are established. As the stand ages, many small trees die. The remaining trees grow larger as the population density drops (figure 22.22).

A

Figure 22.22

This population of lodgepole pine seedlings consists of a large number of individuals very close to one another (*A*). As the trees grow, many of the weaker trees will die, the distance between individuals will increase, and the population density will be reduced (*B*).

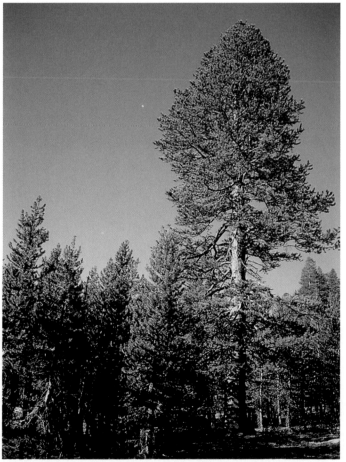

B

THE POPULATION GROWTH CURVE

Because most species of organisms have a high reproductive capacity, there is a tendency for populations to grow if environmental conditions permit. For example, if the usual litter size for a pair of mice is four, the four would produce eight, which in turn would produce sixteen, and so forth. Figure 22.23 shows a graph of change in population size over time known as a **population growth curve.** This kind of curve is typical for situations where a species is introduced into a previously unutilized area.

The change in the size of a population depends on the rate at which new organisms enter the population compared to the rate at which they leave. The number of new individuals added to the population by reproduction per thousand individuals is called **natality.** The number of individuals leaving the population by death per thousand individuals is called **mortality.** When a small number of organisms (two mice) first invades an area, there is a period of time before reproduction takes place when the population remains small and relatively constant. This part of the population growth curve is known as the **lag phase.** During the lag phase both natality and mortality are low. The lag phase occurs because reproduction is not an instantaneous event. Even after animals enter an area they must mate and produce young. This may take days or years depending on the animal.

In organisms that take a long time to mature and produce young, such as elephants, deer, and many kinds of plants, the lag phase may be measured in years. With the mice in our example, it will be measured in weeks. The first litter of young will be able to reproduce in a matter of weeks. Furthermore, the original parents will probably produce an additional litter or two during this time period. Now we have several pairs of mice reproducing more than just once. With several pairs of mice reproducing, natality increases while mortality remains low; therefore, the population begins to grow at an ever-increasing (accelerating) rate. This portion of the population growth curve is known as the **exponential growth phase.**

The number of mice (or any other organism) cannot continue to increase at a faster and faster rate because, eventually, something in the environment will become limiting and cause an increase in the number of deaths. For animals, food, water, or nesting sites may be in short supply, or predators or disease may kill many individuals. Plants may lack water, soil nutrients, or sunlight. Eventually, the number of individuals entering the population by reproduction or immigration will come to equal the number of individuals leaving it by death or migration, and the population size becomes stable. Often there is both a decrease in natality and an increase in mortality at this point. This portion of the population growth curve is known as the **stable**

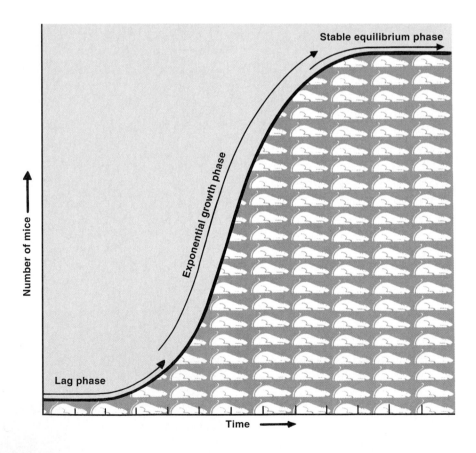

Figure 22.23

In this mouse population, the period of time in which there is little growth is known as the lag phase. This is followed by a rapid increase in population as the offspring of the originating population begin to reproduce themselves; this is known as the exponential growth phase. Eventually the population reaches a stable equilibrium phase, during which the birthrate equals the deathrate.

equilibrium phase. It is important to recognize that this is still a population with births, deaths, migration, and a changing mix of individuals; however, the size of the population is stable.

POPULATION-SIZE LIMITATIONS

Populations cannot continue to increase indefinitely; eventually, some factor or set of factors acts to limit the size of a population, leading to the development of a stable equilibrium phase or even to a reduction in population size. The identifiable factors that prevent unlimited population growth are known as **limiting factors.** All of the different limiting factors that act on a population are collectively known as **environmental resistance,** and the maximum population that an area can support is known as the **carrying capacity** of the area. In general, organisms that are small and have short life spans tend to have fluctuating populations and do not reach a carrying capacity, whereas large organisms that live a long time tend to reach an optimum population size that can be sustained over an extended period (figure 22.24). A forest ecosystem contains populations of many insect species that fluctuate widely and rarely reach a carrying capacity, but the number of specific tree species or large animals such as owls or deer is relatively constant.

However, the carrying capacity is not an inflexible number. Often environmental changes, such as climate variations, disease epidemics, forest fires, or floods can change the carrying capacity of an area for specific species. In aquatic ecosystems one of the major factors that determines the carrying capacity is the amount of nutrients in the water. In areas where nutrients are abundant, the numbers of various kinds of organisms are high. Often nutrient levels fluctuate with changes in current or runoff from the land, and plant and animal populations fluctuate as well. In addition, a change that negatively affects the carrying capacity for one species may increase the carrying capacity for another. For example, the cutting down of a mature forest followed by the growth of young trees increases the carrying capacity for deer and rabbits, which use the new growth for food, but decreases the carrying capacity for squirrels, which need mature, fruit-producing trees as a source of food and old, hollow trees for shelter.

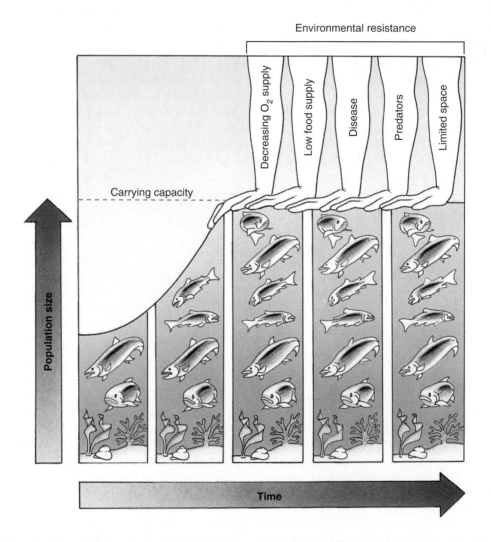

Figure 22.24

A number of factors in the environment, such as food, oxygen supply, diseases, predators, and space determine the number of organisms that can survive in a given area—the carrying capacity of that area. The environmental factors that limit populations are collectively known as environmental resistance.

The actions of government can have a significant impact on the population growth patterns of nations. Some countries have policies that encourage couples to have children. The U.S. tax code indirectly encourages births by providing tax advantages to the parents of children. Some countries in Europe are concerned about the lack of working-age people in the future and are considering ways to encourage births.

China and India are the two most populous countries in the world. Both have over 1 billion people. However, China has taken steps to control its population and now has a total fertility rate of 1.8 children per woman while India has a total fertility rate of 3.2. The total fertility rate is the average number of children born to a woman during her lifetime. The differences in total fertility rates between these two countries are the result of different policy decisions over the last fifty years. The history of China's population policy is an interesting study of how government policy affects reproductive activity among its citizens. When the People's Republic of China was established in 1949, the official policy of the government was to encourage births because more Chinese would be able to produce more goods and services, and production was the key to economic prosperity. The population grew from 540 million to 614 million between 1949 and 1955 while economic progress was slow. Consequently, the government changed its policy and began to promote population control.

The first family-planning program began in 1955, as a means of improving maternal and child health. Birthrates fell, but other social changes resulted in widespread famine and increased deathrates and low birthrates in the late 1950s and early 1960s.

The present family-planning policy began in 1971 with the launching of the *wan xi shao* campaign. Translated, this phrase means "later" (marriages), "longer" (intervals between births), and "fewer" (children). As part of this program the legal ages for marriage were raised. For women and men in rural areas, the ages were raised to 23 and 25, respectively; for women and men in urban areas, the ages were raised to 25 and 28, respectively. These policies resulted in a reduction of birthrates by nearly 50 percent between 1970 and 1979.

An even more restrictive "one child" campaign was begun in 1978–1979. The program offered incentives for couples to restrict their family size to one child. Couples enrolled in the program would receive free medical care, cash bonuses for their work, special housing treatment, and extra old-age benefits. Those who broke their pledge were penalized by the loss of these benefits as well as other economic penalties. By the mid-1980s less than 20 percent of the eligible couples were signing up for the program. Rural couples, particularly, desired more than one child. In fact, in a country where over 60 percent of the population is rural, the rural total fertility rate was 2.5 children per woman. In 1988 a second child was sanctioned for rural couples if their first child was a girl, which legalized what had been happening anyway.

The programs appear to have had an effect because the current total fertility rate has fallen to 1.8 children per woman. Replacement fertility, the total fertility rate at which the population would eventually stabilize, is 2.1 children per woman per lifetime. Furthermore, over 80 percent of couples use contraception. Abortion is also an important aspect of this program, with a ratio of more than 600 abortions per 1,000 live births.

By contrast, during the same fifty years India has had little success in controlling its population. In 2000 a new plan was unveiled which includes the goal of bringing the total fertility rate from its current 3.2 children per woman to 2 (replacement rate) by 2010. In the past the emphasis of government programs was on meeting goals of sterilization and contraceptive use but this has not been successful. Today less than 50 percent of couples use contraceptives. This new plan will emphasize improvements in the quality of life of the people. The major thrusts will be to reduce infant and maternal death, immunize children against preventable disease, and encourage girls to attend school. It is hoped that improved health will remove the perceived need for large numbers of births. Currently, less than 50 percent of the women in India can read and write. The emphasis on improving the educational status of women is related to the experiences of other developing countries. In many other countries it has been shown that an increase in the education level of women has been linked to lower fertility rates.

It seems overly optimistic to think that the total fertility rate can be reduced from 3.2 to 2 in just ten years, but programs that emphasize improvements in maternal and child health and increasing the educational level of women have been very effective in reducing total fertility in other countries.

Source: Data from H. Yuan Tien, "China's Demographic Dilemmas," in *Population Bulletin, 1992*, Population Referene Bureau, Inc., Washington, D.C., and Natural Family Planning Commission of China; and more recent data taken from Population Reference Bureau, Inc.

LIMITING FACTORS TO HUMAN POPULATION GROWTH

Today we hear differing opinions about the state of the world's human population. On one hand we hear that the population is growing rapidly. By contrast we hear that some countries are afraid that their populations are shrinking. Other countries are concerned about the aging of their populations because birthrates and deathrates are low. In magazines and on television we see that there are starving people in the world. At the same time we hear discussions about the problem of food surpluses and obesity in many countries. Some have even said that the most important problem in the world today is the rate at which the human population is growing; others maintain that the growing population will provide markets for goods and be an economic boon. How do we reconcile this mass of conflicting information?

It is important to realize that human populations follow the same patterns of growth and are acted upon by the same kinds of limiting factors as are populations of other organisms. When we look at the curve of population growth over the past several thousand years, estimates are that the human population remained low and constant for thousands of years but has increased rapidly in the past few hundred years (figure 22.25).

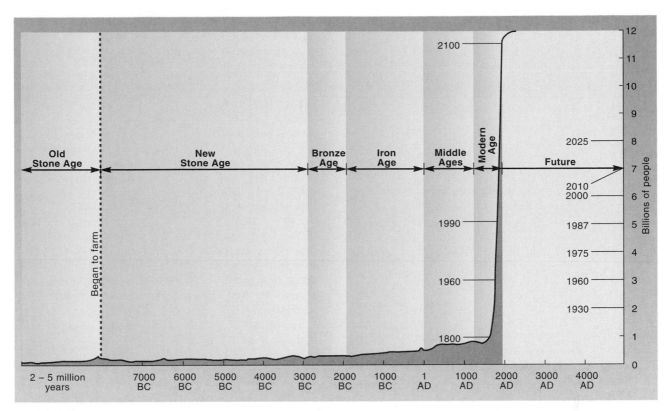

Figure 22.25

The number of humans doubled from A.D. 1800 to 1930 (from 1 billion to 2 billion), doubled again by 1975 (4 billion), and is projected to double again by the year 2025. How long can the human population continue to double before the earth's ultimate carrying capacity is reached?

Source: Data from Jean Van Der Tak et al., "Our Population Predicament: A New Look," in *Population Bulletin*, vol. 34, no. 5, December 1979, Population Reference Bureau, Washington, D.C.

For example, it has been estimated that when Columbus discovered America, the Native American population was about 1 million and was at or near the carrying capacity. Today, the population of the U.S. and Canada is about 300 million people. Does this mean that humans are different from other animal species? Can the human population continue to grow forever?

The human species is no different from other animals. It has an upper limit set by the carrying capacity of the environment, but the human population has been able to increase astronomically because technological changes and displacement of other species have allowed us to shift the carrying capacity upward. Much of the exponential growth phase of the human population can be attributed to the removal of diseases, improvement in agricultural methods, and replacement of natural ecosystems with artificial agricultural ecosystems. But even these conditions have their limits. There will be some limiting factors that eventually cause a leveling off of our population growth curve. We cannot increase beyond our ability to get raw materials and energy, nor can we ignore the waste products we produce or the other organisms with which we interact.

Humans are different from most other organisms in a fundamental way: We are able to predict the outcome of a specific course of action. Current technology and medical knowledge are available to control human population and improve the health and well-being of the people of the world. Why then does the human population continue to grow, resulting in human suffering and stressing the environment in which we live? Because we are social animals that have freedom of choice, we frequently do not do what is considered "best" from an unemotional, unselfish, biological point of view. People make decisions based on historical, social, cultural, ethical, and personal considerations. What is best for the population as a whole may be bad for you as an individual. The biggest problems associated with control of the human population are not biological problems but rather require the efforts of philosophers, theologians, politicians, and sociologists. If our population increases, so will political, social, and biological problems; there will be less individual freedom, there will be intense competition for resources, and famine and starvation will become even more common. The knowledge and technology necessary to control the human population are available, but the will is not. What will eventually limit the size of our population? Will it be lack of resources, lack of energy, accumulated waste products, competition among ourselves, or rational planning of family size?

SUMMARY

Ecology is the study of how organisms interact with their environment. The *environment* consists of *biotic* and *abiotic* components that are interrelated in an *ecosystem*. All *ecosystems* must have a constant *input of energy* from the sun. *Producer organisms* are capable of trapping the sun's energy and converting it into *biomass*. *Herbivores* feed on producers and are in turn eaten by *carnivores*, which may be eaten by *other carnivores*. Each level in the *food chain* is known as a *trophic level*. Other kinds of organisms involved in food chains are *omnivores*, which eat both plant and animal food, and *decomposers*, which break down dead organic matter and waste products. All *ecosystems* have a large *producer* base with successively smaller amounts of energy at the *herbivore*, *primary carnivore*, and *secondary carnivore trophic levels*. This is because each time *energy* passes from *one trophic level* to the next, about 90 percent of the energy is lost from the *ecosystem*. A *community* consists of the interacting *populations* of *organisms* in an area. The organisms are interrelated in many ways in food chains that interlock to create *food webs*. Because of this interlocking, changes in one part of the *community* can have effects elsewhere.

Major land-based regional *ecosystems* are known as *biomes*. The *temperate deciduous forest, northern coniferous forest, temperate rainforest, tropical rainforest, desert, savanna,* and *tundra* are examples of biomes. Aquatic communities can be divided into marine, freshwater, and estuarine communities.

Each organism in a community occupies a specific space known as its *habitat* and has a specific functional role to play known as its *niche*. An *organism's habitat* is usually described in terms of some conspicuous element of its surroundings. The *niche* is difficult to describe because it involves so many interactions with the physical environment and other living things.

Interactions between organisms fit into several categories. *Predation* is one organism benefiting (*predator*) at the expense of the organism killed and eaten (*prey*). *Parasitism* is one organism benefiting (*parasite*) by living in or on another organism (*host*) and deriving nourishment from it. *Commensal* relationships exist when one organism is helped but the other is not affected. *Mutualistic* relationships benefit both organisms. *Competition* causes harm to both of the organisms involved, although one may be harmed more than the other and may become extinct, evolve into a different niche, or be forced to migrate.

Many atoms are cycled through ecosystems. The *carbon atoms* of living things are trapped by *photosynthesis*, passed from organism to organism as food, and released to the atmosphere by *respiration*. Nitrogen originates in the atmosphere, is trapped by *nitrogen-fixing bacteria*, passes through a series of organisms, and is ultimately released to the atmosphere by *denitrifying bacteria*.

A *population* is a group of organisms of the same species in a particular place at a particular time. *Populations* differ from one another in *gene frequency, age distribution, sex ratio,* and *population density*. A typical population growth curve consists of a *lag phase* in which population rises very slowly, followed by an *exponential growth phase* in which the population increases at an accelerating rate, followed by a leveling off of the population in a *stable equilibrium phase* as the *carrying capacity* of the environment is reached.

Humans as a species have the same limits and influences that other organisms do. However, humans can reason and predict, thus offering the possibility of population control through *conscious population limitation.*

KEY TERMS

abiotic factors (p. **482**)
benthic (p. **491**)
biodiversity (p. **486**)
biomes (p. **487**)
biotic factors (p. **482**)
carbon cycle (p. **498**)
carnivore (p. **483**)
carrying capacity (p. **505**)
commensalism (p. **495**)
community (p. **482**)
competition (p. **498**)
consumer (p. **483**)
decomposers (p. **484**)
ecology (p. **482**)
ecosystem (p. **482**)
environment (p. **482**)
environmental resistance (p. **505**)
estuary (p. **491**)
exponential growth phase (p. **504**)
food chain (p. **483**)
free-living nitrogen-fixing bacteria (p. **501**)
freshwater community (p. **491**)
habitat (p. **493**)
herbivore (p. **483**)
keystone species (p. **487**)
lag phase (p. **504**)

limiting factors (p. **505**)
marine community (p. **491**)
mortality (p. **504**)
mutualism (p. **497**)
natality (p. **504**)
niche (p. **493**)
nitrogen cycle (p. **501**)
omnivore (p. **483**)
organism (p. **482**)
parasitism (p. **495**)
pelagic (p. **491**)
plankton (p. **491**)
population (p. **501**)
population growth curve (p. **504**)
predation (p. **495**)
primary carnivore (p. **483**)
primary consumer (p. **483**)
producer (p. **483**)
secondary carnivore (p. **483**)
secondary consumer (p. **483**)
stable equilibrium phase (p. **504**)
symbiotic nitrogen-fixing bacteria (p. **501**)
trophic level (p. **483**)

APPLYING THE CONCEPTS

1. As energy is passed from one trophic level to the next in a food chain, about _____ percent is lost at each transfer.
   (a) 10
   (b) 50
   (c) 75
   (d) 90

2. The primary factor that determines whether a geographic area will support temperate deciduous forest or prairie is
   (a) the amount of rainfall.
   (b) the severity of the winters.
   (c) the depth of the soil.
   (d) the kinds of animals present.

3. Which one of the following organisms is most likely to be at the second trophic level?
   (a) a maple tree
   (b) a snake
   (c) a fungus
   (d) an elephant

4. Tree → insect → spider → frog → fish → human
   In the food chain above, the spider is
   (a) a tertiary consumer.
   (b) a herbivore.
   (c) at the third trophic level.
   (d) a producer.

5. If two species of organisms occupy the same niche,
   (a) mutualism will result.
   (b) competition will be very intense.
   (c) both organisms will become extinct.
   (d) both will need to enlarge their habitat.

6. If nitrogen-fixing bacteria were to become extinct
   (a) life on earth would stop immediately since there would be no source of nitrogen.
   (b) life on earth would continue indefinitely.
   (c) life on earth would slowly dwindle as useful nitrogen became less available.
   (d) life on earth would be unchanged except that proteins would be less common.

7. Nitrogen of the organic molecules in your body comes from
   (a) the air you breath.
   (b) the water you drink.
   (c) carbohydrates in the food you eat.
   (d) proteins in the food you eat.

8. Many plants (flowers) provide nectar for insects. The insects, in turn, pollinate the flower. This relationship between the insect and plant represents
   (a) parasitism.
   (b) commensalism.
   (c) mutualism.
   (d) predation.

9. As population density increases, which one of the following is likely to occur?
   (a) Natality will increase.
   (b) Mortality will decrease.
   (c) The population will experience exponential growth.
   (d) Individuals will migrate from the area of highest density.

10. A population made up primarily of prereproductive individuals will
    (a) increase rapidly in the future.
    (b) become extinct.
    (c) rarely occur.
    (d) remain stable for several generations.

11. Listed below are the sex ratios for four populations. All other things being equal (including current population size), which population should experience the greatest future growth?
    (a) 1 male: 1 female
    (b) 2 male: 1 female
    (c) 1 male: 2 female
    (d) 3 male: 1 female

12. The current human population of the world is experiencing
    (a) a population decline.
    (b) slow steady growth.
    (c) stable equilibrium.
    (d) exponential growth.

Answers
1. d 2. a 3. d 4. c 5. b 6. c 7. d 8. c 9. d 10. a 11. c 12. d

QUESTIONS FOR THOUGHT

1. Describe the flow of energy through an ecosystem.
2. What role does each of the following play in an ecosystem: sunlight, plants, the second law of thermodynamics, consumers, decomposers, herbivores, carnivores, and omnivores?
3. Why is there usually a larger herbivore biomass than a carnivore biomass?
4. List a predominant abiotic factor in each of the following biomes: temperate deciduous forest, coniferous forest, desert, tundra, tropical rainforest, and savanna.
5. Can energy be recycled through an ecosystem? Explain why or why not.
6. Describe your niche.
7. What do parasites, commensal organisms, and mutualistic organisms have in common? How are they different?
8. Trace the flow of carbon atoms through a community that contains plants, herbivores, decomposers, and parasites.
9. Describe four different roles played by bacteria in the nitrogen cycle.
10. List four kinds of limiting factors that help to set the carrying capacity for a species.
11. If competition is intense how can both kinds of organisms continue to exist?
12. How does the population growth curve of humans compare with that of other kinds of animals?

Human Biology: Sex and Nutrition

This is a photograph of human sperm cells on the surface of an egg cell. In humans and many other organisms that reproduce sexually, sperm cells are the result of a complex cell division process known as meiosis.

O U T L I N E

SEXUAL REPRODUCTION

Nearly all organisms have a method of shuffling and exchanging genetic information. This involves organisms that have two sets of genetic data, one inherited from each of their parents **Sexual reproduction** is the formation of a new individual by the union of two sex cells. Before sexual reproduction can occur, each parent must form sex cells. The sex cells produced by male organisms are called **sperm** and those produced by females are called **eggs.** A general term sometimes used to refer to either eggs or sperm is the term **gamete** (sex cell). During this process, the two sets of genetic information found in each body cell must be reduced to one set. The cellular process that is responsible for generating gametes is called *gametogenesis.* This is somewhat similar to shuffling a deck of cards and dealing out hands; the shuffling and dealing assure that each hand will be different. An organism with two sets of chromosomes can produce many combinations of chromosomes when it produces sex cells, just as many different hands can be dealt from one pack of cards. When a sperm unites with an egg (*fertilization*), a new organism containing two sets of genetic information is formed. This new organism's genetic information might have survival advantages over the information found in either parent; this is the value of sexual reproduction.

The life cycle of a sexually reproducing organism begins with the fertilization of an egg by a sperm. This new cell called a *zygote* divides by mitosis as do the resulting cells and growth occurs. When the organism reaches sexual maturity it is able to produces gametes by *meiosis* and the life cycle is complete.

Notice in figure 23.1 that the zygote and its descendants have two sets of chromosomes. However, the male gamete and the female gamete each contain only one set of chromosomes. These sex cells are said to be **haploid.** The haploid number of chromosomes is noted as *n*. A zygote contains two sets and is said to be **diploid.**

Meiosis
Sex cell formation

Single adult cell ⟶ Single gamete (sperm or egg)
Contains 2 sets of chromosomes ⟶ Contains 1 set of chromosomes
2n ⟶ n

The diploid number of chromosomes is noted as $2n$ ($n + n = 2n$). Diploid cells have two sets of chromosomes, one set from each parent. Remember, a chromosome is composed of two chromatids, each containing double-stranded DNA. These two chromatids are attached to each other at a point called the *centromere.* In a diploid nucleus, the chromosomes occur as *homologous chromosomes*—a pair of chromosomes in a diploid cell that contain similar genes throughout their length. One of the chromosomes of a homologous pair was donated by the father, the other by the mother (figure 23.2).

It is necessary for organisms that reproduce sexually to form gametes having only one set of chromosomes. If gametes contained two sets of chromosomes, the zygote resulting from their union would have four sets of chromosomes. The number of chromosomes would continue to double with each new generation. However, this does not usually happen; the number of chromosomes remains constant generation after generation. How are sperm and egg cells formed so that they get only half the chromosomes of the diploid cell? The answers lie in the process of **meiosis,** the specialized pair of cell divisions that reduce the chromosome number from diploid ($2n$) to haploid (n). One of the major functions of meiosis is to produce cells that have one set of genetic information. Therefore, when fertilization occurs between a haploid sperm (n) and a haploid egg (n) the resulting zygote will have two sets of chromosomes ($n + $ n $= 2n$) as did each parent.

Not every cell goes through the process of meiosis. Only specialized organs are capable of producing haploid cells (figure 23.3). In animals, the organs in which meiosis occurs are called **gonads.** The female gonads that produce eggs are called **ovaries.** The male gonads that produce sperm are called **testes.** Organs that produce gametes are also found in algae and plants. Some of these are very simple. In algae such as *Spirogyra*, individual cells become specialized for gamete production. In plants, the structures are very complex. In flowering plants, the **pistil** produces eggs or ova, and the **anther** produces pollen, which contains sperm.

The Mechanics of Meiosis

The whole point of meiosis is to distribute the chromosomes and the genes they carry so that each daughter cell gets one member of each homologous pair. In this way, each daughter cell gets one complete set of genetic information. Meiosis con-

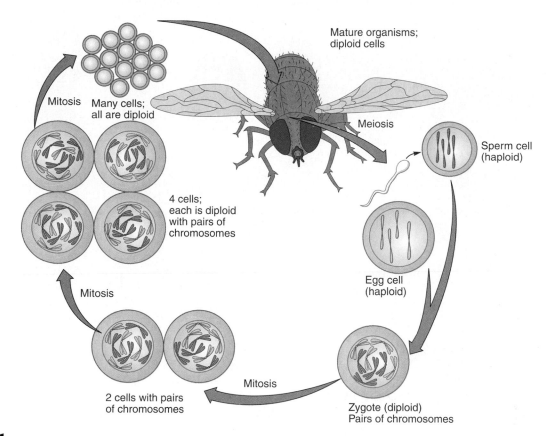

Figure 23.1

The cells of this adult fruit fly have eight chromosomes in their nuclei. In preparation for sexual reproduction, the number of chromosomes must be reduced by half so that fertilization will result in the original number of eight chromosomes in the new individual. The offspring will grow and produce new cells by mitosis, completing the life cycle.

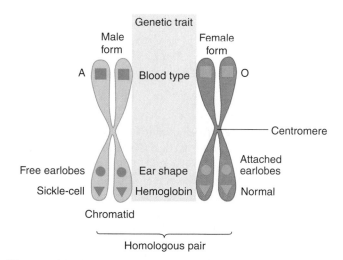

Figure 23.2

A pair of chromosomes of similar size and shape that have genes for the same traits are said to be homologous. Notice that the genes may not be identical but code for the same type of information. Homologous chromosomes are of the same length, have the same types of genes in the same sequence, and have their centromeres in the same location—one came from the male parent and the other was contributed by the female parent.

sists of two divisions known as meiosis I and meiosis II. During meiosis I the number of chromosomes is reduced from the diploid number to the haploid number. Therefore it is often referred to as a **reduction division.** The sequence of events in meiosis I is artificially divided into four phases: prophase I, metaphase I, anaphase I, and telophase I. Meiosis II includes

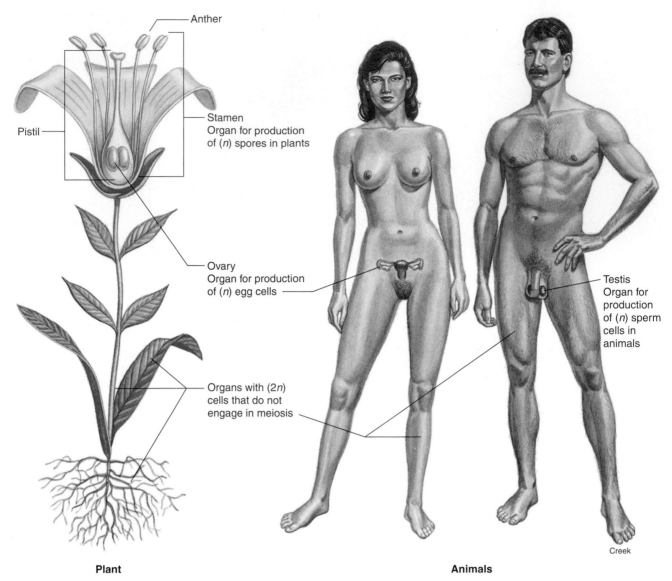

Anther

Stamen
Organ for production
of (*n*) spores in plants

Pistil

Ovary
Organ for production
of (*n*) egg cells

Organs with (2*n*)
cells that do not
engage in meiosis

Testis
Organ for
production
of (*n*) sperm
cells in
animals

Creek

Plant

Animals

Figure 23.3

Both plants and animals produce cells with a haploid number of chromosomes. The male anther in plants and the testes in animals produce haploid male cells, sperm. In both plants and animals, while the ovaries produce haploid female cells, eggs, most other cells are diploid.

four phases: prophase II, metaphase II, anaphase II, and telophase II. The two daughter cells formed during meiosis I continue through meiosis II, so that four cells usually result from the two divisions. Table 23.1 is a review of the stages of meiosis as they would occur in a cell with diploid number of chromosomes equal to 8, (i.e., $n = 4$).

Chromosomes and Sex Determination

You already know that there are several different kinds of chromosomes, that each chromosome carries genes unique to it, and that these genes are found at specific places on the chromosome. Furthermore, diploid organisms have homologous pairs of chromosomes. Genes determine sexual characteristics in the same manner as other types of characteristics. In many organisms, sex-determining genes are located on specific chromosomes known as **sex chromosomes.** All other chromosomes not involved in determining the sex of an individual are known as **autosomes.** In humans and all other mammals, and in some other organisms (e.g., fruit flies), the sex of an individual is determined by the presence of a certain chromosome combination. The genes that determine maleness are located on a small chromosome known as the *Y chromosome.* This Y chromosome behaves as if it and another larger chromosome, known as the *X chromosome,* were homologs, and pairs up with it during meiosis even though they are not identical. Most males have one X and one Y chromosome. Most females have two X chromosomes. Some animals have their sex determined in a completely different way. In bees, for example, the females are diploid and the males are haploid. Other plants and animals have still other chromosomal mechanisms for determining their sex.

Table 23.1 A Review of the Stages of Meiosis

Prophase I

During prophase I, the cell is preparing itself for division. The chromosomes are composed of two chromatids and coil so that they become shorter and thicker. Homologous chromosomes pair with each other. This pairing is called **synapsis.** While homologous chromosomes are paired they can exchange equivalent sections of DNA. This process is called **crossing-over** and is important in mixing up the genes that will be divided into separate gametes. Long fibers known as **spindle fibers** become attached to the chromosomes.

Metaphase I

The synapsed homologous chromosomes move to the equator of the cell as single units. How they are arranged on the equator (which one is on the left and which one is on the right) is determined by chance. In the drawing, three green chromosomes from the father and one purple chromosome from the mother are lined up on the left. Similarly, one green chromosome from the father and three purple chromosomes from the mother are on the right. They could have aligned themselves in several other ways.

Anaphase I

During this stage, homologous chromosomes separate and their number is reduced from diploid to haploid. The two members of each pair of homologous chromosomes move away from each other toward opposite poles. The direction each takes is determined by how each pair was originally arranged on the spindle. Each chromosome is independently attached to a spindle fiber at its centromere. Each chromosome still consists of two chromatids. Because the homologous chromosomes and the genes they carry are being separated from one another, this process is called **segregation.** The way in which a single pair of homologous chromosomes segregates does not influence how other pairs of homologous chromosomes segregate. That is, each pair segregates independently of other pairs. This is known as **independent assortment** of chromosomes.

Telophase I

The two newly forming daughter cells are now haploid (*n*) since each contains only one of each pair of homologous chromosomes.

Prophase II

The chromosomes of each of the two haploid daughter cells shorten and become attached to spindle fibers by their centromeres.

Metaphase II

Chromosomes move to the equator of the cell.

Anaphase II

Centromeres divide, allowing the chromatids to separate toward the poles.

Telophase II

Four haploid (*n*) cells are formed from the division of the two meiosis I cells. Each of these sex cells is genetically different from all the others. These cells become the sex cells (egg or sperm) of a higher organism.

It is possible to examine cells and count chromosomes. Among the easiest cells to view are white blood cells. They are dropped onto a microscope slide so that the cells are broken open and the chromosomes are separated. Photographs are taken of chromosomes from cells in the metaphase stage of mitosis. The chromosomes in the pictures can then be cut and arranged for comparison to known samples (box figure 23.1). This picture of an individual's chromosomal makeup is referred to as that person's *karyotype.*

In the normal process of meiosis, diploid cells have their number of chromosomes reduced to haploid. This involves segregating homologous chromosomes into separate cells during the first meiotic division. Occasionally, a pair of homologous chromosomes does not segregate properly during gametogenesis and both chromosomes of a pair end up in the same gamete. This kind of division is known as *nondisjunction.* One example of the effects of nondisjunction is the condition known as Down syndrome. If an abnormal gamete with two number 21 chromosomes has been fertilized by another containing the typical one copy of chromosome number 21, the resulting zygote would have 47 chromosomes (e.g., 24 from the female plus 23 from the male parent) (box figure 23.2). The child who developed from this fertilization would have 47 chromosomes in every cell of his or her body as a result of mitosis, and would have the symptoms characteristic of Down syndrome. These may include thickened eyelids, some mental impairment, and faulty speech (box figure 23.3). Premature aging is probably the most significant impact of this genetic disease. On the other hand, a child born with only one chromosome 21 rarely survives.

It was thought that the mother's age at childbirth played an important part in the occurrence of *trisomies* (three of one of the kinds

A

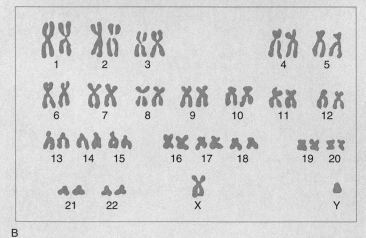

B   C

Box Figure 23.1

Spermatogenesis

One of the biological reasons for sexual activity is the production of offspring. The process of producing gametes includes meiosis and is called **gametogenesis** (gamete formation) (figure 23.4). The term **spermatogenesis** is used to describe gametogenesis that takes place in the testes of males. The two bean-shaped testes are composed of many small sperm-producing tubes, or **seminiferous tubules,** and collecting ducts that store sperm. These are held together by a thin covering membrane. The seminiferous tubules join together and eventually become the epididymis, a long, narrow convoluted tube in which sperm cells are stored and mature before ejaculation (figure 23.5).

Box Figure 23.2

Box Figure 23.3

of chromosomes instead of the normal two) such as Down syndrome. In women, gametogenesis begins early in life, but cells destined to become eggs are put on hold during meiosis I. Beginning at puberty and ending at menopause, one of these cells completes meiosis I monthly. This means that cells released for fertilization later in life are older than those released earlier in life. Therefore, it was believed that the chances of abnormalities such as nondisjunction increase as the age of the mother increases. However, the evidence no longer supports this age-egg link. Currently, the increase in frequency of trisomies with age has been correlated with a decrease in the activity of a woman's immune system. As she ages, her immune system is less likely to recognize the difference between an abnormal and a normal embryo. This means that she is more likely to carry an abnormal fetus to full term.

Box figure 23.4 illustrates the frequency of occurrence of Down syndrome at different ages in women. Notice that the frequency increases very rapidly after age 37. For this reason, many physicians encourage couples to have their children in their early to mid-twenties and not in their late thirties or early forties. Physicians normally encourage older women who are pregnant to have the cells of their fetus checked to see if they have the normal chromosome number. It is important to know that the male parent can also contribute the extra chromosome 21. However, it appears that this occurs less than 30 percent of the time.

Sometimes a portion of chromosome 14 may be cut out and joined to chromosome 21. The transfer of a piece of one nonhomologous chromosome to another is called a chromosomal *translocation.* A person with this 14/21 translocation is monosomic and has only 45 chromosomes; one 14 and one 21 are missing and replaced by the translocated 14/21. Statistically, about 15 percent of the children of carrier mothers inherit the 14/21 chromosome and have Down syndrome. Fewer of the children born to fathers with the 14/21 translocation inherit the abnormal chromosome and are Downic.

Whenever an individual is born with a chromosomal abnormality such as a monosomic or a trisomic condition, it is recommended that both parents have a karyotype evaluation in an attempt to identify the possible source of the problem. This is not to fix blame but to provide information on the likelihood that a next pregnancy would also result in a child with a chromosomal abnormality.

Box Figure 23.4

Leading from the epididymis is the vas deferens, or sperm duct; this empties into the urethra, which conducts the sperm out of the body through the *penis* (figure 23.6). Before puberty, the seminiferous tubules are packed solid with diploid cells called spermatogonia. These cells, which are found just inside the tubule wall, undergo *mitosis* and produce more spermatogonia. Beginning about age eleven, some of the spermatogonia specialize and begin the process of *meiosis,* while others continue to divide by mitosis, assuring a constant and continuous supply of spermatogonia. Once spermatogenesis begins, the seminiferous tubules become hollow and can transport the mature sperm.

Spermatogenesis involves several steps. It begins when some of the spermatogonia in the walls of the seminiferous tubules differentiate and enlarge to become *primary spermatocytes.* These diploid cells undergo the first meiotic division,

Connections . . .

Why Am I Different?

With the possible exception of identical twins, every human that has ever been born is genetically unique. The formation of haploid cells by meiosis and the combination of these sex cells to form a diploid offspring by sexual reproduction results in variety in the offspring. The five factors that influence genetic variation in offspring are mutations, crossing-over, segregation and independent assortment, and fertilization.

Mutations

Two types of mutations will be discussed in chapter 25: point mutations and chromosomal mutations. In point mutations, a change in a DNA nucleotide results in the production of a different protein. In chromosomal mutations, genes are rearranged when parts of chromosomes are duplicated, deleted, or inverted. By causing the production of different proteins, both types of mutations increase variation.

Crossing-Over

When crossing-over occurs during gametogenesis, pieces of genetic material are exchanged between the chromosomes you received from your mother and father. This exchange results in a chromosome with a new combination of genes, combinations of traits that were probably never before found in the family.

Segregation and Independent Assortment

The process of segregation relates to how the chromosomes donated by your father line up or assort themselves in pairs with those donated by your mother. How one pair of chromosomes lines up during metaphase I does not affect how the other pair lines up (i.e., they assort themselves independently of the others). If all those donated by your father line up on the right side of the cell during metaphase I and all those donated by your mother line up on the left, when the chromosomes separate or *segregate*, the resulting gametes will be of two kinds: half of your gametes will contain only your father's chromosomes and the other half of the gametes you form will contain only your mother's chromosomes. If however, some of the chromosomes from your father line up on the right side and some of your mother's line up on the left, the gametes you form will contain some of your mother's and some of your father's chromosomes. There are millions of ways in which these chromosomes can line up and divide.

Fertilization

Because of the large number of possible gametes resulting from independent assortment, segregation, mutations, and crossing-over, an incredibly large number of types of offspring can result. Because human males can produce millions of genetically different sperm and females can produce millions of genetically different eggs, the number of kinds of offspring possible is infinite for all practical purposes.

---

which produces two haploid *secondary spermatocytes.* The secondary spermatocytes go through the second meiotic division, resulting in four haploid *spermatids,* which lose much of their cytoplasm and develop long tails. These cells are then known as *sperm.* The sperm have only a small amount of food reserves. Therefore, once they are released and become active swimmers, they live no more than 72 hours. However, if the temperature is lowered drastically, the sperm become deactivated, freeze, and can live for years outside the testes. This has led to the development of sperm banks. Artificial insemination of cattle, horses, and other domesticated animals with sperm from sperm banks is common.

Spermatogenesis in human males takes place continuously throughout a male's reproductive life, although the number of sperm produced decreases as a man ages. Sperm counts can be taken and used to determine the probability of successful fertilization. For reasons not totally understood, a man must be able to release at least 100 million sperm at one insemination to be fertile. It appears that enzymes in the head of sperm are needed to digest through mucus and protein found in the female reproductive tract. Millions of sperm contribute in this way to the process of fertilization, but only one is involved in fertilizing the egg. A healthy male probably releases about 300 million sperm during each act of *sexual intercourse,* also known as *coitus* or *copulation.*

Oogenesis

The term **oogenesis** refers to the production of egg cells. This process starts during prenatal development of the ovary, when certain cells cease dividing by mitosis and enlarge to become *primary oocytes.* All of the primary oocytes that a woman will ever have are already formed prior to her birth. At this time they number approximately 2 million, but that number is reduced by cell death to between 300,000 to 400,000 cells by the time of puberty.

At puberty and on a regular basis thereafter, the sex hormones stimulate a primary oocyte to continue its maturation process, and it goes through the first meiotic division. But in telophase I, the two cells that form receive unequal portions of cytoplasm. You might think of it as a lopsided division (see figure 23.4). The smaller of the two cells is called a *polar body,* and the larger haploid cell is the *secondary oocyte* (people commonly refer to this cell as an egg or ovum although technically it is not). The other primary oocytes remain in the ovary. Ovulation begins when the soon-to-be released secondary oocyte, encased in a saclike structure known as a *follicle,* grows and moves near the surface of the ovary. When this maturation is complete, the follicle erupts and the secondary oocyte is released. It is swept into the oviduct (fallopian tube) by ciliated cells and travels toward the uterus (figure 23.7). The rest of the follicle develops

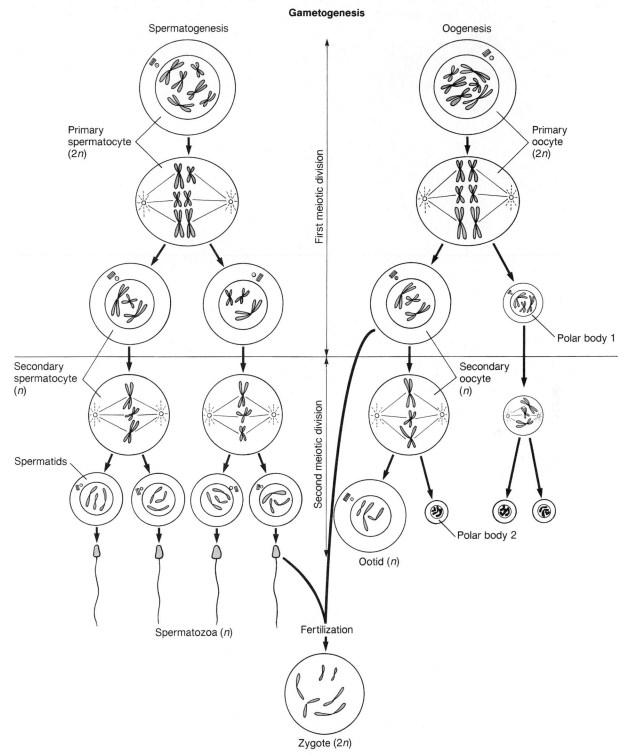

Gametogenesis

Spermatogenesis

Oogenesis

Primary spermatocyte (2n)

Primary oocyte (2n)

First meiotic division

Secondary spermatocyte (n)

Secondary oocyte (n)

Polar body 1

Second meiotic division

Spermatids

Ootid (n)

Polar body 2

Spermatozoa (n)

Fertilization

Zygote (2n)

Figure 23.4

This diagram illustrates the process of gametogenesis in human males and females. Not all of the 46 chromosomes are shown. Carefully follow the chromosomes as they segregate, recalling the details of the process of meiosis explained previously.

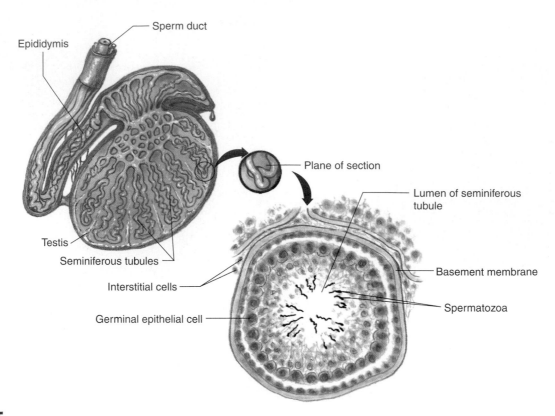

Figure 23.5

The testis consists of many tiny tubes called seminiferous tubules. The walls of the tubes consist of cells that continually divide, producing large numbers of sperm. The sperm leave the seminiferous tubules and enter the epididymis, where they are stored prior to ejaculation through the sperm duct.

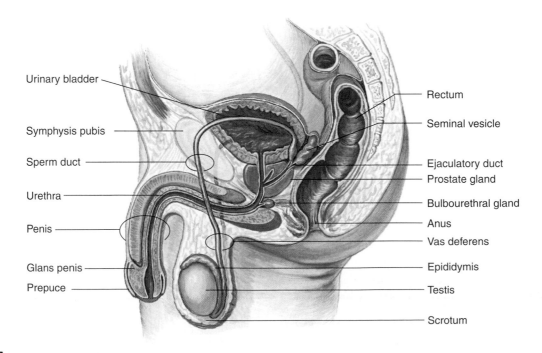

Figure 23.6

The male reproductive system consists of two testes that produce sperm, ducts that carry the sperm, and various glands. Muscular contractions propel the sperm through the vas deferens past the seminal vesicles, prostate gland, and bulbourethral gland, where most of the liquid of the semen is added. The semen passes through the urethra of the penis to the outside of the body.

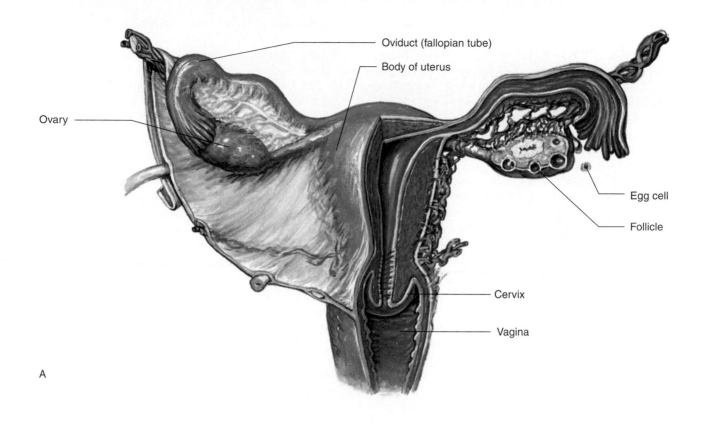

Oviduct (fallopian tube)

Body of uterus

Ovary

Egg cell

Follicle

Cervix

Vagina

A

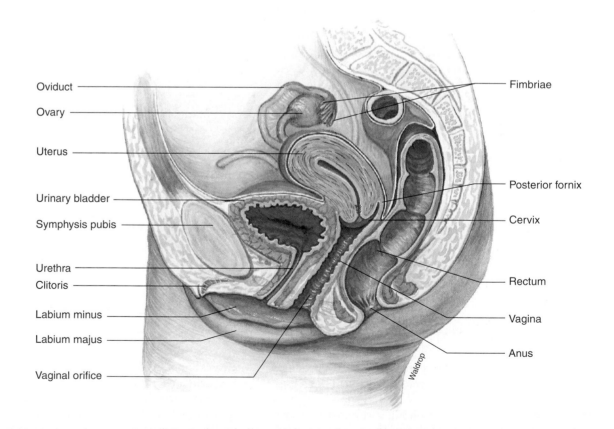

Oviduct

Fimbriae

Ovary

Uterus

Urinary bladder

Posterior fornix

Symphysis pubis

Cervix

Urethra

Clitoris

Rectum

Labium minus

Vagina

Labium majus

Anus

Vaginal orifice

B

Figure 23.7

In the human female reproductive system (*A*), after ovulation the cell travels down the oviduct to the uterus. If it is not fertilized, it is shed when the uterine lining is lost during menstruation. (*B*) The human female reproductive system, side view.

Diseases currently referred to as *sexually transmitted diseases (STDs)* were formerly called *venereal diseases (VDs)*. The term *venereal* is derived from the name of the Roman goddess for love, Venus. Although these kinds of illnesses are most frequently transmitted by sexual activity, many can also be spread by other methods of direct contact, by hypodermic needles, by blood transfusions, and by blood-contaminated materials. Currently, the Centers for Disease Control and Prevention (CDC) in Atlanta, Georgia, recognize over twenty diseases as being sexually transmitted (table 23.A)

The United States has the highest rate of sexually transmitted diseases among industrially developed countries—65 million people (nearly one-fifth of the population) in the United States have incurable sexually transmitted diseases. The CDC estimate there are 15 million new cases of sexually transmitted diseases each year, nearly 4 million among teenagers. Table 23.B lists the most common STDs and estimates of the number of new cases each year. The portions of the public that are most at risk are teenagers, minorities, and women. Some of the most important STDs are described here because of their high incidence in the population and our inability to bring some of them under control. For example, there is no known cure for the HIV virus that is responsible for AIDs. There has also been a sharp rise in the number of gonorrhea cases in the United States caused by a form of the bacterium *Neisseria gonorrhoeae* that has become resistant to the drug penicillin by producing an enzyme that actually destroys the antibiotic. However, most of the infectious agents can be controlled if diagnosis occurs early and treatment programs are carefully followed by the patient.

The spread of STDs during sexual intercourse is significantly diminished by the use of condoms. Other types of sexual contact (i.e., hand, oral, anal) and congenital transmission (i.e., from the mother to the fetus during pregnancy) help to maintain some of these diseases in the population at high enough levels to warrant attention by public health officials, the U.S. Public Health Service, the CDC, and state and local public health agencies. All of these agencies are involved in attempts to raise the general public health to a higher level. Their investigations have resulted in the successful control of many diseases and the identification of special problems, such as those associated with the STDs. Because the United States has an incidence rate of STDs that is 50 to 100 times higher than other industrially developed countries, there is still much that needs to be done.

All public health agencies are responsible for warning members of the public about things that may be dangerous to them. In order to meet these obligations when dealing with sexually transmitted diseases, such as AIDS and syphilis, they encourage the use of one of their most potent weapons, sex education. Individuals must know about their own sexuality if they are to understand the transmission and nature of STDs. Then it will be possible to alter their behavior in ways that will prevent the spread of these diseases. The intent is to present people with biological facts, not to scare them. Public health officials do not have the luxury of advancing their personal opinions when it comes to their jobs. The biological nature of sexual behavior is not a moral issue, but biological facts are needed if people are to make intelligent decisions relating to their sexual behavior. It is hoped that through education, people will alter their high-risk sexual behaviors and avoid situations where they could become infected with one of the STDs.

High-risk behaviors associated with contracting STDs include: sex with multiple partners and failing to use condoms. While some STDs are simply inconvenient or annoying, others severely compromise health and can result in death. As one health official stated, "We should be knowledgeable enough about our own sexuality and the STDs to answer the question, Is what I'm about to do worth dying for?"

Table 23.A

Sexually Transmitted Diseases

Disease	Agent
Genital herpes	Virus
Gonorrhea	Bacterium
Syphilis	Bacterium
Acquired immunodeficiency syndrome (AIDS)	Virus
Candidiasis	Yeast
Chancroid	Bacterium
Genital warts	Virus
Gardnerella vaginalis	Bacterium
Genital *Chlamydia* infection	Bacterium
Genital cytomegalovirus infection	Virus
Genital *Mycoplasma* infection	Bacterium
Group B *Streptococcus* infection	Bacterium
Nongonococcal urethritis	Bacterium
Pelvic inflammatory disease (PID)	Bacterium
Molluscum contagiosum	Virus
Crabs	Body lice
Scabies	Mite
Trichomoniasis	Protozoan
Hepatits B	Virus
Gay bowel syndrome	Variety of agents

Table 23.B

Yearly Estimates of the Number of New Cases of Sexually Transmitted Diseases*

Disease	New Cases Each Year (Estimate)
Genital warts (Human papillomavirus)	5.5 million
Trichomoniasis	5 million
Chlamydia	3 million
Genital herpes	1 million
Gonorrhea	650,000
Hepatitis B	120,000
Syphilis	70,000
Human immunodeficieny virus (HIV) (AIDS)	40,000

* Data from the Centers for Disease Control and Prevention publication, *Tracking the Hidden Epidemics Trends in STDs in the United States 2000*.

into a glandlike structure, the *corpus luteum,* which produces hormones (progesterone and estrogen) that prevent the release of other eggs.

If the secondary oocyte is fertilized, it completes meiosis with the sperm DNA inside and becomes the zygote. If the secondary oocyte is not fertilized, it passes through the *vagina* to the outside during menstruation. During her lifetime, a female releases about 300 to 500 secondary oocytes. Obviously, few of these cells are fertilized.

One of the characteristics to note here is the relative age of the sex cells. In males, sperm production is continuous throughout life. Sperm do not remain in the tubes of the male reproductive system for very long. They are either released shortly after they form or die and are harmlessly absorbed. In females,

The Human Genome Project was first proposed in 1986 by the U.S. Department of Energy (DOE), and cosponsored soon after by the National Institutes of Health (NIH). These agencies were the main research agencies within the U.S. government responsible for developing and planning the project. Later, a private U.S. corporation, Celera Genomics, joined the effort as a competitor. It is one of the most ambitious projects ever undertaken in the biological sciences. The goal was nothing less than the complete characterization of the genetic makeup of humans. The project was completed early in 2001 when the nucleotide sequence of all 23 pairs of human chromosomes was determined. With this in hand, scientists will now be able to produce a map of each of the chromosomes that will show the names and places of all our genes. This international project involving about one hundred laboratories worldwide required only sixteen years to complete. Work began in many of these labs in 1990. Powerful computers are used to store and share the enormous amount of information derived from the analyses of human DNA. To get an idea of the size of this project, consider this: A human Y chromosome (one of the smallest of the human chromosomes) is estimated to be composed of 28 million nitrogenous bases. The larger X chromosome may be composed of 160 million nitrogenous base pairs!

Two kinds of work progressed simultaneously. First, *physical maps* were constructed by determining the location of specific "markers" (known sequences of bases) and their closeness to genes (Box Figure 23.5). A kind of chromosome map already exists that pictures patterns of colored bands on chromosomes, a result of chromosome-staining procedures. Using these banded chromosomes, the markers were then related to these colored bands on a specific region of a chromosome. Work is continuing on the Human Genome Project to identify the location of specific genes. Each year a more complete picture is revealed. The second kind of work was for labs to determine the exact order of nitrogenous bases of the DNA for each chromosome. Techniques exist for determining base sequences, but it is a time-consuming job to sort out the several million bases that may be found in any one chromosome. Coming from behind with new, speedier techniques, Celera Genomics was able to catch up to NIH labs and completed their sequencing at almost the same time. The benefit of having these two organizations as competitors is that when finished they could compare and contrast results. Amazingly, the discrepancies between their findings were declared insignificant. It was originally estimated, for example, that there were between 100,000 and 140,000 genes in the human genome. However, when the results were compared the evidence from both organizations indicated that there are only 30,000 to 40,000 genes. Knowing this information provides insights into the evolution of humans and the mutation rates of males versus females. This knowledge will make future efforts to work with the genome through bioengineering much easier.

When the physical maps are finally completed for all of the human genes, it will be possible to examine a person's DNA and identify genetic abnormalities. This could be extremely useful in diagnosing diseases and providing genetic counseling to those considering having children. This kind of information would also create possibilities for new gene therapies. Once it is known where an abnormal gene is located and how it differs in base sequence from the normal DNA sequence, steps could be taken to correct the abnormality. However, there is also a concern that, as knowledge of our genetic makeup becomes easier to determine, some people may attempt to use this information for profit or political power. This is a real concern because some health insurance companies refuse to insure people with "preexisting conditions" or those at "genetic risk" for certain abnormalities. They fail to realize that between 5 and 50 such "conditions" or mutations are normally found in each individual. Refusing to provide coverage would save these companies the expense of future medical bills incurred by "less-than-perfect" people. Another fear is that attempts may be made to "breed out" certain genes and people from the human population in order to create a "perfect race."

Here are some other intriguing findings from the human genome and the genome identification projects of other organisms:

- Human genes are not scattered at random among the human chromosomes. "Forests" or "clusters" of genes are found on certain chromosomes separated by "deserts" of genes. For example, chromosomes 17 and 19 are forested with thousands of genes while chromosome 18 has many fewer genes.
- Rice appears to have about 50,000 genes.
- Roundworms have about 26,000 genes.
- Fruits flies contain an estimated 13,600 genes.
- Yeast cells have about 6,241 genes.

Peutz-Jeghers syndrome
Diabetes mellitus, insulin resistant
Eye color, green/blue
Mannosidosis
Alzheimer's disease, late onset

Mullerian duct syndrome
Lymphoid leukemia

Atherosclerosis
Familial hypercholesterolemia

Familial Hemiplegic Migraine
Prostate-specific antigen

Immunodeficiency, HLA (II)

Multiple epiphysial dysplasia
Pseudoachondroplasia

Central core disease
Malignant hyperthermia

Polio susceptibility

Xeroderma pigmentosum, D
DNA ligase I deficiency

Hemolytic anemia
Congenital Nephrotic Syndrome

Maple syrup urine disease
Hyperlipoproteinemia (IIIb, II)
Myotonic dystrophy
Hypogonadism

Glutaricacidurea, IIB

Box Figure 23.5

Genes Known to Be on Human Chromosome Number 19. The gene map shows the appropriate positions of several genes known to be on human chromosome number 19.

—Continued top of next page

- There are numerous and virtually identical genes found in many organisms that appear to be very distantly related—for example, mice, humans, and yeasts.

- Genes jump around (transposons) within the chromosomes more than scientists ever thought.
- The mutation rate of male humans is about twice that of females.

- Humans are about 99.9 percent identical at the DNA level! Scientists believe that there is virtually no basis for race since there is much greater variation within a so-called race than there is between the so-called races.

meiosis begins before birth, but the oogenesis process is not completed, and the cell is not released for many years. A secondary oocyte released when a woman is 37 years old began meiosis 37 years before! During that time, the cell was exposed to many influences, a number of which may have damaged the DNA or interfered with the meiotic process. The increased risk of abnormal births in older mothers may be related to the age of the eggs. Such alterations are less likely to occur in males because new gametes are being produced continuously. Also, defective sperm appear to be much less likely to be involved in fertilization.

Hormones control the cycle of changes in breast tissue, in the ovaries, and in the uterus. In particular, estrogen and progesterone stimulate milk production by the breasts and cause the lining of the uterus to become thicker and filled with blood vessels prior to ovulation. This ensures that if fertilization occurs, the resultant embryo will be able to attach itself to the wall of the uterus and receive nourishment. If the cell is not fertilized, the lining of the uterus is shed. This is known as *menstruation, menstrual flow,* the *menses,* or a *period.* Once the wall of the uterus has been shed, it begins to build up again. This continual building up and shedding of the wall of the uterus is known as the menstrual cycle.

At the same time that hormones are regulating ovulation and the menstrual cycle, changes are taking place in the breasts. The same hormones that prepare the uterus to receive the embryo also prepare the breasts to produce milk. These changes in the breasts, however, are relatively minor unless pregnancy occurs.

Sexual Maturation of Young Adults

Following birth, sexuality plays only a small part in physical development for several years. Culture and environment shape the responses that the individual will come to recognize as normal behavior. During **puberty,** normally between 12 and 14 years of age, increasing production of sex hormones causes major changes as the individual reaches sexual maturity. Generally females reach puberty six months to a year before males. After puberty, humans are sexually mature and have the capacity to produce offspring.

The Maturation of Females

Female children typically begin to produce quantities of sex hormones from the hypothalamus, pituitary gland, ovaries, and adrenal glands at 8 to 13 years of age. This marks the onset of

puberty. The **hypothalamus** controls the functioning of many other glands throughout the body, including the **pituitary gland.** At puberty, the pituitary gland in girls begins to produce **follicle-stimulating hormone (FSH).** The increasing amount of this hormone circulating in the blood causes the ovaries to grow and, under the influence of another hormone, luteinizing hormone, begin producing larger quantities of **estrogen.** The increasing supply of estrogen is responsible for the many changes in sexual development that can be noted at this time. These changes include breast growth, changes in the walls of the uterus and vagina, increased blood supply to the clitoris, and changes in the pelvic bone structure.

Estrogen also stimulates the female adrenal gland to produce **androgens,** male sex hormones. The androgens are responsible for the production of pubic hair and they seem to have an influence on the female sex drive. The adrenal gland secretions may also be involved in the development of acne. Those features that are not primarily involved in sexual reproduction but are characteristic of a sex are called **secondary sex characteristics.** In women, the distribution of body hair, patterns of fat deposits, and a higher voice are examples.

A major development during this time is the establishment of the **menstrual cycle.** This involves the periodic growth and shedding of the lining of the uterus. These changes are under the control of a number of hormones produced by the ovaries. The ovaries are stimulated to release their hormones by the pituitary gland, which is in turn influenced by the ovarian hormones. Follicle-stimulating hormone (FSH) and luteinizing hormone (LH) are both produced by the pituitary gland. FSH causes the maturation and development of the ovaries, and LH is important in causing ovulation and in maintaining the menstrual cycle. In addition, secretions of the hypothalamus at puberty bring about the development of a menstrual cycle (table 23.2). Associated with the menstrual cycle is the periodic release of sex cells from the surface of the ovary, called **ovulation** (figure 23.8). Initially, these two cycles, menstruation and ovulation, may be irregular, which is normal during puberty. Eventually, hormone production becomes regulated so that ovulation and menstruation take place on a regular monthly basis in most women, although normal cycles may vary from 21 to 45 days.

As girls progress through puberty, curiosity about the changing female body form and new feelings leads to self-investigation. Studies have shown that sexual activity such as manipulation of the clitoris, which causes pleasurable sensations, is performed by a large percentage of young women. Self-

Table 23.2 Human Reproductive Hormones

Hormone	Production Site	Target Organ	Function
1. Prolactin (lactogenic or luteotropic hormone)	Pituitary gland	Breasts, ovary	Stimulates milk production; also helps maintain normal ovarian cycle
2. Follicle-stimulating hormone	Pituitary gland	Ovary, testis	Stimulates ovary and testis development; stimulates egg production in females and sperm production in males
3. Luteinizing hormone (interstitial cell-stimulating hormone)	Pituitary gland	Ovary, testis	Stimulates ovulation in females and sex-hormone (estrogen and testosterone) production in both males and females
4. Estrogen	Follicle of the ovary	Entire body	Stimulates development of female reproductive tract and secondary sex characteristics
5. Testosterone	Testes	Entire body	Stimulates development of male reproductive tract and secondary sex characteristics
6. Progesterone	Ovaries	Uterus, breasts	Causes uterine thickening and maturation; maintains pregnancy
7. Oxytocin	Pituitary gland	Breasts, uterus	Causes uterus to contract and breasts to release milk
8. Androgens	Testes, adrenal glands	Entire body	Stimulate development of male reproductive tract and secondary sex characteristics in males and females

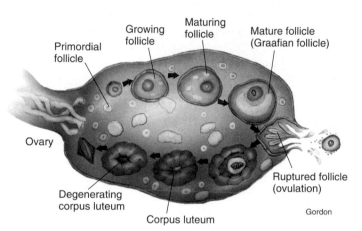

Gordon

Figure 23.8

In the ovary, the egg begins development inside a sac of cells known as a follicle. Each month, one of these follicles develops and releases its product. This release through the wall of the ovary is known as ovulation.

stimulation, frequently to orgasm, is a common result. This stimulation is termed *masturbation,* and it should be stressed that it is considered a normal part of sexual development. *Orgasm* is a complex response to mental and physical stimulation that causes rhythmic contractions of the muscles of the reproductive organs and an intense frenzy of excitement.

The Maturation of Males

Males typically reach puberty about two years later than females, (ages 10 to 15), but puberty in males also begins with a change in hormone levels. At puberty, follicle-stimulating hormone (FSH), the same pituitary-gland secretion produced by females, increases. This is the primary growth stimulator of the testes and ovaries. FSH produced by the male during embryology was responsible for the embryonic development of testes. FSH also stimulates the production of luteinizing hormone, which is often known as *interstitial cell-stimulating hormone (ICSH)* in males, which stimulates the testes to produce **testosterone,** the primary sex hormone in males. The testosterone produced by the embryonic testes caused the differentiation of internal and external genital anatomy in the male embryo. Testosterone is also important in the maturation and production of sperm.

The major changes during puberty include growth of the testes and scrotum, pubic-hair development, and increased size of the penis. Secondary sex characteristics begin to become apparent at age 13 or 14. Facial hair, underarm hair, and chest hair are some of the most obvious. The male voice changes as the larynx (voice box) begins to change shape. Body contours also change and a growth spurt increases height. In addition, the proportion of the body that is muscle increases, while the proportion of body fat decreases. At this time, a boy's body begins to take on the characteristic adult male shape, with broader shoulders and heavier muscles.

In addition to these external changes, the FSH released from the pituitary causes the production of seminal fluid by the *seminal vesicles,* prostate gland, and the bulbourethral glands (figure 23.6). Later, FSH stimulates the production of sperm cells. The release of sperm cells and seminal fluid also begins during puberty and is termed *ejaculation.* This release is generally accompanied by the pleasurable sensations of orgasm. The sensations associated with ejaculation may lead to self-stimulation, or masturbation. Masturbation is a common and normal activity as a boy goes through puberty. Studies of sexual behavior have shown that nearly all men masturbate at some time during their lives.

NUTRITION

Organisms maintain themselves by constantly processing molecules to provide building blocks for new living material and energy to sustain themselves. Autotrophs can manufacture organic molecules from inorganic molecules, but heterotrophs must consume organic molecules to get what they need. All molecules required to support living things are called **nutrients.** Some nutrients are inorganic molecules such as calcium, iron, or potassium, others are organic molecules such as carbohydrates, proteins, fats, and vitamins. All heterotrophs obtain the nutrients they need from food and each kind of heterotroph has particular nutritional requirements. This section deals with the nutritional requirements of humans.

Living Things as Chemical Factories: Matter and Energy Manipulators

The word **nutrition** is used in two related contexts. First of all nutrition is a branch of science that seeks to understand food, its nutrients, how the nutrients are used by the body, and how inappropriate combinations or quantities of nutrients lead to ill health. The word nutrition is also used in a slightly different context to refer to all the processes by which we take in food and utilize it, including *ingestion, digestion, absorption,* and *assimilation. Ingestion* involves the process of taking food into the body through eating. *Digestion* involves the breakdown of complex food molecules to simpler molecules. *Absorption* involves the movement of simple molecules from the digestive system to the circulatory system for dispersal throughout the body. *Assimilation* involves the modification and incorporation of absorbed molecules into the structure of the organism.

Many of the nutrients that enter living cells undergo chemical changes before they are incorporated into the body. These interconversion processes are ultimately under the control of the genetic material, DNA. It is DNA that codes the information necessary to manufacture the enzymes required to extract energy from chemical bonds and to convert raw materials (nutrients) into the structure (anatomy) of the organism.

The food and drink consumed from day to day constitutes a person's **diet.** It must contain the minimal nutrients necessary to manufacture and maintain the body's structure (bones, skin, tendons, muscles, etc.) and regulatory molecules (enzymes and hormones), and to supply the energy (ATP) needed to run the body's machinery. If the diet is deficient in nutrients, or if a person's body cannot process nutrients efficiently, a dietary deficiency and ill health may result. A good understanding of nutrition can promote good health and involves an understanding of the energy and nutrient content in various foods.

Kilocalories, Basal Metabolism, and Weight Control

The unit used to measure the amount of energy in foods is the **kilocalorie (kcal).** One kilocalorie is the amount of heat needed to raise the temperature of one *kilogram* of water one degree Celsius. Remember that the prefix *kilo-* means "1,000 times" the value listed. Therefore, a kilocalorie is 1,000 times more heat energy than a **calorie,** which is the amount of heat needed to raise the temperature of one *gram* of water one degree Celsius. However, the amount of energy contained in food is usually called a Calorie with a capital C. This is unfortunate because it is easy to confuse a Calorie, which is really a kilocalorie, with a calorie. Most books on nutrition and dieting use the term Calorie to refer to *food Calories.* The energy requirements in kilocalories for a variety of activities are listed in table 23.3.

Significant energy expenditure is required for muscular activity. However, even at rest, energy is required to maintain breathing, heart rate, and other normal body functions. The rate at which the body uses energy when at rest is known as the **basal metabolic rate (BMR).** The basal metabolism of most people requires more energy than their voluntary muscular activity. Much of this energy is used to keep your body temperature constant. A true measurement of basal metabolic rate requires a measurement of oxygen used over a specific period under controlled conditions.

Since few of us rest for twenty-four hours a day, we normally require more than the energy needed for basal metabolism. One of these requirements is the amount of energy needed to process the food that we eat. This is called **specific dynamic action (SDA)** and is equal to approximately 10 percent of your total daily kilocalorie intake. In addition to basal metabolism and specific dynamic action, the activity level of a person determines the number of kilocalories needed.

Depending on how obesity is defined, obesity is a problem for 30 percent to 40 percent of the population of the United

---

Table 23.3 Typical Energy Requirements for Common Activities			
Light Activities (120–150 kcal/h)	**Light to Moderate Activities (150–300 kcal/h)**	**Moderate Activities (300–400 kcal/h)**	**Heavy Activities (420–600 kcal/h)**
Dressing	Sweeping floors	Pulling weeds	Chopping wood
Typing	Painting	Walking behind lawnmower	Shoveling snow
Studying	Store clerking	Walking 3.5–4 mi/h on level surface	Walking 5 mi/h
Standing	Bowling	Calisthenics	Walking up hills
Slow walking	Walking 2–3 mi/h	Canoeing 4 mi/h	Cross-country skiing
Sitting activities	Canoeing 2.5–3 mi/h	Doubles tennis	Swimming
	Bicycling on level surface at 5.5 mi/h	Volleyball	Jogging 5 mi/h
		Golf (no cart)	Bicycling 11–12 mi/h or in hills

Connections . . .

Measuring the Caloric Value of Foods

A *bomb calorimeter* is an instrument that is used to determine the heat of combustion or caloric value of foods. To operate the instrument, a small sample is formed into a pellet and sealed inside the bomb, a strong container. The bomb is filled with 30 atmospheres of oxygen and then placed in a surrounding jacket that is filled with water. The sample is electrically ignited and as it aerobically reacts with the oxygen, the bomb releases heat to the surrounding water jacket. As the heat is released the temperature of the water rises. Recall that a calorie is the amount of energy (or heat) needed to increase the temperature of one gram of water one degree Celsius. Therefore, if one gram of food results in a water temperature increase of 13°C for each gram of water, that food has the equivalent of 13 calories.

States. Why is weight control a problem for such a large portion of the population? There are several metabolic pathways that convert carbohydrates (glucose) or proteins to fat. Stored body fat was very important for our prehistoric ancestors, because it allowed them to survive periods of food scarcity. In periods of food scarcity the stored body fat can be used to supply energy. The glycerol portion of the fat molecule can be converted to a small amount of glucose, which can supply energy for red blood cells and nervous tissue that must have glucose. The fatty acid portion of the molecule can be metabolized by most other tissues directly to produce ATP. Today, however, for most of us food scarcity is not a problem, and even small amounts of excess food consumed daily tend to add to our fat stores.

Although energy doesn't weigh anything, the nutrients that contain the energy do. Weight control is a matter of balancing the kilocalories ingested as a result of dietary intake with the kilocalories of energy expended by normal daily activities and exercise. There is a limit to the rate at which a moderately active human body can use fat as an energy source. At the most, one or two pounds of fat tissue per week are lost by an average person when dieting. Since one pound of fatty tissue contains about 3,500 kilocalories, decreasing your kilocalorie intake by 500 to 1,000 kilocalories per day while maintaining a balanced diet (including proteins, carbohydrates, and fats) will result in fat loss of one to two pounds per week. (A pound of pure fat contains about 4,100 kilocalories, but fat tissue contains other materials besides fat, such as water.)

For those who need to gain weight, increasing kilocalorie intake by 500 to 1,000 kilocalories per day will result in an increase of one to two pounds per week, provided the low weight is not the result of a health problem.

If, like millions of others, you feel that you are overweight, you have probably tried numerous diet plans. Not all of these plans are the same, and not all are suitable to your particular situation. If a diet plan is to be valuable in promoting good health, it must satisfy your needs in several ways. It must provide you with needed kilocalories, protein, fat, and carbohydrates. It should also contain readily available foods from all the basic food groups, and it should provide enough variety to prevent you from becoming bored with the plan and going off the diet too soon. A diet should not be something you follow only for a while, then abandon and regain the lost weight.

Amounts and Sources of Nutrients

In an attempt to give people some guidelines for planning a diet that provides adequate amounts of the six classes of nutrients, in the 1970s nutritional scientists and government officials in the United States and many other countries developed nutrient standards. In the United States, these guidelines are known as the **Recommended Dietary Allowances, or RDAs.** RDAs are dietary recommendations, not requirements or minimum standards. They are based on what is thought to be the needs of a healthy person already eating an adequate diet. RDAs do not apply to persons with medical problems who are under stress or suffering from malnutrition. The amount of each nutrient specified by the RDAs has been set relatively high so that most of the population eating those quantities will be meeting their nutritional needs. Keep in mind that since everybody is different, eating the RDA amounts may not meet your personal needs. You may have a special need for additional amounts of a particular nutrient if you have an unusual metabolic condition.

The U.S. RDAs are used when preparing product labels. The federal government requires by law that labels list ingredients from the greatest to the least in quantity. The volume in the package must be stated along with the weight, and the name of the manufacturer or distributor. If any nutritional claim is made, it must be supported by factual information.

A product label that proclaims, for example, that a serving of cereal provides 25 percent of the RDA for vitamin A means that you are getting at least one-fourth of your RDA of vitamin A from a single serving of that cereal. To figure your total RDA of vitamin A, consult a published RDA table for

CONCEPTS APPLIED
At Home with the RDAs

The next time you're in the kitchen or grocery store, read and compare some labels. Compare the percentage of RDAs found in one product from each of the basic food groups to determine which has the highest amount of your chosen nutrient. Now select three foods from the same food group and make a similar comparison.

RDAs and the Glycemic Index—A Reexamination

Recently, the recommendation to reduce the amount of fat and increase the amount of carbohydrates in your diet has been brought under question. Many have thought that all fats are bad and that by avoiding them, it would be possible to prevent weight gain. However, it is now known that certain types of lipids are beneficial if not necessary for optimum health (see chapter 18, "A Closer Look: Omega Fatty Acids and Your Diet").

It is also known that the consumption of large amounts of certain carbohydrates can be detrimental. Carbohydrates are ranked on the *glycemic index* (GI). The GI is a scale that relates to how fast carbohydrates can cause an increase in blood glucose levels. Glucose, also known as blood sugar, is at the top of the GI. Certain complex carbohydrates (e.g., potatoes) are also high on the GI since they are rapidly broken down into glucose by cooking or digestion. Such foods stimulate the production of insulin. With an increase in insulin, there is a greater potential for the development of prob-

lems such as insulin resistance, type II diabetes, and obesity. *Insulin resistance* is a reduced sensitivity to the action of insulin (i.e., Insulin causes glucose to move from the blood into the cells where it can be metabolized.) When this occurs, the body attempts to overcome this resistance by secreting more insulin from the pancreas. The symptoms include abnormal cholesterol and/or triglyceride levels, high blood pressure, and upper body obesity, all risk factors for cardiac disease.

The development of type II diabetes occurs when the pancreas fails to sustain this increase in insulin secretion. Along with high blood levels of insulin, there is an increase in appetite. Increasing your intake of low GI foods (below 55) would be more beneficial. Stay away from high GI foods (i.e., foods with a GI of more than 70). This new knowledge, makes it clear that the Food Guide Pyramid needs to be reexamined.

Estimated GI of Some Foods[1]

Food	GI	Food	GI	Food	GI
Tofu frozen dessert non-dairy	164	Refined white bread	93	Sweet corn	55
Dates	146	Cantaloupe	93	Apple	54
Parsnips	139	High fructose corn syrup	89	Milk—skimmed	46
Potato—baked	121	Ice cream	87	Converted rice	44
French fries	107	Kellogg's corn flakes	84	Snickers candy	41
Pumpkin	107	Pita bread	82	Pasta—spaghetti	38
Graham crackers	106	Popcorn	79	Milk—whole	39
Potato—boiled/mashed	104	Ice cream, low fat	71	Yogurt low fat	33
Glucose	*100*	Chocolate	70	Soy beans	18
Parsnips	97	Baked beans	69	Peanuts	14
Whole wheat bread	97	Oat bran bread	68		
Taco shell	97	Oatmeal	55		

[1]many of these values are for the raw, unprepared food item. Realize that pumpkin pie will not have the same GI as raw pumpkin.

Source: Modified from *Integrated Science*, Second ed, Tillery, Enger, and Ross, The McGraw-Hill Companies, Inc.

adults. It tells you that an adult male requires 1,000 and a female 800 micrograms (μg) of vitamin A per day. Of this, 25 percent is 250 and 200 μg, respectively—the amount you are getting in a serving of that cereal. You will need to get the additional amounts (750 for men and 600 for women) by having more of that cereal or eating other foods that contain vitamin A. If a product claims to have 100 percent of the RDA of a particular nutrient, that amount must be present in the product. However, restricting yourself to that one product will surely deprive you of many of the other nutrients necessary for good health. Ideally, you should eat a variety of complex foods containing a variety of nutrients to ensure that all your health requirements are met.

The Food Guide Pyramid with Five Food Groups

Using RDAs and product labels is a pretty complicated way for a person to plan a diet. Planning a diet around basic food groups is generally easier. The four basic food groups first developed and introduced in 1953 have been modified and updated several times to serve as guidelines in maintaining a balanced diet (figure 23.9). In May 1992, the U.S. Department of Agriculture released the results of its most recent study on how best to educate the public about daily nutrition. The federal government adopted the **Food Guide Pyramid** of the Department of Agriculture as one of its primary tools to help the

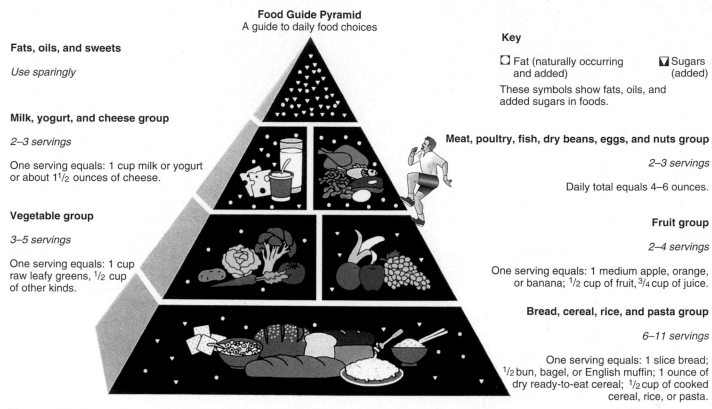

Food Guide Pyramid
A guide to daily food choices

Fats, oils, and sweets

Use sparingly

Milk, yogurt, and cheese group

2–3 servings

One serving equals: 1 cup milk or yogurt or about 1½ ounces of cheese.

Vegetable group

3–5 servings

One serving equals: 1 cup raw leafy greens, ½ cup of other kinds.

Key

☐ Fat (naturally occurring and added)   ◣ Sugars (added)

These symbols show fats, oils, and added sugars in foods.

Meat, poultry, fish, dry beans, eggs, and nuts group

2–3 servings

Daily total equals 4–6 ounces.

Fruit group

2–4 servings

One serving equals: 1 medium apple, orange, or banana; ½ cup of fruit, ¾ cup of juice.

Bread, cereal, rice, and pasta group

6–11 servings

One serving equals: 1 slice bread; ½ bun, bagel, or English muffin; 1 ounce of dry ready-to-eat cereal; ½ cup of cooked cereal, rice, or pasta.

Figure 23.9

In May 1992, the Department of Agriculture released a new guide to good eating. This Food Guide Pyramid suggests that we eat particular amounts of five different food groups while decreasing our intake of fats and sugars. This guide should simplify our menu planning and help ensure that we get all the recommended amounts of basic nutrients. In order to be healthy, exercising on a regular basis is essential.

Source: U.S. Department of Agriculture, 1992.

general public plan for good nutrition. The Food Guide Pyramid contains five basic groups of foods with guidelines for the amounts one needs daily from each group for ideal nutritional planning. The Food Guide Pyramid differs from previous federal government information in than it encourages a reduction in the amount of fats and sugars in the diet while increasing our daily servings of fruits and vegetables. In addition, the new guidelines suggest significantly increasing the amount of grain products we eat each day.

Grain Products Group

Grains include vitamin-enriched or whole-grain cereals and products such as breads, bagels, buns, crackers, dry and cooked cereals, pancakes, pasta, and tortillas. Items in this group are typically dry and seldom need refrigeration. They should provide most of your kilocalorie requirements in the form of complex carbohydrates like starch, which is the main ingredient in most grain products. These foods help you feel you have satisfied your appetite, and many of them are very low in fat.

Fruits Group

You probably can remember discussions of whether a tomato is a vegetable or a fruit. This controversy arises from the fact that the term *vegetable* is not scientifically precise but means a plant material eaten during the main part of the meal. *Fruit,* on the

other hand, is a botanical term for the structure that is produced from the female part of the flower to protect and nourish the ripening seeds. Although, botanically, green beans, peas, and corn are all fruits, nutritionally speaking, they are placed in the vegetable category because they are generally eaten during the main part of the meal. Nutritionally speaking, fruits include such sweet plant products as melons, berries, apples, oranges, and bananas. The Food Guide Pyramid suggests two to four servings of fruit per day. However, since these foods tend to be high in natural sugars, consumption of large amounts of fruits can add significant amounts of kilocalories to your diet. A small apple, half a grapefruit, a half cup of grapes, or six ounces of fruit juice is considered a serving. Fruits provide fiber, carbohydrate, water, and certain vitamins, particularly vitamin C.

Vegetables Group

The Food Guide Pyramid suggests three to five servings from this group each day. Items in this group include nonsweet plant materials, such as broccoli, carrots, cabbage, corn, green beans, tomatoes, potatoes, lettuce, and spinach. A serving is considered one cup of raw leafy vegetables or a half cup of other types. It is wise to include as much variety as possible in this group. If you eat only carrots, several cups each day can become very boring. There is increasing evidence indicating that cabbage, broccoli, and cauliflower can provide some protection from certain types of cancers. This is a good reason to include these foods in your diet.

Foods in this group provide vitamins A and C as well as water and minerals. They also provide fiber, which assists in the proper functioning of the digestive tract.

Dairy Products Group

All of the cheeses, ice cream, yogurt, and milk are in this group. Two servings from this group are recommended each day. Each of these servings should be about 1 cup, or 2 ounces of cheese. Using product labels will help you determine the appropriate serving size of individual items. This group provides minerals, such as calcium, in your diet, but also provides water, vitamins, carbohydrates, and protein. You must also remember that many cheeses contain large amounts of cholesterol and fat for each serving. Low-fat dairy products are increasingly common as manufacturers seek to match their products with the desire of the public for less fat in the diet.

Meat, Poultry, Fish, and Dry Beans Group

This group contains most of the things we eat as a source of protein; for example, nuts, peas, tofu, and eggs are considered a part of this group. It is recommended that we include 4 to 6 ounces (100–170 grams) of these items in our daily diet. In the past, people in the economically developed world ate at least this quantity and frequently much more. Since many sources of protein also include significant fat, and health recommendations suggest reduced fat, more attention is being paid to the quantity of the protein-rich foods in the diet. Not only have we decreased our intake of items from this group, but we also have shifted from the high-fat-content foods, such as beef and pork, to foods that are high in protein but lower in fat content, such as fish and poultry. Beans (except for the oil-rich soybean) are also excellent ways to get needed protein without unwanted fat. Modern food preparers tend to use smaller portions and cook foods in ways that reduce the fat content. Broiled fish, rather than fried, and baked, skinless portions of chicken or turkey (the fat is attached to the skin) are seen more and more often on restaurant menus and on dining-room tables at home.

Remember that the recommended daily portion from this group is only 4 to 6 ounces. This means that one double cheeseburger greatly exceeds this recommendation for the daily intake. Eating excessive amounts of protein can stress the kidneys by causing higher concentrations of calcium in the urine, increase the demand for water to remove toxic keto acids produced from the breakdown of amino acids, and lead to weight gain because of the fat normally associated with many sources of protein. It should be noted, however, that vegetarians must pay particular attention to acquiring adequate sources of protein because they have eliminated a major source from their diet. Although nuts and soybeans are high in protein, they should not be consumed in large quantities because they are also high in fats.

Eating Disorders

The three most common eating disorders are obesity, bulimia, and anorexia nervosa. All three disorders are related to the pre-vailing perceptions and values of the culture in which we live. In many cases there is a strong psychological component as well. Some studies have shown that there is also a shared familial predisposition to these disorders. It has been suggested that genetic factors probably influence the risk for eating disorders and contribute to their co-occurrence with major depression.

Obesity

Overweight and obesity have reached nationwide epidemic proportions. Over 60 million Americans are considered to be obese. Adults who gain a great deal of unnecessary weight and are 15 percent to 20 percent above their ideal weight or have a *body mass index* (BMI = weight/height^2) above 30 are defined as obese. **Obesity** is the condition of being overweight to the extent that a person's health and life span are adversely affected. Obesity occurs when people consistently take in more food energy than is necessary to meet their daily requirements.

On the surface it would appear that obesity is a simple problem to solve. To lose weight all that people must do is consume fewer kilocalories, exercise more, or do both. While all obese people have an imbalance between their need for kilocalories and the amount of food they eat, the reasons for this imbalance are complex and varied.

Bulimia

Bulimia ("hunger of an ox" in Greek) is a disease condition in which the person has a cycle of eating binges followed by purging the body of the food by inducing vomiting or the use of laxatives. Many bulimics also use diuretics that cause the body to lose water and therefore reduce weight. It is often called the silent killer because it is difficult to detect. Bulimics are usually of normal body weight or overweight. The cause is thought to be psychological, stemming from depression, low self-esteem, displaced anger, a need to be in control of one's body, or a personality disorder. Many bulimics have other compulsive behaviors such as drug abuse and are often involved in incidences of theft and suicide.

Vomiting may be induced physically or by the use of some nonprescription drugs. Case studies have shown that bulimics may take forty to sixty laxatives a day to rid themselves of food. For some, the laxative becomes addictive. The binge-purge cycle and associated use of diuretics results in a variety of symptoms that can be deadly.

Anorexia Nervosa

Anorexia nervosa is a nutritional deficiency disease characterized by severe, prolonged weight loss as a result of a voluntary severe restriction in food intake. An anorexic person's fear of becoming overweight is so intense that even though weight loss occurs, it does not lessen the fear of obesity, and the person continues to diet, often even refusing to maintain the optimum body weight for his or her age, sex, and height. This nutritional deficiency disease is thought to stem from sociocultural factors. Our society's preoccupation with weight loss and the desirability of being thin strongly influences this disorder.

Just turn on your television or radio, or look at newspapers, magazines, or billboards, and you can see how our culture encourages people to be thin. Male and female models are thin.

Myths or Misunderstandings About Diet and Nutrition

Myth or Misunderstanding	Reality
1. Exercise burns calories.	You don't burn calories, but you do oxidize (burn) the fuels to provide yourself with the energy needed to perform various activities.
2. Active people who are increasing their fitness need more protein.	The amount of protein needed is very small. Most people get many times the amount of protein required.
3. Vitamins give you energy.	Vitamins are involved with enzymes in the release of energy from food items, but they are not sources of energy.
4. Large amounts of protein are needed to build muscle.	A person can build only a few grams of new muscle per day. Therefore, consuming large amounts of protein will not increase the rate of muscle growth.
5. Large quantities (mega doses) of vitamins will fight disease, build strength, and increase length of life.	Quantities of vitamins that greatly exceed the Recommended Dietary Allowances (RDAs) have not been shown to be beneficial. Large doses of some vitamins are toxic (vitamins A, D, B_3).
6. Special protein supplements are more quickly absorbed and can build muscle faster.	There is adequate protein in nearly all diets. The supplements may be absorbed faster but that does not mean that they will be incorprated into muscle mass faster.

Muscular bodies are considered healthy and any stored body fat unhealthy. Unless you are thin, so the advertisements imply, you will never be popular, get a date, or even marry. Our culture's constant emphasis on being thin has influenced many people to become anorexic and lose too much weight. Anorexic individuals frequently starve themselves to death. Individuals with anorexia are mostly adolescent and preadolescent females, although the disease does occur in males.

Deficiency Diseases

Without minimal levels of the essential amino acids in the diet, a person may develop health problems that could ultimately lead to death. In many parts of the world, large populations of people live on diets that are very high in carbohydrates and fats but low in complete protein. This is easy to understand, since carbohydrates and fats are inexpensive to grow and process in comparison to proteins. For example, corn, rice, wheat, and barley are all high-carbohydrate foods. Corn and its products (meal, flour) contain protein, but it is an incomplete protein that has very low amounts of the amino acids tryptophan and lysine. Without enough of these amino acids, many necessary enzymes cannot be made in sufficient amounts to keep a person healthy. One protein-deficiency disease is called *kwashiorkor,* and the symptoms are easily seen (figure 23.10). A person with this deficiency has a distended belly, slow growth, slow movement, and is emotionally depressed. If the disease is caught in time, brain damage may be prevented and death averted. This requires a change in diet that includes expensive protein, such as poultry, fish, beef, shrimp, or milk. As the world food problem increases, these expensive foods will be in even shorter supply and will become more and more costly.

A Closer Look   Nutritional Needs Associated with Pregnancy and Lactation

Risk-management practices that help in avoiding chronic adult diseases become even more important when planning pregnancy. Studies have shown that an inadequate supply of the essential nutrients can result in infertility, spontaneous abortion, and abnormal fetal development. The period of pregnancy and milk production (lactation) requires that special attention be paid to the diet to ensure proper fetal development, a safe delivery, and a healthy milk supply.

The daily amount of essential nutrients must be increased, as should the kilocaloric intake. Kilocalories must be increased by 300 per day to meet the needs of an increased basal metabolic rate (BMR); the development of the uterus, breasts, and placenta; and the work required for fetal growth. Some of these kilocalories can be obtained by drinking milk, which simultaneously supplies calcium needed for fetal bone development. In those individuals who cannot tolerate milk, supplementary sources of calcium should be used. In addition, the daily intake of protein should be at least 65 grams per day. As was mentioned earlier, most people consume much more than this amount of protein per day. Two essential nutrients, folic acid and iron, should be obtained through prenatal supplements because they are so essential to cell division and development of the fetal blood supply.

The mother's nutritional status affects the developing baby in several ways. If she is under 15 years of age or has had three or more pregnancies in a two-year period, her nutritional stores are inadequate to support a successful pregnancy. The use of drugs such as alcohol, caffeine, nicotine, and "hard" drugs (e.g., heroin) can result in decreased nutrient exchange between the mother and fetus. In particular, heavy smoking can result in low birth weights, and alcohol abuse is responsible for fetal alcohol syndrome (FAS).

Figure 23.10

This starving child shows the symptoms of kwashiorkor, a protein-deficiency disease. If treated with a proper diet containing all amino acids, the symptoms can be reduced.

Starvation is also a common problem in many parts of the world. Very little carbohydrate is stored in the body. If you starve yourself, this small amount will last as a stored form of energy only for about two days. Even after a few hours of fasting, the body begins to use its stored fat deposits as a source of energy; and as soon as the carbohydrates are gone proteins will begin to be used to provide a source of glucose. Some of the keto acids produced during the breakdown of fats and amino acids are released in the breath and can be detected as an unusual odor. People who are fasting, anorexic, diabetic, or have other metabolic problems often have this "ketone breath." During the early stages of starvation, the amount of fat in the body steadily decreases, but the amount of protein drops only slightly (20–30 grams per day) (figure 23.11). This can continue only up to a certain point. After several weeks of fasting, so much fat has been lost from the body that proteins are no longer protected, and cells begin to use them as a primary source of energy (as much as 125 grams per day). This results in a loss of proteins from the cells that prevents them from carrying out their normal functions. When starvation gets to this point it is usually fatal.

The lack of a particular vitamin in the diet can result in a **vitamin-deficiency disease.** A great deal has been said about the need for vitamin and mineral supplements in diets. Some people claim that supplements are essential, while others claim that a well-balanced diet provides adequate amounts of vitamins and minerals. Supporters of vitamin supplements have even claimed that extremely high doses of certain vitamins can prevent poor health or even create "supermen." It is very difficult to substantiate many of these claims, however. Since the function of vitamins and minerals and their regulation in the body is not completely understood, the RDAs are best estimates by experts who have looked at the data from a variety of studies. In fact, the minimum daily requirement of a number of vitamins has not been determined. Vitamin-deficiency diseases that show recognizable symptoms are extremely rare except in cases of extremely poor nutrition.

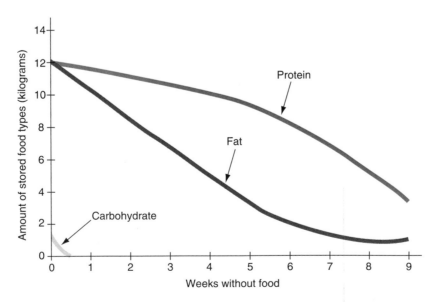

Figure 23.11

Starving yourself results in a very selective loss of the kinds of nutrients stored in the body. Notice how the protein level in the body has the slowest decrease of the three nutrients. This protein-conservation mechanism enables the body to preserve essential amounts of enzymes and other vital proteins.

SUMMARY

Meiosis is a specialized process of cell division resulting in the production of four separate cells, each of which has the *haploid number* of *chromosomes.* The total process involves two sequential divisions during which one *diploid cell* reduces to four *haploid cells.* Since the *chromosomes* act as carriers for *genetic information, genes* separate into different sets during meiosis.

Specific chromosome combinations determine the sex of many kinds of organisms. In humans, *females* have *two X chromosomes,* while *males* have *an X and a Y chromosome.* At *puberty, hormones* influence the development of *secondary sex characteristics* and the functioning of *gonads.*

Sexual reproduction involves the production of *gametes* by *meiosis* in the *ovaries and testes.* In males, each cell that undergoes *spermatogenesis* results in four *sperm;* in females, each cell that undergoes *oogenesis* results in one *oocyte.*

To maintain good health, people must receive *nutrient molecules* that can enter the cells and function in the metabolic processes. The *proper quantity and quality of nutrients* are essential to good health. An important measure of the amount of energy required to sustain a human at rest is the *basal metabolic rate.* To meet this and all additional requirements, the United States has established the *RDAs, Recommended Dietary Allowances,* for each nutrient. Should there be metabolic or psychological problems associated with a person's *normal metabolism,* a variety of disorders may occur, including *obesity, anorexia nervosa, bulimia, kwashiorkor,* and *vitamin-deficiency diseases.*

KEY TERMS

androgens (p. **524**)
anorexia nervosa (p. **530**)
anther (p. **512**)
autosomes (p. **514**)
basal metabolic rate (BMR)
   (p. **526**)
bulimia (p. **530**)
calorie (p. **526**)
crossing-over (p. **515**)
diet (p. **526**)
diploid (p. **512**)
egg (p. **512**)
estrogen (p. **524**)
follicle-stimulating hormone
   (FSH) (p. **524**)
Food Guide Pyramid (p. **528**)
gamete (p. **512**)
gametogenesis (p. **516**)
gonad (p. **512**)
haploid (p. **512**)
hypothalamus (p. **524**)
independent assortment
   (p. **515**)
kilocalorie (p. **526**)
meiosis (p. **512**)
menstrual cycle (p. **524**)
nutrients (p. **525**)
nutrition (p. **526**)

obesity (p. **530**)
oogenesis (p. **518**)
ovaries (p. **512**)
ovulation (p. **524**)
pistil (p. **512**)
pituitary gland (p. **524**)
puberty (p. **524**)
Recommended Dietary
   Allowances (RDAs)
   (p. **527**)
reduction division (p. **513**)
secondary sex characteristics
   (p. **524**)
segregation (p. **515**)
seminiferous tubules (p. **516**)
sex chromosomes (p. **514**)
sexual reproduction (p. **512**)
specific dynamic action
   (SDA) (p. **526**)
sperm (p. **512**)
spermatogenesis (p. **516**)
spindle fibers (p. **515**)
synapsis (p. **515**)
testes (p. **512**)
testosterone (p. **525**)
vitamin-deficiency disease
   (p. **532**)

APPLYING THE CONCEPTS

1. The exchange of genetic material (genes) between segments of homologous chromosomes results in
   (a) new gene combinations.
   (b) zygotes.
   (c) diploid cells.
   (d) segregation of genes.

2. A process that occurs during prophase I is
   (a) segregation.
   (b) synapsis.
   (c) reduction division.
   (d) independent assortment.

3. The fact that each homologous pair of chromosomes in humans separates and moves to the poles without being influenced by the other pairs is
   (a) segregation.
   (b) disintegration.
   (c) independent assortment.
   (d) fertilization.

4. An organism having a diploid number of 12, forms gametes having
   (a) 6 chromosomes.
   (b) 12 chromosomes.
   (c) 18 chromosomes.
   (d) 24 chromosomes.

5. Gametogenesis produces
   (a) sex cells.
   (b) gonads.
   (c) zygotes.
   (d) testes.

6. Ovaries and testes differ from other kinds of organs because
   (a) their cells divide.
   (b) their cells live for only a short time.
   (c) meiosis occurs in them.
   (d) they grow continuously.

7. A complete protein contains
   (a) all the Recommended Dietary Allowances.
   (b) essential fatty acids.
   (c) the suggested kilocalorie amount.
   (d) all the essential amino acids.

8. Converting a nutrient into a molecule in your body's cells is
   (a) digestion.
   (b) assimilation.
   (c) carbohydrate loading.
   (d) protein-sparing.

9. RDAs are easily met by
   (a) eating breakfast cereal.
   (b) using vitamin supplements when you are not feeling well.
   (c) eating a well-balance, varied diet.
   (d) consuming more fiber.

10. Carbohydrates that are not digested in the human digestive tract are
    (a) fibers.
    (b) minerals.
    (c) vitamins.
    (d) fatty acids.
11. The Calories listed on package labels are equal to _____ calories.
    (a) 1
    (b) 100
    (c) 1,000
    (d) 10,000
12. Osteoporosis results from a _____ deficiency.
    (a) vitamin
    (b) calcium
    (c) protein
    (d) energy

Answers

1. a 2. b 3. a 4. a 5. a 6. c 7. d 8. b 9. c 10. a 11. c 12. b

QUESTIONS FOR THOUGHT

1. How do haploid cells differ from diploid cells?
2. What is unique about prophase I?
3. Why is meiosis necessary in organisms that reproduce sexually?
4. Diagram the metaphase I stage of a cell with the diploid number of 8.
5. Diagram fertilization as it would occur between a sperm and an egg with the haploid number of 3.
6. Describe the processes that cause about 50 percent of babies to be born male and 50 percent female.
7. What structures are associated with the human female reproductive system? What are their functions?
8. What structures are associated with the human male reproductive system? What are their functions?
9. What are the differences between oogenesis and spermatogenesis in humans?
10. How are ovulation and menses related to each other?
11. Name the five basic food groups and give two examples of each.
12. What do the initials RDA stand for?

Human Biology: Materials Exchange and Control Mechanisms

Because many different systems each perform a specific set of tasks in the human body, it is necessary to coordinate these various activities with one another.

O U T L I N E

All cells must obtain nutrients and get rid of waste products through their cell membranes. In single-celled organisms the cell membrane is in contact with the cell's surroundings so diffusion and other processes accomplish this exchange. However, large, multicellular organisms consist of trillions of individual cells, most of which are not located near the surface of the organism, yet must exchange materials in the same manner as in single-celled creatures. Therefore, large organisms consist of many different systems that assist in the exchange of materials such as food, wastes, and gases between the cells of the organisms and the external environment. The primary systems involved are the circulatory, respiratory, digestive, and excretory systems.

These systems must be coordinated and regulated in such a way that they supply the needs of the individual cells. Two other systems, the nervous and endocrine systems, are involved in managing these functions. This chapter will look at the systems involved in the exchange of materials and coordination of activities within the human body.

EXCHANGING MATERIALS: BASIC PRINCIPLES

Cells are highly organized units that require a constant flow of energy in order to maintain themselves. The energy they require is provided in the form of nutrient molecules that enter the cell. Oxygen is required for the efficient release of energy from the large organic molecules that serve as fuel. Inevitably, as oxidation takes place, waste products form that are useless or toxic. These must be removed from the cell. All these exchanges of food, oxygen, and waste products must take place through the cell surface.

The ability to transport materials into or out of a cell is determined by its surface area, while its metabolic demands are determined by its volume. So, the larger a cell becomes, the more difficult it is to satisfy its needs. Some cells overcome this handicap by having highly folded cell membranes that substantially increase their surface area. This is particularly true of cells that line the intestine or are involved in the transport of large numbers of nutrient molecules. These cells have many tiny, folded extensions of the cell membrane called *microvilli* (figure 24.1).

In addition to the limitation that surface area presents to the transport of materials, large cells also have a problem with diffusion. The molecular process of diffusion is quite rapid over short distances but becomes very slow over longer distances. Diffusion is generally insufficient to handle the needs of cells if it must take place over a distance of more than 1 millimeter. The center of the cell would die before it received the molecules it needed if the distance were greater. Because of this, it is understandable that the basic unit of life, the cell, must remain small.

All single-celled organisms are limited to a small body size because they handle the exchange of molecules through their cell membranes. Large, multicellular organisms consist of a multitude of cells, many of which are located far from the surface of the organism. Each cell within a multicellular organism must solve the same materials-exchange problems as single-celled organisms. Large organisms have several interrelated systems that are involved in the exchange and transport of materials so that each cell can meet its metabolic needs.

Microvilli →

A

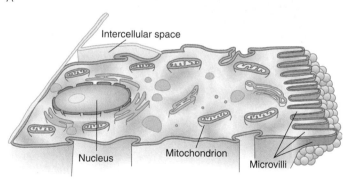

Intercellular space

Nucleus

Mitochondrion

Microvilli

B

Figure 24.1

Intestinal cells that are in contact with the food in the gut have highly folded surfaces. The tiny projections of these cells are called microvilli. These can be clearly seen in the photomicrograph in (*A*). The drawing in (*B*) shows that only one surface has these projections.

TRANSPORTING MATERIALS: THE CIRCULATORY SYSTEM

Large, multicellular organisms like humans consist of trillions of cells. Since many of these cells are buried within the organism far from the body surface, there must be some sort of distribution system to assist them in solving their materials-exchange problems. The primary mechanism used is the circulatory system.

The circulatory system consists of several fundamental parts. **Blood** is the fluid medium that assists in the transport of materials and heat. The **heart** is a pump that forces the fluid blood from one part of the body to another. The heart pumps blood into **arteries,** which distribute blood to organs. It flows into successively smaller arteries until it reaches tiny vessels called **capillaries,** where materials are exchanged between the blood and tissues through the walls of the capillaries. The blood flows from the capillaries into **veins** that combine into larger veins that ultimately return the blood to the heart from the tissues.

The Nature of Blood

Blood is a fluid that consists of several kinds of cells suspended in a watery matrix called **plasma.** This fluid plasma also contains many kinds of dissolved molecules. The primary function of the blood is to transport molecules, cells, and heat from one part of the body to another. The major kinds of molecules that are distributed by the blood are respiratory gases (oxygen and carbon dioxide), nutrients of various kinds, waste products, disease-fighting antibodies, and chemical messengers (hormones). Blood has special characteristics that allow it to distribute respiratory gases very efficiently. Although little oxygen is carried as free, dissolved oxygen in the plasma, *red blood cells* (*RBCs*) contain **hemoglobin,** an iron-containing molecule, to which oxygen molecules readily bind. This allows for much more oxygen to be carried than could be possible if it were simply dissolved in the blood.

Table 24.1 lists the variety of cells found in blood. While the red, hemoglobin-containing *erythrocytes* serve in the transport of oxygen and carbon dioxide, the *white blood cells* (*WBCs*) carried in the blood are involved in defending against harmful agents. These cells help the body resist many diseases. They constitute the core of the *immune system.* The various WBCs participate in providing immunity in several ways. First of all, immune system cells are able to recognize cells and molecules that are dangerous to the body. If a molecule is recognized as dangerous, certain white blood cells called *lymphocytes* produce *antibodies* (*immunoglobulins*) that attach to the dangerous materials. The dangerous molecules that stimulate the production of antibodies are called *antigens* (*immunogens*). When harmful microorganisms (e.g., bacteria, viruses, fungi), or toxic molecules enter the body or cancer cells are produced, other WBCs (1) recognize, (2) boost their abilities to respond to, (3) move toward, and (4) destroy the problem causers. *Neutrophils,* and *monocytes* are specific kinds of WBCs capable of engulfing dan-

gerous material, a process called phagocytosis. Thus they are often called *phagocytes*. While most can move from the bloodstream into the surrounding tissue, monocytes undergo such a striking increase in size that they are given a different name—*macrophages.* Macrophages can be found throughout the body and are the most active of the phagocytes. *Eosinophils* are primarily involved in destroying multicellular parasites such as worms, and *basophils* appear to be primarily involved in the inflammation response, which results in increased blood flow to an injured area, causing swelling and reddening.

Another kind of cellular particle in the blood is the *platelet.* These are fragments of specific kinds of white blood cells and are important in blood clotting. They collect at the site of a wound where they break down, releasing specific molecules, which begin a series of reactions that results in the formation of fibers that trap bloods cells and form a plug in the opening of the wound.

The Heart

Blood can perform its transportation function only if it moves. The organ responsible for providing the energy to pump the blood is the heart. The heart is a muscular pump that provides the pressure necessary to propel the blood throughout the body. It must continue its cycle of contraction and relaxation, or blood stops flowing and body cells are unable to obtain nutrients or get rid of wastes. Some cells, such as brain cells, are extremely sensitive to having their flow of blood interrupted because they require a constant supply of glucose and oxygen. Others, such as muscle cells or skin cells, are much better able to withstand temporary interruptions of blood flow.

The hearts of humans, other mammals, and birds consist of four chambers and four sets of valves that work together to ensure that blood flows in one direction only. Two of these chambers, the right and left *atria* (singular, *atrium*), are relatively thin walled structures that collect blood from the major veins and empty it into the larger, more muscular ventricles (figure 24.2).

The right and left *ventricles* are chambers that have powerful muscular walls whose contraction forces blood to flow through the arteries to all parts of the body. The valves between the atria and ventricles, known as *atrioventricular valves,* are important one-way valves that allow the blood to flow from the atria to the ventricles but prevent flow in the opposite direction. Similarly, there are valves in the aorta and pulmonary artery, known as *semilunar valves.* The **aorta** is the large artery that carries blood from the left ventricle to the body, and the **pulmonary artery** carries blood from the right ventricle to the lungs. The semilunar valves prevent blood from flowing back into the ventricles. If the atrioventricular or semilunar valves are damaged or function improperly, the efficiency of the heart as a pump is diminished, and the person may develop an enlarged heart or other symptoms.

The right and left sides of the heart have slightly different jobs, because they pump blood to different parts of the body. The right side of the heart receives blood from the general body and pumps it through the pulmonary arteries to the lungs,

Table 24.1 The Composition of Blood

Component	Quantity Present	Function
Plasma	55%	Maintain fluid nature of blood
Water	91.5%	
Protein	7.0%	
Other materials	1.5%	
Cellular material	45%	Carry oxygen and carbon dioxide
Red blood cells (erythrocytes)	4.3–5.8 million/mm^3	
White blood cells (leukocytes)	5–9 thousand/mm^3	Immunity
Lymphocytes	25%–30% of white cells present	
Monocytes	3%–7% of white cells present	
Neutrophils	57%–67% of white cells present	
Eosinophils	1%–3% of white cells present	
Basophils	Less than 1% of white cells present	
Platelets	250–400 thousand/mm^3	Clotting

Neutrophils   Eosinophils   Basophils

Lymphocytes   Monocytes   Platelets   Erythrocytes

where exchange of oxygen and carbon dioxide takes place and the blood returns from the lungs to the left atrium. This is called **pulmonary circulation.** The larger, more powerful left side of the heart receives blood from the lungs, delivers it through the aorta to all parts of the body, and returns it to the right atrium by way of large veins known as the *superior vena cava* and *inferior vena cava.* This is known as **systemic circulation.** Both circulatory pathways are shown in figure 24.3. The systemic circulation is responsible for gas, nutrient, and waste exchange in all parts of the body except the lungs.

Arteries, Veins, and Capillaries

Arteries and veins are the tubes that transport blood from one place to another within the body. Figure 24.4 compares the structure and function of arteries and veins. Arteries carry blood away from the heart, because it is under considerable pressure from the contraction of the ventricles. The contraction of the walls of the ventricles increases the pressure in the arteries. A typical pressure recorded in a large artery while the heart is contracting is about 120 millimeters of mercury. This is known as

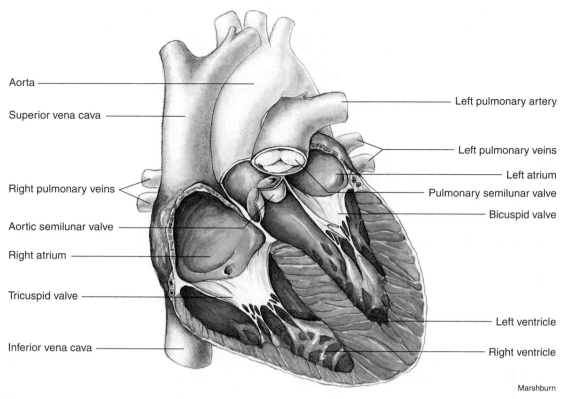

Figure 24.2

The heart consists of two thin-walled chambers called atria that contract to force blood into the two ventricles. When the ventricles contract, the atrioventricular valves (bicuspid and tricuspid) close, and blood is forced into the aorta and pulmonary artery. Semilunar valves in the aorta and pulmonary artery prevent the blood from flowing back into the ventricles when they relax.

the **systolic blood pressure.** The pressure recorded while the heart is relaxing is about 80 millimeters of mercury. This is known as the **diastolic blood pressure.** A blood pressure reading includes both of these numbers and is recorded as 120/80.

The walls of arteries are relatively thick and muscular, yet elastic. Healthy arteries have the ability to expand as blood is pumped into them and return to normal as the pressure drops. This ability to expand absorbs some of the pressure and reduces the peak pressure within the arteries, thus reducing the likelihood that they will burst. If arteries become hardened and less resilient, the peak blood pressure rises and they are more likely to rupture. The elastic nature of the arteries is also responsible for assisting the flow of blood. When they return to normal from their stretched condition they give a little push to the blood that is flowing through them.

Blood is distributed from the large aorta through smaller and smaller blood vessels to millions of tiny capillaries. Some of the smaller arteries, called **arterioles,** may contract or relax to regulate the flow of blood to specific parts of the body. Major parts of the body that receive differing amounts of blood, depending on need, are the digestive system, muscles, and skin.

Veins collect blood from the capillaries and return it to the heart. The pressure in these blood vessels is very low. Some of the largest veins may have a blood pressure of 0.0 mmHg for brief periods. Since pressure in veins is so low, muscular move-

ments of the body are important in helping to return blood to the heart. When muscles of the body contract, they compress nearby veins, and this pressure pushes blood along in the veins. Because valves in the veins allow blood to flow only toward the heart, this activity acts as an additional pump to help return blood to the heart.

Although the arteries are responsible for distributing blood to various parts of the body and arterioles regulate where blood goes, it is the function of capillaries to assist in the exchange of materials between the blood and cells. Capillaries are tiny, thin-walled tubes that receive blood from arterioles. They are so small that red blood cells must go through them in single file. They are so numerous that each cell in the body has a capillary located near it. It is estimated that there are about 1,000 m^2 of surface area represented by the capillary surface in a typical human. Each capillary wall consists of a single layer of cells and therefore presents only a thin barrier to the diffusion of materials between the blood and cells. It is also possible for liquid to flow through tiny spaces between the individual cells of most capillaries (figure 24.5). The flow of blood through these smallest blood vessels is relatively slow. This allows time for the diffusion of such materials as oxygen, glucose, and water from the blood to surrounding cells, and for the movement of such materials as carbon dioxide, lactic acid, and ammonia from the cells into the blood.

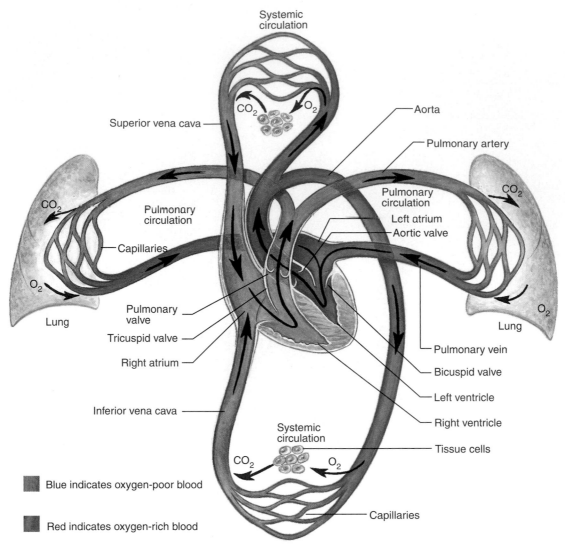

Systemic circulation

Superior vena cava

Aorta

Pulmonary artery

Pulmonary circulation

Pulmonary circulation

CO_2 O_2

Left atrium

Aortic valve

Capillaries

CO_2

O_2

Pulmonary valve

Tricuspid valve

Right atrium

Lung

Lung

CO_2

O_2

Pulmonary vein

Bicuspid valve

Left ventricle

Inferior vena cava

Right ventricle

Tissue cells

Systemic circulation

CO_2 O_2

Capillaries

■ Blue indicates oxygen-poor blood

■ Red indicates oxygen-rich blood

Figure 24.3

The right ventricle pumps blood that is poor in oxygen to the two lungs by way of the pulmonary arteries, where it receives oxygen and turns bright red. The blood is then returned to the left atrium by way of four pulmonary veins. This part of the circulatory system is known as pulmonary circulation. The left ventricle pumps oxygen-rich blood by way of the aorta to all parts of the body except the lungs. This blood returns to the right atrium, depleted of its oxygen, by way of the superior vena cava from the head region and the inferior vena cava from the rest of the body. This portion of the circulatory system is known as systemic circulation.

> **C O N C E P T S   A P P L I E D**
> ### Go with the Flow
>
> Look at figure 24.3. Start with blood in the left lung and trace the flow of blood until it arrives back at the left lung. Make a list of the structures encountered.

In addition to molecular exchange by diffusion, considerable amounts of water and dissolved materials leak through the small holes in the capillaries. This liquid is known as **lymph.** Lymph is produced when the blood pressure forces water and some small dissolved molecules through the walls of the capil-

laries. Lymph bathes the cells but must eventually be returned to the circulatory system by lymph vessels or swelling will occur. Return is accomplished by the *lymphatic system,* a collection of thin-walled tubes that branch throughout the body. These tubes collect lymph that left the circulatory system and ultimately empty it into major blood vessels near the heart. Figure 24.6 shows the structure of the lymphatic system.

EXCHANGING GASES: THE RESPIRATORY SYSTEM

The **lungs** are organs of the body that allow gas exchange to take place between the air and blood. Associated with the lungs is a set of tubes that conducts air from outside the body to the lungs.

Artery

Vein

Epithelial lining

Connective tissue

Elastic tissue

Muscle layers

Valve

Blood flow

Figure 24.4

The walls of arteries are much thicker than the walls of veins. (The pressure in arteries is much higher than the pressure in veins.) The pressure generated by the ventricles of the heart forces blood through the arteries. Veins often have very low pressure. The valves in the veins prevent the blood from flowing backward, away from the heart.

Precapillary sphincter

Arteriole

Blood flow

Artery

Lymph vessel

Capillaries

Venule

Blood flow

Vein

Figure 24.5

Capillaries are tiny blood vessels. Exchange of cells and molecules can occur between blood and tissues through their thin walls. Molecules diffuse in and out of the blood, and cells such as monocytes can move from the blood through the thin walls into the surrounding tissue. There is also a flow of liquid through holes in the capillary walls. This liquid, called lymph, bathes the cells and eventually enters small lymph vessels that return lymph to the circulatory system near the heart.

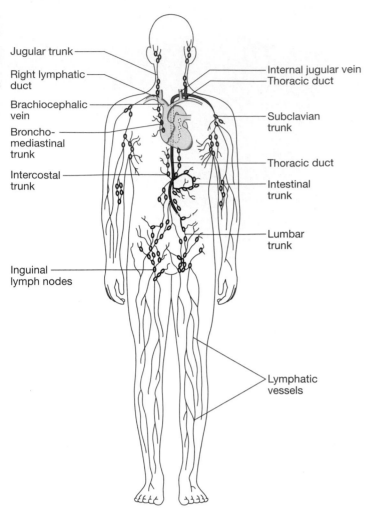

Jugular trunk

Right lymphatic duct

Brachiocephalic vein

Broncho-mediastinal trunk

Intercostal trunk

Inguinal lymph nodes

Internal jugular vein
Thoracic duct

Subclavian trunk

Thoracic duct

Intestinal trunk

Lumbar trunk

Lymphatic vessels

Figure 24.6

The lymphatic system consists of many ducts that transport lymph fluid back toward the heart. Along the way the lymph is filtered in the lymph nodes, and bacteria and other foreign materials are removed before the lymph is returned to the circulatory system near the heart.

The single large-diameter **trachea** is supported by rings of cartilage that prevent its collapse. It branches into two major **bronchi** (singular, *bronchus*) that deliver air to smaller and smaller branches. Bronchi are also supported by cartilage. The smallest tubes, known as **bronchioles,** contain smooth muscle and are therefore capable of constricting. Finally, the bronchioles deliver air to clusters of tiny sacs, known as **alveoli** (singular, *alveolus*), where the exchange of gases takes place between the air and blood.

The nose, mouth, and throat are also important parts of the air-transport pathway because they modify the humidity and temperature of the air and clean the air as it passes. The lining of the trachea contains cells that have cilia that beat in a direction that moves mucus and foreign materials from the lungs. The foreign matter may then be expelled by swallowing, coughing, or other means. Figure 24.7 illustrates the various parts of the respiratory system.

Breathing is the process of moving air in and out of the lungs. It is accomplished by the movement of a muscular organ known as the **diaphragm,** which separates the chest cavity and the lungs from the abdominal cavity. In addition, muscles located between the ribs (*intercostal* muscles) are attached to the ribs in such a way that their contraction causes the chest wall to move outward and upward, which increases the size of the chest cavity. During inhalation, the diaphragm moves downward and the external intercostal muscles of the chest wall contract, causing the volume of the chest cavity to increase. This results in a lower pressure in the chest cavity compared to the outside air pressure. Consequently, air flows from the outside higher-pressure area through the trachea, bronchi, and bronchioles to the alveoli. During normal relaxed breathing, exhalation is accomplished by the chest wall and diaphragm simply returning to their normal, relaxed position. Muscular contraction is not involved (figure 24.8).

Exercising causes an increase in the amount of carbon dioxide in the blood because muscles are oxidizing glucose more rapidly. This lowers the pH of the blood. Certain brain cells and specialized cells in the aortic arch and carotid arteries are sensitive to changes in blood pH. When they sense a lower blood pH, nerve impulses are sent more frequently to the diaphragm and intercostal muscles. These muscles contract more rapidly and more forcefully, resulting in more rapid, deeper breathing. Because more air is being exchanged per minute, carbon dioxide is lost from the lungs more rapidly. When exercise stops, blood pH rises, and breathing eventually returns to normal. Bear in mind, however, that moving air in and out of the lungs is of no value unless oxygen is diffusing into the blood and carbon dioxide is diffusing out.

OBTAINING NUTRIENTS: THE DIGESTIVE SYSTEM

All cells must have a continuous supply of nutrients that provides the energy they require and the building blocks needed to construct the macromolecules typical of living things. This section will deal with the processing and distribution of different kinds of nutrients.

The digestive system consists of a muscular tube with several specialized segments. In addition, there are glands that secrete digestive juices into the tube. Several different kinds of activities are involved in getting nutrients to the cells that need them.

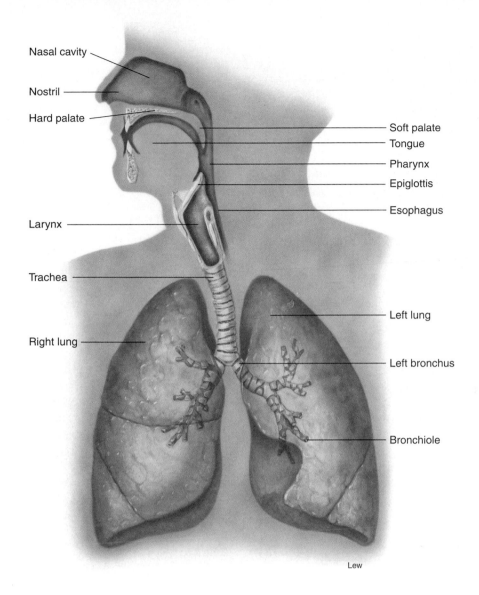

Labels on figure:
Nasal cavity
Nostril
Hard palate
Larynx
Trachea
Right lung

Soft palate
Tongue
Pharynx
Epiglottis
Esophagus
Left lung
Left bronchus
Bronchiole

Lew

Figure 24.7
Although the alveoli of the lungs are the places where gas exchange takes place, there are many other important parts of the respiratory system. The nasal cavity cleans, warms, and humidifies the air entering the lungs. The trachea is also important in cleaning the air going to the lungs.

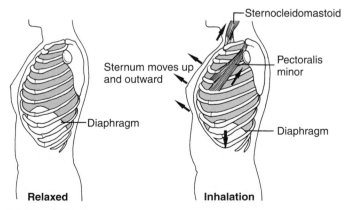

Labels: Sternocleidomastoid, Sternum moves up and outward, Pectoralis minor, Diaphragm, Diaphragm, Relaxed, Inhalation

Figure 24.8
During inhalation, the diaphragm and external intercostal muscles between the ribs contract, causing the volume of the chest cavity to increase. During a normal exhalation, these muscles relax, and the chest volume returns to normal.

Mechanical and Chemical Processing

The digestive system is designed as a disassembly system. Its purpose is to take large chunks of food and break them down to smaller molecules that can be taken up by the circulatory system and distributed to cells. The first step in this process is mechanical processing.

It is important to grind large particles into small pieces by chewing in order to increase their surface area and allow for more efficient chemical reactions. It is also important to add water to the food, which further disperses the particles and provides the watery environment needed for these chemical reactions. Materials must also be mixed, so that all the molecules that need to interact with one another have a good chance of doing so. The mouth and the stomach are the major body regions involved in reducing the size of food particles. The teeth are involved in cutting and grinding food to increase its surface area.

A Closer Look Cigarette Smoking and Your Health

Cigarette smoking is becoming less and less acceptable in our society. The banning of smoking in public buildings and on domestic air flights attests to this fact. Yet in spite of social pressure to quit smoking, research linking smoking with lung and heart disease, and evidence that even secondhand smoke can be harmful, over one-fifth of American adults are smokers (a smoker in this case is defined as someone who has smoked one hundred or more cigarettes in his or her lifetime). Among Americans with a high school education or less, smoking is even more prevalent.

Hazards of Cigarette Smoking

Bronchitis. Cigarette smoking is the leading cause of chronic bronchitis, which involves the inflammation of the bronchi. A common symptom of bronchitis is a harsh cough that expels a greenish-yellow mucus.

Emphysema. Emphysema is a progressive disease in which some of the alveoli are lost. People afflicted with this disease have less and less respiratory surface area and experience greater difficulty getting adequate oxygen, even though they may be breathing more rapidly. It may be caused by cigarette smoke and other air pollutants that damage alveoli. This damage reduces the capacity of the lungs to exchange gases with the bloodstream. A common symptom of emphysema is difficulty exhaling. Several years of an emphysema sufferer's forced breathing can increase the size of the chest and give it a barrel appearance.

Asthma. Cigarette smoke is one of many environmental factors that may trigger an asthma attack. Asthma is an allergic reaction that results in the narrowing of the lungs' air passages and the excess production of fluids that limit the amount of air that can enter the lungs. Symptoms of asthma include coughing, wheezing, and difficulty breathing.

Lung Cancer. Lung cancer develops twenty times more frequently in heavy smokers than in nonsmokers. Typically, lung cancer starts in the bronchi. Cigarette smoke and other pollutants cause cells below the surface of the bronchi to divide at an abnormally high rate. This malignant growth may spread through the lung and move into other parts of the body. Occurrence of cancers of the mouth, larynx, esophagus, pancreas, and bladder are also significantly greater in smokers than in nonsmokers.

Pneumonia. Cigarette smokers have an increased risk of developing pneumonia. Pneumonia involves the infection or inflammation of alveoli, which leads to fluid filling the alveolar sacs. Pneumonia is typically caused by the bacterium *Streptococcus pneumoniae* but in some cases is caused by other bacteria, fungi, protists, or viruses.

Smoking During Pregnancy. Cigarette smoking during pregnancy has been linked to low birth weight and higher rates of fetal and infant death. Children of mothers who smoke during pregnancy have a higher incidence of heart abnormalities, cleft palate, and sudden infant death syndrome (SIDS). Nursing infants of smoking mothers have higher than normal rates of intestinal problems, and infants exposed to secondhand smoke have an increased incidence of respiratory disorders.

Heart Disease. Smoking is a major contributor to heart disease. The action of nicotine from cigarette smoke results in constriction of blood vessels and the reduction of blood flow.

The watery mixture that is added to the food in the mouth is known as *saliva*, and the three pairs of glands that produce saliva are known as *salivary glands*. Saliva contains the enzyme *salivary amylase*, which initiates the chemical breakdown of starch. Saliva also lubricates the mouth and helps to bind food before swallowing. Figure 24.9 describes and summarizes the function of the structures of the digestive system.

Once the food has been chewed, it is swallowed and passes down the esophagus to the stomach. The process of swallowing involves a complex series of events. First, a ball of food, known as a *bolus*, is formed by the tongue and moved to the back of the mouth cavity. Here it stimulates the walls of the throat, also known as the *pharynx*. Nerve endings in the lining of the pharynx are stimulated, causing a reflex contraction of the walls of the esophagus, which transports the bolus to the stomach. Since both food and air pass through the pharynx it is important to prevent food from getting into the lungs. During swallowing, the larynx is pulled upward. This causes a flap of tissue called the *epiglottis* to cover the opening to the trachea and prevent food from entering the trachea. In the stomach, additional liquid, called *gastric juice,* is added to the food. Gastric juice contains enzymes and hydrochloric acid. The major enzyme of the stomach is *pepsin,* which initiates the chemical breakdown of protein. The pH of gastric juice is very low, generally around pH 2. The entire mixture is churned by the contractions of the three layers of muscle in the stomach wall. The combined activities of enzymatic breakdown, chemical breakdown by hydrochloric acid, and mechanical processing by muscular movement result in a thoroughly mixed liquid called *chyme.* Chyme eventually leaves the stomach through a valve known as the pyloric sphincter and enters the small intestine.

The first part of the **small intestine** is known as the **duodenum.** In addition to producing enzymes, the duodenum secretes several kinds of hormones that regulate the release of food from the stomach and the release of secretions from the pancreas and liver. The **pancreas** produces a number of different digestive enzymes and also secretes large amounts of bicarbonate ions, which neutralize stomach acid so that the pH of the duodenum is about pH 8. The **liver** is a large organ in the upper abdomen that performs several functions. One of its functions is the secretion of *bile*. When bile leaves the liver it is stored in the *gallbladder* prior to being released into the duodenum. When bile is released from the gallbladder, it assists mechanical mixing by breaking large fat globules into smaller particles. This process is called *emulsification*.

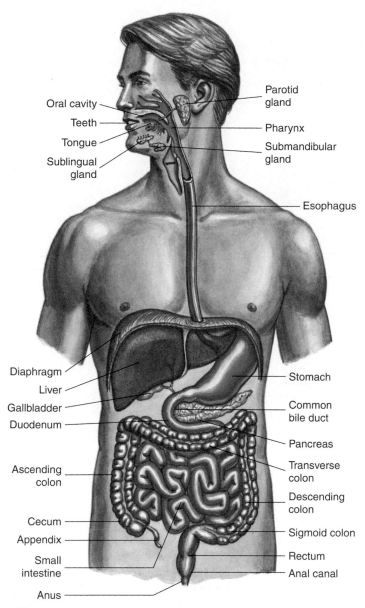

Figure 24.9

In the digestive system, the teeth, tongue, and enzymes from the salivary glands modify the food before it is swallowed. The stomach adds acid and enzymes and further changes the texture of the food. The food is eventually emptied into the duodenum, where the liver and pancreas add their secretions. The small intestine also adds enzymes and is involved in absorbing nutrients. The large intestine is primarily involved in removing water.

Along the length of the intestine, additional watery juices are added until the mixture reaches the **large intestine.** The large intestine is primarily involved in reabsorbing the water that has been added to the food tube along with saliva, gastric juice, bile, pancreatic secretions, and intestinal juices. The large intestine is also home to a variety of different kinds of bacteria. Most live on the undigested food that makes it through the small intestine. Some provide additional benefit by producing vitamins that can be absorbed from the large intestine. A few kinds may cause disease.

Nutrient Uptake

The process of digestion has resulted in a variety of simple organic molecules that are available for absorption from the tube of the gut into the circulatory system. The small intestine is a very long tube; the longer the tube, the greater the internal surface area. In a typical adult human it is about 3 meters long. In addition to length, the lining of the intestine consists of millions of tiny projections called *villi,* which increase the surface area. When we examine the cells that make up the villi, we find that they also have folds in their surface membranes. All of these characteristics increase the surface area available for the transport of materials from the gut into the circulatory system (figure 24.10).

Figure 24.10

The surface area of the intestinal lining is increased by the many fingerlike projections known as villi. Within each villus are capillaries and lacteals. Most kinds of materials enter the capillaries, but most fat-soluble substances enter the lacteals, giving them a milky appearance. Lacteals are part of the lymphatic system. Because the lymphatic system empties into the circulatory system, fat-soluble materials also eventually enter the circulatory system. The close relationship between these vessels and the epithelial lining of the villus allows for efficient exchange of materials from the intestinal cavity to the circulatory system.

Scientists estimate that the cumulative effect of all of these features produces a total intestinal surface area of about 250 m². That is equivalent to about the area of a basketball court.

The surface area by itself would be of little value if it were not for the intimate contact of the circulatory system with this lining. Each villus contains several capillaries and a branch of the lymphatic system called a *lacteal.* The close association between the intestinal surface and the circulatory and lymphatic systems allows for the efficient uptake of nutrients from the cavity of the gut into the circulatory system.

Several different kinds of processes are involved in the transport of materials from the intestine to the circulatory system. Some molecules, such as water and many ions, simply diffuse through the wall of the intestine into the circulatory system. Other materials, such as amino acids and simple sugars, are assisted across the membrane by carrier molecules. Fatty acids and glycerol are absorbed into the intestinal lining cells where they are resynthesized into fats and enter lacteals in the villi. Since the lacteals are part of the lymphatic system, which eventually empties its contents into the circulatory system, fats are also transported by the blood. They just reach the blood by a different route.

WASTE DISPOSAL:
THE EXCRETORY SYSTEM

Because cells are modifying molecules during metabolic processes, harmful waste products are constantly being formed. Urea is a common waste; many other toxic materials must be eliminated as well. Among these are large numbers of hydrogen ions produced by metabolism. This excess of hydrogen ions must be removed from the bloodstream. Other molecules, such as water and salts, may be consumed in excessive amounts and must be removed. The primary organs involved in regulating the level of toxic or unnecessary molecules are the **kidneys.** The urine they produce flows to the urinary bladder through tubes known as *ureters.* The urine leaves the bladder through a tube known as the urethra. (figure 24.11).

The kidneys consist of about 2.4 million tiny units called **nephrons.** At one end of a nephron is a cup-shaped structure called *Bowman's capsule,* which surrounds a knot of capillaries known as a *glomerulus.* In addition to Bowman's capsule, a nephron consists of three distinctly different regions: the *proximal convoluted tubule,* the *loop of Henle,* and the *distal convoluted tubule.* The distal convoluted tubule of a nephron is connected to a collecting duct that transports fluid to the ureters, and ultimately to the urinary bladder, where it is stored until it can be eliminated (figure 24.12).

As in the other systems, the excretory system involves a close connection between the circulatory system and a surface. In this case the large surface is provided by the walls of the millions of nephrons, which are surrounded by capillaries. Three major activities occur at these surfaces: filtration, reabsorption, and secretion. The glomerulus presents a large surface for

Inferior vena cava

Kidney

Aorta

Ureter

Urinary bladder

Urethra

Creek

Figure 24.11
The primary organs involved in removing materials from the blood are the kidneys. The urine produced by the kidneys is transported by the ureters to the urinary bladder. From the bladder, the urine is emptied to the outside of the body by way of the urethra.

the filtering of material from the blood to Bowman's capsule. Blood that enters the glomerulus is under pressure from the muscular contraction of the heart. The capillaries of the glomerulus are quite porous and provide a large surface area for the movement of water and small dissolved molecules from the blood into Bowman's capsule. Normally, only the smaller molecules, such as glucose, amino acids, and ions, are able to pass through the glomerulus into Bowman's capsule at the end of the nephron. The various kinds of blood cells and larger molecules like proteins do not pass out of the blood into the nephron. This physical filtration process allows many kinds of molecules to leave the blood and enter the nephron. The volume of material filtered in this way is about 7.5 liters per hour. Since your entire blood supply is about 5 to 6 liters, there must be some method of recovering much of this fluid. The proximal convoluted tubule reabsorbs most of the useful molecules as the material passes through the nephron. The loop of Henle is involved in the reabsorption of water and the distal convoluted tubule regulates ions, such as hydrogen ions and potassium ions.

Some molecules that pass through the nephron are relatively unaffected by the various activities going on in the kidney. One of these is urea, which is filtered through the glomerulus into Bowman's capsule. As it passes through the nephron, much of it stays in the tubule and is eliminated in

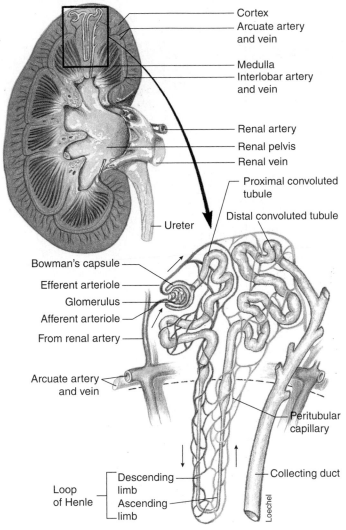

Cortex
Arcuate artery and vein
Medulla
Interlobar artery and vein
Renal artery
Renal pelvis
Renal vein
Proximal convoluted tubule
Distal convoluted tubule
Ureter
Bowman's capsule
Efferent arteriole
Glomerulus
Afferent arteriole
From renal artery
Arcuate artery and vein
Peritubular capillary
Descending limb
Loop of Henle
Ascending limb
Collecting duct
Loechel

Figure 24.12

The nephron and the closely associated blood vessels create a system that allows for the passage of materials from the circulatory system to the nephron by way of the glomerulus and Bowman's capsule. Materials are added to and removed from the fluid in the nephron via the tubular portions of the nephron.

the urine. Many other kinds of molecules, such as minor metabolic waste products and some drugs, are also treated in this manner.

CONTROL MECHANISMS

Living things are complex machines with many parts that must function in a coordinated fashion. All systems are integrated and affect one another in many ways. For example, when you run up a hill, your leg and arm muscles move in a coordinated way to provide power. They burn fuel (glucose) for energy and produce carbon dioxide and lactic acid as waste products, which tend to lower the pH of the blood. Your heart beats faster to provide oxygen and nutrients to the muscles, you breathe faster to supply the muscles with oxygen and get rid of carbon dioxide, and

the blood vessels in the muscles dilate to allow more blood to flow to them. As you run, you generate excess heat. As a result, more blood flows to the skin to get rid of the heat and sweat glands begin to secrete, thus cooling the skin. All of these automatic internal adjustments help the body maintain a constant level of oxygen, carbon dioxide, and glucose in the blood; constant pH; and constant body temperature. These processes can be summed up in the concept of *homeostasis*. **Homeostasis** is the process of maintaining a constant internal environment as a result of monitoring and modifying the functioning of various systems.

There are many kinds of sense organs located within organs and on the surface of the body that respond to specific kinds of stimuli. A **stimulus** is any change in the environment that the organism can detect. Some stimuli, like light or sound, are typically external to the organism; others, like the pain generated by an infection, are internal. The reaction of the organism to a stimulus is known as a **response** (figure 24.13).

The nervous and endocrine systems are the major systems of the body that integrate stimuli and generate appropriate responses necessary to maintain homeostasis. The *nervous system* consists of a network of cells with fibrous extensions that carry information throughout the body. The *endocrine system* consists of a number of glands that communicate with one another and with other tissues through chemicals distributed throughout the organism. **Glands** are organs that manufacture molecules that are secreted either through ducts or into surrounding tissue, where they are picked up by the circulatory system. *Endocrine glands* have no ducts and secrete their products, called **hormones,** into the circulatory system; other glands, such as the digestive glands and sweat glands, empty their contents through ducts. These kinds of glands are called *exocrine glands.*

Although the functions of the nervous and endocrine systems can overlap and be interrelated, these two systems have quite different methods of action. The nervous system functions very much like a computer. A message is sent along established pathways from a specific initiating point to a specific end point, and the transmission is very rapid. The endocrine system functions in a manner analogous to a radio broadcast system. Radio stations send their signals in all directions, but only those radio receivers that are tuned to the correct frequency can receive the message. Messenger molecules (hormones) are typically distributed throughout the body by the circulatory system, but only those cells that have the proper receptor sites can receive and respond to the molecules.

The Structure of the Nervous System

The basic unit of the nervous system is a specialized cell called a *neuron,* or *nerve cell.* A typical neuron consists of a central body called the *soma,* or *cell body,* that contains the nucleus and several long, protoplasmic extensions called nerve fibers. There are two kinds of fibers: *axons,* which carry information away from the cell body, and *dendrites,* which carry information toward the

Stimulus	Integration	Response

Figure 24.13

A stimulus is any detectable change in the surroundings of an organism. When an organism receives a stimulus, it processes the information and may ignore the stimulus or generate a response to it.

cell body (figure 24.14). Most nerve cells have one axon and several dendrites.

Neurons are arranged into two major systems. The *central nervous system,* which consists of the brain and spinal cord, is surrounded by the skull and the vertebrae of the spinal column. It receives input from sense organs, interprets information, and generates responses. The *peripheral nervous system* is located outside the skull and spinal column and consists of bundles of long axons and dendrites called nerves. There are two different sets of neurons in the peripheral nervous system. *Motor neurons* carry messages from the central nervous system to muscles and glands, and *sensory neurons* carry input from sense organs to the central nervous system. Motor neurons typically have one long axon that runs from the spinal cord to a muscle or gland, while sensory neurons have long dendrites that carry input from the sense organs to the central nervous system.

The Nature of the Nerve Impulse

Since most nerve cells have long fibrous extensions, it is possible for information to be passed along the nerve cell from one end to the other. The message that travels along a neuron is known as a **nerve impulse.** A nerve impulse is not a simple electric current but involves a specific sequence of chemical events involving activities at the cell membrane.

Since all cell membranes are differentially permeable, it is difficult for some ions to pass through the membrane, and the combination of ions inside the membrane is different from that on the outside. Cell membranes also contain proteins that ac-

tively transport specific ions from one side of the membrane to the other. Active transport involves the cell's use of ATP to move materials from one side of the cell membrane to the other. Because ATP is required, this is an ability that cells lose when they die. One of the ions that is actively transported from cells is the sodium ion (Na^+). At the same time sodium ions (Na^+) are being transported out of cells, potassium ions (K^+) are being transported into the normal resting cells. However, there are more sodium ions (Na^+) transported out than potassium ions (K^+) transported in.

Because a normal resting cell has more positively charged Na^+ ions on the outside of the cell than on the inside, a small but measurable voltage exists across the membrane of the cell. The voltage difference between the inside and outside of a cell membrane is about 70 millivolts (0.07 volt). The two sides of the cell membrane are, therefore, polarized in the same sense that a battery is polarized, with a positive and negative pole. A resting neuron has its positive pole on the outside of the cell membrane and its negative pole on the inside of the membrane (figure 24.15).

When a cell is stimulated at a specific point on the cell membrane, the cell membrane changes its permeability and lets sodium ions (Na^+) pass through it from the outside to the inside and potassium ions (K^+) diffuse from the inside to the outside. The membrane is thus *depolarized;* it loses its difference in charge as sodium ions (Na^+) diffuse into the cell from the outside. Sodium ions diffuse into the cell because initially they are in greater concentration outside the cell than inside. When the membrane becomes more permeable, they are able to diffuse

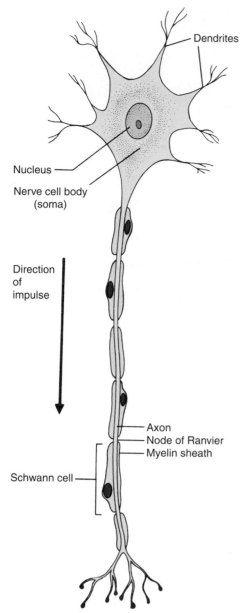

Figure 24.14

Nerve cells consist of a nerve-cell body that contains the nucleus and several fibrous extensions. The shorter, more numerous fibers that carry impulses to the nerve cell body are dendrites. The long fiber that carries the impulse away from the cell body is the axon.

Labels on figure:
- Dendrites
- Nucleus
- Nerve cell body (soma)
- Direction of impulse
- Axon
- Node of Ranvier
- Myelin sheath
- Schwann cell

lease of a molecule that stimulates depolarization of the next neuron in the chain.

Activities at the Synapse

Between the fibers of adjacent neurons in a chain is a space called the **synapse.** Many chemical events occur in the synapse that are important in the function of the nervous system. When a neuron is stimulated, an impulse passes along its length from one end to the other. When the impulse reaches a synapse, a molecule called a **neurotransmitter** is released into the synapse from the axon. It diffuses across the synapse and binds to specific receptor sites on the dendrite of the next neuron. When enough neurotransmitter molecules have bound to the second neuron, an impulse is initiated in it as well. Several kinds of neurotransmitters are produced by specific neurons. These include dopamine, epinephrine, acetylcholine, and several other molecules. The first neurotransmitter identified was *acetylcholine.*

As long as a neurotransmitter is bound to its receptor it continues to stimulate the nerve cell. Thus if acetylcholine continues to occupy receptors, the neuron continues to be stimulated again and again. An enzyme called *acetylcholinesterase* destroys acetylcholine and prevents this from happening. The breakdown products of the acetylcholine can be used to remanufacture new acetylcholine molecules at the end of the axon. The destruction of acetylcholine allows the second neuron in the chain to return to normal. Thus it will be ready to accept another burst of acetylcholine from the first neuron a short time later. Neurons must also constantly manufacture new acetylcholine molecules or they will exhaust their supply and be unable to conduct an impulse across a synapse.

Caffeine and nicotine are considered stimulants because they make it easier for impulses to pass through the synapse. The nerve cells are sensitized and will respond to smaller amounts of acetylcholine than normal. Drinking coffee, taking caffeine pills, or smoking will increase the sensitivity of the nervous system to stimulation. However, as with most kinds of drugs, continual use or abuse tends to lead to a loss of the effect as the nervous system adapts to the constant presence of the drugs.

Because of the way the synapse works, impulses can go in only one direction: Only axons secrete acetylcholine, and only dendrites have receptors. This explains why there are sensory and motor neurons to carry messages to and from the central nervous system.

Endocrine System Function

As mentioned previously, the endocrine system is basically a broadcasting system in which glands secrete messenger molecules, called hormones, which are distributed throughout the body by the circulatory system (figure 24.17). However, each kind of hormone attaches only to appropriate receptor molecules on the surfaces of certain cells. The cells that receive the message typically respond in one of three ways: (1) Some cells release products that have been previously manufactured, (2) other cells are stimulated to synthesize molecules or to

into the cell, toward the area of lower concentration. The depolarization of one point on the cell membrane causes the adjacent portion of the cell membrane to change its permeability as well, and it also depolarizes. Thus, a wave of depolarization passes along the length of the neuron from one end to the other (figure 24.16). The depolarization and passage of an impulse along any portion of the neuron is a momentary event. As soon as a section of the membrane has been depolarized, potassium ions diffuse back into the cell. This reestablishes the original polarized state, and the membrane is said to be *repolarized.* When the nerve impulse reaches the end of the axon, it stimulates the re-

Figure 24.15

All cells, including nerve cells, have an active transport mechanism that pumps Na$^+$ out of cells and simultaneously pumps K$^+$ into them. The end result is that there are more Na$^+$ ions outside the cell and more K$^+$ ions inside the cell. In addition, negative ions such as Cl$^-$ are more numerous inside the cell. Consequently, the outside of the cell is positive (+) compared to the inside, which is negative (−).

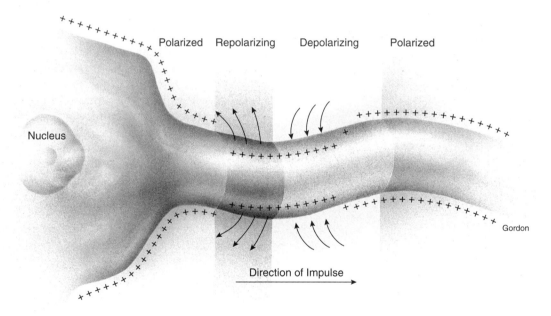

Figure 24.16

When a nerve cell is stimulated, a small portion of the cell membrane depolarizes as Na$^+$ flows into the cell through the membrane. This encourages the depolarization of an adjacent portion of the membrane, and it depolarizes a short time later. In this way a wave of depolarization passes down the length of the nerve cell. Shortly after a portion of the membrane is depolarized, the ionic balance is reestablished. It is repolarized and ready to be stimulated again.

begin metabolic activities, and (3) some are stimulated to divide and grow.

These different kinds of responses mean that some endocrine responses are relatively rapid, while others are very slow. For example, the release of the hormones *epinephrine* and *norepinephrine* from the adrenal medulla, located near the kidney, causes a rapid change in the behavior of an organism. The heart rate increases, blood pressure rises, blood is shunted to muscles, and the breathing rate increases. You have certainly experienced

this reaction many times in your lifetime as when you nearly have an automobile accident or slip and nearly fall.

Another hormone, called *antidiuretic hormone (ADH)*, acts more slowly. It is released from the posterior pituitary gland at the base of the brain and regulates the rate at which the body loses water through the kidneys. It does this by encouraging the reabsorption of water. The effects of this hormone can be noticed in a matter of minutes to hours. Insulin is another hormone whose effects are quite rapid. Insulin is produced by the pancreas, located near the stomach, and stimulates cells—particularly muscle, liver, and fat cells—to take up glucose from the blood. After a meal that is high in carbohydrates, the level of glucose in the blood begins to rise, stimulating the pancreas to release insulin. The increased insulin causes glucose levels to fall as the sugar is taken up by cells. People with diabetes have insufficient or improperly acting insulin or lack the receptors to respond to the insulin, and therefore have difficulty regulating glucose levels in their blood.

The responses that result from the growth of cells may take weeks or years to occur. For example, *growth-stimulating hormone (GSH)* is produced by the anterior pituitary gland over a period of years and results in typical human growth. After sexual maturity, the amount of this hormone generally drops to very low levels and body growth stops. Sexual development is also largely the result of the growth of specific tissues and organs. The male sex hormone *testosterone*, produced by the testes, causes the growth of male sex organs and a change to the adult body form. The female counterpart, *estrogen*, results in the development of female sex organs and body form. In all of these cases, it is the release of hormones over long periods, continually stimulating the growth of sensitive tissues, that results in the normal developmental pattern. The absence or inhibition of any of these hormones early in life changes the normal growth process.

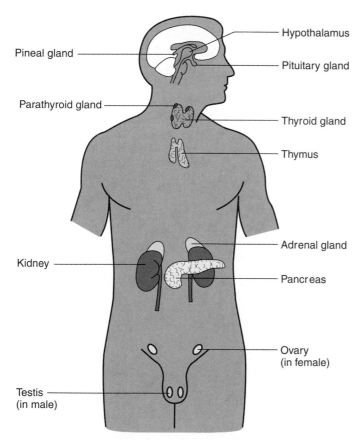

Figure 24.17
The endocrine glands are located at various locations within the body and cause their effects by secreting hormones.

Because the pituitary is constantly receiving information from the hypothalamus of the brain, many kinds of sensory stimuli to the body can affect the functioning of the endocrine system. One example is the way in which the nervous system and endocrine system interact to influence the menstrual cycle. At least three different hormones are involved in the cycle of changes that affect the ovary and the lining of the uterus. It is well documented that stress caused by tension or worry can interfere with the normal cycle of hormones and delay or stop menstrual cycles. In addition, young women living in groups, such as in college dormitories, often find that their menstrual cycles become synchronized. Although the exact mechanism involved in this phenomenon is unknown, it is suspected that input from the nervous system causes this synchronization. (Odors and sympathetic feelings have been suggested as causes.)

Although we still tend to think of the nervous and endocrine systems as being separate and different, it is becoming clear that they are interconnected. These two systems cooperate to bring about appropriate responses to environmental challenges. The nervous system is specialized for receiving and sending short-term messages, whereas activities that require long-term, growth-related actions are handled by the endocrine system.

SENSORY INPUT

The activities of the nervous and endocrine systems are often responses to some kind of input received from the sense organs. Sense organs of various types are located throughout the body. Many of them are located on the surface, where environmental changes can be easily detected. Hearing, sight, and touch are good examples of such senses. Other sense organs are located within the body and indicate to the organism how its various parts are changing. For example, pain and pressure are often used to monitor internal conditions. The sense organs detect changes, but the brain is responsible for perception—the recognition that a stimulus has been received. Sensory abilities involve many different kinds of mechanisms, including chemical recognition, the detection of energy changes, and the monitoring of forces.

Chemical Detection

All cells have receptors on their surfaces that can bind selectively to molecules they encounter. This binding process can cause changes in the cell in several ways. In some cells it causes depolarization. When this happens, the cells can stimulate neurons and cause messages to be sent to the central nervous system, informing it of some change in the surroundings. In other cases, a molecule binding to the cell surface may cause certain genes to be expressed, and the cell responds by changing the molecules it produces. This is typical of the way the endocrine system receives and delivers messages.

Most cells have specific binding sites for particular molecules. Others, such as the taste buds on the tongue, appear to respond to classes of molecules. Traditionally we have distinguished four kinds of tastes: sweet, sour, salt, and bitter.

Connections . . .

The other major chemical sense, the sense of smell, is much more versatile; it can detect thousands of different molecules at very low concentrations. The cells that make up the *olfactory epithelium,* the cells that line the nasal cavity and respond to smells, apparently bind molecules to receptors on their surface. Exactly how this can account for the large number of recognizably different odors is unknown, but the receptor cells are extremely sensitive. In some cases a single molecule of a substance is sufficient to cause a receptor cell to send a message to the brain, where the sensation of odor is perceived. These sensory cells for smell also fatigue rapidly. You have probably noticed that when you first walk into a room, specific odors are readily detected, but after a few minutes you are unable to detect them. Most perfumes and aftershaves are undetectable after 15 minutes of continuous stimulation.

Many internal sense organs also respond to specific molecules. For example, the brain and aorta contain cells that respond to concentrations of hydrogen ions, carbon dioxide, and oxygen in the blood. Remember, too, that the endocrine system relies on the detection of specific messenger molecules to trigger its activities.

Light Detection

The eyes primarily respond to changes in the flow of light energy. The structure of the eye is designed to focus light on a light-sensitive layer of the back of the eye known as the **retina** (figure 24.18). There are two kinds of receptors in the retina of the eye. The cells called *rods* respond to a broad range of wavelengths of light and are responsible for black-and-white vision. Since rods are very sensitive to light, they are particularly useful in dim light. Rods are located over most of the retinal surface except for the area of most acute vision known as the *fovea centralis.* The other receptor cells, called *cones,* are found throughout the retina but are particularly concentrated in the fovea centralis. Cones are not as sensitive to light, but they can detect different wavelengths of light. This combination of receptors gives us the ability to detect color when light levels are high, but we rely on black-and-white vision at night. There are three different varieties of cones: One type responds best to red light, another responds best to green light, and the third responds best to blue light. Stimulation of various combinations of these three kinds of cones allows us to detect different shades of color.

Rods and the three different kinds of cones each contain a pigment that decomposes when struck by light of the proper wavelength and sufficient strength. The pigment found in rods is called *rhodopsin.* This change in the structure of rhodopsin causes the rod to depolarize. Cone cells have a similar mechanism of action, and each of the three kinds of cones has a different pigment. Since rods and cones synapse with neurons, they stimulate a neuron when depolarized and cause a message to be sent to the brain. Thus, the pattern of color and light intensity recorded on the retina is detected by rods and cones and converted into a series of nerve impulses that are received and interpreted by the brain.

However, recently a fifth kind of taste, *umami* (meaty), has been identified that responds to the amino acid glutamate, which is present in many kinds of foods and can be added as a flavor enhancer (monosodium glutamate). The taste buds that give us the sour sensation appear to respond to the presence of hydrogen ions (H^+), because acids taste sour. Sodium chloride stimulates the taste buds that give us the sensation of a salty taste. However, the sensations of sweetness, bitterness, and umami occur when molecules bind to specific surface receptors on the cell. Sweetness can be stimulated by many kinds of organic molecules, including sugars and artificial sweeteners, and also by inorganic lead compounds. When a molecule binds to a sweetness receptor, a molecule is split and its splitting stimulates an enzyme that leads to the depolarization of the cell. The sweet taste of lead salts in old paints partly explains why children sometimes eat paint chips. Because the lead interferes with normal brain development, this behavior can have disastrous results. Many other kinds of compounds of diverse structures give the bitter sensation. The cells that respond to bitter sensations have a variety of receptor molecules on their surface. When a substance binds to one of the receptors, the cell depolarizes. In the case of umami, it is the glutamate molecule that binds to receptors on the cells of the taste buds.

It is also important to understand that much of what we often refer to as taste involves such inputs as temperature, texture, and smell. Cold coffee has a different taste than hot coffee even though they are chemically the same. Lumpy, cooked cereal and smooth cereal have different tastes. If you are unable to smell food, it doesn't taste as it should, which is why you sometimes lose your appetite when you have a stuffy nose. We still have much to learn about how the tongue detects chemicals and the role that other associated senses play in modifying taste.

Figure 24.18

The eye contains a cornea and lens that focus the light on the retina of the eye. The light causes pigments in the rods and cones of the retina to decompose. This leads to the depolarization of these cells and the stimulation of neurons that send messages to the brain.

Labels on figure:
Superior rectus muscle
Sclera
Choroid
Retina
Fovea centralis
Central artery
Central vein
Lew
Optic nerve
Inferior rectus muscle
Conjunctiva
Posterior chamber
Cornea
Anterior chamber
Pupil
Lens
Iris
Posterior chamber
Zonular fibers of suspensory ligament
Ora serrata
Vitreous chamber

Sound Detection

The ears respond to changes in sound waves. Sound is produced by the vibration of molecules. Consequently, the ears are detecting changes in the quantity of energy and the quality of sound waves. Sound has several characteristics. Loudness, or volume, is a measure of the intensity of sound energy that arrives at the ear. Very loud sounds will literally vibrate your body and can cause hearing loss if they are too intense. Pitch is a quality of sound that is determined by the frequency of the sound vibrations. High-pitched sounds have short wavelengths; low-pitched sounds have long wavelengths.

Figure 24.19 shows the anatomy of the ear. The sound that arrives at the ear is first funneled by the external ear to the **tympanum,** also known as the *eardrum.* The cone-shaped nature of the external ear focuses sound on the tympanum and causes it to vibrate at the same frequency as the sound waves reaching it. Attached to the tympanum are three tiny bones known as the *malleus* (hammer), *incus* (anvil), and *stapes* (stirrup). The malleus is attached to the tympanum, the incus is attached to the malleus and stapes, and the stapes is attached to a small, membrane-covered opening called the oval window in a snail-shaped structure known as the cochlea. The vibration of the tympanum causes the tiny bones (malleus, incus, and stapes) to vibrate, and they in turn cause a corresponding vibration in the membrane of the *oval window.*

The *cochlea* of the ear is the structure that detects sound and consists of a snail-shaped set of fluid-filled tubes. When the oval window vibrates, the fluid in the cochlea begins to move, causing a membrane in the cochlea, called the *basilar membrane,* to vibrate. High-pitched, short-wavelength sounds cause the basilar membrane to vibrate at the base of the cochlea near the oval window. Low-pitched, long-wavelength sounds vibrate the basilar membrane far from the oval window. Loud sounds cause the basilar membrane to vibrate more vigorously than do faint sounds. Cells on this membrane depolarize when they are stimulated by its vibrations. Because they synapse with neurons, messages can be sent to the brain.

Because sounds of different wavelengths stimulate different portions of the cochlea, the brain is able to determine the pitch of a sound. Most sounds consist of a mixture of pitches that are heard. Louder sounds stimulate the membrane more forcefully, causing the sensory cells in the cochlea to send more nerve impulses per second. Thus, the brain is able to perceive the loudness of various sounds, as well as the pitch.

Associated with the cochlea are two fluid-filled chambers and a set of fluid-filled tubes called the *semicircular canals.* These structures are not involved in hearing but are involved in maintaining balance and posture. In the walls of these canals and chambers are cells similar to those found on the basilar membrane. These cells are stimulated by movements of the head and by the position of the head with respect to the force of gravity. The constantly changing position of the head results in sensory input that is important in maintaining balance.

Touch

What we normally call the sense of *touch* consists of a variety of different kinds of input. Some receptors respond to pressure, others to temperature, and others, which we call *pain receptors,* usually respond to cell damage. When these receptors are appropriately stimulated, they send a message to the brain.

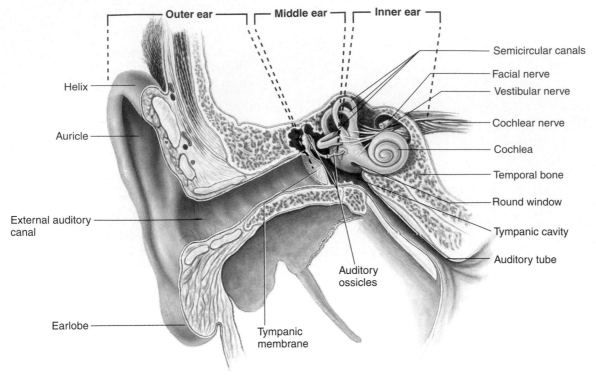

Figure 24.19

The ear consists of an external cone that directs sound waves to the tympanum. Vibrations of the tympanum move the ear bones and vibrate the oval window of the cochlea, where the sound is detected. The semicircular canals monitor changes in the position of the head, helping us maintain balance.

Because receptors are stimulated in particular parts of the body, the brain is able to localize the sensation. However, not all parts of the body are equally supplied with these receptors. The tips of the fingers, lips, and external genitals have the highest density of these nerve endings, while the back, legs, and arms have far fewer receptors.

Some internal receptors, such as pain and pressure receptors, are important in allowing us to monitor our internal activities. Many pains generated by the internal organs are often perceived as if they were somewhere else. For example, the pain associated with heart attack is often perceived to be in the left arm. Pressure receptors in joints and muscles are important in providing information about the degree of stress being placed on a portion of the body. This is also important information to send back to the brain so that adjustments can be made in movements to maintain posture. If you have ever had your foot "go to sleep" because the nerve stopped functioning, you have experienced what it is like to lose this constant input of nerve messages from the pressure sensors to assist in guiding the movements you make. Your movements become uncoordinated until the nerve function returns to normal.

OUTPUT MECHANISMS

The nervous system and endocrine system cause changes in several ways. Both systems can stimulate muscles to contract and glands to secrete. The endocrine system is also able to change metabolism of cells and regulate the growth of tissues. The nervous system acts upon two kinds of organs: muscles and glands. The actions of muscles and glands are simple and direct: muscles contract and glands secrete.

Muscles

The ability to move is one of the fundamental characteristics of animals. Through the coordinated contraction of many muscles, the intricate, precise movements of a dancer, basketball player, or writer are accomplished. It is important to recognize that muscles can pull only by contracting; they are unable to push by lengthening. The work of any muscle is done during its contraction. Relaxation is the passive state of the muscle. There must always be some force available that will stretch a muscle after it has stopped contracting and relaxes. Therefore, the muscles that control the movements of the skeleton are present in antagonistic sets—for every muscle's action there is another muscle that has the opposite action. For example, the biceps muscle causes the arm to flex (bend) as the muscle shortens. The contraction of its antagonist, the triceps muscle, causes the arm to extend (straighten) and at the same time stretches the relaxed biceps muscle (figure 24.20).

There are three major types of muscle: *skeletal, smooth,* and *cardiac.* These differ from one another in several ways. *Skeletal muscle* is voluntary muscle; it is under the control of the nervous system. The brain or spinal cord sends a message to skeletal muscles, and they contract to move the legs, fingers, and other parts of the body. This does not mean that you must make a con-

Connections . . .

Triceps
extends
arm

Biceps
flexes
arm

Creek

Figure 24.20

Because muscles cannot actively lengthen, it is necessary to have sets of muscles that oppose one another. The contraction and shortening of one muscle cause the stretching of a relaxed muscle.

scious decision every time you want to move a muscle. Many of the movements we make are learned initially but become automatic as a result of practice. For example, walking, swimming, or riding a bicycle required a great amount of practice originally, but now you probably perform these movements without thinking about them. They are, however, still considered voluntary actions.

Smooth muscles make up the walls of muscular internal organs, such as the gut, blood vessels, and reproductive organs. They have the property of contracting as a response to

being stretched. Since much of the digestive system is being stretched constantly, the responsive contractions contribute to the normal rhythmic movements associated with the digestive system. These are involuntary muscles; they can contract on their own without receiving direct messages from the nervous system. Although they are involuntary, they can be stimulated to contract by the presence of certain hormones (the hormone oxytocin causes the smooth muscle of the uterus to contract) or their degree of contraction can be modified by nervous stimulation.

Cardiac muscle is the muscle that makes up the heart. It has the ability to contract rapidly like skeletal muscle, but does not require nervous stimulation to do so. Nervous stimulation can, however, cause the heart to speed or slow its rate of contraction. Hormones, such as epinephrine and norepinephrine, also influence the heart by increasing its rate and strength of contraction. Cardiac muscle also has the characteristic of being unable to stay contracted. It will contract quickly but must have a short period of relaxation before it will be able to contract a second time. This makes sense in light of its continuous, rhythmic, pumping function.

Glands

The glands of the body are of two different kinds. Those that secrete into the bloodstream are called **endocrine glands.** We have already talked about several of these: the pituitary, thyroid, ovary, and testis are examples. The **exocrine glands** are those that secrete to the surface of the body or into one of the tubular organs of the body, such as the gut or reproductive tract. Examples are

Aerobic exercise occurs when the muscles being contracted are supplied with sufficient oxygen to continue aerobic cellular respiration:

$$C_6H_{12}O_6 + 6O_2 \rightarrow 6CO_2 + 6H_2O + \text{energy}$$

This type of exercise involves long periods of activity with elevated breathing and heart rate. It results in strengthened chest muscles, which enable more complete exchange of air during breathing. It also improves the strength of the heart, enabling it to pump more efficiently. Flow of blood to the muscles also improves. All these changes increase endurance.

Anaerobic exercise takes place when insufficient amounts of oxygen reach the contracting muscle cells and they shift to anaerobic cellular respiration:

$$C_6H_{12}O_6 \rightarrow \text{lactic acid} + \text{less energy}$$

The buildup of lactic acid results in muscle pain and eventually prevents further contraction. Anaerobic exercise involves explosive bouts of activity, as in sprints or jumping. This kind of exercise increases muscle strength but does little to improve endurance.

Resistance exercise occurs when muscles contract against an object that does not allow the muscles to move that object. This type of exercise does not improve the ability of your body to deliver oxygen to your muscles, nor does it increase your endurance. However, it does stimulate your muscle cells to manufacture more contractile protein fibers.

Make a list of your activities that would be considered aerobic, anaerobic, or resistance exercise. Which activities would result in weight reduction? Weight gain? Improved cardiovascular fitness?

---

the salivary glands, intestinal mucous glands, and sweat glands. Some of these glands, such as salivary glands and sweat glands, are under nervous control. When stimulated by the nervous system, they secrete their contents.

Many other exocrine glands are under hormonal control. Many of the digestive enzymes of the stomach and intestine are secreted in response to local hormones produced in the gut. These are circulated through the blood to the digestive glands, which respond by secreting the appropriate digestive enzymes and other molecules. Endocrine glands also respond to nervous stimulation or stimulation by hormones.

Growth Responses

The hormones produced by the endocrine system can have a variety of effects. As mentioned earlier, hormones can stimulate smooth muscle to contract and can influence the contraction of cardiac muscle as well. Many kinds of glands, both endocrine and exocrine, are caused to secrete as a result of a hormonal stimulus. However, the endocrine system has one major effect that is not equaled by the nervous system: Hormones regulate growth. *Growth-stimulating hormone (GSH)* is produced over a period of years to bring about the increase in size of most of the structures of the body. A low level of this hormone results in a person with small body size. It is important to recognize that the amount of growth-stimulating hormone (GSH) present varies from time to time. It is present in fairly high amounts throughout childhood and results in steady growth. It also appears to be present at higher levels at certain times, resulting in growth spurts. Finally, as adulthood is reached, the level of this hormone falls, and growth stops.

Similarly, testosterone produced during adolescence influences the growth of bone and muscle to provide men with larger, more muscular bodies than those of women. In addition, there is growth of the penis, growth of the larynx, and increased growth of hair on the face and body. The primary female hormone, estrogen, causes growth of reproductive organs and development of breast tissue. It is also involved, along with other hormones, in the cyclic growth and sloughing of the wall of the uterus.

SUMMARY

The *circulatory, respiratory, digestive,* and *excretory systems* are involved in the exchange of materials across cell membranes. All of these systems have special features that provide *large surface* areas to allow for necessary *exchanges.*

The *circulatory system* consists of a pump, the *heart,* and *blood vessels* that distribute the blood to all parts of the body. The *blood* is a carrier fluid that transports molecules and heat. The exchange of materials between the *blood* and body *cells* takes place through the walls of the *capillaries. Hemoglobin* in *red blood cells* is very important in the transport of *oxygen.*

The *respiratory system* consists of the *lungs* and associated tubes that allow air to enter and leave the lungs. The *diaphragm* and *muscles* of the chest wall are important in the process of *breathing.* In the lungs, tiny sacs called *alveoli* provide a large surface area in association with capillaries, which allows for rapid exchange of oxygen and carbon dioxide.

The *digestive system* is involved in disassembling food molecules. This involves several processes: *grinding* by the teeth and stomach, *emulsification* of fats by bile from the liver, *addition of water* to dissolve molecules, and *enzymatic action* to break complex molecules into simpler molecules for *absorption.* The intestine provides a *large surface area* for the *absorption of nutrients* because it is long and its wall contains many tiny projections that increase surface area.

The *excretory system* is a filtering system of the body. The *kidneys* consist of *nephrons* into which the *circulatory system* filters fluid. Most of this fluid is useful and is reclaimed by the cells that make up the walls of these *tubules.* Materials that are present in excess or those that are harmful are allowed to escape.

A *nerve impulse* is caused by sodium ions entering the cell as a result of a change in the permeability of the cell membrane. Thus, a *wave of depolarization* passes down the length of a *neuron* to the *synapse.* The *axon* of a neuron secretes a *neurotransmitter,* such as *acetylcholine,* into the *synapse,* where these molecules bind to the dendrite of the next cell in the chain, resulting in an impulse in it as well. The *acetylcholinesterase* present in the synapse destroys acetylcholine so that it does not repeatedly stimulate the dendrite.

The body's various *systems* must be integrated in such a way that the *internal environment* stays relatively constant. This concept is called *homeostasis*. Several kinds of *sensory inputs* are possible. Many kinds of chemicals can bind to cell surfaces and be recognized. This is probably how the *sense of taste* and the *sense of smell* function. Light energy can be detected because light causes certain molecules in the *retina* of the eye to decompose and stimulate neurons. *Sound* can be detected because fluid in the *cochlea* of the *ear* is caused to vibrate, and special cells detect this movement and stimulate neurons. The sense of *touch* consists of a variety of receptors that respond to *pressure, cell damage,* and *temperature.*

Glands are of two types: *exocrine glands,* which secrete through ducts into the cavity of an organ or to the surface of the skin, and *endocrine glands,* which release their secretions into the circulatory system. It is becoming clear that the endocrine system and the nervous system are interrelated. Actions of the endocrine system can change how the nervous system functions and the reverse is also true.

KEY TERMS

alveoli (p. **542**)
aorta (p. **537**)
arteries (p. **537**)
arterioles (p. **539**)
blood (p. **537**)
bronchi (p. **542**)
bronchioles (p. **542**)
capillaries (p. **537**)
diaphragm (p. **542**)
diastolic blood pressure
  (p. **539**)
duodenum (p. **544**)
endocrine glands (p. **555**)
exocrine glands (p. **555**)
glands (p. **547**)
heart (p. **537**)
hemoglobin (p. **537**)
homeostasis (p. **547**)
hormones (p. **547**)
kidney (p. **546**)
large intestine (p. **545**)
liver (p. **544**)

lung (p. **540**)
lymph (p. **540**)
nephrons (p. **546**)
nerve impulse (p. **548**)
neurotransmitter (p. **549**)
pancreas (p. **544**)
plasma (p. **537**)
pulmonary artery (p. **537**)
pulmonary circulation
  (p. **538**)
response (p. **547**)
retina (p. **552**)
small intestine (p. **544**)
stimulus (p. **547**)
synapse (p. **549**)
systemic circulation (p. **538**)
systolic blood pressure
  (p. **539**)
trachea (p. **542**)
tympanum (p. **553**)
veins (p. **537**)

APPLYING THE CONCEPTS

1. The primary structures involved in pumping blood are the
   (a) veins.
   (b) atria.
   (c) capillaries.
   (d) ventricles.
2. The fluid portion of the blood that leaves the capillaries and surrounds the cells is
   (a) hemoglobin.
   (b) edema.
   (c) lymph.
   (d) lacteal.
3. Blood is carried through vessels to all parts of the body except the lungs by
   (a) pulmonary circulation.
   (b) the pulmonary artery.
   (c) systemic circulation.
   (d) the lymphatic system.
4. As air passes through the lungs it follows the path:
   (a) trachea → bronchi → bronchioles → alveoli
   (b) trachea → bronchioles → bronchi → alveoli
   (c) bronchi → trachea → alveoli → bronchioles
   (d) bronchioles → alveoli → bronchi → trachea
5. The levels of water, hydrogen ions, salts, and urea in the blood are regulated by the
   (a) liver.
   (b) kidneys.
   (c) bladder.
   (d) rectum.
6. An organism's reaction to a change in the environment is a(n)
   (a) stimulus.
   (b) impulse.
   (c) response.
   (d) perception.
7. A light stimulus is received by the nervous system, which results in growth. This is the result of
   (a) hormones stimulating cells.
   (b) activating muscles.
   (c) the endocrine system stimulating the nervous system.
   (d) increasing nervous activity.
8. When a nerve cell is stimulated,
   (a) acetylcholine is destroyed.
   (b) potassium ions enter the neuron.
   (c) sodium ions enter the neuron.
   (d) calcium attaches to the dendrites.
9. The central nervous system consists of the
   (a) brain only.
   (b) brain and spinal cord.
   (c) brain, spinal cord, and nerves.
   (d) motor neurons and sensory neurons.
10. Olfactory senses detect
    (a) light.
    (b) sound.
    (c) chemicals.
    (d) pain.
11. The chemical messengers secreted by endocrine glands are called
    (a) enzymes.
    (b) hormones.
    (c) neurotransmitters.
    (d) impulses.
12. Large food molecules are chemically broken down to smaller molecules by
    (a) bile salts
    (b) enzymes
    (c) dissolving
    (d) chewing

Answers
1. d 2. c 3. c 4. a 5. b 6. c 7. a 8. c 9. b 10. c 11. b 12. b

QUESTIONS FOR THOUGHT

1. What are the functions of the heart, arteries, veins, arterioles, the blood, and capillaries?
2. Describe three ways in which the digestive system increases its ability to absorb nutrients.
3. Describe the mechanics of breathing.
4. Describe how changing permeability of the cell membrane and the movement of sodium ions cause a nerve impulse.
5. What is the role of acetylcholine in a synapse?
6. What is actually detected by the nasal epithelium, taste buds, cochlea of the ear, and retina of the eye?
7. List the differences between the central and peripheral nervous systems.
8. List two differences in the way the endocrine and nervous systems function.
9. Why do large, multicellular organisms need a circulatory system?
10. How are the following things related: capillaries, nephrons, alveoli, villi?
11. Describe how villi and the length of the small intestine are related to the ability to absorb nutrients.
12. Describe two ways in which smooth muscle and skeletal muscle differ in the way they function.

CHAPTER TWENTY-FIVE

Mendelian and Molecular Genetics

The expression of many genes is influenced by the environment.
The gene for dark hair in the cat is sensitive to temperature and expresses itself only in the parts of the body that stay cool.

OUTLINE

hy do you have a particular blood type or hair color? Why do some people have the same skin color as their parents, while others have a skin color different from that of their parents? Why do flowers show such a wide variety of colors? Why is it that generation after generation of plants, animals, and microbes look so much like members of their own kind? These questions and many others can be better answered if you have an understanding of genetics.

GENETICS, MEIOSIS, AND CELLS

A **gene** is a portion of DNA that determines a characteristic. Through meiosis and reproduction, genes can be transmitted from one generation to another. The study of genes, how genes produce characteristics, and how the characteristics are inherited is the field of biology called **genetics.** The first person to systematically study inheritance and formulate laws about how characteristics are passed from one generation to the next was an Augustinian monk named Gregor Mendel (1822–1884). Mendel's work was not generally accepted until 1900, when three men, working independently, rediscovered some of the ideas that Mendel had formulated more than thirty years earlier. Because of his early work, the study of the pattern of inheritance that follows the laws formulated by Gregor Mendel is often called **Mendelian genetics.**

To understand this chapter, you need to know some basic terminology. One term that you have already encountered is *gene.* Mendel thought of a gene as a particle that could be passed from the parents to the *offspring* (children, descendants, or progeny). Today we know that genes are actually composed of specific sequences of DNA nucleotides. The particle concept is not entirely inaccurate, because a particular gene is located at a specific place on a chromosome called its *locus* (*loci* = pl.; location).

Another important idea to remember is that all sexually reproducing organisms have a diploid ($2n$) stage. Since gametes are haploid (n) and most organisms are diploid, the conversion of diploid to haploid cells during meiosis is an important process.

$$2(n) \rightarrow \text{meiosis} \rightarrow (n) \text{ gametes}$$

The diploid cells have two sets of chromosomes—one set inherited from each parent.

$$n + n \text{ gametes} \rightarrow \text{fertilization} \rightarrow 2n$$

Therefore, they have two chromosomes of each kind and have two genes for each characteristic. When sex cells are produced by meiosis, reduction division occurs, and the diploid number is reduced to haploid. Therefore, the sex cells produced by meiosis have only one chromosome of each of the homologous pairs that were in the diploid cell that began meiosis. Diploid organisms usually result from the fertilization of a haploid egg by a haploid sperm. Therefore, they inherit one gene of each type from each parent. For example, each of us has two genes for earlobe shape: one came with our father's sperm, the other with our mother's egg (figure 25.1).

A   B

Figure 25.1

Whether your earlobe is free (*A*) or attached (*B*) depends on the genes you have inherited. As genes express themselves, their actions affect the development of various tissues and organs. Some people's earlobes do not separate from the sides of their heads in the same manner as do those of others. How genes control this complex growth pattern and why certain genes function differently from others is yet to be clarified.

SINGLE-GENE INHERITANCE PATTERNS

In diploid organisms there may be two different forms of the gene. In fact, there may be *several* alternative forms or **alleles** of each gene within a population. In people, for example, there are two alleles for earlobe shape. One allele produces an earlobe that is fleshy and hangs free, while the other allele produces a lobe that is attached to the side of the face and does not hang free. The type of earlobe that is present is determined by the type of allele (gene) received from each parent and the way in which these alleles interact with one another. Alleles are located on the same pair of homologous chromosomes—one allele on each chromosome. These alleles are also at the same specific location, or locus (figure 25.2).

The **genome** is a set of all the genes necessary to specify an organism's complete list of characteristics. The term genome is used in two ways. It may refer to the diploid ($2n$) or haploid (n) number of chromosomes in a cell. Be sure to clarify how your

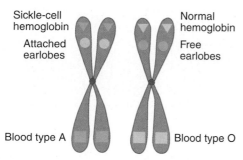

Figure 25.2

Homologous chromosomes contain genes for the same characteristics at the same place. Note that the attached-earlobe allele is located at the ear-shape locus on one chromosome, and the free-earlobe allele is located at the ear-shape locus on the other member of the homologous pair of chromosomes. The other two genes are for hemoglobin structure (alleles for normal and sickled cells) and blood type (alleles for blood types A and O). The examples presented here are for illustrative purposes only. We do not really know if these particular genes are on these chromosomes.

instructor uses this term. The **genotype** of an organism is a listing of the genes present in that organism. It consists of the cell's DNA code; therefore, you cannot see the genotype of an organism. It is not yet possible to know the complete genotype of most organisms, but it is often possible to figure out the genes present that determine a particular characteristic. For example, there are three possible genotypic combinations of the two alleles for earlobe shape. Genotypes are typically represented by upper- and lowercase letters. In the case of the earlobe trait, the allele for free earlobes is designated "E" while that for attached earlobes is "e." A person's genotype could be (1) two alleles for attached earlobes, (2) one allele for attached earlobes and one allele for free earlobes, or (3) two alleles for free earlobes.

How would individuals with each of these three genotypes look? The way each combination of alleles *expresses* (shows) itself is known as the **phenotype** of the organism. The phrase *gene expression* refers to the degree to which a gene goes through protein synthesis to show itself as a physical feature of the individual.

A person with two alleles for attached earlobes will have earlobes that do not hang free. A person with one allele for attached earlobes and one allele for free earlobes will have a phenotype that exhibits free earlobes. An individual with two alleles for free earlobes will also have free earlobes. Notice that there are three genotypes, but only two phenotypes. The individuals with the free-earlobe phenotype can have different genotypes.

Alleles	Genotypes	Phenotypes
E = free earlobes	EE	free earlobes
e = attached earlobes	Ee	free earlobes
	ee	attached earlobes

The expression of some genes is directly influenced by the presence of other alleles. For any particular pair of alleles in an individual, the two alleles from the two parents are either identical or not identical. A person is **homozygous** for a trait when they have the combination of two identical alleles for that par-

ticular characteristic, for example, *EE* and *ee*. A person with two alleles for freckles is said to be homozygous for that trait. A person with two alleles for no freckles is also homozygous. If an organism is homozygous, the characteristic expresses itself in a specific manner. A person homozygous for free earlobes has free earlobes, and a person homozygous for attached earlobes has attached earlobes.

An individual is designated as **heterozygous** when they have two different allelic forms of a particular gene, for example, *Ee*. The heterozygous individual received one form of the gene from one parent and a different allele from the other parent. For instance, a person with one allele for freckles and one allele for no freckles is heterozygous. If an organism is heterozygous, these two different alleles interact to determine a characteristic. A **carrier** is any person who is heterozygous for a trait. In this situation, the recessive allele is hidden, that is, it does not express itself enough to be a phenotype.

Dominant and Recessive Alleles

Often, one allele in the pair expresses itself more than the other. A **dominant allele** masks the effect of other alleles for the trait. For example, if a person has one allele for free earlobes and one allele for attached earlobes, that person has a phenotype of free earlobes. We say the allele for free earlobes is dominant. A **recessive allele** is one that, when present with another allele, has its actions overshadowed by the other; it is masked by the effect of the other allele. Having attached earlobes is the result of having a combination of two recessive characteristics. A person with one allele for free earlobes and one allele for attached earlobes has a phenotype of free earlobes. The expression of recessive alleles is only noted when the organism is homozygous for the recessive alleles. If you have attached earlobes, you have two alleles for that trait. Don't think that recessive genes are necessarily bad. *The term* recessive *has nothing to do with the significance or value of the gene—it simply describes how it can be expressed. Recessive alleles are not less likely to be inherited but must be present in a homozygous condition to express themselves. Also, recessive alleles are not necessarily less frequent in the population.* Sometimes, the physical environment determines whether or not dominant or recessive genes function. For example, in humans genes for freckles do not express themselves fully unless the person's skin is exposed to sunlight (figure 25.3).

Codominance

In cases of dominance and recessiveness, one allele of the pair clearly overpowers the other. Although this is common, it is not always the case. In some combinations of alleles, there is a **codominance.** This is a situation in which both alleles in a heterozygous person express themselves. An example involves the genetic abnormality, *sickle-cell disease.* Having the two recessive alleles for sickle-cell hemoglobin ($Hb^S Hb^S$) can result in abnormally shaped red blood cells. This occurs because the hemoglobin molecules are synthesized with the wrong amino acid sequence. These abnormal hemoglobin molecules tend to attach

Figure 25.3

The expression of many genes is influenced by the environment. The allele for dark hair in the cat is sensitive to temperature and expresses itself only in the parts of the body that stay cool. The allele for freckles expresses itself more fully when a person is exposed to sunlight.

to one another in long, rodlike chains when oxygen is in short supply, that is, with exercise, pneumonia, or emphysema. These rodlike chains distort the shape of the red blood cells into a sickle shape. When these abnormal red blood cells change shape, they clog small blood vessels. The sickled red cells are also destroyed more rapidly than normal cells. This results in a shortage of red blood cells, a condition known as anemia, and an oxygen deficiency in the tissues that have become clogged. People with sickle-cell anemia may experience pain, swelling, and damage to organs such as the heart, lungs, brain, and kidneys.

Sickle-cell anemia can be lethal in the homozygous recessive condition. In the homozygous dominant condition ($Hb^A Hb^A$), the person has normal red blood cells. In the heterozygous condition ($Hb^A Hb^S$), patients produce both kinds of red blood cells. When the amount of oxygen in the blood falls below a certain level, those able to sickle will distort. However, when this occurs, most people heterozygous for the trait do not show severe symptoms. Therefore, these alleles related to one another in a codominant fashion. However, under the right circumstances, being heterozygous can be beneficial. A person with a single sickle-cell allele is more resistant to malaria than a person without this allele.

Genotype	Phenotype
$Hb^A Hb^A$	Normal hemoglobin and nonresistance to malaria
$Hb^A Hb^S$	Normal hemoglobin and resistance to malaria
$Hb^S Hb^S$	Resistance to malaria but death from sickle-cell anemia

Originally, sickle-cell anemia was found at a high frequency in parts of the world where malaria was common, such as tropical regions of Africa and South America. Today, however, this genetic disease can be found anywhere in the world. In the United States, it is most common among black populations whose ancestors came from equatorial Africa.

A classic example of codominance in plants involves the color of the petals of snapdragons. There are two alleles for the color of these flowers. Because neither allele is recessive, we cannot use the traditional capital and small letters as symbols for these alleles. Instead, the allele for white petals is given the

Figure 25.4

The colors of these snapdragons are determined by two alleles for petal color, F^W and F^R. There are three different phenotypes because of the way in which the alleles interact with one another. In the heterozygous condition, neither of the alleles dominates the other.

symbol F^W, and the one for red petals is given the symbol F^R (figure 25.4).

There are three possible combinations of these two alleles:

Genotype	Phenotype
$F^W F^W =$	White flower
$F^R F^R =$	Red flower
$F^R F^W =$	Pink flower

Notice that there are only two different alleles, red and white, but there are three phenotypes, red, white, and pink. Both the red-flower allele and the white-flower allele partially express themselves when both are present, and this results in pink.

X-Linked Genes

Pairs of alleles located on nonhomologous chromosomes separate independently of one another during meiosis when the chromosomes separate into sex cells. Since each chromosome has many genes on it, these genes tend to be inherited as a group. Genes located on the same chromosome that tend to be inherited together are called a *linkage group*. The process of crossing-over, which occurs during prophase I of meiosis I, may split up these linkage groups. Crossing-over happens between homologous chromosomes donated by the mother and the father, and results in a mixing of genes. The closer two genes are to each other on a chromosome, the more probable it is that they will be inherited together.

People and many other organisms have two types of chromosomes. Autosomes (22 pairs) are not involved in sex determination and have the same genes on both members of the homologous pair of chromosomes. *Sex chromosomes* are a pair of chromosomes that control the sex of an organism. In humans, and some other animals, there are two types of sex chromosomes—the X chromosome and the Y chromosome. The Y chromosome is much shorter than the X chromosome and has few genes for traits found on the X chromosome. One genetic trait that is located on the Y chromosome contains the testis-determining gene—SRY. Females are normally produced when two X chromosomes are present. Males are usually produced when one X chromosome and one Y chromosome are present.

Genes found together on the X chromosomes are said to be **X-linked.** Because the Y chromosome is shorter than the X chromosome, it does not have many of the alleles that are found on the comparable portion of the X chromosome. Therefore, in a man, the presence of a single allele on his only X chromosome will be expressed, regardless of whether it is dominant or recessive. A Y-linked trait in humans is the SRY gene. This gene controls the differentiation of the embryonic gonad to a male testis. By contrast, more than 100 genes are on the X chromosome. Some of these X-linked genes can result in abnormal traits such as *color deficiency* (formerly referred to as color blindness), *hemophilia, brown teeth,* and at least two forms of *muscular dystrophy* (Becker's and Duchenne's).

MENDEL'S LAWS OF HEREDITY

Heredity problems are concerned with determining which alleles are passed from the parents to the offspring and how likely it is that various types of offspring will be produced. The first person to develop a method of predicting the outcome of inheritance patterns was Mendel, who performed experiments concerning the inheritance of certain characteristics in sweet pea plants. From his work, Mendel concluded which traits were dominant and which were recessive. Some of his results are shown in table 25.1.

What made Mendel's work unique was that he studied only one trait at a time. Previous investigators had tried to follow numerous traits at the same time. When this was attempted, the total set of characteristics was so cumbersome to work with that no clear idea could be formed of how the offspring inherited traits. Mendel used traits with clear-cut alternatives, such as purple or white flower color, yellow or green seed pods, and tall or dwarf pea plants. He was very lucky to have chosen pea plants in his study because they naturally self-pollinate. When self-pollination occurs in pea plants over many generations, it is possible to develop a population of plants that is homozygous for a number of characteristics. Such a population is known as a *pure line*.

Mendel took a pure line of pea plants having purple flower color, removed the male parts (anthers), and discarded them so that the plants could not self-pollinate. He then took anthers from a pure-breeding white-flowered plant and pollinated the antherless purple flower. When the pollinated flowers produced seeds, Mendel collected, labeled, and planted them. When these seeds germinated and grew, they eventually produced flowers.

You might be surprised to learn that all of the plants resulting from this cross had purple flowers. One of the prevailing hypotheses of Mendel's day would have predicted that the purple and white colors would have blended, resulting in flowers that were lighter than the parental purple flowers. Another hypothesis would have predicted that the offspring would have had a mixture of white and purple flowers. The unexpected result—all of the offspring produced flowers like those of one parent and no flowers like those of the other—caused Mendel to examine other traits as well and formed the basis for much of the rest of his work. He repeated his experiments using pure strains for other traits. Pure-breeding tall plants were crossed with pure-breeding dwarf plants. Pure-breeding plants with yellow pods were crossed with pure-breeding plants with green pods. The results were all the same: the offspring showed the characteristic of one parent and not the other.

Next, Mendel crossed the offspring of the white-purple cross (all of which had purple flowers) with each other to see what the third generation would be like. Had the characteristic of the original white-flowered parent been lost completely? This second-generation cross was made by pollinating these purple flowers that had one white parent among themselves. The seeds produced from this cross were collected and grown. When these plants flowered, three-fourths of them produced purple flowers and one-fourth produced white flowers.

Table 25.1 Dominant and Recessive Traits in Pea Plants

Characteristic	Dominant Allele	Recessive Allele
Plant height	Tall	Dwarf
Pod shape	Full	Constricted
Pod color	Green	Yellow
Seed surface	Round	Wrinkled
Seed color	Yellow	Green
Flower color	Purple	White

One in every twenty persons has a defective allele that causes cystic fibrosis. *Cystic fibrosis* (*CF*) is among the most common lethal genetic disorders affecting most Caucasians in the United States and affects nearly 30,000 people in North America. About 5 percent of American Caucasians are asymptomatic carriers of the recessive allele. About 1,000 new cases are identified in the United States each year. The gene for CF occurs on chromosome 7 and is responsible for the manufacture of cystic fibrosis transmembrane regulator (CFTR) protein. The most common of dozens of possible mutations in this gene is a deletion mutation responsible for the loss of a single amino acid in CFTR. As a result, CFTR is unable to control the movement of chloride ions through chloride-permeable cell membrane channels. This results in the malfunction of sweat glands in the skin and secretion of excess chloride ions, which causes a salty taste of the skin of children with CF. It is also responsible for other symptoms including:

1. Mucus filling bronchioles that results in blocked breathing and frequent upper-respiratory infections
2. Bile duct clogging that interferes with digestion and liver function
3. Pancreas ducts clogged with mucus that prevents the flow of digestive enzymes into the intestinal tract
4. Bowel obstructions caused by thickened stools
5. Sterility in males due to absence of vas deferens development and, on occasion, female sterility due to dense mucus that blocks sperm from reaching eggs

Until recently, the test used to detect this gene in people with a history of cystic fibrosis was only 50 percent accurate and could not detect the gene in persons with no family history of the disease. A new test is 90 percent accurate in detecting people with the gene regardless of their family history. This new test enables physicians to screen prospective parents to better determine the probability that they will have an offspring with cystic fibrosis. Confirmation of CF requires additional testing.

After analyzing his data, Mendel formulated several genetic laws to describe how characteristics are passed from one generation to the next and how they are expressed in an individual.

Mendel's **law of dominance**—When an organism has two different alleles for a given trait, the allele that is expressed, overshadowing the expression of the other allele, is said to be dominant. The gene whose expression is overshadowed is said to be recessive.

Mendel's **law of segregation**—When gametes are formed by a diploid organism, the alleles that control a trait separate from one another into different gametes, retaining their individuality.

Mendel's **law of independent assortment**—Members of one gene pair separate from each other independently of the members of other gene pairs.

At the time of Mendel's research, biologists knew nothing of chromosomes or DNA or of the processes of mitosis and meiosis. Mendel assumed that each gene was separate from other genes. It was fortunate for him that most of the characteristics he picked to study were found on separate chromosomes. If two or more of these genes had been located on the same chromosome (*linked genes*), he probably would not have been able to formulate his laws. The discovery of chromosomes and DNA has led to modifications in Mendel's laws, but it was Mendel's work that formed the foundation for the science of genetics.

STEPS IN SOLVING HEREDITY PROBLEMS: SINGLE-FACTOR CROSSES

The first type of problem we will consider is the easiest type, a single-factor cross. A **single-factor cross** (sometimes called a monohybrid cross: *mono* = one; *hybrid* = combination) is a genetic cross or mating in which a single characteristic is followed from one generation to the next.

For example, in humans, the allele for Tourette syndrome (TS) is inherited as an autosomal dominant allele. For centuries, people displaying this genetic disorder were thought to be possessed by the devil since they displayed such unusual behaviors. These motor and verbal behaviors or tics are involuntary and range from mild (e.g., leg tapping, eye blinking, face twitching) to the more violent forms such as the shouting of profanities, head jerking, spitting, compulsive repetition of words, or even barking like a dog. The symptoms result from an excess production of the brain messenger, dopamine. The most widely used drug to control elevated dopamine levels in Tourette syndrome patients is haloperidol. The penetrance of this trait is different for males and females. Penetrance refers to the all-or-none expression of a given genotype in heterozygous individuals in a population. Understanding penetrance helps answer the question, "If all these people have this gene, why don't I see more people with the condition?" For example, heterozygous males show a 90 percent penetrance of TS; that is, 90 percent of men with TS express the trait enough to be diagnosed with TS. There is only 60 to 70 percent penetrance in the female population. If both parents are heterozygous (have one allele for Tourette syndrome and one allele for no Tourette syndrome) what is the probability that they can have a child without Tourette syndrome? with Tourette syndrome?

In solving a heredity problem there are five basic steps:

Step 1: Assign a symbol for each allele.
Usually a capital letter is used for a dominant allele and a small letter for a recessive allele. Use the symbol *T* for Tourette and *t* for no Tourette.

Allele	Genotype	Phenotype
T = Tourette	*TT*	Tourette syndrome
t = normal	*Tt*	Tourette syndrome
	tt	Normal

Step 2: Determine the genotype of each parent and indicate a mating. Because both parents are heterozygous, the male genotype is *Tt*. The female genotype is also *Tt*. The × between them is used to indicate a mating.

$$Tt \times Tt$$

Step 3: Determine all the possible kinds of gametes each parent can produce.

Remember that gametes are haploid; therefore, they can have only one allele instead of the two present in the diploid cell. Because the male has both the Tourette syndrome allele and the normal allele, half of his gametes will contain the Tourette syndrome allele and the other half will contain the normal allele. Because the female has the same genotype, her gametes will be the same as his.

For genetic problems, a *Punnett square* is used. A **Punnett square** is a box figure that allows you to determine the probability of genotypes and phenotypes of the progeny of a particular cross. Remember, because of the process of meiosis, each gamete receives only one allele for each characteristic listed. Therefore, the male will produce sperm with either a *T* or a *t*; the female will produce ova with either a *T* or a *t*. The possible gametes produced by the male parent are listed on the left side of the square and the female gametes are listed on the top. In our example, the Punnett square would show a single dominant allele and a single recessive allele from the male on the left side. The alleles from the female would appear on the top.

Female genotype
Tt

Male genotype
Tt

Possible female gametes
T and *t*

Possible male gametes
T and *t*

	T	*t*
T		
t		

Step 4: Determine all the gene combinations that can result when these gametes unite.

To determine the possible combinations of alleles that could occur as a result of this mating, simply fill in each of the empty squares with the alleles that can be donated from each parent. Determine all the gene combinations that can result when these gametes unite.

	T	*t*
T	*TT*	*Tt*
t	*Tt*	*tt*

CONCEPTS APPLIED
Are You a 'Taster'?

Some humans inherit the ability to taste the chemical phenylthiocarbamide (PTC), and others are unable to taste PTC. Ask your instructor to furnish you with a supply of paper impregnated with PTC. Place a piece of this paper on your tongue. If you experience a bitter taste, you are a taster. If you do not, you are a nontaster. Take enough strips to test members of your family you can readily contact. These may include your grandparents, parents, siblings, children, aunts, and uncles. After you have tested the members of your family, construct a pedigree as described below. Is it possible for you to determine the genotypes of your family members? Is tasting PTC a dominant or a recessive trait? Could it be a case of lack of dominance? Might it be linked to another trait?

In a pedigree, circles represent females and squares represent males. Symbols of parents are connected by a horizontal mating line, and the offspring are shown on a horizontal line below the parents. Individuals who are tasters are represented by a solid symbol, nontasters by an open symbol. A typical pedigree might look something like the one below, which shows three generations. Create your own pedigree. For this analysis, you can use only "blood" (biological) relatives.

A Closer Look — Blame That Trait on Your Mother!

Within a eukaryotic cell, the bulk of DNA is located in the nucleus. It is this genetic material that controls the majority of biochemical processes of the cell. It has long been recognized that the mitochondria also contain DNA, mtDNA. This genetic material works in conjunction with that located in the nucleus. However, an interesting thing happens when a sperm is formed as a result of meiosis. All the mitochondria are packed into the "necks" and not the "heads" of the sperm. When a sperm penetrates a secondary oocyte, the head enters but the majority of the mitochondria in the neck remain outside. Therefore, fathers rarely if ever contribute mitochondrial genetic information to their children. Should a mutation occur in the mother's mitochondrial DNA, she will pass the abnormality on to her children. Should a mutation occur in the father's mitochondrial DNA, he will not transmit the abnormality. This unusual method of transmission is called *mitochondrial* (or *maternal*) *inheritance*.

Most abnormalities transmitted through mitochondrial genes are associated with muscular weakness because the mitochondria are the major source of ATP in eukaryotic cells. People with one form of this rare type of genetic abnormality, *Leber's hereditary optic neuropathy*, show symptoms of sudden loss of central vision due to optic nerve death in young adults with onset at about 20 years of age. Both males and females may be affected.

Because most genes are comprised of thousands of nucleotide base pairs, there are many different changes in these sequences that can result in the formation of multiple "bad" gene products and therefore, an abnormal phenotype. The same gene can have many different "bad" forms. The fact that a phenotypic characteristic can be determined by many different alleles for a particular characteristic is called *genetic heterogeneity.* For example, two of the best-known forms of *muscular dystrophy* (*MD*) are Duchenne's and Becker's. Duchenne's (DMD) is characterized by a severe, progressive weakening of the muscles, the appearance of a false muscle atrophy (degeneration) in the calves of the legs, onset in early childhood, and a high likelihood of death in the thirties. DMD is caused by a mutation in the dystrophin gene located on the X chromosome and is a recessive trait. Becker's (BMD) is caused by a mutation in the same dystrophin gene but at a different location. BMD is a more mild form of MD.

Step 5: Determine the phenotype of each possible gene combination. In this problem, three of the offspring, *TT, Tt,* and *Tt,* have Tourette syndrome. One progeny, *tt,* is normal. Therefore, the answer to the problem is that the probability of having offspring with Tourette syndrome is 3/4; for no Tourette syndrome, it is 1/4.

Take the time to learn these five steps. All single-factor problems can be solved using this method; the only variation in the problems will be the types of alleles and the number of possible types of gametes the parents can produce.

ALTERNATIVE INHERITANCE SITUATIONS

So far we have considered a characteristic that is determined by simple dominance and recessiveness between two alleles. Other situations, however, may not fit this pattern. Some genetic characteristics are determined by more than two alleles; moreover, some traits are influenced by gene interactions and some traits are inherited differently, depending on the sex of the offspring.

Multiple Alleles

So far we have discussed only traits that are determined by two alleles, for example, *A, a.* However, there can be more than two different alleles for a single trait. All the various forms of the same gene (alleles) that control a particular trait are referred to as **multiple alleles.** However, one person can have only a maximum of two of the alleles for the characteristic. A good example of a characteristic that is determined by multiple alleles is the ABO blood type. There are three alleles for blood type:

Allele*

I^A = blood has type A antigens on red blood cell surface

I^B = blood has type B antigens on red blood cell surface

i = blood type O has neither type A nor type B antigens on red blood cell surface

In the ABO system, A and B show *codominance* when they are together in the same individual, but both are dominant over the O allele. These three alleles can be combined as pairs in six different ways, resulting in four different phenotypes:

Genotype		Phenotype
$I^A I^A$	=	Blood type A
$I^A i$	=	Blood type A
$I^B I^B$	=	Blood type B
$I^B i$	=	Blood type B
$I^A I^B$	=	Blood type AB
ii	=	Blood type O

Multiple-allele problems are worked as single-factor problems. Some examples are in the practice problems at the end of this chapter.

Polygenic Inheritance

Thus far we have considered phenotypic characteristics that are determined by alleles at a specific, single place on homologous chromosomes. However, some characteristics are determined by the interaction of genes at several different loci (on different chromosomes or at different places on a single chromosome). This is called **polygenic inheritance.** A number of different pairs of alleles may combine their efforts to determine a characteristic. Skin color in humans is a good example of this inheritance pattern. According to some experts, genes for skin color are located at a minimum of three loci. At each of these loci, the allele for dark skin is dominant over the allele for light skin. Therefore, a wide variety of skin colors is possible depending on how many dark-skin alleles are present (figure 25.5).

Polygenic inheritance is very common in determining characteristics that are quantitative in nature. In the skin-color example, and in many others as well, the characteristics cannot be categorized in terms of *either/or,* but the variation in phenotypes can be classified as *how much* or *what amount.*

For instance, people show great variations in height. There are not just tall and short people—there is a wide range. Some people are as short as one meter, and others are taller than two meters. This quantitative trait is probably determined by a number of different genes. Intelligence also varies significantly, from those who are severely retarded to those who are geniuses. Many of these traits may be influenced by outside environmental factors such as diet, disease, accidents, and social factors. These are just a few examples of polygenic inheritance patterns.

---

*The symbols I^A and I^B stand for the technical term for the antigenic carbohydrates attached to red blood cells, the *Immunogens.* These alleles are located on human chromosome 9. The ABO system is not the only one used to type blood. Others include the Rh, MNS, and Xg systems.

Locus 1	d^1d^1	d^1D^1	d^1D^1	D^1D^1	D^1d^1	D^1d^1	D^1D^1
Locus 2	d^2d^2	d^2d^2	d^2D^2	D^2d^2	D^2d^2	D^2D^2	D^2D^2
Locus 3	d^3d^3	d^3d^3	d^3d^3	d^3d^3	D^3D^3	D^3D^3	D^3D^3
Total number of dark-skin genes	0	1	2	3	4	5	6
	Very light			Medium			Very dark

Figure 25.5

Skin color in humans is an example of polygenic inheritance. The darkness of the skin is determined by the number of dark-skin genes a person inherits from his or her parents.

Pleiotropy

Even though a single gene produces only one type of mRNA during transcription, it often has a variety of effects on the phenotype of the person. This is called *pleiotropy*. **Pleiotropy** (*pleio-* = changeable) is a term used to describe the multiple effects that a gene may have on the phenotype. A good example of pleiotropy is *Marfan syndrome* (figure 25.6), a disease suspected to have occurred in former U.S. President Abraham Lincoln. Marfan is a disorder of the body's connective tissue but can also have effects in many other organs, including the eyes, heart, blood, skeleton, and lungs. Symptoms generally appear as a tall, lanky body with long arms and spider-fingers, scoliosis (curva-

ture of the spine), and depression of the chest wall. In many cases these nearsighted people also show dislocation of the lens of the eye. The white of the eye (sclera) may appear bluish. Heart problems include dilation of the aorta and abnormalities of the heart valves. Death may be caused by a dissection (tear) in the aorta from the rupture in a weakened and dilated area of the aorta called an aortic *aneurysm*.

Another example of pleiotropy is the genetic abnormality PKU. In this example, a single gene affects many different chemical reactions that depend on the way a cell metabolizes the amino acid phenylalanine, commonly found in many foods (refer to figure 25.7).

A

B

C

Figure 25.6

It is estimated that about 40,000 (the incidence is 1 out of 10,000) people in the United States have Marfan syndrome, an autosomal dominant abnormality. Notice the lanky appearance to the body and face of this person with Marfan syndrome (*A*). Photos (*B*) and (*C*) illustrate their unusually long fingers.

Connections . . .

The Inheritance of Eye Color

It is commonly thought that eye color is inherited in a simple dominant/recessive manner. Brown eyes are considered to be dominant over blue eyes. The real pattern of inheritance, however, is considerably more complicated than this. Eye color is determined by the amount of a brown pigment, known as melanin, present in the iris of the eye. If there is a large quantity of melanin present on the anterior surface of the iris, the eyes are dark. Black eyes have a greater quantity of melanin than brown eyes.

If there is not a large amount of melanin present on the anterior surface of the iris, the eyes will appear blue, not because of a blue pigment but because blue light is returned from the iris (see box figure 25.1). The iris appears blue for the same reason that deep bodies of water tend to appear blue. There is no blue pigment in the water, but blue wavelengths of light are returned to the eye from the water. People appear to have blue eyes because the blue wavelengths of light are reflected from the iris.

Just as black and brown eyes are determined by the amount of pigment present, colors such as green, gray, and hazel are produced by the various amounts of melanin in the iris. If a very small amount of brown melanin is present in the iris, the eye tends to appear green, whereas relatively large amounts of melanin produce hazel eyes.

Several different genes are probably involved in determining the quantity and placement of the melanin and, therefore, in determining eye color. These genes interact in such a way that a wide range of eye color is possible. Eye color is probably determined by polygenic inheritance, just as skin color and height are. Some newborn babies have blue eyes that later become brown. This is because they have not yet begun to produce melanin in their irises at the time of birth.

Box Figure 25.1

Figure 25.7

Phenylketonuria (PKU) is an autosomal recessive disorder located on chromosome 12. This diagram shows how the normal pathways work (these are shown in black). If the enzyme phenylalanine hydroxylase is not produced because of a mutated gene, the amino acid phenylalanine cannot be broken down, and is converted into phenylpyruvic acid, which accumulates in body fluids. There are three major results: (1) mental retardation because phenylpyruvic acid kills nerve cells, (2) abnormal body growth because less of the growth hormone thyroxine is produced, and (3) pale skin pigmentation because less melanin is produced (abnormalities are shown in color). It should also be noted that if a woman who has PKU becomes pregnant, her baby is likely to be born retarded. While the embryo may not have the genetic disorder, the phenylpyruvic acid produced by the pregnant mother will damage the developing brain cells. This is called maternal PKU.

Inheritance Pattern of Spina Bifida

Many polygenic traits are influenced by outside environmental factors, such as diet, drugs, disease, accidents, and social factors. The degree to which the environment impacts the genes can determine the nature of phenotype. Patterns of inheritance that are the result of polygenic traits influenced by environmental factors are called multifactorial inheritance. In fact, most polygenic traits are multifactorial. *Multifactorial inheritance* is not yet completely understood and is hard to trace since it does not follow the typical single-gene inheritance pattern. The first generation (children) to follow individuals with such diseases (parents) shows a markedly high risk of acquiring the gene, whereas the second generation (grandchildren) has a much lower risk. Examples of multifactorial traits include neural tube defects (e.g., spina bifida), insulin-dependent diabetes mellitus, cleft palate, different fingerprints in identical twins, and congenital heart disease. In the case of spina bifida the arches of the vertebrae fail to enclose the spinal cord (typically in the lower back region) and in some cases the cord and nerves may be exposed at birth. This abnormality is a leading cause of still births and death during early infancy and, if unable to be surgically corrected, is a handicap in survivors. One factor that appears to influence this disease is the amount of the vitamin folic acid the mother has in her system prior to conception. Other factors include chromosome disorders, single-gene mutations, and teratogens (chemicals that cause developmental abnormalities).

ENVIRONMENTAL INFLUENCES ON GENE EXPRESSION

Maybe you assumed that the dominant allele would always be expressed in a heterozygous individual. It is not so simple! Here, as in other areas of biology, there are exceptions. For example, the allele for six fingers (*polydactylism*) is dominant over the allele for five fingers in humans. Some people who have received the allele for six fingers have a fairly complete sixth finger; in others, it may appear as a little stub. In another case, a dominant allele causes the formation of a little finger that cannot be bent as a normal little finger. However, not all people who are believed to have inherited that allele will have a stiff little finger. In some cases, this dominant characteristic is not expressed or perhaps only shows on one hand. Thus, there may be variation in the degree to which an allele expresses itself in an *individual*. Geneticists refer to this as *variable expressivity*. A good example of this occurs in the genetic abnormality *neurofibromatosis type I* (NF1) (figure 25.8). In some cases it may not be expressed in the *population* at all. This is referred to as a *lack of penetrance*. Other genes may be interacting with these dominant alleles, causing the variation in expression.

Both internal and external environmental factors can influence the expression of genes. For example, at conception, a male receives genes that will eventually determine the pitch of his voice. However, these genes are expressed differently after puberty. At puberty, male sex hormones are released. This internal environmental change results in the deeper male voice. A male who does not produce these hormones retains a higher-pitched voice in later life. Comparable changes can occur in females when an abnormally functioning adrenal gland causes the release of large amounts of male hormones.

Diet is an external environmental factor that can influence the phenotype of an individual. *Diabetes mellitus*, a metabolic disorder in which glucose is not properly metabolized and is passed out of the body in the urine, has a genetic

Figure 25.8

Neurofibromatosis I is seen in many forms including benign fibromatous skin tumors, "Café-au-lait" spots, nodules in the iris, and possible malignant tumors. It is extremely variable in its expressivity, i.e., the traits may be almost unnoticeable or extensive. An autosomal dominant trait, it is the result of a mutation and the production of a protein (neurofibromin) that normally would suppress the activity of a gene that causes tumor formation.

basis. Some people who have a family history of diabetes are thought to have inherited the trait for this disease. Evidence indicates that they can delay the onset of the disease by reducing the amount of sugar in their diet. This change in the external environment influences gene expression in much the same way that sunlight affects the expression of freckles in humans (see figure 25.3).

WHAT IS A GENE AND HOW DOES IT WORK?

As scientists began to understand the chemical makeup of the **nucleic acids,** an attempt was made to understand how DNA and RNA relate to inheritance, cell structure, and cell activities. The concept that resulted is known as the *central dogma*. The belief can be written in this form:

DNA ⟵ (replication) ⟵ DNA ⟶ (transcription) ⟶ RNA ⟶ (translation) ⟶ Proteins ⟶ Structural / Carrier / Enzymatic/hormonal

What this concept map says is that at the center of it all is DNA, the genetic material of the cell and (going to the left) it is capable of reproducing itself, a process called **DNA replication.** Going to the right, DNA is capable of supervising the manufacture of RNA (a process known as **transcription**), which in turn is involved in the production of protein molecules, a process known as **translation.**

DNA replication occurs in cells in preparation for the cell division processes of mitosis and meiosis. Without replication, daughter cells would not receive the library of information required to sustain life. The transcription process results in the formation of a strand of RNA that is a copy of a segment of the DNA on which it is formed. Some of the RNA molecules become involved in various biochemical processes, while others are used in the translation of the RNA information into proteins. Structural proteins are used by the cell as building materials (feathers, collagen, hair); while others are used to direct and control chemical reactions (enzymes or hormones) or carry molecules from place to place (hemoglobin).

It is the processes of transcription and translation that result in the manufacture of all enzymes. Each unique enzyme molecule is made from a blueprint in the form of a DNA nucleotide sequence, or *gene.* Some of the thousands of enzymes manufactured in the cell are the tools required for transcription and translation to take place. *The enzymes made by the process carry out the process of making more enzymes!*

Tools are made to make more tools! The same is true for DNA replication. Enzymes made from the DNA blueprints by transcription and translation are used as tools to make exact copies of the genetic material! More blueprints are made so that future generations of cells will have the genetic materials necessary for them to manufacture their own regulatory and structural proteins. Without DNA, RNA, and enzymes functioning in the proper manner, life as we know it would not occur.

DNA has four properties that enable it to function as genetic material. It is able to (1) *replicate* by directing the manufacture of copies of itself; (2) *mutate,* or chemically change, and transmit these changes to future generations; (3) *store* information that determines the characteristics of cells and organisms; and (4) use this information to *direct* the synthesis of structural and regulatory proteins essential to the operation of the cell or organism.

The Structure of DNA and RNA

Nucleic acid molecules are enormous and complex polymers made up of monomers called **nucleotides.** Each nucleotide is composed of a sugar molecule (S) containing five carbon atoms, a phosphate group (P), and a molecule containing nitrogen that will be referred to as a *nitrogenous base* (B) (figure 25.9). It is possible to classify nucleic acids into two main groups based on the kinds of sugars and nitrogenous bases used in the nucleotides (i.e., DNA and RNA).

In cells, DNA is the nucleic acid that functions as the original blueprint for the synthesis of proteins. It contains the sugar *deoxyribose;* phosphates; and the nitrogenous bases adenine, guanine, cytosine, and thymine (A, G, C, T). RNA is a type of nucleic acid that is directly involved in the synthesis of protein. It contains the sugar *ribose;* phosphates; and adenine, guanine, cytosine, and uracil (A, G, C, U). There is no thymine in RNA and no uracil in DNA.

DNA and RNA differ in one other respect. DNA is actually a double molecule. It consists of two flexible strands held together between their protruding bases. The two strands are twisted about each other in a coil or double helix (figure 25.10). The two strands of the molecule are held together because they "fit" each other like two jigsaw puzzle pieces that interlock with one another and are stabilized by weak chemical forces, hydrogen bonds. The four kinds of bases always pair in a definite way: adenine (A) with thymine (T), and guanine (G) with cytosine (C). Notice that the large molecules (A and G) pair with the small ones (T and C), thus keeping the two complementary (matched) strands parallel. The bases that pair are said to be **complementary bases.**

$$\text{and} \quad \begin{array}{c} \text{G : C} \\ \text{A : T} \end{array}$$

You can "write" a message in the form of a stable DNA molecule by combining the four different DNA nucleotides (A, T, G, C) in particular sequences. The four DNA nucleotides are being used as an alphabet to construct three-letter words. In order to make sense out of such a code, it is necessary to read in one direction. Reading the sequence in reverse does not always make sense, just as reading this paragraph in reverse would not make sense.

The genetic material of humans and other eukaryotic organisms are *strands* of coiled *double-stranded DNA,* which has histone proteins attached along its length. When eukaryotic chromatin fibers coil into condensed, highly knotted bodies, they are seen easily through a microscope after staining with dye. Condensed like this, a chromatin fiber is referred to as a **chromosome** (figure 25.11B). The genetic material in bacteria is also double-stranded (duplex) DNA, but the ends of the molecule are connected to form a *loop* and they do not form condensed chromosomes (figure 25.12). However, prokaryotic cells have an attached protein called *HU protein.*

Each chromatin DNA strand is different because each strand has a different chemical code. Coded DNA serves as a central cell library. Tens of thousands of messages are in this storehouse of information. This information tells the cell such things as (1) how to produce enzymes required for the digestion of nutrients, (2) how to manufacture enzymes that will metabolize the nutrients and eliminate harmful wastes, (3) how to repair and assemble cell parts, (4) how to reproduce healthy

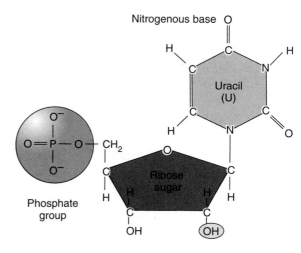

A DNA nucleotide

B RNA nucleotide

C The four nitrogenous bases that occur in DNA

Figure 25.9

(A) The nucleotide is the basic structural unit of all nucleic acid molecules. A thymine nucleotide of DNA is comprised of phosphate, deoxyribose sugar, and the nitrogenous base, thymine (T). Notice in the nucleotides that the phosphate group is written in "shorthand" form as a P inside a circle. (B) The RNA uracil nucleotide is comprised of a phosphate, ribose sugar, and the nitrogenous base, uracil (U). Notice the difference between the sugars and how the bases differ from one another. (C) Using these basic components (phosphate, sugars, and bases) the cell can construct eight common types of nucleotides. Can you describe all eight?

offspring, (5) when and how to react to favorable and unfavorable changes in the environment, and (6) how to coordinate and regulate all of life's essential functions. If any of these functions are not performed properly, the cell may die. The importance of maintaining essential DNA in a cell becomes clear when we consider cells that have lost it. For example, human red blood cells lose their nuclei as they become specialized for carrying oxygen and carbon dioxide throughout the body. Without DNA they are unable to manufacture the essential cell components needed to sustain themselves. They continue to exist for about 120 days, functioning only on enzymes manufactured earlier in their lives. When these enzymes are gone, the cells die. Because these specialized cells begin to die the moment they lose their DNA, they are more accurately called *red blood corpuscles (RBCs)*: "little dying red bodies."

DNA Replication

Because all cells must maintain a complete set of genetic material, there must be a doubling of DNA in order to have enough to pass on to the offspring. DNA replication is the process of

duplicating the genetic material prior to its distribution to daughter cells. When a cell divides into two daughter cells, each new cell must receive a complete copy of the parent cell's genetic information, or it will not be able to manufacture all the proteins vital to its existence. Accuracy of duplication is essential in order to guarantee the continued existence of that type of cell. Should the daughters not receive exact copies, they would most likely die.

1. The DNA replication process begins as an enzyme breaks the attachments between the two strands of DNA. In eukaryotic cells, this occurs in hundreds of different spots along the length of the DNA (figure 25.13A).
2. Moving along the DNA, the enzyme "unzips" the halves of the DNA (figure 25.13B and C). Hydrogen bonds hold each in position (AT, GC) while the new nucleotide is covalently bonded between the sugar and phosphate of the new backbone.
3. Proceeding in opposite directions on each side, the enzyme *DNA polymerase* moves down the length of the DNA, attaching new DNA nucleotides into position (figure 25.13C).
4. The enzyme that speeds the addition of new nucleotides to the growing chain works along with another enzyme to make sure

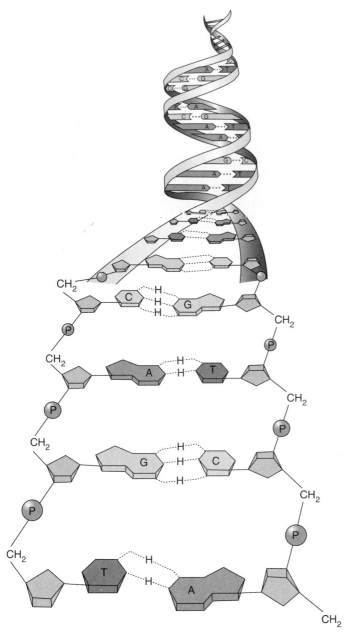

Figure 25.10

DeoxyriboNucleic Acid is a helical molecule. While the parts of each strand are held together by covalent bonds, the two parallel strands are interlinked by nitrogenous bases like jigsaw puzzle pieces. Hydrogen bonds help hold the two strands together.

that no mistakes are made. If the wrong nucleotide appears to be headed for a match, the enzyme will reject it in favor of the correct nucleotide. If a mistake is made and a wrong nucleotide is paired into position, specific enzymes have the ability to replace it with the correct one (figure 25.13D).

5. Replication proceeds in both directions, appearing as "bubbles" (figure 25.13E).

6. The complementary molecules (AT, GC) pair with the exposed nitrogenous bases of both DNA strands by forming new hydrogen bonds (figure 25.13F).

7. Once properly aligned, a bond is formed between the sugars and phosphates of the newly positioned nucleotides. A strong sugar and phosphate backbone is formed in the process (figure 25.13G).

8. This process continues until all the replication "bubbles" join (figure 25.13H).

A new complementary strand of DNA forms on each of the old DNA strands, resulting in the formation of two double-stranded DNA molecules. In this way, the exposed nitrogenous bases of the original DNA serve as a *template,* or pattern, for the formation of the new DNA. As the new DNA is completed, it twists into its double-helix shape.

The completion of the DNA replication process yields two double helices that are identical in their nucleotide sequences. Half of each is new, half is the original parent DNA molecule. The DNA replication process is highly accurate. It has been estimated that there is only one error made for every 2×10^9 nucleotides. A human cell contains 46 chromosomes consisting of about 3,000,000,000 (3 billion) base pairs. This averages to about 1.5 errors per cell! Don't forget that this figure is an estimate. While some cells may have five errors per replication, others may have more, and some may have no errors at all. It is also important to note that some errors may be major and deadly, while others are insignificant. Because this error rate is so small, DNA replication is considered by most to be essentially error-free. Following DNA replication, the cell now contains twice the amount of genetic information and is ready to begin the process of distributing one set of genetic information to each of its two daughter cells.

The distribution of DNA involves splitting the cell and distributing a set of genetic information to the two new daughter cells. In this way, each new cell has the necessary information to control its activities. The mother cell ceases to exist when it divides its contents between the two smaller daughter cells. A cell does not really die when it reproduces itself; it merely starts over again.

DNA Transcription

DNA functions in the manner of a reference library that does not allow its books to circulate. Information from the originals must be copied for use outside the library. The second major function of DNA is the making of these single-stranded, complementary RNA copies of DNA. This operation is called transcription (*scribe* = to write), which means to transfer data from one form to another. In this case, the data are copied from DNA language to RNA language. The same base-pairing rules that control the accuracy of DNA replication apply to the process of transcription. Using this process, the genetic information stored as a DNA chemical code is carried in the form of an RNA copy to other parts of the cell. It is RNA that is used to guide the assembly of amino acids into structural and regulatory proteins. Without the process of transcription, genetic information would be useless in directing cell functions. Although many types of RNA are synthesized from the genes, the three most

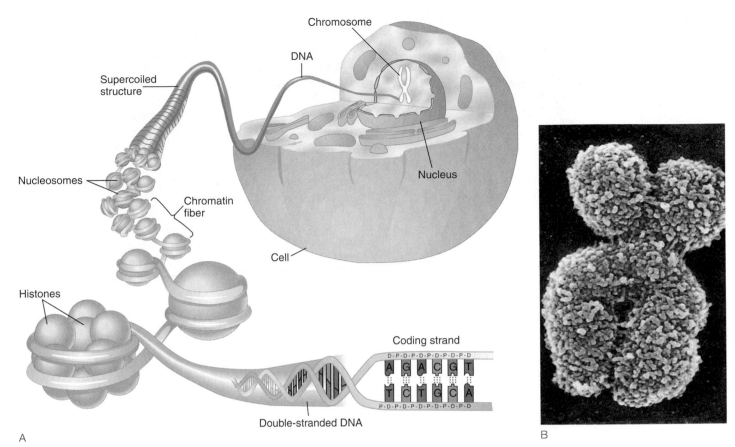

A

B

Figure 25.11

(*A*) Eukaryotic cells contain double-stranded DNA in their nuclei, which takes the form of a three-dimensional helix. One strand is a chemical code (the coding strand) that contains the information necessary to control and coordinate the activities of the cell. The two strands fit together and are bonded by weak hydrogen bonds formed between the complementary, protruding nitrogenous bases according to the base-pairing rule. The length of a DNA molecule is measured in numbers of "base pairs"— the number of rungs on the ladder. (*B*) During certain stages in the reproduction of a eukaryotic cell, the nucleoprotein coils and "supercoils," forming tightly bound masses. When stained, these are easily seen through the microscope. In their supercoiled form, they are called chromosomes, meaning colored bodies.

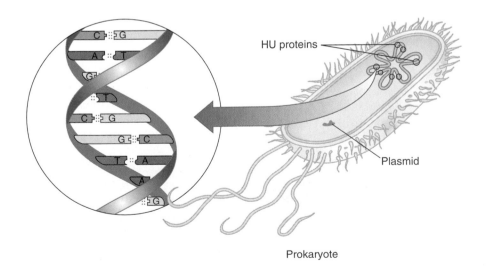

Prokaryote

Figure 25.12

The nucleic acid of prokaryotic cells (the bacteria) does not have the histone protein; rather, it has proteins called HU proteins. In addition, the ends of the giant nucleoprotein molecule overlap and bind with one another to form a loop. The additional small loop of DNA is the plasmid, which contains genes that are not essential to the daily life of the cell.

Figure 25.13

These illustrations summarize the basic events that occur during the replication of double-stranded DNA.

important are *messenger RNA (mRNA), transfer RNA (tRNA),* and *ribosomal RNA (rRNA).* Refer to figure 18.24 for an overview of this process.

Transcription begins in a way that is similar to DNA replication. The DNA is separated by an enzyme, exposing the nitrogenous-base sequences of the two strands. However, unlike DNA replication, transcription occurs only on one of the two DNA strands, which serves as a template, or pattern, for the synthesis of RNA (see figure 25.14). This side is also referred to as the genetic *coding strand* of the DNA. But which strand is copied? Where does it start and when does it stop? Where along the sequence of thousands of nitrogenous bases does the chemical code for the manufacture of a particular enzyme begin and where does it end? If transcription begins randomly, the resulting RNA may not be an accurate copy of the code, and the enzyme product may be useless or deadly to the cell. To answer these questions, it is necessary to explore the nature of the genetic code itself.

RNA polymerase

Noncoding DNA strand

Coding DNA strand

Newly forming RNA

RNA

Figure 25.14

This summary illustrates the basic events that occur during the transcription of one side (the coding strand) of double-stranded DNA. The enzyme attaches to the DNA at a point that allows it to separate the complementary strands. As this enzyme, RNA polymerase, moves down the DNA, new complementary RNA nucleotides are base-paired on one of the exposed strands and linked together, forming a new strand that is complementary to the nucleotide sequence of the DNA. The newly formed (transcribed) RNA is then separated from its DNA complement. Depending on the DNA segment that has been transcribed, this RNA molecule may be a messenger RNA (mRNA), a transfer RNA (tRNA), a ribosomal RNA (rRNA), or an RNA molecule used for other purposes within the cell.

We know that genetic information is in chemical-code form in the DNA molecule. When the coded information is used or *expressed,* it guides the assembly of particular amino acids into structural and regulatory polypeptides and proteins. If DNA is molecular language, then each nucleotide in this language can be thought of as a letter within a four-letter alphabet. Each word, or code, is always three letters (nucleotides) long, and only three-letter words can be written. A **DNA code** is a triplet nucleotide sequence that codes for one of the twenty common amino acids. The number of codes in this language is limited because there are only four different nucleotides, which are used only in groups of three. The order of these three letters is just as important in DNA language as it is in our language. We recognize that CAT is not the same as TAC. If all the possible three-letter codes were written using only the four DNA nucleotides for letters, there would be a total of 64 combinations.

$$4^3 = 4 \times 4 \times 4 = 64$$

When codes are found at a particular place along a coding strand of DNA, and the sequence has meaning, the sequence is a gene. "Meaning" in this case refers to the fact that the gene can be transcribed into an RNA molecule, which in turn may control the assembly of individual amino acids into a polypeptide.

When a gene is transcribed into RNA, the double-stranded DNA is "unzipped," and an enzyme known as *RNA polymerase* attaches to the DNA at the promoter region. It is from this region that the enzymes will begin to assemble RNA nucleotides into a complete, single-stranded copy of the gene, including initiation and termination codes. Triplet RNA nucleotide sequences complementary to DNA codes are called **codons.** Remember that there is no thymine in RNA molecules; it is replaced with uracil. Therefore, the initiation code in DNA (TAC) would be base-paired by RNA polymerase to form the RNA codon AUG. When transcription is complete, the newly assembled RNA is separated from its DNA template and made available for use in the cell; the DNA recoils into its original double-helix form.

In summary:

1. The process begins as one portion of the enzyme RNA polymerase breaks the attachments between the two strands of DNA; the enzyme "unzips" the two strands of the DNA.

2. A second portion of the enzyme RNA polymerase attaches at a particular spot on the DNA called the start codon. It proceeds in one direction along one of the two DNA strands, attaching new RNA nucleotides into position until it reaches a stop codon. The enzymes then assemble RNA nucleotides into a complete, single-stranded RNA copy of the gene. There is no thymine in RNA molecules; it is replaced by uracil. Therefore, the start codon in DNA (TAC) would be paired by RNA polymerase to form the RNA codon AUG.

3. The enzyme that speeds the addition of new nucleotides to the growing chain works along with another enzyme to make sure that no mistakes are made.

4. When transcription is complete, the newly assembled RNA is separated from its DNA template and made available for use in the cell; the DNA recoils into its original double-helix form.

As previously mentioned, three general types of RNA are produced by transcription: messenger RNA, transfer RNA, and ribosomal RNA. Each kind of RNA is made from a specific gene and performs a specific function in the synthesis of polypeptides from individual amino acids at ribosomes. **Messenger RNA (mRNA)** is a mature, straight-chain copy of a gene that describes the exact sequence in which amino acids should be bonded together to form a polypeptide.

Transfer RNA (tRNA) molecules are responsible for picking up particular amino acids and transferring them to the ribosome for assembly into the polypeptide. All tRNA molecules are shaped like a cloverleaf. This shape is formed when they fold and some of the bases form hydrogen bonds that hold the molecule together. One end of the tRNA is able to attach to a specific amino acid. Toward the midsection of the molecule, a triplet nucleotide sequence can base-pair with a codon on mRNA. This triplet nucleotide sequence on tRNA that is complementary to a codon of mRNA is called an **anticodon.** Ribosomal RNA **(rRNA)** is a highly coiled molecule and is used, along with protein molecules, in the manufacture of all ribosomes, the cytoplasmic organelles where tRNA, mRNA, and rRNA come together to help in the synthesis of proteins.

Translation or Protein Synthesis

The mRNA molecule is a coded message written in the biological world's universal nucleic acid language. The code is read in one direction starting at the initiator. The information is used to assemble amino acids into protein by a process called translation. The word *translation* refers to the fact that nucleic acid language is being changed to protein language. To translate mRNA language into protein language, a dictionary is necessary. Remember, the four letters in the nucleic acid alphabet yield 64 possible three-letter words. The protein language has 20 words in the form of 20 common amino acids (table 25.2). Thus, there are more than enough nucleotide words for the 20 amino acid molecules because each nucleotide triplet codes for an amino acid.

Table 25.3 is an amino acid-nucleic acid dictionary. Notice that more than one codon may code for the same amino acid. Some would contend that this is needless repetition, but such "synonyms" can have survival value. If, for example, the gene or the mRNA becomes damaged in a way that causes a particular nucleotide base to change to another type, the chances are still good that the proper amino acid will be read into its proper position. But not all such changes can be compensated for by the codon system, and an altered protein may be produced. Changes can occur that cause great harm. Some damage is so extensive that the entire strand of DNA is broken, resulting in improper protein synthesis or a total lack of synthesis. Any change in DNA is called a **mutation.**

The construction site of the protein molecules (i.e., the translation site) is on the ribosome, a cellular organelle that serves as the meeting place for mRNA and the tRNA that is carrying amino acid building blocks. Ribosomes can be found free in the cytoplasm or attached to the endoplasmic reticulum.

Table 25.2 The Twenty Common Amino Acids and Their Codes

Amino Acid	Three-Letter Code
glycine	Gly
alanine	Ala
valine	Val
leucine	Leu
isoleucine	Ile
methionine	Met
phenylalanine	Phe
tryptophan	Trp
proline	Pro
serine	Ser
threonine	Thr
cysteine	Cys
tyrosine	Tyr
asparagine	ASN
glutamine	Gln
aspartic acid	Asp
glutamic acid	Glu
lysine	Lys
arginine	Arg
histidine	His

These are the twenty common amino acids used in the protein synthesis operation of a cell. Each has a known chemical structure.

Thus, the mRNA moves through the ribosomes, its specific codon sequence allowing for the chemical bonding of a specific sequence of amino acids. Remember that the sequence was originally determined by the DNA. Figure 25.15 shows a possible result of protein synthesis. After transcribing the DNA code, the mRNA delivers its message to a ribosome, where a protein is made. In the case of figure 25.15, the amino acid sequence contains phenylalanine, serine, lysine, and arginine.

Each protein has a specific sequence of amino acids that determines its three-dimensional shape. This shape determines the activity of the protein molecule. The protein may be a structural component of a cell or a regulatory protein, such as an enzyme. Any changes in amino acids or their order changes the action of the protein molecule. The protein insulin, for example, has a different amino acid sequence than the digestive enzyme trypsin. Both proteins are essential to human life and must be produced constantly and accurately. The amino acid sequence of each is determined by a different gene. Each gene is a particular sequence of DNA nucleotides. Any alteration of that sequence can directly alter the protein structure and, therefore, the survival of the organism.

Alterations of DNA

Several kinds of changes to DNA may result in mutations. Phenomena that are either known or suspected causes of DNA damage are called **mutagenic agents.** Agents known to cause damage to DNA are certain viruses (e.g., papillomavirus), weak

Table 25.3 Amino Acid–mRNA Nucleic Acid Dictionary

Second letter

First letter	U	C	A	G	Third letter
U	UUU UUC } Phe UUA UUG } Leu	UCU UCC UCA UCG } Ser	UAU UAC } Tyr UAA Stop UAG Stop	UGU UGC } Cys UGA Stop UGG Try	U C A G
C	CUU CUC CUA CUG } Leu	CCU CCC CCA CCG } Pro	CAU CAC } His CAA CAG } Gln	CGU CGC CGA CGG } Arg	U C A G
A	AUU AUC AUA } Ile AUG Met or start	ACU ACC ACA ACG } Thr	AAU AAC } ASN AAA AAG } Lys	AGU AGC } Ser AGA AGG } Arg	U C A G
G	GUU GUC GUA GUG } Val	GCU GCC GCA GCG } Ala	GAU GAC } Asp GAA GAG } Glu	GGU GGC GGA GGG } Gly	U C A G

1. An mRNA molecule is placed in the smaller of the two portions of a ribosome so that six nucleotides (two codons) are locked into position.

2. The larger ribosomal unit is added to the ribosome/mRNA combination.

3. A tRNA with bases that match the second mRNA codon attaches to the mRNA. The tRNA is carrying a specific amino acid. Once attached, a second tRNA carrying another specific amino acid moves in and attaches to its complementary mRNA codon right next to the first tRNA/amino acid complex.

4. The two tRNAs properly align their two amino acids so that the amino acids may be chemically attached to one another (box figure 25.2).

5. Once the two amino acids are connected to one another by a covalent peptide bond, the first tRNA detaches from its amino acid and mRNA codon.

6. The ribosome moves along the mRNA to the next codon (the first tRNA is set free to move through the cytoplasm to attach to and transfer another amino acid).

7. The next tRNA/amino acid unit enters the ribosome and attaches to its codon next to the first set of amino acids.

8. The tRNAs properly align their amino acids so that they may be chemically attached to one another, forming a chain of three amino acids.

9. Once three amino acids are connected to one another, the second tRNA is released from its amino acid and mRNA (this tRNA is set free to move through the cytoplasm to attach to and transfer another amino acid).

10. The ribosome moves along the mRNA to the next codon, and the fourth tRNA arrives (box figure 25.3).

11. This process repeats until all the amino acids needed to form the protein have attached to one another in the proper sequence. This amino acid sequence was encoded by the DNA gene.

12. Once the final amino acid is attached to the growing chain of amino acids, all the molecules (mRNA, tRNA, and newly formed protein) are released from the ribosome. The "stop" mRNA codon signals this action.

13. The ribosome is again free to become involved in another protein-synthesis operation.

14. The newly synthesized chain of amino acids (the new protein) leaves the ribosome to begin its work. However, the protein may need to be altered by the cell before it will be ready for use (box figure 25.4).

Box Figure 25.2

Box Figure 25.3

Box Figure 25.4

DNA: Cover (complementary to gene)	DNA: Gene	mRNA: Codons (complementary to gene)	tRNA: Anticodons (complementary to mRNA)	Amino acids specified	Protein
T	A	U	A		
T	A	U	A	Phe	Phenylalanine
T	A	U	A		
T	A	U	A		
C	G	C	G	Ser	Serine
T	A	U	A		
A	T	A	U		
A	T	A	U	Lys	Lysine
A	T	A	U		
C	G	C	G		
G	C	G	C	Arg	Arginine
T	A	U	A		

Figure 25.15

There are several steps involved in protein synthesis. (1) mRNA is manufactured from a DNA molecule in the transcription operation; (2) the mRNA enters the cytoplasm and attaches to ribosomes; (3) the tRNA carries amino acids to the ribosome and positions them in the order specified by the mRNA in the translation operation; (4) the amino acids are combined chemically to form a protein.

or "fragile" spots in the DNA, X radiation (X rays) and chemicals found in foods and other products, such as chemicals from burning tobacco. All have been studied extensively, and there is little doubt that they cause mutations. *Chromosomal aberrations* is the term used to describe major changes in DNA. Four types of aberrations include inversions, translocations, duplications, and deletions. An *inversion* occurs when a chromosome is broken and this piece becomes reattached to its original chromosome but in reverse order. It has been cut out and flipped around. A *translocation* occurs when one broken segment of DNA becomes integrated into a different chromosome. *Duplications* occur when a portion of a chromosome is replicated and attached to the original section in sequence. *Deletion* aberrations result when the broken piece becomes lost or is destroyed before it can be reattached.

In some individuals, a single nucleotide of the gene may be changed. This type of mutation is called a *point mutation*. An example of the effects of altered DNA may be seen in human red blood cells. Red blood cells contain the oxygen-transport molecule, hemoglobin. Normal hemoglobin molecules are composed of 150 amino acids in four chains—two alpha and two beta. The nucleotide sequence of the gene for the beta chain is known, as is the amino acid sequence for this chain. In normal individuals, the sequence begins like this:

Val-His-Leu-Thr-Pro-Glu-Glu-Lys . . .

The result is a new amino acid sequence in all the red blood cells:

Val-His-Leu-Thr-Pro-Val-Glu-Lys . . .

This single nucleotide change (known as a *missense point mutation*), which causes a single amino acid to change, may seem minor. However, it is the cause of **sickle-cell anemia,** a disease that affects the red blood cells by changing them from a circular to a sickle shape when oxygen levels are low (figure 25.16). When this sickling occurs, the red blood cells do not flow smoothly through capillaries. Their irregular shapes cause them to clump, clogging the blood vessels. This prevents them from delivering their oxygen load to the oxygen-demanding tissues. A number of physical disabilities may result, including physical weakness, brain damage, pain and stiffness of the joints, kidney damage, rheumatism, and, in severe cases, death.

Changes in the structure of DNA may have harmful effects on the next generation if they occur in the sex cells. Some damage to DNA is so extensive that the entire strand of DNA is broken, resulting in the synthesis of abnormal proteins or a total lack of protein synthesis. A number of experiments indicate that many street drugs, such as LSD (lysergic acid diethylamide), are mutagenic agents and cause DNA to break. Abnormalities have also been identified that are the result of changes in the number or sequence of bases. One way to illustrate these various kinds of mutations is seen in table 25.4.

A powerful new science of gene manipulation, **biotechnology,** suggests that in the future, genetic diseases may be controlled or cured. Since 1953, when the structure of the DNA molecule was first described, there has been a rapid succession of advances in the field of genetics. It is now possible to transfer the DNA from one organism to another. This has made possible the manufacture of human genes and gene products by bacteria.

A

B

Figure 25.16

(A) A normal red blood cell is shown in comparison with (B) a cell having the sickle shape. This sickling is the result of a single amino acid change in the hemoglobin molecule.

Table 25.4 Types of Chromosomal Mutations

Normal Sequence	THE ONE BIG FLY HAD ONE RED EYE
KIND OF MUTATION	**SEQUENCE CHANGE**
Missense	THQ ONE BIG FLY HAD ONE RED EYE
Nonsense	THE ONE BIG
Frameshift:	THE ONE QBI GFLYHA DON ERE DEY
Deletion	THE ONE BIG HAD ONE RED EYE
Duplication	THE ONE BIG FLY FLY HAD ONE RED EYE
Insertion	THE ONE BIG WET FLY HAD ONE RED EYE
Expanding mutation:	
Parents	THE ONE BIG FLY HAD ONE RED EYE
Children	THE ONE BIG FLY FLY FLY HAD ONE RED EYE
Grandchildren	THE ONE BIG FLY FLY FLY FLY FLY FLY HAD ONE RED EYE

Based on R. Lewis, *Human Genetics, Concepts and Applications,* 2nd ed., Dubuque, IA: Wm. C. Brown Publishers, 1997.

A sentence composed of three-letter words can provide an analogy to the effect of mutations on a gene's nucleotide sequence.

GENETIC ENGINEERING

The field of **bioengineering,** also referred to as genetic engineering, is advancing as quickly as the electronics industry. The first bioengineering efforts focused on developing genetically modified (GM) crops that had improvements over past varieties, such as increased resistance to infectious plant disease. This was primarily accomplished through selective breeding and irradiation of cells to produce desirable mutations. The second wave of research involved directly manipulating DNA by using the more sophisticated techniques of recombinant DNA technology, such as the PCR, genetic fingerprinting, and cloning. (The PCR (polymerase chain reaction) is a technique used to amplify or replicate a gene resulting in tens of thousands of copies.) Genetic engineers identify and isolate sequences of nucleotides from a living or dead cell and install them into another living cell. Once these new genes have been installed, they begin to undergo transcription resulting in the production of a protein "foreign" to that organism and undertake DNA replication passing that "foreign" gene down through the generations. There are several steps involved in generating GM organisms: (1) locating the desired gene in a donor organism, (2) isolating that gene, (3) modifying that gene to a more desirable form if necessary, (4) amplifying or replicating that gene using PCR techniques, and (5) introducing the gene into the recipient cell. Genetic modifications have resulted in many changes in food plants. For example the modification of genes in tomatoes causes slower ripening which makes it easier to handle and process them. Currently, crops are being genetically manipulated to manufacture large quantities of specialty chemicals such as antibiotics, steroids, and other biologically useful organic chemicals. In addition, genes for insecticide production and herbicide resistance are also being inserted into select food crops. This practice has led to controversies about the safety of these plants by concerned groups in Mexico, Africa and the European Union where genetically modified corn and other food crops are not accepted.

Although some of these chemicals have been produced in small amounts from genetically engineered microorganisms, crops such as turnips, rice, soybeans, potatoes, cotton, corn, and tobacco can generate tens or hundreds of kilograms of specialty chemicals per year. Many of these GM crops also have increased nutritional value and yet can be cultivated using traditional methods. Such crops have the potential of supplying the essential amino acids, fatty acids, and other nutrients now lacking in the diets of people in underdeveloped or developing na-

A Closer Look   The PCR and Genetic Fingerprinting

In 1989, the American Association for the Advancement of Science named DNA polymerase Molecule of the Year. The value of this enzyme in the polymerase chain reaction (PCR) is so great that it could not be ignored. Just what is the PCR, how does it work, and what can you do with it?

The PCR is a laboratory procedure for copying selected segments of DNA. A single cell can provide enough DNA for analysis and identification! Having a large number of copies of a "target sequence" of nucleotides enables biochemists to more easily work with DNA. This is like increasing the one "needle in the haystack" to such large numbers (100 billion in only a matter of hours) that they're not hard to find, recognize, and work with. The types of specimens that can be used include semen, hair, blood, bacteria, protozoa, viruses, mummified tissue, and frozen cells. The process requires the DNA specimen, free DNA nucleotides, synthetic "primer" DNA, DNA polymerase, and simple lab equipment, such as a test tube and a source of heat.

Having decided which target sequence of nucleotides (which "needle") is to be replicated, scientists heat the specimen of DNA to separate the coding and noncoding strands. Molecules of synthetic "primer" DNA are added to the specimen. These primer molecules are specifically designed to attach to the ends of the target sequence. Next, a mixture of triphosphorylated nucleotides is added so that they can become the newly replicated DNA. The presence of the primer, attached to the DNA and added nucleotides, serves as the substrate for the DNA polymerase. Once added, the polymerase begins making its way down the length of the DNA from one attached primer end to the other. The enzyme bonds the new DNA nucleotides to the strand, replicating the molecule as it goes. It stops when it reaches the other end, having produced a new copy of the target sequence. Because the enzyme will continue to operate as long as enzymes and substrates are available, the process continues, and in a short time there are billions of small pieces of DNA, all replicas of the target sequence.

So what, you say? Well, consider the following. Using the PCR, scientists have been able to:

1. more accurately diagnose such diseases as sickle-cell anemia, cancer, Lyme disease, AIDS, and Legionnaires disease;
2. perform highly accurate tissue typing for matching organ-transplant donors and recipients;
3. help resolve criminal cases of rape, murder, assault, and robbery by matching suspect DNA to that found at the crime scene;
4. detect specific bacteria in environmental samples;
5. monitor the spread of genetically engineered microorganisms in the environment;
6. check water quality by detecting bacterial contamination from feces;
7. identify viruses in water samples;
8. identify disease-causing protozoa in water;
9. determine specific metabolic pathways and activities occurring in microorganisms;
10. determine races, distribution patterns, kinships, migration patterns, evolutionary relationships, and rates of evolution of long-extinct species;
11. accurately settle paternity suits;
12. confirm identity in amnesia cases;
13. identify a person as a relative for immigration purposes;
14. provide the basis for making human antibodies in specific bacteria;
15. possibly provide the basis for replicating genes that could be transplanted into individuals suffering from genetic diseases; and
16. identify nucleotide sequences peculiar to the human genome (an application currently underway as part of the Human Genome Project).

tions. Researchers have also shown, for example, that turnips can produce interferon (an antiviral agent), tobacco can create antibodies to fight human disease, oilseed rape plants can serve as a source of human brain hormones, and potatoes can synthesize human serum albumin that is indistinguishable from the genuine human blood protein.

Another genetic engineering accomplishment has been *genetic fingerprinting*. Using this technique it is possible to show the nucleotide sequence differences among individuals since no two people have the same nucleotide sequences. While this sounds like an easy task, the presence of many millions of base pairs in a person's chromosomes makes this process impractical. Therefore, scientists don't really do a complete genetic fingerprint but focus on certain short sequences that contain many repeats of the same series of bases. They focus on specific short sequences that are known to vary a great deal among individuals. By focusing on the differences found in these specific pieces of DNA, it is possible to determine if two samples are from the same source. The more similar the sequences, the more likely the two samples are from the same person. The less similar the sequences, the less likely the two samples are from the same person. In criminal cases, DNA samples from the crime site can be compared to those taken from suspects. If there is a high number of short, repeating sequence matches, it is highly probable that the suspect was at the scene of the crime and may be the guilty party. This same procedure can also be used to confirm the identity of a person as in cases of amnesia, murder, or accidental death.

Connections . . .

Agarose Gel Electrophoresis

Agarose gel electrophoresis is a process used to separate molecules. Agarose is a complex polysaccharide derived from red algae. When mixed in solution it forms a gel similar to "finger Jello." In the case of DNA, molecules of different size can be separated based on the fact that DNA carries an overall negative electric charge. It is possible to separate a mixture of various lengths of DNA using this method because the heavier segments will move slower than the lighter segments. The various segments will move or migrate through a sheet of gelatin from where the DNA is concentrated to-ward a positive pole. The DNA sample is placed at one end of an agarose gel tray, and a weak voltage is generated in the gel, usually no more than 5 volts per centimeter to the gel. The negatively charged DNA will then migrate slowly to the positive end. It becomes a race from one end of the tray to the other, with the smaller segments always winning. When the power source is shut off, the location of the DNA segments is determined by viewing the gel under ultraviolet light. The gel is then stained with ethidium bromide and the dark "blotches" in the gel can be cut out for purification and analysis.

Pouring agarose   Loading a lane   Ready to run

Box Figure 25.5

Box Figure 25.6

SUMMARY

Genes are *units of heredity* composed of specific *lengths of DNA* that determine the *characteristics* an organism displays. *Specific genes* are at *specific loci* on *specific chromosomes.* The *phenotype* displayed by an organism is the result of the effect of the environment on the ability of the genes to express themselves.

Diploid organisms have *two genes* for each characteristic. The *alternative forms of genes* for a characteristic are called *alleles.* There may be many *different* alleles for a *particular characteristic.* Organisms with *two identical alleles* are *homozygous* for a characteristic; those with *different alleles* are *heterozygous.* Some alleles are *dominant* over other alleles that are said to be *recessive.*

Sometimes *two alleles both express themselves,* and often a gene has more than one recognizable effect on the phenotype of the organism. Some characteristics may be determined by several different *pairs of alleles.*

The successful operation of a *living cell* depends on its ability to *accurately reproduce genes and control chemical reactions. DNA replication* results in an exact *doubling* of the *genetic material.* The process virtually guarantees that *identical* strands of DNA will be passed on to the next generation of cells.

The *enzymes* are responsible for the *efficient control of a cell's metabolism.* However, the *production of protein molecules* is under the con-trol of the *nucleic acids,* the *primary control molecules* of the cell. The structure of the nucleic acids *DNA and RNA* determine the *structure of the proteins,* whereas the *structure of the proteins* determines their *function* in the cell's life cycle. *Protein synthesis* involves the *decoding* of the DNA into specific protein molecules and the use of the *intermediate molecules,* mRNA and tRNA, at the *ribosome.* Errors in any of the *codons* of these molecules may produce observable *changes* in the cell's functioning and lead to cell *death.*

Methods of *manipulating DNA* have led to the *controlled transfer of genes* from one kind of organism to another. This has made it possible for *bacteria* to produce a number of *human* gene products.

KEY TERMS

alleles (p. **560**)
anticodon (p. **576**)
bioengineering (p. **580**)
biotechnology (p. **579**)
carrier (p. **561**)
chromosome (p. **570**)
codominance (p. **561**)
codon (p. **576**)
complementary base (p. **570**)
DNA code (p. **576**)

DNA replication (p. **570**)
dominant allele (p. **561**)
gene (p. **560**)
genetics (p. **560**)
genome (p. **560**)
genotype (p. **561**)
heterozygous (p. **561**)
homozygous (p. **561**)
law of dominance (p. **564**)

APPLYING THE CONCEPTS

1. An example of a phenotype is
   (a) AB type blood.
   (b) an allele for type A and B blood.
   (c) a sperm with an allele for A blood.
   (d) lack of iron in the diet causing anemia.

2. Body size is determined by the interaction of numerous alleles. This is an example of
   (a) autosomes.
   (b) pleiotropy.
   (c) single-factor crosses.
   (d) polygenic inheritance.

3. In order for a recessive X-linked trait to appear in a female, she must inherit a recessive allele from
   (a) neither parent.
   (b) both parents.
   (c) her father only.
   (d) her mother only.

4. It was noticed that in certain flowers, if a flower had red petals they were usually small petals. If the petals were white, pink, or orange they were usually large petals. This could be the result of
   (a) lack of dominance.
   (b) pleiotropy.
   (c) linkage groups.
   (d) polygenic inheritance.

5. A woman with blood type O and a man with blood type AB have a child together. What are the possible bloodtypes of this child?
   (a) AB or O
   (b) A, B, or O
   (c) A or B
   (d) AB, A, B, or O

6. "She has really long fingers and toes and is exceptionally tall." This is a statement of
   (a) genotype.
   (b) phenotype.
   (c) monohybridization.
   (d) locus placement.

7. One difference between preRNA and mature RNA is:
   (a) kinds of nucleotide components.
   (b) presence of extra RNA in preRNA.
   (c) the part of DNA from which they were coded.
   (d) only the age of the molecules.

8. While one strand of duplex DNA is being transcribed to mRNA:
   (a) the complementary strand makes tRNA.
   (b) the complementary strand is inactive.
   (c) the complementary strand at this point is replicating.
   (d) mutations are impossible during this short period.

9. A major difference between the genetic data of prokaryotic and eukaryotic cells is that in prokaryotes the
   (a) genes are RNA not DNA.
   (b) histones are arranged differently.
   (c) double-stranded DNA is circular.
   (d) double-stranded DNA is absent in bacteria.

10. If the DNA gene strand has the base sequence CCA-TAT-TCG, the complimentary DNA strand will have the sequence:
    (a) CCA-TAT-TCG
    (b) GGU-AUA-AGC
    (c) CCA-UAU-UCG
    (d) GGT-ATA-AGC

11. A DNA gene strand with the base sequence CCA-TAT-TCG codes for the amino acid sequence: (Consult the amino acid–nucleic acid dictionary in your text, table 25.3)
    (a) proline-tyrosine-serine
    (b) glycine-isoleucine-threonine
    (c) proline-tyrosine-threonine
    (d) glycine-isoleucine-serine

12. The mRNA codon CAU will form temporary bonds with the
    (a) mRNA anticodon CAU.
    (b) mRNA codon GUA.
    (c) tRNA anticodon GUA.
    (d) tRNAcodon CAU.

Answers
1. a **2.** d **3.** b **4.** c **5.** c **6.** b **7.** b **8.** b **9.** c **10.** d **11.** d **12.** c

QUESTIONS FOR THOUGHT

1. What are the differences between DNA and RNA?
2. List the sequence of events that takes place when a DNA message is translated into protein.
3. Chromosomal and point mutations both occur in DNA. In what ways do they differ? How is this related to recombinant DNA?
4. How does DNA replication differ from the manufacture of an RNA molecule?
5. If a DNA nucleotide sequence is CATAAAGCA, what is the mRNA nucleotide sequence that would base-pair with it?
6. What amino acids would occur in the protein chemically coded by the sequence of nucleotides in question 5?
7. What is the probability of each of the following sets of parents producing the given genotypes in their offspring?

Parents			Offspring	Genotype
(a) AA	×	aa		Aa
(b) Aa	×	Aa		Aa
(c) Aa	×	Aa		aa
(d) AaBb	×	AaBB		AABB
(e) AaBb	×	AaBB		AaBb
(f) AaBb	×	AaBb		AABB

8. If an offspring has the genotype Aa, what possible combinations of parental genotypes can exist?

9. In humans, the allele for albinism is recessive to the allele for normal skin pigmentation.
   (a) What is the probability that a child of a heterozygous mother and father will be an albino?
   (b) If a child is normal, what is the probability that it is a carrier of the recessive albino allele?

10. Dwarfism (achondroplasia) is the result of a mutation in the gene for the production of fibroblast growth factor. It is inherited as an autosomal dominant disorder and often occurs as a new mutation. What might be the explanation if two people of normal height and body stature have five children, one who displays achondroplasia?

11. A woman with color-deficient vision marries a man with normal vision. They have ten children—six boys and four girls.
    (a) How many are expected to have normal vision?
    (b) How many are expected to have color deficiency?

12. Hemophilia is a disease that prevents the blood from clotting normally. It is caused by a recessive allele located on the X chromosome. A boy has the disease; neither his parents nor his grandparents have the disease. What are the genotypes of his parents and grandparents?

Mathematical Review

WORKING WITH EQUATIONS

Many of the problems of science involve an equation, a shorthand way of describing patterns and relationships that are observed in nature. Equations are also used to identify properties and to define certain concepts, but all uses have well-established meanings, symbols that are used by convention, and allowed mathematical operations. This appendix will assist you in better understanding equations and the reasoning that goes with the manipulation of equations in problem-solving activities.

Background

In addition to a knowledge of rules for carrying out mathematical operations, an understanding of certain quantitative ideas and concepts can be very helpful when working with equations. Among these helpful concepts are (1) the meaning of inverse and reciprocal, (2) the concept of a ratio, and (3) fractions.

The term *inverse* means the opposite, or reverse, of something. For example, addition is the opposite, or inverse, of subtraction, and division is the inverse of multiplication. A *reciprocal* is defined as an inverse multiplication relationship between two numbers. For example, if the symbol n represents any number (except zero), then the reciprocal of n is $1/n$. The reciprocal of a number $(1/n)$ multiplied by that number (n) always gives a product of 1. Thus, the number multiplied by 5 to give 1 is $1/5$ $(5 \times 1/5 = 5/5 = 1)$. So $1/5$ is the reciprocal of 5, and 5 is the reciprocal of $1/5$. Each number is the *inverse* of the other.

The fraction $1/5$ means 1 divided by 5, and if you carry out the division it gives the decimal 0.2. Calculators that have a $1/x$ key will do the operation automatically. If you enter 5, then press the $1/x$ key, the answer of 0.2 is given. If you press the $1/x$ key again, the answer of 5 is given. Each of these numbers is a reciprocal of the other.

A *ratio* is a comparison between two numbers. If the symbols m and n are used to represent any two numbers, then the ratio of the number m to the number n is the fraction m/n. This expression means to divide m by n. For example, if m is 10 and n is 5, the ratio of 10 to 5 is $10/5$, or 2:1.

Working with *fractions* is sometimes necessary in problem-solving exercises, and an understanding of these operations is needed to carry out unit calculations. It is helpful in many of these operations to remember that a number (or a unit) divided by itself is equal to 1, for example,

$$\frac{5}{5} = 1 \qquad \frac{\text{inch}}{\text{inch}} = 1 \qquad \frac{5 \text{ inches}}{5 \text{ inches}} = 1$$

When one fraction is divided by another fraction, the operation commonly applied is to "invert the denominator and multiply." For example, 2/5 divided by 1/2 is

$$\frac{\dfrac{2}{5}}{\dfrac{1}{2}} = \frac{2}{5} \times \frac{2}{1} = \frac{4}{5}$$

What you are really doing when you invert the denominator of the larger fraction and multiply is making the denominator (1/2) equal to 1. Both the numerator (2/5) and the denominator (1/2) are multiplied by 2/1, which does not change the value of the overall expression. The complete operation is

$$\frac{\dfrac{2}{5}}{\dfrac{1}{2}} \times \frac{\dfrac{2}{1}}{\dfrac{2}{1}} = \frac{\dfrac{2}{5} \times \dfrac{2}{1}}{\dfrac{1}{2} \times \dfrac{2}{1}} = \frac{\dfrac{4}{5}}{\dfrac{2}{2}} = \frac{\dfrac{4}{5}}{1} = \frac{4}{5}$$

Symbols and Operations

The use of symbols seems to cause confusion for some students because it seems different from their ordinary experiences with arithmetic. The rules are the same for symbols as they are for numbers, but you cannot do the operations with the symbols until you know what values they represent. The operation signs, such as $+$, $\div$, $\times$, and $-$, are used with symbols to indicate the operation that you *would* do if you knew the values. Some of the mathematical operations are indicated several ways. For example, $a \times b$, $a \cdot b$, and ab all indicate the same thing, that a is to be multiplied by b. Likewise, $a \div b$, a/b, and $a \times 1/b$ all indicate that a is to be divided by b. Since it is not possible to carry out the operations on symbols alone, they are called *indicated operations*.

Operations in Equations

An equation is a shorthand way of expressing a simple sentence with symbols. The equation has three parts: (1) a left side, (2) an equal sign $(=)$, which indicates the equivalence of the two sides, and (3) a right side. The left side has the same value and units as the right side, but the two sides may have a very different appearance. The two sides may also have the symbols that indicate mathematical operations $(+, -, \times,$ and so forth) and may be in certain forms that indicate operations $(a/b, ab,$ and so forth). In any case, the equation is a complete expression that states the left side has the same value and units as the right side.

Equations may contain different symbols, each representing some unknown quantity. In science, the expression "solve the equation" means to perform certain operations with one symbol (which represents some variable) by itself on one side of the equation. This single symbol is usually, but not necessarily, on the left side and is not present on the other side. For example, the equation $F = ma$ has the symbol F on the left side. In science, you would say that this equation is solved for F. It could also be solved for m or for a, which will be considered shortly. The equation $F = ma$ is solved for F, and the *indicated operation* is to multiply m by a because they are in the form ma, which means the same thing as $m \times a$. This is the only indicated operation in this equation.

A solved equation is a set of instructions that has an order of indicated operations. For example, the equation for the relationship between a Fahrenheit and Celsius temperature, solved for °C, is $C = 5/9$ $(F - 32)$. A list of indicated operations in this equation is as follows:

1. Subtract 32° from the given Fahrenheit temperature.
2. Multiply the result of (1) by 5.
3. Divide the result of (2) by 9.

Why are the operations indicated in this order? Because the bracket means 5/9 of the *quantity* $(F - 32°)$. In its expanded form, you can see that $5/9(F - 32°)$ actually means $5/9(F) - 5/9(32°)$. Thus, you cannot multiply by 5 or divide by 9 until you have found the quantity of $(F - 32°)$. Once you have figured out the order of operations, finding the answer to a problem becomes almost routine as you complete the needed operations on both the numbers and the units.

Solving Equations

Sometimes it is necessary to rearrange an equation to move a different symbol to one side by itself. This is known as solving an equation for an unknown quantity. But you cannot simply move a symbol to one side of an equation. Since an equation is a statement of equivalence, the right side has the same value as the left side. If you move a symbol, you must perform the operation in a way that the two sides remain equivalent. This is accomplished by "canceling out" symbols until you have the unknown on one side by itself. One key to understanding the canceling operation is to remember that a fraction with the same number (or unit) over itself is equal to 1. For example, consider the equation $F = ma$, which is solved for F. Suppose you are considering a problem in which F and m are given, and the unknown is a. You need to solve the equation for a so it is on one side by itself. To eliminate the m, you do the *inverse* of the indicated operation on m, dividing both sides by m. Thus,

$$F = ma$$

$$\frac{F}{m} = \frac{ma}{m}$$

$$\frac{F}{m} = a$$

Since m/m is equal to 1, the a remains by itself on the right side. For convenience, the whole equation may be flipped to move the unknown to the left side,

$$a = \frac{F}{m}$$

Thus, a quantity that indicated a multiplication (ma) was removed from one side by an inverse operation of dividing by m.

Consider the following inverse operations to "cancel" a quantity from one side of an equation, moving it to the other side:

If the Indicated Operation of the Symbol You Wish to Remove Is:	Perform This Inverse Operation on Both Sides of the Equation
multiplication	division
division	multiplication
addition	subtraction
subtraction	addition
squared	square root
square root	square

SIGNIFICANT FIGURES

The numerical value of any measurement will always contain some uncertainty. Suppose, for example, that you are measuring one side of a square piece of paper as shown in figure A.1. You could say that the paper is *about* 3.5 cm wide and you would be correct. This measurement, however, would be unsatisfactory for many purposes. It does not approach the true value of the length and contains too much uncertainty. It seems clear that the paper width is larger than 3.4 cm but shorter than 3.5 cm. But how much larger than 3.4 cm? You cannot be certain if the paper is 3.44, 3.45, or 3.46 cm wide. As your best estimate, you might say that the paper is 3.45 cm wide. Everyone would agree that you can be certain about the first two numbers (3.4) and they should be recorded. The last number (0.05) has been estimated and is not certain. The two certain numbers, together with one uncertain number, represent the greatest accuracy possible with the ruler being used. The paper is said to be 3.45 cm wide.

A *significant figure* is a number that is believed to be correct with some uncertainty only in the last digit. The value of the width of the paper, 3.45 cm, represents three significant figures. As you can see, the number of significant figures can be determined by the degree of accuracy of the measuring instrument being used. But suppose you need to calculate the area of the paper. You would multiply 3.45 cm × 3.45 cm and the product for the area would be 11.9025 cm². This is a greater accuracy than you were able to obtain with your measuring instrument. The result of a calculation can be no more accurate than the values being treated. Because the measurement had only three significant figures (two certain, one uncertain), then the answer can have only three significant figures. The area is correctly expressed as 11.9 cm².

There are a few simple rules that will help you determine how many significant figures are contained in a reported measurement:

1. All digits reported as a direct result of a measurement are significant.

2. Zero is significant when it occurs between nonzero digits. For example, 607 has three significant figures, and the zero is one of the significant figures.

3. In figures reported as *larger than the digit one,* the digit zero is not significant when it follows a nonzero digit to indicate place. For example, in a report that "23,000 people attended the rock concert," the digits 2 and 3 are significant but the zeros are not significant. In this situation the 23 is the measured part of the figure, and the three zeros tell you an estimate of how many attended the concert, that is, 23 thousand. If the figure is a measurement rather than an estimate, then it is written *with a decimal point after the last zero* to indicate that the zeros *are* significant. Thus 23,000 has *two* significant figures (2 and 3), but 23,000. has *five* significant figures. The figure 23,000 means "about 23 thousand," but 23,000. means 23,000. and not 22,999 or 23,001.

4. In figures reported as *smaller than the digit 1,* zeros after a decimal point that come before nonzero digits *are not* significant and serve only as place holders. For example, 0.0023 has two significant figures, 2 and 3. Zeros alone after a decimal point or zeros after a nonzero digit indicate a measurement, however, so these zeros *are* significant. The figure 0.00230, for example, has three significant figures since the 230 means 230 and not 229 or 231. Likewise, the figure 3.000 cm has four significant figures because the presence of the three zeros means that the measurement was actually 3.000 and not 2.999 or 3.001.

Multiplication and Division

When multiplying or dividing measurement figures, the answer may have no more significant figures than the *least* number of significant figures in the figures being multiplied or divided. This simply means that an answer can be no more accurate than the least accurate measurement entering into the calculation, and that you cannot improve the accuracy of a measurement by doing a calculation. For example, in multiplying 54.2 mi/h × 4.0 h to find out the total distance traveled, the first figure (54.2) has three significant figures but the second (4.0) has only two significant figures. The answer can contain only two significant figures since this is the weakest number of those involved in the calculation. The correct answer is therefore 220 mi, not 216.8 mi. This may seem strange since multiplying the two numbers together gives the answer of 216.8 mi. This answer, however, means a greater accuracy than is possible, and the accuracy cannot be improved over the weakest number involved in the calculation. Since the weakest number (4.0) has only two significant figures the answer must also have only two significant figures, which is 220 mi.

The result of a calculation is *rounded* to have the same least number of significant figures as the least number of a measurement involved in the calculation. When rounding numbers the last significant figure is increased by 1 if the number after it is 5 or larger. If the number after the last significant figure is 4 or less, the nonsignificant figures are simply dropped. Thus, if two significant figures are called for in the answer of the previous example, 216.8 is rounded up to 220 because the last number after the two significant figures is 6 (a number larger than 5). If the calculation result had been 214.8, the rounded number would be 210 miles.

Note that *measurement figures* are the only figures involved in the number of significant figures in the answer. Numbers that are counted or **defined** are not included in the determination of significant figures in an answer. For example, when dividing by 2 to find an average, the 2 is ignored when considering the number of significant figures. Defined numbers are defined exactly and are not used in significant figures. Since 1 kilogram is *defined* to be exactly 1,000 grams, such a conversion is not a measurement.

Addition and Subtraction

Addition and subtraction operations involving measurements, as multiplication and division, cannot result in an answer that implies greater accuracy than the measurements had before the calculation. Recall that the last digit to the right in a measurement is uncertain, that is, it is the result of an estimate. The answer to an addition or subtraction calculation can have this uncertain number *no farther from the decimal place than it was in the weakest number involved in the calculation*. Thus, when 8.4 is added to 4.926, the weakest number is 8.4, and the uncertain number is .4, one place to the right of the decimal. The sum of 13.326 is therefore rounded to 13.3, reflecting the placement of this weakest doubtful figure.

The rules for counting zeros tell us that the numbers 203 and 0.200 both have three significant figures. Likewise, the numbers 230 and 0.23 only have two significant figures. Once you remember the rules, the counting of significant figures is straightforward. On the other hand, sometimes you find a number that seems to make it impossible to follow the rules. For example, how would you write 3,000 with two significant figures? There are several special systems in use for taking care of problems such as this, including the placing of a little bar over the last significant digit. One of the convenient ways of showing significant figures for difficult numbers is to use scientific notation, which is also discussed elsewhere in this appendix. The convention for writing significant figures is to display one digit to the left of the decimal. The exponents are not considered when showing the number of significant figures in scientific notation. Thus if you want to write three thousand showing one significant figure, you would write 3×10^3. To show two significant figures it is 3.0×10^3 and for three significant figures it becomes 3.00×10^3. As you can see, the correct use of scientific notation leaves little room for doubt about how many significant figures are intended.

C O N C E P T S   A P P L I E D
Significant Figures

In an example concerning percentage error, an experimental result of 511 Hz was found for a tuning fork with an accepted frequency value of 522 Hz. The error calculation is

$$\frac{(522 \text{ Hz} - 511 \text{ Hz})}{522 \text{ Hz}} \times 100\% = 2.1\%$$

Since $522 - 511$ is 11, the least number of significant figures of measurements involved in this calculation is *two*. Note that the "100" does not enter into the determination since it is not a measurement number. The calculated result (from a calculator) is 2.1072797, which is rounded off to have only two significant figures, so the answer is recorded as 2.1%.

CONVERSION OF UNITS

The measurement of most properties results in both a numerical value and a unit. The statement that a glass contains 50 cm³ of a liquid conveys two important concepts—the numerical value of 50 and the referent unit of cubic centimeters. Both the numerical value and the unit are necessary to communicate correctly the volume of the liquid.

When working with calculations involving measurement units, *both* the numerical value and the units are treated mathematically. As in other mathematical operations, there are general rules to follow.

1. Only properties with *like units* may be added or subtracted. It should be obvious that adding quantities such as 5 dollars and 10 dimes is meaningless. You must first convert to like units before adding or subtracting.
2. Like or unlike units may be multiplied or divided and treated in the same manner as numbers. You have used this rule when dealing with area (length $\times$ length = length², for example, cm $\times$ cm = cm²) and when dealing with volume (length $\times$ length $\times$ length = length³, for example, cm $\times$ cm $\times$ cm = cm³).

You can use these two rules to create a *conversion ratio* that will help you change one unit to another. Suppose you need to convert 2.3 kg to grams. First, write the relationship between kilograms and grams:

$$1,000 \text{ g} = 1 \text{ kg}$$

Next, divide both sides by what you wish to convert *from* (kilograms in this example):

$$\frac{1,000 \text{ g}}{1 \text{ kg}} = \frac{1 \text{ kg}}{1 \text{ kg}}$$

One kilogram divided by 1 kg equals 1, just as 10 divided by 10 equals 1. Therefore, the right side of the relationship becomes 1 and the equation is:

$$\frac{1,000 \text{ g}}{1 \text{ kg}} = 1$$

The 1 is usually understood, that is, not stated, and the operation is called *canceling*. Canceling leaves you with the fraction 1,000 g/1 kg, which is a conversion ratio that can be used to convert from kilograms to grams. You simply multiply the conversion ratio by the numerical value and unit you wish to convert:

$$= 2.3 \text{ kg} \times \frac{1,000 \text{ g}}{1 \text{ kg}}$$

$$= \frac{2.3 \times 1,000}{1} \frac{\text{kg} \times \text{g}}{\text{kg}}$$

$$= \boxed{2,300 \text{ g}}$$

The kilogram units cancel. Showing the whole operation with units only, you can see how you end up with the correct unit of grams:

$$\text{kg} \times \frac{\text{g}}{\text{kg}} = \frac{\text{kg} \cdot \text{g}}{\text{kg}} = \text{g}$$

Since you did obtain the correct unit, you know that you used the correct conversion ratio. If you had blundered and used an inverted conversion ratio, you would obtain

$$2.3 \text{ kg} \times \frac{1 \text{ kg}}{1,000 \text{ g}} = .0023 \frac{\text{kg}^2}{\text{g}}$$

which yields the meaningless, incorrect units of kg^2/g. Carrying out the mathematical operations on the numbers and the units will always tell you whether or not you used the correct conversion ratio.

SCIENTIFIC NOTATION

Most of the properties of things that you might measure in your everyday world can be expressed with a small range of numerical values together with some standard unit of measure. The range of numerical values for most everyday things can be dealt with by using units (1s), tens (10s), hundreds (100s), or perhaps thousands (1,000s). But the actual universe contains some objects of incredibly large size that require some very big numbers to describe. The sun, for example, has a mass of about 1,970,000,000,000,000,000,000,000,000,000 kg. On the other hand, very small numbers are needed to measure the size and parts of an atom. The radius of a hydrogen atom, for example, is about 0.00000000005 m. Such extremely large and small numbers are cumbersome and awkward since there are so many zeros to keep track of, even if you are successful in carefully counting all the zeros. A method does exist to deal with extremely large or small numbers in a more condensed form. The method is called *scientific notation*, but it is also

sometimes called *powers of ten* or *exponential notation,* since it is based on exponents of 10. Whatever it is called, the method is a compact way of dealing with numbers that not only helps you keep track of zeros but provides a simplified way to make calculations as well.

In algebra you save a lot of time (as well as paper) by writing ($a \times a \times a \times a \times a$) as a^5. The small number written to the right and above a letter or number is a superscript called an *exponent.* The exponent means that the letter or number is to be multiplied by itself that many times, for example, a^5 means a multiplied by itself five times, or $a \times a \times a \times a \times a$. As you can see, it is much easier to write the exponential form of this operation than it is to write it out in the long form. Scientific notation uses an exponent to indicate the power of the base 10. The exponent tells how many times the base, 10, is multiplied by itself. For example,

$$10,000 = 10^4$$
$$1,000 = 10^3$$
$$100 = 10^2$$
$$10 = 10^1$$
$$1 = 10^0$$
$$0.1 = 10^{-1}$$
$$0.01 = 10^{-2}$$
$$0.001 = 10^{-3}$$
$$0.0001 = 10^{-4}$$

This table could be extended indefinitely, but this somewhat shorter version will give you an idea of how the method works. The symbol 10^4 is read as "ten to the fourth power" and means $10 \times 10 \times 10 \times 10$. Ten times itself four times is 10,000, so 10^4 is the scientific notation for 10,000. It is also equal to the number of zeros between the 1 and the decimal point, that is, to write the longer form of 10^4 you simply write 1, then move the decimal point four places to the *right;* 10 to the fourth power is 10,000.

The power of ten table also shows that numbers smaller than 1 have negative exponents. A negative exponent means a reciprocal:

$$10^{-1} = \frac{1}{10} = 0.1$$

$$10^{-2} = \frac{1}{100} = 0.01$$

$$10^{-3} = \frac{1}{1000} = 0.001$$

To write the longer form of 10^{-4}, you simply write 1 then move the decimal point four places to the *left;* 10 to the negative fourth power is 0.0001.

Scientific notation usually, but not always, is expressed as the product of two numbers: (1) a number between 1 and 10 that is called the *coefficient* and (2) a power of ten that is called the *exponent.* For example, the mass of the sun that was given in long form earlier is expressed in scientific notation as

$$1.97 \times 10^{30} \text{ kg}$$

and the radius of a hydrogen atom is

$$5.0 \times 10^{-11} \text{ m}$$

Scientific Notation Expression

What is 26,000,000 in scientific notation?

Answer

Count how many times you must shift the decimal point until one digit remains to the left of the decimal point. For numbers larger than the digit 1, the number of shifts tells you how much the exponent is increased, so the answer is

$$2.6 \times 10^7$$

which means the coefficient 2.6 is multiplied by 10 seven times.

Scientific Notation Calculation

Solve the following problem concerning scientific notation:

$$\frac{(2 \times 10^4) \times (8 \times 10^{-6})}{8 \times 10^4}$$

Answer

First, separate the coefficients from the exponents,

$$\frac{2 \times 8}{8} \times \frac{10^4 \times 10^{-6}}{10^4}$$

then multiply and divide the coefficients and add and subtract the exponents as the problem requires,

$$2 \times 10^{\{[(4) + (-6)] - [4]\}}$$

Solving the remaining additions and subtractions of the coefficients gives

$$2 \times 10^{-6}$$

In these expressions, the coefficients are 1.97 and 5.0, and the power of ten notations are the exponents. Note that in both of these examples, the exponent tells you where to place the decimal point if you wish to write the number all the way out in the long form. Sometimes scientific notation is written without a coefficient, showing only the exponent. In these cases the coefficient of 1.0 is understood, that is, not stated. If you try to enter a scientific notation in your calculator, however, you will need to enter the understood 1.0, or the calculator will not be able to function correctly. Note also that 1.97×10^{30} kg and the expressions 0.197×10^{31} kg and 19.7×10^{29} kg are all correct expressions of the mass of the sun. By convention, however, you will use the form that has one digit to the left of the decimal.

It was stated earlier that scientific notation provides a compact way of dealing with very large or very small numbers, but it provides a simplified way to make calculations as well. There are a few mathematical rules that will describe how the use of scientific notation simplifies these calculations.

To *multiply* two scientific notation numbers, the coefficients are multiplied as usual, and the exponents are *added* algebraically. For example, to multiply (2×10^2) by (3×10^3), first separate the coefficients from the exponents,

$$(2 \times 3) \times (10^2 \times 10^3),$$

then multiply the coefficients and add the exponents,

$$6 \times 10^{(2 + 3)} = 6 \times 10^5$$

Adding the exponents is possible because $10^2 \times 10^3$ means the same thing as $(10 \times 10) \times (10 \times 10 \times 10)$, which equals $(100) \times (1,000)$, or 100,000, which is expressed as 10^5 in scientific notation. Note that two negative exponents add algebraically, for example $10^{-2} \times 10^{-3} = 10^{[(-2) + (-3)]} = 10^{-5}$. A negative and a positive exponent also add algebraically, as in $10^5 \times 10^{-3} = 10^{[(+5) + (-3)]} = 10^2$.

If the result of a calculation involving two scientific notation numbers does not have the conventional one digit to the left of the decimal, move the decimal point so it does, changing the exponent according to which way and how much the decimal point is moved. Note that the exponent increases by one number for each decimal point moved to the left. Likewise, the exponent decreases by one number for each decimal point moved to the right. For example, $938. \times 10^3$ becomes 9.38×10^5 when the decimal point is moved two places to the left.

To *divide* two scientific notation numbers, the coefficients are divided as usual and the exponents are *subtracted*. For example, to divide (6×10^6) by (3×10^2), first separate the coefficients from the exponents,

$$(6 \div 3) \times (10^6 \div 10^2)$$

then divide the coefficients and subtract the exponents,

$$2 \times 10^{(6 - 2)} = 2 \times 10^4$$

Note that when you subtract a negative exponent, for example, $10^{[(3) - (-2)]}$, you change the sign and add, $10^{(3 + 2)} = 10^5$.

Solubilities Chart

	Acetate	Bromide	Carbonate	Chloride	Fluoride	Hydroxide	Iodide	Nitrate	Oxide	Phosphate	Sulfate	Sulfide
Aluminum	S	S	—	S	s	i	S	S	i	i	S	d
Ammonium	S	S	S	S	S	S	S	S	—	S	S	S
Barium	S	S	i	S	s	S	S	S	S	i	i	d
Calcium	S	S	i	S	i	s	S	S	s	i	s	d
Copper(I)	—	s	i	s	i	—	i	—	i	—	d	i
Copper(II)	S	S	i	S	S	i	S	S	i	i	S	i
Iron(II)	S	S	i	S	s	i	S	S	i	i	S	i
Iron(III)	S	S	i	S	s	i	S	S	i	i	S	d
Lead	S	s	i	s	i	i	s	S	i	i	i	i
Magnesium	S	S	i	S	i	i	S	S	i	i	S	d
Mercury(I)	s	i	i	i	d	d	i	S	i	i	i	i
Mercury(II)	S	s	i	S	d	i	i	S	i	i	i	i
Potassium	S	S	S	S	S	S	S	S	S	S	S	i
Silver	s	i	i	i	S	—	i	S	i	i	i	i
Sodium	S	S	S	S	S	S	S	S	d	S	S	S
Strontium	S	S	s	S	i	s	S	S	—	i	i	i
Zinc	S	S	i	S	S	i	S	S	i	i	S	i

S = soluble
i — insoluble
s = slightly soluble
d = decomposes

Relative Humidity Table

Dry-Bulb Temperature (°C)	Difference Between Wet-Bulb and Dry-Bulb Temperatures (°C)																			
	1	2	3	4	5	6	7	8	9	10	11	12	13	14	15	16	17	18	19	20
0	81	64	46	29	13															
1	83	66	49	33	17															
2	84	68	52	37	22	7														
3	84	70	55	40	26	12														
4	86	71	57	43	29	16														
5	86	72	58	45	33	20	7													
6	86	73	60	48	35	24	11													
7	87	74	62	50	38	26	15													
8	87	75	63	51	40	29	19	8												
9	88	76	64	53	42	32	22	12												
10	88	77	66	55	44	34	24	15	6											
11	89	78	67	56	46	36	27	18	9											
12	89	78	68	58	48	39	29	21	12											
13	89	79	69	59	50	41	32	23	15	7										
14	90	79	70	60	51	42	34	26	18	10										
15	90	80	71	61	53	44	36	27	20	13	6									
16	90	81	71	63	54	46	38	30	23	15	8									
17	90	81	72	64	55	47	40	32	25	18	11									
18	91	82	73	65	57	49	41	34	27	20	14	7								
19	91	82	74	65	58	50	43	36	29	22	16	10								
20	91	83	74	66	59	51	44	37	31	24	18	12	6							
21	91	83	75	67	60	53	46	39	32	26	20	14	9							
22	92	83	76	68	61	54	47	40	34	28	22	17	11	6						
23	92	84	76	69	62	55	48	42	36	30	24	19	13	8						
24	92	84	77	69	62	56	49	43	37	31	26	20	15	10	5					
25	92	84	77	70	63	57	50	44	39	33	28	22	17	12	8					
26	92	85	78	71	64	58	51	46	40	34	29	24	19	14	10	5				
27	92	85	78	71	65	58	52	47	41	36	31	26	21	16	12	7				
28	93	85	78	72	65	59	53	48	42	37	32	27	22	18	13	9	5			
29	93	86	79	72	66	60	54	49	43	38	33	28	24	19	15	11	7			
30	93	86	79	73	67	61	55	50	44	39	35	30	25	21	17	13	9	5		
31	93	86	80	73	67	61	56	51	45	40	36	31	27	22	18	14	11	7		
32	93	86	80	74	68	62	57	51	46	41	37	32	28	24	20	16	12	9	5	
33	93	87	80	74	68	63	57	52	47	42	38	33	29	25	21	17	14	10	7	
34	93	87	81	75	69	63	58	53	48	43	39	35	30	28	23	19	15	12	8	5
35	94	87	81	75	69	64	59	54	49	44	40	36	32	28	24	20	17	13	10	7

Solutions for Group A Parallel Exercises

CHAPTER 1

1.1 Answers will vary but should have the relationship of 100 cm in 1 m, for example, 178 cm = 1.78 m.

1.2 Since mass density is given by the relationship $\rho = m/V$, then

$$\rho = \frac{m}{V} = \frac{272 \text{ g}}{20.0 \text{ cm}^3}$$

$$= \frac{272}{20.0} \frac{\text{g}}{\text{cm}^3}$$

$$= \boxed{13.6 \frac{\text{g}}{\text{cm}^3}}$$

1.3 The volume of a sample of lead is given and the problem asks for the mass. From the relationship of $\rho = m/V$, solving for the mass (m) tells you that the mass density (ρ) times the volume (V), or $m = \rho V$. The mass density of lead, 11.4 g/cm^3, can be obtained from table 1.4, so

$$\rho = \frac{m}{V}$$

$$V\rho = \frac{m\cancel{V}}{\cancel{V}}$$

$$m = \rho V$$

$$m = \left(11.4 \frac{\text{g}}{\text{cm}^3} \right) (10.0 \text{ cm}^3)$$

$$11.4 \times 10.0 \frac{\text{g}}{\text{cm}^3} \times \text{cm}^3$$

$$114 \frac{\text{g} \cdot \cancel{\text{cm}^3}}{\cancel{\text{cm}^3}}$$

$$= \boxed{114 \text{ g}}$$

1.4 Solving the relationship $\rho = m/V$ for volume gives $V = m/\rho$, and

$$\rho = \frac{m}{V}$$

$$V\rho = \frac{m\cancel{V}}{\cancel{V}}$$

$$\frac{V\cancel{\rho}}{\cancel{\rho}} = \frac{m}{\rho}$$

$$V = \frac{m}{\rho}$$

$$V = \frac{600 \text{ g}}{3.00 \frac{\text{g}}{\text{cm}^3}}$$

$$= \frac{600}{3.00} \frac{\text{g}}{1} \times \frac{\text{cm}^3}{\text{g}}$$

$$= 200 \frac{\cancel{\text{g}} \cdot \text{cm}^3}{\cancel{\text{g}}}$$

$$= \boxed{200 \text{ cm}^3}$$

1.5 A 50.0 cm^3 sample with a mass of 34.0 grams has a density of

$$\rho = \frac{m}{V} = \frac{34.0 \text{ g}}{50.0 \text{ cm}^3}$$

$$= \frac{34.0}{50.0} \frac{\text{g}}{\text{cm}^3}$$

$$= \boxed{0.680 \frac{\text{g}}{\text{cm}^3}}$$

According to table 1.4, 0.680 g/cm^3 is the mass density of gasoline, so the substance must be gasoline.

1.6 The problem asks for a mass and gives a volume, so you need a relationship between mass and volume. Table 1.4 gives the mass density of water as 1.00 g/cm³, which is a density that is easily remembered. The volume is given in liters (L), which should first be converted to cm³ because this is the unit in which density is expressed. The relationship of ρ = m/V solved for mass is ρV, so the solution is

$$\rho = \frac{m}{V} \therefore m = \rho V$$

$$m = \left(1.00\frac{g}{cm^3}\right)(40,000 \text{ cm}^3)$$

$$= 1.00 \times 40,000 \frac{g}{cm^3} \times cm^3$$

$$= 40,000 \frac{g \cdot cm^3}{cm^3}$$

$$= 40,000 \text{ g}$$

$$= \boxed{40 \text{ kg}}$$

1.7 From table 1.4, the mass density of aluminum is given as 2.70 g/cm³. Converting 2.1 kg to the same units as the density gives 2,100 g. Solving ρ = m/V for the volume gives

$$V = \frac{m}{\rho} = \frac{2,100 \text{ g}}{2.70\frac{g}{cm^3}}$$

$$= \frac{2,100}{2.70}\frac{g}{1} \times \frac{cm^3}{g}$$

$$= 777.78 \frac{g \cdot cm^3}{g}$$

$$= \boxed{780 \text{ cm}^3}$$

1.8 The length of one side of the box is 0.1 m. Reasoning: Since the density of water is 1.00 g/cm³, then the volume of 1,000 g of water is 1,000 cm³. A cubic box with a volume of 1,000 cm³ is 10 cm (since 10 × 10 × 10 = 1,000). Converting 10 cm to m units, the cube is 0.1 m on each edge.

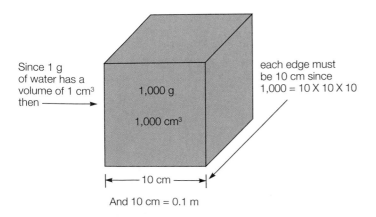

Since 1 g of water has a volume of 1 cm³ then ⟶

1,000 g

1,000 cm³

each edge must be 10 cm since 1,000 = 10 X 10 X 10

⟵ 10 cm ⟶

And 10 cm = 0.1 m

1.9 The relationship between mass, volume, and density is ρ = m/V. The problem gives a volume, but not a mass. The mass, however, can be assumed to remain constant during the compression of the bread so the mass can be obtained from the original volume and density, or

$$\rho = \frac{m}{V} \therefore m = \rho V$$

$$m = \left(0.2\frac{g}{cm^3}\right)(3,000 \text{ cm}^3)$$

$$= 0.2 \times 3,000 \frac{g}{cm^3} \times cm^3$$

$$= 600 \frac{g \cdot cm^3}{cm^3}$$

$$= 600 \text{ g}$$

A mass of 600 g and the new volume of 1,500 cm³ means that the new density of the crushed bread is

$$\rho = \frac{m}{V}$$

$$= \frac{600 \text{ g}}{1,500 \text{ cm}^3}$$

$$= \frac{600}{1,500}\frac{g}{cm^3}$$

$$= \boxed{0.4 \frac{g}{cm^3}}$$

1.10 According to table 1.4, lead has a density of 11.4 g/cm³. Therefore, a 1.00 cm³ sample of lead would have a mass of

$$\rho = \frac{m}{V} \therefore m = \rho V$$

$$m = \left(11.4\frac{g}{cm^3}\right)(1.00 \text{ cm}^3)$$

$$= 11.4 \times 1.00 \frac{g}{cm^3} \times cm^3$$

$$= 11.4 \frac{g \cdot cm^3}{cm^3}$$

$$= 11.4 \text{ g}$$

Also according to table 1.4, copper has a density of 8.96 g/cm³. To balance a mass of 11.4 g of lead, a volume of this much copper would be required:

$$\rho = \frac{m}{V} \therefore V = \frac{m}{\rho}$$

$$V = \frac{11.4 \text{ g}}{8.96\frac{g}{cm^3}}$$

$$= \frac{11.4}{8.96}\frac{g}{1} \times \frac{cm^3}{g}$$

$$= 1.27 \frac{g \cdot cm^3}{g}$$

$$= \boxed{1.27 \text{ cm}^3}$$

CHAPTER 2

2.1 Listing the quantities with their symbols, we can see the problem involves the quantities found in the definition of average speed:

$$\bar{v} = 350.0 \text{ m/s}$$

$$t = 5.00 \text{ s}$$

$$d = ?$$

$$\bar{v} = \frac{d}{t} \therefore d = \bar{v}t$$

$$d = \left(350.0 \frac{\text{m}}{\text{s}}\right)(5.00 \text{ s})$$

$$= (350.0)(5.00) \frac{\text{m}}{\text{s}} \times \text{s}$$

$$= \boxed{1{,}750 \text{ m}}$$

2.2 The initial velocity, final velocity, and time are known and the problem asked for the acceleration. Listing these quantities with their symbols, we have

$$v_i = 0 \text{ m/s}$$

$$v_f = 15.0 \text{ m/s}$$

$$t = 10.0 \text{ s}$$

$$a = ?$$

These are the quantities involved in the acceleration equation, which is already solved for the unknown:

$$a = \frac{v_f - v_i}{t}$$

$$a = \frac{15.0 \text{ m/s} - 0 \text{ m/s}}{10.0 \text{ s}}$$

$$= \frac{15.0}{10.0} \frac{\text{m}}{\text{s}} \times \frac{1}{\text{s}}$$

$$= \boxed{1.50 \frac{\text{m}}{\text{s}^2}}$$

2.3 The distance (d) and the time (t) quantities are given in the problem, and

$$\bar{v} = \frac{d}{t}$$

$$= \frac{285 \text{ mi}}{5.0 \text{ h}}$$

$$= \frac{285}{5.0} \frac{\text{mi}}{\text{h}}$$

$$= \boxed{57 \text{ mi/h}}$$

The units cannot be simplified further. Note two significant figures in the answer, which is the least number of significant figures involved in the division operation.

2.4 Listing the known and unknown quantities:

$$m = 40.0 \text{ kg}$$

$$a = 2.4 \text{ m/s}^2$$

$$F = ?$$

These are the quantities found in Newton's second law of motion, $F = ma$, which is already solved for force (F). Thus,

$$F = (40.0 \text{ kg})\left(2.4 \frac{\text{m}}{\text{s}^2}\right)$$

$$= 40.0 \times 2.4 \frac{\text{kg} \cdot \text{m}}{\text{s}^2}$$

$$= \boxed{96 \text{ N}}$$

2.5 List the known and unknown quantities for the first situation, using an unbalanced force of 18.0 N to give the object an acceleration of 3 m/s²:

$$F_1 = 18 \text{ N}$$

$$a_1 = 3 \text{ m/s}^2$$

For the second situation, we are asked to find the force needed for an acceleration of 10 m/s²:

$$a_2 = 10 \text{ m/s}^2$$

$$F = ?$$

These are the quantities of Newton's second law of motion, $F = ma$, but the mass appears to be missing. The mass can be found from

$$F = ma \therefore m_1 = \frac{F_1}{a_1}$$

$$= \frac{18 \frac{\text{kg} \cdot \text{m}}{\text{s}^2}}{3 \frac{\text{m}}{\text{s}^2}}$$

$$= \frac{18}{3} \frac{\text{kg} \cdot \text{m}}{\text{s}^2} \times \frac{\text{s}^2}{\text{m}}$$

$$= 6 \text{ kg}$$

Now that we have the mass, we can easily find the force needed for an acceleration of 10 m/s²:

$$F_2 = m_2 a_2$$

$$= (6 \text{ kg})\left(10 \frac{\text{m}}{\text{s}^2}\right)$$

$$= 6 \times 10 \frac{\text{kg} \cdot \text{m}}{\text{s}^2}$$

$$= \boxed{60 \text{ N}}$$

2.6 Listing the known and unknown quantities:

$$F = 100 \text{ N}$$

$$a = 0.5 \text{ m/s}^2$$

$$m = ?$$

These are the quantities found in Newton's second law of motion, $F = ma$, which must be first solved for m. Thus,

$$F = ma \therefore m = \frac{F}{a}$$

$$= \frac{100 \frac{\text{kg·m}}{\text{s}^2}}{0.5 \frac{\text{m}}{\text{s}^2}}$$

$$= \frac{100}{0.5} \frac{\text{kg·m}}{\text{s}^2} \times \frac{\text{s}^2}{\text{m}}$$

$$= \boxed{200 \text{ kg}}$$

2.7 Listing the known and unknown quantities:

$$m = 100 \text{ kg}$$

$$v = 6 \text{ m/s}$$

$$p = ?$$

These are the quantities found in the equation for momentum, $p = mv$, which is already solved for momentum (p). Thus,

$$p = mv$$

$$= (100 \text{ kg})\left(6 \frac{\text{m}}{\text{s}}\right)$$

$$= \boxed{600 \frac{\text{kg·m}}{\text{s}}}$$

(Note the lowercase p is the symbol used for momentum. This is one of the few cases where the English letter does not provide a clue about what it stands for. The units for momentum are also somewhat unusual for metric units since they do not have a name or single symbol to represent them.)

2.8 Listing the known and unknown quantities:

$$w = 13{,}720 \text{ N}$$

$$v = 91 \text{ km/h}$$

$$p = ?$$

The equation for momentum is $p = mv$, which is already solved for momentum (p). The weight unit must be first converted to a mass unit:

$$w = mg \therefore m = \frac{w}{g}$$

$$= \frac{13{,}720 \frac{\text{kg·m}}{\text{s}^2}}{9.8 \frac{\text{m}}{\text{s}^2}}$$

$$= \frac{13{,}720}{9.8} \frac{\text{kg·m}}{\text{s}^2} \times \frac{\text{s}^2}{\text{m}}$$

$$= 1{,}400 \text{ kg}$$

The km/h unit should next be converted to m/s. Using the conversion factor from inside the front cover:

$$\frac{0.2778 \frac{\text{m}}{\text{s}}}{1 \frac{\text{km}}{\text{h}}} \times 91 \frac{\text{km}}{\text{h}}$$

$$0.2778 \times 91 \frac{\text{m}}{\text{s}} \times \frac{\text{h}}{\text{km}} \times \frac{\text{km}}{\text{h}}$$

$$25 \frac{\text{m}}{\text{s}}$$

Now, listing the converted known and unknown quantities:

$$m = 1{,}400 \text{ kg}$$

$$v = 25 \text{ m/s}$$

$$p = ?$$

and solving for momentum (p),

$$p = mv$$

$$= (1{,}400 \text{ kg})\left(25 \frac{\text{m}}{\text{s}}\right)$$

$$= \boxed{35{,}000 \frac{\text{kg·m}}{\text{s}}}$$

2.9 Listing the known and unknown quantities:

Bullet → $m = 0.015$ kg   Rifle → $m = 6$ kg

Bullet → $v = 200$ m/s   Rifle → $v = ?$ m/s

Note the mass of the bullet was converted to kilograms. This is a conservation of momentum question, where the bullet and rifle can be considered as a system of interacting objects:

$$\text{Bullet momentum} = -\text{rifle momentum}$$

$$(mv)_b = -(mv)_r$$

$$(mv)_b - (mv)_r = 0$$

$$(0.015 \text{ kg})\left(200 \frac{\text{m}}{\text{s}}\right) - (6 \text{ kg})v_r = 0$$

$$\left(3 \text{ kg} \cdot \frac{\text{m}}{\text{s}}\right) - (6 \text{ kg} \cdot v_r) = 0$$

$$\left(3 \text{ kg} \cdot \frac{\text{m}}{\text{s}}\right) = (6 \text{ kg} \cdot v_r)$$

$$v_r = \frac{3 \text{ kg} \cdot \dfrac{\text{m}}{\text{s}}}{6 \text{ kg}}$$

$$= \frac{3}{6} \frac{\text{kg}}{1} \times \frac{1}{\text{kg}} \times \frac{\text{m}}{\text{s}}$$

$$= \boxed{0.5 \frac{\text{m}}{\text{s}}}$$

The rifle recoils with a velocity of 0.5 m/s.

2.10 A unit conversion is needed:

$$\left(90.0 \frac{\text{km}}{\text{h}}\right)\left(0.2778 \frac{\dfrac{\text{m}}{\text{s}}}{\dfrac{\text{km}}{\text{h}}}\right) = 25.0 \text{ m/s}$$

a.

$$F = ma \therefore m = \frac{F}{a} \text{ and } a = \frac{v_f - v_i}{t}, \text{ so}$$

$$m = \frac{F}{\dfrac{v_f - v_i}{t}} = \frac{5{,}000.0 \dfrac{\text{kg} \cdot \text{m}}{\text{s}^2}}{\dfrac{25.0 \text{ m/s} - 0}{5.0 \text{ s}}}$$

$$= \frac{5{,}000.0 \dfrac{\text{kg} \cdot \text{m}}{\text{s}^2}}{5.0 \dfrac{\text{m}}{\text{s}^2}}$$

$$= \frac{5{,}000.0}{5.0} \frac{\text{kg} \cdot \text{m}}{\text{s}^2} \times \frac{\text{s}^2}{\text{m}}$$

$$= 1{,}000 \frac{\text{kg} \cdot \text{m} \cdot \text{s}^2}{\text{m} \cdot \text{s}^2}$$

$$= \boxed{1.0 \times 10^3 \text{ kg}}$$

b.

$$w = mg$$

$$= (1.0 \times 10^3 \text{ kg})\left(9.8 \frac{\text{m}}{\text{s}^2}\right)$$

$$= (1.0 \times 10^3)(9.8) \text{ kg} \times \frac{\text{m}}{\text{s}^2}$$

$$= 9.8 \times 10^3 \frac{\text{kg} \cdot \text{m}}{\text{s}^2}$$

$$= \boxed{9.8 \times 10^3 \text{ N}}$$

2.11

$$F = \frac{mv^2}{r}$$

$$= \frac{(0.20 \text{ kg})\left(3.0 \dfrac{\text{m}}{\text{s}}\right)^2}{1.5 \text{ m}}$$

$$= \frac{(0.20 \text{ kg})\left(9.0 \dfrac{\text{m}^2}{\text{s}^2}\right)}{1.5 \text{ m}}$$

$$= \frac{0.20 \times 9.0}{1.5} \frac{\text{kg} \cdot \text{m}^2}{\text{s}^2} \times \frac{1}{\text{m}}$$

$$= 1.2 \frac{\text{kg} \cdot \text{m} \cdot \text{m}}{\text{s}^2 \cdot \text{m}}$$

$$= \boxed{1.2 \text{ N}}$$

2.12 a. Newton's laws of motion consider the resistance to a change of motion, or mass, and not weight. The astronaut's mass is

$$w = mg \therefore m = \frac{w}{g} = \frac{1{,}960.0 \dfrac{\text{kg} \cdot \text{m}}{\text{s}^2}}{9.8 \dfrac{\text{m}}{\text{s}^2}}$$

$$= \frac{1{,}960.0}{9.8} \frac{\text{kg} \cdot \text{m}}{\text{s}^2} \times \frac{\text{s}^2}{\text{m}} = 200 \text{ kg}$$

b. From Newton's second law of motion, you can see that the 100 N rocket gives the 200 kg astronaut an acceleration of:

$$F = ma \therefore a = \frac{F}{m} = \frac{100 \dfrac{\text{kg} \cdot \text{m}}{\text{s}^2}}{200 \text{ kg}}$$

$$= \frac{100 \text{ kg} \cdot \text{m}}{200 \text{ s}^2} \times \frac{1}{\text{kg}} = 0.5 \text{ m/s}^2$$

c. An acceleration of 0.5 m/s² for 2.0 s will result in a final velocity of

$$a = \frac{v_f - v_i}{t} \therefore v_f = at + v_i$$

$$= (0.5 \text{ m/s}^2)(2.0 \text{ s}) + 0 \text{ m/s}$$

$$= \boxed{1 \text{ m/s}}$$

CHAPTER 3

3.1 Listing the known and unknown quantities:

$$F = 200 \text{ N}$$
$$d = 3 \text{ m}$$
$$W = ?$$

These are the quantities found in the equation for work, $W = Fd$, which is already solved for work (W). Thus,

$$W = Fd$$
$$= \left(200 \, \frac{\text{kg·m}}{\text{s}^2} \right)(3 \text{ m})$$
$$= (200)(3) \text{ N·m}$$
$$= \boxed{600 \text{ J}}$$

3.2 Listing the known and unknown quantities:

$$F = 440 \text{ N}$$
$$d = 5.0 \text{ m}$$
$$w = 880 \text{ N}$$
$$W = ?$$

These are the quantities found in the equation for work, $W = Fd$, which is already solved for work (W). As you can see in the equation, the force exerted and the distance the box was moved are the quantities used in determining the work accomplished. The weight of the box is a different variable, and one that is not used in this equation. Thus,

$$W = Fd$$
$$= \left(440 \, \frac{\text{kg·m}}{\text{s}^2} \right)(5.0 \text{ m})$$
$$= 2,200 \text{ N·m}$$
$$= \boxed{2,200 \text{ J}}$$

3.3 Note that 10.0 kg is a mass quantity, and not a weight quantity. Weight is found from $w = mg$, a form of Newton's second law of motion. Thus the force that must be exerted to lift the backpack is its weight, or (10.0 kg) × (9.8 m/s²), which is 98 N. Therefore, a force of 98 N was exerted on the backpack through a distance of 1.5 m, and

$$W = Fd$$
$$= \left(98 \, \frac{\text{kg·m}}{\text{s}^2} \right)(1.5 \text{ m})$$
$$= 147 \text{ N·m}$$
$$= \boxed{150 \text{ J}}$$

3.4 Weight is defined as the force of gravity acting on an object, and the greater the force of gravity the harder it is to lift the object. The force is proportional to the mass of the object, as the equation $w = mg$ tells you. Thus the force you exert when lifting is $F = w = mg$, so the work you do on an object you lift must be $W = mgh$.

You know the mass of the box and you know the work accomplished. You also know the value of the acceleration due to gravity, g, so the list of known and unknown quantities is:

$$m = 102 \text{ kg}$$
$$g = 9.8 \text{ m/s}^2$$
$$W = 5,000 \text{ J}$$
$$h = ?$$

The equation $W = mgh$ is solved for work, so the first thing to do is to solve it for h, the unknown height in this problem (note that height is also a distance):

$$W = mgh \therefore h = \frac{W}{mg}$$

$$= \frac{5,000 \, \dfrac{\text{kg·m}}{\text{s}^2} \times \text{m}}{(102 \text{ kg})\left(9.8 \, \dfrac{\text{m}}{\text{s}^2} \right)}$$

$$= \frac{5,000.0}{102 \times 9.8} \, \frac{\text{kg·m}}{\text{s}^2} \times \frac{\text{m}}{1} \times \frac{1}{\text{kg}} \times \frac{\text{s}^2}{\text{m}}$$

$$= \frac{5,000}{999.6} \text{ m}$$

$$= \boxed{5 \text{ m}}$$

3.5 A student running up the stairs has to lift herself, so her weight is the required force needed. Thus the force exerted is $F = w = mg$, and the work done is $W = mgh$. You know the mass of the student, the height, and the time. You also know the value of the acceleration due to gravity, g, so the list of known and unknown quantities is:

$$m = 60.0 \text{ kg}$$
$$g = 9.8 \text{ m/s}^2$$
$$h = 5.00 \text{ m}$$
$$t = 3.92 \text{ s}$$
$$P = ?$$

The equation $P = \dfrac{mgh}{t}$ is already solved for power, so:

a.

$$P = \frac{mgh}{t}$$

$$= \frac{(60.0 \text{ kg})\left(9.8 \frac{m}{s^2}\right)(5.00 \text{ m})}{3.92 \text{ s}}$$

$$= \frac{(60.0)(9.8)(5.00)}{(3.92)} \frac{\left(\frac{\text{kg·m}}{s^2}\right) \times \text{m}}{s}$$

$$= \frac{2940}{3.92} \frac{\text{N·m}}{s}$$

$$= 750 \frac{J}{s}$$

$$= \boxed{750 \text{ W}}$$

b. A power of 750 watts is almost one horsepower.

3.6 Listing the known and unknown quantities:

$$m = 2{,}000 \text{ kg}$$

$$v = 72 \text{ km/h}$$

$$KE = ?$$

These are the quantities found in the equation for kinetic energy, $KE = 1/2mv^2$, which is already solved. However, note that the velocity is in units of km/h, which must be changed to m/s before doing anything else (it must be m/s because all energy and work units are in units of the joule (J). A joule is a newton-meter, and a newton is a kg·m/s^2). Using the conversion factor from inside the front cover of your text,

$$\frac{0.2778 \frac{m}{s}}{1.0 \frac{\text{km}}{h}} \times 72 \frac{\text{km}}{h}$$

$$(0.2778)\,(72) \frac{m}{s} \times \frac{h}{\text{km}} \times \frac{\text{km}}{h}$$

$$20 \frac{m}{s}$$

and

$$KE = \frac{1}{2} mv^2$$

$$= \frac{1}{2}(2{,}000 \text{ kg})\left(20 \frac{m}{s}\right)^2$$

$$= \frac{1}{2}(2{,}000 \text{ kg})\left(400 \frac{m^2}{s^2}\right)$$

$$= \frac{1}{2} \times 2{,}000 \times 400 \frac{\text{kg·m}^2}{s^2}$$

$$= 400{,}000 \frac{\text{kg·m}}{s^2} \times \text{m}$$

$$= 400{,}000 \text{ N·m}$$

$$= \boxed{4 \times 10^5 \text{ J}}$$

Scientific notation is used here to simplify a large number and to show one significant figure.

3.7 Recall the relationship between work and energy—that you do work on an object when you throw it, giving it kinetic energy, and the kinetic energy it has will do work on something else when stopping. Because of the relationship between work and energy you can calculate (1) the work you do, (2) the kinetic energy a moving object has as a result of your work, and (3) the work it will do when coming to a stop, and all three answers should be the same. Thus you do not have a force or a distance to calculate the work needed to stop a moving car, but you can simply calculate the kinetic energy of the car. Both answers should be the same.

Before you start, note that the velocity is in units of km/h, which must be changed to m/s before doing anything else (it must be m/s because all energy and work units are in units of the joule (J). A joule is a newton-meter, and a newton is a kg·m/s^2). Using the conversion factor from inside the front cover,

$$\frac{0.2778 \frac{m}{s}}{1.0 \frac{\text{km}}{h}} \times 54.0 \frac{\text{km}}{h}$$

$$0.2778 \times 54.0 \frac{m}{s} \times \frac{h}{\text{km}} \times \frac{\text{km}}{h}$$

$$15.0 \frac{m}{s}$$

and

$$KE = \frac{1}{2} mv^2$$

$$= \frac{1}{2}(1{,}000.0 \text{ kg})\left(15.0 \frac{m}{s}\right)^2$$

$$= \frac{1}{2}(1{,}000.0 \text{ kg})\left(225 \frac{m^2}{s^2}\right)$$

$$= \frac{1}{2} \times 1{,}000.0 \times 225 \frac{\text{kg·m}^2}{s^2}$$

$$= 112{,}500 \frac{\text{kg·m}}{s^2} \times \text{m}$$

$$= 112{,}500 \text{ N·m}$$

$$= \boxed{1.13 \times 10^5 \text{ J}}$$

Scientific notation is used here to simplify a large number and to easily show three significant figures. The answer could likewise be expressed as 113 kJ.

3.8 **a.** How much energy was used by a 1,000 kg car climbing a hill 51.02 m high is answered by how much work the car did. In this case $W = Fd$ and the force exerted is the weight of the car, $w = mgh$. Thus,

$$w = mgh$$

$$= (1{,}000 \text{ kg})\left(9.8\,\frac{\text{m}}{\text{s}^2}\right)(51.02 \text{ m})$$

$$= 1{,}000 \times 9.8 \times 51.02 \text{ kg} \times \frac{\text{m}}{\text{s}^2} \times \text{m}$$

$$= 499{,}996\,\frac{\text{kg}\cdot\text{m}}{\text{s}^2} \times \text{m}$$

$$= 500{,}000 \text{ N}\cdot\text{m}$$

$$= 5 \times 10^5 \text{ J (or 500 kJ)}$$

b. How much potential energy the car has is found in the potential energy equation, $PE = mgh$. Note the potential energy is path independent, that is, depends only on the vertical height of the hill. Thus,

$$PE = mgh$$

$$= (1{,}000 \text{ kg})\left(9.8\,\frac{\text{m}}{\text{s}^2}\right)(51.02 \text{ m})$$

$$= 1{,}000 \times 9.8 \times 51.02 \text{ kg} \times \frac{\text{m}}{\text{s}^2} \times \text{m}$$

$$= 499{,}996\,\frac{\text{kg}\cdot\text{m}}{\text{s}^2} \times \text{m}$$

$$= 500{,}000 \text{ N}\cdot\text{m}$$

$$= \boxed{5 \times 10^5 \text{ J (or 500 kJ)}}$$

As you can see, the potential energy of the car is exactly the same as the amount of energy used to climb the hill.

3.9 **a.**

$$W = Fd$$

$$= (10 \text{ lb})(5 \text{ ft})$$

$$= (10)(5) \text{ ft} \times \text{lb}$$

$$= 50 \text{ ft}\cdot\text{lb}$$

b. The distance of the bookcase from some horizontal reference level did not change, so the gravitational potential energy does not change.

3.10 The force (F) needed to lift the book is equal to the weight (w) of the book, or $F = w$. Since $w = mg$, then $F = mg$. Work is defined as the product of a force moved through a distance, or $W = Fd$. The work done in lifting the book is therefore $W = mgd$, and:

a.

$$W = mgd$$

$$= (2.0 \text{ kg})(9.8 \text{ m/s}^2)(2.00 \text{ m})$$

$$= (2.0)(9.8)(2.00)\,\frac{\text{kg}\cdot\text{m}}{\text{s}^2} \times \text{m}$$

$$= 39.2\,\frac{\text{kg}\cdot\text{m}^2}{\text{s}^2}$$

$$= 39.2 \text{ J} = \boxed{39 \text{ J}}$$

b.

$$PE = mgh = \boxed{39 \text{ J}}$$

c.

$$PE_{\text{lost}} = KE_{\text{gained}} = mgh = \boxed{39 \text{ J}}$$

(or)

$$v = \sqrt{2gh} = \sqrt{(2)(9.8 \text{ m/s}^2)(2.00 \text{ m})}$$

$$= \sqrt{39.2 \text{ m}^2/\text{s}^2} \quad \text{(Note)}$$

$$= 6.26 \text{ m/s}$$

$$KE = \frac{1}{2}mv^2 = \left(\frac{1}{2}\right)(2.0 \text{ kg})(6.26 \text{ m/s})^2$$

$$= \left(\frac{1}{2}\right)(2.0 \text{ kg})(39.2 \text{ m}^2/\text{s}^2)$$

$$= (1.0)(39.2)\,\frac{\text{kg}\cdot\text{m}^2}{\text{s}^2}$$

$$= \boxed{39 \text{ J}}$$

3.11

$$KE = \frac{1}{2}mv^2$$

$$= \frac{1}{2}(60.0\text{ kg})\left(2.0\,\frac{\text{m}}{\text{s}}\right)^2$$

$$= \frac{1}{2}(60.0\text{ kg})\left(4.0\,\frac{\text{m}^2}{\text{s}^2}\right)$$

$$= 30.0 \times 4.0\text{ kg} \times \left(\frac{\text{m}^2}{\text{s}^2}\right)$$

$$= \boxed{120\text{ J}}$$

$$KE = \frac{1}{2}mv^2$$

$$= \frac{1}{2}(60.0\text{ kg})\left(4.0\,\frac{\text{m}}{\text{s}}\right)^2$$

$$= \frac{1}{2}(60.0\text{ kg})\left(16\,\frac{\text{m}^2}{\text{s}^2}\right)$$

$$= 30.0 \times 16\text{ kg} \times \left(\frac{\text{m}^2}{\text{s}^2}\right)$$

$$= \boxed{480\text{ J}}$$

Thus, doubling the speed results in a four-fold increase in kinetic energy.

3.12 a. The force needed is equal to the weight of the student. The English unit of a pound is a force unit, so

$$W = Fd$$

$$= (170.0\text{ lb})(25.0\text{ ft})$$

$$= \boxed{4{,}250\text{ ft·lb}}$$

b. Work (W) is defined as a force (F) moved through a distance (d), or $W = Fd$. Power (P) is defined as work (W) per unit of time (t), or $P = W/t$. Therefore,

$$P = \frac{Fd}{t}$$

$$= \frac{(170.0\text{ lb})(25.0\text{ ft})}{10.0\text{ s}}$$

$$= \frac{(170.0)(25.0)}{10.0}\,\frac{\text{ft·lb}}{\text{s}}$$

$$= 425\,\frac{\text{ft·lb}}{\text{s}}$$

One hp is defined as $550\,\dfrac{\text{ft·lb}}{\text{s}}$ and

$$\frac{425\text{ ft·lb/s}}{550\,\dfrac{\text{ft·lb/s}}{\text{hp}}} = \boxed{0.77\text{ hp}}$$

(Note that the student's power rating (425 ft·lb/s) is less than the power rating defined as 1 horsepower (550 ft·lb/s). Thus, the student's horsepower must be *less* than 1 horsepower. A simple analysis such as this will let you know if you inverted the ratio or not.)

CHAPTER 4

4.1 Listing the known and unknown quantities:

body temperature $T_F = 98.6°$

$$T_C = ?$$

These are the quantities found in the equation for conversion of Fahrenheit to Celsius, $T_C = \frac{5}{9}(T_F - 32°)$, where T_F is the temperature in Fahrenheit and T_C is the temperature in Celsius. This equation describes a relationship between the two temperature scales and is used to convert a Fahrenheit temperature to Celsius. The equation is already solved for the Celsius temperature, T_C. Thus,

$$T_C = \frac{5}{9}(T_F - 32°)$$

$$= \frac{5}{9}(98.6° - 32°)$$

$$= \frac{333°}{9}$$

$$= \boxed{37°\text{ C}}$$

4.2

$$Q = mc\Delta T$$

$$= (221\text{ g})\left(0.093\,\frac{\text{cal}}{\text{gC°}}\right)(38.0°C - 20.0°C)$$

$$= (221)(0.093)(18.0)\text{g} \times \frac{\text{cal}}{\text{gC°}} \times °C$$

$$= 370\,\frac{\text{g·cal·°C}}{\text{gC°}}$$

$$= \boxed{370\text{ cal}}$$

4.3 First, you need to know the energy of the moving bike and rider. Since the speed is given as 36.0 km/h, convert to m/s by multiplying times 0.2778 m/s per km/h:

$$\left(36.0\,\frac{\text{km}}{\text{h}}\right)\left(0.2778\,\frac{\text{m/s}}{\text{km/h}}\right)$$

$$= (36.0)(0.2778)\,\frac{\text{km}}{\text{h}} \times \frac{\text{h}}{\text{km}} \times \frac{\text{m}}{\text{s}}$$

$$= 10.0\text{ m/s}$$

Then,

$$KE = \frac{1}{2}mv^2$$

$$= \frac{1}{2}(100.0\text{ kg})(10.0\text{ m/s})^2$$

$$= \frac{1}{2}(100.0\text{ kg})(100\text{ m}^2/\text{s}^2)$$

$$= \frac{1}{2}(100.0)(100)\,\frac{\text{kg·m}^2}{\text{s}^2}$$

$$= 5{,}000\text{ J}$$

Second, this energy is converted to the calorie heat unit through the mechanical equivalent of heat relationship, that 1.0 kcal = 4,184 J, or that 1.0 cal = 4.184 J. Thus,

$$\frac{5,000 \text{ J}}{4,184 \, \frac{\text{J}}{\text{kcal}}}$$

$$1.195 \, \frac{\text{J}}{1} \times \frac{\text{kcal}}{\text{J}}$$

$$\boxed{1.20 \text{ kcal}}$$

4.4 First, you need to find the energy of the falling bag. Since the potential energy lost equals the kinetic energy gained, the energy of the bag just as it hits the ground can be found from

$$PE = mgh$$

$$= (15.53 \text{ kg})(9.8 \text{ m/s}^2)(5.50 \text{ m})$$

$$= (15.53)(9.8)(5.50) \, \frac{\text{kg·m}}{\text{s}^2} \times \text{m}$$

$$= 837 \text{ J}$$

In calories, this energy is equivalent to

$$\frac{837 \text{ J}}{4,184 \, \frac{\text{J}}{\text{kcal}}} = 0.200 \text{ kcal}$$

Second, the temperature change can be calculated from the equation giving the relationship between a quantity of heat (Q), mass (m), specific heat of the substance (c), and the change of temperature:

$$Q = mc\Delta T \therefore \Delta T = \frac{Q}{mc}$$

$$= \frac{0.200 \text{ kcal}}{(15.53 \text{ kg})\left(0.200 \, \frac{\text{kcal}}{\text{kgC°}}\right)}$$

$$= \frac{0.200}{(15.53)(0.200)} \, \frac{\text{kcal}}{1} \times \frac{1}{\text{kg}} \times \frac{\text{kgC°}}{\text{kcal}}$$

$$= 0.064 \, \frac{\text{kcal·kgC°}}{\text{kcal·kg}}$$

$$= \boxed{6.4 \times 10^{-2} \text{°C}}$$

4.5 The Calorie used by dietitians is a kilocalorie; thus 250.0 Cal is 250.0 kcal. The mechanical energy equivalent is 1 kcal = 4,184 J, so (250.0 kcal)(4,184 J/kcal) = 1,046,250 J. Since $W = Fd$ and the force needed is equal to the weight (mg) of the person, $W = mgh = (75.0 \text{ kg})(9.8 \text{ m/s}^2) (10.0 \text{ m}) = 7,350$ J for each stairway climbed.

A total of 1,046,250 J of energy from the french fries would require 1,046,250 J/7,350 J per climb, or 142.3 trips up the stairs.

4.6 For unit consistency,

$$T_C = \frac{5}{9}(T_F - 32°) = \frac{5}{9}(68° - 32°) = \frac{5}{9}(36°) = 20°C$$

$$= \frac{5}{9}(32° - 32°) = \frac{5}{9}(0°) = 0°C$$

Glass Bowl:

$$Q = mc\Delta T$$

$$= (0.5 \text{ kg})\left(0.2 \, \frac{\text{kcal}}{\text{kg°C}}\right)(20°C)$$

$$= (0.5)(0.2)(20) \, \frac{\text{kg}}{1} \times \frac{\text{kcal}}{\text{kgC°}} \times \frac{°C}{1}$$

$$= \boxed{2 \text{ kcal}}$$

Iron Pan:

$$Q = mc\Delta T$$

$$= (0.5 \text{ kg})\left(0.11 \, \frac{\text{kcal}}{\text{kgC°}}\right)(20°C)$$

$$= (0.5)(0.11)(20) \text{ kg} \times \frac{\text{kcal}}{\text{kgC°}} \times °C$$

$$= \boxed{1 \text{ kcal}}$$

4.7 Note that a specific heat expressed in cal/g has the same numerical value as a specific heat expressed in kcal/kg because you can cancel the k units. You could convert 896 cal to 0.896 kcal, but one of the two conversion methods is needed for consistency with other units in the problem.

$$Q = mc\Delta T \therefore m = \frac{Q}{c\Delta T}$$

$$= \frac{896 \text{ cal}}{\left(0.056 \, \frac{\text{cal}}{\text{gC°}}\right)(80.0°C)}$$

$$= \frac{896}{(0.056)(80.0)} \, \frac{\text{cal}}{1} \times \frac{\text{gC°}}{\text{cal}} \times \frac{1}{\text{C}}$$

$$= 200 \text{ g}$$

$$= \boxed{0.20 \text{ kg}}$$

4.8 Since a watt is defined as a joule/s, finding the total energy in joules will tell the time:

$$Q = mc\Delta T$$

$$= (250.0\ \text{g})\left(1.00\ \frac{\text{cal}}{\text{gC}^\circ}\right)(60.0^\circ\text{C})$$

$$= (250.0)(1.00)(60.0)\ \text{g} \times \frac{\text{cal}}{\text{gC}^\circ} \times {}^\circ\text{C}$$

$$= 1.50 \times 10^4\ \text{cal}$$

This energy in joules is $(1.50 \times 10^4\ \text{cal})\left(4.184\ \dfrac{\text{J}}{\text{cal}}\right) = 62{,}800\ \text{J}$

A 300 watt heater uses energy at a rate of $300\ \dfrac{\text{J}}{\text{s}}$, so $\dfrac{62{,}800\ \text{J}}{300\ \text{J/s}}$

$= 209\ \text{s}$ is required, which is $\dfrac{209\ \text{s}}{60\ \dfrac{\text{s}}{\text{min}}} = 3.48\ \text{min}$, or

$$\boxed{\text{about } 3\tfrac{1}{2}\ \text{min}}$$

4.9

$$Q = mc\Delta \therefore c = \frac{Q}{m\Delta T}$$

$$= \frac{60.0\ \text{cal}}{(100.0\ \text{g})(20.0^\circ\text{C})}$$

$$= \frac{60.0}{(100.0)(20.0)}\ \frac{\text{cal}}{\text{gC}^\circ}$$

$$= \boxed{0.0300\ \frac{\text{cal}}{\text{gC}^\circ}}$$

4.10 To change water at 80.0°C to steam at 100.0°C requires two separate quantities of heat that can be called Q_1 and Q_2. The quantity Q_1 is the amount of heat needed to warm the water from 80.0°C to the boiling point, which is 100.0°C at sea level pressure ($\Delta T = 20.0^\circ\text{C}$). The relationship between the variable involved is $Q = mc\Delta T$. The quantity Q_2 is the amount of heat needed to take 100.0° water through the phase change to steam (water vapor) at 100.0°C. The phase change from a liquid to a gas (or gas to liquid) is concerned with the latent heat of vaporization. For water, the latent heat of vaporization is given as 540.0 cal/g.

$m = 250.0\ \text{g}$

$L_{v\,(\text{water})} = 540.0\ \text{cal/g}$

$Q = ?$

$$Q_1 = mc\Delta T$$

$$= (250.0\ \text{g})\left(1.00\ \frac{\text{cal}}{\text{gC}^\circ}\right)(20.0^\circ\text{C})$$

$$= (250.0)(1.00)(20.0)\ \text{g} \times \frac{\text{cal}}{\text{gC}^\circ} \times {}^\circ\text{C}$$

$$= 5{,}000\ \frac{\text{g}\cdot\text{cal}\cdot{}^\circ\text{C}}{\text{gC}^\circ}$$

$$= 5{,}000\ \text{cal}$$

$$= 5.00\ \text{kcal}$$

$$Q_2 = mL_v$$

$$= (250.0\ \text{g})\left(540.0\ \frac{\text{cal}}{\text{g}}\right)$$

$$= 250.0 \times 540.0\ \frac{\text{g}\cdot\text{cal}}{\text{g}}$$

$$= 135{,}000\ \text{cal}$$

$$= 135.0\ \text{kcal}$$

$$Q_{\text{Total}} = Q_1 + Q_2$$

$$= 5.00\ \text{kcal} + 135.0\ \text{kcal}$$

$$= \boxed{140.0\ \text{kcal}}$$

4.11 To change 20.0°C water to steam at 125.0°C requires three separate quantities of heat. First, the quantity Q_1 is the amount of heat needed to warm the water from 20.0°C to 100.0°C ($\Delta T = 80.0^\circ\text{C}$). The quantity Q_2 is the amount of heat needed to take 100.0°C water to steam at 100.0°C. Finally, the quantity Q_3 is the amount of heat needed to warm the steam from 100.0° to 125.0°C. According to table 4.3, the c for steam is 0.480 cal/g°C.

$m = 100.0\ \text{g}$

$\Delta T_{\text{water}} = 80.0^\circ\text{C}$

$\Delta T_{\text{steam}} = 25.0^\circ\text{C}$

$L_{v(\text{water})} = 540.0\ \text{cal/g}$

$c_{\text{steam}} = 0.480\ \text{cal/gC}^\circ$

$$Q_1 = mc\Delta T$$

$$= (100.0\ \text{g})\left(1.00\ \frac{\text{cal}}{\text{gC}^\circ}\right)(80.0^\circ\text{C})$$

$$= (100.0)(1.00)(80.0)\ \text{g} \times \frac{\text{cal}}{\text{gC}^\circ} \times {}^\circ\text{C}$$

$$= 8{,}000\ \frac{\text{g}\cdot\text{cal}\cdot{}^\circ\text{C}}{\text{gC}^\circ}$$

$$= 8{,}000\ \text{cal}$$

$$= 8.00\ \text{kcal}$$

$$Q_2 = mL_v$$

$$= (100.0\ \text{g})\left(540.0\ \frac{\text{cal}}{\text{g}}\right)$$

$$= 100.0 \times 540.0\ \frac{\text{g}\cdot\text{cal}}{\text{g}}$$

$$= 54{,}000\ \text{cal}$$

$$= 54.00\ \text{kcal}$$

$$Q_3 = mc\Delta T$$

$$= (100.0\ \text{g})\left(0.480\ \frac{\text{cal}}{\text{gC}^\circ}\right)(25.0^\circ\text{C})$$

$$= (100.0)(0.480)(25.0)\ \text{g} \times \frac{\text{cal}}{\text{gC}^\circ} \times {}^\circ\text{C}$$

$$= 1{,}200\ \frac{\text{g}\cdot\text{cal}\cdot{}^\circ\text{C}}{\text{g}\cdot\text{C}^\circ}$$

$$= 1{,}200\ \text{cal}$$

$$= 1.20\ \text{kcal}$$

$$Q_{total} = Q_1 + Q_2 + Q_3$$
$$= 8.00 \text{ kcal} + 54.00 \text{ kcal} + 1.20 \text{ kcal}$$
$$= \boxed{63.20 \text{ kcal}}$$

4.12 a. Step 1: Cool water from 18.0°C to 0°C.

$$Q_1 = mc\Delta T$$
$$= (400.0 \text{ g})\left(1.00 \frac{cal}{gC°}\right)(18.0°C)$$
$$= (400.0)(1.00)(18.0)\text{g} \times \frac{cal}{gC°} \times °C$$
$$= 7{,}200 \frac{g \cdot cal \cdot °C}{gC°}$$
$$= 7{,}200 \text{ cal}$$
$$= 7.20 \text{ kcal}$$

Step 2: Find the energy needed for the phase change of water at 0°C to ice at 0°C.

$$Q_2 = mL_f$$
$$= (400.0 \text{ g})\left(80.0 \frac{cal}{g}\right)$$
$$= 400.0 \times 80.0 \frac{g \cdot cal}{g}$$
$$= 32{,}000 \text{ cal}$$
$$= 32.0 \text{ kcal}$$

Step 3: Cool the ice from 0°C to ice at −5.00°C.

$$Q_3 = mc\Delta T$$
$$= (400.0 \text{ g})\left(0.50 \frac{cal}{gC°}\right)(5.00°C)$$
$$= 400.0 \times 0.50 \times 5.00 \text{ g} \times \frac{cal}{gC°} \times °C$$
$$= 1{,}000 \frac{g \cdot cal \cdot °C}{gC°}$$
$$= 1{,}000 \text{ cal}$$
$$= 1.0 \text{ kcal}$$
$$Q_{total} = Q_1 + Q_2 + Q_3$$
$$= 7.20 \text{ kcal} + 32.0 \text{ kcal} + 1.0 \text{ kcal}$$
$$= \boxed{40.2 \text{ kcal}}$$

CHAPTER 5

5.1
$$v = f\lambda$$
$$= \left(10 \frac{1}{s}\right)(0.50 \text{ m})$$
$$= 5 \frac{m}{s}$$

5.2 The distance between two *consecutive* condensations (or rarefactions) is one wavelength, so $\lambda = 3.00$ m and

$$v = f\lambda$$
$$= \left(112.0 \frac{1}{s}\right)(3.00 \text{ m})$$
$$= 336 \frac{m}{s}$$

5.3 a. One complete wave every 4 s means that $T = 4.00$ s. (Note that the symbol for the *time of a cycle* is T. Do not confuse this symbol with the symbol for temperature.)

b.
$$f = \frac{1}{T}$$
$$= \frac{1}{4.0 \text{ s}}$$
$$= \frac{1}{4.0} \frac{1}{s}$$
$$= 0.25 \frac{1}{s}$$
$$= \boxed{0.25 \text{ Hz}}$$

5.4 The distance from one condensation to the next is one wavelength, so

$$v = f\lambda \therefore \lambda = \frac{v}{f}$$
$$= \frac{330 \frac{m}{s}}{260 \frac{1}{s}}$$
$$= \frac{330}{260} \frac{m}{s} \times \frac{s}{1}$$
$$= \boxed{1.3 \text{ m}}$$

5.5 a. $v = f\lambda = \left(256 \frac{1}{s}\right)(1.34 \text{ m})$ $= \boxed{343 \text{ m/s}}$

b. $= \left(440.0 \frac{1}{s}\right)(0.780 \text{ m})$ $= \boxed{343 \text{ m/s}}$

c. $= \left(750.0 \frac{1}{s}\right)(0.457 \text{ m})$ $= \boxed{343 \text{ m/s}}$

d. $= \left(2{,}500.0 \frac{1}{s}\right)(0.1372 \text{ m})$ $= \boxed{343 \text{ m/s}}$

5.6 The speed of sound and time are given and you are looking for a distance.

$$v = 1,100 \text{ ft/s}$$

$$t = 4.80 \text{ s}$$

$$d = ?$$

Calculating the total distance the sound traveled,

$$v = \frac{d}{t} \therefore d = vt$$

$$= \left(1,100 \, \frac{\text{ft}}{\text{s}}\right)(4.80 \text{ s})$$

$$= (1,100)(4.80) \frac{\text{ft}}{\text{s}} \times \text{s}$$

$$= 4,800 \text{ ft}$$

Since the sound travels from you to the cliff, then back to you, the cliff must be half the total distance:

$$\frac{4,800 \text{ ft}}{2} = \boxed{2,400 \text{ ft}}$$

5.7 The speed of the sound and the time between the lightning and thunder are given:

$$v = 1,140 \text{ ft/s}$$

$$t = 4.63 \text{ s}$$

$$d = ?$$

The distance that a sound with this velocity travels in the given time is

$$v = \frac{d}{t} \therefore d = vt$$

$$= \left(1,140 \, \frac{\text{ft}}{\text{s}}\right)(4.63 \text{ s})$$

$$= (1,140)(4.63) \frac{\text{ft}}{\text{s}} \times \text{s}$$

$$= \boxed{5,280 \text{ ft or 1 mile}}$$

5.8 a.

$$v = f\lambda \therefore \lambda = \frac{v}{f}$$

$$= \frac{1,125 \, \dfrac{\text{ft}}{\text{s}}}{440 \, \dfrac{1}{\text{s}}}$$

$$= \frac{1,125}{440} \frac{\text{ft}}{\text{s}} \times \frac{\text{s}}{1}$$

$$= 2.56 \, \frac{\text{ft} \cdot \cancel{\text{s}}}{\cancel{\text{s}}}$$

$$= \boxed{2.6 \text{ ft}}$$

b.

$$v = f\lambda \therefore \lambda = \frac{v}{f}$$

$$= \frac{5,020}{440} \frac{\text{ft}}{\text{s}} \times \frac{\text{s}}{1}$$

$$= 11.4 \text{ ft} = \boxed{11 \text{ ft}}$$

CHAPTER 6

6.1 First, recall that a negative charge means an excess of electrons. Second, the relationship between the total charge (q), the number of electrons (n), and the charge of a single electron (e) is $q = ne$. The fundamental charge of a single ($n = 1$) electron (e) is 1.60×10^{-19} C. Thus

$$q = ne \therefore n = \frac{q}{e}$$

$$= \frac{1.00 \times 10^{-14} \text{ C}}{1.60 \times 10^{-19} \, \dfrac{\text{C}}{\text{electron}}}$$

$$= \frac{1.00 \times 10^{-14}}{1.60 \times 10^{-19}} \frac{\text{C}}{1} \times \frac{\text{electron}}{\text{C}}$$

$$= 6.25 \times 10^4 \, \frac{\cancel{\text{C}} \cdot \text{electron}}{\cancel{\text{C}}}$$

$$= \boxed{6.25 \times 10^4 \text{ electron}}$$

6.2

$$\frac{\text{electric}}{\text{current}} = \frac{\text{charge}}{\text{time}}$$

or

$$I = \frac{q}{t}$$

$$= \frac{6.00 \text{ C}}{2.00 \text{ s}}$$

$$= 3.00 \, \frac{\text{C}}{\text{s}}$$

$$= \boxed{3.00 \text{ A}}$$

6.3

$$R = \frac{V}{I}$$

$$= \frac{120.0 \text{ V}}{4.00 \text{ A}}$$

$$= 30.0 \, \frac{\text{V}}{\text{A}}$$

$$= \boxed{30.0 \, \Omega}$$

6.4

$$R = \frac{V}{I} \therefore I = \frac{V}{R}$$

$$= \frac{120.0 \text{ V}}{60.0 \, \dfrac{\text{V}}{\text{A}}}$$

$$= \frac{120.0}{60.0} \cancel{\text{V}} \times \frac{\text{A}}{\cancel{\text{V}}}$$

$$= \boxed{2.00 \text{ A}}$$

6.5 **a.**
$$R = \frac{V}{I} \therefore V = IR$$

$$= (1.20 \text{ A})\left(10.0 \frac{V}{A}\right)$$

$$= \boxed{12.0 \text{ V}}$$

b. Power = (current)(potential difference)

or

$$P = IV$$

$$= \left(1.20 \frac{C}{s}\right)\left(12.0 \frac{J}{C}\right)$$

$$= (1.20)(12.0) \frac{C}{s} \times \frac{J}{C}$$

$$= 14.4 \frac{J}{s}$$

$$= \boxed{14.4 \text{ W}}$$

6.6 Note that there are two separate electrical units that are rates: (1) the amp (coulomb/s) and (2) the watt (joule/s). The question asked for a rate of using energy. Energy is measured in joules, so you are looking for the power of the radio in watts. To find watts ($P = IV$), you will need to calculate the current (I) since it is not given. The current can be obtained from the relationship of Ohm's law:

$$I = \frac{V}{R}$$

$$= \frac{3.00 \text{ V}}{15.0 \frac{V}{A}}$$

$$= 0.200 \text{ A}$$

$$P = IV$$

$$= (0.200 \text{ C/s})(3.00 \text{ J/C})$$

$$= \boxed{0.600 \text{ W}}$$

6.7
$$\frac{(1,200 \text{ W})(0.25 \text{ h})(\$0.10/\text{kWh})}{1,000 \frac{W}{kW}}$$

$$\frac{(1,200)(0.25)(0.10)}{1,000} \frac{W}{1} \times \frac{h}{1} \times \frac{\$}{kWh} \times \frac{kW}{W}$$

$$\boxed{\$0.30} \text{ (3 cents)}$$

6.8 The relationship between power (P), current (I), and volts (V) will provide a solution. Since the relationship considers power in watts, the first step is to convert horsepower to watts. One horsepower is equivalent to 746 watts, so:

$$(746 \text{ W/hp})(2.00 \text{ hp}) = 1,492 \text{ W}$$

$$P = IV \therefore I = \frac{P}{V}$$

$$= \frac{1,492 \frac{J}{s}}{12.0 \frac{J}{C}}$$

$$= \frac{1,492}{12.0} \frac{J}{s} \times \frac{C}{J}$$

$$= 124.3 \frac{C}{s}$$

$$= \boxed{124 \text{ A}}$$

6.9 **a.** The rate of using energy is joule/s, or the watt. Since 1.00 hp = 746 W,

inside motor: (746 W/hp)(1/3 hp) = 249 W

outside motor: (746 W/hp)(1/3 hp) = 249 W

compressor motor: (746 W/hp)(3.70 hp) = 2,760 W

249 W + 249 W + 2,760 W = $\boxed{3,258 \text{ W}}$

b. $\dfrac{(3,258 \text{ W})(1.00 \text{ h})(\$0.10/\text{kWh})}{1,000 \text{ W/kW}} = \$0.33$ per hour

c. ($\$0.33/\text{h}$)(12h/day)(30 day/mo)=$\boxed{\$118.80}$

6.10 The solution is to find how much current each device draws and then to see if the total current is less or greater than the breaker rating:

$$\text{Toaster: } I = \frac{V}{R} = \frac{120 \text{ V}}{15 \text{ V/A}} = 8.0 \text{ A}$$

Motor: (0.20 hp)(746 W/hp) = 150 W

$$I = \frac{P}{V} = \frac{150 \text{ J/s}}{102 \text{ J/C}} = 1.3 \text{ A}$$

Three 100 W bulbs: 3 × 100 W = 300 W

$$I = \frac{P}{V} = \frac{300 \text{ J/s}}{120 \text{ J/C}} = 2.5 \text{ A}$$

$$\text{Iron: } I = \frac{P}{V} = \frac{600 \text{ J/s}}{120 \text{ J/C}} = 5.0 \text{ A}$$

The sum of the currents is 8.0A + 1.3A + 2.5A + 5.0A = 16.8A, so the total current is greater than 15.0 amp and the circuit breaker will trip.

6.11 a. $V_p = 1{,}200$ V

$N_p = 1$ loop

$N_s = 200$ loops

$V_s = ?$

$$\frac{V_p}{N_p} = \frac{V_s}{N_s} \therefore V_s = \frac{V_p N_s}{N_p}$$

$$V_s = \frac{(1{,}200 \text{ V})(200 \text{ loops})}{1 \text{ loop}}$$

$$= \boxed{240{,}000 \text{ V}}$$

b. $I_p = 40$ A $\quad V_p I_p = V_s I_s \therefore I_s = \frac{V_p I_p}{V_s}$

$I_s = ?$ $\qquad I_s = \frac{1{,}200 \text{ V} \times 40 \text{ A}}{240{,}000 \text{ V}}$

$$= \frac{1{,}200 \times 40}{240{,}000} \frac{\text{V·A}}{\text{V}}$$

$$= \boxed{0.2 \text{ A}}$$

6.12 a. $V_s = 12$ V $\qquad \frac{V_p}{N_p} = \frac{V_s}{N_s} \therefore \frac{N_p}{N_s} = \frac{V_p}{V_s}$

$I_s = 0.5$ A

$V_p = 120$ V $\qquad \frac{N_p}{N_s} = \frac{120 \text{ V}}{12 \text{ V}} = \frac{10}{1}$

$\frac{N_p}{N_s} = ?$ $\qquad\qquad$ or

$\qquad\qquad\qquad$ $\boxed{10 \text{ primary to } 1 \text{ secondary}}$

b. $I_p = ?$ $\qquad V_p I_p = V_s I_s \therefore I_p = \frac{V_s I_s}{V_p}$

$$I_p = \frac{(12 \text{ V})(0.5 \text{ A})}{120 \text{ V}}$$

$$= \frac{12 \times 0.5}{120} \frac{\text{V·A}}{\text{V}}$$

$$= \boxed{0.05 \text{ A}}$$

c. $P_s = ?$ $\qquad P_s = I_s V_s$

$$= (0.5 \text{ A})(12 \text{ V})$$

$$= 0.5 \times 12 \frac{\text{C}}{\text{s}} \times \frac{\text{J}}{\text{C}}$$

$$= 6 \frac{\text{J}}{\text{s}}$$

$$= \boxed{6 \text{ W}}$$

CHAPTER 7

7.1 The relationship between the speed of light in a transparent material (v), the speed of light in a vacuum ($c = 3.00 \times 10^8$ m/s), and the index of refraction (n) is $n = c/v$. According to table 7.1, the index of refraction for water is $n = 1.33$ and for ice is $n = 1.31$.

a. $c = 3.00 \times 10^8$ m/s

$n = 1.33$

$v = ?$

$$n = \frac{c}{v} \therefore v = \frac{c}{n}$$

$$v = \frac{3.00 \times 10^8 \text{ m/s}}{1.33}$$

$$= \boxed{2.26 \times 10^8 \text{ m/s}}$$

b. $c = 3.00 \times 10^8$ m/s

$n = 1.31$

$v = ?$

$$v = \frac{3.00 \times 10^8 \text{ m/s}}{1.31}$$

$$= \boxed{2.29 \times 10^8 \text{ m/s}}$$

7.2 $d = 1.50 \times 10^8$ km

$= 1.50 \times 10^{11}$ m

$c = 3.00 \times 10^8$ m/s

$t = ?$

$$v = \frac{d}{t} \therefore t = \frac{d}{v}$$

$$t = \frac{1.50 \times 10^{11} \text{ m}}{3.00 \times 10^8 \frac{\text{m}}{\text{s}}}$$

$$= \frac{1.50 \times 10^{11}}{3.00 \times 10^8} \text{ m} \times \frac{\text{s}}{\text{m}}$$

$$= 5.00 \times 10^2 \frac{\text{m·s}}{\text{m}}$$

$$= \frac{5.00 \times 10^2 \text{ s}}{60.0 \frac{\text{s}}{\text{min}}}$$

$$= \frac{5.00 \times 10^2}{60.0} \text{ s} \times \frac{\text{min}}{\text{s}}$$

$$= \boxed{8.33 \text{ min}}$$

7.3

$$d = 6.00 \times 10^9 \text{ km}$$

$$= 6.00 \times 10^{12} \text{ m}$$

$$c = 3.00 \times 10^8 \text{ m/s}$$

$$t = ?$$

$$v = \frac{d}{t} \therefore t = \frac{d}{v}$$

$$t = \frac{6.00 \times 10^{12} \text{ m}}{3.00 \times 10^8 \frac{\text{m}}{\text{s}}}$$

$$= \frac{6.00 \times 10^{12}}{3.00 \times 10^8} \cancel{\text{m}} \times \frac{\text{s}}{\cancel{\text{m}}}$$

$$= 2.00 \times 10^4 \text{ s}$$

$$= \frac{2.00 \times 10^4 \text{ s}}{3{,}600 \frac{\text{s}}{\text{h}}}$$

$$= \frac{2.00 \times 10^4}{3.600 \times 10^3} \cancel{\text{s}} \times \frac{\text{h}}{\cancel{\text{s}}}$$

$$= \boxed{5.56 \text{ h}}$$

7.4 From equation 7.1, note that both angles are measured from the normal and that the angle of incidence (θ_i) equals the angle of reflection (θ_r), or

$$\theta_i = \theta_r \therefore \boxed{\theta_i = 10°}$$

7.5

$$v = 2.20 \times 10^8 \text{ m/s}$$

$$c = 3.00 \times 10^8 \text{ m/s}$$

$$n = ?$$

$$n = \frac{c}{v}$$

$$= \frac{3.00 \times 10^8 \frac{\text{m}}{\text{s}}}{2.20 \times 10^8 \frac{\text{m}}{\text{s}}}$$

$$= 1.36$$

According to table 7.1, the substance with an index of refraction of 1.36 is $\boxed{\text{ethyl alcohol.}}$

7.6 **a.** From equation 7.3:

$$\lambda = 6.00 \times 10^{-7} \text{ m} \qquad c = \lambda f \therefore f = \frac{c}{\lambda}$$

$$c = 3.00 \times 10^8 \text{ m/s}$$

$$f = ?$$

$$f = \frac{3.00 \times 10^8 \frac{\text{m}}{\text{s}}}{6.00 \times 10^{-7} \text{m}}$$

$$= \frac{3.00 \times 10^8}{6.00 \times 10^{-7}} \frac{\cancel{\text{m}}}{\text{s}} \times \frac{1}{\cancel{\text{m}}}$$

$$= 5.00 \times 10^{14} \frac{1}{\text{s}}$$

$$= \boxed{5.00 \times 10^{14} \text{ Hz}}$$

b. From equation 7.4:

$$f = 5.00 \times 10^{14} \text{ Hz}$$

$$h = 6.63 \times 10^{-34} \text{ J·s}$$

$$E = ?$$

$$E = hf$$

$$= (6.63 \times 10^{-34} \text{ J·s})\left(5.00 \times 10^{14} \frac{1}{\text{s}}\right)$$

$$= (6.63 \times 10^{-34})(5.00 \times 10^{14}) \text{J·}\cancel{\text{s}} \times \frac{1}{\cancel{\text{s}}}$$

$$= \boxed{3.32 \times 10^{-19} \text{ J}}$$

7.7 First, you can find the energy of one photon of the peak intensity wavelength (5.60×10^{-7} m) by using equation 7.3 to find the frequency, then equation 7.4 to find the energy:

Step 1: $c = \lambda f \therefore f = \frac{c}{\lambda}$

$$= \frac{3.00 \times 10^8 \frac{\text{m}}{\text{s}}}{5.60 \times 10^{-7} \text{ m}}$$

$$= 5.36 \times 10^{14} \text{ Hz}$$

Step 2: $E = hf$

$$= (6.63 \times 10^{-34} \text{ J·s})(5.36 \times 10^{14} \text{ Hz})$$

$$= 3.55 \times 10^{-19} \text{J}$$

Step 3: Since one photon carries an energy of 3.55×10^{-19} J and the overall intensity is 1,000.0 W, each square meter must receive an average of

$$\dfrac{1,000.0\,\dfrac{J}{s}}{3.55 \times 10^{-19}\,\dfrac{J}{photon}}$$

$$\dfrac{1.000 \times 10^{3}}{3.55 \times 10^{-19}}\,\dfrac{\cancel{J}}{s} \times \dfrac{photon}{\cancel{J}}$$

$$\boxed{2.82 \times 10^{21}\,\dfrac{photon}{s}}$$

7.8 **a.** $f = 4.90 \times 10^{14}$ Hz $\qquad c = \lambda f \therefore \lambda = \dfrac{c}{f}$

$c = 3.00 \times 10^{8}$ m/s

$\lambda = ?$

$$\lambda = \dfrac{3.00 \times 10^{8}\,\dfrac{m}{s}}{4.90 \times 10^{14}\,\dfrac{1}{s}}$$

$$= \dfrac{3.00 \times 10^{8}}{4.90 \times 10^{14}}\,\dfrac{m}{\cancel{s}} \times \dfrac{\cancel{s}}{1}$$

$$= \boxed{6.12 \times 10^{-7}\,\text{m}}$$

b. According to table 7.2, this is the frequency and wavelength of orange light.

7.9 $f = 5.00 \times 10^{20}$ Hz

$h = 6.63 \times 10^{-34}$ J·s

$E = ?$

$E = hf$

$$= (6.63 \times 10^{-34}\,\text{J·s})\left(5.00 \times 10^{20}\,\dfrac{1}{s}\right)$$

$$= (6.63 \times 10^{-34})(5.00 \times 10^{20})\,\text{J·}\cancel{s} \times \dfrac{1}{\cancel{s}}$$

$$= \boxed{3.32 \times 10^{-13}\,\text{J}}$$

7.10 $\lambda = 1.00$ mm

$= 0.001$ m

$f = ?$

$c = 3.00 \times 10^{8}$ m/s

$h = 6.63 \times 10^{-34}$ J·s

$E = ?$

Step 1: $c = \lambda f \therefore f = \dfrac{v}{\lambda}$

$$f = \dfrac{3.00 \times 10^{8}\,\dfrac{m}{s}}{1.00 \times 10^{-3}\,m}$$

$$= \dfrac{3.00 \times 10^{8}}{1.00 \times 10^{-3}}\,\dfrac{\cancel{m}}{s} \times \dfrac{1}{\cancel{m}}$$

$$= 3.00 \times 10^{11}\,\text{Hz}$$

Step 2: $E = hf$

$$= (6.63 \times 10^{-34}\,\text{J·s})\left(3.00 \times 10^{11}\,\dfrac{1}{s}\right)$$

$$= (6.63 \times 10^{-34})(3.00 \times 10^{11})\,\text{J·}\cancel{s} \times \dfrac{1}{\cancel{s}}$$

$$= \boxed{1.99 \times 10^{-22}\,\text{J}}$$

CHAPTER 8

8.1 Energy is related to the frequency and Planck's constant in equation 8.1, $E = hf$.

For $n = 6$, $E_H = 6.05 \times 10^{-20}$ J

For $n = 2$, $E_L = 5.44 \times 10^{-19}$ J

$E = ?$ J

$E = E_H - E_L$

$$= (-6.05 \times 10^{-20}\,\text{J}) - (-5.44 \times 10^{-19}\,\text{J})$$

$$= \boxed{4.84 \times 10^{-19}\,\text{J}}$$

8.2 For $n = 6$, $E_H = -6.05 \times 10^{-20}$ J

For $n = 2$, $E_L = -5.44 \times 10^{-19}$ J

$h = 6.63 \times 10^{-34}$ J·s

$f = ?$

$$hf = E_H - E_L \therefore f = \dfrac{E_H - E_L}{h}$$

$$f = \dfrac{(-6.05 \times 10^{-20}\,\text{J}) - (-5.44 \times 10^{-19}\,\text{J})}{6.63 \times 10^{-34}\,\text{J·s}}$$

$$= \dfrac{4.84 \times 10^{-19}}{6.63 \times 10^{-34}}\,\dfrac{\cancel{J}}{\cancel{J}\text{·s}}$$

$$= 7.29 \times 10^{14}\,\dfrac{1}{s}$$

$$= \boxed{7.29 \times 10^{14}\,\text{Hz (violet)}}$$

8.3 $(n = 1) = -13.6$ eV

Since the energy of the electron is -13.6 eV, it will require 13.6 eV (or 2.17×10^{-18} J) to remove the electron.

8.4 $q/m = -1.76 \times 10^{11}$ C/kg

$q = -1.60 \times 10^{-19}$ C

$m = ?$

$$\text{mass} = \frac{\text{charge}}{\text{charge/mass}}$$

$$= \frac{-1.60 \times 10^{-19}\,\text{C}}{1.76 \times 10^{11}\,\frac{\text{C}}{\text{kg}}}$$

$$= \frac{-1.60 \times 10^{-19}}{-1.76 \times 10^{11}}\,\cancel{\text{C}} \times \frac{\text{kg}}{\cancel{\text{C}}}$$

$$= \boxed{9.09 \times 10^{-31}\,\text{kg}}$$

8.5
a. Boron: $1s^2 2s^2 2p^1$
b. Aluminum: $1s^2 2s^2 2p^6 3s^2 3p^1$
c. Potassium: $1s^2 2s^2 2p^6 3s^2 3p^6 4s^1$

8.6
a. Boron is atomic number 5 and there are 5 electrons.
b. Aluminum is atomic number 13 and there are 13 electrons.
c. Potassium is atomic number 19 and there are 19 electrons.

8.7
a. Argon: $1s^2 2s^2 2p^6 3s^2 3p^6$
b. Zinc: $1s^2 2s^2 2p^6 3s^2 3p^6 4s^2 3d^{10}$
c. Bromine: $1s^2 2s^2 2p^6 3s^2 3p^6 4s^2 3d^{10} 4p^5$

CHAPTER 9

9.1 Recall that the number of outer energy level electrons is the same as the family number for the representative elements:
a. Li: 1   d. Cl: 7
b. N: 5   e. Ra: 2
c. F: 7   f. Be: 2

9.2 The same information that was used in question 1 can be used to draw the dot notation:

(a) $\dot{\underset{\cdot}{\text{B}}}\cdot$ (c) Ca: (e) $\cdot\ddot{\underset{\cdot\cdot}{\text{O}}}\cdot$

(b) $\cdot\ddot{\underset{\cdot\cdot}{\text{Br}}}:$ (d) K$\cdot$ (f) $\cdot\ddot{\underset{\cdot\cdot}{\text{S}}}:$

9.3 The charge is found by identifying how many electrons are lost or gained in achieving the noble gas structure:
a. Boron 3+
b. Bromine 1−
c. Calcium 2+
d. Potassium 1+
e. Oxygen 2−
f. Nitrogen 3−

9.4 The name of some common polyatomic ions are in box table 9.2. Using this table as a reference, the names are
a. hydroxide
b. sulfite
c. hypochlorite
d. nitrate
e. carbonate
f. perchlorate

9.5 The Roman numeral tells you the charge on the variable-charge elements. The charges for the polyatomic ions are found in box table 9.2. The charges for metallic elements can be found in tables 9.2 and 9.3. Using these resources and the crossover technique, the formulas are as follows:
a. $Fe(OH)_3$ d. NH_4NO_3
b. $Pb_3(PO_4)_2$ e. $KHCO_3$
c. $ZnCO_3$ f. K_2SO_3

9.6 Box table 9.3 has information about the meaning of prefixes and stem names used in naming covalent compounds. (a), for example, asks for the formula of carbon tetrachloride. Carbon has no prefixes, so there is one carbon atom, and it comes first in the formula because it comes first in the name. The "tetra-" prefix means four, so there are four chlorine atoms. The name ends in "-ide," so you know there are only two elements in the compound. The symbols can be obtained from the list of elements on the inside back cover of this text. Using all this information from the name, you can think out the formula for carbon tetrachloride. The same process is used for the other compounds and formulas:
a. CCl_4 d. SO_3
b. H_2O e. N_2O_5
c. MnO_2 f. As_2S_5

9.7 Again using information from box table 9.3, this question requires you to reverse the thinking procedure you learned in question 6.
a. carbon monoxide
b. carbon dioxide
c. carbon disulfide
d. dinitrogen monoxide
e. tetraphosphorus trisulfide
f. dinitrogen trioxide

9.8
a. $2\,SO_2 + O_2 \rightarrow 2\,SO_3$
b. $4\,P + 5\,O_2 \rightarrow 2\,P_2O_5$
c. $2\,Al + 6\,HCl \rightarrow 2\,AlCl_3 + 3\,H_2$
d. $2\,NaOH + H_2SO_4 \rightarrow Na_2SO_4 + 2\,H_2O$
e. $Fe_2O_3 + 3\,CO \rightarrow 2\,Fe + 3\,CO_2$
f. $3\,Mg(OH)_2 + 2\,H_3PO_4 \rightarrow Mg_3(PO_4)_2 + 6\,H_2O$

9.9
a. General form of $XY + AZ \rightarrow XZ + AY$ with precipitate formed: Ion exchange reaction.
b. General form of $X + Y \rightarrow XY$: Combination reaction.
c. General form of $XY \rightarrow X + Y + \ldots$: Decomposition reaction.
d. General form of $X + Y \rightarrow XY$: Combination reaction.
e. General form of $XY + A \rightarrow AY + X$: Replacement reaction.
f. General form of $X + Y \rightarrow XY$: Combination reaction.

9.10 **a.** $C_5H_{12(g)} + 8\,O_{2(g)} \rightarrow 5\,CO_{2(g)} + 6\,H_2O_{(g)}$

b. $HCl_{(aq)} + NaOH_{(aq)} \rightarrow NaCl_{(aq)} + H_2O_{(l)}$

c. $2\,Al_{(s)} + Fe_2O_{3(s)} \rightarrow Al_2O_{3(s)} + 2\,Fe_{(l)}$

d. $Fe_{(s)} + CuSO_{4(aq)} \rightarrow FeSO_{4(aq)} + Cu_{(s)}$

e. $MgCl_{(aq)} + Fe(NO_3)_{2(aq)} \rightarrow$ No reaction (all possible compounds are soluble and no gas or water was formed).

f. $C_6H_{10}O_{5(s)} + 6\,O_{2(g)} \rightarrow 6\,CO_{2(g)} + 5\,H_2O_{(g)}$

CHAPTER 11

11.1 **a.** cobalt-60: 27 protons, 33 neutrons

b. potassium-40: 19 protons, 21 neutrons

c. neon-24: 10 protons, 14 neutrons

d. lead-208: 82 protons, 126 neutrons

11.2 **a.** $^{60}_{27}Co$

b. $^{40}_{19}K$

c. $^{24}_{10}Ne$

d. $^{204}_{82}Pb$

11.3 **a.** cobalt-60: Radioactive because odd numbers of protons (27) and odd numbers of neutrons (33) are usually unstable.

b. potassium-40: Radioactive, again having an odd number of protons (19) and an odd number of neutrons (21).

c. neon-24: Stable, because even numbers of protons and neutrons are usually stable.

d. lead-208: Stable, because even numbers of protons and neutrons *and* because 82 is a particularly stable number of nucleons.

11.4 **a.** $^{56}_{26}Fe \rightarrow {}^{0}_{-1}e + {}^{56}_{27}Co$

b. $^{7}_{4}Be \rightarrow {}^{0}_{-1}e + {}^{7}_{5}B$

c. $^{64}_{29}Cu \rightarrow {}^{0}_{-1}e + {}^{64}_{30}Zn$

d. $^{24}_{11}Na \rightarrow {}^{0}_{-1}e + {}^{24}_{12}Mg$

e. $^{214}_{82}Pb \rightarrow {}^{0}_{-1}e + {}^{214}_{83}Bi$

f. $^{32}_{15}P \rightarrow {}^{0}_{-1}e + {}^{32}_{16}S$

11.5 **a.** $^{235}_{92}U \rightarrow {}^{4}_{2}He + {}^{231}_{90}Th$

b. $^{226}_{88}Ra \rightarrow {}^{4}_{2}He + {}^{222}_{86}Rn$

c. $^{239}_{94}Pu \rightarrow {}^{4}_{2}He + {}^{235}_{92}U$

d. $^{214}_{83}Bi \rightarrow {}^{4}_{2}He + {}^{210}_{81}Tl$

e. $^{230}_{90}Th \rightarrow {}^{4}_{2}He + {}^{226}_{88}Ra$

f. $^{210}_{84}Po \rightarrow {}^{4}_{2}He + {}^{206}_{82}Pb$

11.6 Thirty-two days is four half-lives. After the first half-life (8 days), 1/2 oz will remain. After the second half-life (8 + 8, or 16 days), 1/4 oz will remain. After the third half-life (8 + 8 + 8, or 24 days), 1/8 oz will remain. After the fourth half-life (8 + 8 + 8 + 8, or 32 days), 1/16 oz will remain, or 6.3×10^{-2} oz.

11.7 The Fe-56 nucleus has a mass of 55.9206 u, but the individual masses of the nucleons are

$$26 \text{ protons} \times 1.00728\text{ u} = 26.1893\text{ u}$$

$$30 \text{ neutrons} \times 1.00867\text{ u} = \underline{30.2601\text{ u}}$$
$$56.4494\text{ u}$$

The mass defect is thus

$$56.4494\text{ u}$$
$$\underline{-\ 55.9206\text{ u}}$$
$$0.5288\text{ u}$$

The atomic mass unit (u) is equal to the mass of a mole (g), therefore 0.5288 u = 0.5288 g. The mass defect is equivalent to the binding energy according to $E = mc^2$. For a molar mass of Fe-56, the mass defect is

$$E = (5.29 \times 10^{-4}\text{ kg})\left(3.00 \times 10^8\,\frac{\text{m}}{\text{s}}\right)^2$$

$$= (5.29 \times 10^{-4}\text{ kg})\left(9.00 \times 10^{16}\,\frac{\text{m}^2}{\text{s}^2}\right)$$

$$= 4.76 \times 10^{13}\,\frac{\text{kg·m}^2}{\text{s}^2}$$

$$= 4.76 \times 10^{13}\text{ J}$$

For a single nucleus,

$$\frac{4.76 \times 10^{13}\text{ J}}{6.02 \times 10^{23}\text{ nuclei}} = 7.90 \times 10^{-11}\text{ J/nuclei}$$

A P P E N D I X   E

Problem Solving

Students are sometimes apprehensive when assigned problem exercises. This apprehension comes from a lack of experience in problem solving and not knowing how to proceed. This reading is concerned with a basic approach and procedures that you can follow to simplify problem-solving exercises. Thinking in terms of quantitative ideas is a skill that you can learn. Actually, no mathematics beyond addition, subtraction, multiplication, and division is needed. What is needed is knowledge of certain techniques. If you follow the suggested formatting procedures and seek help from the mathematical review appendix as needed, you will find that problem solving is a simple, fun activity that helps you to learn to think in a new way. Here are some more considerations that will prove helpful.

1. Read the problem carefully, perhaps several times, to understand the problem situation. If possible, make a sketch to help you visualize and understand the problem in terms of the real world.
2. Be alert for information that is not stated directly. For example, if a moving object "comes to a stop," you know that the final velocity is zero, even though this was not stated outright. Likewise, questions about "how far?" are usually asking a question about distance, and questions about "how long?" are usually asking a question about time. Such information can be very important in procedure step 1 (see procedure steps in right column), the listing of quantities and their symbols. Overlooked or missing quantities and symbols can make it difficult to identify the appropriate equation.
3. Understand the meaning and concepts that an equation represents. An equation represents a relationship that exists between variables. Understanding the relationship helps you to identify the appropriate equation or equations by inspection of the list of known and unknown quantities (see procedure step 2). You will find a list of the equations being considered at the end of each chapter. Information about the meaning and the concepts that an equation represents is found within each chapter.
4. Solve the equation *before* substituting numbers and units for symbols (see procedure step 3). A helpful discussion of the mathematical procedures required, with examples, is in appendix A.
5. Note if the quantities are in the same units. A mathematical operation requires the units to be the same; for example, you cannot add nickels, dimes, and quarters until you first convert them all to the same unit of money. Likewise, you cannot correctly solve a problem if one time quantity is in seconds and another time quantity is in hours. The quantities must be converted to the same units before anything else is done (see procedure step 4). There is a helpful section on how to use conversion ratios in appendix A.
6. Perform the required mathematical operations on the numbers and the units as if they were two separate problems (see procedure step 6). You will find that following this step will facilitate problem-solving activities because the units you obtain will tell you if you have worked the problem correctly. If you just write the units that you *think* should appear in the answer, you have missed this valuable self-check.
7. Be aware that not all learning takes place in a given time frame and that solutions to problems are not necessarily arrived at "by the clock." If you have spent a half an hour or so unsuccessfully trying to solve a particular problem, move on to another problem or do something entirely different for awhile. Problem solving often requires time for something to happen in your brain. If you move on to some other activity, you might find that the answer to a problem that you have been stuck on will come to you "out of the blue" when you are not even thinking about the problem. This unexpected revelation of solutions is common to many real-world professions and activities that involve thinking.

Until you are comfortable in a problem-solving situation, you should follow a formatting procedure that will help you organize your thinking. Here is an example of such a formatting procedure:

Step 1: List the quantities involved together with their symbols on the left side of the page, including the unknown quantity with a question mark.

Step 2: Inspect the given quantities and the unknown quantity as listed, and *identify* the equation that expresses a relationship between these quantities. A list of the equations that express the relationships discussed in each chapter is found at the end of that chapter. *Write* the identified equation on the right side of your paper, opposite the list of symbols and quantities.

Step 3: If necessary, *solve* the equation for the variable in question. This step must be done before substituting any numbers or units in the equation. This simplifies things and keeps down the confusion that may otherwise result. If you need help solving an equation, see the section on this topic in appendix A.

Step 4: If necessary, *convert* any unlike units so they are all the same. If the equation involves a time in seconds, for example, and a speed in kilometers per hour, you should convert the km/h to m/s. Again, this step should be done at this point to avoid confusion and incorrect operations in a later step. If you need help converting units, see the section on this topic in appendix A.

Step 5: *Substitute* the known quantities in the equation, replacing each symbol with both the number value and units represented by the symbol.

Step 6: *Perform* the required mathematical operations on the numbers and on the units. This performance is less confusing if you first separate the numbers and units, as shown in the following example and in the examples throughout this text, and then perform the mathematical operations on the numbers and units as separate steps, showing all work.

Step 7: *Draw a box* around your answer (numbers and units) to communicate that you have found what you were looking for. The box is a signal that you have finished your work on this problem.

EXAMPLE PROBLEM

Mercury is a liquid metal with a mass density of 13.6 g/cm³. What is the mass of 10.0 cm³ of mercury?

SOLUTION

The problem gives two known quantities, the mass density (ρ) of mercury and a known volume (V), and identifies an unknown quantity, the mass (m) of that volume. Make a list of these quantities:

$$\rho = 13.6 \text{ g/cm}^3$$

$$V = 10.0 \text{ cm}^3$$

$$m = ?$$

The appropriate equation for this problem is the relationship between mass density (ρ), mass (m), and volume (V):

$$\rho = \frac{m}{V}$$

The unknown in this case is the mass, m. Solving the equation for m, by multiplying both sides by V, gives:

$$V\rho = \frac{mV}{V}$$

$$V\rho = m, \text{ or}$$

$$m = V\rho$$

Now you are ready to substitute the known quantities in the equation and perform the mathematical operations on the numbers and on the units:

$$m = \left(13.6 \, \frac{\text{g}}{\text{cm}^3}\right)(10.0 \text{ cm}^3)$$

$$= (13.6)(10.0)\left(\frac{\text{g}}{\text{cm}^3}\right)(\text{cm}^3)$$

$$= 136 \, \frac{\text{g} \cdot \cancel{\text{cm}^3}}{\cancel{\text{cm}^3}}$$

$$= \boxed{136\text{g}}$$

A

abiotic factors nonliving parts of an organism's environment

absolute humidity a measure of the actual amount of water vapor in the air at a given time—for example, in grams per cubic meter

absolute magnitude a classification scheme to compensate for the distance differences to stars; calculations of the brightness that stars would appear to have if they were all at a defined, standard distance

absolute scale temperature scale set so that zero is at the theoretical lowest temperature possible, which would occur when all random motion of molecules has ceased

absolute zero the theoretical lowest temperature possible, which occurs when all random motion of molecules has ceased

abyssal plain the practically level plain of the ocean floor

acceleration a change in velocity per change in time; by definition, this change in velocity can result from a change in speed, a change in direction, or a combination of changes in speed and direction

accretion disk fat bulging disk of gas and dust from the remains of the gas cloud that forms around a protostar

acetylcholine a neurotransmitter secreted into the synapse by many axons and received by dendrites

acetylcholinesterase an enzyme present in the synapse that destroys acetylcholine

achondrites homogeneously textured stony meteorites

acid any substance that is a proton donor when dissolved in water; generally considered a solution of hydronium ions in water that can neutralize a base, forming a salt and water

acid-base indicator a vegetable dye used to distinguish acid and base solutions by a color change

acquired characteristics characteristics an organism gains during its lifetime that are not genetically determined and therefore cannot be passed on to future generations

active transport use of a carrier molecule to move molecules through a cell membrane in a direction opposite that of the concentration gradient; the carrier requires an input of energy other than the kinetic energy of the molecules

adenine a double-ring nitrogenous-base molecule in DNA and RNA; the complementary base of thymine or uracil

adenosine triphosphate (ATP) a molecule formed from the building blocks of adenine, ribose, and phosphates; it functions as the primary energy carrier in the cell

aerobic cellular respiration the biochemical pathway that requires oxygen and converts food, such as carbohydrates, to carbon dioxide and water; during this conversion, it releases the chemical-bond energy as ATP molecules

air mass a large, more or less uniform body of air with nearly the same temperature and moisture conditions throughout

air mass weather the weather experienced within a given air mass; characterized by slow, gradual changes from day to day

alcohol an organic compound with a general formula of ROH, where R is one of the hydrocarbon groups—for example, methyl or ethyl

aldehyde an organic molecule with the general formula RCHO, where R is one of the hydrocarbon groups—for example, methyl or ethyl

alkali metals members of family IA of the periodic table, having common properties of shiny, low-density metals that can be cut with a knife and that react violently with water to form an alkaline solution

alkaline earth metals members of family IIA of the periodic table, having common properties of soft, reactive metals that are less reactive than alkali metals

alkanes hydrocarbons with single covalent bonds between the carbon atoms

alkenes hydrocarbons with a double covalent carbon-carbon bond

alkyne hydrocarbon with a carbon-carbon triple bond

alleles alternative forms of a gene for a particular characteristic (e.g., attached-earlobe and free-earlobe are alternative alleles for ear shape)

alpha particle the nucleus of a helium atom (two protons and two neutrons) emitted as radiation from a decaying heavy nucleus; also known as an alpha ray

alpine glaciers glaciers that form at high elevations in mountainous regions

alternating current an electric current that first moves one direction, then the opposite direction with a regular frequency

alternation of generations a term used to describe that aspect of the life cycle in which there are two distinctly different forms of an organism; each form is involved in the production of the other and only one form is involved in producing gametes

alveoli tiny sacs that are part of the structure of the lungs where gas exchange takes place

amino acids organic molecules that join to form polypeptides and proteins

amp unit of electric current; equivalent to C/s

ampere full name of the unit amp

amplitude the extent of displacement from the equilibrium condition; the size of a wave from the rest (equilibrium) position

anaphase the third stage of mitosis, characterized by dividing of the centromeres and movement of the chromosomes to the poles

androgens male sex hormones produced by the testes that cause the differentiation of typical internal and external genital male anatomy

angle of incidence angle of an incident (arriving) ray or particle to a surface; measured from a line perpendicular to the surface (the normal)

angle of reflection angle of a reflected ray or particle from a surface; measured from a line perpendicular to the surface (the normal)

angular momentum quantum number in the quantum mechanics model of the atom, one of four descriptions of the energy state of an electron wave; this quantum number describes the energy sublevels of electrons within the main energy levels of an atom

anorexia nervosa a nutritional deficiency disease characterized by severe, prolonged weight loss for fear of becoming obese

anther the sex organ in plants that produces the pollen that contains the sperm

anticline an arch-shaped fold in layered bed rock

anticodon a sequence of three nitrogenous bases on a tRNA molecule capable of forming hydrogen bonds with three complementary bases on an mRNA codon during translation

anticyclone a high-pressure center with winds flowing away from the center; associated with clear, fair weather

antinode region of maximum amplitude between adjacent nodes in a standing wave

aorta the large blood vessel that carries blood from the left ventricle to the majority of the body

apoptosis also known as "programmed cell death"; death of specific cells that has a genetic basis and is not the result of injury

apparent magnitude a classification scheme for different levels of brightness of stars that you see; brightness values range from one to six with the number one (first magnitude) assigned to the brightest star and the number six (sixth magnitude) assigned to the faintest star that can be seen

aquifer a layer of sand, gravel, or other highly permeable material beneath the surface that is saturated with water and is capable of producing water in a well or spring

Archaea the domain in which are found prokaryotic organisms that live in extreme habitats

archaeology the study of past human life and culture from material evidence

area the extent of a surface; the surface bounded by three or more lines

arid dry climate classification; receives less than 25 cm (10 in) precipitation per year

aromatic hydrocarbon organic compound with at least one benzene ring structure; cyclic hydrocarbons and their derivatives

arteries the blood vessels that carry blood away from the heart

arterioles small arteries located just before capillaries that can expand and contract to regulate the flow of blood to parts of the body

artesian term describing the condition where confining pressure forces groundwater from a well to rise above the aquifer

asbestos the common name for any one of several incombustible fibrous minerals that will not melt or ignite and can be woven into a fireproof cloth or used directly in fireproof insulation; about six different commercial varieties of asbestos are used, one of which has been linked to cancer under heavy exposure

assimilation the physiological process that takes place in a living cell as it converts nutrients in food into specific molecules required by the organism

asteroids small rocky bodies left over from the formation of the solar system; most are accumulated in a zone between the orbits of Mars and Jupiter

asthenosphere a plastic, mobile layer of the earth's structure that extends around the earth below the lithosphere

astronomical unit the radius of the earth's orbit is defined as one astronomical unit (AU)

atom the smallest unit of an element that can exist alone or in combination with other elements

atomic mass unit relative mass unit (u) of an isotope based on the standard of the carbon-12 isotope, which is defined as a mass of exactly 12.00 u; one atomic mass unit (1 u) is 1/12 the mass of a carbon-12 atom

atomic number the number of protons in the nucleus of an atom

atomic weight weighted average of the masses of stable isotopes of an element as they occur in nature, based on the abundance of each isotope of the element and the atomic mass of the isotope compared to carbon-12

atria thin-walled sacs of the heart that receive blood from the veins of the body and empty into the ventricles; singular, atrium

atrioventricular valves located between the atria and ventricles of the heart preventing blood from flowing backward from the ventricles into the atria

autosomes chromosomes that typically carry genetic information used by the organism for characteristics other than the primary determination of sex

autotroph an organism that is able to make its food molecules from inorganic raw materials by using basic energy sources such as sunlight

axis the imaginary line about which a planet or other object rotates

axon a neuronal fiber that carries information away from the nerve cell body

B

background radiation ionizing radiation (alpha, beta, gamma, etc.) from natural sources; between 100 and 500 millirems/yr of exposure to natural radioactivity from the environment

Balmer series a set of four line spectra, narrow lines of color emitted by hydrogen atom electrons as they drop from excited states to the ground state

band of stability a region of a graph of the number of neutrons versus the number of protons in nuclei; nuclei that have the neutron to proton ratios located in this band do not undergo radioactive decay

barometer an instrument that measures atmospheric pressure, used in weather forecasting and in determining elevation above sea level

basal metabolic rate (BMR) the amount of energy required to maintain normal body activity while at rest

base any substance that is a proton acceptor when dissolved in water; generally considered a solution that forms hydroxide ions in water that can neutralize an acid, forming a salt and water

basilar membrane a membrane in the cochlea containing sensory cells that are stimulated by the vibrations caused by sound waves

basin a large, bowl-shaped fold in the land into which streams drain; also a small enclosed or partly enclosed body of water

beat rhythmic increases and decreases of volume from constructive and destructive interference between two sound waves of slightly different frequencies

benthic aquatic organisms that live on the bottom

beta particle high-energy electron emitted as ionizing radiation from a decaying nucleus; also known as a beta ray

big bang theory current model of galactic evolution in which the universe was created from an intense and brilliant explosion from a primeval fireball

bile a product of the liver, stored in the gallbladder, which is responsible for the emulsification of fats

binding energy the energy required to break a nucleus into its constituent protons and neutrons; also the energy equivalent released when a nucleus is formed

binomial system of nomenclature a naming system that uses two Latin names, genus and specific epithet, for each type of organism

biodiversity the number of different kinds of organisms found in an area

bioengineering also referred to as *genetic engineering;* the process of purposely altering the genetic makeup of organisms

biogenesis the concept that life originates only from preexisting life

biological species concept the concept that species can be distinguished from one another by their inability to interbreed

biology the science that deals with the study of living things and how living things interact with things around them

biomes large regional communities primarily determined by climate

biotechnology the science of gene manipulation

biotic factors living parts of an organism's environment

black hole the theoretical remaining core of a supernova with an intense gravitational field

blackbody radiation electromagnetic radiation emitted by an ideal material (the blackbody) that perfectly absorbs and perfectly emits radiation

blood the fluid medium consisting of cells and plasma that assists in the transport of materials and heat

Bohr model model of the structure of the atom that attempted to correct the deficiencies of the solar system model and account for the Balmer series

boiling point the temperature at which a phase change of liquid to gas takes place through boiling; the same temperature as the condensation point

boundary the division between two regions of differing physical properties

breaker a wave whose front has become so steep that the top part has broken forward of the wave, breaking into foam, especially against a shoreline

British thermal unit the amount of energy or heat needed to increase the temperature of 1 pound of water 1 degree Fahrenheit (abbreviated Btu)

bronchi major branches of the trachea that ultimately deliver air to bronchioles in the lungs

bronchioles small tubes that deliver air to the alveoli in the lung; they are capable of contracting

bulimia a nutritional deficiency disease characterized by a binge-and-purge cycle of eating

C

calorie the amount of energy (or heat) needed to increase the temperature of 1 gram of water 1 degree Celsius

Calorie the nutritionist's "calorie"; equivalent to 1 kilocalorie

cancer any abnormal growth of cells that has a malignant potential

capillaries tiny blood vessels through the walls of which exchange between cells and the blood takes place

carbohydrates organic compounds that include sugars, starches, and cellulose; carbohydrates are used by plants and animals for structure, protection, and food

carbon cycle the series of processes and organisms through which carbon atoms pass in ecological systems

carbon film a type of fossil formed when the volatile and gaseous constituents of a buried organic structure are distilled away, leaving a carbon film as a record

carbonation in chemical weathering, a reaction that occurs naturally between carbonic acid (H_2CO_3) and rock minerals

carcinogens agents responsible for causing cancer

carnivores animals that eat other animals

carrier (genetic) any individual having a hidden, recessive allele

carrier protein proteins that pick up and deliver molecules from one place in a cell or multicellular organism and transport them to another; for example, certain blood proteins

carrying capacity (ecosystem) the *optimum* population size an area can support over an extended period of time

cast sediments deposited by groundwater in a mold, taking the shape and external features of the organism that was removed to form the mold, then gradually changing to sedimentary rock

cathode rays negatively charged particles (electrons) that are emitted from a negative terminal in an evacuated glass tube

cell the basic structural unit that makes up all living things

cell membrane the outer-boundary membrane of the cell; also known as the *plasma membrane*

cell plate a plant-cell structure that begins to form in the center of the cell and proceeds to the cell membrane, resulting in cytokinesis

cell wall an outer covering on some cells; may be composed of cellulose, chitin, or peptidoglycan depending on the kind of organism

cellulose a polysaccharide abundant in plants that forms the fibers in cell walls that provide structure for plant materials

Celsius scale referent scale that defines numerical values for measuring hotness or coldness, defined as degrees of temperature; based on the reference points of the freezing point of water and the boiling point of water at sea-level pressure, with 100 degrees between the two points

cementation process by which spaces between buried sediment particles under compaction are filled with binding chemical deposits, binding the particles into a rigid, cohesive mass of a sedimentary rock

Cenozoic the most recent geologic era; the time of recent life, meaning the fossils of this era are identical to the life found on the earth today

central nervous system the portion of the nervous system consisting of the brain and spinal cord

centrifugal force an apparent outward force on an object following a circular path that is a consequence of the third law of motion

centrioles organelles composed of microtubules located just outside the nucleus

centripetal force the force required to pull an object out of its natural straight-line path and into a circular path; centripetal means "center seeking"

centromere the unreplicated region where two chromatids are joined

Cepheid variables a bright variable star that can be used to measure distance

cerebellum a region of the brain connected to the medulla oblongata that receives many kinds of sensory stimuli and coordinates muscle movement

cerebrum a region of the brain that surrounds most of the other parts of the brain and is involved in consciousness and thought

chain reaction a self-sustaining reaction where some of the products are able to produce more reactions of the same kind; in a nuclear chain reaction neutrons are the products that produce more nuclear reactions in a self-sustaining series

chemical bond an attractive force that holds atoms together in a compound

chemical change a change in which the identity of matter is altered and new substances are formed

chemical energy a form of energy involved in chemical reactions associated with changes in internal potential energy; a kind of potential energy that is stored and later released during a chemical reaction

chemical equation concise way of describing what happens in a chemical reaction

chemical reaction a change in matter where different chemical substances are created by forming or breaking chemical bonds

chemical weathering the breakdown of minerals in rocks by chemical reactions with water, gases of the atmosphere, or solutions

chemistry the science concerned with the study of the composition, structure, and properties of substances and the transformations they undergo

Chinook a warm wind that has been warmed by compression; also called Santa Ana

chloroplast an energy-converting, membranous, saclike organelle in plant cells containing the green pigment chlorophyll

chondrites subdivision of stony meteorites containing small spherical lumps of silicate minerals or glass

chondrules small spherical lumps of silicate minerals or glass found in some meteorites

chromatid one of two component parts of a chromosome formed by replication and attached at the centromere

chromatin areas or structures within the nucleus of a cell composed of long, loosely arranged molecules of deoxyribonucleic acid (DNA) in association with proteins

chromatin fibers see nucleoproteins

chromosomal aberrations changes in the gene arrangement in chromosomes; for example, translocation and duplication mutations

chromosomes complex tightly coiled structures within the nucleus composed of various kinds of histone proteins and DNA that contains the cell's genetic information

class a group of closely related orders found within a phylum

cleavage furrow an indentation of the cell membrane of an animal cell that pinches the cytoplasm into two parts

climate the general pattern of weather that occurs in a region over a number of years

coalescence process (meteorology) the process by which large raindrops form from the merging and uniting of millions of tiny water droplets

cochlea the part of the ear that converts sound into nerve impulses

coding strand one of two DNA strands that serves as a template, or pattern, for the synthesis of RNA

codominance when an organism has two different alleles for a trait and they both express themselves

codon a sequence of three nucleotides of an mRNA molecule that directs the placement of a particular amino acid during translation

coitus see *sexual intercourse*

cold front the front that is formed as a cold air mass moves into warmer air

combination chemical reaction a synthesis reaction in which two or more substances combine to form a single compound

comets celestial objects originating from the outer edges of the solar system that move about the sun in highly elliptical orbits; solar heating and pressure from the solar wind form a tail on the comet that points away from the sun

commensalism a relationship between two organisms in which one organism is helped and the other is not affected

community all of the kinds of interacting organisms within an ecosystem

compaction the process of pressure from a depth of overlying sediments squeezing together the deeper sediments and squeezing out water

competition a kind of interaction between organisms in which both organisms are harmed to some extent

complementary base a nitrogenous base in nucleic acids that can form hydrogen bonds with another base of a specific nucleotide

complex carbohydrates macromolecules formed as a result of the chemical combining of simple sugars (monomers) to form polysaccharides; for example, starch

compound a pure chemical substance that can be decomposed by a chemical change into simpler substances with a fixed mass ratio

compressive stress a force that tends to compress the surface as the earth's plates move into each other

concentration an arbitrary description of the relative amounts of solute and solvent in a solution; a larger amount of solute makes a concentrated solution, and a small amount of solute makes a dilute concentration

conception fertilization

condensation (sound) a compression of gas molecules; a pulse of increased density and pressure that moves through the air at the speed of sound

condensation (water vapor) when more vapor or gas molecules are returning to the liquid state than are evaporating

condensation nuclei tiny particles such as tiny dust, smoke, soot, and salt crystals that are suspended in the air on which water condenses

condensation point the temperature at which a gas or vapor changes back to a liquid

conduction the transfer of heat from a region of higher temperature to a region of lower temperature by increased kinetic energy moving from molecule to molecule

cones light-sensitive cells in the retina of the eye that respond to different colors of light

constructive interference the condition in which two waves arriving at the same place, at the same time and in phase, add amplitudes to create a new wave

consumers organisms that must obtain energy in the form of organic matter

continental air mass dry air masses that form over large land areas

continental climate a climate influenced by air masses from large land areas; hot summers and cold winters

continental drift a concept that continents shift positions on the earth's surface, moving across the surface rather than being fixed, stationary landmasses

continental glaciers glaciers that cover a large area of a continent; for example, Greenland and the Antarctic

continental shelf a feature of the ocean floor; the flooded margins of the continents that form a zone of relatively shallow water adjacent to the continents

continental slope a feature of the ocean floor; a steep slope forming the transition between the continental shelf and the deep ocean basin

control group the situation used as the basis for comparison in a controlled experiment

control processes mechanisms that ensure that an organism will carry out all metabolic activities in the proper sequence (coordination) and at the proper rate (regulation)

control rods rods inserted between fuel rods in a nuclear reactor to absorb neutrons and thus control the rate of the nuclear chain reaction

controlled experiment an experiment that allows for a comparison of two events that are identical in all but one respect

convection transfer of heat from a region of higher temperature to a region of lower temperature by the displacement of high-energy molecules—for example, the displacement of warmer, less dense air (higher kinetic energy) by cooler, more dense air (lower kinetic energy)

convection cell complete convective circulation pattern; also, slowly turning regions in the plastic asthenosphere that might drive the motion of plate tectonics

convection zone (of a star) part of the interior of a star according to a model; the region directly above the radiation zone where gases are heated by the radiation zone below and move upward by convection to the surface, where they emit energy in the form of visible light, ultraviolet radiation, and infrared radiation

convergent boundaries boundaries that occur between two plates moving toward each other

copulation see *sexual intercourse*

core (of a star) dense, very hot region of a star where nuclear fusion reactions release gamma and X-ray radiation

core (of the earth) the center part of the earth, which consists of a solid inner part and liquid outer part; makes up about 15 percent of the earth's total volume and about one-third of its mass

Coriolis effect the apparent deflection due to the rotation of the earth; it is to the right in the Northern Hemisphere

corpus luteum a glandlike structure produced in the ovary that produces hormones (progesterone and estrogen) that prevent the release of other eggs

correlation the determination of the equivalence in geologic age by comparing the rocks in two separate locations

coulomb unit used to measure quantity of electric charge; equivalent to the charge resulting from the transfer of 6.24 billion particles such as the electron

Coulomb's law relationship between charge, distance, and magnitude of the electrical force between two bodies

covalent bond a chemical bond formed by the sharing of a pair of electrons

covalent compound chemical compound held together by a covalent bond or bonds

creep the slow downhill movement of soil down a steep slope

crest the high mound of water that is part of a wave; also refers to the condensation, or high-pressure part, of a sound wave

critical angle limit to the angle of incidence when all light rays are reflected internally

critical mass mass of fissionable material needed to sustain a chain reaction

crossing-over the process that occurs when homologous chromosomes exchange equivalent sections of DNA during meiosis

crude oil petroleum pumped from the ground that has not yet been refined into usable products

crust the outermost part of the earth's interior structure; the thin, solid layer of rock that rests on top of the Mohorovicic discontinuity

curie unit of nuclear activity defined as 3.70×10^{10} nuclear disintegrations per second

cycle a complete vibration

cyclone a low-pressure center where the winds move into the low-pressure center and are forced upward; a low-pressure center with clouds, precipitation, and stormy conditions

cytokinesis division of the cytoplasm of one cell into two new cells

cytoplasm the more fluid portion of the protoplasm that surrounds the nucleus

cytosine a single-ring nitrogenous-base molecule in DNA and RNA; complementary to guanine

D

dark energy a recently discovered mystery force that is apparently accelerating the expanding universe

dark matter missing matter of the universe that is believed to exist, but is invisible and unseen

data measurement information used to describe something

data points points that may be plotted on a graph to represent simultaneous measurements of two related variables

daughter cells two cells formed by cell division

daughter chromosomes chromosomes produced by DNA replication that contain identical genetic information; formed after chromosome division in anaphase

daughter nuclei two nuclei formed by mitosis

decibel scale a nonlinear scale of loudness based on the ratio of the intensity level of a sound to the intensity at the threshold of hearing

decomposers organisms that use dead organic matter as a source of energy

decomposition chemical reaction a chemical reaction in which a compound is broken down into the elements that make up the compound, into simpler compounds, or into elements and simpler compounds

deflation the widespread removal of base materials from the surface by the wind

delta a somewhat triangular deposit at the mouth of a river formed where a stream flowing into a body of water slowed and lost its sediment-carrying ability

denature a change in the chemical and physical properties of a molecule as a result of the breaking of chemical bonds within the molecule; for example, the change in egg white as a result of cooking

dendrites neuronal fibers that receive information from axons and carry it toward the nerve-cell body

denitrifying bacteria bacteria capable of converting nitrite (NO_2^-) to nitrogen gas (N_2), which is released into the atmosphere

density the compactness of matter described by a ratio of mass per unit volume

deoxyribonucleic acid (DNA) a polymer of nucleotides that serves as genetic information; in prokaryotic cells, it is a double-stranded DNA loop and contains attached HU proteins and in eukaryotic cells, it is found in strands with attached histone proteins—when tightly coiled, it is known as a chromosome

deoxyribose a 5-carbon sugar molecule; a component of DNA

depolarized having lost the electrical difference existing between two points or objects

destructive interference the condition in which two waves arriving at the same point at the same time out of phase add amplitudes to create zero total disturbance

dew condensation of water vapor into droplets of liquid on surfaces

dew point temperature the temperature at which condensation begins

diaphragm a muscle separating the lung cavity from the abdominal cavity that is involved in exchanging the air in the lungs

diastolic blood pressure the blood pressure recorded while the heart is relaxing

diastrophism all-inclusive term that means any and all possible movements of the earth's plates, including drift, isostatic adjustment, and any other process that deforms or changes the earth's surface by movement

diet the food and drink consumed by a person from day to day

differentiation the process of forming specialized cells within a multicellular organism

diffuse reflection light rays reflected in many random directions, as opposed to the parallel rays reflected from a perfectly smooth surface such as a mirror

diffusion net movement of a kind of molecule from an area of higher concentration to an area of lesser concentration

diploid having two sets of chromosomes: one set from the maternal parent and one set from the paternal parent

direct current an electrical current that always moves in one direction

direct proportion when two variables increase or decrease together in the same ratio (at the same rate)

disaccharides two monosaccharides joined together with the loss of a water molecule; examples of disaccharides are sucrose (table sugar), lactose, and maltose

dispersion the effect of spreading colors of light into a spectrum with a material that has an index of refraction that varies with wavelength

divergent boundaries boundaries that occur between two plates moving away from each other

divide line separating two adjacent watersheds

DNA code a sequence of three nucleotides of a DNA molecule

DNA polymerase an enzyme that bonds DNA nucleotides together when they base pair with an existing DNA strand

DNA replication the process by which the genetic material (DNA) of the cell reproduces itself prior to its distribution to the next generation of cells

domain a classification group above that of kingdoms

dome a large, upwardly bulging, symmetrical fold that resembles a dome

dominant allele an allele that expresses itself and masks the effects of other alleles for the trait

Doppler effect an apparent shift in the frequency of sound or light due to relative motion between the source of the sound or light and the observer

double bond covalent bond formed when two pairs of electrons are shared by two atoms

dune a hill, low mound, or ridge of wind-blown sand or other sediments

duodenum the first part of the small intestine, which receives food from the stomach and secretions from the liver and pancreas

E

earthquake a quaking, shaking, vibrating, or upheaval of the earth's surface

earthquake epicenter point on the earth's surface directly above an earthquake focus

earthquake focus place where seismic waves originate beneath the surface of the earth

echo a reflected sound that can be distinguished from the original sound, which usually arrives 0.1 s or more after the original sound

eclipse when the shadow of a celestial body falls on the surface of another celestial body

ecology the branch of biology that studies the relationships between organisms and their environment

ecosystem an interacting unit consisting of a collection of organisms and abiotic factors

egg the haploid sex cell produced by sexually mature females

ejaculation the release of sperm cells and seminal fluid through the penis

El Niño changes in the atmospheric pressure systems, ocean currents, water temperatures, and wind patterns that seem to be linked to worldwide changes in the weather

elastic rebound the sudden snap of stressed rock into new positions; the recovery from elastic strain that results in an earthquake

elastic strain an adjustment to stress in which materials recover their original shape after a stress is released

electric charge a fundamental property of electrons and protons; electrons have a negative electric charge and protons have a positive electric charge

electric circuit consists of a voltage source that maintains an electrical potential, a continuous conducting path for a current to follow, and a device where work is done by the electrical potential; a switch in the circuit is used to complete or interrupt the conducting path

electric current the flow of electric charge

electric field force field produced by an electrical charge

electric field lines a map of an electric field representing the direction of the force that a test charge would experience; the direction of an electric field shown by lines of force

electric generator a mechanical device that uses wire loops rotating in a magnetic field to produce electromagnetic induction in order to generate electricity

electrical energy a form of energy from electromagnetic interactions; one of five forms of energy—mechanical, chemical, radiant, electrical, and nuclear

electrical force a fundamental force that results from the interaction of electrical charge and is billions and billions of times stronger than the gravitational force; sometimes called the "electromagnetic force" because of the strong association between electricity and magnetism

electrical insulators electrical nonconductors, or materials that obstruct the flow of electric current

electrical nonconductors materials that have electrons that are not moved easily within the material—for example, rubber; electrical nonconductors are also called electrical insulators

electrolyte water solution of ionic substances that conducts an electric current

electromagnet a magnet formed by a solenoid that can be turned on and off by turning the current on and off

electromagnetic force one of four fundamental forces; the force of attraction or repulsion between two charged particles

electromagnetic induction process in which current is induced by moving a loop of wire in a magnetic field or by changing the magnetic field

electron subatomic particle that has the smallest negative charge possible and usually found in an orbital of an atom, but gained or lost when atoms become ions

electron configuration the arrangement of electrons in orbitals and suborbitals about the nucleus of an atom

electron dot notation notation made by writing the chemical symbol of an element with dots around the symbol to indicate the number of outer orbital electrons

electron pair a pair of electrons with different spin quantum numbers that may occupy an orbital

electrostatic charge an accumulated electric charge on an object from a surplus or deficiency of electrons; also called "static electricity"

element a pure chemical substance that cannot be broken down into anything simpler by chemical or physical means; there are over 100 known elements, the fundamental materials of which all matter is made

endocrine glands secrete chemical messengers into the circulatory system

endocrine system a number of glands that communicate with one another and other tissues through chemical messengers transported throughout the body by the circulatory system

endoplasmic reticulum (ER) folded membranes and tubes throughout the eukaryotic cell that provide a large surface upon which chemical activities take place

endosymbiotic theory a theory suggesting that some organelles found in eukaryotic cells may have originated as free-living prokaryotes

energy the ability to do work

English system a system of measurement that originally used sizes of parts of the human body as referents

entropy the measure of disorder in thermodynamics

environment the surroundings; anything that affects an organism during its lifetime

environmental resistance the collective set of factors that limit population growth

enzymes protein molecules, produced by organisms, that are able to control the rate at which chemical reactions occur

eon the largest blocks of time in the earth's geologic history—the Hadean, Archean, Proterozoic, and Phanerozoic eons

epinephrine a hormone produced by the adrenal medulla that increases heart rate, blood pressure, and breathing rate

epiphyte a plant that lives on the surface of another plant without doing harm

epochs subdivisions of geologic periods

equation a statement that describes a relationship in which quantities on one side of the equal sign are identical to quantities on the other side

eras the major blocks of time in the earth's geologic history; the Cenozoic, Mesozoic, Paleozoic, and Precambrian

erosion the process of physically removing weathered materials, for example, rock fragments are physically picked up by an erosion agent such as a stream or a glacier

essential amino acids those amino acids that cannot be synthesized by the human body and must be part of the diet (e.g., lysine, tryptophan, and valine)

esters class of organic compounds with the general structure of RCOOR', where R is one of the hydrocarbon groups—for example,

methyl or ethyl; esters make up fats, oils, and waxes and some give fruit and flowers their taste and odor

estrogen one of the female sex hormones that cause the differentiation of typical female internal and external genital anatomy; responsible for the changes in breasts, vagina, uterus, clitoris, and pelvic bone structure at puberty

estuary a special category of aquatic ecosystem that consists of shallow, partially enclosed areas where freshwater enters the ocean

ether class of organic compounds with the general formula ROR', where R is one of the hydrocarbon groups—for example, methyl or ethyl; mostly used as industrial and laboratory solvents

Eubacteria the domain in which are found organisms commonly known as bacteria

Eucarya the domain in which are found all eukaryotic organisms

eukaryote an organism composed of cells possessing a nuclear membrane and other membranous organelles

eukaryotic cells one of the two major types of cells; characterized by cells that have a true nucleus, as in plants, fungi, protists, and animals

euphotic zone the upper layer of the ocean, where the sun's rays penetrate

evaporation process of more molecules leaving a liquid for the gaseous state than returning from the gas to the liquid; can occur at any given temperature from the surface of a liquid

evolution the continuous genetic adaptation of a population of organisms to its environment over time

excited states as applied to an atom, describes the energy state of an atom that has electrons in a state above the minimum energy state for that atom; as applied to a nucleus, describes the energy state of a nucleus that has particles in a state above the minimum energy state for that nuclear configuration

exfoliation the fracturing and breaking away of curved, sheet-like plates from bare rock surfaces via physical or chemical weathering, resulting in dome-shaped hills and rounded boulders

exocrine glands use ducts to secrete to the surface of the body or into hollow organs of the body, for example, sweat glands or pancreas

exosphere the outermost layer of the atmosphere, where gas molecules merge with the diffuse vacuum of space

experiment a re-creation of an event in a way that enables a scientist to gain valid and reliable empirical evidence

experimental group the group in a controlled experiment that is identical to the control group in all respects but one

exponential growth phase a period of time during population growth when the population increases at an accelerating rate

external energy the total potential and kinetic energy of an everyday-sized object

external parasite a parasite that lives on the outside of its host

F

facilitated diffusion diffusion assisted by carrier molecules

Fahrenheit scale referent scale that defines numerical values for measuring hotness or coldness, defined as degrees of temperature; based on the reference points of the freezing point of water and the boiling point of water at sea-level pressure, with 180 degrees between the two points

family (elements) vertical columns of the periodic table consisting of elements that have similar properties

family (organisms) a group of closely related genera within an order

fats organic compounds of esters formed from glycerol and three long-chain carboxylic acids that are also called triglycerides; called fats in animals and oils in plants

fatty acid one of the building blocks of true fats and phospholipids; composed of a long-chain carbon skeleton with a carboxylic acid functional group at one end; for example, linoleic acid

fault a break in the continuity of a rock formation along which relative movement has occurred between the rocks on either side

fault plane the surface along which relative movement has occurred between the rocks on either side; the surface of the break in continuity of a rock formation

ferromagnesian silicates silicates that contain iron and magnesium; examples include the dark-colored minerals olivine, augite, hornblende, and biotite

fertilization the joining of haploid nuclei, usually from an egg and a sperm cell, resulting in a diploid cell called the zygote

fiber natural (plant) or industrially produced polysaccharides that are resistant to hydrolysis by human digestive enzymes

first law of motion every object remains at rest or in a state of uniform straight-line motion unless acted on by an unbalanced force

first law of thermodynamics a statement of the law of conservation of energy in the relationship between internal energy, work, and heat

floodplain the wide, level floor of a valley built by a stream; the river valley where a stream floods when it spills out of its channel

fluids matter that has the ability to flow or be poured; the individual molecules of a fluid are able to move, rolling over or by one another

folds bends in layered bed rock as a result of stress or stresses that occurred when the rock layers were in a ductile condition, probably under considerable confining pressure from deep burial

foliation the alignment of flat crystal flakes of a rock into parallel sheets

follicle the saclike structure near the surface of the ovary that encases the soon-to-be-released secondary oocyte

follicle-stimulating hormone (FSH) the pituitary secretion that causes the ovaries to begin to produce larger quantities of estrogen and to develop the follicle and prepare the egg for ovulation

food chain a sequence of organisms that feed on one another, resulting in a flow of energy from a producer through a series of consumers

Food Guide Pyramid a tool developed by the U.S. Department of Agriculture to help the general public plan for good nutrition; guidelines for required daily intake from each of five food groups

food web a system of interlocking food chains

force a push or pull capable of changing the state of motion of an object; since a force has magnitude (strength) as well as direction, it is a vector quantity

formula describes what elements are in a compound and in what proportions

formula weight the sum of the atomic weights of all the atoms in a chemical formula

fossil any evidence of former life

fossil fuels organic fuels that contain the stored radiant energy of the sun converted to chemical energy by plants or animals that lived millions of years ago; coal, petroleum, and natural gas are the common fossil fuels

Foucault pendulum a heavy mass swinging from a long wire that can be used to provide evidence about the rotation of the earth

fovea centralis the area of sharpest vision on the retina, containing only cones, where light is sharply focused

fracture strain an adjustment to stress in which materials crack or break as a result of the stress

free fall when objects fall toward the earth with no forces acting upward; air resistance is neglected when considering an object to be in free fall

free-living nitrogen-fixing bacteria soil bacteria that provide nitrogen compounds that can be taken up by plants through their roots

freezing point the temperature at which a phase change of liquid to solid takes place; the same temperature as the melting point for a given substance

frequency the number of cycles of a vibration or of a wave occurring in one second, measured in units of cycles per second (hertz)

freshwater water that is not saline and is fit for human consumption

freshwater communities aquatic communities that have low salt concentrations

front the boundary, or thin transition zone, between air masses of different temperatures

frost ice crystals formed by water vapor condensing directly from the vapor phase; frozen water vapor that forms on objects

frost wedging the process of freezing and thawing water in small rock pores and cracks that become larger and larger, eventually forcing pieces of rock to break off

fuel rod long zirconium alloy tubes containing fissionable material for use in a nuclear reactor

full moon the moon phase when the earth is between the sun and the moon and the entire side of the moon facing the earth is illuminated by sunlight

functional group the atom or group of atoms in an organic molecule that is responsible for the chemical properties of a particular class or group of organic chemicals

fundamental charge smallest common charge known; the magnitude of the charge of an electron and a proton

fundamental frequency the lowest frequency (longest wavelength) that can set up standing waves in an air column or on a string

fundamental properties a property that cannot be defined in simpler terms other than to describe how it is measured; the fundamental properties are length, mass, time, and charge

fungus the common name for the kingdom Fungi; examples, yeast and mold

G

g symbol representing the acceleration of an object in free fall due to the force of gravity

galactic clusters gravitationally bound subgroups of as many as 1,000 stars that move together within the Milky Way galaxy

galaxy group of billions and billions of stars that form the basic unit of the universe, for example, the earth is part of the solar system, which is located in the Milky Way galaxy

gallbladder an organ attached to the liver that stores bile

galvanometer a device that measures the size of an electric current from the size of the magnetic field produced by the current

gamete a haploid sex cell

gametogenesis the generating of gametes; the meiotic cell-division process that produces sex cells; oogenesis and spermatogenesis

gametophyte stage a life-cycle stage in plants in which a haploid sex cell is produced by mitosis

gamma ray very short wavelength electromagnetic radiation emitted by decaying nuclei

gases a phase of matter composed of molecules that are relatively far apart moving freely in a constant, random motion and have weak cohesive forces acting between them, resulting in the characteristic indefinite shape and indefinite volume of a gas

gasohol solution of ethanol and gasoline

gastric juice the secretions of the stomach that contain enzymes and hydrochloric acid

Geiger counter a device that indirectly measures ionizing radiation (beta and/or gamma) by detecting "avalanches" of electrons that are able to move because of the ions produced by the passage of ionizing radiation

gene a unit of heredity located on a chromosome and composed of a sequence of DNA nucleotides

gene flow the movement of genes from one generation to another or from one place to another

gene frequency a measure of how often a specific gene for a characteristic shows up in the individuals of a population

generative processes actions that increase the size of an individual organism (growth) or increase the number of individuals in a population (reproduction)

genetic recombination the gene mixing that occurs as a result of sexual reproduction

genetics the study of genes, how genes produce characteristics, and how the characteristics are inherited

genome a set of all the genes necessary to specify an organism's complete list of characteristics

genotype the catalog of genes of an organism, whether or not these genes are expressed

genus (plural, genera) a group of closely related species within a family

geographic barriers geographic features that keep different portions of a species from exchanging genes

geographic isolation a condition in which part of the gene pool is separated by geographic barriers from the rest of the population

geologic time scale a "calendar" of geologic history based on the appearance and disappearance of particular fossils in the sedimentary rock record

geomagnetic time scale time scale established from the number and duration of magnetic field reversals during the past 6 million years

giant planets the large outer planets of Jupiter, Saturn, Uranus, and Neptune that all have similar densities and compositions

glacier a large mass of ice on land that formed from compacted snow and slowly moves under its own weight

gland an organ that manufactures and secretes a material either through ducts or directly into the circulatory system

globular clusters symmetrical and tightly packed clusters of as many as a million stars that move together as subgroups within the Milky Way galaxy

glycerol an alcohol with three hydroxyl groups per molecule; for example, glycerin (1,2,3 propanetriol)

glycogen a highly branched polysaccharide synthesized by the human body and stored in the muscles and liver; serves as a direct reserve source of energy

glycol an alcohol with two hydroxyl groups per molecule; for example, ethylene glycol that is used as an antifreeze

Golgi apparatus a stack of flattened, smooth, membranous sacs; the site of synthesis and packaging of certain molecules in eukaryotic cells

gonad a generalized term for organs in which meiosis occurs to produce gametes; ovary or testis

granite light-colored, coarse-grained igneous rock common on continents; igneous rocks formed by blends of quartz and feldspars, with small amounts of micas, hornblende, and other minerals

greenhouse effect the process of increasing the temperature of the lower parts of the atmosphere through redirecting energy back toward the surface; the absorption and reemission of infrared radiation by carbon dioxide, water vapor, and a few other gases in the atmosphere

ground state energy state of an atom with electrons at the lowest energy state possible for that atom

groundwater water from a saturated zone beneath the surface; water from beneath the surface that supplies wells and springs

growth-stimulating hormone (GSH) a hormone produced by the anterior pituitary gland that stimulates tissues to grow

guanine a double-ring nitrogenous-base molecule in DNA and RNA. It is the complementary base of cytosine

gyre the great circular systems of moving water in each ocean

H

habitat the place or part of an ecosystem occupied by an organism

hail a frozen form of precipitation, sometimes with alternating layers of clear and opaque, cloudy ice

hair hygrometer a device that measures relative humidity from changes in the length of hair

half-life the time required for one-half of the unstable nuclei in a radioactive substance to decay into a new element

halogen member of family VIIA of the periodic table, having common properties of very reactive nonmetallic elements common in salt compounds

haploid having a single set of chromosomes resulting from the reduction division of meiosis

hard water water that contains relatively high concentrations of dissolved salts of calcium and magnesium

heart the muscular pump that forces the blood through the blood vessels of the body

heat total internal energy of molecules, which is increased by gaining energy from a temperature difference (conduction, convection, radiation) or by gaining energy from a form conversion (mechanical, chemical, radiant, electrical, nuclear)

heat of formation energy released in a chemical reaction

Heisenberg uncertainty principle you cannot measure both the exact momentum and the exact position of a subatomic particle at the same time—when the more exact of the two is known, the less certain you are of the value of the other

hemoglobin an iron-containing molecule found in red blood cells, to which oxygen molecules bind

herbivores animals that feed directly on plants

hertz unit of frequency; equivalent to one cycle per second

Hertzsprung-Russell diagram diagram to classify stars with a temperature-luminosity graph

heterotroph an organism that requires a source of organic material from its environment; it cannot produce its own food

heterozygous describes a diploid organism that has two different alleles for a particular characteristic

high short for high-pressure center (anticyclone), which is associated with clear, fair weather

high latitudes latitudes close to the poles; those that for a period of time during the winter months receive no solar radiation at noon

high-pressure center another term for anticyclone

homeostasis the process of maintaining a constant internal environment as a result of constant monitoring and modification of the functioning of various systems

homologous chromosomes a pair of chromosomes in a diploid cell that contain similar genes at corresponding loci throughout their length

homozygous describes a diploid organism that has two identical alleles for a particular characteristic

hormones chemical messengers secreted by endocrine glands to regulate other parts of the body

horsepower measurement of power defined as a power rating of 550 ft·lb/s

host an organism that a parasite lives in or on

hot spots sites on the earth's surface where plumes of hot rock materials rise from deep within the mantle

humid moist climate classification; receives more than 50 cm (20 in) precipitation per year

humidity the amount of water vapor in the air; see *relative humidity*

hurricane a tropical cyclone with heavy rains and winds exceeding 120 km/h

hydration the attraction of water molecules for ions; a reaction that occurs between water and minerals that make up rocks

hydrocarbon an organic compound consisting of only the two elements hydrogen and carbon

hydrocarbon derivatives organic compounds that can be thought of as forming when one or more hydrogen atoms on a hydrocarbon have been replaced by an element or a group of elements other than hydrogen

hydrogen bonding a strong bond that occurs between the hydrogen end of one molecule and the fluorine, oxygen, or nitrogen end of another molecule

hydrologic cycle water vapor cycling into and out of the atmosphere through continuous evaporation of liquid water from the surface and precipitation of water back to the surface

hydronium ion a molecule of water with an attached hydrogen ion, H_3O^+

hypothalamus a region of the brain located in the floor of the thalamus and connected to the pituitary gland that is involved in sleep and arousal; emotions such as anger, fear, pleasure, hunger, sexual response, and pain; and automatic functions such as temperature, blood pressure, and water balance

hypothesis a tentative explanation of a phenomenon that is compatible with the data and provides a framework for understanding and describing that phenomenon

I

ice-crystal process a precipitation-forming process that brings water droplets of a cloud together through the formation of ice crystals

ice-forming nuclei small, solid particles suspended in air; ice can form on the suspended particles

igneous rocks rocks that formed from magma, which is a hot, molten mass of melted rock materials

immune system a system of white blood cells specialized to provide the body with resistance to disease; there are two types, antibody-mediated immunity and cell-mediated immunity

impulse a change of motion is brought about by an impulse; the product of the size of an applied force and the time the force is applied

incandescent matter emitting visible light as a result of high temperature; for example, a lightbulb, a flame from any burning source, and the sun are all incandescent sources because of high temperature

incident ray line representing the direction of motion of incoming light approaching a boundary

incus the ear bone that is located between the malleus and the stapes

independent assortment the segregation, or assortment, of one pair of homologous chromosomes independently of the segregation, or assortment, of any other pair of chromosomes

index fossils distinctive fossils of organisms that lived only a brief time; used to compare the age of rocks exposed in two different locations

index of refraction the ratio of the speed of light in a vacuum to the speed of light in a material

inertia a property of matter describing the tendency of an object to resist a change in its state of motion; an object will remain in unchanging motion or at rest in the absence of an unbalanced force

inferior vena cava a major vein that returns blood to the heart from the lower body

infrasonic sound waves having too low a frequency to be heard by the human ear; sound having a frequency of less than 20 Hz

initiation code the code on DNA with the base sequence TAC that begins the process of transcription

inorganic chemistry the study of all compounds and elements in which carbon is not the principal element

insulators materials that are poor conductors of heat—for example, heat flows slowly through materials with air pockets because the molecules making up air are far apart; also, materials that are poor conductors of electricity, for example, glass or wood

intensity a measure of the energy carried by a wave

interference phenomenon of light where the relative phase difference between two light waves produces light or dark spots, a result of light's wavelike nature

intermediate-focus earthquakes earthquakes that occur in the upper part of the mantle, between 70 to 350 km below the surface of the earth

intermolecular forces forces of interaction between molecules

internal energy sum of all the potential energy and all the kinetic energy of all the molecules of an object

internal parasite a parasite that lives inside its host

interphase the stage between cell divisions in which the cell is engaged in metabolic activities

interstitial cell-stimulating hormone(ICSH) the chemical messenger molecule released from the pituitary that causes the testes to produce testosterone, the primary male sex hormone; same as follicle-stimulating hormone in females

intertropical convergence zone a part of the lower troposphere in a belt from 10°N to 10°S of the equator where air is heated, expands, and becomes less dense and rises around the belt

intrusive igneous rocks coarse-grained igneous rocks formed as magma cools slowly deep below the surface

inverse proportion the relationship in which the value of one variable increases while the value of the second variable decreases at the same rate (in the same ratio)

inversion a condition of the troposphere when temperature increases with height rather than decreasing with height; a cap of cold air over warmer air that results in trapped air pollution

ion an atom or a particle that has a net charge because of the gain or loss of electrons; polyatomic ions are groups of bonded atoms that have a net charge

ion exchange reaction a reaction that takes place when the ions of one compound interact with the ions of another, forming a solid that comes out of solution, a gas, or water

ionic bond chemical bond of electrostatic attraction between negative and positive ions

ionic compounds chemical compounds that are held together by ionic bonds—that is, bonds of electrostatic attraction between negative and positive ions

ionization process of forming ions from molecules

ionization counter a device that measures ionizing radiation (alpha, beta, gamma, etc.) by indirectly counting the ions produced by the radiation

ionized an atom or a particle that has a net charge because it has gained or lost electrons

ionosphere refers to that part of the atmosphere—parts of the thermosphere and upper mesosphere—where free electrons and ions reflect radio waves around the earth and where the northern lights occur

iron meteorites meteorite classification group whose members are composed mainly of iron

island arcs curving chains of volcanic islands that occur over belts of deep-seated earthquakes; for example, the Japanese and Indonesian islands

isomers chemical compounds with the same molecular formula but different molecular structure; compounds that are made from the same numbers of the same elements but have different molecular arrangements

isotope atoms of an element with identical chemical properties but with different masses; isotopes are atoms of the same element with different numbers of neutrons

J

jet stream a powerful, winding belt of wind near the top of the troposphere that tends to extend all the way around the earth, moving generally from the west in both hemispheres at speeds of 160 km/h or more

joint a break in the continuity of a rock formation without a relative movement of the rock on either side of the break

joule metric unit used to measure work and energy; can also be used to measure heat; equivalent to newton-meter

K

Kelvin scale a temperature scale that does not have arbitrarily assigned referent points, and zero means nothing; the zero point on the Kelvin scale (also called absolute scale) is the lowest limit of temperature, where all random kinetic energy of molecules ceases

ketone an organic compound with the general formula RCOR', where R is one of the hydrocarbon groups; for example, methyl or ethyl

keystone species species in an ecosystem that affects many aspects of the ecosystem and whose removal causes significant alteration of the ecosystem

kidneys the primary organs involved in regulating blood levels of water, hydrogen ions, salts, and urea

kilocalorie the amount of energy required to increase the temperature of 1 kilogram of water 1 degree Celsius; equivalent to 1,000 calories

kilogram the fundamental unit of mass in the metric system of measurement

kinetic energy the energy of motion; can be measured from the work done to put an object in motion, from the mass and velocity of the object while in motion, or from the amount of work the object can do because of its motion

kinetic molecular theory the collection of assumptions that all matter is made up of tiny atoms and molecules that interact physically, that explain the various states of matter, and that have an average kinetic energy that defines the temperature of a substance

kingdom a category within a domain used in the classification of organisms

Kuiper Belt a disk-shaped region of small icy bodies some 30 to 100 AU from the sun; the source of short-period comets

kwashiorkor a protein-deficiency disease common in malnourished children caused by prolonged protein starvation leading to reduced body size, lethargy, and low mental ability

L

lack of dominance the condition of two unlike alleles both expressing themselves, neither being dominant

lag phase a stage of a population growth curve during which both natality and mortality are low

lake a large inland body of standing water

large intestine the last portion of the food tube; primarily involved in reabsorbing water

last quarter the moon phase between the full phase and the new phase when the moon is perpendicular to a line drawn through the earth and the sun; one half of the lighted moon can be seen from the earth, so this phase is called the last quarter

latent heat refers to the heat "hidden" in phase changes

latent heat of fusion the heat absorbed when 1 gram of a substance changes from the solid to the liquid phase, or the heat released by 1 gram of a substance when changing from the liquid phase to the solid phase

latent heat of vaporization the heat absorbed when 1 gram of a substance changes from the liquid phase to the gaseous phase, or the heat released when 1 gram of gas changes from the gaseous phase to the liquid phase

lava magma, or molten rock, that is forced to the surface from a volcano or a crack in the earth's surface

law of conservation of energy energy is never created or destroyed; it can only be converted from one form to another as the total energy remains constant

law of conservation of mass same as law of conservation of matter; mass, including single atoms, is neither created nor destroyed in a chemical reaction

law of conservation of momentum the total momentum of a group of interacting objects remains constant in the absence of external forces

law of dominance when an organism has two different alleles for a trait, the allele that is expressed and overshadows the expression of the other allele is said to be dominant; the allele whose expression is overshadowed is said to be recessive

law of independent assortment members of one allelic pair will separate from each other independently of the members of other allele pairs

law of segregation when gametes are formed by a diploid organism, the alleles that control a trait separate from one another into different gametes, retaining their individuality

light ray model using lines to show the direction of motion of light to describe the travels of light

light-year the distance that light travels through empty space in one year, approximately 9.5×10^{12} km

limiting factors the identifiable factors that prevent unlimited population growth

limnetic zone the region in a lake that does not have rooted vegetation

line spectrum narrow lines of color in an otherwise dark spectrum; these lines can be used as "fingerprints" to identify gases

linear scale a scale, generally on a graph, where equal intervals represent equal changes in the value of a variable

lines of force lines drawn to make an electric field strength map, with each line originating on a positive charge and ending on a negative charge; each line represents a path on which a charge would experience a constant force and lines closer together mean a stronger electric field

linkage group genes located on the same chromosome that tend to be inherited together

lipid large, nonpolar, organic molecules that do not easily dissolve in polar solvents such as water; there are three different types of lipids: *true fats* (pork chop fat or olive oil), *phospholipids* (the primary component of cell membranes), and *steroids* (most hormones)

liquids a phase of matter composed of molecules that have interactions stronger than those found in a gas but not strong enough to keep the molecules near the equilibrium positions of a solid, resulting in the characteristic definite volume but indefinite shape of a liquid

liter a metric system unit of volume usually used for liquids

lithosphere solid layer of the earth's structure that is above the asthenosphere and includes the entire crust, the Moho, and the upper part of the mantle

littoral zone the region in a lake with rooted vegetation

liver an organ of the body responsible for secreting bile, filtering the blood, detoxifying molecules, and modifying molecules absorbed from the gut

locus the spot on a chromosome where an allele is located

loess very fine dust or silt that has been deposited by the wind over a large area

longitude angular distance of a point east or west from the prime meridian on a parallel

longitudinal wave a mechanical disturbance that causes particles to move closer together and farther apart in the same direction that the wave is traveling

longshore current a current that moves parallel to the shore, pushed along by waves that move accumulated water from breakers

loudness a subjective interpretation of a sound that is related to the energy of the vibrating source, related to the condition of the transmitting medium, and related to the distance involved

low latitudes latitudes close to the equator; those that receive vertical solar radiation at noon during some part of the year

luminous an object or objects that produce visible light; for example, the sun, stars, lightbulbs, and burning materials are all luminous

lunar highlands light-colored mountainous regions of the moon

lung a respiratory organ in which air and blood are brought close to one another and gas exchange occurs

L-wave seismic waves that move on the solid surface of the earth much as water waves move across the surface of a body of water

lymph liquid material that leaves the circulatory system to surround cells

lymphatic system a collection of thin-walled tubes that collects, filters, and returns lymph from the body to the circulatory system

lysosome a specialized, submicroscopic organelle that holds a mixture of hydrolytic enzymes

M

macromolecule very large molecule, with a molecular weight of thousands or millions of atomic mass units, that is made up of a combination of many smaller, similar molecules

magma a mass of molten rock material either below or on the earth's crust from which igneous rock is formed by cooling and hardening

magnetic field model used to describe how magnetic forces on moving charges act at a distance

magnetic poles the ends, or sides, of a magnet about which the force of magnetic attraction seems to be concentrated

magnetic quantum number from the quantum mechanics model of the atom, one of four descriptions of the energy state of an electron wave; this quantum number describes the energy of an electron orbital as the orbital is oriented in space by an external magnetic field, a kind of energy sub-sublevel

magnetic reversal the flipping of polarity of the earth's magnetic field as the north magnetic pole and the south magnetic pole exchange positions

main sequence stars normal, mature stars that use their nuclear fuel at a steady rate; stars on the Hertzsprung-Russell diagram in a narrow band that runs from the top left to the lower right

malleus the ear bone that is attached to the tympanum

manipulated variable in an experiment, a quantity that can be controlled or manipulated; also known as the independent variable

mantle middle part of the earth's interior; a 2,870 km (about 1,780 mile) thick shell between the core and the crust

maria smooth dark areas on the moon

marine climate a climate influenced by air masses from over an ocean, with mild winters and cool summers compared to areas farther inland

marine communities aquatic communities that have a high salt content

maritime air mass a moist air mass that forms over the ocean

mass a measure of inertia, which means a resistance to a change of motion

mass defect the difference between the sum of the masses of the individual nucleons forming a nucleus and the actual mass of that nucleus

mass movement erosion caused by the direct action of gravity

mass number the sum of the number of protons and neutrons in a nucleus defines the mass number of an atom; used to identify isotopes; for example, uranium 238

masturbation stimulation of one's own sex organs

matter anything that occupies space and has mass

meanders winding, circuitous turns or bends of a stream

measurement the process of comparing a property of an object to a well-defined and agreed-upon referent

mechanical energy the form of energy associated with machines, objects in motion, and objects having potential energy that results from gravity

mechanical weathering the physical breaking up of rocks without any changes in their chemical composition

medulla oblongata a region of the more primitive portion of the brain connected to the spinal cord that controls such automatic functions as blood pressure, breathing, and heart rate

meiosis the specialized pair of cell divisions that reduces the chromosome number from diploid ($2n$) to haploid (n)

melting point the temperature at which a phase change of solid to liquid takes place; the same temperature as the freezing point for a given substance

Mendelian genetics the pattern of inheriting characteristics that follows the laws formulated by Gregor Mendel

menstrual cycle (menses, menstrual flow, period) the repeated building up and shedding of the lining of the uterus

Mesozoic a geologic era; the time of middle life, meaning some of the fossils for this time period are similar to the life found on the earth today, but many are different from anything living today

messenger RNA (mRNA) a molecule composed of ribonucleotides that functions as a copy of the gene and is used in the cytoplasm of the cell during protein synthesis

metabolic processes the total of all chemical reactions within an organism; for example, nutrient uptake and processing, and waste elimination

metal matter having the physical properties of conductivity, malleability, ductility, and luster

metamorphic rocks previously existing rocks that have been changed into a distinctly different rock by heat, pressure, or hot solutions

metaphase the second stage in mitosis, characterized by alignment of the chromosomes at the equatorial plane

meteor the streak of light and smoke that appears in the sky when a meteoroid is made incandescent by friction with the earth's atmosphere

meteor shower event when many meteorites fall in a short period of time

meteorite the solid iron or stony material of a meteoroid that survives passage through the earth's atmosphere and reaches the surface

meteoroids remnants of comets and asteroids in space

meteorology the science of understanding and predicting weather

meter the fundamental metric unit of length

metric system a system of referent units based on invariable referents of nature that have been defined as standards

microclimate a local, small-scale pattern of climate; for example, the north side of a house has a different microclimate than the south side

microvilli tiny projections from the surfaces of cells that line the intestine

middle latitudes latitudes equally far from the poles and equator; between the high and low latitudes

mineral (geology) a naturally occurring, inorganic solid element or chemical compound with a crystalline structure

minerals (biology) inorganic elements that cannot be manufactured by the body but are required in low concentrations; essential to metabolism

miscible fluids two or more kinds of fluids that can mix in any proportion

mitochondrion a membranous organelle resembling a small bag with a larger bag inside that is folded back on itself; serves as the site of aerobic cellular respiration

mitosis a process that results in equal and identical distribution of replicated chromosomes into two newly formed nuclei

mixture matter composed of two or more kinds of matter that has a variable composition and can be separated into its component parts by physical means

model a mental or physical representation of something that cannot be observed directly that is usually used as an aid to understanding

moderator a substance in a nuclear reactor that slows fast neutrons so the neutrons can participate in nuclear reactions

mold the preservation of the shape of an organism by the dissolution of the remains of a buried organism, leaving an empty space where the remains were

mole an amount of a substance that contains Avogadro's number of atoms, ions, molecules, or any other chemical unit; a mole is thus 6.02×10^{23} atoms, ions, or other chemical units

molecule from the chemical point of view: a particle composed of two or more atoms held together by an attractive force called a chemical bond; from the kinetic theory point of view: smallest particle of a compound or gaseous element that can exist and still retain the characteristic properties of a substance

momentum the product of the mass of an object times its velocity

monomer individual, repeating units or segments of complex molecules that chemically combine to form long chainlike molecules called polymers; for example, monosaccharides link to form polysaccharides

monosaccharides simple sugars containing 3 to 6 carbon atoms; the most common kinds are 6-carbon molecules such as glucose and fructose

moraines deposits of rocks and other mounded materials bulldozed into position by a glacier and left behind when the glacier melted

morphological species concept the concept that species can be distinguished from one another by structural characteristics

mortality the number of individuals leaving the population by death per thousand individuals in the population

motor neurons those neurons that carry information from the central nervous system to muscles or glands

motor unit all of the muscle cells stimulated by a single neuron

mountain a natural elevation of the earth's crust that rises above the surrounding surface

multiple alleles a term used to refer to conditions in which there are several different alleles for a particular characteristic, not just two

multiregional hypothesis states that *Homo erectus* evolved into *Homo sapiens.*

mutagenic agent anything that causes permanent change in DNA

mutation any change in the genetic information of a cell

mutualism a relationship between two organisms in which both organisms benefit

N

natality the number of individuals entering the population by reproduction per thousand individuals in the population

natural frequency the frequency of vibration of an elastic object that depends on the size, composition, and shape of the object

natural selection the processes that encourage the passage of beneficial genes and discourage the passage of harmful or unfavorable genes from one generation to the next

nebulae a diffuse mass of interstellar clouds of hydrogen gas or dust that may develop into a star

negative electric charge one of the two types of electric charge; repels other negative charges and attracts positive charges

negative ion atom or particle that has a surplus, or imbalance, of electrons and, thus, a negative charge

nephrons millions of tiny tubular units that make up the kidneys, which are responsible for filtering the blood

nerve cell see *neuron*

nerve impulse a series of changes that take place in the neuron, resulting in a wave of depolarization that passes from one end of the neuron to the other

nerves bundles of neuronal fibers

nervous system a network of neurons that carry information from sense organs to the central nervous system and from the central nervous system to muscles and glands

net force the resulting force after all forces have been added; if a net force is zero, all the forces have canceled each other and there is not an unbalanced force

neuron the cellular unit consisting of a cell body and fibers that makes up the nervous system; also called nerve cell

neurotransmitter a molecule released by the axons of neurons that stimulates other cells

neutralized acid or base properties have been lost through a chemical reaction

neutron neutral subatomic particle usually found in the nucleus of an atom

neutron star very small superdense remains of a supernova with a center core of pure neutrons

new crust zone zone of a divergent boundary where new crust is formed by magma upwelling at the boundary

newton a unit of force defined as $kg \cdot m/s^2$; that is, a 1 newton force is needed to accelerate a 1 kg mass 1 m/s^2

niche the functional role of an organism

nitrifying bacteria bacteria that can convert ammonia (NH_3) into nitrite-containing (NO_2^-) compounds, which in turn can be converted into nitrate-containing (NO_3^-) compounds

nitrogenous base a category of organic molecules found as components of the nucleic acids. There are five common types: thymine, guanine, cytosine, adenine, and uracil

noble gas members of family VIII of the periodic table, having common properties of colorless, odorless, chemically inert gases; also known as rare gases or inert gases

node regions on a standing wave that do not oscillate

noise sounds made up of groups of waves of random frequency and intensity

nonelectrolytes water solutions that do not conduct an electric current; covalent compounds that form molecular solutions and cannot conduct an electric current

nonferromagnesian silicates silicates that do not contain iron or magnesium ions; examples include the minerals of muscovite (white mica), the feldspars, and quartz

nonmetal an element that is brittle (when a solid), does not have a metallic luster, is a poor conductor of heat and electricity, and is not malleable or ductile

nonsilicates minerals that do not have the silicon-oxygen tetrahedra in their crystal structure

norepinephrine a hormone produced by the adrenal medulla that increases heart rate, blood pressure, and breathing rate

normal a line perpendicular to the surface of a boundary

normal fault a fault where the hanging wall has moved downward with respect to the foot wall

north pole (of a magnet) short for "north seeking"; the pole of a magnet that points northward when it is free to turn

nova a star that explodes or suddenly erupts and increases in brightness

nuclear energy the form of energy from reactions involving the nucleus, the innermost part of an atom

nuclear fission nuclear reaction of splitting a massive nucleus into more stable, less massive nuclei with an accompanying release of energy

nuclear force one of four fundamental forces, a strong force of attraction that operates over very short distances between subatomic particles; this force overcomes the electric repulsion of protons in a nucleus and binds the nucleus together

nuclear fusion nuclear reaction of low-mass nuclei fusing together to form more stable and more massive nuclei with an accompanying release of energy

nuclear membrane the structure surrounding the nucleus that separates the nucleoplasm from the cytoplasm

nuclear reactor steel vessel in which a controlled chain reaction of fissionable materials releases energy

nucleic acids complex molecules that store and transfer genetic information within a cell; constructed of fundamental monomers known as nucleotides; the two common forms are DNA and RNA

nucleoli (singular, *nucleolus*) nuclear structures composed of completed or partially completed ribosomes and the specific parts of chromosomes that contain the information for their construction

nucleons name used to refer to both the protons and neutrons in the nucleus of an atom

nucleoproteins the double-stranded DNA with attached proteins; also called chromatin fibers

nucleosomes histone clusters with their encircling DNA

nucleotide the building block of the nucleic acids, composed of a 5-carbon sugar, a phosphate, and a nitrogenous base

nucleus (atom) tiny, relatively massive and positively charged center of an atom containing protons and neutrons; the small, dense center of an atom

nucleus (cell) the central part of a cell that contains the genetic material

numerical constant a constant without units; a number

nutrients molecules required by the body for growth, reproduction, or repair

nutrition collectively, the processes involved in taking in, assimilating, and utilizing nutrients

O

obese a term describing a person who gains a great deal of unnecessary weight and is 15% to 20% above his or her ideal weight

obligate intracellular parasites infectious particles (viruses) that can function only when inside a living cell

observed lapse rate the rate of change in temperature compared to change in altitude

occluded front a front that has been lifted completely off the ground into the atmosphere, forming a cyclonic storm

ocean the single, continuous body of salt water on the surface of the earth

ocean basin the deep bottom of the ocean floor, which starts beyond the continental slope

ocean currents streams of water within the ocean that stay in about the same path as they move over large distances; steady and continuous onward movement of a channel of water in the ocean

ocean wave a moving disturbance that travels across the surface of the ocean

oceanic ridges long, high, continuous, suboceanic mountain chains; for example, the Mid-Atlantic Ridge in the center of the Atlantic Ocean Basin

oceanic trenches long, narrow, deep troughs with steep sides that run parallel to the edge of continents

octet rule a generalization that helps keep track of the valence electrons in most representative elements; atoms of the representative elements (A families) attempt to acquire an outer orbital with eight electrons through chemical reactions

offspring descendants of a set of parents

ohm unit of resistance; equivalent to volts/amps

Ohm's law the electric potential difference is directly proportional to the product of the current times the resistance

oil field petroleum accumulated and trapped in extensive porous rock structure or structures

oils organic compounds of esters formed from glycerol and three long-chain carboxylic acids that are also called triglycerides; called fats in animals and oils in plants

olfactory epithelium the cells of the nasal cavity that respond to chemicals

omnivores animals that are carnivores at some times and herbivores at others

oogenesis the specific name given to the gametogenesis process that leads to the formation of eggs

Oort cloud a spherical "cloud" of small, icy bodies from 30,000 AU out to a light-year from the sun; the source of long-period comets

opaque materials that do not allow the transmission of any light

orbital the region of space around the nucleus of an atom where an electron is likely to be found

order a group of closely related families within a class

ore mineral mineral deposits with an economic value

organ a structure composed of two or more kinds of tissues that perform a particular function

organ system a group of organs that performs a particular function

organelles cellular structures that perform specific functions in the cell; the function of an organelle is directly related to its structure

organic acids organic compounds with a general formula of RCOOH, where R is one of the hydrocarbon groups; for example, methyl or ethyl

organic chemistry the study of compounds in which carbon is the principal element

organism an independent living unit

orgasm a complex series of responses to sexual stimulation that results in intense frenzy of sexual excitement

osmosis the net movement of water molecules through a selectively permeable membrane

osteoporosis a disease condition resulting from the demineralization of the bone, resulting in pain, deformities, and fractures; related to a loss of calcium

out-of-Africa hypothesis states that modern humans (*Homo sapiens*) originated in Africa, as had several other hominid species, and migrated from Africa to Asia and Europe and displaced species such as *H. erectus* and *H. heidelbergensis* that had migrated into these areas previously

oval window the membrane-covered opening of the cochlea, to which the stapes is attached

ovaries the female sex organs that produce haploid sex cells—the eggs or ova

overtones higher resonant frequencies that occur at the same time as the fundamental frequency, giving a musical instrument its characteristic sound quality

oviduct the tube (fallopian tube) that carries the oocyte to the uterus

ovulation the release of a secondary oocyte from the surface of the ovary

oxbow lake a small body of water, or lake, that formed when two bends of a stream came together and cut off a meander

oxidation the process of a substance losing electrons during a chemical reaction; a reaction between oxygen and the minerals making up rocks

oxidation-reduction reaction a chemical reaction in which electrons are transferred from one atom to another; sometimes called "redox" for short

oxidizing agents substances that take electrons from other substances

oxidizing atmosphere an atmosphere that contains molecular oxygen

oxytocin a hormone released from the posterior pituitary that causes contraction of the uterus

ozone shield concentration of ozone in the lower portions of the stratosphere that absorbs potentially damaging ultraviolet radiation, preventing it from reaching the surface of the earth

P

paleontology the science of discovering fossils, studying the fossil record, and deciphering the history of life

Paleozoic a geologic era; time of ancient life, meaning the fossils from this time period are very different from anything living on the earth today

pancreas an organ of the body that secretes many kinds of digestive enzymes into the duodenum

parasite an organism that lives in or on another organism and derives nourishment from it

parasitism a relationship which involves one organism living in or on another living organism from which it derives nourishment

parts per billion concentration ratio of parts of solute in every one billion parts of solution (ppb); could be expressed as ppb by volume or as ppb by weight

parts per million concentration ratio of parts of solute in every one million parts of solution (ppm); could be expressed as ppm by volume or as ppm by weight

Pauli exclusion principle no two electrons in an atom can have the same four quantum numbers; thus, a maximum of two electrons can occupy a given orbital

pelagic describes aquatic organisms that float or swim actively

penis the portion of the male reproductive system that deposits sperm in the female reproductive tract

pepsin an enzyme produced by the stomach that is responsible for beginning the digestion of proteins

percent by volume the volume of solute in 100 volumes of solution

percent by weight the weight of solute in 100 weight units of solution

perception recognition by the brain that a stimulus has been received

perigee when the moon's elliptical orbit brings the moon closest to the earth

perihelion the point at which an orbit comes closest to the sun

period (geologic time) subdivisions of geologic eras

period (periodic table) horizontal rows of elements with increasing atomic numbers; runs from left to right on the element table

period (wave) the time required for one complete cycle of a wave

periodic law similar physical and chemical properties recur periodically when the elements are listed in order of increasing atomic number

peripheral nervous system the fibers that communicate between the central nervous system and other parts of the body

permeability the ability to transmit fluids through openings, small passageways, or gaps

permineralization the process that forms a fossil by alteration of an organism's buried remains by circulating groundwater depositing calcium carbonate, silica, or pyrite

petroleum oil that comes from oil-bearing rock, a mixture of hydrocarbons that is believed to have formed from ancient accumulations of buried organic materials such as remains of algae

pH scale scale that measures the acidity of a solution with numbers below 7 representing acids, 7 representing neutral, and numbers above 7 representing bases

phagocytosis the process by which the cell wraps around a particle and engulfs it

pharynx the region at the back of the mouth cavity; the throat

phase change the action of a substance changing from one state of matter to another; a phase change always absorbs or releases internal potential energy without a temperature change

phases of matter the different physical forms that matter can take as a result of different molecular arrangements, resulting in characteristics of the common phases of a solid, liquid, or gas

phenotype the physical, chemical, and behavioral expression of the genes possessed by an organism

phospholipids a class of complex water-insoluble organic molecules that resemble fats but contain a phosphate group (PO_4) in their structure; a component of cellular membranes

photoelectric effect the movement of electrons in some materials as a result of energy acquired from absorbed light

photon a quantum of energy in a light wave; the particle associated with light

phylogeny the science that explores the evolutionary relationships among organisms and seeks to reconstruct evolutionary history

phylum a subdivision of a kingdom

physical change a change of the state of a substance but not the identity of the substance

phytoplankton planktonic organisms that carry on photosynthesis

pistil the sex organ in plants that produces eggs or ova

pitch the frequency of a sound wave

pituitary gland the gland at the base of the brain that controls the functioning of other glands throughout the organism

placenta an organ made up of tissues from the embryo and the uterus of the mother that allows for the exchange of materials between the mother's bloodstream and the embryo's bloodstream; it also produces hormones

Planck's constant proportionality constant in the relationship between the energy of vibrating molecules and their frequency of vibration; a value of 6.63×10^{-34} Js

plankton aquatic organisms that are so small and weakly swimming that they are simply carried by currents.

plasma (biology) the watery matrix that contains the molecules and cells of the blood

plasma (physics) a phase of matter; a very hot gas consisting of electrons and atoms that have been stripped of their electrons because of high kinetic energies

plasma membrane the outer-boundary membrane of the cell; see *cell membrane.*

plastic strain an adjustment to stress in which materials become molded or bent out of shape under stress and do not return to their original shape after the stress is released

plate tectonics the theory that the earth's crust is made of rigid plates that float on the asthenosphere

pleiotropy the multiple effects that a gene may have on the phenotype of an organism

point mutation a change in the DNA of a cell as a result of a loss or change in a single nitrogenous-base

polar air mass cold air mass that forms in cold regions

polar body the smaller of two cells formed by unequal meiotic division during oogenesis

polar climate zone climate zone of the high latitudes; average monthly temperatures stay below 10°C (50°F), even during the warmest month of the year

polarized light whose constituent transverse waves are all vibrating in the same plane; also known as plane-polarized light

Polaroid a film that transmits only polarized light

polyatomic ion ion made up of many atoms

polygenic inheritance the concept that a number of different pairs of alleles may combine their efforts to determine a characteristic

polymers huge, chainlike molecules made of hundreds or thousands of smaller repeating molecular units called monomers

polypeptide chain polymers of amino acids; sometimes called proteins

polysaccharides polymers consisting of monosaccharide units joined together in straight or branched chains; starches, glycogen, or cellulose

pond a small body of standing water, smaller than a lake

pons a region of the brain immediately anterior to the medulla oblongata that connects to the cerebellum and higher regions of the brain and controls several sensory and motor functions of the head and face

population a group of organisms of the same species located in an area

population growth curve a graph of the change in population size over time

porosity the ratio of pore space to the total volume of a rock or soil sample, expressed as a percentage; freely admitting the passage of fluids through pores or small spaces between parts of the rock or soil

positive electric charge one of the two types of electric charge; repels other positive charges and attracts negative charges

positive ion atom or particle that has a net positive charge due to an electron or electrons being torn away

potential energy energy due to position; energy associated with changes in position (e.g., gravitational potential energy) or changes in shape (e.g., compressed or stretched spring)

power the rate at which energy is transferred or the rate at which work is performed; defined as work per unit of time

Precambrian the time before the time of ancient life, meaning the rocks for this time period contain very few fossils

precipitation water that falls to the surface of the earth in the solid or liquid form

predation a relationship between two organisms that involves the capturing, killing, and eating of one by the other

predator an organism that captures, kills, and eats another animal

pressure defined as force per unit area; for example, pounds per square inch (lb/in^2)

prey an organism captured, killed, and eaten by a predator

primary carnivores carnivores that eat herbivores and are therefore on the third trophic level

primary coil part of a transformer; a coil of wire that is connected to a source of alternating current

primary consumers organisms that feed directly on plants—herbivores

primary loop part of the energy-converting system of a nuclear power plant; the closed pipe system that carries heated water from the nuclear reactor to a steam generator

primary oocyte the diploid cell of the ovary that begins to undergo the first meiotic division in the process of oogenesis

primary spermatocyte the diploid cell in the testes that undergoes the first meiotic division in the process of spermatogenesis

prime meridian the referent meridian (0°) that passes through the Greenwich Observatory in England

principle of uniformity a frame of reference that the same processes that changed the landscape in the past are the same processes you see changing the landscape today

principal quantum number from the quantum mechanics model of the atom, one of four descriptions of the energy state of an electron wave; this quantum number describes the main energy level of an electron in terms of its most probable distance from the nucleus

prions proteins that can be passed from one individual to another and cause disease

probability the chance that an event will happen, expressed as a percent or fraction

producers organisms that produce new organic material from inorganic material with the aid of sunlight

prokaryote an organism composed of cells that lack a nuclear membrane and other membranous organelles

prokaryotic cells one of the two major types of cells; they do not have a typical nucleus bound by a nuclear membrane and lack many of the other membranous cellular organelles; for example, members of the Eubacteria and Archaea

promoter a region of DNA at the beginning of each gene, just ahead of an initiator code

proof a measure of ethanol concentration of an alcoholic beverage; proof is double the concentration by volume; for example, 50 percent by volume is 100 proof

properties qualities or attributes that, taken together, are usually unique to an object; for example, color, texture, and size

prophase the first phase of mitosis during which individual chromosomes become visible

proportionality constant a constant applied to a proportionality statement that transforms the statement into an equation

protein synthesis the process whereby the tRNA utilizes the mRNA as a guide to arrange the amino acids in their proper sequence according to the genetic information in the chemical code of DNA

proteins macromolecular polymers made of smaller molecules of amino acids, with molecular weight from about six thousand to fifty million; proteins are amino acid polymers with roles in biological structures or functions; without such a function, they are known as polypeptides

proton subatomic particle that has the smallest possible positive charge, usually found in the nucleus of an atom

protoplanet nebular model a model of the formation of the solar system that states that the planets formed from gas and dust left over from the formation of the sun

protoplasm the living portion of a cell as distinguished from the nonliving cell wall

protostar an accumulation of gases that will become a star

pseudoscience use of the appearance of science to mislead; the assertions made are not valid or reliable

psychrometer a two-thermometer device used to measure the relative humidity

puberty a time in the life of a developing individual characterized by the increasing production of sex hormones, which cause it to reach sexual maturity

pulmonary artery the major blood vessel that carries blood from the right ventricle to the lungs

pulmonary circulation the flow of blood through certain chambers of the heart and blood vessels to the lungs and back to the heart

pulsars the source of regular, equally spaced pulsating radio signals believed to be the result of the magnetic field of a rotating neutron star

Punnett square a method used to determine the probabilities of allele combinations in an offspring

pure substance materials that are the same throughout and have a fixed definite composition

pure tone sound made by very regular intensities and very regular frequencies from regular repeating vibrations

P-wave a pressure, or compressional, wave in which a disturbance vibrates materials back and forth in the same direction as the direction of wave movement

P-wave shadow zone a region on the earth between 103° and 142° of arc from an earthquake where no P-waves are received; believed to be explained by P-waves being refracted by the core

pyloric sphincter a valve located at the end of the stomach that regulates the flow of food from the stomach to the duodenum

Q

quad one quadrillion Btu (10^{15} Btu); used to describe very large amounts of energy

quanta fixed amounts; usually referring to fixed amounts of energy absorbed or emitted by matter ("quanta" is plural and "quantum" is singular)

quantities measured properties; includes the numerical value of the measurement and the unit used in the measurement

quantum mechanics model of the atom based on the wave nature of subatomic particles, the mechanics of electron waves; also called wave mechanics

quantum numbers numbers that describe energy states of an electron; in the Bohr model of the atom, the orbit quantum numbers could be any whole number 1, 2, 3, and so on out from the nucleus; in the quantum mechanics model of the atom, four quantum numbers are used to describe the energy state of an electron wave

R

rad a measure of radiation received by a material (radiation absorbed dose)

radiant energy the form of energy that can travel through space; for example, visible light and other parts of the electromagnetic spectrum

radiation the transfer of heat from a region of higher temperature to a region of lower temperature by greater emission of radiant energy from the region of higher temperature

radiation zone part of the interior of a star according to a model; the region directly above the core where gamma and X rays from the core are absorbed and reemitted, with the radiation slowly working its way outward

radioactive decay the natural spontaneous disintegration or decomposition of a nucleus

radioactive decay constant a specific constant for a particular isotope that is the ratio of the rate of nuclear disintegration per unit of time to the total number of radioactive nuclei

radioactive decay series series of decay reactions that begins with one radioactive nucleus that decays to a second nucleus that decays to a third nucleus and so on until a stable nucleus is reached

radioactivity spontaneous emission of particles or energy from an atomic nucleus as it disintegrates

radiometric age age of rocks determined by measuring the radioactive decay of unstable elements within the crystals of certain minerals in the rocks

range the geographical distribution of a species

rarefaction a thinning or pulse of decreased density and pressure of gas molecules

ratio a relationship between two numbers, one divided by the other; the ratio of distance per time is speed

real image an image generated by a lens or mirror that can be projected onto a screen

recessive allele an allele that, when present with a dominant allele, does not express itself and is masked by the effect of the dominant allele

recombinant DNA DNA that has been constructed by inserting new pieces of DNA into the DNA of another organism, such as a bacterium

Recommended Dietary Allowances (RDA) U.S. dietary guidelines for a healthy person that focus on the amounts of foods desired from six classes of nutrients

red giant stars one of two groups of stars on the Hertzsprung-Russell diagram that have a different set of properties than the main sequence stars; bright, low-temperature giant stars that are enormously bright for their temperature

redox reaction short name for oxidation-reduction reaction

reducing agent supplies electrons to the substance being reduced in a chemical reaction

reduction division (also meiosis) a type of cell division in which daughter cells get only half the chromosomes from the parent cell

referent referring to or thinking of a property in terms of another, more familiar object

reflected ray a line representing direction of motion of light reflected from a boundary

reflection the change when light, sound, or other waves bounce backwards off a boundary

refraction a change in the direction of travel of light, sound, or other waves crossing a boundary

regulator protein proteins that help determine the activities that will occur in a cell or multicellular organism; for example, enzymes and some hormones

relative dating dating the age of a rock unit or geological event relative to some other unit or event

relative humidity ratio (times 100 percent) of how much water vapor is in the air to the maximum amount of water vapor that could be in the air at a given temperature

reliable a term used to describe results that remain consistent over successive trials

rem measure of radiation that considers the biological effects of different kinds of ionizing radiation

replacement (chemical reaction) reaction in which an atom or polyatomic ion is replaced in a compound by a different atom or polyatomic ion

replacement (fossil formation) process in which an organism's buried remains are altered by circulating groundwaters carrying elements in solution; the removal of original materials by dissolutions and the replacement of new materials an atom or molecule at a time

representative elements name given to the members of the A-group families of the periodic table; also called the main-group elements

reservoir a natural or artificial pond or lake used to store water, control floods, or generate electricity; a body of water stored for public use

resonance when the frequency of an external force matches the natural frequency of a material and standing waves are set up

response the reaction of an organism to a stimulus

responsive processes those abilities to react to external and internal changes in the environment; for example, irritability, individual adaptation, and evolution

retina the light-sensitive region of the eye

reverberation apparent increase in volume caused by reflections, usually arriving within 0.1 second after the original sound

reverse fault a fault where the hanging wall has moved upward with respect to the foot wall

revolution the motion of a planet as it orbits the sun

rhodopsin a light-sensitive pigment found in the rods of the retina

ribonucleic acid (RNA) a polymer of nucleotides formed on the template surface of DNA by transcription. Three forms that have been identified are mRNA, rRNA, and tRNA

ribose a 5-carbon sugar molecule that is a component of RNA

ribosomal RNA (rRNA) a globular form of RNA; a part of ribosomes

ribosomes small structures composed of two protein and ribonucleic acid subunits; involved in the assembly of proteins from amino acids

Richter scale expresses the intensity of an earthquake in terms of a scale with each higher number indicating 10 times more ground movement and about 30 times more energy released than the preceding number

ridges long, rugged mountain chains rising thousands of meters above the abyssal plains of the ocean basin

rift a split or fracture in a rock formation, land formation, or in the crust of the earth

rip current strong, brief current that runs against the surf and out to sea

RNA polymerase an enzyme that attaches to the DNA at the promoter region of a gene and assists in combining RNA nucleotides when the genetic information is transcribed into RNA

rock a solid aggregation of minerals or mineral materials that have been brought together into a cohesive solid

rock cycle understanding of igneous, sedimentary, or metamorphic rock as a temporary state in an ongoing transformation of rocks into new types; the process of rocks continually changing from one type to another

rock flour rock pulverized by a glacier into powdery, silt-sized sediment

rods light-sensitive cells in the retina of the eye that respond to low-intensity light but do not respond to different colors of light

rotation the spinning of a planet on its axis

runoff water moving across the surface of the earth as opposed to soaking into the ground

S

salinity a measure of dissolved salts in seawater, defined as the mass of salts dissolved in 1,000 g of solution

salivary amylase an enzyme present in saliva that breaks starch molecules into smaller molecules

salivary glands glands that produce saliva

salt any ionic compound except one with hydroxide or oxide ions

San Andreas fault in California, the boundary between the North American Plate and the Pacific Plate that runs north-south for some 1,300 km (800 miles) with the Pacific Plate moving northwest and the North American Plate moving southeast

saturated air air in which an equilibrium exists between evaporation and condensation; the relative humidity is 100 percent

saturated molecule an organic molecule that has the maximum number of hydrogen atoms possible

saturated solution the apparent limit to dissolving a given solid in a specified amount of water at a given temperature; a state of equilibrium that exists between dissolving solute and solute coming out of solution

scientific law a relationship between quantities, usually described by an equation in the physical sciences; is more important and describes a wider range of phenomena than a scientific principle

scintillation counter a device that indirectly measures ionizing radiation (alpha, beta, gamma, etc.) by measuring the flashes of light produced when the radiation strikes a phosphor

sea a smaller part of the ocean with characteristics that distinguish it from the larger ocean

sea breeze cool, dense air from over water moving over land as part of convective circulation

seafloor spreading the process by which hot, molten rock moves up from the interior of the earth to emerge along mid-oceanic rifts, flowing out in both directions to create new rocks and spread apart the seafloor

seamounts steep, submerged volcanic peaks on the abyssal plain

second the standard unit of time in both the metric and English systems of measurement

second law of motion the acceleration of an object is directly proportional to the net force acting on that object and inversely proportional to the mass of the object

second law of thermodynamics a statement that the natural process proceeds from a state of higher order to a state of greater disorder

secondary carnivores carnivores that feed on primary carnivores and are therefore at the fourth trophic level

secondary coil part of a transformer, a coil of wire in which the voltage of the original alternating current in the primary coil is stepped up or down by way of electro-magnetic induction

secondary consumers animals that eat other animals—carnivores

secondary loop part of a nuclear power plant; the closed pipe system that carries steam from a steam generator to the turbines, then back to the steam generator as feedwater

secondary oocyte the larger of the two cells resulting from the unequal cytoplasmic division of a primary oocyte in meiosis I of oogenesis

secondary sex characteristics characteristics of the adult male or female, including the typical shape that develops at puberty: broader shoulders, heavier long-bone muscles, development of facial hair, axillary hair, and chest hair, and changes in the shape of the larynx in the male; rounding of the pelvis and breasts and changes in deposition of fat in the female

secondary spermatocyte cells in the seminiferous tubules that go through the second meiotic division, resulting in four haploid spermatids

sedimentary rocks rocks formed from particles or dissolved minerals from previously existing rocks that were deposited from air or water

sediments accumulations of silt, sand, or gravel that settled out of the atmosphere or out of water

segregation the separation and movement of homologous chromosomes to the opposite poles of the cell

seismic waves vibrations that move as waves through any part of the earth, usually associated with earthquakes, volcanoes, or large explosions

seismograph an instrument that measures and records seismic wave data

selecting agents specific environmental factors that favor the passage of certain characteristics from one generation to the next and discourage others

selectively permeable the property of a membrane that allows certain molecules to pass through it, but interferes with the passage of others

semen the sperm-carrying fluid produced by the seminal vesicles, prostate glands, and bulbourethral glands of males

semiarid climate classification between arid and humid; receives between 25 and 50 cm (10 and 20 in) precipitation per year

semicircular canals a set of tubular organs associated with the cochlea that sense changes in the movement or position of the head

semiconductors elements that have properties between those of a metal and those of a nonmetal, sometimes conducting an electric current and sometimes acting like an electrical insulator depending on the conditions and their purity; also called metalloids

semilunar valves pulmonary artery and aorta valves that prevent the flow of blood backward into the ventricles

seminal vesicle a part of the male reproductive system that produces a portion of the semen

seminiferous tubules sperm-producing tubes in the testes

sensory neurons those neurons that send information from sense organs to the central nervous system

sex chromosomes a pair of chromosomes that determines the sex of an organism; X and Y chromosomes

sex-determining chromosome the chromosomes X and Y that are primarily responsible for determining if an individual will develop as a male or female

sexual intercourse the mating of male and female; the deposition of the male sex cells, or sperm cells, in the reproductive tract of the female; also known as coitus or copulation

sexual reproduction the propagation of organisms involving the union of gametes from two parents

sexuality a term used in reference to the totality of the aspects—physical, psychological, and cultural—of our sexual nature

shallow-focus earthquakes earthquakes that occur from the surface down to 70 km deep

shear stress produced when two plates slide past one another or by one plate sliding past another plate that is not moving

shell model of the nucleus model of the nucleus that has protons and neutrons moving in energy levels or shells in the nucleus (similar to the shell structure of electrons in an atom)

shells the layers that electrons occupy around the nucleus

shield volcano a broad, gently sloping volcanic cone constructed of solidified lava flows

shock wave a large, intense wave disturbance of very high pressure; the pressure wave created by an explosion, for example

sickle-cell anemia a disease caused by a point mutation; this malfunction produces sickle-shaped red blood cells

silicates minerals that contain silicon-oxygen tetrahedra either isolated or joined together in a crystal structure

sill a tabular-shaped intrusive rock that formed when magma moved into the plane of contact between sedimentary rock layers

simple harmonic motion the vibratory motion that occurs when there is a restoring force opposite to and proportional to a displacement

single bond covalent bond in which a single pair of electrons is shared by two atoms

single-factor cross a genetic study in which one characteristic is followed from the parental generation to the offspring

small intestine the portion of the digestive system immediately following the stomach; responsible for digestion and absorption

soil a mixture of unconsolidated weathered earth materials and humus, which is altered, decay-resistant organic matter

solenoid a cylindrical coil of wire that becomes electromagnetic when a current runs through it

solids a phase of matter with molecules that remain close to fixed equilibrium positions due to strong interactions between the molecules, resulting in the characteristic definite shape and definite volume of a solid

solstices time when the sun is at its maximum or minimum altitude in the sky, known as the summer solstice and the winter solstice

solubility dissolving ability of a given solute in a specified amount of solvent, the concentration that is reached as a saturated solution is achieved at a particular temperature

solute the component of a solution that dissolves in the other component; the solvent

solution a homogeneous mixture of ions or molecules of two or more substances

solvent the component of a solution present in the larger amount; the solute dissolves in the solvent to make a solution

soma the cell body of a neuron, which contains the nucleus

sonic boom sound waves that pile up into a shock wave when a source is traveling at or faster than the speed of sound

sound quality characteristic of the sound produced by a musical instrument; determined by the presence and relative strengths of the overtones produced by the instrument

south pole (of a magnet) short for "south seeking"; the pole of a magnet that points southward when it is free to turn

speciation the process of generating new species

species a group of organisms that can interbreed naturally to produce fertile offspring

specific dynamic action (SDA) the amount of energy needed to process the food that we eat

specific heat each substance has its own specific heat, which is defined as the amount of energy (or heat) needed to increase the temperature of 1 gram of a substance 1 degree Celsius

speed a measure of how fast an object is moving—the rate of change of position per change in time; speed has magnitude only and does not include the direction of change

sperm the haploid sex cells produced by sexually mature males

spermatids haploid cells produced by spermatogenesis that change into sperm

spermatogenesis the specific name given to the gametogenesis process that leads to the formation of sperm

spin quantum number from the quantum mechanics model of the atom, one of four descriptions of the energy state of an electron wave; this quantum number describes the spin orientation of an electron relative to an external magnetic field

spinal cord the portion of the central nervous system located within the vertebral column, which carries both sensory and motor information between the brain and the periphery of the body

spindle fibers an array of microtubules extending from pole to pole; used in the movement of chromosomes

spontaneous generation the theory that living organisms arose from nonliving material

spontaneous mutation a change in the DNA of an organism for which there is no known cause

stable equilibrium phase a stage in a population growth curve following rapid growth in which there is both a decrease in natality and an increase in mortality so that the size of the population is stable

standard atmospheric pressure the average atmospheric pressure at sea level, which is also known as normal pressure; the standard pressure is 29.92 inches or 760.0 mm of mercury (1,013.25 millibar)

standard unit a measurement unit established as the standard upon which the value of the other referent units of the same type are based

standing waves condition where two waves of equal frequency traveling in opposite directions meet and form stationary regions of maximum displacement due to constructive interference and stationary regions of zero displacement due to destructive interference

stapes the ear bone that is attached to the oval window

starch a group of complex carbohydrates (polysaccharides) that plants use as a stored food source and that serves as an important source of food for animals

stationary front occurs when the edge of a weather front is not advancing

steam generator part of a nuclear power plant; the heat exchanger that heats feedwater from the secondary loop to steam with the very hot water from the primary loop

step-down transformer a transformer that decreases the voltage

step-up transformer a transformer that increases the voltage

steroids a group of lipid molecules characterized by an arrangement of interlocking rings of carbon; they serve as hormones that aid in regulating body processes

stimulus any change in the internal or external environment of an organism that it can detect

stony meteorites meteorites composed mostly of silicate minerals that usually make up rocks on the earth

stony-iron meteorites meteorites composed of silicate minerals and metallic iron

storm a rapid and violent weather change with strong winds, heavy rain, snow, or hail

strain adjustment to stress; a rock unit might respond to stress by changes in volume, changes in shape, or by breaking

stratosphere the layer of the atmosphere above the troposphere where temperature increases with height

stream body of running water

stress a force that tends to compress, pull apart, or deform rock; stress on rocks in the earth's solid outer crust results as the earth's plates move into, away from, or alongside each other

structural protein protein molecules whose function is to provide support and shape to a cell or multicellular organism; for example, muscle protein fibers

structural similarities one of the characteristics of living things; describes the fact that all living things are composed of cells, either prokaryotic or eukaryotic

subduction zone the region of a convergent boundary where the crust of one plate is forced under the crust of another plate into the interior of the earth

sublimation the phase change of a solid directly into a vapor or gas

submarine canyons a feature of the ocean basin; deep, steep-sided canyons that cut through the continental slopes

subspecies regional groups within a species that are significantly different structurally, physiologically, or behaviorally, yet are capable of exchanging genes by interbreeding with other members of the species

superconductors some materials in which, under certain conditions, the electrical resistance approaches zero

supercooled water in the liquid phase when the temperature is below the freezing point

superior vena cava a major vein that returns blood to the heart from the head and upper body

supernova a rare catastrophic explosion of a star into an extremely bright, but short-lived phenomenon

supersaturated containing more than the normal saturation amount of a solute at a given temperature

surf the zone where breakers occur; the water zone between the shoreline and the outermost boundary of the breakers

surface wave a seismic wave that moves across the earth's surface, spreading across the surface like water waves spread on the surface of a pond from a disturbance

S-wave a sideways, or shear wave in which a disturbance vibrates materials from side to side, perpendicular to the direction of wave movement

S-wave shadow zone a region of the earth more than 103° of arc away from the epicenter of an earthquake where S-waves are not recorded; believed to be the result of the core of the earth that is a liquid, or at least acts like a liquid

swell regular groups of low-profile, long-wavelength waves that move continuously

symbiotic nitrogen-fixing bacteria bacteria that live in the roots of certain kinds of plants, where they convert nitrogen gas molecules into compounds that the plants can use to make amino acids and nucleic acids

synapse the space between the axon of one neuron and the dendrite of the next, where chemicals are secreted to cause an impulse to be initiated in the second neuron

synapsis the condition in which the two members of a pair of homologous chromosomes come to lie close to one another

syncline a trough-shaped fold in layered bedrock

systemic circulation the flow of blood through certain chambers of the heart and blood vessels to the general body and back to the heart

systolic blood pressure the blood pressure recorded in a large artery while the heart is contracting

T

talus steep, conical or apron-like accumulations of rock fragments at the base of a slope

taxonomy the science of classifying and naming organisms

telophase the last phase in mitosis characterized by the formation of daughter nuclei

temperate climate zone climate zone of the middle latitudes; average monthly temperatures stay between 10°C and 18°C (50°F and 64°F) throughout the year

temperature how hot or how cold something is; a measure of the average kinetic energy of the molecules making up a substance

tensional stress the opposite of compressional stress; occurs when one part of a plate moves away from another part that does not move

termination code the DNA nucleotide sequence at the end of a gene with the code ATT, ATC, or ACT that signals "stop here"

terrestrial planets planets Mercury, Venus, Earth, and Mars that have similar densities and compositions as compared to the outer giant planets

testes the male sex organs that produce haploid cells called sperm

testosterone the male sex hormone produced in the testes that controls male sexual development

thalamus a region of the brain that relays information between the cerebrum and lower portions of the brain, providing some level of awareness in that it determines pleasant and unpleasant stimuli and is involved in sleep and arousal

theory a broad, detailed explanation that guides the development of hypotheses and interpretations of experiments in a field of study

theory of natural selection the idea that some individuals within a population will have favorable combinations of genes that make it very likely that those individuals will survive, reproduce, and pass their genes to the next generation

thermometer a device used to measure the hotness or coldness of a substance

third law of motion whenever two objects interact, the force exerted on one object is equal in size and opposite in direction to the force exerted on the other object; forces always occur in matched pairs that are equal and opposite

thrust fault a reverse fault with a low-angle fault plane

thunderstorm a brief, intense electrical storm with rain, lightning, thunder, strong winds, and sometimes hail

thymine a single-ring nitrogenous-base molecule in DNA but not in RNA; it is complementary to adenine

thyroid-stimulating hormone (TSH) a hormone secreted by the pituitary gland that stimulates the thyroid to secrete thyroxine

thyroxine a hormone produced by the thyroid gland that speeds up the metabolic rate

tidal bore a strong tidal current, sometimes resembling a wave, produced in very long, very narrow bays as the tide rises

tidal currents a steady and continuous onward movement of water produced in narrow bays by the tides

tides periodic rise and fall of the level of the sea from the gravitational attraction of the moon and sun

tissue a group of specialized cells that work together to perform a specific function

tornado a long, narrow, funnel-shaped column of violently whirling air from a thundercloud that moves destructively over a narrow path when it touches the ground

total internal reflection condition where all light is reflected back from a boundary between materials; occurs when light arrives at a boundary at the critical angle or beyond

trachea a major tube supported by cartilage that carries air to the bronchi; also known as the windpipe

transcription the process of manufacturing RNA from the template surface of DNA; three forms of RNA that may be produced are mRNA, rRNA, and tRNA

transfer RNA (tRNA) a molecule composed of ribonucleic acid. It is responsible for transporting a specific amino acid into a ribosome for assembly into a protein

transform boundaries in plate tectonics, boundaries that occur between two plates sliding horizontally by each other along a long, vertical fault; sudden jerks along the boundary result in the vibrations of earthquakes

transformer a device consisting of a primary coil of wire connected to a source of alternating current and a secondary coil of wire in which electromagnetic induction increases or decreases the voltage of the source

transition elements members of the B-group families of the periodic table

translation the assembly of individual amino acids into a polypeptide

transparent term describing materials that allow the transmission of light; for example, glass and clear water are transparent materials

transportation the movement of eroded materials by agents such as rivers, glaciers, wind, or waves

transverse wave a mechanical disturbance that causes particles to move perpendicular to the direction that the wave is traveling

triglyceride organic compound of esters formed from glycerol and three long-chain carboxylic acids; also called fats in animals and oils in plants

triple bond covalent bond formed when three pairs of electrons are shared by two atoms

trophic level a step in the flow of energy through an ecosystem

tropical air mass a warm air mass from warm regions

tropical climate zone climate zone of the low latitudes; average monthly temperatures stay above 18°C (64°F), even during the coldest month of the year

tropical cyclone a large, violent circular storm that is born over the warm, tropical ocean near the equator; also called hurricane (Atlantic and eastern Pacific) and typhoon (in western Pacific)

troposphere layer of the atmosphere from the surface to where the temperature stops decreasing with height

trough the low mound of water that is part of a wave; also refers to the rarefaction, or low-pressure part of a sound wave

true fats also known as *neutral fats;* a category of lipids composed of glycerol and fatty acids; for example, pork chop fat or olive oil

tsunami very large, fast, and destructive ocean wave created by an undersea earthquake, landslide, or volcanic explosion; a seismic sea wave

tumor a mass of undifferentiated cells not normally found in a certain portion of the body

turbidity current a muddy current produced by underwater landslides

tympanum the eardrum

typhoon the name for hurricanes in the western Pacific

U

ultrasonic sound waves too high in frequency to be heard by the human ear; frequencies above 20,000 Hz

unconformity a time break in the rock record

undertow a current beneath the surface of the water produced by the return of water from the shore to the sea

unit in measurement, a well-defined and agreed-upon referent

universal law of gravitation every object in the universe is attracted to every other object with a force directly proportional to the

product of their masses and inversely proportional to the square of the distance between the centers of the two masses

unpolarized light light consisting of transverse waves vibrating in all conceivable random directions

unsaturated molecule an organic molecule that does not contain the maximum number of hydrogen atoms; a molecule that can add more hydrogen atoms because of the presence of double or triple bonds

uracil a single-ring nitrogenous-base molecule in RNA but not in DNA; it is complementary to adenine

uterus the organ in female mammals in which the embryo develops

V

vacuole a large sac within the cytoplasm of a cell, composed of a single membrane

vagina the passageway between the uterus and outside of the body; the birth canal

valence the number of covalent bonds an atom can form

valence electrons electrons of the outermost shell; the electrons that determine the chemical properties of an atom and the electrons that participate in chemical bonding

valid a term used to describe meaningful data that fit into the framework of scientific knowledge

Van Allen belts belts of radiation caused by cosmic-ray particles becoming trapped and following the earth's magnetic field lines between the poles

vapor the gaseous state of a substance that is normally in the liquid state

variables changing quantities usually represented by a letter or symbol

veins the blood vessels that return blood to the heart

velocity describes both the speed and direction of a moving object; a change in velocity is a change in speed, in direction of travel, or both

ventifacts rocks sculpted by wind abrasion

ventricles the powerful muscular chambers of the heart whose contractions force blood to flow through the arteries to all parts of the body

vibration a back-and-forth motion that repeats itself

villi tiny fingerlike projections in the lining of the intestine that increase the surface area for absorption

viroids infectious particles that are composed solely of a small, single strand of RNA

virtual image an image where light rays appear to originate from a mirror or

lens; this image cannot be projected on a screen

virus a nucleic acid particle coated with protein that functions as an obligate intracellular parasite

vitamin-deficiency disease poor health caused by the lack of a certain vitamin in the diet; for example, scurvy for lack of vitamin C

vitamins organic molecules that cannot be manufactured by the body but are required in very low concentrations for good health

volcanism volcanic activity; the movement of magma

volcano a hill or mountain formed by the extrusion of lava or rock fragments from a mass of magma below

volt the ratio of work done to move a quantity of charge

voltage source a device that does work in moving a quantity of charge

volume how much space something occupies

vulcanism volcanic activity; the movement of magma

W

warm front the front that forms when a warm air mass advances against a cool air mass

water table the boundary below which the ground is saturated with water

watershed the region or land area drained by a stream; a stream drainage basin

watt metric unit for power; equivalent to J/s

wave a disturbance or oscillation that moves through a medium

wave equation the relationship of the velocity of a wave to the product of the wavelength and frequency of the wave

wave front a region of maximum displacement in a wave; a condensation in a sound wave

wave height the vertical distance of an ocean wave between the top of the wave crest and the bottom of the next trough

wave mechanics alternate name for quantum mechanics derived from the wavelike properties of subatomic particles

wave period the time required for two successive crests or other successive parts of the wave to pass a given point

wavelength the horizontal distance between successive wave crests or other successive parts of the wave

weak acid acids that only partially ionize because of an equilibrium reaction with water

weak base a base only partially ionized because of an equilibrium reaction with water

weathering slow changes that result in the breaking up, crumbling, and destruction of any kind of solid rock

white dwarf stars one of two groups of stars on the Hertzsprung-Russell diagram that have a different set of properties than the main sequence stars; faint, white-hot stars that are very small and dense

wind a horizontal movement of air that moves along or parallel to the ground, sometimes in currents or streams

wind abrasion the natural sand-blasting process that occurs when wind-driven particles break off small particles of rock and polish the rock they strike

wind chill factor a factor that compares heat loss from bodies in still air with those in moving air; moving air removes heat more rapidly and causes a person to feel that the air is colder than its actual temperature; the cooling power of wind

work the magnitude of applied force times the distance through which the force acts; can be thought of as the process by which one form of energy is transformed to another

X

X chromosome the chromosome in a human female egg (and in one-half of sperm cells) that is associated with the determination of sexual characteristics

X-linked gene a gene located on one of the sex-determining X chromosomes

Y

Y chromosome the sex-determining chromosome in one-half of the sperm cells of human males responsible for determining maleness

Z

zone of saturation zone of sediments beneath the surface in which water has collected in all available spaces

zooplankton small, weakly swimming animals of many kinds

zygote a diploid cell that results from the union of an egg and a sperm; a fertilized egg

Photographs

Front Matter *Page xi:* © Vol. 223/Corbis.

Chapter 1 *Opener:* © Werner H. Muller/Peter Arnold, Inc. *1.1:* © Spencer Grant/Photo Researchers, Inc. *1.2:* © Rafael Macia/Photo Researchers, Inc. *1.3:* © Bill W. Tillery. *1.13:* © Chad Slattery/Stone/Getty Images. *1.14A:* © Barrington Brown/Photo Researchers, Inc. *1.14B:* © Hank Morgan/Photo Researchers, Inc. *1.14C:* © David M. Phillips/Visual Unlimited. *1.15A:* © David R. Frazier/Photo Researchers, Inc. *1.17 left:* Corbis/Bettmann Archive. *1.17 right:* © Bob Coyle/Red Diamond Photos.

Chapter 2 *Opener:* © Douglas Faulkner/Photo Researchers, Inc. *2.1:* © Darrell Wong/Stone/Getty Images. *2.7:* © Bill W. Tillery. *2.9:* © Scala/Art Resource, NY. *Box Fig. 2.1:* © Eunice Harris/Photo Researchers, Inc. *2.16:* Arizona State University Media Relations Office. *2.18:* © Aldo Torelli/Stone/Getty Images. *2.22:* Arizona State University Media Relations Office. *Box Fig. 2.2:* The Image Bank/Getty Images. *2.24:* NASA.

Chapter 3 *Opener:* © Hank Morgan/Science Source/Photo Researchers, Inc. *3.1, 3.9, 3.10A:* © Bill W. Tillery. *3.10B:* © Bill Beatty/Visuals Unlimited. *3.11:* Arizona Public Service Company. *3.13:* Arizona Public Service Company. *Box Fig. 3.1:* © Bill W. Tillery.

Chapter 4 *Opener:* © Lawrence Livermore Nat. Laboratory/SPL/Photo Researchers, Inc. *4.1:* Bethlehem Steel Corporation. *4.2:* Courtesy of John C. Wheatly, Center for High Resolution Electron Microscopy. *4.6:* Reprinted with permission from Honeywell, Inc. *4.14:* Manville Company. *4.20:* © Bill W. Tillery.

Chapter 5 *Opener:* © Clyde H. Smith/Peter Arnold, Inc. *5.1:* © Patrick Docklin/Stone/Getty Images. *5.12:* © The McGraw-Hill Companies, Inc./Bob Coyle, photographer.

Chapter 6 *Opener:* © Keith Kent/Peter Arnold, Inc. *6.1, 6.11A, 6.11B, 6.11C:* © Bill W. Tillery. *Box Fig. 6.2A:* ©1989 Paul Silverman/Fundamental Photographs. *6.12, Box Fig. 6.5A, Box Fig. 6.5B, 6.26A, 6.26B:* © Bill W. Tillery.

Chapter 7 *Opener:* © James L. Amos/Peter Arnold, Inc. *7.1:* © John Kieffer/Peter Arnold, Inc. *7.7:* © Bill W. Tillery. *7.8A, 7.8B:* 1987 Libbey-Owens-Ford Co. *7.16:* © Bill W. Tillery.

Chapter 8 *Opener:* © Astrid & Hanns-Frieder Michler/SPL/Photo Researchers, Inc. *8.1:* © Ken Eward/Science Source/Photo Researchers, Inc.

Chapter 9 *Opener:* © David Weintraub/Photo Researchers, Inc. *9.1:* © Bill W. Tillery. *9.3A, 9.3B:* © Leonard Lessin/Peter Arnold, Inc. *9.4:* © Bill W. Tillery. *9.5:* © Richard Megna/Fundamental Photographs. *9.9:* © Charles M. Falco/Photo Researchers, Inc. *9.11, 9.12:* © Bill W. Tillery. *9.13:*

© Paul Sisul/Stone /Getty Images. *9.15, 9.16, Box Fig. 9.2, 9.18:* © Bill W. Tillery.

Chapter 10 *Opener:* © Blair Seitz/Photo Researchers, Inc. *10.1:* © Craig Tuttle/CORBIS. *10.11A, 10.11B, 10.12, 10.14, Box Fig. 10.1:* © Bill W. Tillery.

Chapter 11 *Opener:* © Yann Arthus-Bertrand/Peter Arnold, Inc. *11.1:* Arizona Public Service Company. *11.2A, 11.2B:* © Fundamental Photographs. *Box Fig. 11.1A:* © SIU Med/Photo Researchers, Inc. *Box Fig. 11.1B:* © Steve Grand/SPL/Photo Researchers, Inc. *11.14A, 11.14B, 11.16, 11.17:* Arizona Public Service Company. *Box Fig. 11.2:* © Bill W. Tillery.

Chapter 12 *Opener:* NASA. *12.1A:* Photo © UC Regents/Lick Observatory. *12.2:* © J. Fuller/Visuals Unlimited. *12.9, 12.11:* © UC Regents/Lick Observatory. *Box Fig. 12.1:* © Hencoup Enterprises/SPL/Photo Researchers, Inc. *12.13:* © UC Regents/Lick Observatory.

Chapter 13 *Opener:* NASA. *13.1, 13.4, 13.5:* NASA. *13.6:* © UC Regents/Lick Observatory. *13.7, 13.8, 13.10A, 13.10B, 13.11:* NASA. *13.12A, 13.12B:* © Space Telescope Science Institute/NASA/SPL/Photo Researchers, Inc. *13.13, 13.14:* NASA. *13.17:* Sky and Telescope. *13.18:* © UC Regents/Lick Observatory. *13.20A, 13.20B:* Center for Meteorite Studies, Arizona State University.

Chapter 14 *Opener:* © Brian Parker/Tom Stack & Associates. *14.1:* © Bill W. Tillery. *14.4:* © Charles Falco/Photo Researchers, Inc. *14.5, 14.8, 14.9:* © Bill W. Tillery. *Box Fig. 14.1, 14.10:* Photo by C.C. Plummer.

Chapter 15 *Opener:* © Jeff Foot/Tom Stacks & Associates. *15.1:* Photo by A. Post, USGS Photo Library, Denver, CO. *15.2:* © Bill W. Tillery. *15.4A:* Robert W. Northrop, Photographer/Illustrator. *15.4B:* © Bill W. Tillery. *15.5:* Photo by C.C. Plummer. *15.8B:* Photo by Frank M. Hanna. *15.12D:* Photo by University of Colorado, Courtesy National Geophysical Data Center, Boulder, Colorado. *15.15:* NASA. *Fig. 15.16B:* © John S. Shelton. *15.18:* © Bill W. Tillery. *15.20:* National Park Service, Photo by Wm. Belnap, Jr. *15.22A:* © A.J. Copley/Visuals Unlimited. *15.22B:* © L. Linkhard/Visuals Unlimited. *15.23:* © Ken Wagner/Visuals Unlimited. *15.24A, 15.24B:* © Bill W. Tillery. *15.26:* © William J. Weber/Visuals Unlimited. *15.27:* © Doug Sherman/Geofile. *15.29:* Photo by Burton A. Amundson. *15.30A:* Photo by D.A. Rahn, courtesy of Rahn Memorial Collections, Western Washington University. *15.31:* Photo by C.C. Plummer.

Chapter 16 *Opener:* © Galen Rowell/Mountain Light Photography. *16.1:* © Bill W. Tillery. *Box Fig. 16.1:* NOAA. *16.16, 16.18A, 16.18B, 16.19A, 16.19B, 16.19C, 16.19D, 16.19E, 16.19F:* © Bill W. Tillery. *16.20, 16.22:* NOAA. *16.31:* © Telegraph Herald/Photo by Patti Carr. *16.32, 16.33:* NOAA. *16.35:* © David Parker/SPL/Photo Researchers, Inc. *16.36:*

© Bill W. Tillery. *16.39:* Dr. Charles Hogue, Curator of Entomology, Los Angeles County Museum of Natural History. *16.40:* Photo by Bob Wallen. *16.41:* Photo by Elizabeth Wallen.

Chapter 17 *Opener:* © James H. Karales/Peter Arnold, Inc. *17.1:* Salt River Project. *17.9:* City of Tempe, Arizona. *17.10:* Salt River Project. *17.13, 17.14:* © Bill W. Tillery. *17.18:* © John S. Shelton.

Chapter 18 *Opener:* Kenneth Edward/BioGrafx-Science Source/Photo Researchers, Inc. *18.1, 18.9:* American Petroleum Institute. *18.14:* © Bill W. Tillery. *18.17A:* © R.J. Erwin 1988/Photo Researchers, Inc. *18.17B:* Image obtained at www.kurtstueber.de

Chapter 19 *Opener:* M. Sameni and B.F. Sloane, Wayne State University School of Medicine, Detroit, MI. *19.1 upper left:* © Science VU/Visuals Unlimited. *19.1 upper right:* © A.M. Siegelman/Visuals Unlimited. *19.1 lower left:* © John Garrett/Stone/Getty Images. *19.1 lower right:* © Ray Coleman/Photo Researchers, Inc. *19.2A:* © Stock Montage. *19.2B:* © Dr. Jeremy Burgess/SPL/Photo Researchers, Inc. *19.9 top:* © K. G. Murti/Visuals Unlimited. *19.9 middle:* © Warren Rosenberg/Biological Photo Service. *19.9 lower right:* © Richard Rodewald/Biological Photo Service. *19.9 lower left:* © David M. Phillips/Visuals Unlimited. *19.10:* Fred Ross. *19.12B:* © Dr. Keith Porter. *19.13:* © Don Fawcett/Photo Researchers, Inc. *19.14A:* M. Sameni and B.F. Sloane, Wayne State University School of Medicine, Detroit, MI. *19.17:* From William Jensen and R.B. Park, Cell Ultrastructure, 1967, p. 57, © Brooks/Cole Publishing, a division of International Thompson Publishing, Inc.

Chapter 20 *20.1:* © 1999 M.D. Escher/Cordon Art-Baarn-Holland. All rights reserved. *20.5:* © Ted Levin/Animals Animals/Earth Scenes. *20.6:* © D.O.E./Science Source/Photo Researchers, Inc. *20.7:* © John Cunningham/Visuals Unlimited. *20.10:* © Thomas Kitchin/Tom Stack & Associates. *20.12A:* © Walt Anderson/Visuals Unlimited. *20.12B:* Eldon Enger. *20.12C:* © William J. Weber/Visuals Unlimited. *20.13 left, 20.13 right:* Tom & Pat Leeson.

Chapter 21 *Opener:* Courtesy Eldon Enger. *21.2:* Photo by Bob Wallen. *21.6:* © Bill W. Tillery. *21.7:* © W.B. Saunders/Biological Photo Service. *21.9:* Robert W. Northrop, Photographer/Illustrator. *21.10:* U.S. Geological Survey. *21.15A:* © Tom McHugh/Photo Researchers, Inc. *21.15B:* © Alvin E. Staffan/Photo Researchers, Inc. *21.16:* © Stock Montage. *21.17A:* © A.M. Siegelman/Visuals Unlimited. *21.17B:* © John Gerlach/Tom Stack & Associates. *21.17C:* © Paul W. Johnson/Biological Photo Service. *21.17D:* Eldon Enger. *21.17E:* © Glenn Oliver/Visuals Unlimited. *21.18A:* © Tom Stack/Tom Stack & Associates. *21.18B:* © Don & Pat Valenti/Tom Stack & Associates. *21.20 upper left, 21.20 lower left, 21.20 middle:* Eldon Enger. *21.20 upper right:* © Claudia E. Mills/Biological

Photo Service. *21.20 bottom:* © P. Nuridsany/Photo Researchers, Inc.

Chapter 22 *Opener:* Eldon Enger. *22.1A:* © Gregory K. Scott/Photo Researchers, Inc. *22.1B, page 488:* Eldon Enger. *Page 489 top left:* © Pat & Tom Leeson/Photo Researchers, Inc. *Page 489 bottom left:* © C.P. Hickman/Visuals Unlimited. *Page 489 top right and bottom right:* Eldon Enger. *Page 490 left:* © Peter K. Ziminski/Visuals Unlimited. *Page 490 top right:* © Michael Giannenchini/Photo Researchers, Inc. *Page 490 bottom:* Eldon Enger. *22.11A:* © John D. Cunningham/Visuals Unlimited. *22.11B:* Eldon Enger. *22.12A:* © David Waters/Envision. *22.12B:* © Stephen Dalton/Photo Researchers, Inc. *22.13A:* © J.H. Robinson/Photo Researchers, Inc. *22.13B:* © Gary Milburn/Tom Stack & Associates. *22.13C:* © Stephen J. Kraseman/Peter Arnold, Inc. *22.14A:* © Douglas Faulkner/Sally Faulkner Collection. *22.14B:* © Kjell Sanved/Visuals Unlimited. *22.15:* Eldon Enger. *22.16:* © Susanna Pashko/Envision. *22.21:* © G.R. Higbee/Photo Researchers, Inc. *22.22A:* © Stephen J. Krasemann/Photo Researchers, Inc. *22.22B:* © Walt Anderson/Visuals Unlimited.

Chapter 23 *Opener:* © David Scharf/Peter Arnold, Inc. *Box Fig. 23.3:* © M. Coleman/Visuals Unlimited. *Page 527:* LECO Corporation. *23.10:* © Paul A. Souders.

Chapter 24 *24.1A:* © Dr. Keith Porter.

Chapter 25 *Opener:* © Renee Lynn/Photo Researchers, Inc. *25.1A, 25.1B:* © The McGraw-Hill Companies, Inc./Bob Coyle, photographer. *25.3 left:* © Renee Lynn/Photo Researchers, Inc. *25.3 middle, 25.3 right:* Courtesy Mary Drapeau. *25.4:* © John D. Cunningham/Visuals Unlimited. *25.6A, 25.6B, 25.6C:* Courtesy Jeanette Navia. *25.8:* Neurofibromatosis, Inc. Lanham, MD. *25.11B:* © BioPhoto Associates/Photo Researchers, Inc. *25.16A, 25.16B:* © Stanley Flegler/Visuals Unlimited. *Box Fig. 25.5 all:* Photo courtesy R. A. Bowen. *Box Fig. 25.6:* Photo courtesy R. A. Bowen.

Illustrations

Chapter 3 *3.17:* Source: www.eia.doc.gov/emeu/sep/us/frame.html.

Chapter 13 *Table 13.3:* Source: Data from NASA.

Chapter 14 *14.15:* From Charles C. Plummer and David McGeary, *Physical Geology,* 6th ed., © 1993 Wm. C. Brown Communications, Inc., Dubuque, Iowa. Reproduced with permission of The McGraw-Hill Companies. *14.16:* From Charles C. Plummer and David McGeary, *Physical Geology,* 6th ed., © 1993 Wm. C. Brown Communications, Inc., Dubuque, Iowa. Reproduced with permission of The McGraw-Hill Companies. *14.21:* From W.C. Pitman, III; R.L. Larson; & E.M. Herron; with permission of the publisher, the Geological Society of America, Boulder, Colorado, USA. Copyright ©1974 Geological Society of America. *14.22:* From Carla W. Montgomery, *Physical Geology,* 3rd ed., ©1993 Wm. C. Brown Communications, Inc., Dubuque, Iowa. Reproduced with permission of The McGraw-Hill Companies. *14.23:* Source: After W. Hamilton, U.S. Geological Survey. *14.28A–B:*

From Charles C. Plummer and David McGeary, *Physical Geology,* 6th ed., © 1993 Wm. C. Brown Communications, Inc., Dubuque, Iowa. Reproduced with permission of The McGraw-Hill Companies.

Chapter 15 *15.3:* From Carla W. Montgomery, *Physical Geology,* 3rd ed., ©1993 Wm. C. Brown Communications, Inc., Dubuque, Iowa. Reproduced with permission of The McGraw-Hill Companies. *15.6:* From Carla W. Montgomery, *Physical Geology,* 3rd ed., © 1993 Wm. C. Brown Communications, Inc., Dubuque, Iowa. Reproduced with permission of The McGraw-Hill Companies. *15.8A:* From David McGeary and Charles C. Plummer, Earth Revealed, 2nd ed., ©1994 Wm. C. Brown Communications, Inc., Dubuque, Iowa. Reproduced with permission of The McGraw-Hill Companies *15.9A–C:* From Charles C. Plummer and David McGeary, *Physical Geology,* 6th ed., ©1993 Wm. C. Brown Communications, Inc., Dubuque, Iowa. Reproduced with permission of The McGraw-Hill Companies. *15.10A–B:* From Charles C. Plummer and David McGeary, *Physical Geology,* 6th ed., ©1993 Wm. C. Brown Communications, Inc., Dubuque, Iowa. Reproduced with permission of The McGraw-Hill Companies. *15.12A–C:* From Charles C. Plummer and David McGeary, *Physical Geology,* 6th ed., ©1993 Wm. C. Brown Communications, Inc., Dubuque, Iowa. Reproduced with permission of The McGraw-Hill Companies. *15.13:* From Carla W. Montgomery, *Physical Geology,* 3rd ed., ©1993 Wm. C. Brown Communications, Inc., Dubuque, Iowa. Reproduced with permission of The McGraw-Hill Companies.*15.14A–B:* From Carla W. Montgomery, *Physical Geology,* 3rd ed., ©1993 Wm. C. Brown Communications, Inc., Dubuque, Iowa. Reproduced with permission of The McGraw-Hill Companies. *Table 15.1:* Source: National Oceanic and Atmospheric Administration. *15.17:* From Carla W. Montgomery, *Physical Geology,* 3rd ed., © 1993 Wm. C. Brown Communications, Inc., Dubuque, Iowa. Reproduced with permission of The McGraw-Hill Companies. *15.21A–B:* From Carla W. Montgomery, *Physical Geology,* 3rd ed., ©1993 Wm. C. Brown Communications, Inc., Dubuque, Iowa. Reproduced with permission of The McGraw-Hill Companies. *15.28A–C:* From Charles C. Plummer and David McGeary, *Physical Geology,* 6th ed., ©1993 Wm. C. Brown Communications, Inc., Dubuque, Iowa. Reproduced with permission of The McGraw-Hill Companies. *15.32:* From Charles C. Plummer and David McGeary, *Physical Geology,* 5th ed., ©1991 Wm. C. Brown Communications, Inc., Dubuque, Iowa. Reproduced with permission of The McGraw-Hill Companies.

Chapter 16 *Box Figure 16.1:* From www.nws.noaa.gov/om/windfall/html

Chapter 17 *Pages 359–360, A Closer Look: Water Quality and Wastewater Treatment; Box Figure 17.1; and Box Figure 17.2:* Drawings, and some text, from *How Wastewater Treatment Works . . . The Basics,* The Environmental Protectional Agency, Office of Water, www.epa.gov/owm/featinfo.htm

Chapter 19 *19.17A–B:* From John W. Hole, Jr., *Human Anatomy And Physiology,* 5th ed., ©1995 Wm. C. Brown Communications, Inc., Dubuque,

Iowa. Reproduced with permission of The McGraw-Hill Companies.

Chapter 20 *20.8:* Source: Data from Georghiou, University of California at Riverside.

Chapter 21 *21.3:* From Carla W. Montgomery, *Physical Geology,* 3rd ed., ©1993 Wm. C. Brown Communications, Inc., Dubuque, Iowa. Reproduced with permission of The McGraw-Hill Companies. *21.4:* From Carla W. Montgomery, *Physical Geology,* 3rd ed., ©1993 Wm. C. Brown Communications, Inc., Dubuque, Iowa. Reproduced with permission of The McGraw-Hill Companies. *21.11:* From Carla W. Montgomery, *Physical Geology,* 3rd ed., ©1993 Wm. C. Brown Communications, Inc., Dubuque, Iowa. Reproduced with permission of The McGraw-Hill Companies.

Chapter 22 *22.6:* Figure from "Biotic Communities of the Aspen Parkland of Central Canada" by Ralph D. Bird from *Ecology* 11 (2): 410. Reprinted by permission of the Ecological Society of America. *22.20:* Source: U.S. Bureau of the Census and the United Nations, as reported in Joseph A. McFalls, Jr., "Population: A Lively Introduction," *Population Bulletin,* vol. 46, no. 2, Washington, D.C.: Population Reference Bureau, Inc., October 1991. *22.25:* Source: Data from Jean Van Der Tak et al., "Our Population Predicament: A New Look" in *Population Bulletin,* Vol. 34, No. 5, December 1979, Population Reference Bureau, Washington, D.C.

Chapter 23 *23.5:* From John W. Hole, Jr., *Human Anatomy and Physiology,* 4th ed., ©1993 Wm. C. Brown Communications, Inc., Dubuque, Iowa. Reproduced with permission of The McGraw-Hill Companies. *23.6:* From John W. Hole, Jr., *Human Anatomy and Physiology,* 4th ed., ©1993 Wm. C. Brown Communications, Inc., Dubuque, Iowa. Reproduced with permission of The McGraw-Hill Companies. *23.7A:* Reprinted with permission from *Concepts of Human Anatomy and Physiology,* 5th Edition by Kent M. Van De Graaff and Stuart Ira Fox. *23.7B:* From John W. Hole, Jr., *Human Anatomy and Physiology,* 4th ed., ©1993 Wm. C. Brown Communications, Inc., Dubuque, Iowa. Reproduced with permission of The McGraw-Hill Companies. *Table 23.A:* Data from the Centers for Disease Control and Prevention publication, *Tracking the Hidden Epidemics, Trends in STDs in the United States 2000. Table 23.B:* Data from the Centers for Disease Control and Prevention publication, *Tracking the Hidden Epidemics, Trends in Stds in the United States 2000. Table 23.2:* Reprinted with permission from *Concepts of Human Anatomy and Physiology,* 5th Edition by Kent M. Van De Graaff and Stuart Ira Fox.

Chapter 24 *24.8:* From John W. Hole, Jr., *Human Anatomy and Physiology,* 5th ed., ©1995 Wm. C. Brown Communications, Inc., Dubuque, Iowa. Reproduced with permission of The McGraw-Hill Companies.

Chapter 25 *Table 25.4:* Based on R. Lewis, *Human Genetics, Concepts and Applications,* 2nd ed., Dubuque, Iowa: Wm. C. Brown Publishers, 1997.

Table of Atomic Weights (Based on Carbon-12)

Name	Symbol	Atomic Number	Atomic Weight	Name	Symbol	Atomic Number	Atomic Weight
Actinium	Ac	89	(227)	Mendelevium	Md	101	258.10
Aluminum	Al	13	26.9815	Mercury	Hg	80	200.59
Americium	Am	95	(243)	Molybdenum	Mo	42	95.94
Antimony	Sb	51	121.75	Neodymium	Nd	60	144.24
Argon	Ar	18	39.948	Neon	Ne	10	20.179
Arsenic	As	33	74.922	Neptunium	Np	93	(237)
Astatine	At	85	(210)	Nickel	Ni	28	58.71
Barium	Ba	56	137.34	Niobium	Nb	41	92.906
Berkelium	Bk	97	(247)	Nitrogen	N	7	14.0067
Beryllium	Be	4	9.0122	Nobelium	No	102	259.101
Bismuth	Bi	83	208.980	Osmium	Os	76	190.2
Bohrium	Bh	107	264	Oxygen	O	8	15.9994
Boron	B	5	10.811	Palladium	Pd	46	106.4
Bromine	Br	35	79.904	Phosphorus	P	15	30.9738
Cadmium	Cd	48	112.40	Platinum	Pt	78	195.09
Calcium	Ca	20	40.08	Plutonium	Pu	94	244.064
Californium	Cf	98	242.058	Polonium	Po	84	(209)
Carbon	C	6	12.0112	Potassium	K	19	39.098
Cerium	Ce	58	140.12	Praseodymium	Pr	59	140.907
Cesium	Cs	55	132.905	Promethium	Pm	61	144.913
Chlorine	Cl	17	35.453	Protactinium	Pa	91	(231)
Chromium	Cr	24	51.996	Radium	Ra	88	(226)
Cobalt	Co	27	58.933	Radon	Rn	86	(222)
Copper	Cu	29	63.546	Rhenium	Re	75	186.2
Curium	Cm	96	(247)	Rhodium	Rh	45	102.905
Dubnium	Db	105	(262)	Rubidium	Rb	37	85.468
Dysprosium	Dy	66	162.50	Ruthenium	Ru	44	101.07
Einsteinium	Es	99	(254)	Rutherfordium	Rf	104	(261)
Erbium	Er	68	167.26	Samarium	Sm	62	150.35
Europium	Eu	63	151.96	Scandium	Sc	21	44.956
Fermium	Fm	100	257.095	Seaborgium	Sg	106	(266)
Fluorine	F	9	18.9984	Selenium	Se	34	78.96
Francium	Fr	87	(223)	Silicon	Si	14	28.086
Gadolinium	Gd	64	157.25	Silver	Ag	47	107.868
Gallium	Ga	31	69.723	Sodium	Na	11	22.989
Germanium	Ge	32	72.59	Strontium	Sr	38	87.62
Gold	Au	79	196.967	Sulfur	S	16	32.064
Hafnium	Hf	72	178.49	Tantalum	Ta	73	180.948
Hassium	Hs	108	(269)	Technetium	Tc	43	(99)
Helium	He	2	4.0026	Tellurium	Te	52	127.60
Holmium	Ho	67	164.930	Terbium	Tb	65	158.925
Hydrogen	H	1	1.0079	Thallium	Tl	81	204.37
Indium	In	49	114.82	Thorium	Th	90	232.038
Iodine	I	53	126.904	Thulium	Tm	69	168.934
Iridium	Ir	77	192.2	Tin	Sn	50	118.69
Iron	Fe	26	55.847	Titanium	Ti	22	47.90
Krypton	Kr	36	83.80	Tungsten	W	74	183.85
Lanthanum	La	57	138.91	Uranium	U	92	238.03
Lawrencium	Lr	103	260.105	Vanadium	V	23	50.942
Lead	Pb	82	207.19	Xenon	Xe	54	131.30
Lithium	Li	3	6.941	Ytterbium	Yb	70	173.04
Lutetium	Lu	71	174.97	Yttrium	Y	39	88.905
Magnesium	Mg	12	24.305	Zinc	Zn	30	65.38
Manganese	Mn	25	54.938	Zirconium	Zr	40	91.22
Meitnerium	Mt	109	(268)				

Note: A value given in parentheses denotes the number of the longest-lived or best-known isotope.